W9-AQA-782

THE 68000 MICROPROCESSOR FAMILY

Architecture, Programming, and Applications

Second Edition

Michael A. Miller

DeVry Institute of Technology—Phoenix

Merrill, an imprint of
Macmillan Publishing Company
New York

Maxwell Macmillan Canada
Toronto

Maxwell Macmillan International
New York Oxford Singapore Sydney

Cover: Jook C. Leung/SPG International
Editor: Dave Garza
Developmental Editor: Suzanne Murchland
Production Editor: Constantina Geldis
Cover Designer: Robert Vega
Production Buyer: Patricia A. Tonneman

This book was set in Times Roman by Publication Services, Inc., and was printed and bound by Book Press, Inc., a Quebecor America Book Group Company. The cover was printed by Lehigh Press, Inc.

Printed in the United States of America.

Earlier edition, entitled *The 68000 Microprocessor: Architecture, Programming, and Applications,* copyright © 1988 by Merrill Publishing Company.

Macmillan Publishing Company
866 Third Avenue
New York, NY 10022

Macmillan Publishing Company is part of the
Maxwell Communication Group of Companies.

Maxwell Macmillan Canada, Inc.
1200 Eglington Avenue East, Suite 200
Don Mills, Ontario M3C 3N1

Library of Congress Cataloging-in-Publication Data

Miller, Michael A.
The 68000 microprocessor family : architecture, programming, and applications / Michael A. Miller.
p. cm.
Rev. ed. of: The 68000 microprocessor.c1988.
Includes index.
ISBN 0-02-381560-4
1. Motorola 68000 (Microprocessor) I. Miller, Michael A. 68000 microprocessor. II. Title. III. Title: Sixty-eight thousand microprocessor family.
QA76.8.M6895M553 1992
004.16–dc20 91-21616
CIP

Printing: 1 2 3 4 5 6 7 8 9 Year: 2 3 4 5

This text is dedicated to Ann, who understands very little of its contents, but who has had the patience and fortitude to put up with me while I was creating it.

MERRILL'S INTERNATIONAL SERIES IN ENGINEERING TECHNOLOGY

ADAMSON	*Applied Pascal for Technology,* 0-675-20771-1
	The Complete Dictionary of Electronics, 0-02-300820-2
	Microcomputer Repair, 0-02-300825-3
	Structured BASIC Applied to Technology, 0-675-20772-X
	Structured C for Technology, 0-675-20993-5
	Structured C for Technology (w/disks), 0-675-21289-8
ANTONAKOS	*The 68000 Microprocessor: Hardware and Software Principles and Applications,* 0-675-21043-7
ASSER/ STIGLIANO/ BAHRENBURG	*Microcomputer Servicing: Practical Systems and Troubleshooting,* 0-675-20907-2
	Microcomputer Theory and Servicing, 0-675-20659-6
	Lab Manual to accompany Microcomputer Theory and Servicing, 0-675-21109-3
ASTON	*Principles of Biomedical Instrumentation and Measurement,* 0-675-20943-9
BATESON	*Introduction to Control System Technology, Third Edition,* 0-675-21010-0
BEACH/ JUSTICE	*DC/AC Circuit Essentials,* 0-675-20193-4
BERLIN	*Experiments in Electronic Devices to accompany Floyd's Electronic Devices and Electronic Devices: Electron Flow Version, Third Edition,* 0-02-308422-7
	The Illustrated Electronics Dictionary, 0-675-20451-8
BERLIN/ GETZ	*Experiments in Instrumentation and Measurement,* 0-675-20450-X
	Fundamentals of Operational Amplifiers and Linear Integrated Circuits, 0-675-21002-X
	Principles of Electronic Instrumentation and Measurement, 0-675-20449-6
BERUBE	*Electronic Devices and Circuits Using MICRO-CAP II,* 0-02-309160-6
BOGART	*Electronic Devices and Circuits, Second Edition,* 0-675-21150-6
BOGART/ BROWN	*Experiments in Electronic Devices and Circuits, Second Edition,* 0-675-21151-4
BOYLESTAD	*DC/AC: The Basics,* 0-675-20918-8
	Introductory Circuit Analysis, Sixth Edition, 0-675-21181-6

BOYLESTAD/ KOUSOUROU	*Experiments in Circuit Analysis, Sixth Edition,* 0-675-21182-4 *Experiments in DC/AC Basics,* 0-675-21131-X
BREY	*Microprocessors and Peripherals: Hardware, Software, Interfacing, and Applications, Second Edition,* 0-675-20884-X *The Intel Microprocessors—8086/8088, 80186, 80286, 80386, and 80486—Architecture, Programming, and Interfacing, Second Edition,* 0-675-21309-6
BROBERG	*Lab Manual to accompany Electronic Communication Techniques, Second Edition,* 0-675-21257-X
BUCHLA	*Digital Experiments: Emphasizing Systems and Design, Second Edition,* 0-675-21180-8 *Experiments in Electric Circuits Fundamentals, Second Edition,* 0-675-21409-2 *Experiments in Electronics Fundamentals: Circuits, Devices and Applications, Second Edition,* 0-675-21407-6
BUCHLA/ McLACHLAN	*Applied Electronic Instrumentation and Measurement,* 0-675-21162-X
CICCARELLI	*Circuit Modeling: Exercises and Software, Second Edition,* 0-675-21152-2
COOPER	*Introduction to VersaCAD,* 0-675-21164-6
COX	*Digital Experiments: Emphasizing Troubleshooting, Second Edition,* 0-675-21196-4
CROFT	*Getting a Job: Resume Writing, Job Application Letters, and Interview Strategies,* 0-675-20917-X
DAVIS	*Technical Mathematics,* 0-675-20338-4 *Technical Mathematics with Calculus,* 0-675-20965-X *Study Guide to Accompany Technical Mathematics,* 0-675-20966-8 *Study Guide to Accompany Technical Mathematics with Calculus,* 0-675-20964-1
DELKER	*Experiments in 8085 Microprocessor Programming and Interfacing,* 0-675-20663-4
FLOYD	*Digital Fundamentals, Fourth Edition,* 0-675-21217-0 *Electric Circuits Fundamentals, Second Edition,* 0-675-21408-4 *Electronic Devices, Third Edition,* 0-675-22170-6 *Electronic Devices: Electron Flow Version,* 0-02-338540-5 *Electronics Fundamentals: Circuits, Devices, and Applications, Second Edition,* 0-675-21310-X *Fundamentals of Linear Circuits,* 0-02-338481-6 *Principles of Electric Circuits, Electron Flow Version, Second Edition,* 0-675-21292-8 *Principles of Electric Circuits, Third Edition,* 0-675-21062-3
FULLER	*Robotics: Introduction, Programming, and Projects,* 0-675-21078-X
GAONKAR	*Microprocessor Architecture, Programming, and Applications with the 8085/8080A, Second Edition,* 0-675-20675-8 *The Z80 Microprocessor: Architecture, Interfacing, Programming, and Design,* 0-675-20540-9
GILLIES	*Instrumentation and Measurements for Electronic Technicians,* 0-675-20432-1
GOETSCH	*Industrial Supervision: In the Age of High Technology,* 0-675-22137-4
GOETSCH/ RICKMAN	*Computer-Aided Drafting with AutoCAD,* 0-675-20915-3
GOODY	*Programming and Interfacing the 8086/8088 Microprocessor,* 0-675-21312-6
HUBERT	*Electric Machines: Theory, Operation, Applications, Adjustment, and Control,* 0-675-21136-0
HUMPHRIES	*Motors and Controls,* 0-675-20235-3

HUTCHINS	*Introduction to Quality: Management, Assurance and Control,* 0-675-20896-3
KEOWN	*PSpice and Circuit Analysis,* 0-675-22135-8
KEYSER	*Materials Science in Engineering, Fourth Edition,* 0-675-20401-1
KIRKPATRICK	*The AutoCAD Book: Drawing, Modeling and Applications, Second Edition,* 0-675-22288-5
	Industrial Blueprint Reading and Sketching, 0-675-20617-0
KRAUT	*Fluid Mechanics for Technicians,* 0-675-21330-4
KULATHINAL	*Transform Analysis and Electronic Networks with Applications,* 0-675-20765-7
LAMIT/ LLOYD	*Drafting for Electronics,* 0-675-20200-0
LAMIT/ WAHLER/ HIGGINS	*Workbook in Drafting for Electronics,* 0-675-20417-8
LAMIT/ PAIGE	*Computer-Aided Design and Drafting,* 0-675-20475-5
LAVIANA	*Basic Computer Numerical Control Programming, Second Edition,* 0-675-21298-7
MACKENZIE	*The 8051 Microcontroller,* 0-02-373650-X
MARUGGI	*Technical Graphics: Electronics Worktext, Second Edition,* 0-675-21378-9
	The Technology of Drafting, 0-675-20762-2
	Workbook for the Technology of Drafting, 0-675-21234-0
McCALLA	*Digital Logic and Computer Design,* 0-675-21170-0
McINTYRE	*Study Guide to accompany Electronic Devices, Third Edition,* and *Electronic Devices, Electron Flow Version,* 0-02-379296-5
	Study Guide to accompany Electronics Fundamentals, Second Edition, 0-675-21406-8
MILLER	*The 68000 Microprocessor Family: Architecture, Programming, and Applications, Second Edition,* 0-02-381560-4
MONACO	*Essential Mathematics for Electronics Technicians,* 0-675-21172-7
	Introduction to Microwave Technology, 0-675-21030-5
	Laboratory Activities in Microwave Technology, 0-675-21031-3
	Preparing for the FCC General Radiotelephone Operator's License Examination, 0-675-21313-4
	Student Resource Manual to accompany Essential Mathematics for Electronics Technicians, 0-675-21173-5
MONSSEN	*PSpice with Circuit Analysis,* 0-675-21376-2
MOTT	*Applied Fluid Mechanics, Third Edition,* 0-675-21026-7
	Machine Elements in Mechanical Design, Second Edition, 0-675-22289-3
NASHELSKY/ BOYLESTAD	*BASIC Applied to Circuit Analysis,* 0-675-20161-6
PANARES	*A Handbook of English for Technical Students,* 0-675-20650-2
PFEIFFER	*Proposal Writing: The Art of Friendly Persuasion,* 0-675-20988-9
	Technical Writing: A Practical Approach, 0-675-21221-9
POND	*Introduction to Engineering Technology,* 0-675-21003-8
QUINN	*The 6800 Microprocessor,* 0-675-20515-8
REIS	*Digital Electronics Through Project Analysis,* 0-675-21141-7
	Electronic Project Design and Fabrication, Second Edition, 0-02-399230-1
	Laboratory Manual for Digital Electronics Through Project Analysis, 0-675-21254-5
ROLLE	*Thermodynamics and Heat Power, Third Edition,* 0-675-21016-X
ROSENBLATT/ FRIEDMAN	*Direct and Alternating Current Machinery, Second Edition,* 0-675-20160-8

ROZE *Technical Communication: The Practical Craft,* 0-675-20641-3

SCHOENBECK *Electronic Communications: Modulation and Transmission, Second Edition,* 0-675-21311-8

SCHWARTZ *Survey of Electronics, Third Edition,* 0-675-20162-4

SELL *Basic Technical Drawing,* 0-675-21001-1

SMITH *Statistical Process Control and Quality Improvement,* 0-675-21160-3

SORAK *Linear Integrated Circuits: Laboratory Experiments,* 0-675-20661-8

SPIEGEL/LIMBRUNNER *Applied Statics and Strength of Materials,* 0-675-21123-9

STANLEY, B.H. *Experiments in Electric Circuits, Third Edition,* 0-675-21088-7

STANLEY, W.D. *Operational Amplifiers with Linear Integrated Circuits, Second Edition,* 0-675-20660-X

SUBBARAO *16/32-Bit Microprocessors: 68000/68010/68020 Software, Hardware, and Design Applications,* 0-675-21119-0

TOCCI *Electronic Devices: Conventional Flow Version, Third Edition,* 0-675-20063-6

Fundamentals of Pulse and Digital Circuits, Third Edition, 0-675-20033-4

Introduction to Electric Circuit Analysis, Second Edition, 0-675-20002-4

TOCCI/OLIVER *Fundamentals of Electronic Devices, Fourth Edition,* 0-675-21259-6

WEBB *Programmable Logic Controllers: Principles and Applications, Second Edition,* 0-02-424970-X

WEBB/GRESHOCK *Industrial Control Electronics,* 0-675-20897-1

WEISMAN *Basic Technical Writing, Sixth Edition,* 0-675-21256-1

WOLANSKY/AKERS *Modern Hydraulics: The Basics at Work,* 0-675-20987-0

WOLF *Statics and Strength of Materials: A Parallel Approach,* 0-675-20622-7

PREFACE

This text presents concepts and applications for the 68000 microprocessor family. Because it is an upper-level text in digital and microprocessor technology, any student using this text should have background in the following areas of digital electronics:

- Number systems and binary codes
- Logic gates and Boolean algebra
- Registers, counters, and sequential logic
- Memories and memory systems
- Analog to/from digital conversion
- Basic microprocessor computer concepts

The student also should have knowledge of bipolar and field-effect transistors as well as basic circuit analysis.

Most chapters begin with objectives and a glossary of important terms used in that chapter. Throughout each chapter, examples are presented to clarify concepts and programs are developed to enhance the understanding and use of the 68000 and its associated circuitry. I have attempted to present a balanced approach to hardware and software concepts. Because of the nature of these concepts, however, some places in the text are more weighted toward one or the other of these areas.

Chapter 1 is an introduction to 16- and 32-bit microprocessor theory, Chapter 2 introduces the 68000 from a hardware perspective. A basic system is developed using the 68000 as the central processing unit (CPU). I use this system throughout the text primarily to illustrate the step-by-step requirements for designing such a system. This approach is different from a circuit analysis view of an *existing* system, and thus is a more versatile way to describe system design concepts.

The 68000-based system developed in the text is intended to be primarily a pedagogically helpful tool for the student to understand the design process rather than to be a practical piece of equipment.

Chapter 3 introduces the 68000 instruction set and processing capabilities. The system developed in Chapter 2 is used to present 68000 programming concepts.

Chapter 4 continues the discussion of programming concepts and includes timing diagrams and signal interrelationships.

Interfacing the 68000 to existing 8-bit parallel peripheral devices is discussed in Chapter 5. Motorola selected the 68000's control signals so that Motorola's 6800 family byte-size devices, in pairs, can provide 16-bit parallel buffering to the outside world. Although Motorola has developed some newer devices to be interfaced directly with the 68000, these devices are still 8-bit oriented and thus must be used in pairs. Chapter 6 explores serial interface devices used with the 68000. Both the 8-bit ACIA and 16-bit DUART are discussed.

In Chapter 7 Motorola's two primary training assemblers, TUTOR 1.3 and EXORmacs, are discussed. Exception processing and programs are covered in Chapter 8. Included in this discussion are interrupts, error routines, and software traps.

Chapter 9 ties together the first eight chapters and presents some application programs. Chapter 10 discusses two other processors in the 68000 family, the MC68008 and MC8010/12 microprocessors. Specific 16-bit support devices are described as well.

The MC68020, Motorola's 32-bit microprocessor, is introduced in Chapter 11. Additional concepts relating to this chip, such as on-board instruction cache and dynamic bus sizing, are explained. Chapter 12 covers coprocessors (using the 68000 family floating-point coprocessor, the MC68881) and modular programming support.

In this edition, two new chapters are added to cover material on Motorola's 32-bit processors, the MC68030 (Chapter 13) and the MC68040 (Chapter 14). Memory management concepts are expanded upon in the discussion of the 68030, and floating point operations are discussed in relation to the 68040.

The appendices contain the instruction set for the 68000, a set of timing diagrams for the 68000, the schematic of the system developed throughout the text, and a set of data sheets for the devices discussed in the text. The bibliography lists the references used in the book, as well as those on data communications and support devices.

Acknowledgments

I am grateful to the engineers at Motorola who designed the devices discussed within this text. Motorola's training staff provided the basic information and training needed for understanding the theories upon which these devices are based.

Finally, I would like to express my appreciation to the editors at Merrill, an imprint of Macmillan Publishing Company, for their guidance and belief in this project, and to the following reviewers, whose suggestions and comments proved valuable throughout the process of writing the first edition: Mike Bachelder, South Dakota School of Mines and Technology; Gary Boyington, Chemeketa Community College, Salem, Oregon; Barry Brey, DeVry Institute of Technology, Columbus, Ohio; Robert A. Feugate, Flagstaff, Arizona; Steven Friedman, High Tech Institute, Phoenix, Arizona; Frank Getz, Delaware Technical and Community College; Gene Jacot, Lansing Community College, Lansing, Michigan; Ed Kimble, ITT Technical Institute, Ft. Wayne, Indiana; Victor Michael, Williamsport Community College, Williamsport, Pennsylvania; Michael Naughton, Minneapolis Technical Institute; Larry Oliver, Columbus Technical Institute, Columbus, Ohio; John C. Skroder, Texas A & M University; Melvin Smith, Greenville Technical College, Greer, South Carolina; Joseph G. Tront, Virginia Polytechnic Institute; Larry Welles, ITT Technical Institute, Indianapolis, Indiana; and Roy A. Wilson, Monrovia, California. I also thank the reviewers of this second edition: Kenneth Dennis, San Jacinto College; Ed Serwon, Erie Community College; Tim Sheer, South Dakota School of Mines and Technology; Curtis Johnson, University of Houston—University Park; Dave Cummings, Metropolitan State College; and T. Damarla, University of Kentucky.

CONTENTS

1

FUNDAMENTALS OF 16-BIT MICROPROCESSOR-BASED COMPUTERS

CHAPTER OBJECTIVES

Chapter 1 is an overall review of microprocessor principles based on a general model of a 16-bit processor. The glossary defines important terms in this chapter. Terminology used in this chapter reflects usage by the staff of Motorola, Inc., who designed and produced the processors discussed in this text. Architecture concepts are then discussed, followed by specific software applications using a microprocessor-based system.

1.1 GLOSSARY

address Binary value of a memory location where data is stored.
address mode How the location of data used by an instruction is determined.
assembler program Software that translates source code to object code.
buffer Device used to isolate one circuit from another with the purpose of providing increased drive to the load circuit.
bus Group of wires that carry similar types of signals.
counter Electronic circuit that counts numerically.
CPU Central Processing Unit. Section of the computer that performs data manipulation and arithmetic and logic operations.
execute To perform an operation as directed by an instruction.
flag Status indicator.
input/output (I/O) devices Peripheral devices such as keyboards, displays, and disk drives, used to input to or take data from a computer.
interface Buffer between a system's buses and the devices serviced by them.
interrupt Hardware signal that informs the processor that a user, through some peripheral device, wants to stop the current program and begin execution of a different program.

label Name, in a program, associated with a specific memory address.
machine code Binary ones and zeroes interpreted by a computer as instructions and data.
microprocessor Single integrated circuit (IC) central processing unit.
multiple-precision arithmetic Addition or subtraction of numbers larger than system's basic data size.
multiplexing Timesharing of a common function or bus.
opcode Operation portion of an instruction.
operand Information portion of an instruction.
peripheral devices External devices used with a computer, such as printers and tape drives.
pointer Register that directs data to or from a memory location.
port Data entry or exit point for an I/O device.
program Orderly list of instructions that are executed by a computer.
pseudo op Assembler directive that assigns labels and/or data to a memory location.
RAM Random-access read/write memory.
register Set of data latches that temporarily store information.
read-only memory (ROM) Permanent memory that holds data and/or instructions.
subroutine Short program integrated into a larger program.
syntax Formatting rules for instructions and operands.

1.2 MICROPROCESSOR-BASED COMPUTERS

Today, many people are familiar with personal computers and the fact that **microprocessors** are included in them. An operator can perform many procedures with these computers simply by following the instructions presented on the screen, which is as much as many people will ever need or want to know about a computer. The rest of us, through interest or technical need, desire to know more about how the electronics perform the computing processes.

The microprocessor-based computer works by interpreting a set of instructions (**opcodes**) designed specifically for the central processing unit (**CPU**). These opcodes compose the instruction set of the microprocessor. A *process* is performed when the system **executes** a series of consecutive instructions, called a **program.** Instructions must be presented to the processor in the correct order and format for the program to be run successfully. In general, that format is the opcode followed by an **operand,** if one is required. The opcode defines the instruction or process which the computer is to do. The operand defines the data or location of the data for the process to use when executing the instruction. To understand how the microprocessor performs, we first need to take a look at the hardware (architecture) that makes up a basic computer system.

The Architecture: What's in It?

The architecture of a computer is the physical makeup of the device or, in essence, what's inside the device. Three basic blocks compose a computing system. These blocks, illustrated in Figure 1.1, are (1) the central processing unit, also known as the microprocessing unit (MPU); (2) the memory; and (3) input/output (I/O) interfacing to peripheral devices, such as keyboards, printers and monitors.

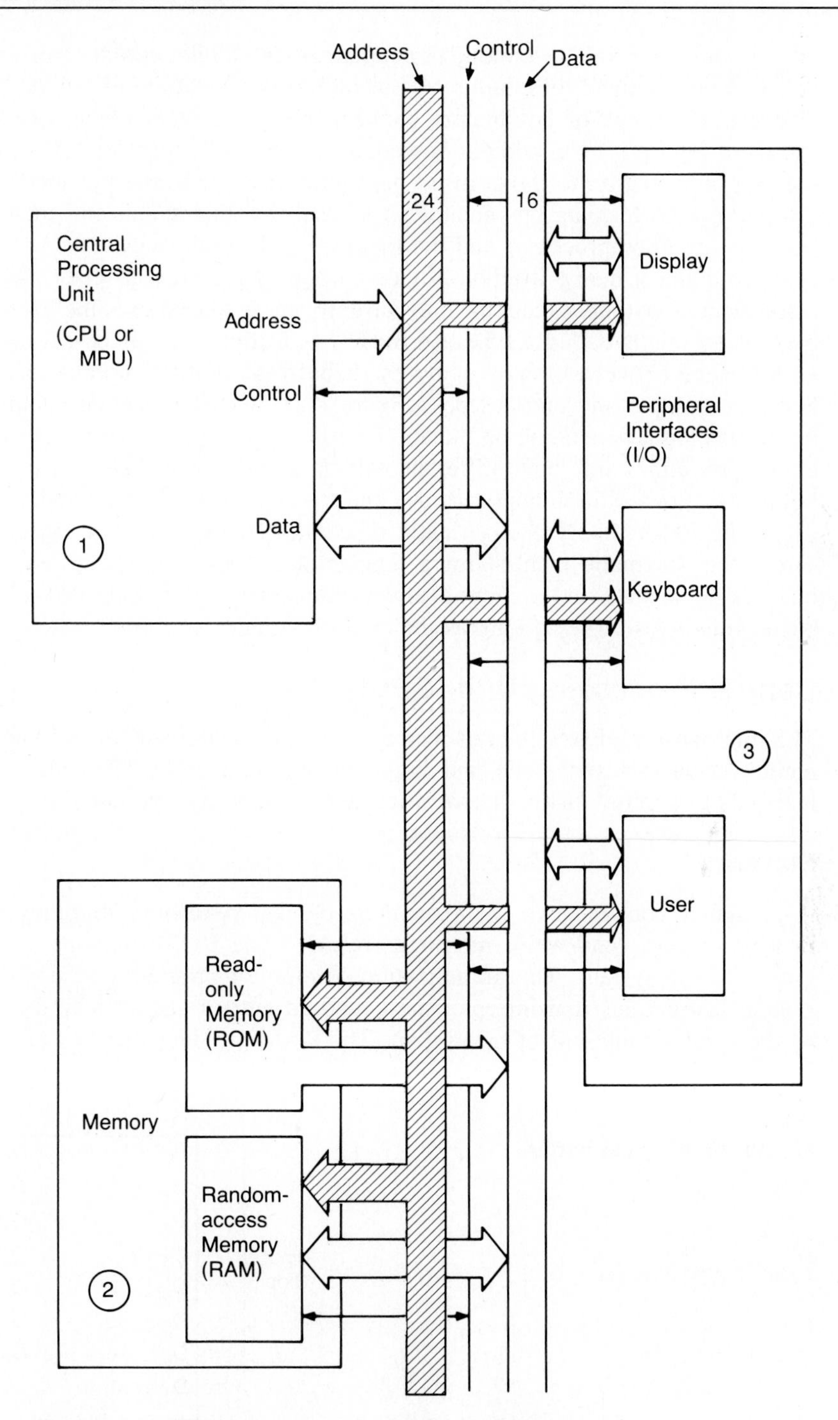

Figure 1.1 Block diagram of a basic computer

Groups of wires or conductors called **buses** handle similar information or signals and interconnect these units. A bus allows data containing several bits to travel along these groups of conductors between many different devices or circuits. As shown in Figure 1.1, an address bus composed of 24 conductors allows memory addressing information to move from the microprocessor to the memory or I/O chips. Similarly, a 16-bit data bus creates an electrical path for data information to move between the microprocessor and the memory and I/O devices.

To connect many different device outputs to a common set of lines, one must disconnect all outputs except the one currently being placed onto the lines. The device used to do this is called a **tri-state buffer.** A **buffer** is an electrical circuit used as an **interface** between circuits or buses. It buffers the differing electrical conditions between its input and output sides (Figure 1.2). A buffer may be latching (holding) or non-latching, depending on the application. Specifically, a buffer is used when the circuits on either side have different voltage or current requirements for temporary holding and/or isolation applications. The tri-state buffer has the additional capacity of having its output side electrically disconnected by a control signal. Figure 1.3 shows the schematic symbol for a regular and a tri-state buffer. With the regular buffer, a logic state on the input is always transferred to the output. With the tri-state buffer this is also true as long as the control signal is in the on state.

Central Processing Unit (CPU)

The microprocessor or CPU has many functions, including program and data movement control and arithmetic and logic computations. The CPU also supplies and responds to control signals between it and the other system units.

Memory

A computer contains two main memory blocks: **read-only memory (ROM)** and **random-access read/write memory (RAM)**. The ROM contains a program that initializes the system. This initialization or monitor program contains the operating system instructions that interpret the keyboard entries and display data as required by the various functions of the system. The RAM provides memory area for the user

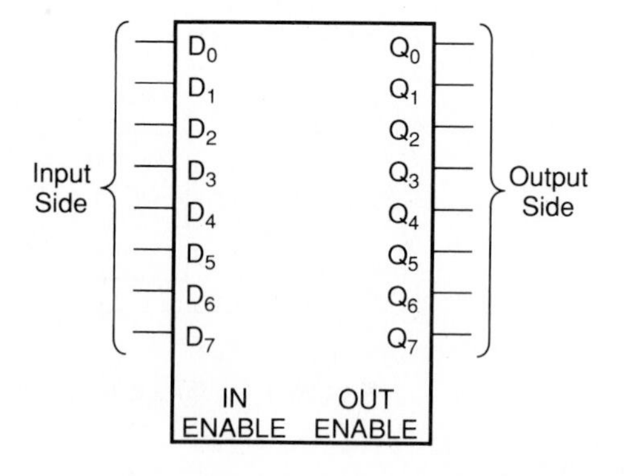

Figure 1.2 **Typical buffer**

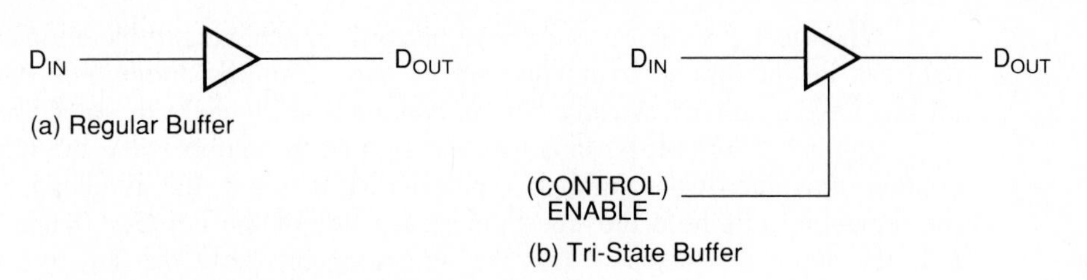

Figure 1.3 **Schematic symbols for buffers**

to enter data and program system instructions, as well as a general scratch-pad area for the initialization program stored in ROM.

Peripheral Interfaces (I/O)

The peripheral interface area contains buffers between the computer buses and the system I/O and **peripheral devices.** These devices allow the human to interface with and use the computer.

1.3 A 16-BIT MICROPROCESSOR

System Data

To use any system, you must know some important characteristics. For any microprocessor, they include bus sizes, the instruction set, and the specific architecture of the system. Most 16-bit microprocessors are so designated because their external data buses are 16 bits (two bytes) wide. That is, a data size presented by the processor to the remaining circuitry contains 16 binary bits of information. **Address** bus sizes vary from 16 to 32 bits wide depending on the specific device. Figure 1.4 is an illustration of a generic 16-bit microprocessor with a 16-bit data bus and a 24-bit address bus. This allows the microprocessor to directly access 2^{24} or 16 megabytes of memory using the 24-bit address bus.

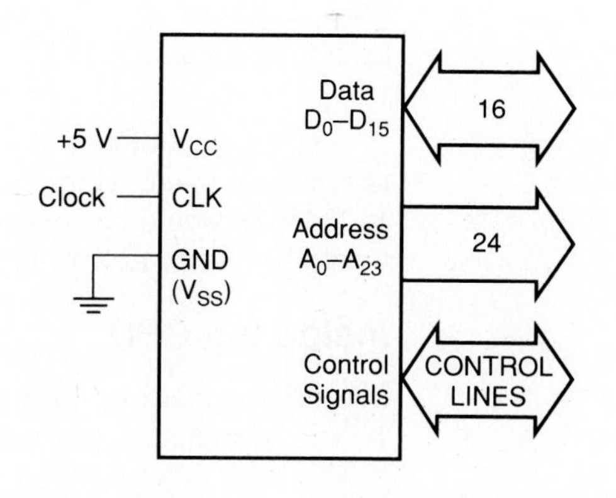

Figure 1.4 **Generic 16-bit microprocessor**

When using a microprocessor to perform mathematical operations, the type of data can be interpreted as having one of two meanings: *magnitude only* (no sign: all bits have a numerical value) or *2's complement signed*. For signed numbers, the *most significant bit* (MSB) indicates the sign of the number and the remaining bits contain the value of the number. When the MSB is a 0, the number is positive and the remaining bits hold the true numerical value of the number. When the MSB is a 1, the number is negative and the remaining bits hold the negative value in 2's complement form. To obtain the actual numerical value, you must manually convert the data to its positive equivalent by using a 2's complement process or by using additional programming following the arithmetic that produced the negative result. This process also can be performed by the computer if it is programmed to do the operation. Given these restrictions, the range of numerical values for a 16-bit data word are

Magnitude Only:	\$0000–\$FFFF or 0 to 65,535 base ten
Positive Numbers (MSB = 0):	\$0000–\$7FFF or 0 to 32,767
Negative Numbers (MSB = 1):	\$FFFF–\$8000 or −1 to −32,768

A dollar sign (\$) preceding a number is used by Motorola to indicate that the number is a hexadecimal number. Other microprocessor manufacturers, such as Intel and Zilog, use a trailing H to designate a hexadecimal (HEX) value. Documentation by those companies would show \$7FFF as 7FFFH.

EXAMPLE 1.1

After a completed addition process, the result of the operation produced \$B300. What are the possible HEX and decimal equivalent values that this data can represent?

Solution:

Considering the data as an unsigned, magnitude-only number, the HEX value is \$B300 and its decimal equivalent is 45,824. Treating the number as a signed number, note that its value is negative (MSB = 1). The number may be taken as shown: that is, \$B300, or complemented and presented as −\$4D00, which is −19,712 in decimal. The actual interpretation of the number is up to the user and depends on the application.

Throughout this text, as numbers get larger, a four-digit convention will be used to represent them. This is done to make the numbers easier to read. As an example, \$12345678 would be written as \$1234 5678 and the binary value 100111010101110 would be shown as 100 1110 1010 1110.

Inside the CPU

The CPU is made up of many sections. Using the functional block diagram (Figure 1.5) as a reference, we will explore these sections. Within the block diagram is a

Figure 1.5 Functional block diagram

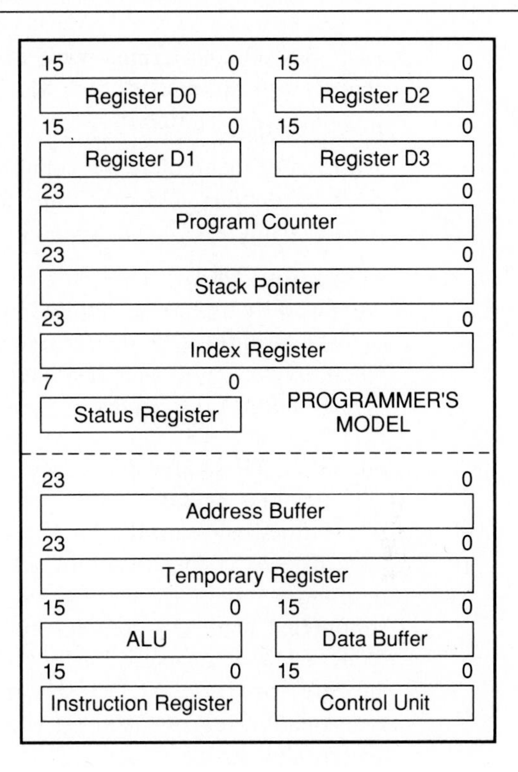

section called the programmer's model; it is a functional block diagram of the device from a user's (programmer's) viewpoint.

Program Counter (PC) The 24-bit program counter is a pre-settable binary up counter used to point to memory locations where the program is stored. As an instruction is executed (acted upon), the program counter is incremented so that it points to the next instruction to be fetched (retrieved) from memory.

Address Buffer The address buffer is a 24-bit parallel load buffer register that temporarily holds memory address information to be applied to the address bus to access a memory location. The system we will explore in this chapter uses a one address per one byte memory scheme. That is, even though the data bus is 2 bytes wide, each byte of data in memory has its own address. When accessing data from location $00 2000, for instance, the processor is actually fetching a byte from location $00 2000 and a byte from location $00 2001. The high byte of the data is stored at the even address ($00 2000) and the low byte at the odd address ($00 2001). This convention is used to facilitate byte access in a 16-bit system. Again, please note that the orientation of high and low byte parts of the 16-bit data is Motorola's and that both Intel and Zilog use a reverse convention. That is, the low byte of a 16-bit word is stored at the lower address ($00 2000) and the high byte at the higher address ($00 2001). This book will retain the Motorola convention throughout. This chapter will not discuss byte access, but will reserve that as part of Chapter 2.

Index Register This 24-bit register is used as a memory **pointer.** Instructions are available to manipulate the contents of this register to redirect the pointer to different areas in memory. Data then are accessed from the memory pointed to by the index register using specific index address mode instructions.

Computing Units The computing side of the CPU contains the *arithmetic logic unit* (ALU) with inputs supplied by *data* **registers** D0–D3 and the *temporary* (TEMP) register. Additionally, data registers D0–D3 perform the function of an *accumulator* when they are used as a destination register. An accumulator has the task of holding (accumulating) the results of an arithmetic operation.

Registers D0–D3 receive data as directed by a microprocessor instruction or operation. These operations may cause information to be transferred between registers and memory, between the registers themselves or cause the data in the registers to be modified and returned to a destination register or memory location.

The ALU does the real work of arithmetic computation and logic manipulations required by the instruction being executed. It performs 16-bit size addition and subtraction using 2's complement addition. Logic manipulations (AND, OR, NOT, XOR, etc.) are enacted, bit-by-bit, on the 16-bit data word. For example, all 16 bits are affected by an AND instruction with the least significant bit (LSB), or b0, of a destination register ANDed with the b0 of the source, b1 ANDed with b1 of the source, and so forth.

Status Register The real strength of any computer is its ability to make decisions based on the results of an operation. To do this, a microprocessor reacts to these results as monitored by the **flags** of the status register. Figure 1.6 shows a status register containing six flag indicators used by the CPU: X, I, N, Z, V, and C, which are described below.

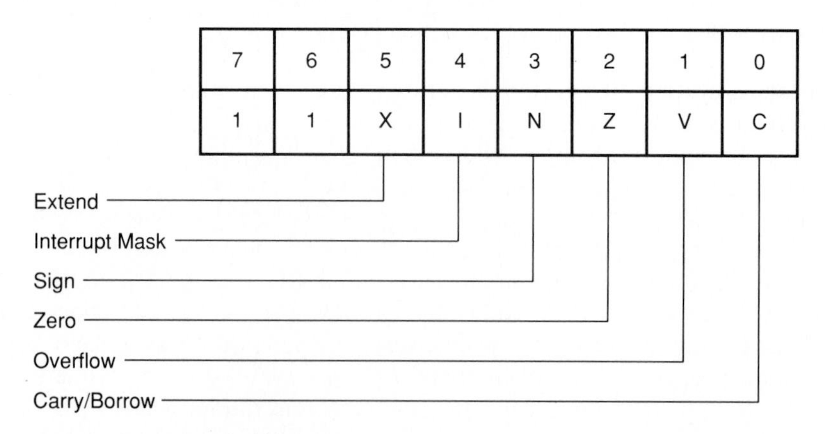

Figure 1.6 **Status register**

Interrupt Flag (I): The operation of a program may be interrupted by the application of an interrupt request by an external source to the interrupt request (IRQ) control pin of the microprocessor. To prevent the interruption from occurring in response to this signal, a flag in the status register, called the *interrupt mask* or I bit, is set. This flag is set by a set I (SEI) instruction and cleared (or reset) to allow interrupt handling by a clear I (CLI) instruction. Interrupt processes are dealt with in detail in later chapters.

Sign Flag (N): The N flag monitors the MSB of the 16-bit data word. If this bit is high (negative number), the N flag is set. If low, the N flag is reset.

Zero Flag (Z): If the result of an operation leaves $0000 in the destination, the Z flag is set. Otherwise, it is reset.

Overflow (V): The overflow flag is set if the result of an operation yields a *signed* number which is too large for the system to handle, that is, a number greater than +32,767 or −32,768. A second way to treat this flag, for arithmetic operations, is to consider it as a sign-error flag for signed number operations. This error occurs when two positive numbers (MSB low) are added directly or by 2's complement addition (for subtraction) and the result is negative (MSB high) or when two negative numbers (MSB high) are summed and the results are positive (MSB low). The actual operation of this flag will become clearer with the examples following the status flag discussion.

Carry/Borrow (C) and Extend (X): These flags are set if a carry was generated by *unsigned* addition or if a borrow was required for an *unsigned* subtraction. The C flag is maintained as an indication of the carry/borrow state while the X flag is used for **multiple-precision arithmetic.** Multiple-precision arithmetic operations allow numbers larger than 16 bits to be summed and subtracted.

Examples illustrate how the flags react to various CPU operations. Assume that the I flag is reset for these examples. An add operation sets or resets each of the remaining flags according to the result of the sum.

EXAMPLE 1.2

In this example, add $3500 to $2700. In binary, $3500 is 0011 0101 0000 0000 and $2700 is 0010 0111 0000 0000. What are the states of the flags after the addition?

Solution:

$$\begin{array}{rcr} 0011\ 0101\ 0000\ 0000 & = & 3500 \\ +\ 0010\ 0111\ 0000\ 0000 & = & +\ 2700 \\ \hline 0101\ 1100\ 0000\ 0000 & = & 5C00 \end{array}$$

As a result of the operation, no carry occurred, so the C and X flags are reset. The answer is positive (the MSB of the data is a logic 0), making the N bit low. The results are not $0000 0000, so the Z flag is reset. For the overflow (V) flag, we need to examine the sign of the numbers which were summed and the result. If an error occurred, the V flag is set. In our case, both added numbers are positive and the result is positive, which is consistent with the problem. The V flag will be reset. The actual interpretation of the overflow result is that the addition result ($5C00) is within the signed number limits of the computer's data word size (here, that limit is $7FFF). The flags of the status register, as shown previously in Figure 1.6, are organized as 11XINZVC. For this example, by substituting the states of the flags into their proper place, the status register contains 1100 0000 or $C0.

EXAMPLE 1.3

For this example, determine the contents of the status register when adding $8000 to $8000, which produces a unique set of results.

Solution:

$8000 is 1000 0000 0000 0000 in binary and the sum is

```
  1000 0000 0000 0000  =    8000
+ 1000 0000 0000 0000  =  + 8000
 10000 0000 0000 0000  =    0000 with a carry.
```

The 16-bit result is $0000, causing the Z bit to be set reflecting the zero-value result. A carry was generated, setting the C and X flags. Zero is a positive number (MSB = 0), making N = 0. Now comes the overflow bit. First, recognize that $8000 is a negative number (MSB = 1); thus two negative numbers are being summed. We expect a negative result, but instead we get a positive one, causing the V bit to be set. Why was there an incorrect result? Essentially, the sum produced a value exceeding the signed value limit of the system. Here the decimal equivalent sum of −32,768 and −32,768 is −65,536 which exceeds the negative value limit of −32,768.

Note that when considering the numbers as unsigned, the most significant bit of the result does not fit into the 16-bit data size. This bit has a dual meaning: first, as a carry bit, which is an indication of an unsigned overflow and, second, as the unsigned numerical value, 131,072. The status register contains $D7 for this example.

EXAMPLE 1.4

Now try one more addition example.

$6B13 + $7C2C

Solution:

$6B13 + $7C2C = $E73F. The result is negative (N = 1), not $0000 (Z = 0), and no carry is generated (C and X = 0). Two positive numbers were summed and the result was negative, meaning that we exceeded the positive limit ($7FFF) of the system, causing V = 1. The contents of the status register for this example are $CA.

For subtraction, the system performs 2's complement addition and the flags are affected as with regular addition, except for the carry and extend flags. The C and X flags indicate a borrow, which is an active low carry in 2's complement addition. This is verified by example.

EXAMPLE 1.5

What are the contents of the status register as a result of subtracting $8C00 from $3700?

Solution:

Subtraction				2's Complement Addition		
0011 0111 0000 0000	=	3700	=	0011 0111 0000 0000	=	3700
− 1000 1100 0000 0000	=	− 8C00	=	+ 0111 0100 0000 0000	=	+ 7400
1010 1011 0000 0000	=	AB00	=	1010 1011 0000 0000		AB00

First, notice that a borrow is required to subtract the larger $8C00 from $3700. The subtraction is actually performed by taking the 2's complement of the subtrahend ($8C00) and adding it to the principal value ($3700). During the addition, no carry is generated. This relates to the need for a borrow indication to show that the borrow is an active low carry when performing subtraction by 2's complement addition. The C and X flags are set to indicate the borrow condition. Next, the results are negative (N = 1) and non-zero (Z = 0). The overflow is determined as before, by examining the sign conditions of the sum and result. We are adding two positive numbers ($3700 and $7400—the 2's complement of $8C00) and getting a negative result, causing the V flag to be set. The status register contents are $DB as a result of this subtraction.

Another way to look at subtraction, for the overflow bit, is a bit more confusing, but no less true. Subtracting a negative number from a positive number produces a positive answer. Try some decimal examples to feel confident of this. Since our example produced a negative answer, the overflow bit is set.

Keep in mind that the computer makes no distinction between types of data. As far as the computer is concerned, it reacts to the conditions of the data bits as a result of an operation, leaving the interpretation and use of the flag indicators up to the discretion of the user. For you, the user, to make use of these flag conditions during the decision-making process of programming, you must clearly understand the data types as applied to the flags.

The sign flag (N), for instance, has no meaning if we treat the data as unsigned numbers. The N flag indicates if the number is negative or positive, requiring the MSB of the data to be interpreted as a sign bit and have no numerical value. Similarly, the overflow (V) flag is an indication of sign error or oversized signed number results. It, too, has no meaning for unsigned numbers.

In contrast, the carry (C) and extend (X) flags indicate when a carry or borrow is generated from the sum or addition of numbers in which all 16 bits have numerical value. The zero flag (Z) indicates when the result is 0 regardless of the type of data used.

Instruction Register and Control Unit Program instructions are coded into groups of binary bits called operation codes: opcodes, for short. Opcodes are fetched from memory as pointed to by the program counter and loaded into the instruction register of the microprocessor. They are decoded and sent to a control unit which interprets the decoded opcodes, combines them with timing signals, and produces the necessary control signals to operate the system. These control signals enable register outputs onto the buses, cause registers to be loaded from the buses, cause the program counter to be incremented, and cause the ALU to perform the operations required by the decoded instruction.

Stack Pointer Register The stack pointer register is a second 24-bit register. It points to an area in memory called the *stack*. The stack saves the contents of the program counter and/or user registers whenever a **subroutine** or interrupt occurs. A subroutine is a short program that usually contains a specific task or set of tasks to be performed at a recurring period within a program. The subroutine is called and executed by the computer, after which program execution is returned to the main program at the point after the subroutine was called. When a subroutine is called, the contents of the program counter are sent to the stack. Table 1.1 illustrates the events that occur during a subroutine call.

The subroutine call instruction (A) is pointed to by the program counter and fetched into the CPU. The program counter, in turn, now points to the next instruction

Table 1.1 Subroutine Events

Memory Location	Example Data
Location A	Subroutine Call Instruction
Location B	Next Regular Instruction
Location C	First Subroutine Instruction
Location D	Return from Subroutine

Program Counter		Activity	Stack
Location A	←	Fetch Subroutine CALL	Empty
Location B	→	Push onto Stack	Location B
Location C ⋮ Location D	←	DO Subroutine Return from Subroutine	↓
Location B	←	Pull from Stack	Location B
		Resume Normal Program	Empty

(B); the program counter contents are stored (pushed) at the location pointed to by the stack pointer register. The stack pointer is decremented so that it always points to the next available location on the stack. The address of the first instruction (C) of the subroutine is then loaded into the program counter and execution of the subroutine program begins. The subroutine program must end with a return from subroutine (RTS) instruction. The only process an RTS performs is to recover the program counter information from the stack.

The stack pointer is first incremented; then, the program counter is loaded from the location pointed to by the stack pointer. Notice that this is the same location where the original program contents were saved as a result of the subroutine call. This operation is called a *pull* or *pop* of the stack contents. The difference in terminology arises from several manufacturers' preferences. Motorola, the manufacturer of the 68000 family of 16/32-bit microprocessors, selected the term *pull,* while Intel, which makes the 8086 line of 16/32-bit processors, and Zilog, which produces the Z8000, chose the term *pop* to describe the extracting of data from the stack. This text, based on Motorola's devices, uses the term *pull*.

Since the program counter is a 3-byte register, the actual push and pull operations involve two memory stack locations. During the push, the program counter's two low bytes are saved first. The stack pointer is decremented by two for both bytes and then the third byte of the program counter is pushed onto the stack. Because we are dealing with a 16-bit data bus, the upper byte during this push is $00. During the pull sequence for the program counter, the first location is pulled and the upper byte of the program counter is transferred. The upper byte of the data bus is ignored by the processor. The stack pointer is incremented by two and the lower two bytes are stored into the program counter. Figure 1.7 illustrates the process of pushing and pulling the program counter.

An **interrupt** is a process by which a currently running program is stopped or interrupted by a request from an external source. A new program is begun and is terminated by a *return from interrupt* (RTI) instruction. The RTI returns the CPU operation to the original program at the point where it was interrupted. This requires

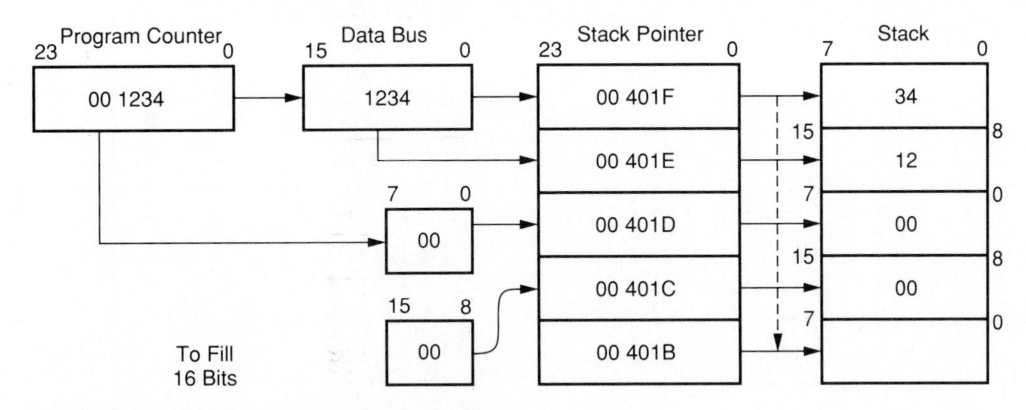

Figure 1.7 Pushing the PC onto the stack

the saving of the original register contents at the point of interruption and the restoration of this information when the main program execution resumes. Program counter information is pushed and pulled automatically by the interrupt and RTI processes. The remaining registers of the CPU must be pushed onto the stack and pulled from the stack using specific *push* and *pull* instructions.

1.4 CONTROL SIGNALS

Figure 1.8 shows a typical pin out diagram for a microprocessor chip, indicating its control signal and bus complement. The 16-bit data bus and 24-bit address bus convey data and address information between the processor and other system components.

Data Lines (D_0–D_{15}) and Address Lines (A_0–A_{23})

The data bus is bi-directional in that data can be both read and written by the processor. The address bus, in contrast, is unidirectional since address information

Figure 1.8 **Pin out diagram**

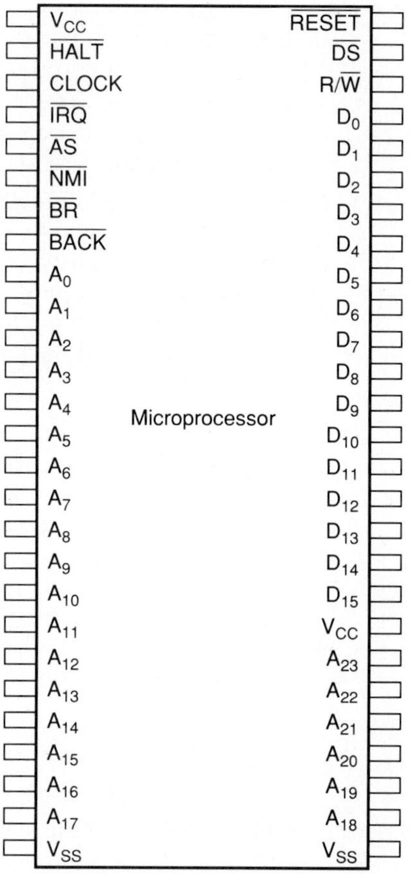

Microprocessor (left pins)	Microprocessor (right pins)
V_{CC}	$\overline{RESET}$
$\overline{HALT}$	$\overline{DS}$
CLOCK	$R/\overline{W}$
$\overline{IRQ}$	D_0
$\overline{AS}$	D_1
$\overline{NMI}$	D_2
$\overline{BR}$	D_3
$\overline{BACK}$	D_4
A_0	D_5
A_1	D_6
A_2	D_7
A_3	D_8
A_4	D_9
A_5	D_{10}
A_6	D_{11}
A_7	D_{12}
A_8	D_{13}
A_9	D_{14}
A_{10}	D_{15}
A_{11}	V_{CC}
A_{12}	A_{23}
A_{13}	A_{22}
A_{14}	A_{21}
A_{15}	A_{20}
A_{16}	A_{19}
A_{17}	A_{18}
V_{SS}	V_{SS}

is supplied by the processor to external devices. The remaining leads handle timing and control of data and address movement.

$\overline{\text{RESET}}$ (Input)

Throughout this text, an active low signal is indicated by an overbar above the signal designation, such as $\overline{\text{RESET}}$. This is an active low reset requiring the line to be brought to a logic 0 to cause the reset operation to take place. While the $\overline{\text{RESET}}$ line is held active, the address and data bus, as well as many of the control signals, is placed into a high-impedance (tri-state) condition. Internally, the interrupt mask (I) of the status register is reset to prevent interrupt requests from being serviced and the address buffers are cleared to $00 0000. Once the $\overline{\text{RESET}}$ signal is removed, the contents of the address buffers are placed on the address bus and then the contents of location $00 0000 are fetched and loaded into the program counter. The processor performs an opcode fetch and begins to process the initialization program which is stored at the location pointed to by the program counter. This method of retrieving program counter information is called *vectoring*. Location $00 0000 is called a vector address, while the contents of that location form a vectored address which points to the first instruction of a program (in this case, the initialization program).

The initialization program is written to initialize peripheral devices, such as keyboards and monitors, so that they can be used by the user. A logo or message is usually displayed to inform the user that the system is ready for use. This program resides in (permanent) read-only memory (ROM) address space so that each time the system is reset, the initialization program will be there to be fetched and executed. Since this generic processor vectors to $00 0000 upon a reset, ROM devices are located in the lower portion of the memory space (or map) of systems using this processor.

Address Strobe ($\overline{\text{AS}}$) (Output)

Address strobe ($\overline{\text{AS}}$) is an output signal which the microprocessor uses to tell other devices, such as memory or interfaces, that the address information on the address bus is valid and true.

Read/Write (R/$\overline{\text{W}}$) (Output)

R/$\overline{\text{W}}$ deciphers data transfer direction between the microprocessor and RAM or peripheral interfaces. A high on R/$\overline{\text{W}}$ indicates that the microprocessor is reading data from memory or an I/O device. A low allows the microprocessor to store data in memory or the I/O device.

$\overline{\text{HALT}}$ (Input)

The $\overline{\text{HALT}}$ pin is used as a slow memory wait signal. If a device requires more than normal cycle time to digest data or a command from the microprocessor, it can hold this line active until it is ready to let the microprocessor continue. Until it is removed, this signal effectively suspends microprocessor sequencing. No internal changes to the microprocessor take place while the $\overline{\text{HALT}}$ is low.

Interrupt Request ($\overline{IRQ}$) (Input)

Interrupt request ($\overline{IRQ}$) is an input signal from a peripheral device requesting an interruption of the current program to run a different one. The device sets the $\overline{IRQ}$ line low. If the I bit in the status register is set, the microprocessor completes its current instruction and initiates the interrupt process by pushing the program counter contents onto the stack and fetching new program counter contents from vector location $00 0004. This is done in a similar manner as the $\overline{RESET}$ vectoring. At the end of the interrupting program an RTI instruction restores the old program counter contents and the original program is resumed from the point at which it was interrupted. If the I bit is reset upon receipt of an active $\overline{IRQ}$, the interrupt process is not initiated and the processor continues on with its regular program.

Non-Maskable Interrupt Request ($\overline{NMI}$) (Input)

The non-maskable interrupt ($\overline{NMI}$) request is identical to the $\overline{IRQ}$ except that the state of the status register I bit is ignored. If the $\overline{NMI}$ line is made active, the system begins the interrupt process following the completion of the current instruction regardless of the state of the I bit. The vector location for the $\overline{NMI}$ is $00 0008.

Clock Signal (CLK) (Input)

The clock signal is used internally to synchronize all microprocessor activity. Typical clock frequencies for the 16/32-bit processors range from 4 MHz to 25 MHz.

1.5 INPUT/OUTPUT STRUCTURE

For the Motorola line of microprocessors, there are no separate input/output instructions or structures. That is, input/output peripheral interface devices are treated in the same manner as memory devices. They are addressed as well as written to and from as if they were an extension of the memory area. In contrast, Intel and Zilog microprocessors handle transfers of data to and from peripheral devices using IN and OUT instructions along with specific port addressing separate from memory address space. For example, the Intel 8086/8 has a 20-bit memory address bus which allows that processor to access up to 1 Mbyte memory locations. Additionally, when executing IN and OUT instructions, the 8086/8 uses 16 bits to define an I/O (**port**) address. Thus the 8086/8 can access up to 64K ports and 1 Mbyte of memory in one system. An additional signal, M/$\overline{IO}$, is used as part of the address decoding scheme to separate memory and I/O space. This signal is brought low when executing IN and OUT instructions and remains high when accessing memory.

Since Motorola does not provide a separate I/O structure, our generic microprocessor will follow that philosophy. I/O interfaces are accessed in the same way as memory locations and therefore will occupy some space in the memory addressing map. The advantage of this is that the entire instruction set can be used with I/O devices, like keyboard and monitor interfaces, in the same manner as any other memory location. For Intel and Zilog I/O, only IN and OUT instructions directly affect I/O devices. However, if a system designer preferred to connect an I/O as a memory access device in an Intel or Zilog system, it could be done. But many of the I/O interfaces designed to operate in these environments are made to respond specifically to the $M/\overline{IO}$ signal and smaller address size. A drawback to memory-mapping the I/O devices is that when an interface is set to occupy an address, it reduces the available address space for memory devices. For instance, a keyboard interface which uses the addresses $21 0000 and $21 0002 would prevent a memory chip from operating on that page (usually from $21 0000 to $21 FFFF) unless there is some added circuitry to defeat access to the chip when the keyboard is being accessed. With a processor like the one used in this chapter, the loss of a page of memory out of 16 Mbytes is not necessarily worrisome, especially if all existing I/O devices are mapped into that page.

Light-emitting Diode (LED) Display

Two essential peripheral devices are the keyboard, for data entry, and a display to indicate the current status of the system or the results of the program. To demonstrate the principles behind operating these peripherals, we will use a simplistic display and keyboard. A basic display is the 7-segment light-emitting diode (LED) unit illustrated in Figure 1.9. This diagram shows the LED orientation and segment designations

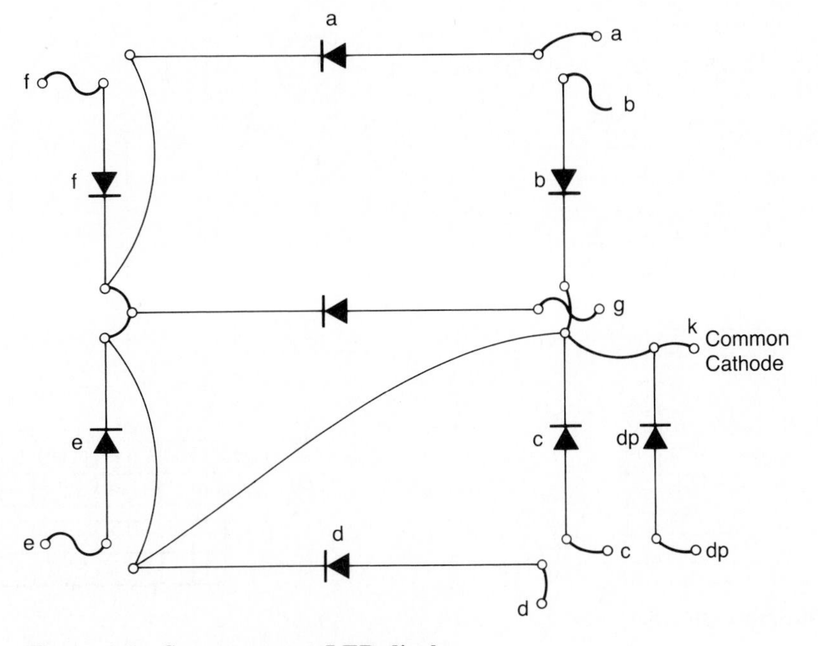

Figure 1.9 Seven-segment LED display

from a to g and a decimal point (dp). Each individual segment is illuminated by placing a positive potential on the anode of that segment and 0 volts on the common cathode of the unit.

Figure 1.10 is a schematic diagram of a microprocessor-driven display system (indicated by the data and port address lines). The 74173 chip is a quad-data flip-flop latch that buffers and holds data from the microprocessor and delivers it to the display system. The data presented to the flip-flops are latched on a negative transition on the input disable lines of the 74173s. These active high-input disables (low enables) are driven by address decoders that output a 0 in response to address

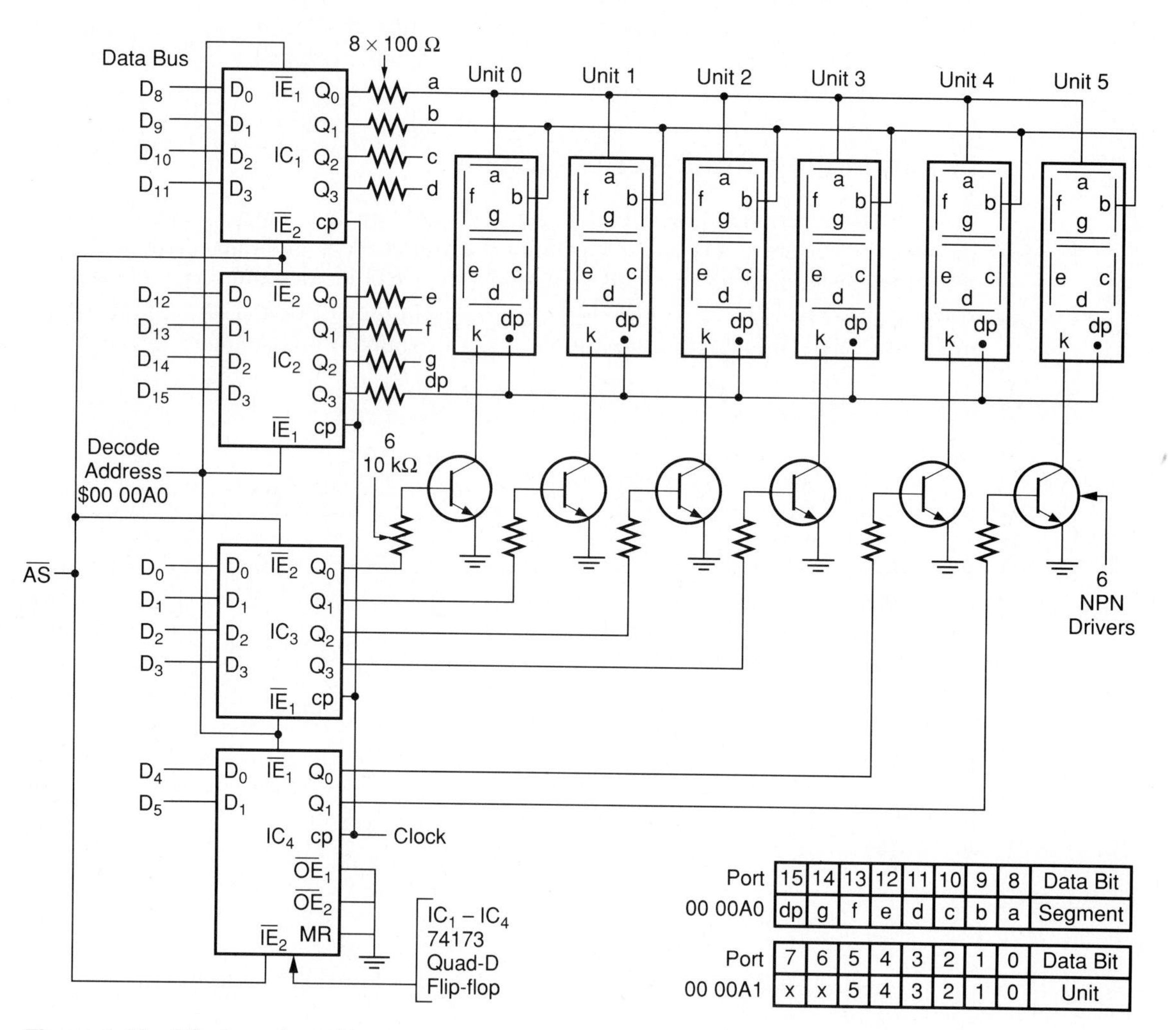

Port	15	14	13	12	11	10	9	8	Data Bit
00 00A0	dp	g	f	e	d	c	b	a	Segment

Port	7	6	5	4	3	2	1	0	Data Bit
00 00A1	x	x	5	4	3	2	1	0	Unit

Figure 1.10 Display schematic

$20 00A0 from the microprocessor. The upper byte on the data bus contains the segment codes needed to place a positive (or logic 1) on selected segment anodes while the lower byte contains the unit code which selects one of six units to receive a low on its common cathode. Data are transferred from the microprocessor by MOVE instructions to port address $20 00A0. Bits which are set high in the upper byte send this positive voltage through a current-limiting resistor to the selected segment. Similarly, a high bit in the lower byte sends a high potential to the base of a selected unit driver. That turns on the driver transistor which supplies a low voltage potential to the common cathode input of the selected unit. The combination of the positive potentials on selected segment anodes and a driven cathode illuminates one unit with the desired character.

EXAMPLE 1.6

What data must be sent to location $20 00A0 to light the character r on units 1 and 4?

Solution:

The upper byte of location $20 00A0 is set to 0101 0000 or $50 to select segments e and g. The lower byte is 0001 0010 or $12 to select units 1 and 4. An instruction which moves the data $2012 to location $20 00A0 causes a lower case r to appear on units 1 and 4.

Just as output information can be sent to a peripheral device by storing data in a given memory location which accesses that device, data can also be read into the system from a designated location connected to an input device such as an *ASCII* (American Standard Code for Information Interchange) keyboard. This type of keyboard sends out a different 7-bit code for every key pressed. The ASCII code (Table 1.2, page 20) defines the 7-bit binary code for the alphabet, numbers, punctuation, and control characters (such as Return, Shift, and Escape) commonly used by today's computers and communications systems. A simple scheme for interconnecting the keyboard to the system is shown in Figure 1.11 on page 21. The seven bits produced by the keyboard are stored in two quad-data flip-flops (74173) whenever a key is pressed, because the keyboard sends out a single signal to indicate a key has been depressed. This signal is used as the enabling input to the 74173s and to set b_7 used by the processor to recognize that a key press has occurred. The output enables of the quad-data flip-flops are connected to an address decoder which generates a signal whenever the microprocessor reads location $20 00A4. When the read occurs, the outputs of the register along with b_7 flip-flop are read. The program must first check b_7 to determine if the data being read are valid key data. If that bit is low, then the data are disregarded. If it is high, then the key data are accepted as valid and

Table 1.2 ASCII Codes

MSB (HEX)	0	1	2	3	4	5	6	7
LSB (HEX)								
0	NUL	DLE	SP	0	@	P	'	p
1	SOH	DC1	!	1	A	Q	a	q
2	STX	DC2	"	2	B	R	b	r
3	ETX	DC3	#	3	C	S	c	s
4	EOT	DC4	$	4	D	T	d	t
5	ENQ	NAK	%	5	E	U	e	u
6	ACK	SYN	&	6	F	V	f	v
7	BEL	ETB	'	7	G	W	g	w
8	BS	CAN	(	8	H	X	h	x
9	HT	EM	)	9	I	Y	i	y
A	LF	SUB	*	:	J	Z	j	z
B	VT	ESC	+	;	K	[	k	{
C	FF	FS	,	<	L	\	l	|
D	CR	GS	–	=	M	]	m	}
E	SO	RS	.	>	N	↑	n	~
F	SI	US	/	?	O	←	o	DEL

MSB = b_6, b_5, b_4
LSB = b_3, b_2, b_1, b_0

used. The reading of \$20 00A4 also causes b_7 to be reset and available for the next key press. Because only seven bits are needed for the ASCII code and one for the key press status, the remaining eight bits to complete the 16-bit data width are held low.

EXAMPLE 1.7

What data are read from location \$20 00A4 following the pressing of the S key?

Solution:

Bits in D_8–D_{15} are permanently low; b_7 is now high, indicating a valid key press and the ASCII code for an S is 101 0011. Put together, the data at \$20 00A4 is \$00D3.

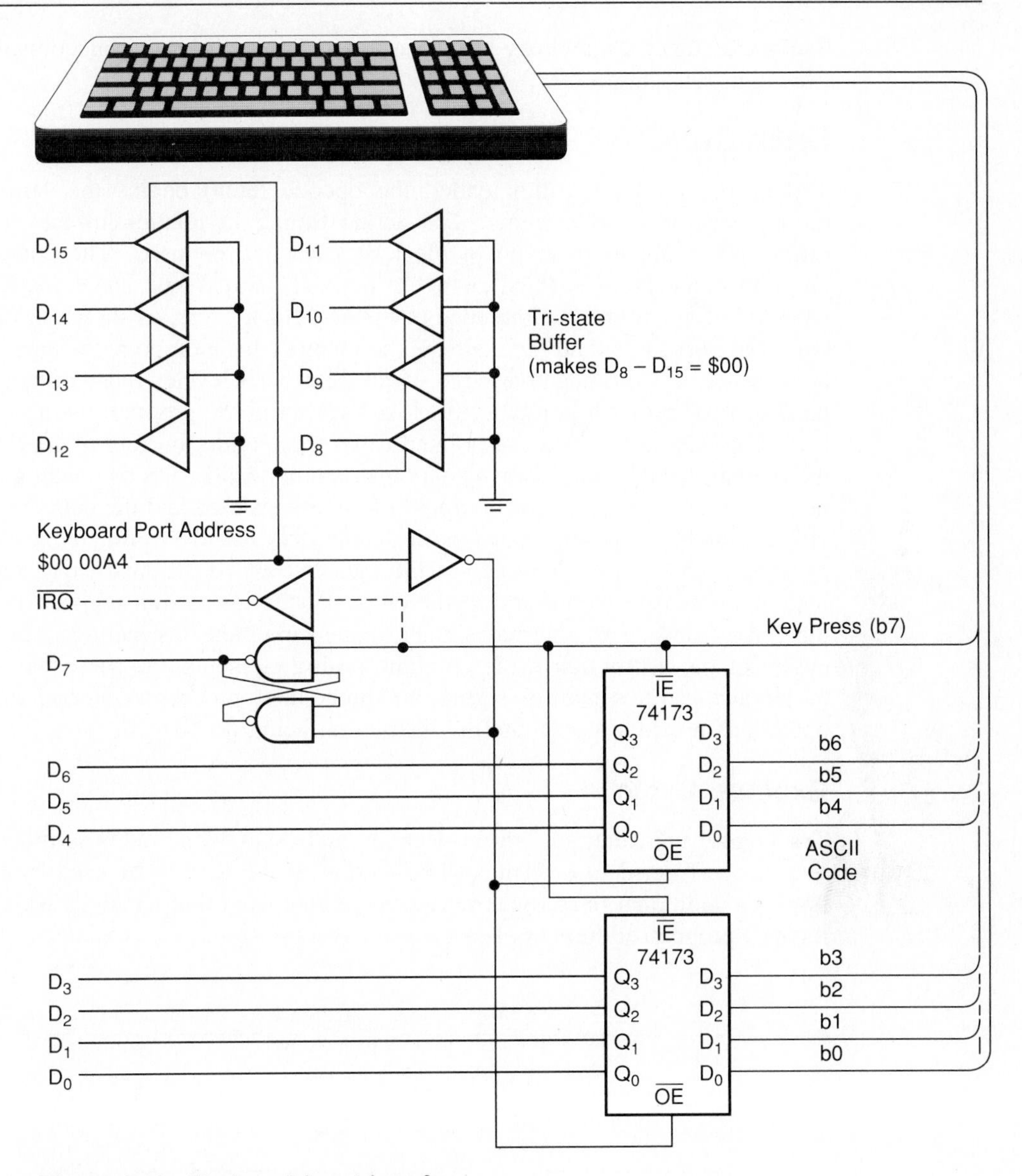

Figure 1.11 Keyboard input interface

1.6 EXECUTING A MICROPROCESSOR SYSTEM PROGRAM

Using the system's keyboard, the user directs the microprocessor to begin a program by entering the starting address into the computer's program counter. The initialization program directs movement of this information as a result of interpreting a RUN command. Once an address has been entered into the program counter, program execution can begin.

Fetch Cycle

The instruction fetch (often called the opcode fetch) begins the same for most microprocessor-based systems. The actual timing in relation to the clock cycles differs according to the type of clock synchronization used. The contents of the program counter are enabled onto the internal address bus and transferred to the address buffers on the leading edge of the first clock cycle, as shown in Figure 1.12. Once the buffers contain the address, the memory location pointed to by the address is accessed. On the negative edge of the first clock cycle, the program counter is incremented so that it points to the next location in the program listing.

The data at the location pointed to by the address buffers is the first word of the program which must always be an instruction opcode. On the leading edge of the next clock cycle, the memory output buffers are enabled and the data are transferred via the data bus to the microprocessor data buffers. On the trailing edge of this clock cycle, the data is transferred from the data buffers to the instruction register. The instruction decoder connected to the output of the instruction register then converts the binary opcode to internal signals representing the instruction. The remaining cycles of the instruction cycle combine timing elements and these internal signals to produce various control signals to direct microprocessor internal and external operations required to execute the instruction being decoded.

Execute Cycle

The control unit interprets the instruction and notes which address mode operation is to be performed to determine the location of the data to be used by the instruction. An additional fetch cycle retrieves the data from that location, which might be the next program address or elsewhere in memory. Again, the contents of the program

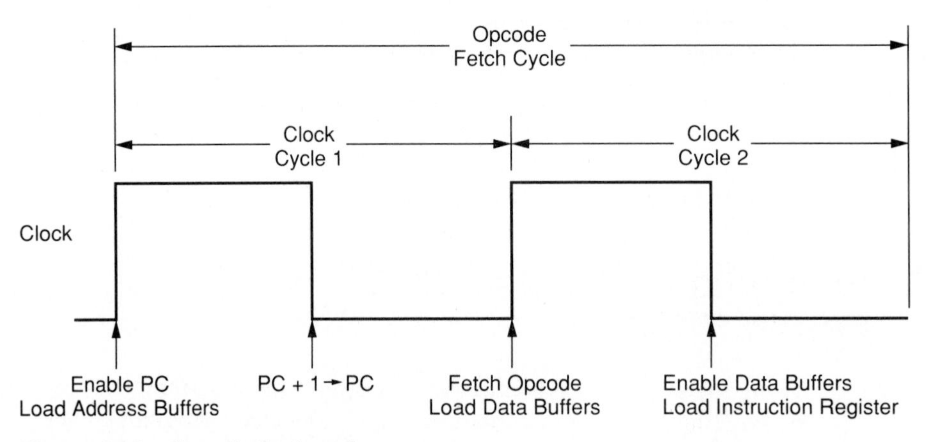

Figure 1.12 Opcode fetch cycle

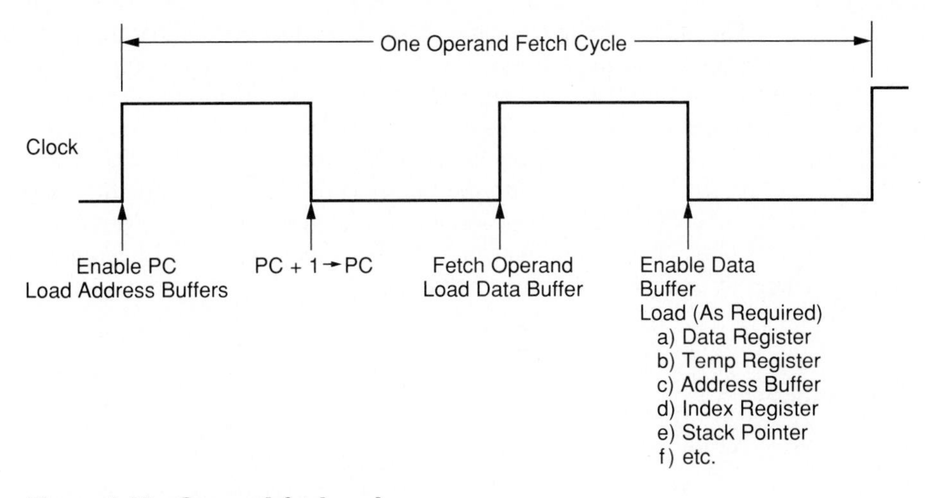

Figure 1.13 Operand fetch cycle

counter are transferred to the address buffers via the CPU's internal bus and the program counter is incremented to point to the next line in the program. This time, the data retrieved from memory do not go to the instruction register because this is an operand and not an opcode fetch. Instead, either the data are loaded into the address buffers to point to the location of the instruction's data or the data fetched are used immediately by the instruction itself (Figure 1.13). In the first instance, the data at the location pointed to by the operand address previously fetched is used by the instruction. The total time to fetch an instruction and execute it is called the *machine cycle* or *instruction cycle*. The program counter is incremented during each fetch of information (opcode or operand) from the program list. It now points to the program location of the opcode of the next instruction. The next opcode is fetched and the process is repeated.

1.7 PROGRAMMING FUNDAMENTALS

The microprocessor-based computer works by deciphering a set of instructions designed specifically for the CPU. These instructions, in the form of binary opcodes, compose the instruction set of the microprocessor. A *program* is a listing of instructions and data for a computer to use to perform various operations. A process or operation is performed when the microprocessor executes consecutive instructions fetched from memory that were presented to it in correct order and format (**syntax**). In general, that format is the opcode followed by the source operand and the destination operand as required. The code defines the process that the computer is to perform. The operands define the source and destination data or location of the data for the process to use when executing the instruction. Table 1.3 on pages 24 and 25 is a list of instructions and their hexadecimal opcodes that are used by our generic 16-bit processor.

The first column in Table 1.3a states the instruction by name and function. The second column contains the instruction syntax as it appears in an **assembler program** listing. Here, mnemonics, a shorthand notation for each instruction, is used to identify each operation. A program, called an assembler, is capable of translating the instructions in this form into **machine** or *object* **code.** The machine code is the ones and zeros that make up the binary code for each instruction. The binary codes for the opcodes of each instruction are shown in the third column of Table 1.3a. Two instructions, the conditional branch (Bcc) and the MOVE, require additional code charts to complete their machine code. These charts are included in Table 1.3a.

Table 1.3 Instruction Set

a) Instructions and Opcodes

Instruction Name	Instruction Syntax	Opcode Binary Code															
		15	14	13	12	11	10	9	8	7	6	5	4	3	2	1	0
Add Decimal	ABCD Ds,Dd	1	1	0	0	←Dd→			1	0	0	0	0	0	←Ds→		
Add Binary	ADD Ds,Dd	1	1	0	1	←Dd→			0	0	1	0	0	0	←Ds→		
Add Binary with Extended	ADDX Ds,Dd	1	1	0	1	←Dd→			1	0	1	0	0	0	←Ds→		
AND Binary	AND Ds,Dd	1	1	0	0	←Dd→			0	0	1	0	0	0	←Ds→		
Conditional Branch	Bcc.S ⟨displace⟩ *see Condition Code chart for Condition Codes	0	1	1	0	←cc→				←8 Bit Displacement→							
Branch Always	BRA.S⟨displace⟩	0	1	1	0	0	0	0	0	←8 Bit Displacement→							
Clear Register	CLR Dd	0	1	0	0	0	0	1	0	0	1	0	0	0	←Dd→		
Compare Registers	CMP Ds,Dd	1	0	1	1	←Dd→			0	0	1	0	0	0	←Ds→		
Jump to Subroutine	JSR ⟨address⟩	0	1	0	0	1	1	1	0	1	0	1	1	1	0	0	0
Move Source to Destination	MOVE ⟨source⟩, ⟨dest⟩ *See mode chart for mode code	0	0	1	1	←Dd→			←Dest→ Mode			←Source→ Mode			←Ds→		
Complement Binary	NEG Dd	0	1	0	0	0	1	0	0	0	1	0	0	0	←Dd→		
Return from Subroutine Return from Interrupt	RTS/RTI	0	1	0	0	1	1	1	0	0	1	1	1	0	1	0	1
Subtract Decimal	SBCD Ds,Dd	1	0	0	0	←Dd→			1	0	0	0	0	0	←Ds→		
Subtract Binary	SUB Ds,Dd	1	0	0	1	←Dd→			0	0	1	0	0	0	←Ds→		
Subtract Binary with Extended	SUBX Ds,Dd	1	0	0	1	←Dd→			1	0	1	0	0	0	←Ds→		

Table 1.3 *(continued)*

b) Condition Code Chart

cc	Name	Test	cc Bits
CC	Carry Clear	C = 0	0 1 0 0
CS	Carry Set	C = 1	0 1 0 1
EQ	Equal (to 0)	Z = 1	0 1 1 1
GE	Greater Than or Equal	$N \oplus V = 0$	1 1 0 0
GT	Greater Than	$N \oplus V + Z = 0$	1 1 1 0
HI	Higher	$C + Z = 0$	0 0 1 0
LE	Less or Equal	$N \oplus V + Z = 1$	1 1 1 1
LS	Lower or Same	$C + Z = 1$	0 0 1 1
LT	Less Than	$N \oplus V = 1$	1 1 0 1
MI	Minus	N = 1	1 0 1 1
NE	Not Equal (to 0)	Z = 0	0 1 1 0
PL	Plus	N = 0	1 0 1 0
VC	Overflow Clear	V = 0	1 0 0 0
VS	Overflow Set	V = 1	1 0 0 1

c) Mode Chart

Address Mode	Mode Code	D_d or D_s Code
Register	0 0 0	D_d or D_s number
Index	1 0 1	0 0 0
Address Short	1 1 1	0 0 0
Address Long	1 1 1	0 0 1
Immediate	1 1 1	1 0 0

ADD and SUBTRACT Instructions

There are three different forms of add and subtract instructions. Fundamental binary add and subtract instructions have the syntax ADD Ds,Dd and SUB Ds,Dd. Ds designates a source data register which can be D0–D3 or a 16-bit immediate data value. Dd designates a destination data register which also can be any data register from D0–D3 or the index register. These instructions perform the operation of add and subtract by using the indicated registers and placing the results into register Dd.

EXAMPLE 1.8

Write the instructions that add and subtract registers D1 and D3.

Solution:

Assuming D1 is the destination register and expecting D3 to be subtracted from D1, the two instructions are as follows:

ADD D3,D1 for addition and SUB D3,D1 for subtraction

ABCD and SBCD instructions use binary coded decimal (BCD) values as data. In essence, these instructions perform decimal addition and subtraction. Binary addition and subtraction (represented in hexadecimal numbers) is followed by an additional operation to yield BCD results. Data that are used must be in BCD form (from 0–9); otherwise the use of the BCD instructions produces meaningless results. After the two numbers are summed, either by straight addition or 2's complement addition for subtraction, the value 6 is added hexadecimally to the result under the following conditions:

1. 6 is added to a digit if it is greater than 9.
2. 6 is added to a digit if the addition in that digit position produces a carry to the next position.

EXAMPLE 1.9

How does the microprocessor produce the decimal result 9500 by adding 3800 to 5700?

Solution:

The hexadecimal sum of $3800 and $5700 yields $8F00. Since $F in the result is greater than 9, a $6 is added to it. The final result is $8F00 + $600 = 9500.

EXAMPLE 1.10

Show how 16745 is produced by adding 7811 to 8934.

Solution:

The HEX result is $10145. A carry was generated by the addition of $8 to $9, so a $6 is added to that digit's result. A carry is also generated by adding $7 + $8 + carry from the previous column. $6 is also added to that column, so the final result is: $10145 + $6600 = 16745. Note that the final carry out is shown by setting the C flag in the status register while the destination register would hold the final BCD sum of 6745.

The last set of add and subtract instructions is the extend types, ADDX and SUBX, used for multiple-precision arithmetic. They are used when it is desirable to add or subtract numbers larger than 16 bits. The ADDX adds two numbers plus the value of the X flag, putting the results in the destination register. The SUBX subtracts the source register and X flag values from the destination register, placing the results in the destination register. To understand the need for this, consider the following decimal addition example.

EXAMPLE 1.11

Add the decimal numbers 678 and 456, paying close attention to any carry conditions created by the process.

Solution:

Addition starts by summing the two least significant digits, 6 and 8. For this sum, there is no carry which precedes it, but there is one produced by the addition which is 4 and carry 1. The next column adds 7 + 5 + the carry from adding 6 + 8. This sum produces 3 and a carry of 1 to the last column which is added to that sum: 6 + 4 + 1 = 1 and final carry of 1. Diagrammatically:

```
 111
  678
+ 456
 1134
```

When adding 16-bit numbers, the intermediate carrys from one digit to another are handled by the instruction so the process can be performed. The final carry out is stored into the C and X flags as described earlier. However, when adding numbers larger than 16 bits, more than one addition process must be performed. For example, adding 32-bit numbers requires two 16-bit additions. The ADDX instruction allows the carry generated by the first 16-bit add to be included in the second 16-bit add.

EXAMPLE 1.12

Show how $1234 5678 is added to $ABCD EF90 using ADD and ADDX.

Solution:

The lower two 16-bit data are added together using the ADD instruction since there is no carry input involved in the addition. The result produces a sum of $4608 and a carry, setting the C and X flag. The second addition is done using the ADDX instruction which adds $1234 + $ABCD + X flag = $CE02. The complete sum is $CE02 4608. The C and X flags are cleared since there was no carry from the second addition.

AND Instruction

The AND instruction is the only logic instruction used in our simplified generic processor. Its syntax is AND and does a binary logic AND of each bit in the source and destination registers on a bit by bit basis. The results are placed in the destination register.

EXAMPLE 1.13

What are the results in D2 if D0 contains $1C34 and D2 contains $5678 following the instruction AND D0,D2?

Solution:

$1C34 = 0001 11000 0011 0100 in binary and $5678 is 0101 0110 0111 1000. ANDing these two values together bit by bit yields the following:

0001 1100 0011 0100
0101 0110 0111 1000
0001 0100 0011 0000

Branch Instruction

A *branch* instruction has the effect of changing the direction of program flow. It does this by modifying the contents of the program counter so that the next instruction fetched comes from a different area in memory from the one which contained the branch instruction. The unconditional branch causes this change of flow to occur whenever the instruction is executed. Its syntax is BRA.S ⟨displacement⟩. BRA stands for branch only and the .S indicates a short branch. The displacement is an 8-bit signed displacement. The displacement is added to the current contents of the program counter after it has fetched the branch instruction. A short branch is a branch that changes the location of the program within +127 and −128 locations from the current program counter value. Long branches use a 16-bit displacement and will not be considered in this chapter.

EXAMPLE 1.14

From what location is the next instruction fetched following a BRA.S $44 if the BRA.S instruction is at address $10 C444?

Solution:

While the processor is fetching the BRA.S $44 instruction, the program counter is incremented twice from $10 C444 to $10 C446. The instruction decoder interprets the BRA.S instruction and adds the displacement value of $44 to the program counter, changing its value to $10 C48A. This is the location from which the next instruction is fetched.

The second form of branch instruction is the conditional branch, Bcc.S ⟨displacement⟩. This is the decision-making instruction. It causes the change of program flow described for the BRA.S instruction if the condition it is testing is true. If the condition is not met, program flow continues on to the next instruction as if the branch instruction were not there. The conditions that are tested are shown in the third part of Table 1.3.

Unsigned Data Flag Tests Unsigned number comparisons are made using branch instructions that react to the status of the carry flag (C). A compare operation is performed by subtracting one number from another to set the flag condition based on the difference between the numbers. The branch tests that could be used are branch if carry clear (BCC) or branch if carry set (BCS). For subtraction, if the carry is clear, then no borrow took place and the value in the destination data register was greater than or equal to the source value. Similarly, the BCS also can be called a branch if less than, since a borrow indicates that the number in the destination was less than the source value.

Two other branch instructions are used for unsigned tests: branch if higher than (BHI) and branch if less than or the same (BLS). These check to see if the results of ORing the C and the Z flags are zero (BHI) or one (BLS). If the number in the destination is higher than the number in the source register being compared to it, the C and the Z flags are reset. If the number in the destination is less than the number being compared to it, the C flag is set as before. If the two numbers are the same, the result of the comparison is 0 and the Z flag is set.

The use of the term *comparison* in the preceding discussion is deliberate. The instruction set lets us subtract without compromising the original numerical value. This is done with the compare instruction CMP Ds,Dd, compare source data to the destination contents. This instruction subtracts source from destination, but discards the results. The original data remain unchanged as desired. Some of the flags are set or reset as a result of the subtraction. In other words, a branch test can be made while maintaining the data in both source and destination registers.

Signed Number Testing Signed number data can be compared by several branch instructions. We recognize these instructions by the use of signed number flags N and V. Branch if minus (BMI) and branch if plus (BPL) operate on the condition of the N flag. Branch if overflow clear (BVC) and branch if overflow set (BVS) are branches that test the condition of overflow (or sign error) due to an operation.

Additional signed number tests use combinations of condition code flags to determine if a number is more positive or more negative when compared with another number. The tests usually use this Boolean combination as a test: $N \oplus V$ (N exclusive OR V).

The value of this expression in relation to the branch instruction can be seen if you look at its truth table (Table 1.4). The N bit tells the computer the condition of the results: negative or positive. However, the overflow tells the computer if the results have a sign error. Line 1 says the results are positive ($N = 0$) and no overflow occurred ($V = 0$). So the results are truly positive. When $N = 1$ and the V flag is still low, the results are truly negative. If the overflow flag is set, the results are incorrect and should be the opposite sign. By exclusive ORing the N and V flag, we observe that when the results should be positive, the exclusive OR yields a logic 0. When the results should be negative, the exclusive OR yields a logic 1.

Table 1.4 Sign Relationships

N	V	$N \oplus V$	Sign Value
0	0	0	Positive
0	1	1	Negative
1	0	1	Negative
1	1	0	Positive

The branch instruction BGE stands for *branch if greater than or equal to the comparing value*. The flag test is $N \oplus V = 0$. The branch is asking if the result of subtracting two numbers is greater than or equal to zero (positive). If they are, then the data at the destination was greater than (more positive) or equal to the source value. BGT expands the flag test to $(N \oplus V) + Z = 0$. This test asks only if the destination value was greater than the source number. BLT and BLE perform the same type of test, but instead of looking for a greater result, they branch if the original value is less than (BLT) or less than or equal to (BLE) the compared value. An example using these branch instructions and the results will help to clarify any confusion.

EXAMPLE 1.15

D1 and D3 hold two values to be compared. They can be signed or unsigned numbers, depending on which branch test you, the programmer, use to determine the data type. Which test is used for each data type to check if the value in D1 is larger than the one in D3?

Solution:

To test the numbers as unsigned, the BHI.S instruction is used following the CMP instruction:

```
CMP D3,D1        * D1–D3 sets flags
BHI.S BIGGER     * If D1 is bigger (C and Z are both 0) then
                   take the branch.
```

To make the same test for signed numbers, replace BHI.S with BGT.S. The results of this test will determine if D1 contains the more positive of the two values. It checks the N, V, and Z flags, and if they satisfy the Boolean expression: $N + V + Z = 0$, then the number in D1 is indeed more positive [(N is 0 and no overflow indicating sign error occurred) and not equal $(Z = 0)$] than the number in D3.

CLEAR Instruction

The CLR Dd instruction causes the destination register to have the value $0000 in it and all the flags in the status register except the I and Z flags to be reset. The I flag is unchanged and the Z flag is set to indicate the zero value contents of the destination register.

COMPARE Instruction

The instruction used to compare two values is CMP Ds,Dd where Ds, the source, can be any of the four data registers or an immediate 16-bit data value; and Dd, the destination, can be any of the four data registers, D0–D3. The operation is performed by subtracting the source from the destination and discarding the results. Status flags are set or reset based on the subtraction and can be used by conditional branch instructions described in previous sections for decision making.

Subroutine Jump Instruction

Subroutines are separate programs designed to perform specific tasks, such as driving a display, reading a keyboard, performing a multiplication or a square root operation. They are essentially complete programs that require the input of some data and/or return some result. Subroutines are written so that they can be used any number of times in a program or by different programs as needed. A main program executes its instructions sequentially until it fetches a branch instruction or a jump to subroutine instruction. The former instruction has already been described. The jump to subroutine has the syntax, JSR ⟨address⟩, where the address is the location of the first instruction of the subroutine program. When the microprocessor fetches and executes this instruction, it pushes the program counter contents onto the stack and loads the program counter with the address operand of the JSR instruction. This diverts program fetches and executions to the subroutine program. The subroutine's last instruction is a return from subroutine (RTS) which causes the stack to be pulled and the original contents of the program counter to be returned to the PC. At that point, the main program resumes its execution with the instruction following the JSR instruction.

2's Complement Instruction

The complement of negate instruction, NEG Dd, performs the 2's complement of the number in Dd. It takes the value in the destination register and returns its negative value. This instruction is valuable when it is desired to produce the true magnitude value of a negative number following an arithmetic operation.

EXAMPLE 1.16

Demonstrate the use of the NEG instruction to produce the true value of the result of subtracting $9311 from $456B.

Solution:

The subtraction is performed by using a SUB instruction. $456B is previously stored in register D0 and then $9311 is subtracted from it using SUB #$9311,D0. The # preceding the data value is Motorola's syntax indicating that the value is to be used as an immediate data value and not to be mistaken for an address or register designation. The subtract operation results in a value of $B25A. The C and X flag are set to indicate that a borrow was required to do the subtraction. The C flag also indicates that the answer is a 2's complement value rather than a true magnitude value. Following the SUB instruction is a BCC instruction to change the program flow around the next instruction. This test checks to see if the subtraction produced a true magnitude result ($C = 0$) or a 2's complement negative result ($C = 1$). The instruction which follows the BCC is bypassed if the branch test is true (results in D0 are positive true value). If the test is false, then the branch is not taken. The NEG D0 instruction is executed, which causes the value in D0 to be complemented to $4DA6, the true magnitude of the result. It is up to the user to signify that this value is really −$4DA6. The short program section for this example is:

```
SUB #$9311,D0      * Do subtraction
BCC $02            * Check carry flag
NEG D0             * 2's complement D0
```

MOVE Instruction

The workhorse of the instruction set is the MOVE instruction. It has the syntax, MOVE ⟨source⟩, ⟨destination⟩, where the source can be any data register, the index register, or any immediate 16-bit data or 16- or 24-bit address value. The destination can be any one of the data registers, the index or stack pointer register or a 16- or 24-bit address value. The purpose of this instruction is to move data from one place to another. Data can be moved between registers and between registers and memory. A mode chart is provided in Table 1.3c to assist in deciphering the machine code given an instruction syntax. Examples of using the MOVE instruction are incorporated in the discussion of the addressing modes used with this processor. With the MOVE instruction it is important to remember that the C, X, and I flags are not affected by the instruction, but Z and N are. The overflow (V) flag is always reset by a MOVE instruction.

1.8 ADDRESSING MODES

An **address mode** is used to designate the location of information to be used by an instruction or to designate the new location for program execution in the case of instructions that change program flow (BRANCH and SUBROUTINES). The address modes used with our 16-bit microprocessor are *immediate, register, address short, address long, relative, inherent,* and *index*.

Immediate Address Mode

The immediate address mode is the easiest to understand. The data used by the instruction is immediately in the operand of the instruction. An immediate operand is indicated by a # sign preceding the operand value.

EXAMPLE 1.17

Write the instruction that immediately loads register D0 with the value $9311.

Solution:

The syntax of the MOVE instruction is MOVE followed by the source ($9311) and destination (D0) operands. A # precedes the source operand to indicate that the data is immediate:

```
MOVE #$9311,D0
```

This instruction moves $9311 into D0.

Register Mode

The register mode is one we have used quite frequently already. It says the data is either located in a register or will be sent to a register. In Example 1.17, the destination mode used is the register mode since the destination of the instruction is data register D0. The registers used for this mode include the four data registers (D0–D3), the index register (IX), and the stack pointer (SP).

Address Short and Address Long

Two addressing modes which allow the use of memory address operands are address short and address long. They are both used when it is desirable to access data directly from memory or to write to memory. Address short is used when the memory location is located on *page 0* of the memory map. Page 0 memory locations are those whose upper byte of the address is 00. Page 0 covers locations $00 0000 to $00 FFFF. Since the upper byte is 0, that byte does not need to be included in the instruction. The system adds the leading zeros as needed. Address long is usable for any address within the 16 Mbyte address range ($00 0000 to $FF FFFF). Address short is preferred because it takes one less fetch cycle to retrieve the operand from the program list. Address short operands can be contained in one 16-bit location. Address long operands require two 16-bit locations; the first holds the upper or page byte and the second holds the lower two bytes of the address. The first location's high byte must be $00 to be a valid address.

EXAMPLE 1.18

Write the instruction that moves data from location $00 1234 to location $22 4C00.

Solution:

The source location is on page 0 so that the address short mode can be used. The destination is on page $22 so address long must be employed. The instruction looks like:

```
MOVE $1234,$224C00
```

Notice that the # sign used in the immediate mode is not used in the address short or long modes. This says that this data is not immediate, but is an address which contains the data. Also note that the source operand $1234 occupies one 16-bit location while $22 4C00 must occupy two which contain $0022 and $4C00.

Address long and short are also used with the JSR instruction to indicate the address of the suboroutine being jumped to.

Relative Address Mode

The relative address mode is used by branch type instructions only. Relative addresses are one byte signed displacements that are added to the program counter to change the flow of program execution. This process was described earlier and examples were given.

Inherent Address Mode

Inherent addressing has no operand. Any data or register used by the instruction is inherent to the instruction, eliminating the need for a separate operand value. The only inherent instruction in our instruction set is return from subroutine (RTS), which uses the stack and stack pointer.

Index Mode

The index register is a 24-bit, memory address pointer register. It can be modified by an ADD, SUB, or MOVE instruction using the index register as a destination. These instructions are not index mode instructions. They are standard instructions that are applied to the index register, treating that unit as a 24-bit register. Once the index register is initialized with address information or modified by one of these preceding instructions, instructions using the index address mode can benefit from the index register's pointer capability.

In the *index address mode,* a memory address is formed by taking the contents of the index register and adding them to the signed 16-bit offset number which is fetched from the operand portion of the instruction. For example, MOVE $6IX,D0 causes register D0 to be loaded with the data from a memory location created by adding $6 to the contents of the index (IX) register. This addition occurs in the TEMP register, leaving the contents of the IX register unaffected by an index address mode use.

EXAMPLE 1.19

Demonstrate the use of the index register as a pointer to add two numbers located at locations $00 0120 and $00 0124, placing the results into location $00 0130.

Solution:

First, load the index register with one of the address values, say $00 0120, using the immediate address mode:

```
MOVE #$0120,IX
```

Next, move the data from that same location and add the data from the next location (4 bytes away) and then store the results at the last location (16 bytes above the reference address of \$00 0120):

```
MOVE $0IX,D0        * IX + 0 = $00 0120
ADD $4IX,D0         * IX + 4 = $00 0124
MOVE D0,$10X        * IX + 16 = $00 0130
```

The power of the index mode is realized when a program must access a bank of data stored in sequential locations. The index register is used to point to the top of the bank. As the program processes the data, the index register is modified to move the pointer down the stack of data.

EXAMPLE 1.20

Write a program segment that adds two banks of data, each ten bytes long, and stores the results at a bank location starting at location \$00 1240. The original data banks start at locations \$00 1200 and \$00 1220.

Solution:

Whenever using the index register, you must start by initializing the IX register:

```
MOVE #$1200,IX      * $00 1200 → IX
```

Data from the first bank is fetched and placed into D0. Next, data from the second bank is added to D0 and the results are stored in the third bank:

```
MOVE $0IX,D0        * Bank 1 → D0
ADD $20X,D0         * Bank 2 + D0 → D0
MOVE D0,$40,D0      * D0 → Bank 3
```

After the first addition has occurred, the index register is incremented twice to move to the next 16-bit data location in the bank. Then, the index register is tested to check if there is additional data to be added. Since ten 16-bit locations are in each bank, the last address in bank 1 is \$00 1212. When the contents of IX reach \$00 1214, we are out of data and the program should stop.

```
ADD #$2,IX                  * $00 0002 + IX → IX
CMP #$1214,IX               * IX − $1214 set/reset Z
BNE ⟨to MOVE $0IX,D0⟩       * IX not = $00 1214?
Stop Program
```

Notice in Example 1.20 how the index register facilitated the manipulating of data in the three banks of information.

1.9 ASSEMBLY LISTINGS

Writing programs in machine language (opcodes and operands) is cumbersome and time-consuming. The first desktop computer, made by Altair, required that actual binary information be entered through the front panel switches. This required the programs to be written using machine language before they were entered. Assembler programs now exist which translate source codes (mnemonic representation of instructions) into machine codes that the microprocessor can understand and use. These assemblers allow the user to create a program using the mnemonic language for the microprocessor. To appreciate the difference, we will create a short program using both techniques. Although hexadecimal numbers are used to represent the machine codes, following and deciphering the program is difficult. Here is the machine code for a sample program:

```
303C 001C D07C 00B7 31C0 2080 60FD
```

What does the program do? The answer becomes clearer when the same program is written in source or mnemonic code:

```
START   MOVE #$1C,D0
        ADD #$B7,D0
        MOVE D0,$2080
DONE    BRA.S DONE
```

Program 1.1

The source code format of Program 1.1 is sectioned into three fields. The first field is the label field, where a label name may be used. These **labels** are used to refer to address and data information in a more readable form. Here the label START indicates the beginning of the program. The next field is the mnemonic or opcode field. It contains the opcode mnemonics (from the instruction set, Table 1.3) and the instruction's operand. A third field, the comment field, is not shown in Program 1.1. Comments are preceded by a delimiter, such as an *, and contain any clarifying comment the user wishes to add to the program listing. A full line of comments may be used by placing an * at the beginning of the line. Comments are messages to the reader and writer of programs but are ignored when the computer translates (assembles) the program into machine code.

Without being fully versed in the instruction language, you can still manage to interpret the program as adding two numbers and saving the result. Imagine trying to write an extensive program using object (machine) code only. If we use the source code, an organized thought process can be developed. Our example program starts by loading register D0 with the number $001C. It then adds $00B7 to that value and saves the result at location $00 2080.

An assembler permits the user to compose the program in source code and, upon the issue of an assemble command, it performs the chore of converting the program to machine code and stores it in a designated memory area. A typical assembly directive (command) causes the computer to execute the following procedure. First, a prompt (or cue) appears on the screen, which signals the user that the system is awaiting a command. The user types COMPOSE followed by a carriage return (⟨cr⟩). Line numbers appear and the user can now type in the source program following the format required by the assembler. The BREAK key is used to exit any mode including the COMPOSE mode. After typing in the program source code, the user presses the BREAK key and types the command word ASSEMBLE followed by a carriage return. The computer then asks for a starting address location for the assembled program to be stored. The user now types in the address and again follows with a return. The assembler program then begins the process of assembling the program and placing the machine code into its proper sequence in memory, starting at the designated starting address. Some assemblers automatically list the fully assembled program upon completion of the process, while others simply return a prompt when done. If we assemble our program example starting at location $00 0120 and use a question mark (?) as a prompt, it appears on the computer's screen as:

```
? COMPOSE ⟨cr⟩
001 START     MOVE     #$1C,D0    ⟨cr⟩
002           ADD      #$B7,D0    ⟨cr⟩
003           MOVE     D0,$2080   ⟨cr⟩
004 DONE      BRA.S    DONE       ⟨cr⟩
005 ⟨break⟩
? ASSEMBLE ⟨cr⟩
STARTING ADDRESS ? 000120 ⟨cr⟩
000120   303C001C      START     MOVE #$1C,D0
000124   D07C00B7                ADD #$B7,D0
000128   31C02080                MOVE D0,$2080
00012C   60FD          DONE      BRA.S DONE
DEFINED LABELS
START 000120    DONE 00012C
?
```

Program 1.2

Two additional fields are added to the assembled listing. The very first field holds the address of the opcode of each instruction in the program. The next field contains the machine or object code of the instruction. This is the opcode (derived from Table 1.3) followed by the operand in hexadecimal notation. The machine code is the language the microprocessor understands, while the source code is the shorthand that programmers can decipher. The remaining fields of the assembled program have the same function as the source listing described after Program 1.1.

Most assemblers perform a two-pass operation. The first pass through the source listing seeks out any labels and assigns an address value to each. In our example, the label START is detected and assigned the value $00 0120 and the label DONE is assigned the value $00 012C. The second pass performs the actual job of converting from the source listing to machine codes. Any labels used in the body of the program are replaced with their address values as operands of each instruction. Because of this, a label can be used in place of an address in the operand portion of an instruction. Since the actual address values are not assigned until the assembly process takes place, programs can be located anywhere in memory without conflict. This is called *address-free* or *address-independent* programming. But once the program is assembled, it resides in a specific area of memory and becomes *address dependent* on that location. Our program has a fixed address reference for the MOVE D0,$2080 instruction. Since address $00 2080 must hold the results of the program, the program itself cannot use that location for anything else. If the operand of the MOVE instruction is changed to a label, then the actual location of the result becomes solely dependent on the starting address at the time of assembly. In that instance, the program can be placed anywhere, including at $00 2080.

1.10 PSEUDO OPERATIONS

Pseudo operations (pseudo ops) facilitate the use of labels. They are not assembled as program instructions and do not occupy program memory space. Instead, they are interpreted during the assembler's first pass and their job chiefly is to match labels and address values. Each assembler program publisher selects and defines its choices for pseudo op codes. The examples discussed in this section are those used by Motorola. One such pseudo op is the define storage command (DEFS), which assigns an address to a label and reserves a preset number of following locations for data storage. The syntax (format) for this pseudo op is LABEL DEFS ⟨number⟩. This pseudo op can be placed anywhere in the listing since it is acted upon during the first pass and has no effect during the second pass. Placing it at the beginning of a source listing causes lower addresses preceding the program to be assigned to the labels. Placing it at the end of a listing causes addresses following a program to be assigned to the labels. Placing a DEFS in the middle of a source list inserts storage space in the middle of a program, which is not advisable. Let's use the DEFS operation for our program and observe the effects of assembling the program.

EXAMPLE 1.14

Place the DEFS statement at the beginning of a source program:

```
001 STORE     DEFS     2
002 START     MOVE     #$1C,D0        * First number
003           ADD      #$B7,D0        * Add second number
004           MOVE     D0,STORE       * Store results
005 DONE      BRA.S    DONE
```

Program 1.3

After this program is assembled starting at location \$00 0120, the result appears as:

```
000122 303C001C START    MOVE   #$1C,D0      * First number
000126 D07C00B7          ADD    #$B7,D0      * Add second number
00012A 31C00120          MOVE   D0,STORE     * Store results
00012E 60FD      DONE    BRA.S  DONE
DEFINED LABELS
STORE 000120    START 000122    DONE 00012E
```

Program 1.4

Notice that the result is not stored at \$00 2080 as in the original program, but, instead, the storage is done at location \$00 0120, which is the address assigned by the assembler directive. Comments are also retained in the listing, but they do not affect label assignment or conversion from source to object code.

Next, place the DEFS statement at the end of the listing instead of at the beginning and observe the differences in the assembled listing.

```
000120 303C001C START  MOVE   #$1C,D0      * First number
000124 D07C00B7        ADD    #$B7,D0      * Add second number
000128 31C0012E        MOVE   $D0,STORE    * Store results
00012C 60FD     DONE   BRA.S  DONE
DEFINED LABELS
START 000120    DONE 00012C    STORE 00012E
```

Program 1.5

Since the assembler assigns addresses on the first pass, it discovers the correct address for STORE at the end of the program and uses that address (\$00 012E) as the value for STORE when it converts the program to machine code.

The original program could not be assembled starting at location \$00 2080 without creating a conflict between a program location and a data location resulting from the MOVE D0,\$2080 instruction. When the program is first run, the result stored at location \$00 2080 destroys the opcode \$303C of the program. Running the program a second time results in some form of failure depending on the number that now resides in location \$00 2080. We can see the benefit of using labels for address-independent programming. Either example listing using the DEFS directive can be assembled starting at location \$00 2080. In the first example, STORE is equated with \$00 2080 and the first instruction begins at \$00 2082 so there is no conflict between data and program locations.

Other General-Nature Pseudo Ops

DEFB The define byte (DEFB) directive assigns an address to a label as before. In addition, a designated value of data is stored at that location. The syntax for this operative is

LABEL DEFB ⟨one byte data⟩

As an example, let's use the label DATA1 and store $4C at the location designated as DATA1. The directive would appear in the source listing as

```
DATA1        DEFB        $4C
```

Assembling this directive starting at location $00 0800 creates the label DATA1 for location $00 0800. The data at location $00 0800 becomes $4C.

DEFW The define word (DEFW) directive is similar to the DEFB pseudo op except the data size handled is 2 bytes instead of 1. The data are stored at two sequential locations, with the first byte at the location associated with the label and the second byte at the next location. The syntax is

LABEL DEFW ⟨two bytes data⟩

If we change the DEFB example to use DEFW, it becomes

```
DATA1        DEFW        4C2E
```

Assembling this directive at $00 0800 again assigns the label DATA1 to location $00 0800. The data at that location are, once again, $4C. This time, however, the data $2E are stored at location $00 0801.

EQU The equate (EQU) directive fixes a given address or data to a label. This pseudo op is used when an address location is set and is the only one usable by a program sequence or when a label must have a fixed data value. For example, when you address decode I/O devices, only one memory location is associated with the device and a label must be precise so that it has the same address for every starting address usable by an assembler program. The syntax for equate is

LABEL EQU ⟨address or data⟩

Retaining the data used in the DEFB example, the EQU pseudo op assigns the label DATA1 to $00 0800:

```
DATA1        EQU        $000800
```

ORG The origin (ORG) statement tells the assembler to relocate the assembling process. Used in the middle of a program, this pseudo op causes program sections to be stored in different areas of memory. Locating subroutines apart from main programs is one use of the origin operation. A label is optional. The syntax for ORG is as follows:

LABEL ORG ⟨address⟩

There are other pseudo ops, but these are the most common. Labels themselves are usually limited to 6 characters and must begin with a letter to differentiate them from numerical operands such as data or addresses. This, of course, creates a corollary: Each numerical operand must begin with a number from 0 to 9 to be distinguished from a label. To accommodate operands such as $B700, an assembler allows a leading zero to satisfy this requirement. The assembler promptly ignores the leading zero when it performs the conversion to machine code. The correct syntax for our ADD example becomes ADD $0B700 to satisfy the assembler.

Labels are used in all following discussions involving program examples since most are in source format. To avoid confusion with actual numeric data used in these programs, we will not use the leading zero convention.

1.11 DELAY SUBROUTINE

The first program example is the delay subroutine illustrated in the flowchart of Figure 1.14 on page 42. Often, programmers need a short time delay, for example, when trying to display a message onto a 6-unit set of 7-segment LED displays. These displays will be **multiplexed** (lighted one at a time) to reduce power drain. To do this successfully, each unit remains illuminated for a minimum amount of time. The time delay provides the necessary delay between units of the display. One way to cause a delay is to load a register with a value and continue to decrement (reduce by one) the register until it reaches 0. No other processing is done until the register is fully decremented. Since it takes time to cause the decrement to occur, you can insert a time delay anywhere needed. The program assembly listing is as follows:

```
000300 303C0030 DELAY  MOVE #$30,D0    * Counter value
000304 90470001 LESS   SUB #$1,D0      * Decrement counter
000308 66FA            BNE.S LESS      * Test loop
00030A 4E75            RTS
DEFINED LABELS
DELAY 000300      LESS 000304
```

Program 1.6

The counter is first loaded with a value ($30 in this example) and then decremented until it reaches zero. As long as the results are not 0 (BNE), the program returns to location $00 0304 to repeat the loop portion of the program. It is only when

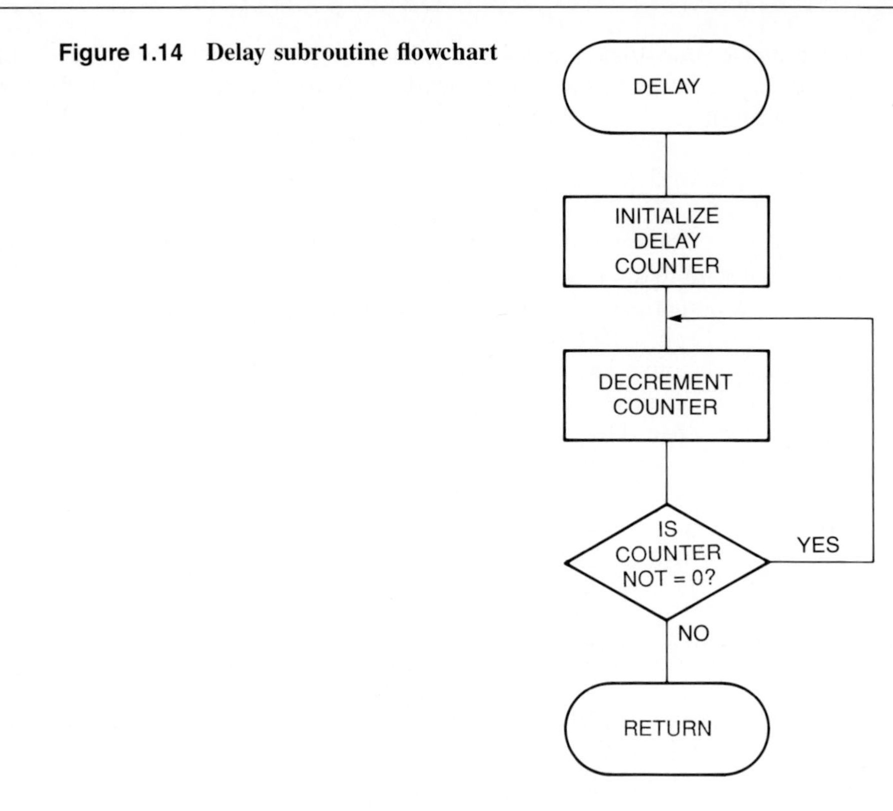

Figure 1.14 Delay subroutine flowchart

D0 reaches 0 that we fall out of the loop and return to the main program. The subroutine yields about 300 μs of delay with a 4-MHz clock.

1.12 INDEX MODE AND MULTIPLE-PRECISION ARITHMETIC

The index address mode is useful when it is desirable to add numbers of unspecified length and is limited only by the availability of RAM space to hold the numbers.

Location COUNT holds the number of bytes each number has (the two numbers are the same length) and location POINT contains the address of the two least significant of number A. Number B starts 64 ($40) locations away from number A, which limits the size of the data for this example to 64 bytes each.

The job of the program is to add these two numbers and store the results beginning at a location 64 bytes from number B. The basic addition process follows the principles in Example 1.12 with several differences. The index register is used as a data pointer to fetch both numbers and store the results correctly in a continuous looping program as long as portions of the numbers remain to be summed. Since all additions are done within the loop, we need to initialize the program carefully as will be illustrated by a program example shortly. The add with carry process in the loop forces the two least significant byte pairs to be summed with the contents of the extend flag. These tow byte pairs, being the first to be added, require the carry flag to be zero to start with for a correct result. The following additions are then carried out as before, with the two numbers summed with the current contents of the carry flag as established by the last sum. The following is the initialization and addition portion of our program:

```
START   CLR DO            * Clear C flag
        MOVE POINT,IX     * Initialize index register
        MOVE COUNT,D1     * Initialize counter
LOOP    MOVE $00IX,D0     * Get number A
        ADDX $40IX,D0     * Add number B to A
        MOVE D0,$80IX     * Save results
```

Program 1.7

The data from number A are fetched by the MOVE \$00IX,D0 instruction from location POINT, plus 0 locations. To this is added the data from location POINT, plus \$40 locations, and the results are stored at address POINT, plus \$80. By incrementing index register twice, all three addresses from numbers A and B and the storage area are increased by 2. Register B, holding the count, is decremented after each addition (pass through the loop). To ensure that there are more data to work with, the counter (register B) is tested for a non-zero condition. If the test passes, more data are present and the loop is repeated. If the test fails, we have run out of data and fall out of the loop. The carry flag is then tested for a final carry condition, the result of which is stored in the next location following the most significant sum byte. The second half of the program, putting these elements together, is as follows:

```
        ADD #$2,1X        * Increment index register
        SUB #$1,D1        * Decrement D1
        BNE.S LOOP        * More data?
        BCC.S NULL        * No carry out?
        MOVE #$1,$801X    * Store carry out data
DONE    BRA.S DONE
NULL    CLR $801X         * Clear carry out
        BRA.S DONE
```

Program 1.8

1.13 LETTER GRADE PROGRAM

Now we will write a program that stores, in location \$00 2080, the letter (A, B, C, D, or F) equivalent to a numerical grade found in location \$00 2082. The statement of the problem is simple and direct. The first task for our solution is to determine the level of the grade by comparing it to the limits of letter grades A–F. We have some unknown value and would like to determine if that value is greater than, say, 59. If we compare 59 with the value in $00 2082, one of three results occurs. The number could be greater than 59, equal to 59, or less than 59. The computer tells which of these cases is true by using the status flags, specifically the zero and carry flags since this is an unsigned number comparison (grade values are never negative). We can see, immediately, that if the number is equal to 59, the result of the subtraction is 0 and the Z flag is set. For the other two cases, the Z flag is reset. To determine if the number is less than or greater than 59, we must use the C flag as a borrow indicator. If the number is less than 59, then a borrow is required to do the subtraction function of the comparison instruction. This is indicated by the setting of the C flag. If the number is greater than 59, no borrow is required and the C flag is reset. It may be necessary to make several comparisons to determine which level our grade is.

Note that the compare instruction (CMP) performs subtraction without altering the contents of the accumulator. The subtraction of the compare instruction is performed, flags are set or reset based on the result, but the original data is not changed. The subtraction performed is a hexadecimal subtraction, but grades are in decimal form. There is no conflict here because the result we desire is not a determination of how high an A, B, C, D, or F grade is, but rather at what letter level the grade is. In other words, if the results of a comparison produce a hexadecimal value, we don't care. We are not using the subtraction results. As an example, suppose a grade is 76. Comparing that number to 89 will produce a "less than" result. The actual subtraction would yield $76 – $89 = $ED. This number is discarded by the compare instruction. The important flag conditions affected by this instruction are $Z = 0$ and $C = 1$. The result is not 0 and a borrow is needed to do the subtraction. In effect, the result of the compare instruction is that the number 76 is still in the accumulator and we have determined that it is less than 89. Using the BHI instruction [which causes a branch if the carry flag and the zero flag are both reset ($C + Z = 0$)], a branch does not occur following our example. The number in the accumulator can then be compared to the limit for a B grade (79) and tested again. Once more, the BHI does not branch. On the third test, the comparison produces a C and Z flag condition of zero and the branch does occur. The program at the new location stores the letter C into location $00 2080. The data locations needed for this program are as follows:

00 2080	Letter results
00 2082	Original grade

Figure 1.15 is the flowchart for the program. Take a minute to study the diagram. It says what we have described for the program process so far. Writing the actual program is a problem exercise at the end of the chapter.

1.14 MESSAGE APPLICATION PROGRAM

To solve most programming problems, we need a clear statement of the task to be performed and an idea of what is available from the system to help us perform

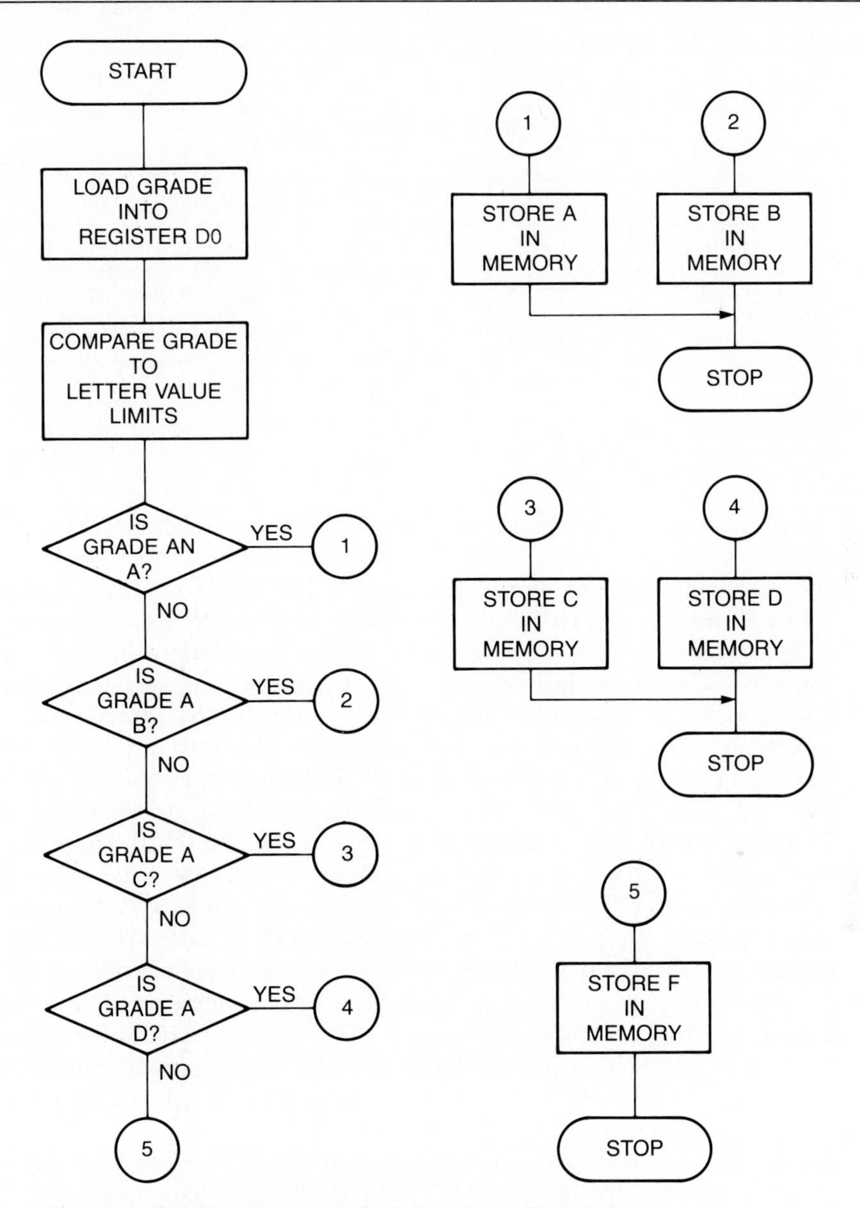

Figure 1.15 Number grade to letter grade flowchart

it. This example involves driving the display system (Figure 1.10) with a message stored in memory.

The problem is to vary the segment data as each unit is selected to display the message. Each character also must be held in its unit for a long enough period to be visible. Changing segment data and unit locations causes a type of multiplexing action. Electrically, only one unit will be illuminated at a time. Each needs to be

Table 1.5 HEX Contents of $ 00 00A0 for HELLO! Message

Character	Unit	Segment Code								Unit Code								Port $ 00 00A0 (HEX)
		dp	g	f	e	d	c	b	a	x	x	5	4	3	2	1	0	
H	0	0	1	1	1	0	1	1	0	0	0	0	0	0	0	0	1	7601
E	1	0	1	1	1	1	0	0	1	0	0	0	0	0	0	1	0	7902
L	2	0	0	1	1	1	0	0	0	0	0	0	0	0	1	0	0	3804
L	3	0	0	1	1	1	0	0	0	0	0	0	0	1	0	0	0	3808
O	4	0	0	1	1	1	1	1	1	0	0	0	1	0	0	0	0	3F10
!	5	1	0	0	0	0	1	1	0	0	0	1	0	0	0	0	0	8620

cycled or multiplexed rapidly enough so that the message appears constant to the viewer.

The following program is developed to light the 7-segment display of Figure 1.10 with the message HELLO!. The process is begun by first developing the seven-segment codes for each of the letters of the message along with their corresponding unit driver codes. For instance, the letter H requires that segments b, c, e, f, and g be illuminated. To do this, data bits 9, 10, 12, 13, and 14 of location $00 00A0 must be set high. This translates to binary data 0111 0110 or $76. Since the H will be lit on unit 0, bit 0 of location $00 00A1 must be set high and the remaining seven bits (1–7) must be set low. Putting the segment and unit driver codes together as a 16-bit data word makes the data sent to port to be $00 00A0 $7601 light an H on unit 0. Table 1.5 lists the segment and unit driver codes as well as the data required to be placed into port $00 00A0 for each of the characters in the message HELLO!. These data are stored in memory locations $00 0040 (letter H) through $00 004A (character !) since port $00 00A0 cannot hold all of them at once.

Essentially, to produce our message, a short display program must be run six times, once for each message character. The data for each of the characters from location $00 0040 through $00 004A must be fetched, each in turn, and sent to port $00 00A0. The DELAY subroutine developed earlier is called between each character to allow some time for that character to be displayed. One pass through the message would not leave the characters displayed long enough to be seen, so the program must be repeated continually to refresh the display message.

Register D0 is used to supply the character and unit driver data to port $00 00A0 within this DISPLAY subroutine:

```
DISPLAY   MOVE D0,$00A0     * Output to port $00 00A0
          JSR DELAY
          RTS
```

Program 1.9

After the data are sent to the port, the DELAY subroutine is called. When the DELAY subroutine is finished, it returns to DISPLAY subroutine which fetches and executes the RTS instruction to return to the program that called the DISPLAY subroutine. This calling program's task is to fetch the data for D0, call the DISPLAY subroutine, point to the next character's data, get it, and recall the DISPLAY subroutine. After fetching the data for all six characters, the calling routine should stop.

The DISPLAY subroutine is called up by a JSR DISPLAY instruction in the calling program. As an example, we will locate this instruction at address $00 0122. The address of the first instruction of the DISPLAY subroutine, which is also the value of the label, DISPLAY, is $00 2800. The use of a subroutine within a subroutine (the DELAY subroutine in the DISPLAY subroutine) is an example of *nesting* subroutines. The system allows subroutines to be called within other subroutines by pairing stack pushes resulting from the JSR instruction, with stack pulls resulting from RTS instructions. During the following discussion, refer to Figure 1.16 on page 48 as an illustration of this point.

The program that calls the DISPLAY subroutine reaches location $ 00 0122 and fetches the JSR opcode. During the fetch cycle, the program counter (PC) is incremented normally to $00 0124. There, the microprocessor fetches the operand of the JSR instruction with the PC advancing to $00 0128. The PC had incremented to $00 0124 when it fetched the 16-bit opcode and to $00 0128 when fetching the 24-bit operand from two 16-bit locations, $00 0124 (holding $0000) and location $00 0126 (holding $2800). This operand is truncated to the 24-bit address $00 2800 which is the address of the DISPLAY subroutine's first instruction.

When the microprocessor decodes the JSR opcode, it stores the operand into the TEMP register and pushes the current contents of the PC ($00 0128) onto the stack. Leading zeros are added to the high byte of the PC contents to form two 16-bit words, $0000 and $0128, which are stored onto the stack. The stack pointer (SP) is decremented 4 times, from an initial value of $00 401F to $00 401B as the PC data is stacked. Following the stacking operation, the contents of the TEMP register are transferred to the PC and an opcode fetch is begun from location $00 2800, to which the PC is now pointing.

The DISPLAY program is fetched and executed normally until it reaches the JSR DELAY instruction at location $00 2804. Once again, the JSR opcode and its operand are fetched and the PC is incremented 6 times to $00 280A. The JSR operand, $00 3100, is moved to the TEMP register; and the current contents of the PC ($00 280A) are pushed onto the stack, causing the SP to decrement to a value of $00 4018. $00 3100 is transferred from the TEMP register into the PC and the DELAY subroutine is fetched and executed.

The last instruction of the DELAY subroutine is a return from subroutine (RTS) instruction at location $00 310C. As the RTS opcode is fetched, the PC is incremented to $00 310E. However, when the RTS opcode is decoded, the CPU is directed to pull two stack locations and store the data into the PC. As the data is pulled from the stack, the SP is incremented back to $00 401B. The data on the stack that is pulled is $00 280A. After the RTS is completed, the CPU starts an opcode fetch cycle

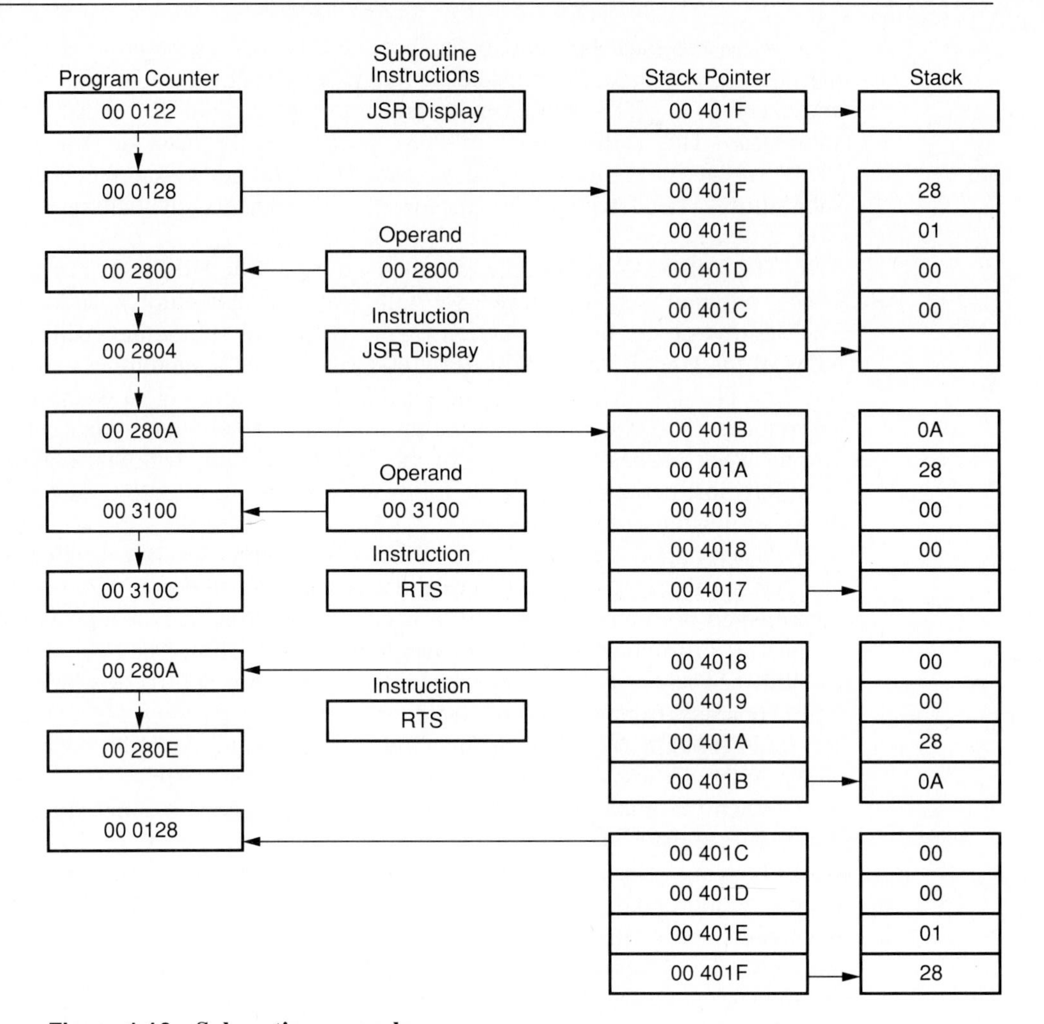

Figure 1.16 Subroutine example

which fetches the instruction at location $00 280A, the location in the DISPLAY subroutine following the JSR DELAY instruction. The instruction at this location is another RTS which, again, causes the stack to be pulled, restoring the SP to $00 401F and the PC to $00 0128. The PC now fetches the instruction which follows the JSR DISPLAY instruction and the original calling program execution is resumed.

The calling program's task is to provide the correct data to the DISPLAY subroutine. Each character's code data are retrieved from memory locations $00 0040 through $00 004A, one at a time. The index address mode is used so that data may be more easily accessed. A counter is also maintained to determine when all six characters have been fetched and used. This program we shall call SHOW and will be written as a subroutine that may be used by any other program at any time.

```
SHOW    MOVE #$0040,IX    * Initialize the index register
        MOVE #$6,D1       * Initialize character counter
MORE    MOVE $00IX,D0     * Get segment and unit codes
        JSR DISPLAY       * Call DISPLAY subroutine
        ADD #$2,IX        * Increment index pointer
        SUB #$1,D1        * Decrement counter
        BNE.S MORE        * Get next character
        RTS
```

Program 1.10

This subroutine does one pass through the six units. In order to make the message visible, it must be called over and over. This can be done by inserting a BRA.S $FB instruction following the JSR SHOW. This causes the program flow to move back to the JSR SHOW instruction upon return from the SHOW subroutine. Other applications may require that the message be displayed only for a short time. In those cases, a counter is maintained between the JSR SHOW and BRANCH back to it. This counter would allow the SHOW subroutine to be called a specified number of times before the program continued. Once the program did continue, the message would no longer be displayed.

```
        MOVE #$20,D3      * Initialize SHOW counter
AGAIN   JSR SHOW
        SUB #$1,D3        * Decrement SHOW counter
        BNE.S AGAIN
        CLR $00A0         * Blank display
```

Program 1.11

This program segment reruns the SHOW subroutine 32 times and then blanks the display (sets all segments and unit drivers off) before it continues on to the next task.

EXAMPLE 1.21

Use the SHOW subroutine to display the sign condition (POS or nEg) of the result of subtracting two numbers found in locations $00 00E0 and $00 00E2.

Solution:

A first step might be to flowchart the program (see Figure 1.17, page 50). This helps you to organize the flow of execution. The flowchart also provides a diagrammatic

Figure 1.17 Display sign flowchart

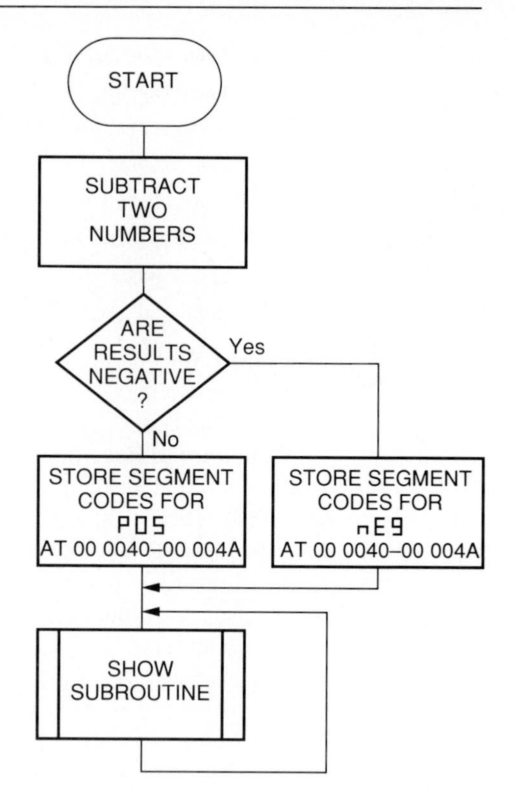

illustration of your program that others can follow—so keep it straightforward and readable to anyone.

To make our program work, data locations holding the two original numbers and the segment codes for POS and nEg must have been previously loaded with the correct data. Once this is done, the data can be fetched and used in the program. The first task is to do the subtraction and then test the result for a negative or positive value:

```
MOVE $00E0,D0       * Get first number
SUB $00E2,D0        * Subtract second number
BMI.S NEG           * Branch if result negative
```

Program 1.12

The segment codes needed for the correct display are then fetched according to the result of the test:

```
        MOVE #$0040,IX      * Initialize index pointer
        MOVE #$7301,D0      * Code for P in unit 0
        MOVE D0,$0IX        * Store unit 0 code
        MOVE #$3F02,D0      * Code for O in unit 1
        MOVE D0,$2IX        * Store unit 1 code
        MOVE #$6D04,D0      * Code for S in unit 2
        MOVE D0,$4IX        * Store unit 2
NULL    CLR $6IX            * Blank unit 3
        CLR $8IX            * Blank unit 4
        CLR $0AIX           * Blank unit 5
AGAIN   JSR SHOW
        BRA.S AGAIN
```

Program 1.13

After the character code data are transferred to $00 0040–$00 004A, the SHOW subroutine is called. Branching back to the SHOW subroutine after it has been run allows the message to be seen by the user. The following segment, branched to by the BMI instruction, selects the nEg codes.

```
        MOVE #$0040,IX      * Initialize index pointer
NEG     MOVE #$5401,D0      * Code for n in unit 0
        MOVE D0,$0IX        * Store unit 0
        MOVE #$7902,D0      * Code for E in unit 1
        MOVE D0,$2IX        * Store unit 1
        MOVE #$6F,D0        * Code for g in unit 2
        MOVE D0,$4IX        * Store unit 2
        BRA.S NULL
```

Program 1.14

1.15 READING THE KEYBOARD

Reading the keyboard interfaced to the microprocessor system is done by loading a data register from port $00 00A4 and checking the state of bit 7. If that bit is high, then a key was pressed; if it is low, no key was pressed. In the case of a key press, bits 6–0 contain the ASCII code for the key. If no key was pressed, then bits 6–0 should all be low. In order to do this, a short routine must be executed within each program to check the keyboard regularly. This routine is:

```
SEARCH   MOVE $00A4,D0    * Read keyboard
         MOVE D0,D3       * Save keyboard data
         AND #$80,D3      * Isolate bit 7
         BNE.S KEYS       * Determine key and action
         RTS              * Return if no key press
KEYS     JSR KEYMRK       * Key decoding
         RTS              * Return after key use
```

Program 1.15

First, the SEARCH routine determines if a key was pressed by isolating bit 7 from the data read from port $00 00A4. This is accomplished by reading the port into register D0 and making a copy of the data in D3. Then an AND instruction forces all the bits of D3 to be low except bit 7. The condition of bit 7 is determined by its state as read from port $00 00A4. This process of isolating a bit in a data word is called *masking*. Since the system adds leading zeros as needed, the actual data being ANDed with D3, in binary, is 0000 0000 1000 0000. ANDing occurs on a bit by bit basis, so all bits except bit 7 are definitely low.

The method used to read the keyboard as described above is called *polling* and is the simplest method to accomplish the task. However, polling does require the current program to stop what it is doing occasionally and scan the keyboard to see if a key is pressed. That is, the SEARCH subroutine must be called on a regular basis to look for a possible key press. Polling is convenient when you wish to "lock out" the keyboard for a period of time. This is done simply by NOT calling the SEARCH program until your program is ready to do so. This process should be done carefully, because if you inadvertently omit the jump to SEARCH subroutine in a critical part of your program, you may be forced to reset the system to terminate program execution.

Keyboard Detection by Interrupt

A second method to achieve keyboard scanning is by using the interrupt request (IRQ) input. This signal, when asserted, causes an interruption of the current program and forces the CPU to divert to a different program. The KEY PRESS lead of Figure 1.11 needs to be connected to the CPU $\overline{\text{IRQ}}$ input through an inverter instead of D7 flip-flop to port $00 00A4. This lead can be left connected to both the inverter and flip-flop if it is desirable to have the option of using polling or interrupts for key scanning.

When any key is pressed, KEY PRESS is asserted, still loading the ASCII code into the 74173s. It is inverted and is detected by the CPU as an interrupt request. If the I bit in the status register is set, the CPU completes its current instruction and then begins the interrupt process. The current program counter contents are pushed onto the stack and the stack pointer is decremented four times. The CPU places vector address \$00 0004 into the address buses and does a memory read, loading the information at that location into the TEMP register. A second read at location \$00 0006 is also sent to the TEMP register. Twenty-four bits of data in the TEMP register are now transferred to the PC and an opcode fetch is started. Thus the first instruction of the interrupt program (the SEARCH routine in this case), is executed. The interrupt program ends with a return from interrupt (RTI) instruction; it is similar to an RTS instruction in that its main purpose is to pull the stack. This restores the contents of the PC with the address of the instruction following the last one executed when the interrupt occurred. An opcode fetch is begun and the original program execution is resumed.

EXAMPLE 1.22

A program receives an interrupt request while fetching an opcode from location \$00 023C. The SEARCH program is stored at location \$00 08A4 with the RTI instruction at location \$00 08C6. Demonstrate the CPU activity that occurs if the stack is located starting at location \$00 401F.

Solution:

Refer to Figure 1.18 on page 54 as the process is described. The CPU detects the active $\overline{\text{IRQ}}$ during the fetching of an opcode at \$00 023C. It completes the fetch of the instruction and its operand, then executes the instruction itself. The PC has been duly incremented during the fetches to \$00 023E. The I bit in the status register is checked and found high so the interrupt process is begun. First, the I bit is automatically set low to prevent further interrupts from occurring. Next, the current contents of the PC (\$00 023E) are pushed onto the stack and the SP is decremented to \$00 401B.

The CPU then places the vector (\$00 0004–\$00 0006) onto the address bus and fetches its contents by the process described earlier, and next loads the PC with \$00 08A4. The CPU then executes the interrupt program that ends by setting the I bit (MOVE to the status register) and executing an RTI at location \$00 08C6. The RTI causes the stack to be pulled and the stack pointer to be incremented and restored to its original value of \$00 401F. \$00 023E is replaced into the PC and the original program is resumed from where it was interrupted.

The advantage of doing key scanning by interrupt is that the user program does not have to pay attention to the keyboard at all. The user program executes as if the keyboard does not exist. Only when a key is pressed is the user program execution disrupted. "Keyboard lock out" can still be accomplished by resetting the I bit in the status register to cause the CPU to ignore the $\overline{\text{IRQ}}$ signal when it is asserted.

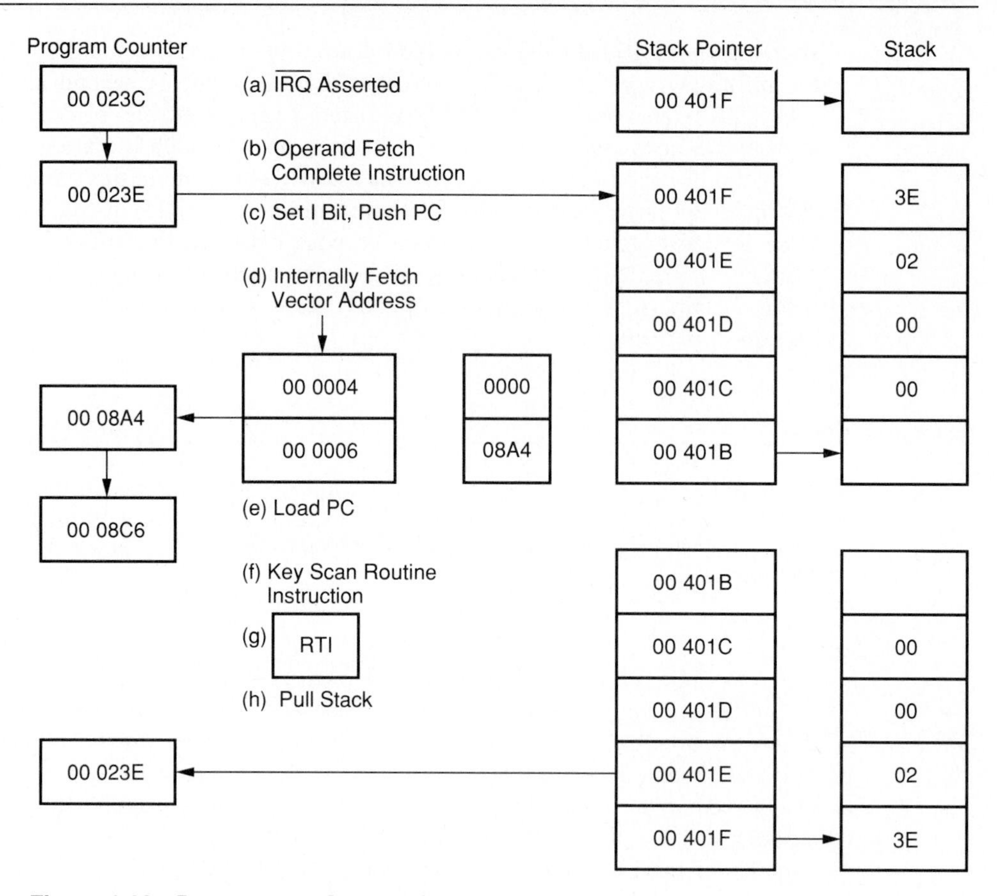

Figure 1.18 Interrupt stack operation

1.16 KEYBOARD APPLICATION PROGRAM: GRADE CONVERSION

Now we are ready for an application program using the keyboard to enter the data. Return to the earlier grades problem (Section 1.13) and, in addition, store in a location called LETTER the letter-grade equivalent to the numerical grade entered from the keyboard. Once again, the compare instruction is used to determine the grade level without altering the numerical value originally entered.

Start by using SEARCH to enter the grade into register D0. Next, compare the value in register D0 with the limits of each letter grade until its equivalent is determined. By using register D1 to keep track of the letter grade, we can develop a fairly short and direct program.

```
START    JSR SEARCH      * Keyboard entry to D0
         MOVE #$0A,D1    * Letter A
         CMP #$89,D0     * Letter B limit
         BLS.S GRADEB    * Grade not an A?
STORE    MOVE D1,LETR    * Store letter grade
DONE     BRA.S DONE
GRADEB   ADD #$1,D1      * Letter B
         CMP #$79,D0     * Letter C limit
         BHI.S STORE     * Grade is a B?
         ADD #$1,D1      * Letter C
         CMP #$69,D0     * Letter D limit
         BHI.S STORE     * Grade is a C?
         ADD #$1,D1      * Letter D
         CMP #$59,D0     * Letter F limit
         BHI.S STORE     * Grade is a D?
         ADD #$2,D0      * Letter F
         BRA.S STORE     * Grade is an F
```

Program 1.16

As a matter of practice, work this program out with several numerical grade values until you are satisfied that it stores the correct letter grade each time. Notice the use of the branch if less than or the same (BLS) and the branch if higher (BHI) instructions to test the results of the compare instructions. Their use is needed because we are dealing with unsigned (magnitude-only) values.

This program illustrates the "art" in programming; that is, finding different and improved solutions to a programming problem. Reducing memory space to hold programs or shortening execution time, thus making a program more efficient, is a primary objective of any programmer after the initial solution has been found. Attempting to better a program helps you, the student, to understand machine-level programming and gives you an opportunity to become more familiar with a microprocessor's instruction set. Compare this program solution to the original grades program at the beginning of the chapter.

1.17 DISPLAYING A KEY: COMBINING SUBROUTINES

SEARCH is a useful subroutine for a system initialization program. This subprogram, residing in ROM, is the key to making a computer system useful to the user. Another necessary subroutine is some form of display subroutine similar to the SHOW subroutine developed earlier. A system requires some method to display the results of user programs and to show the data as it is entered into its memory. A problem arises in that data entered from the keyboard are not in the same form required by the 7-segment displays. The initialization program requires a *lookup table* that relates, in some fashion, the values of the keys with their 7-segment equivalents. A lookup table is a number of memory locations set aside to hold fixed data that can be accessed (looked up) and used by any program at any time. As an example, this lookup table starts at location $00 0050 with the label CODES.

The problem now is to get the 7-segment code equivalent to a pressed key into the correct location between $00 0040 and $00 004A to be displayed on a selected unit. We will call this new subroutine KEYSEE. Location LOOK ($00 004C) holds lookup table pointer information.

KEYSEE Program

The KEYSEE program is begun by establishing a lookup table pointer at $00 0050, labeled CODES, and saving that pointer at location LOOK. A section of the lookup table is shown in Table 1.5 along with the table addresses used for this program. Additional locations—SHOWPT, to maintain a pointer for display locations $00 0040 through $00 004A, and UNIT, to keep track of display unit codes—are used by KEYSEE.

```
        MOVE #$6,D2           * Display counter
        MOVE #$1,UNIT         * Initialize UNIT
        CLR SHOWPT            * Upper address = $00
        MOVE #$40,SHOWPT+2    * $0040 --> SHOWPT+2
READ    MOVE #$CODES,IX       * Initialize index register
        MOVE IX,LOOK          * Store base value at LOOK
        JSR SEARCH
```

Program 1.17

Key data are in register D0 at the end of SEARCH and have the ASCII value of the key pressed. Location LOOK+2 contains the lower 16 bits of the base address of the lookup table which has the 7-segment codes for each of the keys. By adding the ASCII key value of the key pressed to this base value, we develop a pointer into that table.

```
ADD D0,D0           * Double D0
ADD LOOK+2,D0       * Key value + base --> D0
MOVE D0,LOOK+2      * D0 --> LOOK+2
MOVE IX,LOOK        * Place pointer value into IX
```

Program 1.18

The ASCII code is doubled to ensure that the new value is even. This value is added to the base pointer lower 16 bits at LOOK+2. The even pointer value is stored back to LOOK+2, so that LOOK now contains the modified pointer information for the key pressed. The modified pointer is then loaded into the index register which then points to the table location holding the 7-segment code for the pressed letter. Assuming that the first key press is to be illuminated in unit 0, the value $01 is appended to the 7-segment code and stored at location $00 0040.

```
MOVE $0IX,D0        * Get 7-segment code
ADD UNIT,D0         * Append unit code
MOVE IX,SHOWPT      * Reinitialize IX
MOVE D0,$0IX        * Store character code
```

Program 1.19

EXAMPLE 1.23

How does the 7-segment code for letter A get into location $00 0040?

Solution:

The SEARCH routine fetches ASCII code 41 for the letter A. When the index register contents were stored at LOOK, the data $0000 was stored at $00 004C and the value $0050 at $00 004E. Address LOOK+2 is $00 004E, so that when the doubled contents of D0 ($0082) are added to LOOK+2 ($0050), the result, $00E2 is stored back to $00 004E. The index register is loaded from location LOOK with the modified pointer value. MOVE $0IX,D0 retrieves the 7-segment code ($7700) from location $00 00E2 (see Table 1.5) and appends $01, the unit code, to it with the add instruction, making the data sent to $00 0040, $7701.

Following storage of the data at $00 0040, the value at SHOWPT+2 is incremented twice to point to the next display storage location, UNIT is incremented once to hold the next unit's code value, and the display counter is decremented and if it is not zero, a branch is taken back to READ.

```
ADD #$2,SHOWPT+@     * Adjust display store
ADD #$1,UNIT         * Next display unit
SUB #$1,D2           * Decrement display counter
BNE.S READ           * Get next key code
RTS
```

Program 1.20

Once the six display locations, $00 0040–$00 004A, have been loaded with character and unit codes, the SHOW subroutine can be called to display them on the 7-segment display.

The approach to programming used in this chapter is one of modular design, by developing subroutines to be used as individual sections of an overall program. By using subroutines, larger problems are broken into smaller subunits or subprograms. These subprograms, when used together, can perform larger tasks. The subroutines developed in this chapter are representative of the types of program blocks used in system initialization and other programs.

PROBLEMS

1.1 Explain the function of the program counter.

1.2 Why is it necessary to have a portion of memory dedicated for ROM?

1.3 Which unit receives opcode information during a fetch cycle?

1.4 What is the decimal equivalent of the largest unsigned number that a 16-bit processor can store at a memory location?

1.5 What are the decimal equivalents of the largest positive and largest negative numbers that can be stored at a memory location?

1.6 How many different memory locations can be directly addressed by the microprocessor described in this chapter? Supply answer in both decimal and hexadecimal form.

1.7 Determine if the overflow flag (V) is set or reset by each of these addition problems (all numbers are HEX):
(a) 921C + 8B51 (b) 6CA1 + 5F22 (c) 9999 + 3E02 (d) BB1C + 2CCC

1.8 For each of the problems in Problem 1.7, what are the conditions of the X, N, C, and Z flags?

1.9 Using the order of the status flags in the status register, what is the hexadecimal equivalent of the status register as a result of these subtraction problems (assume unused flags are reset low)? All numbers are HEX:
(a) 952C − 4C11 (b) DD45 − BC1B (c) 1000 − EFFF (d) 8000 − 1000

1.10 List the three main operations of the opcode fetch cycle.

1.11 How many clock cycles does an opcode fetch use?

1.12 Which control signal indicates to memory devices that a valid address is present on the address bus?

1.13 Describe briefly the reset process for the microprocessor used in this chapter.

1.14 What is the difference in operation between interrupt requests $\overline{\text{IRQ}}$ and $\overline{\text{NMI}}$?

1.15 What are the segment and unit codes necessary to illuminate the character H in units 1 and 4 using the display circuit in the text?

All subroutines and circuits developed in this chapter are usable for the program problems below.

1.16 Using the 7-segment display circuit of this chapter, write a program to display J on units 3 and 5.

1.17 There are two ways to blank the display used in this chapter. What are they?

1.18 Write a program which finds the largest unsigned number in a data bank in locations $00 0080 to $00 009E and stores this number in location $00A0.

1.19 What changes have to be made to the program in Problem 1.18 to use signed data? The value sought becomes the most positive number.

1.20 Write a subroutine that will take data in register D0 and display that data on any two consecutive display units selected by storing the first unit number in a memory location labeled UNITPIC. Call the subroutine DATAWORD.

1.21 Extend the grades program to display the grade and its letter value as shown below. Hint: Use the subroutines DATAWORD and SHOW.

Figure 1.19 Problem 1.21

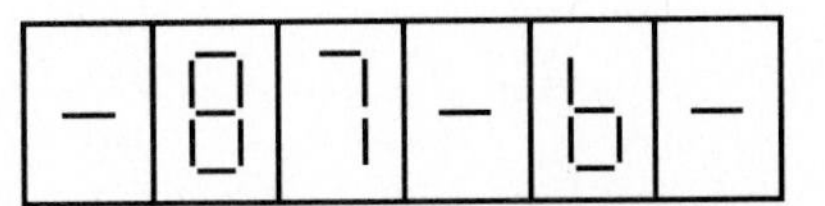

Note: Values for illustration purposes only.

1.22 Write a program which displays a 6-digit binary count on the 7-segment display, starting with 000000 and incrementing to 111111, with a sufficient amount of delay between each count to allow the display to be read. The count returns to 000000 and resumes counting in a loop, repeating the sequence continually.

1.23 Write a program which adds two BCD numbers entered to the system from the keyboard used in this chapter.

1.24 Handshaking is a process whereby the computer and a peripheral device signal each other during a data transfer. Write a program that outputs the value of two key presses from the keyboard to a peripheral connected to port $00 8004. After the second key is pressed, b7 of the port is to be made high, signaling the peripheral that data is available at the port. The program will initially set bit 7 low. Bit 7 is returned low and the program repeated when it senses bit 15 of port $00 8004 being made low by the operator of the peripheral device, who uses the data sent out through the port. Use the subroutines developed in this chapter and the figure below for orientation of the data in port $00 8004:

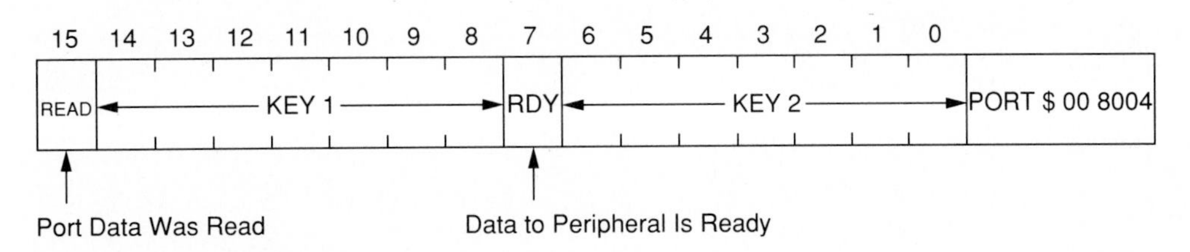

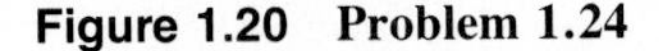

Figure 1.20 Problem 1.24

1.25 Extend the program in Problem 1.24 to include an indication on the display of the control signal activity. Initially, the display shows the message rEAdy until the first key is pressed. After b7 of port $00 8004 is made high, the message SEnd is displayed. When b15 is sensed as low, the program repeats.

2

INTRODUCTION TO THE 68000

CHAPTER OBJECTIVES

The objective of Chapter 2 is to introduce the 68000 microprocessor from a hardware standpoint. The architecture of the device is discussed first, followed by an example system developed around the 68000. Interconnection with ROM and RAM devices establishes this basic system.

2.1 GLOSSARY

assert Make active.
contention Situation in which sets of data vie for the use of a bus or control line.
exception Any programs other than user programs.
long word For the 68000, 32-bit binary data.
mask Disable or omit.
strobe Control signal that activates an action.
supervisor mode Operational mode for execution of exception programs.
trace mode Single-step execution mode for debugging.
user mode Operational mode for execution of user programs.
word For the 68000, 16-bit binary data.

The 8-bit microprocessors were the mainstay of the small-computer industry through the 1970s and early 1980s. However, just as the transistor and small-scale ICs were eventually eclipsed, so was the 8-bit microprocessor. Motorola developed the 68000 to be the start of a new line of microprocessors, sacrificing upward compatibility from their 8-bit microprocessor, the 6800. The principles and programming techniques discussed in Chapter 1 help lay down a foundation for the 68000. Basic addressing modes and memory transfer concepts for the 68000 are similar to those discussed in Chapter 1. New ideas are incorporated and the overall functions of the microprocessor have been expanded.

2.2 OPERATING MODES OF THE 68000

The 68000 family of microprocessors operates in one of two modes, the **user** or **supervisor.** The first of these two, the user mode, is reserved for executing programs written by system users and programmers. The supervisor mode handles all other programs presented to the 68000. These programs are grouped under the heading of **exception** programs; that is, they are exceptions to normal user programs. Examples of these programs are interrupt handling, initialization, traps, and bus error programs. Details on exception programs are discussed in later chapters. The 68000 switches into the supervisor mode automatically whenever an exception program is to be executed. It is returned to the user mode by a return from exception instruction.

2.3 ARCHITECTURE OF THE 68000

Figure 2.1 is the programming model and pin out for the 68000. First, note that there are no separate accumulators or index registers. Instead, there are eight data registers (D0 through D7), seven address registers (A0 through A6), and two stack pointers: a user stack pointer (A7) and a supervisor stack pointer (A7′). The stack pointer registers can be used for most applications as an eighth address register. All 8 data registers can be used as accumulators for data handling, and all 15 data and address registers, as well as the stack pointers, can be used as index registers.

Finally, in the program model, there is a status register. The lower 8 bits of the status register compose the condition control register (CCR) flags, while the upper half composes the supervisor status register and contains some new flags that are not used in the 8-bit microprocessors.

All data and address registers, as well as the program counter, are 32 bits wide, meaning that the size of data that can be handled by the 68000 is 32 bits and not 16. Then why call this device a 16-bit device? The answer lies in the data bus size, which is 16 bits (D_0–D_{15}) To understand this apparent conflict, think back to the index register described in Chapter 1, which is similar to the index register of the 6800 and Zilog Z80 microprocessors. This register contained 24 bits, even though the data bus is 16 bits wide. To load the index register required two memory locations that hold the data to be moved to the index register. Those data were transferred 16 bits at a time—first the upper word was moved to the index register along the data bus followed by the lower word. In the 68000, similar types of transfers will move 32 bits along a 16-bit data bus.

In designing the 68000 family of microprocessors, Motorola elected to incorporate many of the features of its original top-of-the-line microprocessor, the 68020 (discussed in detail in later chapters) into the earlier members of the family. Internally, the 68000 microprocessors are 32-bit devices. That is, the data and address buses, as well as the register complement, all are 32 bits wide. Externally, the 68000 is interfaced with a 16-bit data bus and a 24-bit address bus.

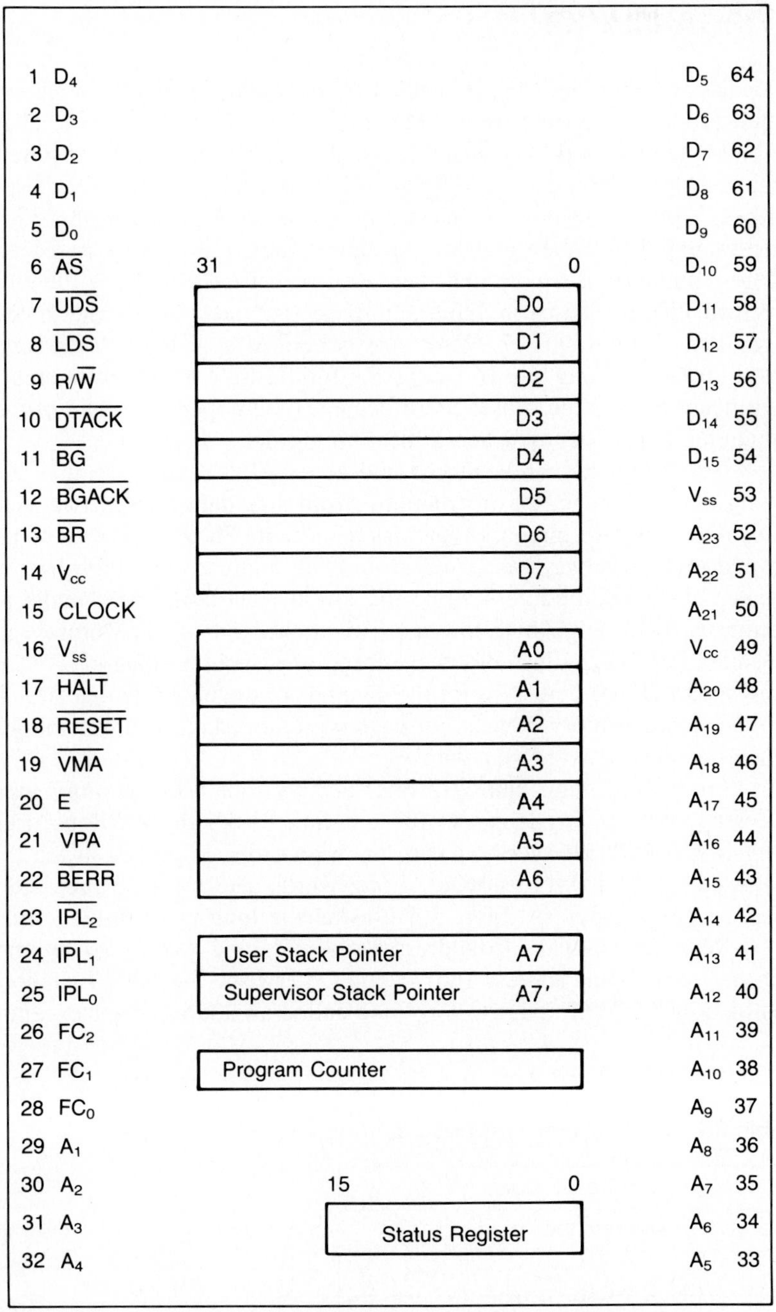

Pin	Signal	Signal	Pin
1	D_4	D_5	64
2	D_3	D_6	63
3	D_2	D_7	62
4	D_1	D_8	61
5	D_0	D_9	60
6	$\overline{AS}$	D_{10}	59
7	$\overline{UDS}$	D_{11}	58
8	$\overline{LDS}$	D_{12}	57
9	$R/\overline{W}$	D_{13}	56
10	$\overline{DTACK}$	D_{14}	55
11	$\overline{BG}$	D_{15}	54
12	$\overline{BGACK}$	V_{ss}	53
13	$\overline{BR}$	A_{23}	52
14	V_{cc}	A_{22}	51
15	CLOCK	A_{21}	50
16	V_{ss}	V_{cc}	49
17	$\overline{HALT}$	A_{20}	48
18	$\overline{RESET}$	A_{19}	47
19	$\overline{VMA}$	A_{18}	46
20	E	A_{17}	45
21	$\overline{VPA}$	A_{16}	44
22	$\overline{BERR}$	A_{15}	43
23	$\overline{IPL_2}$	A_{14}	42
24	$\overline{IPL_1}$	A_{13}	41
25	$\overline{IPL_0}$	A_{12}	40
26	FC_2	A_{11}	39
27	FC_1	A_{10}	38
28	FC_0	A_9	37
29	A_1	A_8	36
30	A_2	A_7	35
31	A_3	A_6	34
32	A_4	A_5	33

31 0
D0
D1
D2
D3
D4
D5
D6
D7

A0
A1
A2
A3
A4
A5
A6

User Stack Pointer A7
Supervisor Stack Pointer A7'

Program Counter

15 0
Status Register

Figure 2.1 68000 program model and pin assignment (Courtesy of Motorola, Inc.)

2.4 DATA ORGANIZATION

The sizes of data that can be handled by the 68000 instructions are bit, byte (8-bit), **word** (16-bit), and **long word** (32-bit) widths. Additionally, special instructions interpret byte data as BCD (binary coded decimal) numbers. All instructions involving data registers affect the condition control register flags, while most operations on address registers do not. To make things less complicated in this text, program applications will use data registers to handle data and address registers to operate as address pointers. But note that they are not restricted to these functions. Data sizes are formatted as shown in Table 2.1. Byte size data occupy the lowest eight bits of a register—b0 through b7. Word size is held in b0–b15 with the lower byte of the word in b0–b7. Long words occupy the full 32-bit contents of the register. The most significant byte of the 4-byte word occupies the upper eight bits of the register. The remaining bytes occupy space in declining order.

Byte size data can be stored and accessed at each individual address by specifying that address in the instruction. Word size data require two bytes of data (16 bits) and, thus, two memory locations to store it. These data must be held in memory so that they are always accessed at an even address. The high byte of the word is located at the even address while the low byte is at the next highest address. This conforms to the high byte-low address orientation of all Motorola microprocessor systems. Table 2.2 illustrates the storage of a word in memory. Memory maps are drawn with 16 bits across, since the data bus of the 68000 is that size. Actually, there are two 8-bit memory devices for each word size data. Both memories are accessed when transferring word size data.

Long word data, being 32 bits, occupy four 8-bit memory locations and are accessed or transferred one word at a time. Like the previous organization, the long word would be accessed with the high-order word first at the lower addresses followed by the low-order word. Long words also must be accessed starting at an even-numbered address. Table 2.3 illustrates a long word in memory.

Remember always to address word and long word data to or from memory starting at an even address (b0 = 0). Otherwise, the 68000 cannot execute your instruction, but will divert to a routine called an address error exception.

Table 2.1 Register Data Organization

Register Bits				
31-24	23-16	15-8	7-0	Data Size
			Byte	Byte*
		High byte	Low byte	Word
High word		Low word		Long word

*Data registers only

Table 2.2 Word Organization in Memory

Even Address	Even Address + 1
High byte (D_8–D_{15})	Low byte (D_0–D_7)

Table 2.3 Long Word Organization in Memory

Even Address	Even Address + 1
Upper byte of the upper word (D_{24}–D_{31})	Lower byte of the upper word (D_{16}–D_{23})

Even Address + 2	Even Address + 3
Upper byte of the lower word (D_8–D_{15})	Lower byte of the lower word (D_0–D_7)

2.5 68000 PIN DESCRIPTIONS

A_1–A_{23} (Outputs)

Twenty-three of the 24 address lines form the address bus to interconnect memories and other devices to the microprocessor. Address information flows from the microprocessor to the devices connected to the bus. Twenty-four address bits allow up to 16,777,216 bits, or 16 megabytes, of direct addressing. Address line A_0 is not available directly from the microprocessor. Instead, it is used in combination with data size information to direct the condition of control lines $\overline{UDS}$, the upper data **strobe,** and $\overline{LDS}$, the lower data strobe. Memory locations are referenced using all address lines (A_0–A_{23}). This requires that $\overline{UDS}$ and $\overline{LDS}$ be used in the address decoding scheme for a 68000-based system. Odd addresses ($A_0 = 1$) and even addresses ($A_0 = 0$) define exact locations for byte size data. Instead of decoding these locations with A_0, the data strobes are used as illustrated later.

D_0–D_{15} (Bidirectional)

The 68000 has a 16-bit data bus that allows up to a word of data to be transferred between the microprocessor and other devices at one time. The data bus is bidirectional and, with the aid of the data strobes, allows transfers of byte size data on both halves of the bus as well as word transfers using the entire 16-bit width of the bus.

$\overline{AS}$ (Address Strobe—Output)

This control signal indicates that a valid address is present on the address bus during a memory transfer operation. This signal can be used as a steering signal or as a strobe input to external devices that require address data to be latched into them.

$\overline{\text{DTACK}}$ (Data Transfer Acknowledge—Input)

External devices must make this line low to complete a memory transfer cycle successfully. In response to $\overline{\text{DTACK}}$ during a memory read cycle, the 68000 latches the information on the data bus into the data buffers. For a memory write cycle, $\overline{\text{DTACK}}$ indicates to the 68000 that the information sent out on the data lines reached the memory device for which it was intended. The 68000 expects either this signal or a $\overline{\text{BERR}}$ (see below) to be returned in response to memory read or write operations. If the microprocessor does not receive either of these signals during a memory transfer cycle, it inserts wait cycles, doing very little, as it patiently waits for $\overline{\text{DTACK}}$. In this manner, slower memory devices can be used within a 68000-based system.

$\overline{\text{RESET}}$ (Bidirectional)

The reset line of the 68000 is a bidirectional control line. As an output lead, it responds to a RESET instruction by going low for 124 clock cycles. In this condition, it can be used to reset external devices under software control. No internal units of the 68000 are affected by the RESET instruction (except the program counter, which is incremented during the fetch of the instruction from memory).

As an input, $\overline{\text{RESET}}$, in combination with the $\overline{\text{HALT}}$ input, puts the 68000 into a master reset condition. The master reset sets the interrupt **mask** to disable all maskable interrupt requests. It also causes the program counter and supervisor stack pointer to be initialized with data from memory. The supervisor stack pointer is loaded with data from locations $00 0000–$00 0003 and the program counter is initialized with data from locations $00 0004–$00 0007. The 68000 is placed into the supervisor mode to execute a $\overline{\text{RESET}}$ exception routine, which is the system's restart or initialization program.

$\overline{\text{HALT}}$ (Bidirectional)

$\overline{\text{HALT}}$ is the other bidirectional control line. As an input by itself, $\overline{\text{HALT}}$ makes all control signals become inactive, places the address and data buses to high impedance, and places the 68000 into an idle state. The 68000 enters the halt state in response to an active $\overline{\text{HALT}}$ at the completion of the current bus cycle. It can be taken out of the halt state by removing the $\overline{\text{HALT}}$ input. Making $\overline{\text{HALT}}$ and $\overline{\text{RESET}}$ active simultaneously generates a master reset in the 68000. In combination with $\overline{\text{BERR}}$ (see next paragraph), the $\overline{\text{HALT}}$ causes the bus cycle where the error occurred to be rerun in hope that the error will no longer be there. As an output line, the microprocessor makes the $\overline{\text{HALT}}$ low if it has been halted as a result of an internal bus failure, such as a double bus fault. A double bus fault occurs when a bus error is experienced while the bus error exception routine is in process.

$\overline{\text{BERR}}$ (Bus Error—Input)

Logic can be designed between the 68000 and external devices to allow the external devices to return a $\overline{\text{BERR}}$ in place of a $\overline{\text{DTACK}}$ in the event of a system problem. The application of a $\overline{\text{BERR}}$ by itself causes a bus error exception routine to be run by the microprocessor. **Asserting** (making the signal active) $\overline{BERR}$ along with $\overline{HALT}$ causes the 68000 to rerun the bus cycle in which the error was detected.

R/$\overline{W}$ (Read/Write—Output)

This is a standard read/write line. It is high when reading data from memory and low when writing data to memory.

$\overline{UDS}$ and $\overline{LDS}$ (Upper and Lower Data Strobe—Output)

These signals are directed by A_0 to select whether data are to be moved on the upper or lower half or both halves of the data bus. Table 2.4 is a truth table of the $\overline{LDS}$, $\overline{UDS}$, and R/$\overline{W}$ signals and their effects on the data bus. When transferring byte size data, A_0 internally selects whether $\overline{LDS}$ (for odd addresses $A_0 = 1$) or $\overline{UDS}$ (for even addresses $A_0 = 0$) is driven active. For word size transfers, addresses always are referenced to even bounds ($A_0 = 0$) and both data strobes are driven active.

The $\overline{UDS}$ and $\overline{LDS}$ are used with address decoders to address memory chips via their chip-select inputs. In this way, 8-bit memories can be used with the 68000. The simplest data transfer to illustrate is a word transfer, since the entire data bus is used. When the 68000 is directed by an instruction to move a word of data between it and memory, both data strobes are made active. This causes memory devices connected to both halves of the data bus to be enabled so that the transfer is completed successfully.

Byte data always are transferred to or from the lowest byte of a data register. This information can be supplied to a memory device connected either to the upper half of the data bus (even address) or to the lower half of the bus (odd address).

Table 2.4 Data Strobe Truth Table

A_0	$\overline{UDS}$	$\overline{LDS}$	R/$\overline{W}$	Data Bus(D_0-D_{15})
0	Low	High	High	Byte read on D_8-D_{15}
1	High	Low	High	Byte read on D_0-D_7
0	Low	Low	High	Word read on D_0-D_{15}
0	Low	Low	Low	Word write on D_0-D_{15}
1	High	Low	Low	Byte write on D_0-D_7
0	Low	High	Low	Byte write on D_8-D_{15}
-	Low	Low	Tri state	No valid data on the bus
-	High	High	-	No valid data on the bus

1. During byte read, data are valid on the selected portion of the data bus only.
2. During byte write, valid data can be present on the entire bus. However, only the portion selected is written into memory.
3. The condition of A_0 is an internal state governed by the odd or even value of the memory location being accessed.

During a byte transfer from the data register to memory, the lowest byte from the register is copied into both halves of the microprocessor's data bus buffers. That is, D_0–D_7 are identical to D_8–D_{15}. The appropriate data strobe ($\overline{UDS}$ for even addresses and $\overline{LDS}$ for odd addresses) is made active, causing the data on the data bus to be loaded into or read from the correctly enabled memory device. The data on the other half of the data bus are not used, so that only the appropriate byte is transferred between the microprocessor and memory. For byte read operations from memory to a data register, the memory device is required only to place data on the appropriate half of the 16-bit data bus in relation to the even or odd address being read. The microprocessor ignores the other half of the bus during the transfer.

A 32-bit long word is transferred one word at a time and requires two memory transfer cycles. The high word is moved first to the data buffers and transferred as a word onto the data bus. The second memory cycle transfers the lower word out to memory. The correct addresses are maintained and pointed to by the address buffers connected to the address lines of the processor.

FC_0, FC_1, and FC_2 (Function Control—Outputs)

The 68000 has two processing modes: user mode and supervisor mode. The user mode is used for normal programming, while the supervisor mode handles any programs except the normal user program. The function control lines indicate what type of bus cycle is being performed by the 68000. Bus cycles are user and supervisor program fetches, and user and supervisor data transfers. They can be used as steering signals to the devices connected to the 68000 buses. Table 2.5 is a truth table for the function control lines.

Table 2.5 Truth Table for Function Control Lines

FC_2	FC_1	FC_0	Cycle Type
0	0	0	Undefined
0	0	1	User data
0	1	0	User program
0	1	1	Undefined
1	0	0	Undefined
1	0	1	Supervisor data
1	1	0	Supervisor program
1	1	1	Interrupt acknowledge

1. Data areas are restricted to data used by a program.
2. Program areas contain actual coded programs.
3. User areas are normal programmer's memory.
4. Supervisor areas hold exception routines.

The function control lines restrict areas of memory to their designated functions through logic connection to various memory chip's chip selects. In this manner, a section of memory dedicated to user program listings can be accessed only during an instruction fetch. FC_2, FC_1, and FC_0 = 010 at that time and are decoded to enable the section of memory dedicated to user program space. Other combinations of the function control lines are generated according to the appropriate action of the microprocessor and are used similarly to restrict access to designated memory areas.

6800 Interfacing Signals

Several signals allow 6800 peripheral devices to be used with the 68000. They are the enable (E), valid memory address (VMA), and valid peripheral address (VPA). They allow the system designer to take advantage of many existing byte size devices such as peripheral interface adapters (PIAs) and asynchronous communications interface adapters (ACIAs).

E (Enable—Output) This output supplies the phase 2 clock input to 6800-family interface chips. It is developed by being held low for 6 clock cycles and high for 4 clock cycles, for a total of 10 clock cycles for each E cycle. As such, a 40 percent duty cycle clock (1/10 the frequency of the 68000 clock input) is continually output on the E line. The E clock is used in place of a system clock to operate the slower 8-bit devices when they are used in a 68000-based system.

$\overline{\text{VMA}}$ (Valid Memory Address—Input) $\overline{\text{VMA}}$ indicates that a valid address is present on the address bus when addressing 6800-type peripheral chips. This signal is synchronized to the E signal and is generated in response to a $\overline{\text{VPA}}$ input to the 68000 from the address decoders. It is used as part of the chip select process to 8-bit peripherals such as Motorola's Peripheral Interface Adapter (PIA) and Asynchronous Communications Interface Adapter (ACIA).

$\overline{\text{VPA}}$ (Valid Peripheral Address—Output) This input to the 68000 indicates that an access is desired between the 68000 and a 6800 or other 8-bit-type peripheral interface chip during the current bus cycle. Since these devices are slower than 68000 microprocessors and support chips, data transfers between the 68000 and these devices are synchronized to the E signal. A $\overline{\text{VPA}}$ input causes the 68000 to generate a $\overline{\text{VMA}}$ and insert wait states until a $\overline{\text{DTACK}}$ is asserted, indicating the transfer has been successfully completed.

The $\overline{\text{VPA}}$ line can also be used with the interrupt handling process of the 68000. If an interrupting device asserts this line active in response to an interrupt acknowledge (FC_2, FC_1, FC_0 = 111), a process called auto-vectoring occurs. Normally, in response to an interrupt acknowledge, the interrupt-requesting device supplies interrupt vector information to the microprocessor on its data bus. This information is used by the microprocessor to develop a corresponding vector address that points to an associated interrupt program. The auto-vector operation causes the microprocessor internally to select fixed vector information associated with the priority level of the interrupt request. Using this information, the microprocessor formulates the vector address.

$\overline{IPL_0}$, $\overline{IPL_1}$, $\overline{IPL_2}$ (Interrupt Requests—Inputs)

The 16-bit microprocessor described in Chapter 1 had two interrupt request inputs, $\overline{IRQ}$ and $\overline{NMI}$. The 68000 uses a different scheme. Three interrupt input lines, $\overline{IPL_0}$, $\overline{IPL_1}$, and $\overline{IPL_2}$, input interrupt requests in a priority-level scheme (IPL stands for Interrupt Priority Level). Each of the 8 combinations placed on these lines represents an interrupt level from 0 (all input lines high) to 7 (all input lines low).

Level 0 is a "no interrupt request" level—all IPL inputs are inactive (not asserted). Level 7, on the other hand, is designated as the non-maskable interrupt. This interrupt request level cannot be **masked** or ignored by the microprocessor under normal operation. The remaining levels (1–6) are serviced if their levels are higher than the level set by the 3 interrupt mask bits (I_0, I_1, and I_2) of the supervisor status register.

Direct Memory Access Arbitration Signals

Direct memory accessing (DMA) is a process by which the microprocessor relinquishes control of the buses (i.e., setting them to a high-impedance or tri-state condition) so that another device can do memory transfers using the buses directly without microprocessor intervention. This type of transfer is much faster since no time is needed for microprocessor instruction fetch cycles. Three control leads described in the following paragraphs are related to DMA operations.

$\overline{BR}$ (Bus Request—Input) An external device sets this lead low to request that the 68000 turn over control of its buses so that the DMA device can use them.

$\overline{BG}$ (Bus Grant—Output) The 68000 informs the device that it is ready to relinquish its bus control by making this output low.

$\overline{BGACK}$ (Bus Grant Acknowledge—Input) In response to the $\overline{BG}$ signal, the device must assert this lead and maintain it active during the entire DMA transfer operation. Upon receipt of this signal, the 68000 sets its data, address, and control buses to high-impedance states, in effect disconnecting itself from these buses. The DMA device can now apply the necessary signals to these buses to perform data transfers between itself and the existing memory and I/O interfaces of the system. The device returns control to the 68000 by allowing this line to go inactive.

V_{cc}, V_{ss} (Gnd), and CLOCK Pins

The last few pins are the V_{cc}, V_{ss} (Gnd), and CLOCK pins. Source voltage (V_{cc}) is nominally 5 V with respect to ground (V_{ss}). Maximum applicable voltages that can be put on V_{cc} or any input pin without damage to the 68000 are specified from −0.3 V to +7.0 V. This is *not* an operating range of voltages, but rather a maximum non-destruct voltage limit. Actual logic voltage levels for input and output pins are shown in Table 2.6. The clock rate depends on the particular version of the 68000 used and ranges from 4 MHz to 12.5 MHz. Typical rise and fall times needed for the clock are 10 ns maximum for all rates except 12.5 MHz, where it is 5 ns. The E signal, since it is $\frac{1}{10}$ of the clock rate, ranges from 400 kHz to 1.25 MHz, which is sufficient to operate 6800 peripheral devices with maximum clock inputs of 2 MHz.

Table 2.6 Operating Logic Voltage Levels

Characteristic	Symbol	Minimum	Maximum
Input high voltage	V_{IH}	2.0 V	V_{CC}
Input low voltage	V_{IL}	V_{SS} - 0.3 V	0.8 V
Output high voltage	V_{OH}	2.4 V	--
Output low voltage	V_{OL}	--	0.5 V

2.6 THE STATUS REGISTER

You are now familiar with status flags in the user condition code register (CCR) portion of the status register (see Section 1.3). The sign (N), zero (Z), overflow (V), and carry/borrow (C) flags are identical in function to the flags of the status register described in Chapter 1.

The supervisor portion of the status register contains the S flag, which indicates that the 68000 is in the supervisor mode of operation, and the T flag, which stands for the **trace** or single-step **mode.** In this mode, the user program instructions are executed one at a time. A trace exception routine, previously written into ROM, is run after each instruction to allow the user to examine the register contents. This is helpful for debugging software problems. The remaining three usable bits of the supervisor status register are the interrupt mask bits: I_0, I_1, and I_2. They are set to select the level above which interrupts will first be serviced. The input level on $\overline{IPL_0}$, $\overline{IPL_1}$, and $\overline{IPL_2}$ must be higher than the level set by the three interrupt mask bits for an interrupt to be serviced. The organization of the status bits is illustrated in Table 2.7. Unused status bits are permanently held low.

2.7 MEMORY CHIPS AND MEMORY MAPPING A 68000 SYSTEM

After selecting a microprocessor to design a system, one must determine what and how much memory to incorporate into the system. Although much time and energy has been spent designing the microprocessors and specialized peripheral devices to operate in a word size environment, unfortunately very little time has been spent

Table 2.7 Status Register

Supervisor Status Register								Condition Code Register							
15	14	13	12	11	10	9	8	7	6	5	4	3	2	1	0
T	0	S	0	0	I_2	I_1	I_0	0	0	0	X	N	Z	V	C

on memory devices. Manufacturers have been content to use 8-bit memories for two reasons: (1) there are plenty of 8-bit devices available; and (2) there has been a rush to perfect impressive microprocessors and coprocessors that are attractive to the customer, leaving little time to develop large memories for these systems. For discussion purposes, two separate erasable, programmable read-only memory (EPROM) devices are used to illustrate the process of interconnecting ROM with the 68000. They are Motorola's 6836, an 8-bit device, and Intel's 27210, a 16-bit device.

M6836E16: ROM

The M6836E16 is a Motorola 16K-byte combination ROM/EPROM. This memory is arranged as a 16,384 × 8 memory chip, meaning that it contains 16,384 directly addressable locations, each 8 bits wide. The lower 2K are bulk-erasable EPROM locations, the remaining 14K are dedicated to ROM memory. The pin out diagram for this chip is shown in Figure 2.2 and includes 8 data lines (DQ_0–DQ_7) that provide a means for information to be programmed into the chip's EPROM or ROM area. The DQ lines also allow data to be transferred to the data bus from any of the ROM's locations. Fourteen address lines allow data access from among the 16,384

Figure 2.2 **6836 ROM/EPROM (Courtesy of Motorola, Inc.)**

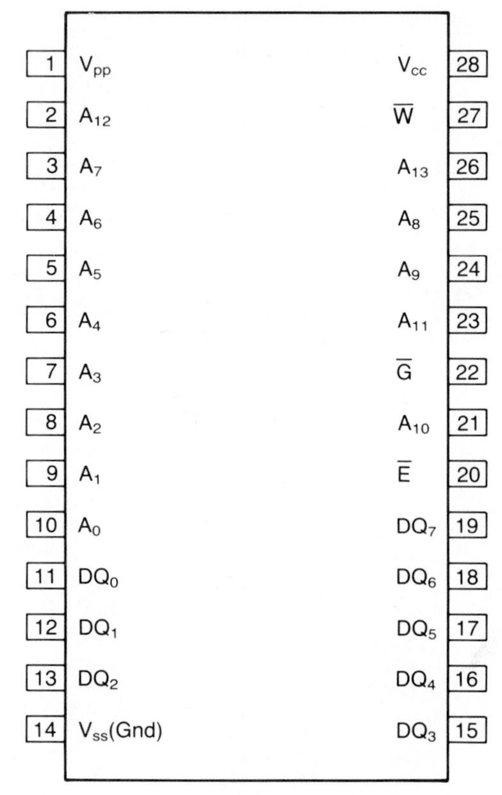

memory locations within the chip. Among the control lines are a chip enable ($\overline{E}$) and an output enable ($\overline{G}$), which are used under normal memory read accesses of the M6836E16. For data to be read from this chip, both the chip and the output enable must be asserted along with a valid address supplied on the address lines.

The manufacturer (Motorola) initially writes data into the mask ROM portion by using a program listing supplied by the system user. Once this program is made into a mask, it is burnt into the ROM; that is, the binary data codes for the program are permanently stored into the chip's memory. This is done by burning a fusible link for one type of binary data (a 0 in the case of the M6836E16) and by leaving the link intact for the opposite type of data bit.

When required, the user writes data into the programmable portion of the chip using the write ($\overline{W}$) and program voltage (V_{pp}) inputs. When the ROM chip is in the system and is being used in a read-only mode, $\overline{W}$ should be (but doesn't have to be) made inactive and V_{pp} is connected to the nominal V_{cc} chip voltage of 5 V. To access the ROM, $\overline{E}$ and $\overline{G}$, the chip and output enables, are asserted along with a valid address.

During the programming operation of the EPROM, the chip is enabled ($\overline{E}$ active) and a 21-V programming voltage is applied to V_{pp}. After correct data and address information have been presented to the chip, $\overline{W}$ is asserted, allowing the data to be written into the chip at the selected location. The output enable ($\overline{G}$) is held inactive during programming to prevent chip data from conflicting with data input to it on the DQ lines.

Vector locations, resulting from the application of a master reset and from exception processing operations, are at the lower-end 68000 address area. These vectors, along with a restart or initialization program, must reside permanently in the system memory, making it necessary to place ROM chips containing this information at the lower address range of the 68000 address map. Since this establishes the lowest address for ROM as $00 0000, what are the expected upper addresses of the first pair of ROM chips? To find this out, we need to create a sample system. Within this system, we will provide for 32K of ROM.

The 16-bit 68000 data bus requires two 8-bit locations for each word of data. The microprocessor allows 8-bit memory chips to be used by using the $\overline{LDS}$ and $\overline{UDS}$ (lower and upper data strobes) control signals. The address line A_0 is not directly available from the 68000 but is used, instead, to guide the data strobes $\overline{LDS}$ and $\overline{UDS}$. For byte accesses, the condition of A_0 determines if the byte data are at an odd address ($\overline{LDS}$ asserted) or an even address ($\overline{UDS}$ asserted). For word and long word data accesses, A_0 directs both data strobes to be asserted simultaneously, causing the full 16-bit data bus to be used. Because of this action, $\overline{UDS}$ and $\overline{LDS}$ become integral parts of the chip-selection process.

The M6836E16 ROM has 14 address lines, which make 16K locations available within it; but our sample system requires 32K locations. Using additional address line inputs to the ROM would increase the number of available locations; however, neither additional address lines nor memory cells are available inside a single M6836E16 ROM. The solution to this dilemma is to use two M6836E16 ROM chips and select them individually, using the data strobes and address lines to assert the chip enables of each ROM.

Essentially, the task ahead is to map out the memory—to allocate memory addresses to the memory chips. There is only one absolute rule to follow at all times when laying out a memory map. That rule, which can be called the rule of non-contention, specifies that *any single address accesses one and only one location*. For example, if address $00 0000 accesses a location in ROM number 1, it *cannot* access any other location in that or any other memory chip. If two memory locations were to be accessed by one address, line **contention** would result. Both sets of data would contend for use of the data bus at the same time. As a result information on the data bus would be scrambled and useless. The devices may be damaged as they try to attain opposite levels (low logic or close to 0 V versus high logic or approximately 4.5 V). Equal consideration must be given to the various data sizes handled by the 68000. In other words, byte data must be accessible individually, while word size data require dual-byte accessibility.

Reread the rule carefully, because there is a subtle corollary. Again, the rule says that any single address can access only a single 8-bit memory location. However, any single memory location can have any number of addresses used to access it *as long as no single address violates the rule of non-contention*. An example will help to bring the full meaning home.

EXAMPLE 2.1

The address $00 0000 accesses the first location in ROM 1. The address $10 0000 also can access this same location. The addresses $01 0000 and $11 0000 can access the same location in another ROM. However, neither $00 0000 nor $10 000 can access the second ROM. Similarly, $01 0000 and $11 0000 do not address ROM 1.

Memory Map

The job of memory mapping is to lay out a system to ensure that the basic rule of non-contention is followed. Actually, the first step is done for you by the nature of the RESET vector locations $00 0000–$00 0007. ROM 1 and ROM 2 must start at locations $00 0000 and $00 0001, respectively. ROM 1 will hold even address bytes and ROM 2 the odd address bytes when accessing word and long word information from the ROM area of memory. At what locations do ROM 1 and ROM 2 end? There are 16,384 locations in each ROM, making a total of 32,768 for both. ROM 1 will be activated by $\overline{UDS}$ and ROM 2, by $\overline{LDS}$. In this respect, ROM 1 contains the even addresses and ROM 2 the odd addresses for the range beginning at $00 0000 and ending 32,767 locations later ($00 0000 is one location). In hexadecimal, 32,767 is $7FFF, making the highest address in ROM 1 $00 7FFE and in ROM 2, $00 7FFF.

This is the first step—to determine the addresses for each chip. The next part of the process is to decide which 68000 lines to connect to the chips to make them operational with the microprocessor. The data lines D_0–D_7 are connected directly between the microprocessor and ROM 2. That is, D_0 of the microprocessor is connected to DQ_0 of ROM 2, designating the lower byte of word size data to the odd addresses of the ROM area in memory. Similarly, data lines D_8–D_{15} are connected to ROM 1. Address lines A_1–A_{14} connect the three chips. The remaining problem involves the rest of the address bus, control lines $\overline{UDS}$ and $\overline{LDS}$, and the chip enables of the two ROMs.

By expanding the address bus for ROM 1 and ROM 2, we develop a scheme as shown in Table 2.8. As usual a 1 indicates a constant high and a 0 a constant low. X indicates those address lines that will be variable and that will be used to select an individual location within a chip. A U indicates connection to $\overline{UDS}$. An L represents $\overline{LDS}$.

There is only one chip enable on the M6836E16, but there are many more address lines to be accounted for. In addition, the address strobe control signal, $\overline{AS}$, is used so that the ROM is accessed only when information on the address bus is valid. An *address decoder* overcomes this shortcoming and provides a way to select different chips by different address combinations. A 74138 "1 of 8" decoder performs the function needed. The pin out diagram for this chip is shown in Figure 2.3. The 74138 sets one of its 8 outputs low for each of 8 binary combinations on its 3 input leads. For instance, if we set A_0, A_1, and A_2 of the 74138 high, output $\overline{O}_7$ goes low. If we set A_2 and A_1 high, but make A_0 low, then output $\overline{O}_6$ goes low and $\overline{O}_7$ is returned high. The decoder chip also has 3 enable inputs: 2 active low and 1 active high.

To start with, the $\overline{AS}$ signal from the microprocessor is connected to an active low enable of the 74138. For any of the decoder's outputs to become active, this line must be low, indicating the presence of a valid address on the address bus.

A method called *partial address decoding* reduces the amount of circuitry for the system by decoding part of the address bus when enabling the system's various chips. In contrast, by using the entire address bus, *full address decoding* is required and every location must be accounted for. Partial address decoding works as long as there is no violation of the rule of non-contention. The address assignment for this system does not involve address lines A_{19}–A_{23}, but address lines A_{15}–A_{18} must be accounted for. Three of these address lines can be handled by one 74138 decoder chip (Figure 2.3, page 76) whose selected outputs are connected through AND gates to the ROM's chip enables, as shown in Figure 2.4 on page 77. The other inputs to the AND gates are connected individually to either $\overline{UDS}$ or $\overline{LDS}$ to complete the chip selection. The fourth address line, A_{15}, is not connected

Table 2.8 ROM 1 and ROM 2 Memory Map

← ADDRESS BUS →																								
23	22	21	20	19	18	17	16	15	14	13	12	11	10	9	8	7	6	5	4	3	2	1	0	DEVICE
					0	0	0		X	X	X	X	X	X	X	X	X	X	X	X	X	X	U	ROM 1
					0	0	0		X	X	X	X	X	X	X	X	X	X	X	X	X	X	L	ROM 2

U = $\overline{UDS}$ L = $\overline{LDS}$ X = Variable

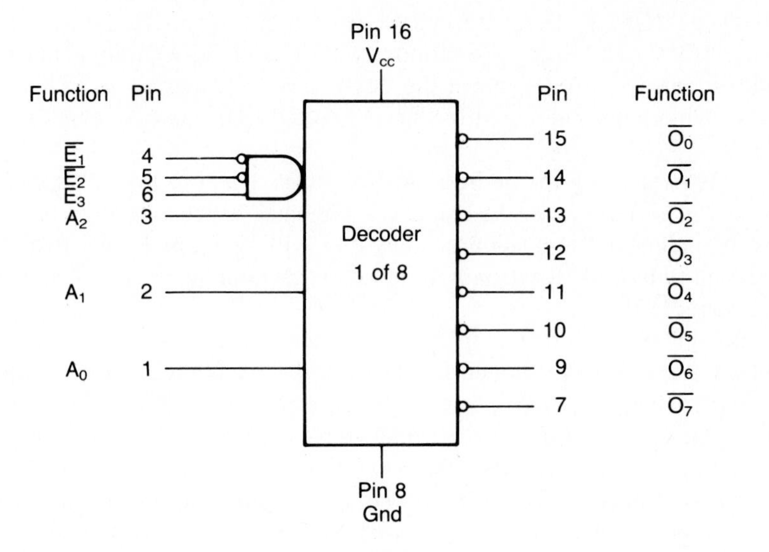

$\overline{E_1}$	$\overline{E_2}$	E_3	A_2	A_1	A_0	$\overline{0}$	$\overline{1}$	$\overline{2}$	$\overline{3}$	$\overline{4}$	$\overline{5}$	$\overline{6}$	$\overline{7}$
1	X	X	X	X	X	← All high or inactive →							
X	1	X	X	X	X								
X	X	0	X	X	X								
0	0	1	0	0	0	0	1	1	1	1	1	1	1
↓	↓	↓	0	0	1	1	0	1	1	1	1	1	1
			0	1	0	1	1	0	1	1	1	1	1
			0	1	1	1	1	1	0	1	1	1	1
			1	0	0	1	1	1	1	0	1	1	1
			1	0	1	1	1	1	1	1	0	1	1
			1	1	0	1	1	1	1	1	1	0	1
			1	1	1	1	1	1	1	1	1	1	0

Figure 2.3 74138 decoder and truth table (Courtesy of Signetics)

or used with the ROM chips. Leaving this line open allows for the RAM chip selection described later in the chapter.

Figure 2.4 illustrates the interconnections for purposes of address decoding between the 68000, the ROM chips, and the decoder chip. Address lines A_{16}, A_{17}, and A_{18} are wired to the decoder chips A_0, A_1, and A_2 inputs, respectively.

Output 0 (made active by the decoder's inputs all being low) is connected to the AND gates. To make this line active, A_{15}, A_{16}, and A_{17} all must be low. Address lines A_1–A_{14} are directly connected between the 68000 and the ROMs. These are the variable addresses that actually access a memory location within a ROM once it

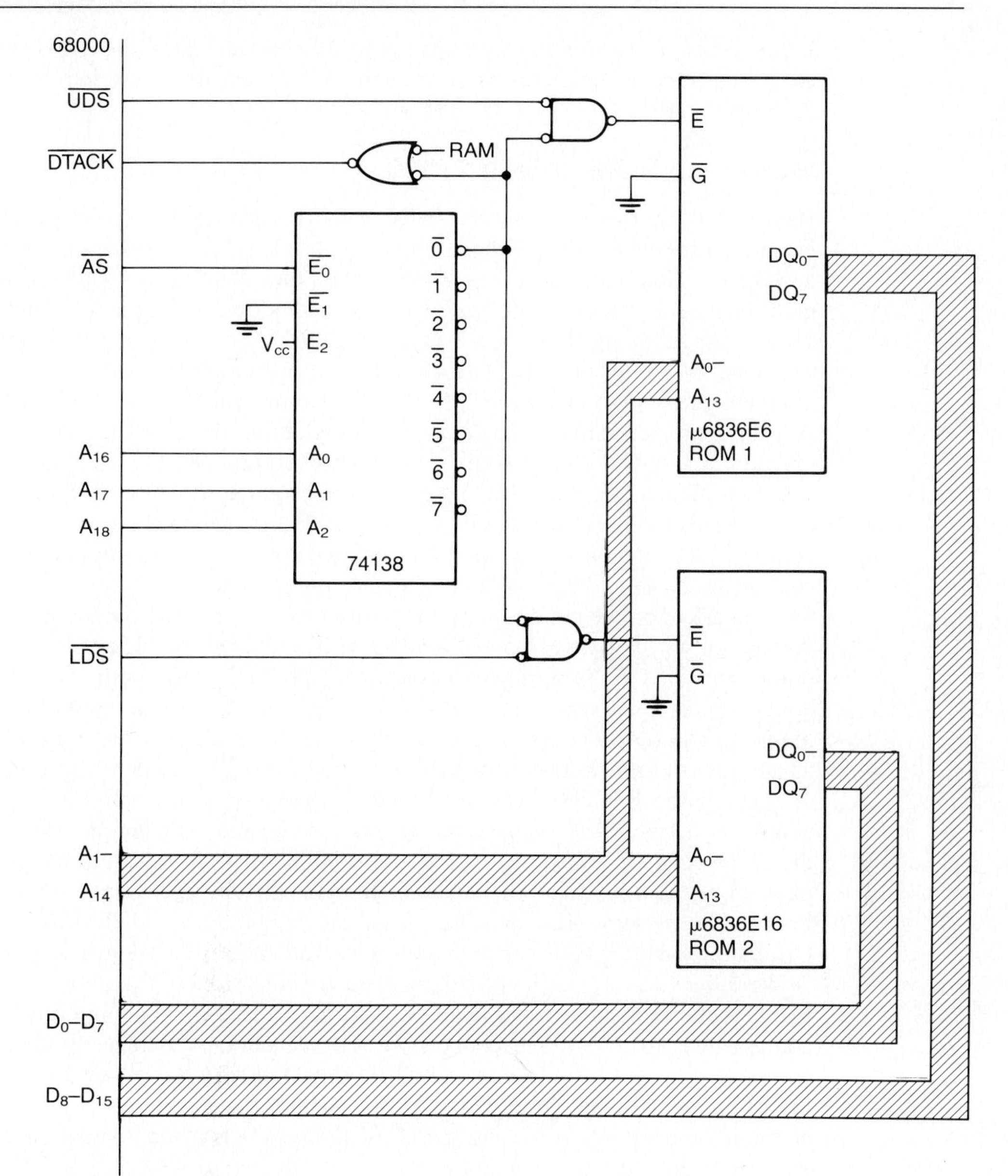

Figure 2.4 ROM decoding

is selected. $\overline{UDS}$ is connected to the AND gate feeding ROM 1, making it accessible by even addresses only, while $\overline{LDS}$, through the other AND gate, is used for ROM 2, restricting its use to odd addresses.

The 68000 data bus is separated into two 8-bit halves, the upper of which is connected to ROM 1 (enabled by $\overline{UDS}$) and the lower, to ROM 2 (enabled by

$\overline{LDS}$). $\overline{DTACK}$ is returned whenever the ROM area (and later, the RAM) is correctly addressed. The address bus is connected as laid out in the system memory map developed earlier.

Word Size ROM: Intel's 27210

Byte size memories were initially used with 16-bit microprocessors like the 68000 because of the availability of those chips. To this date, Motorola has not manufactured a word size ROM or RAM. However, Intel, the manufacturer of the 8086 16-bit microprocessor, has placed several word size memories on the market with its 27xxx-series chips. Among these is a 1-megabit EPROM, the 27210, which is organized as a 64K by 16-bit memory. The 40-pin, dual in-line (DIP) and 44-pin pad pin out diagrams are shown in Figure 2.5. All 32 output data (O_0–O_{15}) and address (A_0–A_{15}) lines are accessible from the device. In addition, there are control leads for chip enable ($\overline{CE}$), output enable ($\overline{OE}$), and program control ($\overline{PGM}$). The operating mode selection (Table 2.9, page 80) is a guide to the functions of the EPROM.

The EPROM is electrically programmable by placing the correct data and addressing information on the data and address lines. The preprogrammed value of each memory cell is a 1. Actually, 0's are programmed into these cells, while 1's have no effect on the condition of a memory cell. During programming, the program voltage on V_{pp} is between 12 V and 13 V. The chip is enabled ($\overline{CE}$ = 0) and the output enable ($\overline{OE}$) is disabled. After the correct data and address information are applied to the appropriate leads, $\overline{PGM}$ is brought low to write a word of data into the memory. The EPROM is erased by applying an intense ultraviolet light (3000–4000 angstroms) to the window on the device. This returns all cells to the 1 state.

After the EPROM is programmed, it operates as any other ROM—data are read from it when it is enabled and properly addressed. For the preceding example, the 27210 can be substituted for the two M6836E16 ROM chips in the memory map. Figure 2.6 on page 80 illustrates the connections necessary to use the 27210 in a 68000 system. The data bus from the 68000 is wired directly (or through buffers to provide additional source current availability to a system) to the EPROM. Address lines A_1–A_{16} of the microprocessor are connected to the address inputs of the EPROM. Once again, partial address decoding is used by attaching three additional address lines to a 74138 decoder chip. The decoder is enabled by an active $\overline{AS}$. Output 0 is, as before, made active when A_{17}, A_{18}, and A_{19} all are low. This output drives the chip enable input of the ROM. A buffer is inserted between the $\overline{CE}$ line and the output enable ($\overline{OE}$) and the $\overline{DTACK}$ input to provide a small delay between the chip enable and these signals.

Whenever a word size read from the ROM is requested by the 68000, the 27210 is enabled and it places a full word of data onto the data bus. As indicated by the truth table for $\overline{UDS}$, $\overline{LDS}$, and R/$\overline{W}$, the full word is read by the microprocessor. During byte accesses, once again, the ROM chip puts a full word onto the data bus. During byte reads, address lines A_1–A_{16} from the 68000 are identical for even or odd addresses. For example, a byte access from location $00 3002 or from location $00 3003 causes those address lines to be the same. One difference is, of course, in the state of $\overline{UDS}$ and $\overline{LDS}$. The other difference is that the microprocessor reads

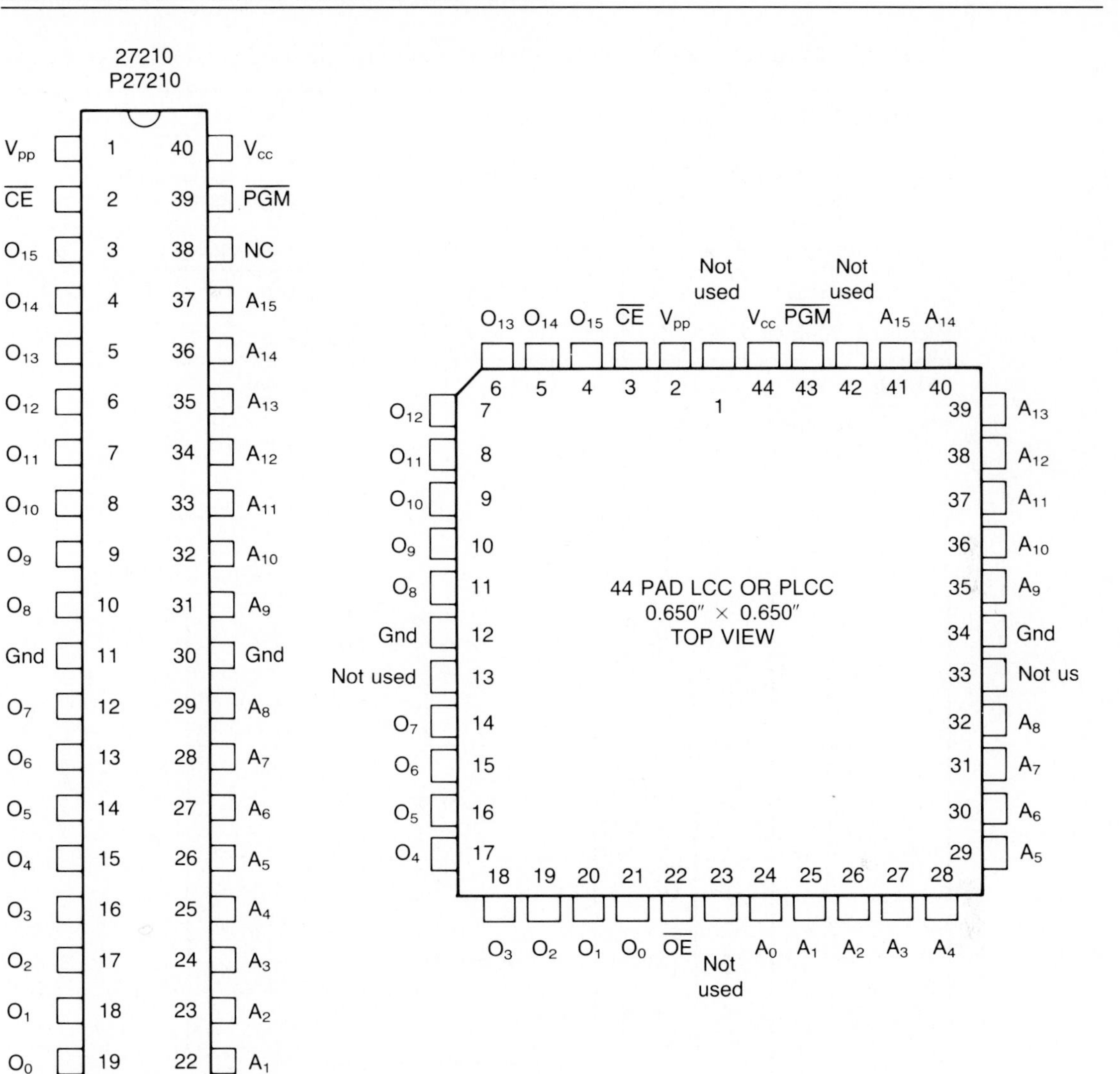

Figure 2.5 27210 pin diagrams (Courtesy of Intel Corporation)

only that half of the data bus being accessed by the byte transfer. Refer to Table 2.4 and note that when an odd address is accessed, the upper byte of the data is invalid and not read. Similarly, during a byte access from an even address, the lower byte on the data bus is considered invalid and not read by the microprocessor. Comparing the two ROM devices, you can see that using the newer 16-bit wide ROM is more straightforward and less complex to include in a 68000-based system.

Table 2.9 27210 Operating Modes

Operating Mode	$\overline{CE}$	$\overline{OE}$	$\overline{PGM}$	V_{PP}	V_{CC}	Data Lines
READ	0	0	X	X	5.0 V	Data out
Output Disable	0	1	X	X	5.0 V	High impedance
Deselected	1	X	X	X	5.0 V	High impedance
Programming	0	1	0	12.5 V	6.0 V	Data in
Program Inhibit	0	1	1	12.5 V	6.0 V	Undefined
Program Verify	0	0	1	12.5 V	6.0 V	Data out

X can be a 1 or 0 (5.0 V or 0.0 V)

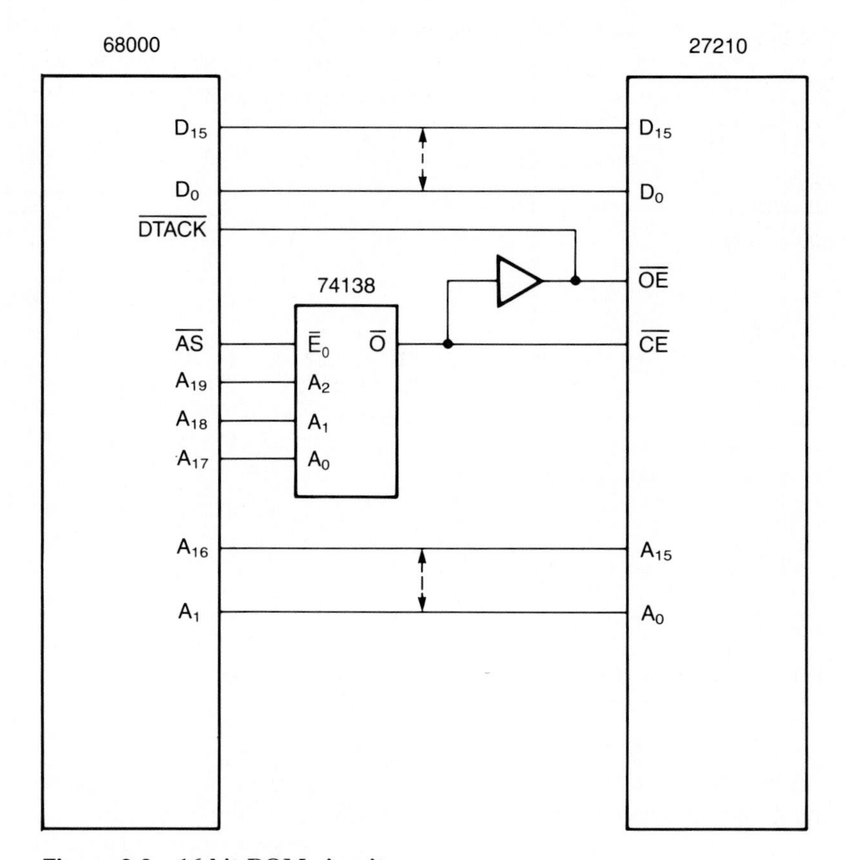

Figure 2.6 **16-bit ROM circuit**

μPD43256 Static RAM

The μPD43256 is a RAM manufactured by NEC Electronics, Inc. It is a complementary metal oxide semiconductor (CMOS), or MIX-MOS (as NEC designates it), static 32K×8 memory whose pin out diagram is shown in Figure 2.7. The designation 32K×8 tells us that there are 32,768 locations, each containing a byte of data. Fifteen address lines are needed to select from the 32K locations. This chip is supplied with a write enable ($\overline{WE}$) control lead to direct data in ($\overline{WE}$ active) or out ($\overline{WE}$ inactive) of the memory. A chip select ($\overline{CS}$) and an output enable ($\overline{OE}$) are provided for address decoding.

The next task is to isolate the ROM address area from the RAM address area using as few connections as possible. To do this isolation, keep the rule of noncontention in mind. Make sure when you address any ROM chip that no RAM chip is accessed and vice versa. One common way to do this is to make the RAM addresses continue sequentially from the last ROM location. This method is particularly adaptable in a small-memory system. In contrast, larger systems require that more space be set aside for future ROM additions. Address line A_{16} is used to separate the ROM and RAM areas. This bit is low for both ROMs and becomes high for all RAM addresses. As a result, the RAM area begins at address \$01 0000 and continues sequentially to the last RAM location.

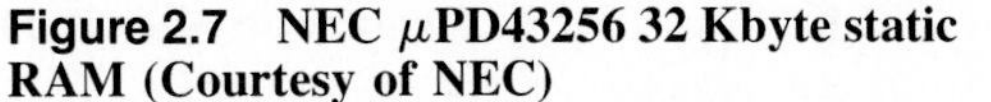

Figure 2.7 NEC μPD43256 32 Kbyte static RAM (Courtesy of NEC)

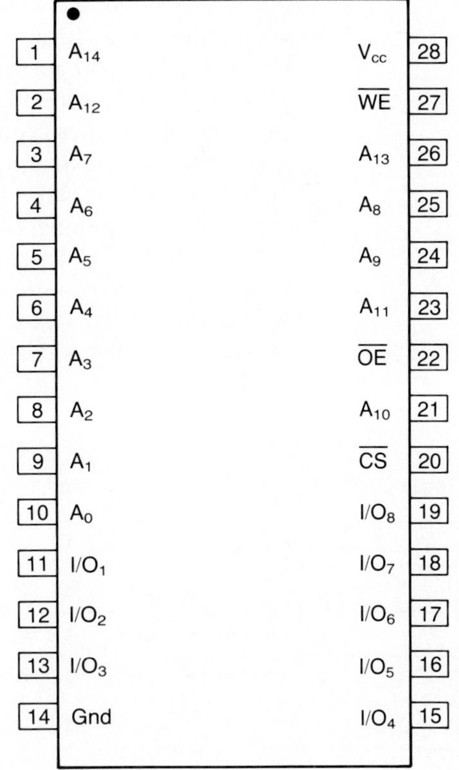

RAM 1 starts at location $01 0000 and contains $7FFF (32K) locations, all of which are even addresses, while RAM 2 is used for odd addresses within the same address range. Output 1 from the decoder chip in Figure 2.4 is connected to 2 AND gates, whose outputs go to the $\overline{\text{CS}}$ inputs of the RAMs in similar fashion to the AND gates for the ROMs. Likewise, the other inputs to each of the AND gates come from either $\overline{\text{UDS}}$ (for RAM 1) or $\overline{\text{LDS}}$ (for RAM 2). The extended address map for the system memory is shown in Table 2.10, and the schematic diagram illustrating the RAM interconnections is shown in Figure 2.8.

The two outputs of the decoder chip now being used are ORed to supply a $\overline{\text{DTACK}}$ back to the 68000 when either RAM or ROM is accessed. $\overline{\text{BERR}}$, which is illustrated and discussed in more detail later, is activated whenever the ROM or RAM is incorrectly addressed. A short delay between the $\overline{\text{AS}}$ and the input to the AND gate driving the $\overline{\text{BERR}}$ signal allows time for the address decoder and OR gate driving $\overline{\text{DTACK}}$ to propagate a successful access signal disabling $\overline{\text{BERR}}$. Data bus lines are connected to the RAMs as directed by $\overline{\text{UDS}}$ and $\overline{\text{LDS}}$. Address lines are used to enable further RAM chip selects according to the memory map, as well as to address the memory cells properly within the memories. The R/$\overline{\text{W}}$ is connected to the $\overline{\text{WE}}$ of each RAM chip to direct data in or out of the RAM being selected. This signal is not needed with the ROMs, since data direction always goes from the ROM to the microprocessor during normal operation.

Note that this is not a complete system. There is no provision for interfacing to the outside world, which will be discussed in Chapters 5 and 6. That discussion involves interface chips and other concepts not yet presented.

It is left to the reader to take a few minutes and apply any address from $00 0000–$FF FFFF to the basic system, based on the memory map developed to this point. For each address you select, only one memory location in one chip is accessed. You will find, of course, quite a bit of address duplication. For instance, $00 0020 and $10 0020 address the same location in ROM 1. That's perfectly all right, as long as they address nothing else. Despite this duplicity of addresses, a system designer will lay out a map using a designated set of addresses from among the duplicate groups for the programmer to use while operating this system. That doesn't mean that the other addresses are unusable; it just means that for clarity and directness, you should use the addresses designated by the system designer.

Table 2.10 8-bit Device Memory Map

23	22	21	20	19	18	17	16	15	14	13	12	11	10	9	8	7	6	5	4	3	2	1	0	Chip
					0	0	0		X	X	X	X	X	X	X	X	X	X	X	X	X	X	U	ROM 1
					0	0	0		X	X	X	X	X	X	X	X	X	X	X	X	X	X	L	ROM 2
					0	0	1	X	X	X	X	X	X	X	X	X	X	X	X	X	X	X	U	RAM 1
					0	0	1	X	X	X	X	X	X	X	X	X	X	X	X	X	X	X	L	RAM 2

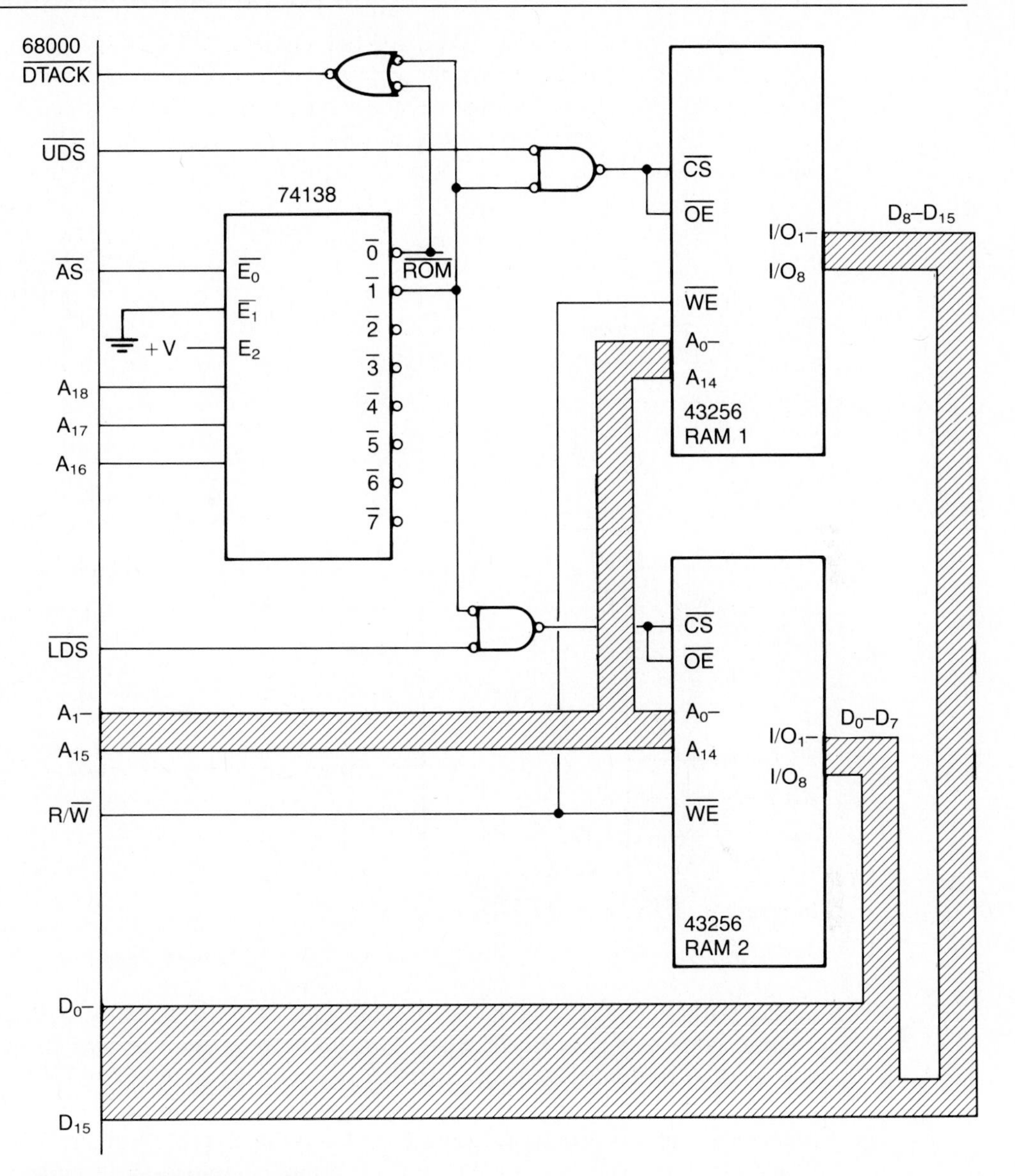

Figure 2.8 RAM decoding

$\overline{\text{DTACK}}$ and $\overline{\text{BERR}}$

(Figure 2.9 on page 84 is a schematic diagram of the $\overline{\text{DTACK}}$ and $\overline{\text{BERR}}$ logic using 8-bit memory devices.) A small amount of logic circuitry is required to enable $\overline{\text{DTACK}}$ whenever a memory access is performed. This is done by ORing all address decoder outputs. If any one of them is active, $\overline{\text{DTACK}}$ is asserted, indicating that

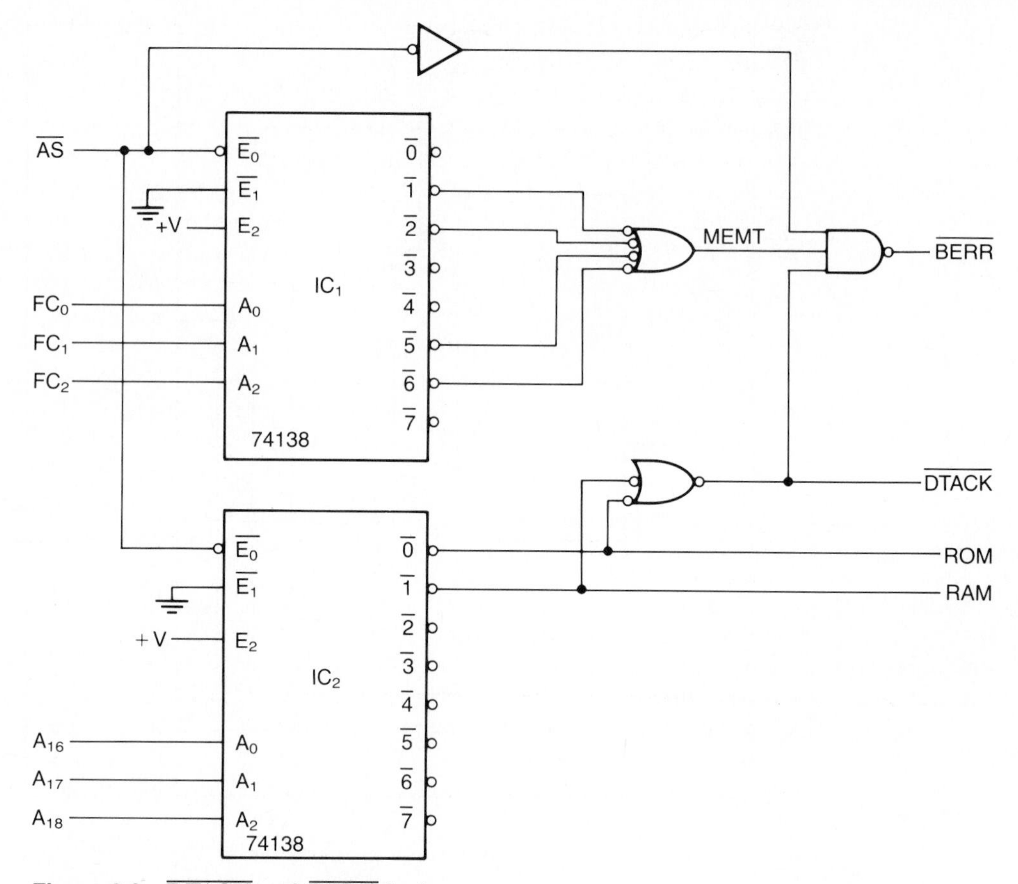

Figure 2.9 **$\overline{\text{DTACK}}$ and $\overline{\text{BERR}}$ logic**

a successful memory transfer is in progress. $\overline{\text{DTACK}}$ actually indicates that the memory area being accessed exists and can be read from or written into has been accessed. A method to generate a $\overline{\text{BERR}}$ is illustrated by ANDing $\overline{\text{DTACK}}$, $\overline{\text{AS}}$, and the function control (FC) outputs to produce a $\overline{\text{BERR}}$ in the event that an unused memory address is accessed during a read or write cycle. In Figure 2.8, this occurs if address lines A_{16}, A_{17}, or A_{18} are set to any combination other than 000 or 100 (A_{16}, A_{17}, and A_{18}, respectively) during a memory access cycle indicated by the decoding of the function control signals.

$\overline{\text{DTACK}}$ is active whenever one of the memory chips is accessed for a read or write operation. Schematically, a $\overline{\text{DTACK}}$ is generated whenever one of the selected outputs from the 74138 address decoder (chip IC_2 of Figure 2.9) is made active (low). These outputs are ORed and sent to the 68000 as $\overline{\text{DTACK}}$.

FC_0, FC_1, and FC_2 (Table 2.5) have 4 combinations of states that indicate that a memory bus cycle is in process. These are used for user program fetches, user data transfers, supervisor program fetches, and supervisor data transfers. One use of the function code signals is to separate areas of memory into designated uses as indicated by those functions. A second use is to detect whenever any memory bus cycle is in process regardless of the memory address involved. IC_1 of Figure 2.9 decodes the function control lines into one of four active states when a memory transfer is being processed by the 68000. These, in turn, are ORed to produce a single signal (MEMT) for memory transfer. This signal is ANDed with the inverted $\overline{AS}$ and the $\overline{DTACK}$ signal produced as described in the preceding paragraph. The $\overline{BERR}$ output from the NAND gate is made active when a memory transfer occurs (MEMT and $\overline{AS}$ are active, indicating a valid address and memory transfer bus cycle) and $\overline{DTACK}$ is not generated. $\overline{DTACK}$ will not be generated if the memory transfer requests an address in memory where no chip is connected. This request is not decoded through IC_2, and the resulting lack of an active output from that decoder prevents $\overline{DTACK}$ from becoming active. A $\overline{BERR}$ returned to the microprocessor in place of a $\overline{DTACK}$ causes the system to cease trying to execute the current instruction and run a bus error exception routine instead. The inclusion of the $\overline{AS}$ into the $\overline{BERR}$ AND gate ensures that a $\overline{BERR}$ is not generated unless a valid memory transfer is being attempted. A high on $\overline{AS}$ is inverted to a low and holds the gate inactive at all other times.

Figure 2.10 on page 86 is the schematic diagram for the system up to this point. In addition to the interconnections between the microprocessor, decoder chips, the memories, and the $\overline{DTACK/BERR}$ logic, there are other required connections and circuitry. The data bus and address buses are shown along with the R/$\overline{W}$ line. There is a resistor-capacitor RESET circuit that produces an active reset for approximately 10 ms at power up or application of the RESET switch. This signal must cause $\overline{RESET}$ and $\overline{HALT}$ to be active simultaneously. The capacitor is initially discharged through lack of power in the system or by the short applied by the switch. When power is applied for the first time or after the switch is open, the capacitor begins to charge. At first, with no voltage across the capacitor, $\overline{RESET}$ and $\overline{HALT}$ are driven low by the buffer connected in the circuit. Eventually, the capacitor charges sufficiently to cause the buffer to switch to the high state, making $\overline{RESET}$ and $\overline{HALT}$ inactive. The voltage across the capacitor reaches approximately 3.2 V after one time constant ($R \times C$) or, in this case, 10 ms. In essence, the capacitor charges to 63% of the supply voltage in this time period (63% of 5 V is approximately 3.2 V). This is sufficiently high to be detected by the buffer as a 1 level. As such, the amount of time that the reset inputs are active is approximately 10 ms.

PROBLEMS

2.1 Starting with the address $00 0120, make a memory list that would hold the data $1234 ABCD.

2.2 List the 4 types of data by size that can be used with the 68000.

2.3 What are the sizes of the data and address buses of the 68000?

2.4 What are the conditions of $\overline{UDS}$, $\overline{LDS}$, R/$\overline{W}$, and the three function control signals during a user program instruction fetch?

2.5 What signals are active to write a word of data to user data RAM?

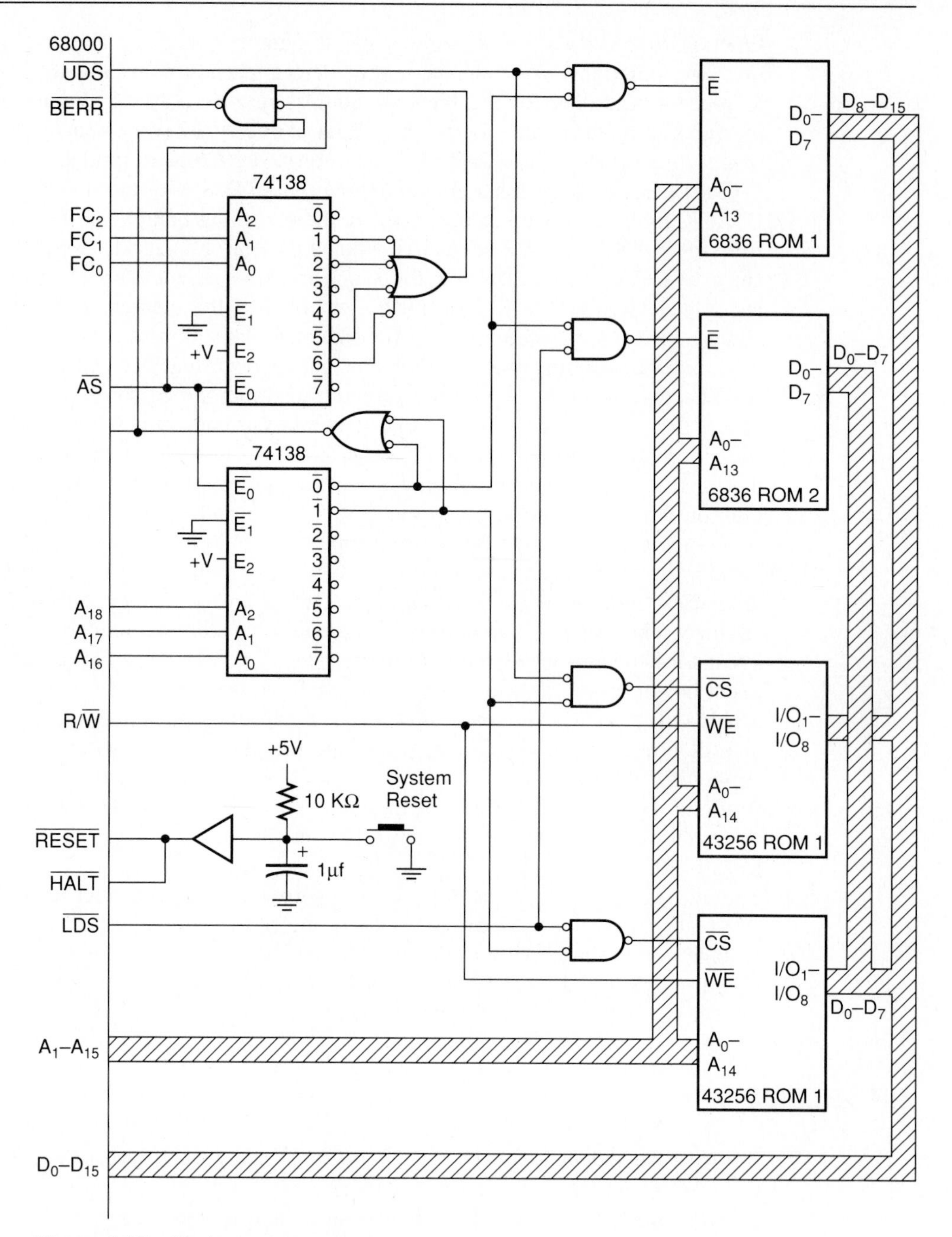

Figure 2.10 Basic system memory

2.6 There are two possible signal responses returned to the 68000 to a memory read bus cycle. Which signals are made active by both of these responses? What does the 68000 do if the response indicates that an unused memory address access was attempted?

2.7 Give the control signals associated with each of the following descriptions:
(a) indicates a valid address during a memory read cycle
(b) requests control of the buses by a DMA device
(c) must remain active during a DMA operation
(d) is the master reset for the 68000
(e) causes a bus error routine to be run
(f) are made active or not active by A_0
(g) indicates a valid address to 6800 peripheral devices
(h) informs a DMA device that it can have control of the bus
(i) is required for a successful completion of a memory cycle
(j) is made active for a period of time as a result of a RESET instruction

2.8 Write two additional lines on the system chart to expand the RAM area to include two additional chips, RAM 3 and RAM 4. What is the address range for this additional RAM area?

2.9 Since the data bus is 16 bits wide, address references for memory are made to even addresses only and include two locations (8 bits each). For example, location \$00 1000 contains \$3C2B. This means that location \$00 1000 contains \$3C and location \$00 1001 contains \$2B. A word data transfer is made from location \$00 1000 using the data already shown. The transfer is made to data register 3 (D3) and does not affect the upper word of D3. If D3 originally contained \$1234 5678, what are its contents after the transfer?

2.10 The extend (X) flag of the user portion of the status register is a new condition flag. What causes it to be set and what is it used for?

2.11 What do the T and S flags of the supervisor status indicate?

3

THE 68000 INSTRUCTION SET

CHAPTER OBJECTIVES

The objective of this chapter is to acquaint the student with the format of the 68000 instructions and addressing modes. MOVE and arithmetic instructions are discussed to exemplify the method used to interpret the instruction set. A basic program example is presented as an illustration of the concepts presented throughout the chapter.

3.1 GLOSSARY

displacement Value added to a pointer register to produce an effective address.
effective address Actual data, source, or destination address used by an instruction.
leading zero Extends a data word to its selected size without altering its value. For example, $C2 as word size data requires two leading zeroes: $00C2.
opword An operation's binary machine code.
prefetch Program opword or operand fetch done during other machine cycle operations.
sign extend To copy the MSB of a data value into the remaining upper bits of the data word. For example, the byte value $C2 sign extended to word size produces $FFC2. Compare this with *leading zero*.

3.2 INTRODUCTION

By the time 16-bit microprocessors became a reality, several instruction set schemes had been used on existing microprocessors. Most companies borrowed liberally from the concepts of their own devices and those of competing manufacturers. For instance, moving data between a register and memory is done by load and store instructions for Motorola's 6800 (LDAA, LDAB, STAA, and STAB—load and store accumulator A or B). For its 8-bit microprocessor, the 8085, Intel chose to use the move mnemonic for this process—MOV A,M or MOV M,A. They still used load and store mnemonics for multiple register data movement—LHLD and SHLD (load and store HL register pair). Zilog, the manufacturer of the Z80 microprocessor, reverted to load mnemonics to perform data movements similar to Intel devices—LD A,C and LD BC,HL. The microprocessors that use both types of instructions limit memory-to-register data accesses to load type instructions and register-to-register transfers to move instructions. Motorola selected the move-type instruction to manipulate data transfers within a 68000-based system. For this case, the MOVE instruction encompasses all types of data accesses: memory-to-memory, register-to-register, and memory-to-register. As microprocessors continue to evolve, we expect to see fewer differences in data movement instruction types.

3.3 THE MOVE INSTRUCTION

The primary instruction type used to move data between registers and/or memory in the 68000 is the MOVE instruction. The basic syntax for the MOVE instruction is

MOVE.⟨data size⟩ ⟨source effective address⟩, ⟨destination effective address⟩

The abbreviation ⟨ea⟩ is a shorthand notation used in place of the words *source* or *destination effective address* and the term *data size* is also shortened to *size*, so the syntax of MOVE now becomes

MOVE.⟨size⟩ ⟨ea⟩, ⟨ea⟩

The **source effective address** is the location that holds data fetched by the 68000 and the **destination effective address** is the location into which the microprocessor stores data. These addresses are determined by the instruction and the address mode. Similarly, 8-bit microprocessors also use effective addresses derived from the instructions and the address modes. In all cases, the effective address is the final value of the memory address used by the instruction.

Size is designated as B for byte, W for word, and L for long word data. The effective address determines the addressing mode that accesses the data used by the instruction.

3.4 REGISTER DIRECT ADDRESSING

Register direct addressing uses data already present in the designated register for the transfer. For example, MOVE.L D2,D1 moves the 32-bit contents of D2 (as the source) into D1 (as the destination). A 32-bit move is the least complex move since all of the destination register's contents are modified by the move. When dealing with data sizes less than 32 bits, care and attention must be given to where data are sent and what happens to the remaining bits within the register as a result of the transfer. For most operations dealing with data registers D0–D7, data moved into them replace the designated data size without affecting the rest of the register. Data are entered, filling the lower register areas. In other words, byte size data transfers affect b0–b7 of a data register; word size, b0–b15; and long words, all 32 bits. For word and long word size data, the lowest byte is held in the lower bit locations. Table 2.1 is a diagram of register data organization. Data formats for address registers are identical, except that byte size data are never used with address registers.

EXAMPLE 3.1

D_0 initially contains $1234 5678 and D_1 holds $ABCD EF09. What are the contents of both registers after a MOVE.W D0,D1?

Solution:

W indicates a word size data transfer. Register D0 is the source of the transfer, and D_1 is the register that receives the new data. The word size indicates that only the lower word data are affected. Thus $5678 is transferred to D1's lower word. All transfers are nondestructive reads of the source register, so D0's contents remain unaffected. D0 contains $1234 5678 and D1 contains $ABCD 5678 at the completion of the instruction. Notice that the upper word of D1 is not affected by the move. Data moves from address registers to data registers are done in a like manner. Since the source is an address register, data sizes are restricted to word and long word.

3.5 ADDRESS REGISTER DIRECT ADDRESSING

When the microprocessor moves data into an address register, data sizes are restricted to word and long word only. Because of this restriction and other differing responses of data movement into address registers, a separate MOVE instruction called MOVEA is used for the address move. Entering data into an address register generates a different response from moving data into a data register. Long word moves present no new problems—all 32 bits are accounted for. The new concern involves word size data moves into address registers. We noted in the data register that the upper word is not affected by a word move. However, word moves to address registers cause the upper word to be modified as well by **sign extension.** The MSB of the word size data transferred to the address register is copied into b16–b31 of the register. Because of these added considerations, Motorola selected a slightly different mnemonic for an address move, MOVEA, although most 68000 assemblers accept the MOVE syntax as well. Assemblers written by Motorola recognize a MOVE instruction to an address register and assemble it with a MOVEA **opword** code automatically.

EXAMPLE 3.2

A0 originally contains $0023 4B32, D0 contains $1234 5678, and D1 holds $34DC ACB5. What are the contents of A0 as a result of the following two instructions?

(a) MOVEA.W D0,A0 (b) MOVEA.W D1,A0

Solution:

Begin as in Example 3.1—move the lower word from the source register to the destination register. For part (a), $5678 replaces A0's lower word. If we look at $5678's MSB, we find it is a 0 ($5 = 0101). A0's upper word is sign extended by filling in its bits with 0's. For the instruction in part (a), A0 = $0000 5678. For part (b), the MSB of D1's lower word is a 1 ($A = 1010). After the lower word is moved from D1 to A0, the upper word of A0 is filled with 1's and A0 = $FFFF ACB5. As before, the contents of the source registers are unaltered by the MOVEA instruction.

3.6 ADDRESS REGISTER INDIRECT

The next effective address to be considered is one in which an address register is used as a memory pointer. This is similar to an index mode, discussed in Chapter 1, without an offset. Additionally, for Intel and Zilog microprocessors, this addressing mode is similar to using a register pair as a data address pointer. Parentheses surrounding an address register as an effective address designate this address mode. For example, MOVE.L (A0),D0 moves 32 bits from memory starting at a location pointed to by register A_0 and puts this data into register D0. Data registers cannot be used as pointers, so there is no data register indirect addressing mode.

Address register indirect can be used as either a source or destination effective address. As such, a memory-to-memory move is done by using this mode for both. MOVE.L (A0),(A1) moves the data starting at the location pointed to by A0 into the memory area starting where A1 points to. This mode also can be used to fetch data for an address register.

The rules of sign extension always apply when the microprocessor loads address registers. A MOVEA.W (A2),A5 loads the lower word of A5 from the location pointed to by A2 and the upper word of A5 is sign extended according to the value of the data moved.

EXAMPLE 3.3

Using the same initial data for each instruction below, what are the contents of each of the destinations?

00 2000 1234	00 2006 F00D	A_0 = 0000 2000
00 2002 5678	D_0 = 1266 6565	A_1 = 0000 2004
00 2004 BEAD	D_1 = AAAB B2DD	D_3 = ABCD 573D
		A_2 = 0000 2002

(a) MOVE.W (A0),D3 (b) MOVE.L (A1),D5 (c) MOVEA.W (A1),A4
(d) MOVE.L D0,(A0) (e) MOVE.W D1,(A1)

Solution:

(a) A0 = \$0000 2000. Location \$00 2000 contains the data \$1234. This word is moved into data register D3 by the first instruction. The upper word of D3 is not affected by the move. D3 = \$ABCD 1234.

(b) A1 contains \$0000 2004, which means it points to address \$00 2004. The move into D5 is a long word done by loading the upper portion of D5 starting with the beginning address of the data. D5 = \$BEAD F00D. Compare the register data organization with the memory organization of data charted in Chapter 2. Notice that the upper byte of data in memory is stored at the lowest address for long word size data. Registers hold the upper byte of long word data in b24–b31. The remaining bytes follow in logical sequence.

(c) This is an address register move using word size data. A1 points to address \$00 2004, which contains the data \$BEAD. This word is moved into A4 and is sign extended into the upper word of that register so that A4 = \$FFFF BEAD.

(d) A0 points to location \$00 2000. This time, data are moved from a register into memory. The contents of D0 are \$1266 6565, and the move is a long word. Following the correct data organization in both memory and registers, we find that the upper word from the register is moved to the lower addresses so that \$00 2000 = \$1266. Actually, address \$00 2000 contains \$12 and address \$00 2001 contains \$66; however, the convention of showing a full 16 bits at even addresses is used when considering 16-bit data buses, but two 8-bit memory locations are still involved. The word \$6565 is stored at location \$00 2002.

(e) Moving a word from D1 to location \$00 2004 (pointed to by A1) causes the lower word of D1 to be stored in memory, so that \$00 2004 = \$B2DD.

EXAMPLE 3.4

Using the same data as in Example 3.3, consider the instruction MOVE.L (A0),(A2). This instruction causes the data starting at location \$00 2000 to be copied into memory starting at location \$00 2002. The choice of overlapping memory is deliberate. Notice that this is a long word move operation, meaning that the data held at locations \$00 2000–\$00 2003 are to be moved to locations \$00 2002–\$00 2005. There is no conflict in the move because the 68000 is equipped to fetch the data and move it correctly. The data from locations \$00 2002 and \$00 2003 are fetched before the new data previously fetched from \$00 2000 and \$00 2001 are stored into those locations. The results are that location \$00 2002 contains \$1234 and location \$00 2004 holds \$5678 at the completion of this instruction.

3.7 ADDRESS REGISTER INDIRECT WITH DISPLACEMENT

The index addressing mode allowed a programmer to create an effective address by combining the contents of the index register, used as a base address pointer, with an offset number. This number, when added to the contents of the index register, formed the address of the data used by the instruction. The 68000 has a similar type of addressing mode called address register indirect with **displacement.** In this case, the contents of an address register, used as the base pointer, are added to a 16-bit, sign extended displacement to form the effective address of the data. The syntax for the MOVE instruction using this mode as a source and data register direct as a destination is

MOVE.⟨size⟩ d16(An),Dn

The term d16 indicates the need for a 16-bit operand with the instruction. The term An stands for any address register, and Dn, the data registers. An example best illustrates what occurs.

EXAMPLE 3.5

A0 contains $0000 0200; D0 contains $1234 5678. The data held in memory starting at location $00 0200 are $BADE, and $00 0202 holds 0AF2. What are the contents of D0 after the instruction MOVE.W $2(A0),D0?

Solution:

The contents of A0 are added to the displacement to form the effective address. The displacement is $2. The first operation the assembler performs is to transfer the contents of the address register used as the data pointer (A0) to the address buffers. The displacement operand is filled with leading 0's to extend it to 16 bits ($0002). The 16-bit number is now sign extended to a full 32 bits ($0000 0002) and added to the contents of the address buffer. The effective address in the address buffers is $00 0202, and the contents of A0 are still $0000 0200. There is no conflict here with the number of bits shown—address registers are full 32-bit registers. However, only 24 of the 32 address buffers that actually supply address information to the address bus and the $\overline{\text{LDS}}/\overline{\text{UDS}}$ logic circuitry are used for addressing purposes. The internal 32-bit sizes were designed to allow upward compatibility with future 68000-family microprocessors. The data fetched from this effective address are loaded into the lower word of D0. As a result, D0 contains $1234 0AF2 at the completion of the instruction. Address register indirect can be used for both source and destination effective addresses for many of the instructions of the 68000.

3.8 ABSOLUTE LONG AND ABSOLUTE SHORT

The address register indirect instruction requires an additional word besides the op-word to be complete. The displacement becomes the operand portion of the two-word instruction. Absolute long and short also involve the use of operands; one word for absolute short and two words for absolute long. The previous instructions involved data either already in a register (register direct) or data at an address pointed to by an address register. Two address modes allow data to be fetched from memory without the aid of an address register as a pointer. These are the absolute modes, in which the address of the data is written as the operands of the instructions. The syntax of the MOVE instruction using these modes as a source is

MOVE.⟨size⟩ ⟨address⟩,⟨ea⟩

These two modes are similar to the direct and extended modes discussed in Chapter 1. The extended mode requires a full 16-bit address as an effective address operand and allows any address to be referenced by an instruction. The direct mode uses a shortened address form called a *page 0* address. Page 0 indicates that all the addresses on that "page" have a value of $00 as their upper byte. The address operand of the instruction references only the lower byte value. The microprocesor is directed by this address mode to formulate the effective address by setting the upper byte to $00 and fetching the operand and including it as the effective address's lower byte. The advantage of this mode is that it requires only a single operand fetch as opposed to the two fetches needed to fetch an extended mode's operand.

Use of a 16-bit address places the instruction in the absolute short mode, which causes the 16-bit address to be sign extended to the full 24-bit address size. For example, MOVE.L $A02C,D3 loads D3 with a 32-bit data word starting from location $FF A02C. Using a full 24-bit address causes the absolute long mode to be selected. Data are transferred during program fetches in word sizes. As such, the absolute long address is fetched as a long word. Once again, the shorter address bus causes that address to be truncated to 24 bits. When writing the instruction, you should habitually restrict address notations to the actual 24-bit address bus size.

EXAMPLE 3.6

Write an absolute instruction to fetch the word size data from location $01 3FC2 and move it to location $00 2000.

Solution:

The absolute address mode is used for both source and destination. The data moved are word size and the instruction becomes

MOVE.W $013FC2,$2000

When this instruction is assembled into its object code, four words of code are produced. First is the opword that defines the operation and the address modes for both source and destination. The next two bytes are the source operands ($0001 and $3FC2), and the fourth word is the destination operand, $2000.

3.9 IMMEDIATE MODE

Every microprocessor provides a method to load registers with data directly written into the program. Almost universally, this mode is called the *immediate* addressing mode. The address concept in this case is the address of the location of the operand portion of the immediate mode instruction. All three major data sizes can be used in this mode, but caution must be exercised. Program fetches from memory are always word size, because all opwords are word size and must be fetched on even address bounds like any word size data transfer. When byte size data are encountered in the program (for example, in an immediate mode instruction), the byte used is the lower byte of the word fetched from memory. In other words, a byte size operand fetch results in a word size data move with the upper byte ignored. The assembler program, when faced with a byte size operand, fills in **leading zeros** to a word size to cause a full word size to be occupied in memory for the operand area of the instruction.

The syntax of an immediate mode MOVE instruction is always

MOVE.⟨size⟩ #⟨data⟩,⟨ea⟩

This mode is never used as a destination because it makes no sense to move data into data. Data can be loaded into a register or memory location, but data cannot be entered into other data.

EXAMPLE 3.7

Write a MOVE instruction that causes address register A3 to be loaded with $00 0020. Write this instruction three different ways, using the immediate mode for all.

Solution:

The trick here is to be fully aware of the requirements for entering data into an address register. First, byte size transfers to address registers are not allowed; we cannot include that method in our three instructions. Keep in mind that address registers are sign extended when loaded with word size data. Here are two of the three forms that an assembler accepts to perform our task:

MOVEA.L #$20,A3 and MOVEA.W #$20,A3

In both cases, the assembler adds leading zeros as far as the data size of the transfer ($0000 0020 for long and $0020 for word). The long word transfer then fills the full 32 bits of the address register. For the word transfer, the data word is sign extended so that, once again, A3 contains $0000 0020.

The third form of the instruction reflects what an assembler does if the size of the data to be transferred is not specified. The instruction becomes

MOVE #$20,A3

The assembler defaults to (automatically selects) word size data when the instruction does not specify a size. In all our previous word size examples, we could have omitted the .W and the instructions would have been correctly assembled and executed.

The remaining address modes introduce complexities into the process of determining where data are located for the instruction using it. They, in return, offer immense versatility in the recovery of data for a program. The first of these is address register indirect with *predecrement* or *postincrement*.

3.10 ADDRESS REGISTER INDIRECT WITH PREDECREMENT

Any address register can be used to perform stack pointer-type operations—that is, to move data from one place (register or memory location) to another and update the pointer(s) by the number of bytes transferred. Two address registers, A7 and A7′, are fixed as user and supervisor stack pointers, respectively. They can be used as regular address registers for all address modes discussed so far. Note that when you change the contents of registers A7 and A7′, any processing involving subroutines or interrupts can get lost if modified pointers cause the program to lose track of the stack locations. Always be careful when you alter the contents of A7 or A7′.

The address register indirect mode with predecrement lets you use the other seven address registers as stack pointers. With this mode, the designated address register is first decremented by one, two, or four, depending on the data size transferred. Its contents are used as the effective address where the data are to be moved to or retrieved from. Syntaxes for the predecrement mode are, as source,

MOVE.⟨size⟩ −(An),⟨ea⟩

and, as destination,

MOVE.⟨size⟩ ⟨ea⟩,−(An)

EXAMPLE 3.8

Address pointer A2 = $00 2044. From what location is D3 loaded with the instruction MOVE.B −(A2),D3?

Solution:

The data size to be moved is a byte, thus only one location is involved. The source is in the predecrement mode, which causes the contents of A2 first to be decremented to $00 2043. The data at that location are then fetched and loaded into D3's lowest byte. The predecrement mode causes both the address pointer (A2) and the data register (D3) in this example to be altered. Since only a byte size transfer is performed, accessing an odd address is permissible. Do not forget that word and long word transfers always must be done starting at an even address.

EXAMPLE 3.9

Using the following memory data and register contents as the initial conditions for each instruction, what are the contents of the address registers, data registers, and/or memory locations of as a result of each operation?

\$00 2000 = 1234	D0 = \$4455 66AA
\$00 2002 = 6789	D1 = \$AABB CCDD
\$00 2004 = BADE	A0 = \$0000 2006
\$00 2006 = ABCD	A1 = \$0000 2004

(a) MOVE −(A1),D0 (b) MOVE.L −(A0),D5 (c) MOVE.W D1,−(A1)
(d) MOVEA −(A0),A2 (e) MOVE.L −(A1),−(A0)

Solution:

(a) The transfer size is a word, so A1 first is decremented twice and the word at location \$00 2002 is fetched and loaded into D0. D0 = \$4455 6789, and A1 = \$0000 2002.

(b) Since a long word is transferred, the address register A0 first is decremented four times to \$0000 2002 and then the data are moved to D5. D5 = \$6789 BADE.

(c) Using predecrement as a destination creates the classic stack push operation, where the stack pointer first is decremented (A1 = \$00 2002, in this case) and then data are pushed onto the stack (transferred to memory). \$00 2002 = \$CCDD.

(d) Several changes occur with this example. Again, it is a word size move, so A0 is decremented to \$0000 2004. The data at this location (\$BADE) are transferred to A2. Since this is an address register load, A2 is sign extended to b31. A2 = \$FFFF BADE.

(e) This is a memory-to-memory transfer. Both address registers are predecremented by 4 for the long word transfer. A0 = \$0000 2002, and A1 = \$0000 2000. The long word starting at location \$00 2000 is moved to memory beginning at location \$00 2002. At the completion of the instruction, these are the contents of memory:

\$00 2000 = \$1234	\$00 2002 = \$1234
\$00 2004 = \$6789	\$00 2006 = \$ABCD

3.11 ADDRESS REGISTER INDIRECT WITH POSTINCREMENT

This address mode is essentially the same as the predecrement, except that the address register used as the pointer is incremented after the data transfer has been completed instead of being decremented before the transfer is initiated. The syntax for the postincrement mode MOVE instruction, as a source, is

MOVE.⟨size⟩ (An)+,⟨ea⟩

and, as a destination,

MOVE.⟨size⟩ ⟨ea⟩,(An)+

The opposite nature of address register modification provides a complement to the predecrement instruction; thus, Motorola has provided a means to use address registers as stack pointers for true stack-type operations. Displacement values, as used with the address register indirect with displacement, are not allowed in conjunction with either the predecrement or postincrement addressing modes.

EXAMPLE 3.10

Use the predecrement and postincrement modes with address register A0 first to push the contents of D0 onto a stack. Then retrieve that information with a pull-type stack operation.

Solution:

Using predecrement to formulate the destination effective address produces the necessary push operation, MOVE.L D0,−(A0). This causes the full contents of D0 to be transferred to memory, starting at a location pointed to by A0 once it has been decremented by four. The instruction for the pull function is MOVE.L (A0)+,D0. This instruction restores the contents of D0 and increments A0 back to its original value.

EXAMPLE 3.11

What does the instruction MOVE.L (A0)+,−(A0) accomplish?

Solution:

Actually, this is a very long *no operation* (NOP). The action fetches the long word from memory and increments A0 four times. Next, A0 is decremented four times by the predecrement function, and the data are stored where they were just fetched from. A0 has the same data before and after the instruction. The data retrieved from memory are replaced in exactly the same locations they were taken from. Since nothing was altered as a result of the instruction, the result was the waste of a number of machine cycles to execute the sequence.

When you use the postincrement or predecrement modes, incrementing and decrementing of pointer information occur in the address buffers. The final value in the buffers is transferred to the designated address register at the end of each part (source and destination) of the instruction. For instance, MOVE.L A0,−(A0) stores the original contents of register A0 starting at a memory location pointed to by the original value, minus 4.

EXAMPLE 3.12

Address register A5 = $0020 403C. What is moved to memory and where in memory is it moved as a result of the instruction MOVE.L A5,−(A5)? What are the contents of A5 at the completion of the instruction?

Solution:

The contents of A5 first are transferred to the address buffers where they are predecremented four times, for the long word move, to $20 4038. The contents of A5 ($0020 403C) then are moved to memory, beginning at $20 4038, so that $20 4038 contains $0020 and $20 403A contains $403C. After the transfer, the contents of the address buffers are moved to register A5, which holds $0020 4038 at the end of the instruction.

EXAMPLE 3.13

Address register A3 = $0030 6010. What are the contents of A3 at the end of the instruction MOVEA.W −(A3),A3?

Solution:

This instruction starts like Example 3.12 in that the contents of the data pointer (A3) are transferred to the address buffer and decremented twice. A memory fetch is performed, loading A3 from location $0030 600E. The data are sign extended in register A3. After the source operation is finished, the contents of the address buffer are transferred into A3 as before. However, the updated information in A3 is replaced by the data fetched to complete the destination process. Thus A3 contains the sign extended data fetched from location $0030 600E at the end of instruction.

3.12 ADDRESS REGISTER WITH INDEX ADDRESSING MODE

Motorola chiefly is responsible for introducing and using index register concepts with microprocessors. The use of an index register is extended to the 68000 in a mode called *address register indirect with index and displacement*. All data and address registers, including stack pointers A7 and A7′, can be used as index registers. The size of the data extracted from the index register can be either a sign extended word or long word. The indexed displacement is restricted to sign extended byte size. The effective address of the data is now calculated by adding together the contents of the address register, the index register, and the displacement The syntax for a MOVE instruction using this mode as a source is

MOVE.⟨size⟩ d8(An,Rn.⟨size⟩),⟨ea⟩

The term d8 indicates the byte size displacement and Rn stands for any data or address register.

EXAMPLE 3.14

From where are the data fetched to be loaded into D_2 by the instruction MOVE.L $6(A2,D5.W),D2? A2 = $0000 2000; D5 = $0123 0046; and D2 = $1234 5678.

Solution:

D2 is loaded with data from (A2)+(D5.W)+6, or

$$\$0000\ 2000 + \$0000\ 0046 + \$0000\ 0006 = \$0000\ 204C$$

Notice that the displacement is sign extended from the byte of \$06 to \$0000 0006. As specified by the instruction, only the lower word of D5 is used and it, too, is sign extended. After the three values are summed, they are truncated by the address buffer limitations to an address of \$00 204C.

3.13 PROGRAM COUNTER WITH DISPLACEMENT MODE

The 6800 and, to a smaller extent, the Z80 8-bit microprocessors used an address mode called *relative addressing* to do some of the program-order modifications caused by jump, branch, or subroutine call operations. In those cases, a displacement value was added to the program counter to alter its contents. This change caused the next instruction to be fetched from a new location in memory. The 68000 also uses this type of address mode to facilitate its conditional and unconditional branching instructions. The mode is renamed *program counter with displacement* or *program counter relative* and works similarly to the relative addressing modes of the 8-bit microprocessors.

Computing the branch address manually is similar to computing relative addresses for either the Motorola 6800 or Zilog Z80. That is, a constant offset displacement of two is added along with the relative address operand to the contents of the program counter. The resulting sum is the address of the next instruction to be executed by the 68000. As an equation, the relationship is expressed as

$$\text{DEST} = \text{BRANCH} + \text{REL} + \$2$$

where DEST is the destination address, BRANCH is the location of the branch instruction's opword, and REL is the value of the relative operand used with the instruction. Relative zero (the location where the displacement would be zero, effectively cancelling the branch result by branching to the place it would get to by normal program execution) is two locations after the branch opword location.

Two forms of the branch instruction are designated by the relative address size. The short branch limits relative addresses to a signed byte size. Using the branch always (BRA) instruction for illustration, the syntax for a short branch is

BRA.S ⟨label⟩

The byte size offset is included in the opword when assembled, making this a one-word instruction. Forward branches are restricted to \$7F or +127, while the maximum negative branch value is −128 or \$80. The long branch allows offsets to be word size, increasing the offset range to ±32K locations (\$7FFF and \$8000). The mnemonic for the long branch, BRA ⟨label⟩, requires an additional word operand to hold the offset.

EXAMPLE 3.15

What is the displacement to move the program execution from location $00 102C, where a BRA is located, to location $00 1150, where a MOVEA is located?

Solution:

The new destination location for the MOVE instruction is $00 1150. From this value, subtract the branch opword location of $00 102C plus 2:

REL = DEST − (BRANCH + 2)
REL = $00 1150 − $00 102E = $00 0122

This value is truncated to $0122 as the branch's displacement operand.

EXAMPLE 3.16

What is the new location branched to with the instruction $00 1122 BRA $34C2?

Solution:

Add the displacement to the opword address plus 2 to arrive at the address of the next instruction:

BRANCH + REL + $2 = DEST
$00 1122 + $00 34C2 + $2 = $00 45E6

Keep in mind that the displacement is sign extended before it is added to the program counter. Sign extension results in displacements which can be negative and cause program execution to return to an earlier instruction. This use of the branching instruction creates program loops.

EXAMPLE 3.17

What is the displacement that causes a program to execute its next instruction at $00 1206 with a branching instruction at $00 243C?

Solution:

The same computation as in Example 3.16 takes place. Two is added to the opword's address value ($00 243C + $2 = $00 243E) and that value is subtracted from the next instruction location and truncated to word size:

$$\$00\ 1206 - \$00\ 243E = \$EDC8$$

The instruction at location $00 243C has a source code of BRA $EDC8.

An extension of the program counter with displacement mode is the *program counter with index*. The index register used with this mode follows the same set of rules as it does with the address indirect mode. That is, the size of the data used in the index register is either word or long word and any register can be used as an index register. The indexed displacement is the sign extended byte included in the instruction opword. The effective address for an instruction that uses this mode is found by adding the contents of the program counter to the index register and the displacement value.

EXAMPLE 3.18

What is the effective address if the mode used is program counter, indexed with a displacement of (a) $20 and (b) $94? The program counter holds $0000 2036 and the index register (D1) holds $0000 0012.

Solution:

The effective address in the program counter with index is the sum of the program counter contents, the index register contents, and the sign extended byte displacement. For (a), the sum is

$$\$0000\ 2036 + \$0000\ 0012 + \$0000\ 0020 = \$0000\ 2068$$

For (b), the computation becomes

$$\$0000\ 2036 + \$0000\ 0012 + \$FFFF\ FF94 = \$0000\ 1FDC$$

As an address, the upper byte is omitted, so the final answers are (a) $00 2068 and (b) $00 1FDC.

3.14 ADDITIONAL MOVE INSTRUCTIONS

So far, the MOVE and MOVEA instructions have illustrated the various 68000 addressing modes. Now let's explore some of the more general operations within this microprocessor's instruction set.

MOVEM (Multiple Move)

The MOVE and MOVEA instructions let us load and store information using the various data and address registers as well as memory locations throughout a 68000-

based system. Often, we want to move data to or from several of the registers and memory. The multiple move (MOVEM) instruction serves this purpose well, but there are a few precautions one must take when using it. The syntax for the MOVEM instruction is

MOVEM.⟨size⟩ ⟨register list⟩,⟨ea⟩

or

MOVEM.⟨size⟩ ⟨ea⟩,⟨register list⟩

The data size that can be transferred using this instruction is word or long word only. All data and address registers can be included in the register list in any order. The assembler software sets the correct order of transfer in a register list mask.

The basic concept behind this instruction is that data are moved from consecutive memory locations to specified registers within the register list or vice versa—from the registers into memory locations. The effective address specifies the starting location of memory involved in the transfer. There are three forms of addressing modes used with the MOVEM instruction. One is the address register indirect with predecrement mode; another, the postincrement mode; and last, all other modes, grouped under the term *control modes*. Table 3.1 shows the effective addresses allowed for source and destination modes of the MOVEM instruction.

For the control modes (all but predecrement and postincrement), the order of the registers involved is the same regardless of the direction of data transfer. That order starts with data registers D0–D7; then address registers A0–A7. For example, suppose we want to load data registers 3, 5, and 6, as well as address registers 2, 3, and 7 using MOVEM. In that case, the order in which the registers would receive data is D3, D5, D6, A2, A3, and A7. The data are retrieved starting at the address specified by the effective address and moving up through higher addresses. The register list itself is specified in one of two ways. First, individual registers are separated by slashes. Using the registers of the example just given, the list appears as

D3/D5/D6/A2/A3/A7

Table 3.1 MOVEM Effective Addresses

Address Mode	Source	Destination
Data register direct	No	No
Address register direct	No	No
Address register indirect	Yes	Yes
Address register indirect with displacement	Yes	Yes
Postincrement	Yes	No
Predecrement	No	Yes
Index	Yes	Yes
Absolute word and long word	Yes	Yes
Immediate	No	No
Program counter with displacement	Yes	No
Program counter with index	Yes	No

The order of the list as source code is unimportant. The assembler sets the order within its masking operation.

The second method for writing the register list involves handling consecutive number registers. For instance, let's load registers D2, D3, D4, D5, D6, A3, and A6 using MOVEM. Because they are consecutive, the data registers can be written with a dash between the highest and lowest value registers: D2–D6. The entire list then becomes

D2–D6/A3/A6

Now let us put all the parts of a MOVEM instruction together.

EXAMPLE 3.19

Write a MOVEM instruction to load registers D2, D5, D7, A1, A2, A3, A4, and A6 from memory, starting at location $00 2000.

Solution:

First, derive the register list:

D2/D5/D7/A1–A4/A6

Although the order isn't critical at this point, it is always best to be organized and keep the registers in correct order. One tends to make fewer mistakes that way. Next, the effective address of the data starts at $00 2000. In this instance, we can use the absolute word address mode for the source effective address. Putting it all together and using long word as the data size to be transferred, the MOVEM source instruction becomes

MOVEM.L $2000,D2/D5/D7/A1–A4/A6

Upon execution of the instruction in Example 3.19, data register D2 is loaded with the data from locations $00 2000–$00 2003. D5 gets its data starting at location $00 2004, and the rest follow in order.

Interchanging the source effective address with the register list causes data to be moved from the registers, starting with the lowest data register, into memory, starting at the effective address value. In Example 3.19, the instruction becomes

MOVEM.L D2/D5/D7/A1–A4/A6,$2000

Data from D2 is now stored at location $00 2000–$00 2003. D5's information is moved to memory starting at location $00 2004 and the rest follow in sequence.

A few precautions are necessary when using the MOVEM instruction. The first precaution involves the use of word size data transfers to registers from memory. To this point, we had to consider only that word size data moves fill the lower word of all registers and that address registers are then sign extended through b31. This still applies to MOVEM operations, but data registers as well as address registers are sign extended. Regular MOVE instructions do not cause anything to happen to data registers' upper words during word size data moves. MOVEM *does* cause all registers to be sign extended when word size data are transferred from memory into the registers.

EXAMPLE 3.20

Using the data given, what are the contents of the registers after the MOVEM instruction shown?

$00 2000 = 1234 $00 2002 = ABCD
$00 2004 = 5566 $00 2006 = DEAF

MOVEM.W $2000,D1/D3/A5/A2

Solution:

The source effective address is $00 2000, and the order of word transfers is to D1, D3, A2, and A5. The lower word of D1 is loaded from $00 2000; then, it is sign extended so that D1 = $0000 1234 as a result of the instruction. D3 receives its word from $00 2002, and it is sign extended to $FFFF ABCD. A2 is loaded from $00 2004 so that A2 = $0000 5566. A5, getting its data from location $00 2006, shows $FFFF DEAF after the execution of this MOVEM instruction.

The second precaution in applying the MOVEM instruction involves the use of the predecrement and postincrement addressing modes. The MOVEM instruction is designed so that address registers can be used as stack pointers. Also, MOVEM allows easy pushing and pulling of registers and memory locations. As such, data must be pulled from a stack in the opposite order from which it was pushed to restore those data into the correct registers. The predecrement address mode is reserved for moving data to memory from the registers in the list, forcing this mode to be used only as a destination reference. The postincrement, performing the pull function of restoring data into registers from memory, is used only as a source effective address.

The third precaution is a continuation of the stack operation concept. Not only must the direction in which the address pointer moves be opposite for the push and pull operation, but also the order in which the registers are affected must be opposite from one operation to the next. For example, if we were to push the data from registers D2, D3, and D4 onto the stack starting with D4 and using the predecrement mode, the memory would look like this:

Lowest location: D2 contents
Next highest set: D3 contents
Highest location: D4 contents

In addition, the address pointer used would have been decremented to the lowest address value. To retrieve the data, the postincrement mode is used. The address pointer begins to climb in value and load the registers from memory starting at the lowest location. Therefore, register D2 must be loaded first; then D3 and D4 are loaded. In effect, the stack created by this action is a last in-first out (LIFO) store and retrieve operation. The MOVEM provides for these complications first by restricting when the post increment and predecrement modes can be used. Second, the address list for these two modes is fixed to be opposite in order. The register list is set as before for the postincrement mode; that is, registers are loaded from memory, starting with the lowest data register and ending with the highest address register. This process is the same as for the control modes. The predecrement register mask causes the registers' contents to be moved to memory in opposite order, starting with the highest address register and continuing through to the lowest data register. Because the pointer is decremented, the resulting stack list has the same order in memory as have the other addressing modes. The lowest data register's data are found at the lowest address, and the highest address register data are found at the highest address.

EXAMPLE 3.21

Registers D1, D2, and A3 are to have word size data pushed onto a stack pointed to by A0. A0 contains $0000 2034 before the operation. Write the instruction and show the contents of A0 and memory after it has been executed.

D1 = $1234 5678
D2 = $ABCD EF02
A3 = $0000 CA33

Solution:

The instruction will be the MOVEM, and the predecrement mode will be used as the destination effective address:

```
MOVEM.W D1/D2/A3,-(A0)
```

Since word size data are being moved, A0 is decremented by 2 first; then the data are stored into memory. The memory list after the instruction is

$00 202E = $5678
$00 2030 = $EF02
$00 2032 = $CA33

Notice that as the pointer register is decremented the register contents are stored starting with A3, then D2, and, finally, D1. A0 contains $0000 202E at the end of this instruction.

EXAMPLE 3.22

Now reverse the process. Take the results of Example 3.21 and apply them to the instruction

MOVEM.W (A0)+,D1/D2/A3

Solution:

This is the pull operation. The stack pointer, A0, is incremented after each data transfer. Data are sent to the registers normally—D1, D2, and, finally, A3, as the pointer increases in value. A0 held $0000 202E at the end of Example 3.21. Data from that location are transferred to D1, which then is sign extended to the value $0000 5678. A0 is incremented twice and now points to location $00 2030. $EF02 is moved to D2 and sign extended to $FFFF EF02. Finally, after A0 again is incremented, the last data word is transferred to A3 and sign extended to $FFFF CA33. Following this last transfer, A0 is incremented twice and contains $0000 2034, the value it began with in Example 3.21.

EXAMPLE 3.23

Compare the control mode with the postincrement when pulling data into a pointer. The problem here is what happens to the contents of the pointer register when it is included in the memory list. The following data are held in memory before the execution of each instruction. What are the contents of the address registers at the completion of the instructions?

00 1000 1122	00 1004 5566	00 1008 99AA
00 1002 3344	00 1006 7788	

(a) MOVE #$1000,A3
MOVEM (A3),A0–A4

(b) MOVE #$1000,A3
MOVEM (A3)+,A0–A4

Solution:

In both cases the contents of A3 ($00 1000) are moved into the address buffers, which are used to do the actual pointing to memory. In example (a), the five registers are loaded from memory as expected. Each, in turn, is sign extended so that it contains the following data:

A0 = 0000 1122
A1 = 0000 3344
A2 = 0000 5566
A3 = 0000 7788
A4 = FFFF 99AA

Nothing more is done to the registers, including A3.

In example (b), the same results take place at first. Additionally, the address buffers are incremented by the postincrement process and contain \$00 100A, as expected. Once the instruction has been completed, the contents of the address buffers are transferred back to A3. The only difference between the two operations is the contents of A3: \$0000 7788 for example (a) and \$0000 100A for example (b). In example (a), the data pointer is not needed after the transfer, and it is not restored into A3. For the postincrement mode, it is needed and therefore restored into A3.

The multiple move instruction is a powerful data movement tool, allowing multiple data transfer by a single instruction. Be careful to assign the correct registers to the register list. Also, since sign extension applies to all registers using the MOVEM instruction, make sure that necessary data in the upper words of registers are not destroyed inadvertently.

MOVEQ (Move Quick)

The move quick (MOVEQ) instruction is an immediate mode only operation that moves a single sign-extended byte of data into a data register. The syntax for this instruction is

MOVEQ #⟨data⟩,Dn

where ⟨data⟩ is any byte of data from 00–\$FF. The data first is sign extended to 32 bits and then transferred into the designated data register. For example, MOVEQ #\$CA,D3 puts \$FFFF FFCA into register D3.

3.15 READING THE INSTRUCTION SET DETAILS

The instruction set in Appendix A contains details about every 68000 instruction. The description of the MOVE instruction in the following paragraphs illustrates how to derive needed information about an instruction from this appendix.

Each instruction is listed alphabetically by the syntax of its operation. In earlier microprocessor data sheets, this syntax was called the instruction's mnemonic. The mnemonic for the MOVE instruction is MOVE. Other MOVE-type instructions, MOVEA, MOVEM, and MOVEQ, are described individually under their respective mnemonics.

Following the mnemonic is a brief description of the instruction's operation, the applicable data sizes, and the general syntax for the entire instruction. The MOVE instruction header is illustrated in Table 3.2 on page 110.

The block following the header (see Table 3.3, page 110) holds the effective address chart detailing which addressing modes can be used for source and destination effective addresses. Also included in this block are the condition code flags affected by the instruction.

Table 3.2 MOVE Instruction Header

MOVE:	Move Data
Operation:	Moves data from the source effective address to the destination effective address
Data Size:	Byte, word, long word
Syntax:	MOVE.⟨size⟩ ⟨ea⟩,⟨ea⟩

Table 3.3 MOVE Effective Addresses and Condition Codes

Address Mode	Source	Destination
Data register direct	Yes	Yes
Address register direct	Yes	No
Address register indirect	Yes	Yes
Address register indirect with displacement	Yes	Yes
Postincrement	Yes	Yes
Predecrement	Yes	Yes
Absolute word and long word	Yes	Yes
Immediate	Yes	No
Program counter with displacement	Yes	No
Program counter with index	Yes	No

Condition Codes:
- N: Set if result is negative; reset otherwise.
- Z: Set if result is zero; reset otherwise.
- V: Reset
- C: Reset
- X: Unaffected

The remaining section (see Table 3.4) deals with the translation from the assembly form to the object code of the instruction. The first word is always the opword followed by operands, if they are required.

The term Rn is a general notation for any register whose number is n. The effective address mode code in Table 3.5 also precedes the instruction set in Appendix

Table 3.4 Opword Encoding Examples

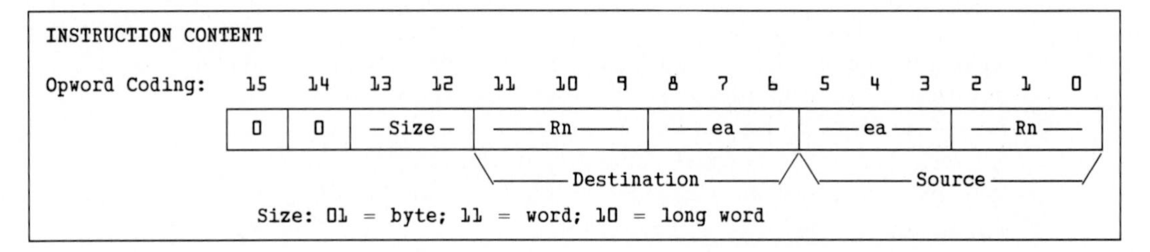

INSTRUCTION CONTENT

Opword Coding:

15	14	13	12	11	10	9	8	7	6	5	4	3	2	1	0
0	0	— Size —		—— Rn ——			—— ea ——			—— ea ——			—— Rn ——		
				Destination						Source					

Size: 01 = byte; 11 = word; 10 = long word

Table 3.5 Effective Address Mode Chart

Address Mode	Syntax	Mode	Register Field
Data register direct	Dn	000	Register number
Address register direct	An	001	Register number
Address register indirect	(An)	010	Register number
Address register indirect with displacement	d(An)	101	Register number
Postincrement	(An)+	011	Register number
Predecrement	−(An)	100	Register number
Index	d(An,Xi)	110	Address register number
Absolute word	Abs.W	111	000
Absolute long word	Abs.L	111	001
Program counter	d(PC)	111	010
Program counter with index	d(PC,Xi)	111	011
Immediate	Imm	111	100

A. There is a 3-bit binary code for each address mode, which is placed in the mode fields of the opword. Additionally, as shown in the table, the program counter and the immediate and absolute modes use a binary code in the register fields when these modes are used.

Immediate, absolute, indirect with displacement, and index modes require operands or word extensions. For the immediate mode, the operand is data and can be either word or long word size. When using byte size data in the immediate mode, the assembler inserts leading zeros to complete a word size operand. When the data are transferred, this upper byte of zeros is ignored.

The operand for absolute mode instructions is either a single word extension for absolute word or two words for absolute long. These operands are the actual address of the data to be used by the instruction.

For the displacement operand, a sign-extended word is used for address register indirect with displacement and program counter with displacement modes. In other words, a word of data is used in the operand. When the instruction is executed, that word is sign extended to 32 bits and added to the address register or program counter specified by the instruction.

Last, the operand required for index mode operations is actually an extension of the opword and is more correctly called an *extension word*. The word includes information about the designated index register and the index displacement used for the instruction. The word is formatted as shown in Table 3.6 on page 112.

Table 3.6 Index Register Extension Word

Bits:	15	14	13	12	11	10	9	8	7	6	5	4	3	2	1	0
	R		Rn		S	0	0	0	Displacement							

R = Register type: 0 = data; 1 = address
S = Register size: 0 = word; 1 = long word
Rn = Register number
Displacement = Sign-extended byte

EXAMPLE 3.24

What is the bit pattern for the instruction MOVE.L $CA(A3),(A1)+?

Solution:

Starting with the opword (see Table 3.4), b_{15} and b_{14} are fixed as lows. b_{13} and b_{12} indicate the size of the data transfer (which is long in this example), so b_{13} is a 1, while b_{12} is low. Putting these upper four bits together makes the upper nibble of the opword 0010 or $2. b_6–b_{11} define the destination effective address, (A1)+. The mode is postincrement, making b_8, b_7, and b_6 equal 011. b_{11}, b_{10}, and b_9 hold the address register number 1 (001). Last, b_0–b_5 show the source effective address: b_5, b_4, and b_3 contain 101 for the address register indirect with displacement mode and b_2, b_1, and b_0 are 011 for address register number 3. Laying out the bit pattern for this instruction altogether becomes

0010 0010 1110 1011 = $22EB

However, we are not done with this instruction. It also requires an operand for the displacement word of the source effective address. The assembler code shows this displacement to be $CA. The displacement is actually a sign-extended word; the assembler will add enough leading zeros to make a full word for the operand portion of the instruction, which becomes $00CA. The full instruction is $22EB00CA.

Now, what happens if both source and destinations have displacement or operand needs? For example, how is MOVE.W $013FC2,$2000 assembled into machine code? First, the opword is assembled in similar manner as in Example 3.24 and becomes $31F9. Recognize that the source effective address is absolute long and the destination is absolute word when verifying the opword as coded. The operand generated by the source follows the opword, and lastly, the operand required by the destination effective address brings up the rear. The entire instruction, when properly encoded, appears as $31F900013FC22000. $31F9 is the opword as shown earlier; 00013FC2 is the absolute long address for the source. When the instruction is executed, the 24-bit limitation of the address bus truncates the operand to $01 3FC2. $2000 is the destination operand for absolute word. During the execution of the instruction, this number is sign extended to $0000 2000 and truncated by the address bus to address $00 2000.

3.16 THE ARITHMETIC GROUP

Now that data can be moved fairly freely with a variety of MOVE instructions, we can introduce instructions that use the moved data. As you read these instructions, remember that a number of addressing modes are applicable to each instruction. You can find which modes can be used in Appendix A under each instruction's header.

Addition

The basic ADD instruction allows data to be fetched using all of the addressing modes as a source and then added to a data register. Limitations are presented when data are added to a destination source. The syntax for the ADD instruction is

ADD.⟨size⟩ ⟨ea⟩,Dn or ADD.⟨size⟩ Dn,⟨ea⟩

Data sizes can be byte, word, or long word. All condition code flags are set or reset based on the result of the addition. With data sizes smaller than long word, the addition does not affect the unused portion of the register. The condition code flags are also based on the data size used, disregarding the remaining data in the register (or memory location) used with the instruction.

EXAMPLE 3.25

D3 = $1234 5678 and D1 = $2233 4455. What do the registers hold after each of these instructions? Also, what are the states of the condition code flags after each operation?

(a) ADD.B D3,D1 (b) ADD.W D3,D1 (c) ADD.L D3,D1

Solution:

In all three cases, the contents of D3 are added to D1, with the result placed into D1. Part (a) uses byte data, causing $78 to be added to $55, yielding $CD. No other data numbers are affected. D3 is totally unchanged, and D1 = $2233 44CD. The result of this byte addition produces a negative number ($CD), so N = 1. The result is not 0, making Z = 0. The X flag matches the C flag and, since no carry is generated by the ADDition, both are reset to 0. Both original numbers ($78 and $55) are positive, but produce a negative sum when a positive sum is expected. As a result, the overflow (V) bit is set.

Instruction (b) is a word addition affecting the lower word of D1 only. $5678 is added to $4455 to produce $9ACE. This sum replaces the lower word of the destination register D1: $2233 9ACE. The condition codes yield the same result as they did for part (a); that is, this sum of two positive numbers ($4455 and $5678) again produces a negative result with no carry out.

Part (c) is a long word addition affecting the full 32-bit contents of D1. $1234 5678 + $2233 4455 = $3467 9ACE. Again, no carry is generated and the result is not 0. However, the result this time is positive, causing the N and V flags to be reset to 0.

ADDA (Add Address) ADDA⟨ea⟩,An is provided for additions using address registers as the destination. The reason behind the separate ADD instruction is that address register destinations are limited to word and long word operations. Additionally, the entire address register is used regardless of data size. To clarify this last statement, first consider adding $7000 to A0 (initially containing $0066 A234). If this is done using the ADD instruction and if $0066 A234 is in D0, the result would be $0066 1234. The C flag is set, indicating that a carry has occurred. However, for ADDA and register A0, the sum is the result of sign extending the word data, $7000 in this example, and adding that number to the full 32-bit contents of A0, producing $0067 1234 in A0. The condition code flags are not affected by an ADDA instruction. Conditional branches done following this instruction will react to conditions created by previous instructions.

EXAMPLE 3.26

What are the results of the following instructions if D0 contains $2C34 C11C and A0 contains $0112 B33C initially?

(a) ADDA.L D0,A0 (b) ADDA.W D0,A0 (c) ADD.W A0,D0

Solution:

Part (a) is a straightforward 32-bit addition with the result placed into A0. D0 is not altered by that instruction and A0 = $2D47 7458. Part (b) is a word add, with A0 as the destination. The lower word of D0 is sign extended to $FFFF C11C and added to A0. Again, D0 is unaltered and A0 = $0112 7458. Part (c) is included as a contrast to part (b). This is a word add with the result placed in D0. In this case, the upper word of D0 is unaffected. The lower word of A_0 is not sign extended, but is added directly to D0's lower word. A0 itself is not changed, and D0 = 2C34 7458.

ADDX (Add Extended) Multiple-precision additions are done using any word size data with the ADDX instruction. This instruction is limited to two modes of operation:

Data register direct ADDX.⟨size⟩ Dy,Dx
Predecrement ADDX.⟨size⟩ −(Ay),−(Ax)

The result in the destination is the sum of the source contents and the destination contents plus the value of the X bit in the condition code portion of the status register.

ADDI (Add Immediate) and ADDQ (Add Quick) Two other ADD instructions specify immediate source data for the add operation. They are ADDI, which uses any of the three data sizes, and ADDQ, which is restricted to the data range of 1–8. These data always are treated as positive integers. If the destination is an address register for ADDQ, the entire register contents are affected by the addition. For other destinations, only the data size specified by the operation is affected. ADDQ is used in place of an increment instruction since the contents of a register can be increased by any amount from 1–8. Similarly, SUBQ can be used in place of a decrement instruction. All condition codes are affected by the instruction so that tests can be made based on the result of the operation.

For ADDI, an operand containing the immediate data is required. Word size operations use one word of extension, while long word data require two words of operand (high word followed by the low word). For byte size data, leading zeros are placed in the upper byte, preceding the byte of data, so that the operand length remains word size. During the actual addition, the leading zeros are ignored.

ABCD (Add Decimal with Extended) The last remaining instruction in the ADD group is add decimal extended (ABCD). This instruction treats the lowest byte of the data register or memory location specified as a BCD number from 00–99. It then performs the multiple-precision sum of the two BCD bytes and the X flag of the condition code register. This instruction is limited to byte size data and to the same two addressing modes as that of the ADDX instruction. The syntax for each of the ADD instructions is summarized in Table 3.7.

Subtraction

The SUBTRACT instructions have the same format and restrictions as the ADD group. The base mnemonic for this group is SUB and the syntaxes are the same as for the ADD group. The subtract operation subtracts the source data from the destination and places the difference in the destination register at the completion of the instruction. As before, where address registers are involved as destination registers, the word source data are first sign extended and then the operation is completed using all 32 bits of the address register.

Table 3.7 ADD Syntax Summary

Instruction	Syntax
Add decimal with extended	ABCD Dy,Dx ABCD -(Ay),-(Ax)
Add binary	ADD.⟨size⟩ ⟨ea⟩,Dn ADD.⟨size⟩ Dn,⟨ea⟩
Add address	ADD.⟨size⟩ ⟨ea⟩,An
Add immediate	ADDI.⟨size⟩ #data,⟨ea⟩
Add quick	ADDQ.⟨size⟩ #data,⟨ea⟩
Add extended	ADDX.⟨size⟩ Dy,Dx ADDX.⟨size⟩ -(Ay),-(Ax)

Multiplication

The 68000 provides two instructions to perform multiplication functions directly. They are multiply unsigned (MULU) and multiply signed (MULS). As indicated, one is used for signed numbers, where the MSB of the data size is an indication of the negative or positive nature of the number. For unsigned numbers, every data bit has a numerical value. Beyond that difference, these instructions operate similarly. Thus, we will concentrate on the MULU in the following discussion.

The MULTIPLY instructions are restricted to word size data and have the following syntax:

MULU ⟨ea⟩,Dn

The lower word of the effective address (memory or register) is multiplied by the lower word of the destination data register. The full 32-bit product is saved in the destination register. The zero and sign flags of the condition code register are affected based on the result of the operation. Since the maximum data value that this instruction can use is $FFFF, the highest product ($FFFF × $FFFF) will not exceed 32 bits. The V and C flags are therefore reset whenever MULU is used. The only mode not usable as a source effective address is address register direct, restricting operation to data registers and/or memory locations.

EXAMPLE 3.27

D0 contains $1234 A678, D1 contains $A012 0002, D2 contains $336B 4105, and D3 contains $B11C A024. What are the results of MULU D0,D1 and MULU D2,D3?

Solution:

The lower words of the registers are multiplied and the result is placed in D1. D0 remains unchanged. D1 = $A678 × $0002 = $0001 4CF0.

Multiplying $4105 by $A024 is a little more difficult to do by hand. However, the result is $28AC 44B4, which means that D3 contains $28AC 44B4 after the second instruction.

Division

The division process is the reverse of multiplication. A DIVIDE instruction yields a 16-bit quotient when a long word is divided by a word size divisor. By experimenting with various data values, you can verify that multiplying two 16-bit numbers does not produce a number greater than 32 bits. Thus, an overflow condition does not exist with multiplication. Unfortunately, dividing a 32-bit number by a 16-bit number sometimes produces a quotient greater than 16 bits, thereby creating an overflow condition. For instance, dividing $4444 4444 by $0002 results in a $2222 2222 quotient. Quotients larger than a word cause the overflow bit in the condition code register to be set and have no effect on destination register Dn.

Just as there are two MULTIPLY instructions, there are also two DIVIDE instructions: divide signed (DIVS) and divide unsigned (DIVU). They perform essentially the same function. We will again concentrate on the unsigned instruction. The syntax for the DIVU is

DIVU ⟨ea⟩,Dn

where the only restriction for the source effective address is the same as for the MULU instruction: You cannot use address register direct as an effective address. The size of data in the destination data register is initially long word, and the data fetched from the effective address location are used as the word divisor. The process is to divide the contents of the destination register by the contents of the effective address and place the result into the destination register. The upper word of the result contains any remainder from the division, while the lower word contains the quotient. If the quotient is larger than a word, the overflow bit is set.

Multiplying by 0 produces a product of 0, but dividing by 0 is undefined. As such, the 68000 causes a divide by 0 exception routine to be run if that condition occurs. This exception routine interrupts the divide process and executes a program at the vector location associated with the divide by 0 exception.

EXAMPLE 3.28

What is the result of DIVU D3,D2 if D2 = $28AC 44B5 and D3 = $1C32 A024 initially?

Solution:

The numbers used in this example are the values used in Example 3.27, but a 1 has been added to the product of that example so that a remainder results from the division. The operation performed divides the contents of D2 ($28AC 44B5) by the lower word of D3 ($A024). The result of the division produces a quotient of $4105 with a remainder of $0001. D2 = $0001 4105 after the instruction is executed.

EXAMPLE 3.29

What are the contents of D5 if the instruction DIVU D4,D5 is performed using the data at the beginning of this section?

Solution:

D5 initially contains $4444 4444 and D4's lower word is $0002. The division process produces an overflow since the quotient result would be a 32-bit value: $2222 2222. The overflow flag is set, and the data in D4 and D5 remain unchanged.

Table 3.8 Instruction Set Summary by Group

Group	Instruction in the Group
Transfer	MOVE, EXCHANGE, BIT Manipulation, PUSH, LOAD, SIGN EXTEND
Arithmetic	ADD, SUBTRACT, MULTIPLY, DIVIDE, COMPARE
Logic	AND, OR, XOR, NEGATE, NOT
Program counter modify	Branch, Decrement and Branch, RETURN, JUMP, Subroutine Calls, TRAPs, LINK
Miscellaneous	CHECK, BIT TEST, NOP

The other condition code flags affected by the DIVIDE instruction reflect the condition of the quotient. The N and Z flags are set if the quotient is negative or 0, respectively. The C flag is always reset to 0 and the X flag is not affected.

This, then, is the arithmetic group of instructions. The instruction set contains other groups which are summarized in Table 3.8. The remainder of this chapter is devoted to the branch group of instructions, leaving the remaining groups for later chapters.

3.17 BRANCH GROUP

Decision making on the 68000 is performed by executing conditional branch instructions that react to the states of the various flags in the status register. The basic syntax of these instructions is

Bcc ⟨relative address or label⟩

where cc is any one of the conditions listed in Table 3.9.

These conditional branch tests are the same group used by the 16-bit microprocessor described in Chapter 1, and their applications are the same. The branch instructions require that a relative address value be specified. This relative address is the signed value of either byte or word size. If the address is byte size, it is included as a portion of the opword. For word size, the byte value in the opword is zero and an operand word is needed to write the relative address value. How that operand is constructed depends on the assembler used. In the most crude form, the user must supply the actual relative address. This is rarely done because of the complexity involved in calculating a correct relative address. A slight error causes the program execution to continue at an address unintended by the user. Also, relative addresses cannot branch to an odd address since all opwords and operands are word length and must be fetched on even address bounds.

Motorola's training board, the education board, assembles the lines of code one at a time. The relative address operand for the branch instruction is translated

Table 3.9 Condition Flag Tests

cc	Name	Flag Conditions
CC	Carry clear	$C = 0$
CS	Carry set	$C = 1$
EQ	Equal to (0)	$Z = 1$
GE	Greater than or equal to	$N \oplus V = 0$
GT	Greater than	$(N \oplus V) + Z = 0$
HI	Higher than	$C + Z = 0$
LE	Less than or equal to	$(N \oplus V) + Z = 1$
LS	Less than or the same as	$C + Z = 1$
LT	Less than	$N \oplus V = 1$
MI	Minus, or negative	$N = 1$
NE	Not equal to (0)	$Z = 0$
PL	Plus, or positive	$N = 0$
VC	No overflow	$V = 0$
VS	Result is too large	$V = 1$

by the assembler from the real address of the destination location. This is easier than computing the relative address yourself. There is a problem in that there are times when you do not know where that destination address is. In those cases, insert a dummy operand when you first write the source program. When you reach your destination location, note its address. Go back later to the branch instruction and insert the correct address.

There is an additional annoying feature of the education board assembler related to address assignments. Recall that the use of labeling facilitates address assignments and allows for address-free (or address-independent) programming. Unfortunately, the education board assembler does not support the use of labels. All addresses must be included within the source listing. Motorola's EXORmacs™ is a full assembler that supports labels as well as a full set of pseudo ops. With this assembler, labels are used in place of the relative address operands when writing branch instructions.

Decrement and Branch

The 68000 does not have a decrement register instruction. The discussion of arithmetic instructions above mentioned that the SUBQ instruction is used in place of a decrement instruction. Under the branch group, there is a special instruction provided because of the usual heavy need for a looping counter branch operation. This operation involves performing a number of operations until either a condition is met or a counter counts out.

This concept can be applied to the problem of a checkbook containing 10 checks. A short program can be written to tell the user if there still is money in the account or if there are no more checks. The first instructions deduct the amount of the check from the account balance and then check if the remaining amount is negative. If not, a counter holding the number of checks is decremented. If there are still more checks, the program loops back to the first instruction, awaiting the next withdrawal. If any withdrawal indicates the account is exhausted or when the user runs out of checks, the program exits the loop to execute instructions to inform the user of the problem. Figure 3.1 is a flowchart of this checkbook problem.

The test condition, decrement and branch (DBcc) instruction performs the test and looping portion of the looping program. First, the specified condition (cc) is checked for a match. If the match does not occur (i.e., the test is false), the lower word of a selected register is decremented. As long as the register contents are not −1 ($FFFF), a branch is executed to a destination location. The loop is exited if either the condition proves true or the counter counts down to −1. The syntax for this instruction is

DBcc Dn,⟨relative address⟩

A label can be used in place of the relative address as long as the assembler in use recognizes labels and does the relative address computation during the assembly process. The term Dn designates any data register. The cc conditions are listed in Table 3.9 along with two additional test condition codes. These condition codes are false (F), meaning that the test is never satisfied, and true (T), where the test is always true or met. The false test creates a condition in which the branch is always taken until the counter reaches −1. In a sense, this creates a basic decrement instruction with a built-in test for when the counter counts out.

The DBT Dn,⟨address⟩ produces an unconditional branch and since the test is always true, a branch is performed immediately. This form is rarely used because the instruction set provides an unconditional branch, BRA ⟨address⟩. Relative addresses for the DBcc instruction are limited to word size only, while both byte and word size can be used by the BRA instruction.

3.18 CHECKBOOK APPLICATION OF THE DBcc INSTRUCTION

The purpose of this first program is to become more familiar with the 68000 instruction format. Let's extend the brief checkbook example of the DBcc instruction into a program. Refer to the flowchart in Figure 3.1 during the following discussion.

Input and output information to a 68000-based system is handled through designated memory locations; these locations are affected by all instructions that involve normal memory operations. Locations designated by the labels ENTRY and EXIT

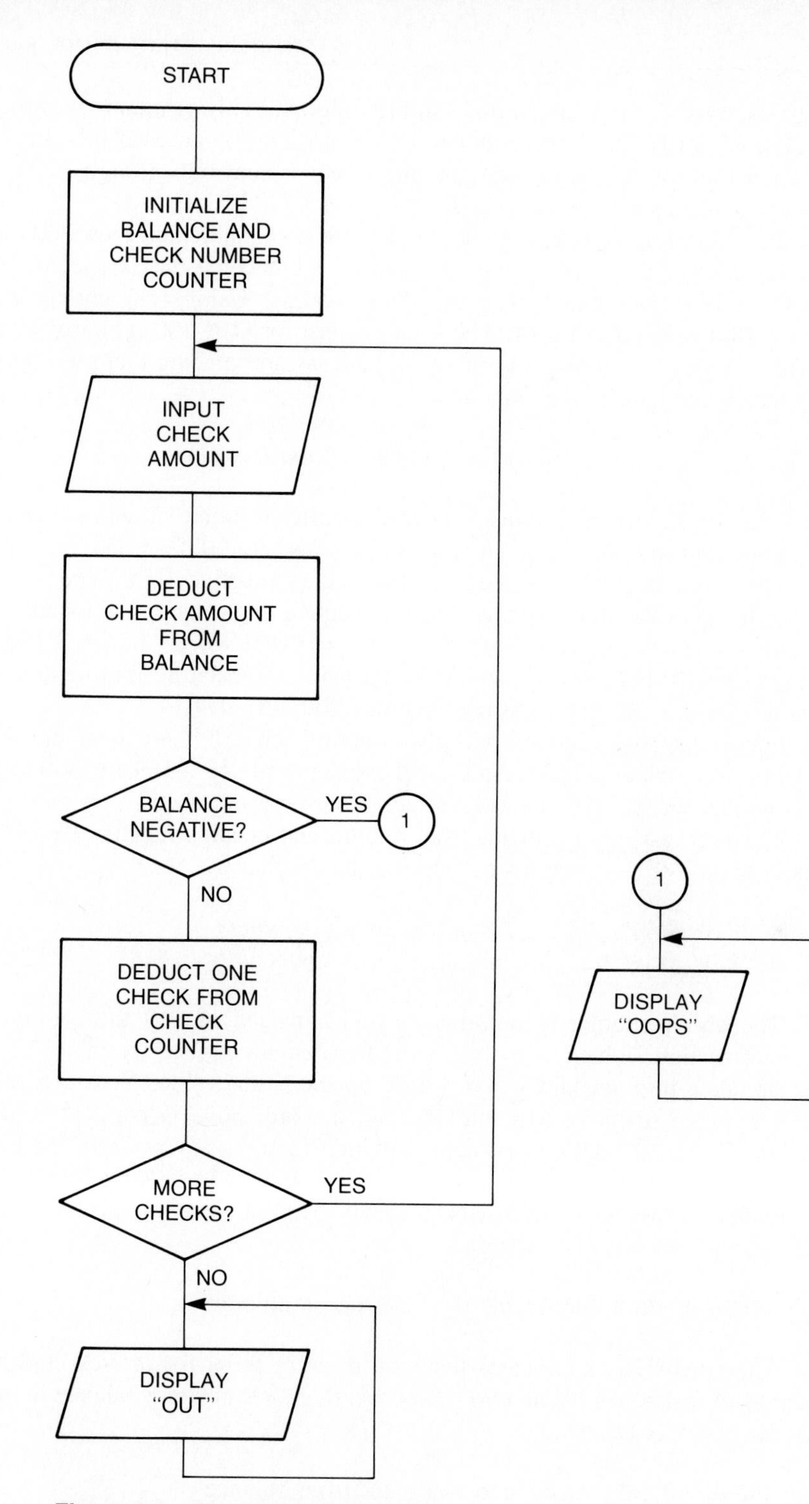

Figure 3.1 Checkbook flowchart

will be used for I/O devices in this sample program. This program takes the amount brought in from ENTRY and deducts it from a preset balance. If the check amount overdraws the account, a message to that effect is sent out through EXIT. We will represent that message by the ASCII characters OOPS.

The American Standard Code for Information Interchange (ASCII) is a 7-bit character code; that is, every combination of 7 bits represents a specific character. (Table 1.2 is a complete listing of ASCII codes.) Using four characters, OOPS requires 28(4×7) bits of data. The code pattern for *O* is 1001111; for *P,* 1010000; and for *S,* 1010011. Combining these bit patterns and placing them in a left-to-right order makes the data for the message

1001111100111110100001010011

Rounding out 32 bits by making the most significant nibble all zeros means that the data, translated into hexadecimal, for the message are $09F3 E853.

The other condition is that the balance is positive, but there are no more checks. In this case, the message OUT is sent through the EXIT location. The bit pattern for OUT (21 bits) is 1001111 for *O,* 1010101 for *U,* and 1010100 for *T,* or 100111110101011010100 for the entire message. Translating into hexadecimal and including leading 0's, the message becomes $0013 EAD4.

Let us say the account initially contains $500.00 and there are 25 checks available. The hexadecimal values for these constants are $1F4 for 500 and $19 for 25. Now that we have the data, let's tackle the program itself.

The first task is to initialize the amount and check counters. We do this with immediate MOVE instructions to selected data registers.

```
MOVE.W #$1F4,D0    * Initialize amount counter
MOVE.W #$18,D1     * Initialize check counter
```

The check counter is initialized to $18 instead of $19 to accommodate the DBcc instruction, which will loop until the counter register reaches −1. The −1 must be taken into account when setting up the count value. Note that only a byte size is required to move $18 into D_1, but a word move is performed. The intent is to use D1 for the DBcc instruction, which requires a word size for the counter register, making a word move to initialize the counter necessary. Next, we retrieve the amount of the check from ENTRY, which, as a memory location, requires an absolute-type MOVE instruction:

```
READ  MOVE.W ENTRY,D2    * Get check amount
```

The label READ indicates the loop re-entry point for the test and branch instruction. A test is needed here to verify that there is a positive balance in the account once the check is cashed:

```
SUB.W D2,D0     * Deduct amount from balance
DBMI D1,READ    * Branch if balance is positive and D1 > -1
```

The DBMI instruction first checks to see if the balance is negative. If it isn't, register D1 (which holds the check count) is decremented. If the number of checks does not reach −1, the loop returns to READ and awaits the next check entry. Once the loop is exited, the program must determine if the cause was a negative balance or an exhaustion of checks. The DBcc instruction does not affect the condition codes as it performs its operation. As such, the flags, particularly the N flag in this case, were set or reset depending on the SUB instruction. Essentially, if the N flag is set at this point, then the loop is exited because of a negative balance. If the sign flag is reset, then the cause of the exit is that all checks have been used. There is, of course, the possibility that both actions have occurred, but since the overdraw is the more serious problem, we will accept the OOPS message for that possibility as well.

```
        BMI OOPS                * Check if overdrawn
        MOVE.L #$13EAD4,EXIT    * Send OUT message
DONE    BRA DONE                * End program
OOPS    MOVE.L #$9F3E853,EXIT   * Send OOPS message
        BRA DONE
```

Program 3.1

When writing data in a text such as this, it is convenient to enter spaces between groups of data for clarity. However, when entering data into a source listing, those spaces must be omitted. The eternal loop on the DONE BRA DONE line is one method to terminate a program. No further instructions can be executed following this instruction since it is stuck in the loop. Exit from the loop is done by hardware means—either a RESET or BREAK key on the keyboard or test set.

```
        MOVE.W #$1F4,D0         * Initialize amount counter
        MOVE.W #$18,D1          * Initialize check counter
READ    MOVE.W ENTRY,D2         * Get check amount
        SUB.W D2,D0             * Deduct from balance
        DBMI D1,READ            * Positive balance loop
        BMI OOPS                * Check if overdrawn
        MOVE.L #$13EAD4,EXIT    * Send OUT message
DONE    BRA DONE                * End program
OOPS    MOVE.L #$9F3E853,EXIT   * Send OOPS message
        BRA DONE
```

Program 3.2

Review the short program we have just developed to get a feel for the source coding. Designating data sizes is a fairly new process for experienced as well as inexperienced programmers. Adjusting to the source and destination formats to indicate the flow of data also requires practice. As with the 16-bit microprocessor described in Chapter 1, labels will be used to indicate addresses within a program. It will depend largely upon the system available to you whether you can extend the label facility to your application. Eventually, you will have systems available to you whose assemblers allow the use of labels. They are easier and more convenient to use than absolute addresses.

PROBLEMS

For all questions involving data, use the following data:

D0 = $1234 5678	A0 = $0000 2000	$00 2000 = $FACE
D1 = $9ABC DEF0	A1 = $0000 2001	$00 2002 = $0AFE
D2 = $1122 3344	A2 = $0000 2002	$00 2004 = $4A63
D3 = $5566 7788	A3 = $0000 2004	$00 2006 = $0002
D4 = $0012 0034	A4 = $0001 1006	$00 2008 = $FFFF
D5 = $C01D FF32	A5 = $0000 200A	$00 200A = $2006
D6 = $AABB CCDD	A6 = $0001 1004	$00 200C = $1010
D7 = $0001 0004	A7 = $0000 4000	$00 200E = $ABCD

3.1 What data are contained in each of the destination registers for the following instructions?
(a) MOVE.L D2,D0 (b) MOVE.W (A0),D3 (c) MOVE.B D6,D1
(d) MOVEA (A5),A2 (e) MOVE.B $3(A3),D4 (f) MOVEA.L A1,A5
(g) MOVEA #$C345,A6

3.2 What is the value of the effective address used as the source of the instruction MOVE.W 3(A1,D7.W),D1? What are the contents of D1 as a result of the instruction?

3.3 What are the contents of the data registers following the instruction MOVEM (A2)+,D0–D5? What are the contents of A2 at the end of that instruction?

3.4 List the memory locations and their contents as a result of executing the instruction MOVEM.L D3/A5/A2/D6,−(A6).

3.5 Write a short program segment that transfers 53 words of data, starting at location $00 2002, to an area starting at location $01 1006. Both memory banks increase in address from the starting locations. Hint: Use DBcc with D4 as the counter register.

3.6 What are the contents of each destination register or location for the instructions below? Assume the X flag is reset.
(a) ADD D0,D3 (b) ADDA D1,A1 (c) ADDQ #$3,(A3)
(d) ADDI #$C235,A2 (e) ABCD D3,D2 (f) SUB.L D2,D1
(g) SUBA.L $2(A3),A0 (h) SUBI.B D6,D5

3.7 What are the contents of the destination registers following each of these instructions:
(a) MULU D2,D7 (b) DIVU 5(A1),D6

3.8 A list of 10 words is already stored starting at location $00 4000. Write a program to store the sum of the LSBs of these words at location $00 5000. Keep in mind that the data at the 10 locations could produce a sum that is either byte or word size. A full word result should therefore be stored at location $00 5000.

3.9 Location \$00 2000 contains the lower word of an address as data. The upper portion of this address is \$00. The next location (\$00 2002) contains initial counter information. Write a program that finds the largest unsigned word in a data bank that starts at the location pointed to by the address data held in location \$00 2000. The number of words in the bank is the data at location \$00 2002. Store this value at location \$00 2004.

3.10 What is wrong with the following instructions? Use the instruction set in Appendix A as your reference.

(a) MOVEA.B D3,A2 (b) ADDQ #\$A,D2 (c) MOVE 3(A0)+,D2
(d) ADD.L \$002003,D6 (e) MOVE.L D4,#\$2000

3.11 Which of the following instructions cause the overflow flag to be set? Again, use the data given at the beginning of the section to answer this question.

(a) ADD D0,D3 (b) DIVU D2,D5 (c) MULU D5,D6
(d) ADD.L D6,D5 (e) DIVU D7,D6 (f) MULU D7,D4

4

68000 PROGRAM APPLICATIONS AND BUS CYCLE TIMING

CHAPTER OBJECTIVES

This chapter discusses the interrelationship between programs and bus cycle timing by using example programs to examine 68000 bus cycle operations. The function control leads, $R/\overline{W}$, and data strobes guide the operation of various portions and devices during program execution. Occasionally, references are made to the control signal truth tables in Chapter 2 and their functional descriptions. After we examine bus cycle timing, we will develop a number of program applications to give you a feel for the 68000 instruction set.

4.1 MEMORY READ CYCLE TIMING

The checkbook program developed in Chapter 3 illustrates some of the 68000 timing requirements. To execute the program, as with every microprocessor-based computer system, the first instruction (MOVE.W #$F4,D0) must be fetched from memory. Looking up the MOVE instruction in Appendix A, we translate the source code of our instruction into the following bit patterns:

Source mode is immediate, so the mode/register pattern is 111 100

Destination mode is data register direct using register D0, making its mode/register pattern 000 000

Word size data are used, making size = 11

Putting the opword pattern together produces 0011 0000 0011 1100 = $303C

Since this is an immediate mode instruction, an operand word extension is required. This word is the actual data used: $01F4. The complete machine code is

\$303C01F4 for this MOVE instruction. The first operation of the program is to fetch the opword from memory. For illustration, this program resides in memory beginning at location \$00 1000. Refer to Figure 4.1 to follow the actual read bus cycle timing discussion.

The initial action of the microprocessor sets the function control leads (FC_2, FC_1, and FC_0) to 010, the combination for a user program function (see Table 2.5). Then a valid address is placed on address lines A_1–A_{23}. At this point, the system has the address and function code information available. If areas of memory must be hardware restricted to their function, that logic process would take place at this time. Chip enables on the memories involved would not yet be asserted until the $\overline{AS}$ control signal goes low.

The address strobe $\overline{AS}$ and data strobes $\overline{LDS}$ and $\overline{UDS}$ are made active simultaneously (2000 ns maximum) after a valid address is placed on A_1–A_{23} by the microprocessor. The address strobe allows the address bus to be decoded and passed to one of a memory device's chip selects. The data strobes also activate another chip select on their designated memories. Since this is a memory read cycle, the $R/\overline{W}$ remains high throughout the operation. The program counter is incremented twice (to \$00 1002) to account for both bytes of the opword.

Once the memory chips are fully selected, valid data must be placed on the data bus within 65 ns of asserting data acknowledge ($\overline{DTACK}$). $\overline{DTACK}$ must be returned

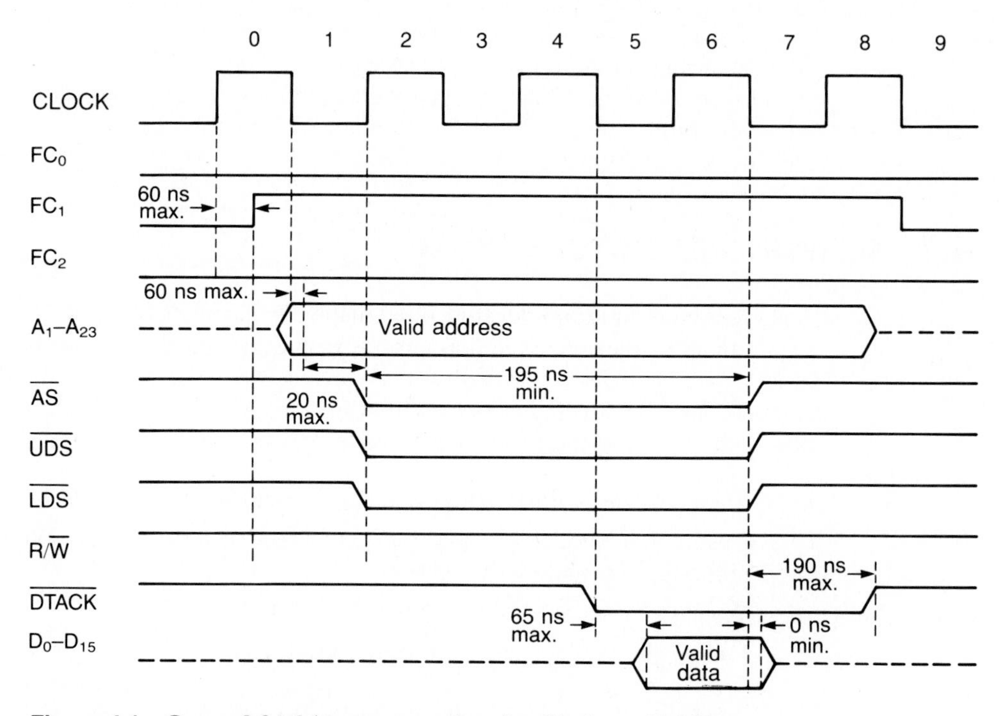

Figure 4.1 Opword fetch/memory read cycle. Clock = 10 MHz

to the microprocessor at some time after the selection process and before the address strobe is removed. This signal informs the microprocessor that a successful data transfer has been completed. Because this is an asynchronous control, a number of clock cycles (or wait states) can occur before $\overline{\text{DTACK}}$ is asserted. Once the memory chips are fully selected, valid data must be placed on the data bus within 65 ns of asserting $\overline{\text{DTACK}}$. $\overline{\text{AS}}$, $\overline{\text{UDS}}$, $\overline{\text{LDS}}$, and $\overline{\text{DTACK}}$ are made inactive at the time shown in Figure 4.1. The bus error and rerun timing for an invalid or unsuccessful data transfer are discussed later in this chapter.

4.2 PREFETCH OPERATION

The 68000 uses a two-word prefetch mechanism to absorb program fetch cycles within execute times. Initially, two words are fetched from the program listing. As each word is used by the microprocessor to execute the program, a new word is fetched from the list. When data fetches from memory are required, the prefetch operation is delayed temporarily to avoid conflict created by multiple use requirements of address and data buses. The program counter points to the last word fetched (in essence, to the location of the second prefetched word) so that it points to the location following the currently used word from the program. This is done so that the program counter always points to the next instruction in a program list. The second word prefetched following a branch opword is not used, since the branching operation modifies the contents of the program counter and causes a new opword to be fetched from memory.

EXAMPLE 4.1

Using the following program segment, trace through the fetch operation of the 68000.

```
004000 307C2000 MOVEA #$2000,A0
004004 601C     BRA.S $1C
004006 D240     ADD D0,D1
004008 3081     MOVE D1,(A0)
```

Solution:

During the execution of the instruction preceding the MOVE, two words are prefetched into the 68000: the opword, $307C, and the operand, $2000. The program counter at this point contains $00 4002, the location of the operand that was prefetched. The opword is transferred to the instruction register and the next word ($D240) is fetched from the program list. The program counter now holds $00 4004. As $2000 is loaded into A0 and the register is sign extended to $0000 2000, the next word from the program ($3081) is prefetched, causing the program counter to be incremented to $00 4008. The branch always (BRA.S) opword ($601C) is placed into the instruction register and is immediately recognized as a program-modifying instruction. The prefetch is not performed. Instead, the relative address of the BRA.S is sign extended and added to the contents of the program counter ($00 4008 + $00 0040 = $00 4048). Two new words are prefetched, starting at location $00 4048, and the operation continues until the program is run completely.

4.3 WAIT STATE INSERTION

The basic memory read cycle is the same for all standard reads. The major difference is in the levels of the function control leads, which are high or low according to the type of memory operation that is in progress. If a slow memory is used (one with a long access time), wait states are inserted between cycles 4 and 5. These wait states are created by delaying the return of $\overline{DTACK}$ until data are available from the memory. The number of wait cycles depends on how long the device maintains the $\overline{DTACK}$ inactive. The memory bus cycle is resumed to completion following the application of $\overline{DTACK}$.

4.4 THE WRITE CYCLE

The next to last instruction in Program 3.1 is a long word move to a memory location called EXIT (see page 123). Assigning consecutive addresses for every byte in the program, we find that this instruction's opword is at program location $00 101C. Going through the same process as with the earlier MOVE instruction, we can construct the machine code for this instruction (use $00 2000 for EXIT):

```
MOVE.L #$09F3E853,EXIT = $2E3C09F3E8532000
```

The first word is the opword, followed by the source operand and, finally, the destination operand. As before, the opword is fetched first (either by prefetch during the previous instruction cycle or as the first fetch in the sequence). While the opword is being acted upon, the first word of the source operand is fetched. The decoding circuitry recognizes the data transfer as a long word immediate mode operation. As such, the internal workings of the microprocessor move the first data word into a temporary register and fetch the second word simultaneously. This second word is also held in the temporary buffer. During the temporary storage of the second data word, the destination origin word is fetched and, being an absolute address mode, is transferred to the address buffers. Since the mode used is absolute short, the upper word of the address buffers is set to zero.

While the address bus is being used to do the transfer to memory, a prefetch is not possible. The first data word is transferred to the address pointed to by the address buffers. The address buffer is incremented twice and the second word is transferred to memory. Immediately following the transfer, the microprocessor enters a memory opword read cycle and fetches the next program word.

Figure 4.2 is the memory write timing diagram used for the memory transfer described in the previous paragraph. To begin with, the function control lines are sent out indicating (in our example) that user data transfer is ready to start. Shortly after they are sent out, the address information is supplied to the address bus within the time frames shown in Figure 4.2. As a result, the address decoders begin the chip-enabling process of the system's memories. The actual decoding occurs with the application of the $\overline{AS}$ signal, a minimum of 20 ns after valid address data are on the address bus. The read/write (R/$\overline{W}$) signal is made low, indicating a memory write operation, and the data are placed on the data bus. A short time of 10 ns, called *setup time,* allows the data lines to settle, after which $\overline{LDS}$ and $\overline{UDS}$, the data strobes, are applied to complete the memory-selection process.

Following the selection of the memory chips for the write and before the process is completed, $\overline{DTACK}$ is returned to the microprocessor to indicate that a successful transfer has occurred. When a reasonable hold time to ensure that the data are written into the memory has passed, the data are removed from the bus lines and the control signals are returned to their inactive states. There is a complete diagram

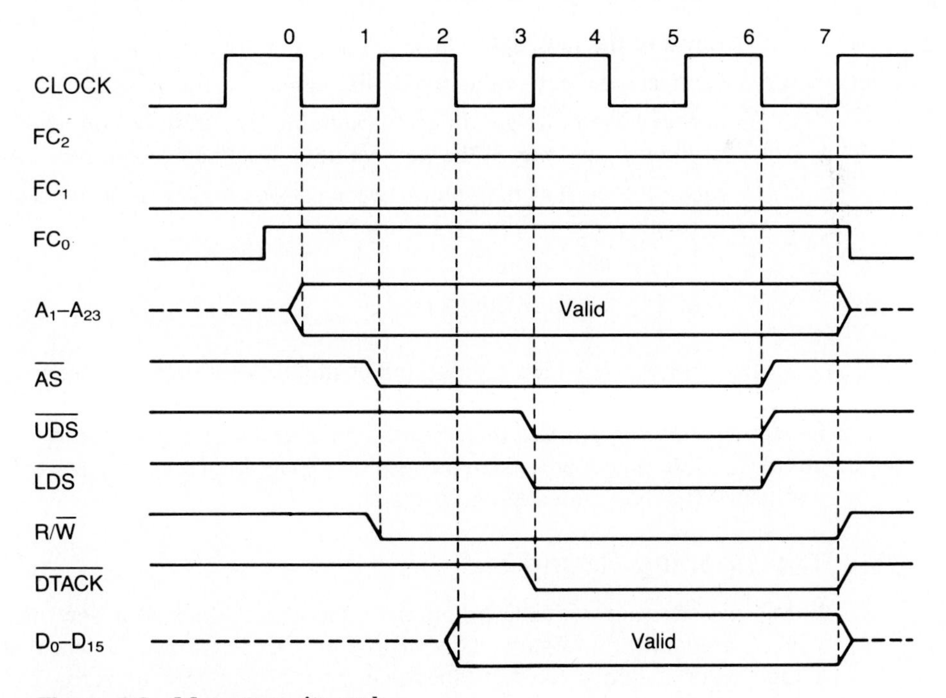

Figure 4.2 Memory write cycle

of the actual times for the occurrence of the signals thus far described in Appendix B, which contains all the timing diagrams for the 68000 microprocessor.

4.5 DATA INSERTION APPLICATION PROGRAM

To illustrate the ways data are manipulated by the 68000, let us develop a data insertion program to insert a new data word into an ordered list of data and to update the word count total to reflect the addition of the word. The ideas behind this program lead to sorting and alphabetizing routines. Address COUNT holds the hexadecimal value of the number of words in the list. This allows the list to contain from 0 values to 65,535 words. The beginning address of the list is contained at location LISTTOP, allowing the list to be anywhere in memory. The word values are unsigned numbers, and the list starts with the lowest value and increases as the addresses that hold the data increase. The new value to be inserted is stored at location VALUE.

To approach any programming problem, assess it thoroughly and decide what actions the program is expected to perform. If we list these requirements for the problem stated, it looks like this:

1. Initialize the word counter and check for no contents.
2. Initialize the data pointer.
3. Retrieve the new data value.
4. Compare the new value to the list values.
5. Set up a loop that continues through the list, from bottom (highest address) to the top, as long as the new value is less than the current list value.
6. Move the current list values two locations higher until the correct location is found for the value to be inserted.
7. Insert the new value.
8. Add 1 to the word counter.

The process by which a programmer initially develops the needs of a program makes a significant difference in the degree of success experienced or effort expended in putting the program together. Some of the tasks in the preceding list are fairly obvious, while others are a bit more subtle. Your skills and abilities will develop as you gain experience in creating programs.

The Opening Round

Before we start out with the program example, let's look at a new instruction that will be used in the program. This instruction, CLR.⟨size⟩⟨ea⟩, sets that portion of a register or memory location specified by the data size to zero. The remaining bits are unaffected by the operation. Address register direct is not allowed with this instruction, that is the data in an address register cannot be set to zeros by using the CLR instruction.

The first part of the program initializes the system to retrieve and use the data. In this example, we must first initialize an address pointer and a word list counter. Additionally, if the new data are to be the first words entered into the list, then the processing will be different.

```
CLR.L D6                * Clear counter register
MOVE.L #LISTTOP,A0      * Initialize data pointer
MOVE.W VALUE,D0         * Get new data word
MOVE.W COUNT,D6         * Retrieve counter value
MOVE.W D6,D5            * Copy count value
ADD.W D5,D5             * Double count value
ADDA.L D5,A0            * Adjust pointer
TST.W D6                * Is word count at D6 zero?
BEQ NEWLIST             * No values now in list?
SUBQ #$1,D6             * Adjust for DBcc instruction
```

Program 4.1

At first, all we do is get the data to the first several locations and transfer them to A0, D0, and D6. The count data are copied into D5 and doubled. This action allows the program to maintain the actual count value in D6 and to create an offset to be added to the address pointer A0. This offset adjusts the pointer so that it points to the bottom of the list regardless of how long the list is. Next, it is necessary to test D6 to find if the word count is zero, indicating that the new word to be added is actually the first word to be entered into the list. This is done using a TST instruction, which subtracts zero from the contents of D6. This causes D6 to remain unchanged, but does set or reset the N and Z flags based on the state of the data in D6. To account for the process of decrementing to -1 by a DBcc instruction, the count value in D6 is reduced by one using the SUBQ instruction. SUBQ allows a quicker subtraction operation than a regular SUB instruction. The source operand for the SUBQ is always an immediate data value from one to eight. This instruction is widely used as decrement instruction since no such instruction other than the DBcc is supplied for the 68000.

Now that the pointer and counter are correctly established, it's time to go through the list to find where the new word will be inserted. This is done with a CHECK loop that compares the value of the new word that was loaded into D_0 with the rest of the list values.

```
CHECK    MOVE.W -(A0),D4     * Get list value
         MOVE.W D4,2(A0)     * Save list value
         CMP D4,D0           * Compare data with list values
         DBHI D6,CHECK       * Loop test,instruction
         MOVE.W D0,2(A0)     * Insert new value
         MOVE.W D4,(A0)      * Reinsert list value
```

Program 4.2

The first instruction in this block retrieves the list value and places it into register D4. The address pointer is predecremented and points to the value just accessed. This value is then moved to the next word location in the list, making the original location available for new data. Next, the data in D0 (the new value) are compared with the data from the list in D4. The DBHI checks to see if the new word is higher than the value in the data list; if it is, then the test is true. A true result causes the program to fall out of the DBcc loop. This is the point at which the new word must be entered into the list. If the new value is lower or the same value as that in the list, the lower word of register D6 is decremented. If that value is greater than -1, the loop is repeated. Notice that the pointer moves as a result of using predecrement with the compare instruction.

If the new word is higher than the data tested, then it is inserted at the previous location. The loop is not executed because the test condition for higher (HI) proves true in this case.

Inserting the Word and Reordering the List

The new word is moved into the location from which the last list value was accessed, writing over the original list value (which has a lower value). Therefore, the original value (D4) must now be inserted into the next available location. The remaining requirements of the program are to increment the list count and to provide for the case where the new value is the only (or first) value in the list.

```
DONE     ADDQ #$1,COUNT      * Increment count value
FINI     BRA.S FINI
NEWLIS   MOVE D0,(A0)        * Insert new first word
         BRA DONE
```

Program 4.3

The word counter is incremented by one to account for the new word added to the list. The program ends in an endless loop (BRA FINI). If the new word is the only word in the list, NEWLIST is branched to. In that case, the new word is saved and the program branches to DONE. The complete listing for this program is shown below.

```
          CLR.L D6            * Clear counter register
          MOVE.L LISTTOP,A0   * Initialize data pointer
          MOVE VALUE,D0       * Get new word
          MOVE COUNT,D6       * Retrieve count value
          MOVE D6,D5          * Copy count value
          ADD D5,D5           * Double count value
          ADDA.L D5,A0        * Adjust pointer
          TST D6              * Is word count at D6 zero?
          BEQ.S NEWLIST       * No values now in list?
          SUBQ #$1,D6         * Adjust for DBcc instruction
CHECK     MOVE -(A0),D4       * Get list value
          MOVE D4,2(A0)       * Save list value
          CMP D4,D0           * Compare data with list values
          DBHI D6,CHECK       * Loop test
          MOVE D0,2(A0)       * Insert new value
          MOVE D4,(A0)        * Reinsert list value
DONE      ADDQ #$1,COUNT      * Increment count value
FINI      BRA.S FINI
NEWLIST   MOVE D0,(A0)        * Insert new first word
          BRA DONE

Note: Size defaults to word (.W)
```

Program 4.4

4.6 THE SWAP INSTRUCTION

The ability to handle different sizes of data is convenient, but a problem arises when the data are not where you need them. Word size data manipulations are performed on the lower word portion of registers or memory locations. Suppose you need to perform an operation on the upper half of a 32-bit word, but not on the lower half. The SWAP instruction easily and rapidly performs a word swap, exchanging the upper and lower word positions in a register. The syntax of the SWAP instruction is

```
SWAP Dn
```

where Dn refers to any data register. In the next program, we will make good use of the SWAP operation.

4.7 OVERFLOW DIVIDE PROBLEM

Both the DIVU and DIVS operations experience overflow problems when confronted with a quotient larger than 16 bits. For example, if you divide $2245 6689 by $0002, the result is $1122 B344, with a remainder of $1. Using the DIVU instruction does not change the contents of the data registers. The overflow flag is set, indicating that the quotient is too large. The possibility of performing such divisions is very likely, so we need to develop a program to handle the situation.

The initial condition for the divide problem is that the divisor is stored manually by the user at location DIVSE. The value to be divided is a 32-bit number, stored in memory starting at location DIVDE. The resulting 32-bit quotient is saved starting at location QUOT. Finally, the remainder from the division is saved at location REMAIN. The task ahead is to split the original dividend into two separate dividends, divide each by the divisor and combine the results. Since each division is performed on a 16-bit dividend rather than a 32-bit one, no overflows result. One thing to consider during this process is the remainder from the division of the upper half of the original 32-bit dividend. It must be appended to the lower half before that portion can be divided by the divisor.

To illustrate the process, it is best to walk through the problem. Separating the original number into two halves produces $2245 and $6689. Dividing the upper half by 2 yields $1122 with a remainder of 1. This remainder must be added as $1 0000 to the lower half number before it can be divided: $1 6689. After the division is performed a second time, the result is $B344 with a remainder of $1. Combining the two results yields the final answers: quotient = $1122 B344 and remainder = $0001. Verify manually that the result is correct. The only difficulty in this example is the remainder from dividing 5 by 2. It is a value of 16 that is added to the next lower digit (6) to make 22. Divide 22 by 2 to get 11 ($B). The remaining divisions are straightforward.

4.8 THE OVERFLOW DIVISION APPLICATION PROGRAM

As before, this program is developed in blocks to clarify their use. DU and DL refer to the upper and lower portions of the dividend in the comments column. Similar notations are used for the quotients (QU and QL) and the remainders (RU and RL). Each represents a word of data.

The first requirement is to initialize the registers used in the program.

```
CLR.L D1            * D1 = 0000 0000
MOVE.L DIVDE,D0     * D0 = DU DL
MOVE.W D0,D1        * D1 = 0000 DL
CLR.W D0            * D0 = DU 0000
SWAP D0             * D0 = 0000 DU
```

Program 4.5

The dividend is retrieved from memory and split into two parts. The CLEAR instructions set portions of the registers to zeros and the SWAP instruction positions the upper dividend portion into the lower word part of D0. Now the first division operation can be performed. After the division, the result is copied into D3 and the remainder is isolated by clearing the lower word of D3. Now the addition of the upper remainder to the lower dividend is performed, followed by the second division.

```
DIVU DIVSE,D0     * D0 = RU QU
MOVE.L D0,D3      * D3 = RU QU
CLR.W D3          * D3 = RU 0000
ADD.L D3,D1       * D1 = RU DL
DIVU DIVSE,D1     * D1 = RL QL
```

Program 4.6

The final result is now in pieces in D0 (QU) and D1 (RL and QL). They must be combined and stored into memory.

```
        SWAP D0             * D0 = QU RU
        CLR.W D0            * D0 = QU 0000
        ADD.W D1,D0         * D0 = QU QL
        MOVE.L D0,QUOT      * Store quotient
        SWAP D1             * D1 = QL RL
        MOVE.W D1,REMAIN    * Store remainder
DONE    BRA.S DONE
```

Program 4.7

Be very careful to consider the size of data each instruction uses and exactly where those are initially. This program is a good exercise in data manipulation. Notice the subtle word ADD that combines the two quotient halves, while leaving the final remainder (RL) unaffected. The last SWAP places that remainder in a position to be transferred to memory. This program can be used for any divisions, even those where no overflow occurs. The result, in that case, places $0000 in the upper word of the quotient. The complete source listing for the divide with overflow program is shown in Program 4.8.

```
         CLR.L D1            * D1 = 0000 0000
         MOVE.L DIVDE,D0     * D0 = DU DL
         MOVE.W D0,D1        * D1 = 0000 DL
         CLR.W D0            * D0 = DU 0000
         SWAP D0             * D0 = 0000 DU
         DIVU DIVSE,D0       * D0 = RU QU
         MOVE.L D0,D3        * D3 = RU QU
         CLR.W D3            * D3 = RU 0000
         ADD.L D3,D1         * D1 = RU DL
         DIVU DIVSE,D1       * D1 = RL QL
         SWAP D0             * D0 = QU RU
         CLR.W D0            * D0 = QU 0000
         ADD.W D1,D0         * D0 = QU QL
         MOVE.L D0,QUOT      * Store quotient
         SWAP D1             * D1 = QL RL
         MOVE.W D1,REMAIN    * Store remainder
DONE     BRA.S DONE
```

Program 4.8

4.9 DECIMAL AVERAGING PROGRAM

For the next program problem, consider a list of student test scores. The problem is to produce the truncated (whole number only) average of the grades. The data for this program are BCD numbers (actual grades and the number of test scores). Before we can begin working on the program itself, we must examine some new instructions.

ABCD and SBCD Instructions

The 68000 adds and subtracts BCD numbers using the special instructions add BCD (ABCD) and subtract BCD (SBCD). These instructions perform the indicated operation interpreting data as BCD numbers. The inclusion of the X flag in the arithmetic operation of both instructions allows multiple-precision additions and subtractions.

If you look up these instructions in the instruction set (Appendix A), you will discover that they have some important restrictions and considerations. The first of these involves the addressing modes that retrieve the data used by the instructions. The source and destination effective addresses must both be in the data direct mode or both be in the predecrement addressing mode. As a result, the acceptable syntaxes for these instructions are

```
ABCD Dy,Dx     ABCD -(Ay),-(Ax)

SBCD Dy,Dx     SBCD -(Ay),-(Ax)
```

where subscripts x and y are any numbers from 0 to 7.

Data size for these two instructions is limited to byte size only. The X bit provides for larger size additions or subtractions, but also requires the reuse of the instruction for each byte portion of the full data number. The process is similar to the multiple-precision arithmetic examples of Chapter 2.

As if these restrictions were not sufficient to cause concern while using these instructions, there are additional considerations involving the condition code flags. The N and V flags are undefined since they have no meaning when using BCD data. They reflect the polarity of signed binary or hexadecimal data only. The X and C flags are affected as expected; they are set by a carry (for ABCD) or borrow (for SBCD) generated by the arithmetic operation. The difficulty appears when dealing with the Z flag. This flag is cleared (reset to zero) if the result of the operation is not zero. This is the normal response of the Z flag. However, if the results *are* zero, the Z flag for these instructions remains unchanged. That is, there is no true indication of the zero condition of the data at the end of either instruction's execution.

TST Instruction

How, then, are the zero and non-zero conditions of the results of the ABCD or SBCD instructions checked? The answer lies in the use of the test (TST) instruction. In essence, the TST instruction subtracts zero from the operand data. This action does not change the original data, but it does cause the Z and N flags to reflect the condition of that data. The Z flag, which was concern with the BCD arithmetic instructions, is now set or reset depending on whether the data are zero or not zero. The N flag reflects the polarity of the data, which, again, has no bearing when using BCD data. The X flag is not affected by the TST instruction, allowing the instruction's use in a multiple-precision program loop. The syntax for the TST instruction is

```
TST.⟨size⟩ ⟨ea⟩
```

Grade Averaging Program

The purpose of the grade averaging program is to produce the truncated average of a set of 50 or fewer student test scores. The number of test scores to be averaged is placed in a memory location labeled COUNT. The resulting average is stored, at the program's end, in location AVERAGE. Finally, the list of grades begins with the first grade at location GRADE.

In analyzing the problem, first establish the actions to be performed by the program. Initially, the grades must be fetched from memory and summed. The sum is divided by the number of test scores to produce the average. The average for this problem will be truncated to keep the program relatively simple. Truncating discards that portion of the data that is less than one.

Since the grades are BCD numbers, the addition is done using the ABCD instruction. The maximum number of tests is 50. If each student scored 100, then the maximum sum is 50 × 100, or 5000. This value can be held in a word (2-byte) size register. Also, because the sum is most likely to be greater than 99 and the ABCD instruction is restricted to byte size data, double precision arithmetic is required.

After the sum has been calculated, it must be divided by the number of test scores to produce the average. There are two divide instructions (DIVU and DIVS), but these operate on binary or hexadecimal forms of data and not on BCD data. The division process will be performed by successive subtractions of the divisor (the number of test scores) from the grade sum. Each time the subtraction is performed, a counter is incremented. This process is repeated until a subtraction produces a borrow. At this point, the count in the counter register is the divided (average) result of the division process.

Initializing the Program

The lower bytes of D0 and D1 hold the BCD sum of the test grades. D5 holds the constant, 1, which increments and decrements BCD counters using the data register direct mode for ABCD and SBCD. D3 and D6 are used for housekeeping chores throughout the program. D2 holds the number of test grades, and A0 functions as the data pointer when retrieving the grades from the grade list. When the program starts, the sum and housekeeping registers are cleared, while the constant, grade counter, and pointers are initialized with their respective data (Program 4.9).

```
CLR D0              * Sum low byte
CLR D1              * Sum high byte
CLR D6              * Constant 0
MOVE #1,D5          * Constant 1
MOVE.B COUNT,D2     * Grade counter
MOVE.L #GRADE,A0    * Initialize pointer
```

Program 4.9

The counter register (D2) is loaded from location COUNT using a byte size data move. This value is a BCD number equal to or less than 50. The data pointer (A0) is initialized using a long word data move in the immediate addressing mode.

Retrieving and Summing the Grades

The first grade is fetched from the list at location GRADE, and the grade counter (D2) is reduced by one. If this process results in the counter reaching zero, indicating only one test grade was in the list, then that test grade is stored as the resulting average and the program terminates.

```
MOVE.B (A0)+,D0        * Fetch first grade
MOVE.B D0,D3           * Copy grade
SBCD D5,D2             * Reduce counter by 1
TST.B D2               * Set/Reset Z flag
BEQ.S RESULT           * Only one grade?
```

Program 4.10

The grade is fetched into D0 and copied into D3. D3 is used later in the program to transfer the average into memory. Using the postincrement addressing mode causes the pointer (A0) to be incremented and point to the next grade data. The SBCD instruction is used to cause the counter register to be decremented by one, accounting for the first data fetch. The subtraction is BCD, forcing the use of the TST instruction to check the condition of the Z flag. If there is only one test score, the result will be zero and program execution moves to location RESULT to store that score as the average. If there is more than one test score, the program continues by fetching the remaining grades and summing them. This is done in a looping program segment.

```
AVE     MOVE.B (A0)+,D3      * Fetch grades
        ABCD D3,D0           * Sum grades
        ABCD D6,D1           * Sums carry (X flag)
        SBCD D5,D2           * Decrement counter
        TST.B D2             * Check counter
        BNE.S AVE            * More grades?
```

Program 4.11

The next grade is fetched from memory and moved to D3, which no longer needs to hold a copy of the first grade. D3 and D0 are added, with the result returned to D0. D6 and D1, which were initially cleared, are added using the next ABCD instruction. The state of the X flag is added to D1 at this time as well. If the first addition produces a carry, then the X flag is set. The second ABCD instruction adds zero to zero (the original contents of D6 and D1) and to one (the X flag). This sum is placed into D1 at the completion of the instruction. This operation is the same as using the add with carry (ADC) instruction of the 68000 followed by the decimal add adjust (DAA) instruction.

EXAMPLE 4.2

Grades 97 and 85 are summed using the preceding program segment. What are the contents of the lower bytes of D0 and D1 after the addition is done?

Solution:

The first ABCD instruction produces 82 in the lower byte of D0 and sets the X flag. The second ABCD instruction produces 01 in D1's lower byte. The result of the addition is 97 + 85 = 182.

Next, the counter register is decremented by one using the SBCD instruction again. The results are checked for zero by the TST instruction. As long as the grade list is not exhausted, the counter will not be zero and the looping instruction (BNE) will return program execution to AVE.

Finding the Average

Once the grades are summed (indicated by the counter reaching zero), the program falls out of the AVE loop and begins the division process (DIV loop). The counter is reinitialized with the number of scores from location COUNT, and the housekeeping register (D3) is cleared. The division is done by subtracting the counter value from the sum until the resulting difference generates a borrow. Each time the subtraction is done, a separate counter, the "sub counter" (D3), is incremented. When the subtraction is completed, D3 contains the dividend (and hence, the average of the grades).

```
        CLR D3
        MOVE.B COUNT,D2     * Retrieve divisor
DIV     SBCD D2,D0          * Successive subtractions
        BCS.S RESULT        * Subtraction done?
        ABCD D5,D3          * Increment sub counter
        BRA.S DIV           * Loop instruction
```

Program 4.12

The first subtraction performs the BCD process, causing the extend (X) flag to be set if a borrow occurs. The second SBCD instruction subtracts zero and the extend flag from the upper byte of the BCD sum results. If this subtraction generates a borrow as well, then the subtraction process is complete. Each time the carry/borrow flag is not set as a result of the second subtraction, the sub counter (D3) is incremented. When the loop is finally exited by the borrow condition of the second SBCD, D3 contains the result of the program.

EXAMPLE 4.3

What are the contents of D0, D1, and D3 if the original sum is 255 and the number of test scores is 3?

Solution:

During the first pass through the DIV loop, 3 is subtracted from 55 (the contents of D0), producing 52 and causing the X flag to be reset. The second SBCD instruction produces 2 when subtracting the contents of D6 (0) and the X flag (also 0) from D1 (2). The X flag, as a result of the second SBCD process, is reset. This allows D3 to increase by 1 (ABCD D5,D3), and the loop resumes by the BRA instruction.

After 18 passes through the loop, the total value has been reduced to 201, with D1 still containing 2. D0 holds 01 and D3 contains 18. The next run through the loop subtracts 3 from the contents of D0, producing 98 and causing the X flag to be set. The second SBCD instruction now subtracts D6 (0) and the X flag (1) from D1, reducing the contents of D1 to 1. The looping process continues until eventually both D1 and D0 are equal to 0. D3 holds 85 at this point, meaning that the loop has been executed 85 times. On the eighty-sixth run through the loop, the first SBCD instruction produces 97 in D0 and sets the X flag. The second subtraction causes D1 to become 99 ($0 - 0 - 1$) and sets the X and C flags. The branch if carry set (BCS) instruction senses the set condition of the carry flag and branches out of the loop. The next instruction is executed at location RESULT. D0 = 97, D1 = 99, and D3 = 85.

Saving the Average in Memory

Notice in the example solution discussion that D3 holds the average of the three grades ($255 \div 3$). Regardless of whether the program enters RESULT from the beginning because there is only one test score or from the BCS instruction in the DIV loop, the expected average will be contained in D3. The only action remaining is to store this result in memory and terminate the program.

```
RESULT    MOVE.B D3,AVERAGE
DONE      BRA.S DONE
```

The complete listing of the grade averaging program is shown in Program 4.13.

```
        CLR D0                 * Sum low byte
        CLR D1                 * Sum high byte
        CLR D6                 * 0 constant
        MOVE #1,D5             * 1 constant
        MOVE.B COUNT,D2        * Initialize grade counter
        MOVEA.L #GRADE,A0      * Initialize grade pointer
        MOVE.B (A0)+,D0        * First grade
        MOVE.B D0,D3           * Copy first grade
        SBCD D5,D2             * Decrement counter
        TST.B D2               * Check counter for 0
        BEQ.S RESULT           * Only one test score?
AVE     MOVE.B (A0)+,D3        * Get grades
        ABCD D3,D0             * Sum grades
        ABCD D6,D1
        SBCD D5,D2             * Decrement counter
        TST.B D2               * Check counter for 0
        BNE.S AVE
        CLR D3                 * Clear average counter
        MOVE.B COUNT,D2        * Restore score count
DIV     SBCD D2,D0             * Start subtraction
        SBCD D6,D1
        BCD.S RESULT           * Division done?
        ABCD D5,D3             * Increment average counter
        BRA.S DIV
RESULT  MOVE.B D3,AVERAGE      * Save program results
DONE    BRA.S DONE             * Terminates program execution
```

Program 4.13

PROBLEMS

4.1 What size data are fetched from memory during each program fetch?

4.2 What is the last control signal applied from the microprocessor to address a particular memory device during a memory read cycle?

4.3 What signal must be returned to the 68000 for a successful data transfer?

4.4 What is meant by prefetch? How does it reduce program execution time?

4.5 What signal is asserted when you use slow memories? What determines the length of the wait state?

4.6 What are the contents of D0 after these instructions? Initially, D0 = \$1234 5678.
(a) SWAP D0 (b) CLR D0 (c) TST D0

4.7 Initially, D1 = \$3CB3 5A27 and D0 = \$1234 0002. What are the contents of D0 and D1 after the instruction DIVU D0,D1?

4.8 Why is it unnecessary to write an overflow multiply program if the 2 numbers to be multiplied are word size?

4.9 What is wrong with this program segment? How can you correct it?

```
       MOVE #5,D0
       MOVE #1,D1
       MOVE #20,D2
MULT   ABCD D2,D2
       SBCD D1,D0
       BNE.S MULT
DONE   BRA.S DONE
```

4.10 What are the contents of D0, D1, and D2 at the end of the program in Problem 4.9 after is has been corrected?

5

PARALLEL INTERFACING THE 68000

CHAPTER OBJECTIVES

The objective of this chapter is to acquaint you with the requirements for using Motorola's 8-bit parallel interface chip, the peripheral interface adapter (PIA), in a 68000-based system. A keyboard (an input device) and an LED display (an output circuit) illustrate the theory behind input/output interfacing.

5.1 GLOSSARY

acknowledge Verification of a signal or a process.
handshake Method of exchanging control information during data transfer between devices or stations.
interrupt program Program run in place of the current program in response to an interrupt request. The current program is resumed once the interrupt program is finished.
interrupt request Signal that causes the current program to be deferred so that a different program can be run.
mask In the context of interrupt handling, prevents the microprocessor from detecting an interrupt request.
port 8-bit access buffer that handles data to and from an I/O device.

5.2 68000 PERIPHERAL PHILOSOPHY

The 68000, like its earlier cousin, the 6800, uses memory-mapped I/O interfacing. There are no special **port** addressing or interface instructions—all transfers are effected by MOVE-type instructions. The biggest advantage of memory mapping an interface is that it is treated as any other memory location and is affected by the full instruction set. The disadvantage is the loss of memory space designated for I/O use. To use a 68000-based system, keyboard and display peripheral devices must be interfaced to the system.

5.3 A 68000 SEVEN-SEGMENT LED DISPLAY

The LED display circuit used in Chapter 1 is limited to six units for simplicity. The display circuitry in this chapter is expanded to ten units to accommodate the size of information to be displayed, which is six hexadecimal characters for addresses and four characters for data. This translates into six units for addresses and four for data. Figure 5.1 is a schematic diagram of the display circuit for this system. Segment location $06 0000 is configured similarly to the 16-bit model in Chapter 1, with each segment of the seven-segment display driven independently by one of the eight bits from that memory location.

In the case of the unit drive, b_0–b_3 drive a BCD decoder that in turn drives the common cathodes of the displays through a current driver. This scheme allows the display units to be selected within the bounds of an 8-bit data size. Location $06 0001 is used as the unit storage location and is labeled UNIT. Location $06 0000, holding the segment code, is labeled SEGMENTS. The word size combination at location $06 0000 holds the data for the segment and unit code so that a unit can be selected and lighted with a character by a single word size move. A segment is selected by placing a high in the corresponding bit position of location SEGMENTS, and a unit is selected with the correct binary information supplied to the decoder from location UNIT.

5.4 THE KEYBOARD INTERFACE

A hexadecimal keyboard, summarized in Figure 5.2 on page 150, is used to input characters 0–F and a word size data location. Key input information is gathered by a single move operation from that location. Location $06 0004, designated KEYHI, inputs data from keys 8–F, while location $06 0005, labeled KEYLO, receives information from keys 0–7. Key input information is high when no keys are pressed and low when a key is actuated. Key data are latched into the debounce buffers placed between KEYHI, KEYLO, and the keyboard.

5.5 USING THE INTERFACES

Driving the display is a matter of selecting the correct segment and unit code and moving these data to $06 0000 and $06 0001. For example, the character C is to be lighted on unit 3. The C requires segments a, d, e, and f to light while the rest remain off. Data into the SEGMENTS location are 0011 1001 ($39) to make the corresponding bits high, which drives the appropriate segment LEDs. Unit 3 is driven on when output $\overline{2}$ of the BCD decoder is asserted low. This is produced by sending the binary 0010 to A_3, A_2, A_1, and A_0 of the decoder from location UNIT. The data in UNIT are therefore $02. Putting the two pieces of data together produces

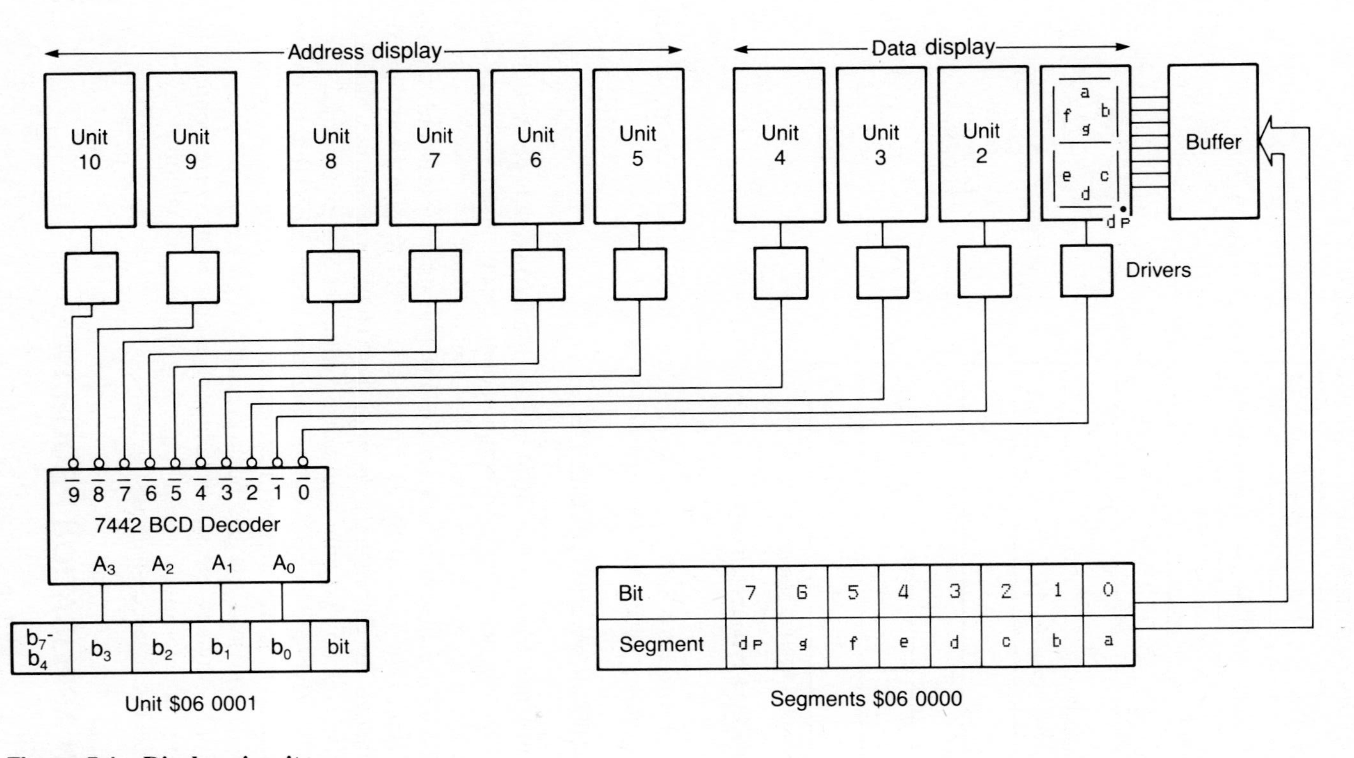

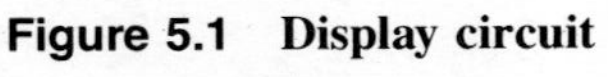

Figure 5.1 Display circuit

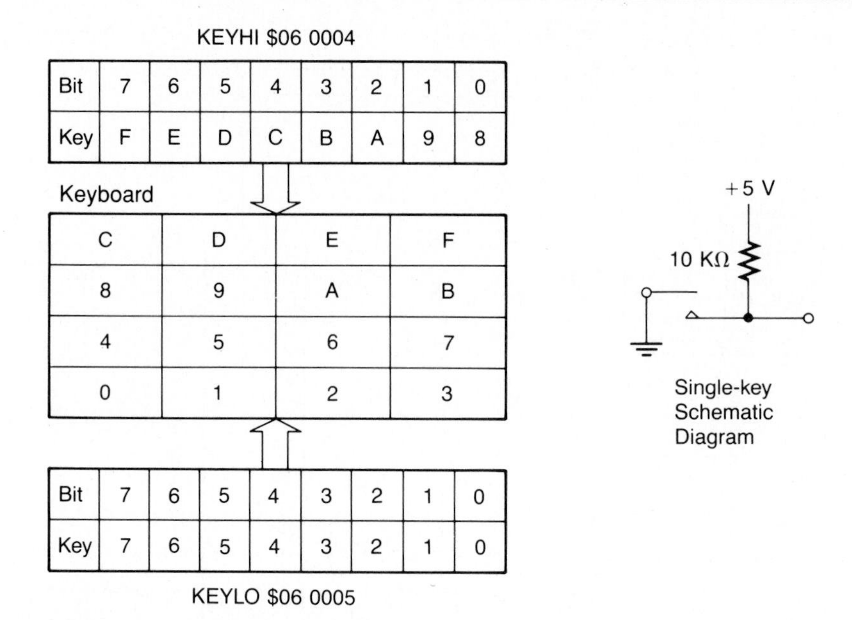

Figure 5.2 Keyboard circuit

the word size data \$3902, which is moved to SEGMENTS to perform the task of lighting unit 3 with a C. The instruction is

```
MOVE.W #$3902,SEGMENTS
```

Since the unit decoder is a BCD circuit, binary inputs beyond 9 (1001) produce no active low output. These bit combinations can be used in UNIT to deselect (not select) any unit and thereby make the display blank.

To light different characters on different units necessitates a looping program that changes the data at SEGMENTS and UNIT with a short time delay between each change. In this manner, as with the SHOW program in Chapter 1, the display is multiplexed so that only one unit is lighted at any given time.

The keyboard is read by moving data from KEYHI into the microprocessor with a MOVE.W KEYHI,D0 instruction. The data in D0 can then be examined to determine which key(s) has been pressed. If no key is pressed, the data received are \$FFFF. Any key pressed is indicated as low in its corresponding bit position. A rotate **mask** determines which single key is pressed. Multiple-key presses are tested with specific data compare instructions.

5.6 KEYSCAN APPLICATION PROGRAM

When you develop a general keyscan program, it is necessary to read in the data from the keyboard and to be concerned with possible key bounce. Using hardware to reduce the effects of the latter condition requires the use of debounce latches. Using a program that responds to key presses following a short delay after the key has been released will further reduce these effects.

```
KEYSCAN   MOVE KEYHI,D0        * Get key data
          NOT D0               * Complement D0
          BEQ KEYSCAN          * No key pressed?
BOUNCE    MOVE KEYHI,D1
          NOT D1
          BNE BOUNCE          * Key released?
```

Program 5.1

The first loop reads the keyboard until a key is pressed. When this occurs, the complemented key data are held in D0. The second loop checks the keyboard and waits until the key has been released. D1 contains $0000 0000 from complementing the data after the key is released. D1 is used as a counter to develop the key value as D0 is rotated and tested for the actual key input data.

```
FIND      ROR #$1,D0        * Move LSB to carry flag
          BCS RETURN        * Carry set?
          ADDQ #$1,D1       * Increment key count
          BRA FIND
RETURN    RTS
```

Program 5.2

Upon the return from this subroutine, D1 contains the actual value of the key pressed. All instructions requiring data sizes in the program default to word size. D1's word is composed of a zero in the upper byte and the key value in the lower byte. Keeping this in mind, you can expand the subroutine into a dual-key press routine. These first two sections form a subroutine labeled KEYSCAN. To get the first key input, this subroutine first is called, returning the value of the first key pressed in D1. Next, these data are transferred to another data register (D2) to allow the KEYSCAN routine to be used a second time. The key data in D2 are shifted into the upper byte of the lower word. KEYSCAN is called again, returning the second key value in D1's lower byte. Adding the two bytes of information results in a full word of data holding the two key entries.

```
KEYWORD   JSR KEYSCAN      * Call keyscan
          MOVE D1,D2       * Copy key value
          ASL #$4,D2       * Shift key value
          JSR KEYSCAN      * Get second key value
          ADD D2,D1        * Combine key values
          RTS
```

Program 5.3

What about expanding to four key entries? This is accomplished in a few instructions. The object is to move the data in the lower word of D1 into the upper word of D1 and then fetch the second pair of key presses.

```
KEYLONG   JSR KEYWORD       * Get two key values
          SWAP D1           * Put values in upper word
          JSR KEYWORD       * Get two more values
          RTS
```

Program 5.4

There are three keyboard interface routines to collect key information: KEYSCAN, for a single key press; KEYWORD, for two consecutive key presses; and KEYLONG, for four consecutive key entries. After you have developed a display subroutine, it can be inserted in the KEYSCAN subroutine to allow each key press to be displayed as the keys are actuated.

5.7 THE DISPLAY APPLICATION PROGRAM

To operate the LED display, we must expand the ideas from the SHOW subroutine developed for the microprocessor in Chapter 1. For this application, 10 segment data locations that hold segment codes for the display character of each unit are reserved in a memory bank headed by location SEGCODE. The subroutine takes the data from the segment code bank and transfers them to SEGMENTS while continually updating the data in location UNIT. A short delay is incorporated between each data change to allow the current character to be displayed on the designated unit.

```
SHOW    MOVEA.L #$SEGCODE.A0    * Initialize pointer
        CLR SEGMENTS            * Blank display
        MOVEQ #$9,D1            * Unit counter
```

Program 5.5

These first three instructions initialize the system to handle display data. By making the lower byte of the unit register all low with the CLR SEGMENT instruction, Unit 1 is selected (see Figure 5.1). The segments themselves are turned off because SEGMENTS has also been cleared. A0 has been selected to act as a pointer to the segment code data bank. Finally, a counter is set to allow use of the DBcc instruction in keeping track of the 10 units. The next section of the program retrieves the segment data and changes the unit pointer at the correct time.

```
NEXT    MOVE.B (A0)+,SEGMENTS    * Get segment data
        MOVE.B D1,UNIT           * Select unit
        MOVE #$30,D0             * Delay counter
DELAY   DBF D0,DELAY
        DBF D1,NEXT              * Unit counter loop
        RTS
```

Program 5.6

The NEXT loop begins by getting the segment code data from the data bank and incrementing that pointer (A0). This is a byte move from one memory location, pointed to by A0, to another, labeled SEGMENTS. No register or other byte of memory is affected by this instruction. The contents of D1 are moved to UNIT to select the unit that will display the character from SEGMENTS. A short delay follows, and the unit data are altered in preparation for the next character to be displayed. After the delay, D1 is decremented. If D1 has not yet reached -1, the DBF instruction loops back to NEXT. Once D1 has reached -1, all units have been illuminated once (for about 100 μs) and the program resumes its normal execution by the RTS instruction. One sweep of the display is not sufficient for the message to be seen by the user. The number of times the SHOW subroutine is to be run is decided by the programmer and is a function of the instructions which follow the jump to the SHOW subroutine.

The interfaces illustrated here are used for direct machine code programming of the 68000. Programming this microprocessor directly is a cumbersome task. Assemblers that make the task a lot easier require a more extensive keyboard that includes letters, punctuation, numbers, and a cathode ray tube (CRT) display. However, the interfaces illustrated in this chapter expand upon the 68000 instruction set and programming concepts.

5.8 THE PERIPHERAL INTERFACE ADAPTOR (PIA)

Motorola's primary 8-bit parallel interface chip is the 6821 PIA. This PIA moves 8 bits of data between the microprocessor via the data bus and an I/O device connected to one of the two 8-bit PIA interface or peripheral ports. Hardware handshaking and interrupts can also be handled through the PIA.

The first aspect of the PIA to consider is the movement of data between I/O devices and the 68000. Pairs of PIAs are used like 8-bit memories in that one handles data transfers on the lower byte of the data bus while the second deals with the upper byte. Motorola, in designing its peripheral interface, made the PIA appear to the microprocessor as another set of address locations. To fully understand how this is accomplished, first you need to examine the architecture of the PIA. Figure 5.3 is a programming and signal model of the 6821 PIA.

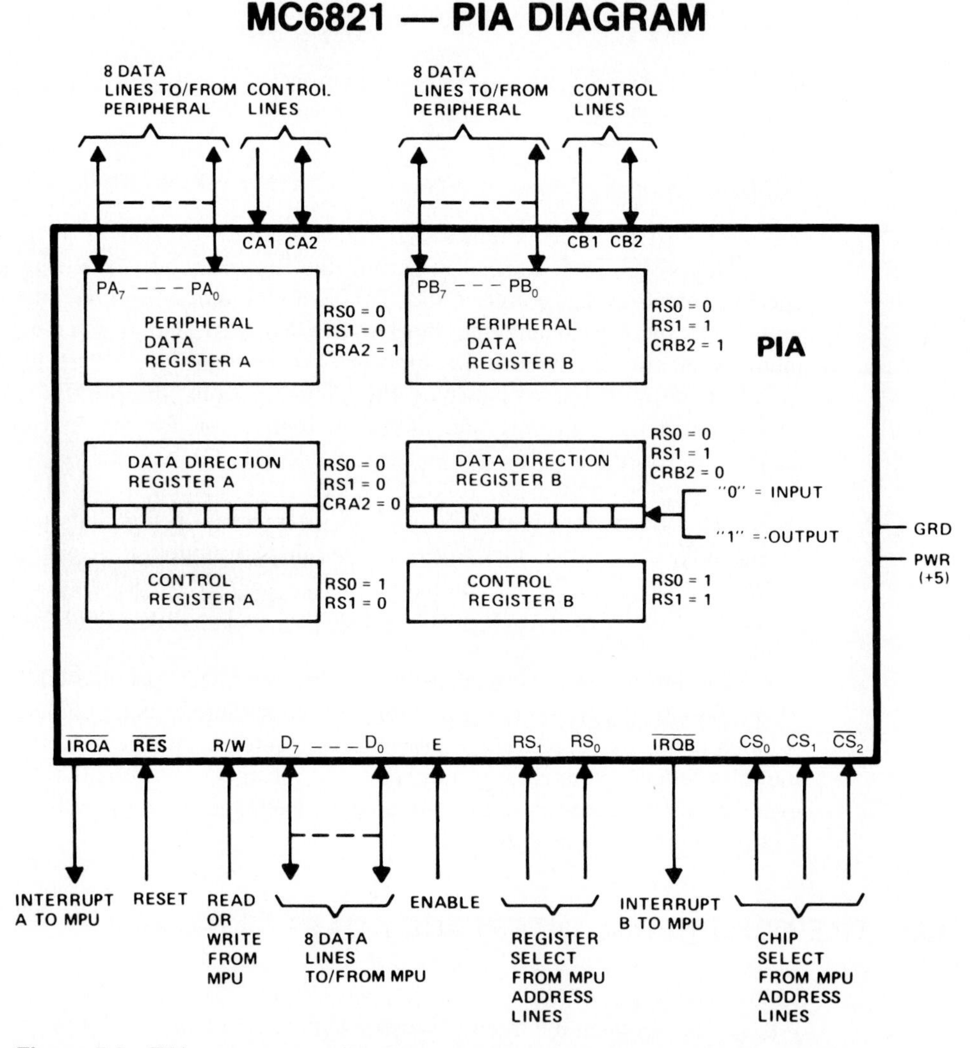

Figure 5.3 **PIA program model (Reproduced by permission of Motorola, Inc.)**

5.9 ARCHITECTURE OF THE PIA

Everything at the top of the PIA in Figure 5.3 is connected to the I/O devices. Similarly, the pins on the lower part of the diagram are connected to the 68000 microprocessor. There are two sets of eight peripheral data leads: one from the A side of the PIA, labeled port A, and one set from port B. These leads carry the actual data information from the PIA to the I/O devices. There are four additional leads for receiving interrupt requests and for data-flow control (handshaking).

On the microprocessor interconnections' side of the PIA, there are eight data lines that pass data back and forth between the PIA's various registers and the microprocessor's data bus. $\overline{IRQA}$ and $\overline{IRQB}$ pass interrupt requests from ports A and B to the microprocessor. $\overline{RESET}$ resets the PIA by setting all PIA registers to zero. The E line synchronizes the internal PIA operations and also connects to the E output line of the 68000.

The three chip selects allow the device to be selected by decoding various address lines in the same manner as the chip selects for the system's memory. Register select inputs RS_0 and RS_1 form the remaining addressing inputs to the PIA and usually are connected to address lines A_1 and A_2, respectively. They select which register is accessed by the corresponding addresses. A truth table and logic diagram for these leads are shown in Table 5.1 and Figure 5.4 (page 156).

Notice in Table 5.1 that there are six PIA registers (a control register, a data register, and a direction register) for each port. Table 5.1 also shows some ambiguity in that the data and direction registers for each side share the same address. A bit in the control register determines which of the two is being accessed at any given time.

The data registers act as buffer registers between the microprocessor buses and the peripheral data leads. Data on the peripheral leads can propagate either from the I/O device to the PIA port or, in the reverse order, from the port to the I/O device. The direction registers determine which way the data propagates. The two control registers direct the information flow within the PIA and set the conditions for passing

Table 5.1 Register Select Truth Table

RS_1	RS_0	Register	Port
0	0	DDR/PDR	A
0	1	Control	A
1	0	DDR/PDR	B
1	1	Control	B

DDR = Data direction register
PDR = Peripheral data register

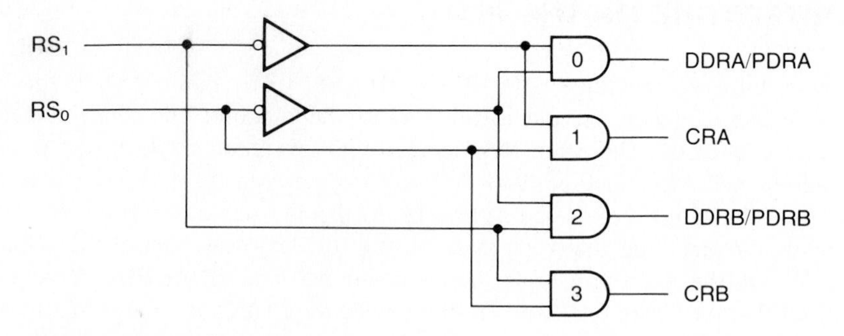

Figure 5.4 Register select logic

interrupt and handshake information between the I/O device, the PIA, and the CPU. Table 5.2 is a diagram of the control register and its functions. As you need to use the various bits in this register, refer to this diagram.

To begin a detailed discussion of the PIA, consider the movement of data from the microprocessor to the I/O device via the PIA. First, the propagation direction is selected by entering data into the direction register. Each of the eight peripheral lines of each port can be selected individually either as an input data lead, bringing data in from the I/O device, or as an output data line, sending data to the I/O device. The selection is done by setting the corresponding bits of the direction register to 1's for output selection or to 0's for an input choice.

5.10 SELECTING DATA FLOW DIRECTION

Refer to Table 5.1 and note that the direction register is accessed when the register select lines (RS_1 and RS_0) are both low for port A and when RS_1 is high and RS_0 is low for port B. The problem is to resolve the apparent conflict with the data registers accessed by the same register select combinations. Within the control register is a

Table 5.2 PIA Control Register

		CA2 Control				CA1 Control	
Bit 7	Bit 6	Bit 5	Bit 4	Bit 3	Bit 2	Bit 1	Bit 0
IRQA1	IRQA2	0 = CA2 input	Transit. 0 = neg. 1 = pos.	CA2 MASK	Register Select	Transition	CA1 MASK
1 = CA1 interrupt	1 = CA2 interrupt	1 = CA2 output	0 = mode	0 = hand-shake 1 = pulse	0 = DDR 1 = PDR	0 = neg. 1 = pos.	0 = mask 1 = not masked
			1 = steady state	0 = low 1 = high			

flag (b_2) that is set aside for the job of selecting which of the two registers is accessed when the register selects are of the combination described at the beginning of this paragraph. If b_2 of the control register is low, the direction register is accessed. A high in b_2 of the control register directs access to the data register.

Before we use an example to illustrate the usage and results of initializing the direction register, relate the SEGMENTS and UNITS addresses in this chapter to the PIA register select lines. Location \$06 0000 is designated for SEGMENTS and \$06 0001 for UNITS. Expanding the lower nibble (last four bits) to their binary equivalents yields 0000 for SEGMENTS and 0001 for UNITS. Pins RS_1 and RS_0 of a PIA are connected to A_2 and A_1, respectively. That makes RS_1 and RS_0 both low for addressing the display. If you consult Table 5.1, you can see that this combination accesses either the data or direction register of port A of each PIA. $\overline{UDS}$ directs the data on the upper half of the data bus to the segment PIA, and $\overline{LDS}$ directs data to the lower byte to the unit PIA. The control register addresses for these PIAs are determined from Table 5.1 and are formed when RS_1 is low and RS_0 is high. Thus, control register A's address for the segment PIA is \$06 0002 and the units control register's address is \$06 0003. Table 5.3 lists the address schemes developed so far, along with appropriate identifying labels.

Each register can be accessed using byte size data. Additionally, both the SEGMENTS and UNIT data registers can be accessed using word size data. The control registers (CONS and CONU) are accessed simultaneously using word size data as well. Be careful when you use long word data with PIAs. Control and data registers, as well as direction registers for any PIA, can only be accessed individually, since the data to or from them must travel along the same microprocessor data lines. Word moves involve two different PIAs to access both SEGMENTS and UNIT ports simultaneously. A long word move, since it moves one word at a time, can be used to transfer data to the PIAs. The upper word accessing the lower addresses is transferred first, then the lower word follows. The result of this type of move is examined in a programming example later in the next section of this chapter.

When the $\overline{RESET}$ is asserted, the PIA clears the six registers. This is one way to make b_2 of the control register low, allowing access to the direction register. Unfortunately, often it is necessary to initialize the PIA without the $\overline{RESET}$ signal. To ensure that b_2 of the control registers is set low, use a software instruction such

Table 5.3 PIA 1 and PIA 2 Port A Display Addresses

Address	$\overline{UDS}$	$\overline{LDS}$	Label
06 0000	0	1	SEGMENTS
06 0001	1	0	UNIT
06 0002	0	1	CONS
06 0003	1	0	CONU

as CLR CONS. This clear instruction, which is the first instruction in a program to configure or set up the PIA as an interface to an I/O device, sets most of the control register bits of both PIAs to zero or low. As interrupt request flags, b_6 and b_7 are unaffected by instructions involving the control register.

Since $06 0000 and $06 0001 drive the seven-segment display, it is necessary to direct the peripheral leads to propagate data from the PIA to the LED display; port A of both PIAs becomes an output port. Because each port lead is an output lead, all bits of the A side direction register must be set high. The following instructions perform that task directly:

```
CLR CONS                 * Sets b2 of the control register low
MOVE #$FFFF,SEGMENTS     * Makes SEGMENTS and UNIT output ports
```

The data and direction registers share the same address, which makes the labels SEGMENTS and UNIT appropriate to access either type of register. A label is related to an address, and its name aids in the description of a location's use. In this case, the labels serve dual roles in that the data are sent to the correct place within the PIA as directed by b_2 in the control register.

5.11 INITIAL DATA OUTPUTTED TO THE DISPLAYS

After the program segment in the preceding section has been executed, port A of both PIAs is set as an 8-bit output port. Data stored into the data registers from the microprocessor are sent immediately to the display. Another effect of clearing the six registers by a $\overline{\text{RESET}}$ signal is to set a port's output registers to $00; thus when the port is selected initially as an output port, the data sent out first are all low. Note that some applications do not use a $\overline{\text{RESET}}$ signal; one should set the data registers low with software. Setting the data registers low will ensure that the display, in this case, will be blank initially (both segments and unit codes are $00). To do this, access to the data register must be made available by making b_2 of the control register high. The required control register data are 0000 0100 or $04. The instructions that set the output lines low are

```
MOVE #$0404,CONS     * Makes b2 high
CLR SEGMENTS         * Sets output lines low
```

Keeping the note about the long word move in mind, you can rewrite the PIA display initialization program this way:

```
CLR CONS                      * Sets b2 low
MOVE.L #$FFFF0404,SEGMENTS    * Selects output ports/sets b2 high
CLR SEGMENTS                  * Sets output data low
```

The long word instruction first transfers the upper word to the direction registers (SEGMENTS location), then the lower word to the CONS location (control registers), so that the registers within the PIAs are accessed in the correct order.

Once the PIAs are configured to handle data, information passed from the microprocessor to the output ports drives the display's numerous LED drives. This is effected by treating SEGMENTS and UNITS as memory locations and storing into them the codes that cause the selected segments and units to light. Once the data are stored, they remain there until new data are stored or until the unit is RESET or power is turned off.

5.12 KEYBOARD INTERFACE

Addresses \$06 0004 and \$06 0005 are connected to the keyboard as described earlier. These addresses access port B of PIAs 1 and 2. The data locations are labeled KEYHI for \$06 0004 and KEYLO for \$06 0005. Port B side control registers are labeled CTLHI and CTLLO. The direction registers share the same address as the data registers and therefore have the same labels. The initialization of the port B registers is similar to that for port A, except that the direction of the data flow is into the PIA from the keyboard instead of out to the display. Direction registers for the B side are set low to make the peripheral data leads act as input leads. Relying on the $\overline{\text{RESET}}$ signal would make the process of initializing these ports easy, since the direction register is set to zero by that signal. The full initialization is accomplished by Program 5.7.

```
CLR CONHI              * Set b2 low
CLR KEYHI              * Select input direction
MOVE #$0404,CONHI      * Set b2 high
```

Program 5.7

Appending this program segment to the one for the display completes that portion of the initialization program that handles the display and keyboard for our basic system. Since b_2 of all control registers is set high, reading or writing to the PIA deals with the actual data to or from the display and keyboard.

Two additional PIA's, 3 and 4, are wired to a terminal strip to allow user application connections, completing our basic system. These PIAs are used as user interface ports, allowing the programmer to connect the computer to any I/O device. The complete set of port addresses and labels is listed in Table 5.4 on page 160.

5.13 ADDING THE PIAs TO THE BASIC SYSTEM

So far, the effects of A_1 and A_2 on selecting the registers within the PIA have been the only addressing considerations. The PIA is initially selected by chip selects enabled by decoding the rest of the address lines. Connecting the PIAs to the basic system is done in the same manner as connecting the RAM and ROM to the microprocessor address bus. In Figure 5.3 notice that there are three chip selects, two active high

Table 5.4 Port Address Summary

Port	Register	Address	Label
PIA1-A	Data/Direction	06 0000	SEGMENTS
PIA2-A	Data/Direction	06 0001	UNIT
PIA1-A	Control	06 0002	CONS
PIA2-A	Control	06 0003	CONU
PIA1-B	Data/Direction	06 0004	KEYHI
PIA2-B	Data/Direction	06 0005	KEYLO
PIA1-B	Control	06 0006	CONHI
PIA2-B	Control	06 0007	CONLO
PIA3-A	Data/Direction	06 0008	PORTAHI
PIA4-A	Data/Direction	06 0009	PORTALO
PIA3-A	Control	06 000A	ACONHI
PIA4-A	Control	06 000B	ACONLO
PIA3-B	Data/Direction	06 000C	PORTBHI
PIA4-B	Data/Direction	06 000D	PORTBLO
PIA3-B	Control	06 000E	BCONHI
PIA4-B	Control	06 000F	BCONLO

and one active low, provided with the chip. It is necessary to use these chip selects to isolate the PIAs from each other as well as from the RAM and ROM that make up the remaining memory map of the system. Table 2.10 is reproduced here as Table 5.5, but is extended to include the PIAs.

The first task is to isolate the selection of the PIAs from the selection of the RAM and ROM. Examine the PIA addresses and notice that all have b_{17} and b_{18} high. These same bits are low for all memory addresses. The solution to this problem is to connect an active high chip select of each PIA to A_{17} and A_{18} and a low chip select to A_{16}.

These address lines are decoded through an address decoder similar to the one that selects the memory chips. The $\overline{\text{VMA}}$ signal, which allows timing of the 6800 devices to be synchronized with the E signal from the 68000 rather than with the system clock, is connected to one of the PIA's chip selects. Output $\overline{6}$ of the address decoder is asserted when A_{18} and A_{17} are high and A_{16} is low. This output is directed to an active low chip select of the PIAs.

The remaining task is to create separate chip selections for each PIA. The differences in the PIA addresses reside in the lower nibble. A_1 and A_2 are reserved for RS_0 and RS_1, while the data strobes $\overline{\text{UDS}}$ and $\overline{\text{LDS}}$, representing A_0, are connected

Table 5.5 System Memory Map

Address Lines																								
23	22	21	20	19	18	17	16	15	14	13	12	11	10	9	8	7	6	5	4	3	2	1	0	Chip
					0	0	0		X	X	X	X	X	X	X	X	X	X	X	X	X	X	U	ROM 1
					0	0	0		X	X	X	X	X	X	X	X	X	X	X	X	X	X	L	ROM 2
					0	0	1	X	X	X	X	X	X	X	X	X	X	X	X	X	X	X	U	RAM 1
					0	0	1	X	X	X	X	X	X	X	X	X	X	X	X	X	X	X	L	RAM 2
					1	1	0													0	R	S	U	PIA 1
					1	1	0													0	R	S	L	PIA 2
					1	1	0													1	R	S	U	PIA 3
					1	1	0													1	R	S	L	PIA 4

U = $\overline{UDS}$ L = $\overline{LDS}$ R = RS_1 S = RS_0

to an active low chip select on each appropriate PIA in the same manner as for the memory chips. Active low chip selects of PIAs 1 and 2 and active high chip selects of PIAs 3 and 4 are connected to A_3 to complete the address selection requirements of isolating each individual PIA according to its addresses.

5.14 CONNECTING THE PIA TO THE 68000

Figure 5.5 on page 162 is an interconnection scheme for PIAs 1 and 2 to the basic system developed in Chapter 4. A_{18}, A_{17}, and A_{16} are connected to the 74138 decoder inputs, allowing output $\overline{6}$ from the decoder to represent A_{18} and A_{17} high and A_{16} low. A_3 is ANDed with $\overline{VMA}$ and passed to a chip select on PIAs 3 and 4. The same address line is inverted, ANDed with $\overline{VMA}$, and directed toward a chip select on PIAs 1 and 2. This selects PIAs 1 and 2 when A_3 is low and PIAs 3 and 4 when it is high. $\overline{VMA}$ is made active once the 68000 receives an active $\overline{VPA}$ during a data transfer cycle. $\overline{VPA}$, in turn, is made active whenever one of the PIAs is selected. The actual timing and sequence is examined later in the chapter. The E line from the 68000 contains the clock signal (described in Chapter 2) and is used as the enabling clock input to the PIAs. RS_0 and RS_1 are connected to A_1 and A_2 to select the internal PIA registers. $\overline{RESET}$ and R/$\overline{W}$ are direct connections between the two devices. The upper data bus and data strobe ($\overline{UDS}$) are brought to PIAs 1 and 3, while the lower data bus and strobe are directed to PIAs 2 and 4. The interrupt request lines ($\overline{IRQA}$ and $\overline{IRQB}$) supply inputs to an encoder that converts each to an interrupt level input to the 68000 $\overline{IPL}$ lines. A line from the address decoder is ORed with a composite signal from the function control lines to supply the $\overline{VPA}$ signal required for successful data transfers when you use 6800 devices with the 68000. Interrupt use and application, as well as the $\overline{VPA}$ response to an interrupt acknowledge, are discussed in Section 8.8.

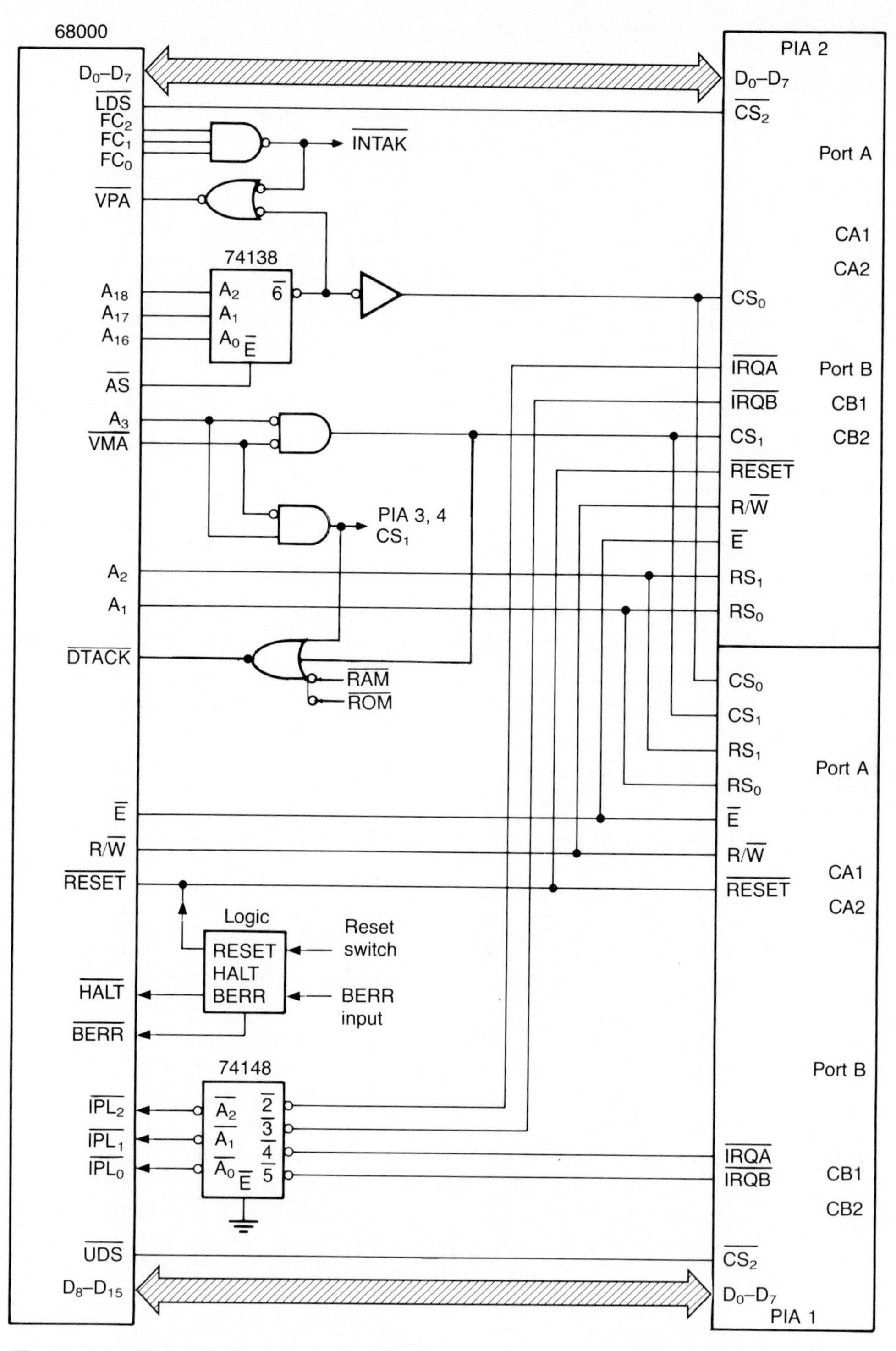

Figure 5.5 **PIA connections to the 68000**

The $\overline{\text{VMA}}$ signal is sent to the PIA to indicate that a valid address for a 6800 interface device is on the address bus, while $\overline{\text{VPA}}$ informs the 68000 that the address of the data transfer between it and the device is a valid peripheral address to a 6800-type interface chip. Besides performing address translation, these signals also handshake during the transfer of information between these chips. They are synchronized with the E clock, which is 1/10 the frequency of the system clock. These differences allow the slower 8-bit devices to be used with the 68000. Figure 5.6 on page 164 is a timing diagram for the control signals used with the 6800 8-bit devices.

$\overline{\text{AS}}$ and the appropriate data strobes are made active along with the function codes at the beginning of the transfer cycle. The address decoder returns an active $\overline{\text{VPA}}$ to the 68000. In response, the microprocessor makes $\overline{\text{VMA}}$ active. These signals are supplied to the PIAs in the form of chip selects; that is, the PIA selected is activated by having its address correctly decoded, the designated data strobe being applied, and $\overline{\text{VMA}}$ being asserted. Returning a $\overline{\text{DTACK}}$ to the 68000 informs it that the address to the PIA is valid and a successful data transfer is underway. The microprocessor inserts wait states, internally, as needed to allow time for the transfer to complete.

5.15 HANDSHAKING AND INTERRUPT HANDLING USING THE PIA

So far, we have developed an address scheme for four PIAs and a method by which they are configured as data carriers. But the PIA is a complicated device. Besides providing a method to transfer data between the computer and I/O devices, the PIA also can direct or control this transfer. The term that covers this idea, in a broad sense, is *handshaking*.

The concept behind any handshake can be reduced to a common everyday example involving a telephone. A telephone ring blares its way into the middle of a conversation, effectively interrupting the current comment. This interrupt can be ignored (**masked**) or attended to (serviced). Have you ever tried to ignore a telephone ring? Some people can do just that, and some cannot. Those who cannot ignore that shrill tone are unable to mask the interrupt, making the ring a nonmaskable interrupt (NMI). But some can treat the ring as a request to interrupt whatever we are doing. This **interrupt request** (IRQ) can be masked (ignored) or serviced (answered). Assume the ring is answered. You might want to save the current conversation in your memory, so that when the phone call is completed, you can resume it. Try it sometime—it's not so easy as you think.

The first word spoken on the phone after answering is usually "hello." This word is not really information because it doesn't inform the caller of anything except that somebody is on the line. In effect, "hello" acts as an **acknowledge** signal telling the caller that you have answered the phone. At that point, the caller probably will ask to speak to someone. Again, these words are not data of any sort because they don't tell us what the caller wants. They are a reply to the "hello" and a request to be connected to someone. Once the conversation is underway, pauses at the end of

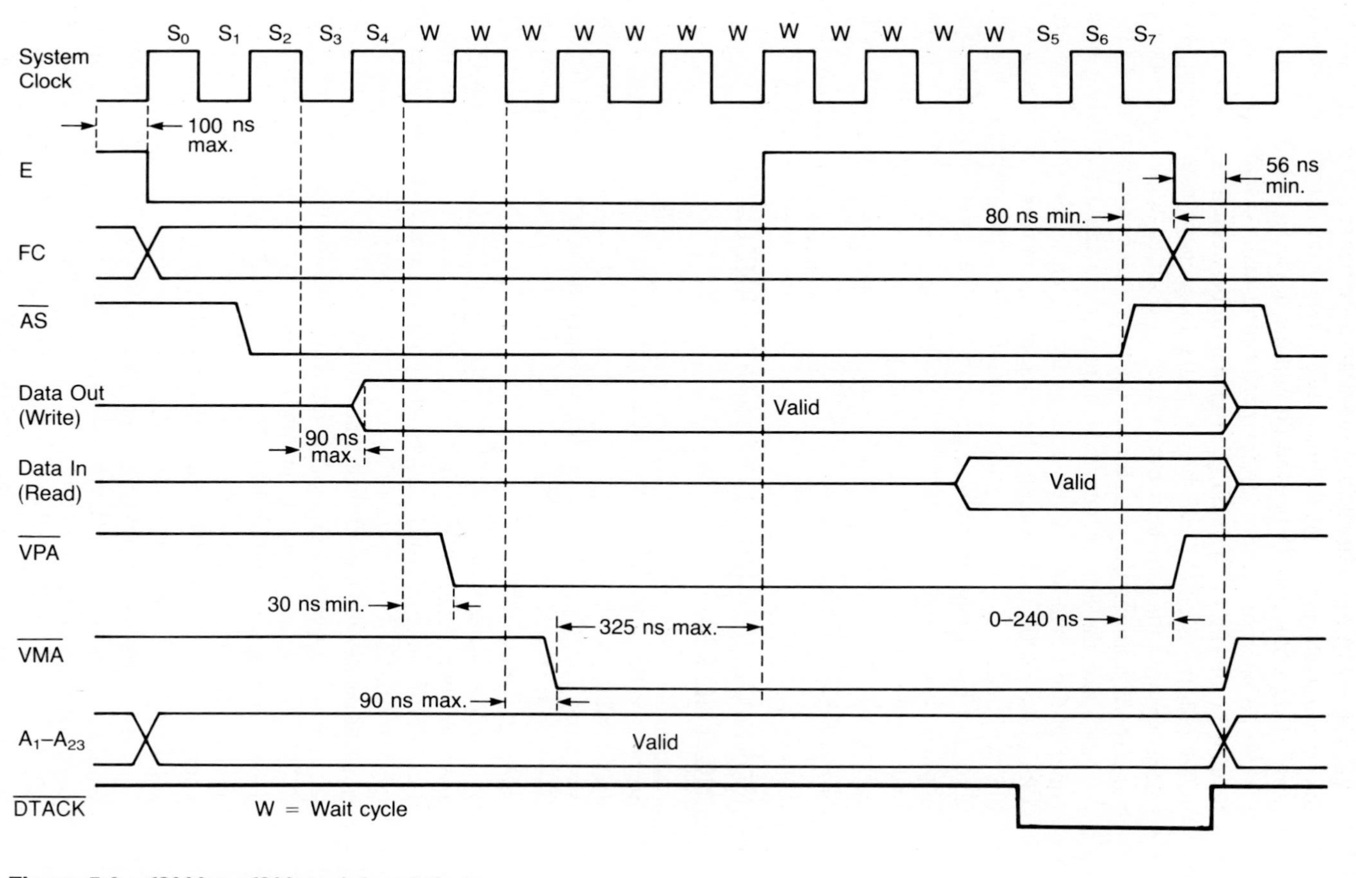

Figure 5.6 **68000 to 6800 peripheral timing**

sentences or thoughts signal the listener that the talker is finished for the moment and that a reply is expected. At the end of the conversation, good-byes are exchanged as a type of sign-off handshake and the phones are hung up, breaking the connection.

The PIA is capable of a similar type of handshake control by using the control leads CA1, CA2, CB1, and CB2. How these handshake signals are configured to perform their tasks is a job for the remaining bits of the control register. Refer to Table 5.2, the diagram of the control register. Much of our discussion involving the control leads will be the same for port A and port B. Differences occur in the pulse and handshake modes, but all other conditions are identical.

CA1 and CB1 Control Lines

CA1 and CB1 are always input control leads; that is, control information is passed from the I/O device to the PIA using these lines. To be specific, these lines are the interrupt request lines to the PIA. Within the control register, there is the option of passing on interrupt requests through the PIA's $\overline{IRQ}$ lines to the $\overline{IPL}$ encoder for the 68000 MPU or preventing (masking) them. b_0 and b_1 of the control register determine the effect of CA1 or CB1, which in turn depends on what control register is used. b_1 selects whether the valid interrupt request signal from the I/O device should be a change (transition) from a low level to a high level or vice versa. If the correct transition does occur, b_7 of the control register is set. This bit is the interrupt request flag for CA1. Many people call this flag a ready flag, indicating that a peripheral device is ready to handle data transfers.

b_0 of the control register decides if the $\overline{IRQA}$ will go low when b_7 is set. The $\overline{IRQA}$ line is an open-collector type of line and can be wired directly with the $\overline{IRQB}$. In turn, it can be connected to a microprocessor's interrupt request line or wired directly to an interrupt level encoder to be used with the 68000. With these connections, an interrupt request can be passed from the PIA to the microprocessor. Figure 5.7 shows a logic representation of what is occurring internally in the PIA

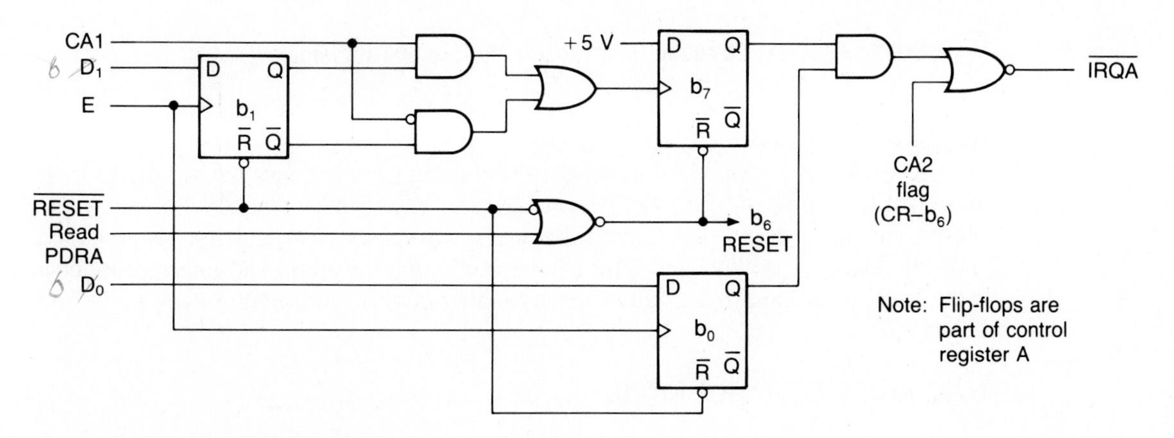

Figure 5.7 CA1 control logic

to this point. CA1 is fed to two AND gates. One has an inverted input so that it responds to CA1 going low, while the other responds to CA1 going high. The Q and $\overline{Q}$ outputs from a flip-flop associated with b1 direct which of the two AND gates will pass the interrupt request along. Note that if b1 is low, the AND gate activated by the low transition passes CA1 information. If b1 is high, then a positive transition is passed along. The outputs from the two AND gates are fed into an OR gate so that the AND gate responsible for passing along the interrupt request will have that information feed b_7 flip-flop's clock of the control register.

b_7 is set only by the receipt of a correct interrupt request on CA1. This flag bit is reset by two methods. In the first method, a reset signal is applied to the $\overline{\text{RESET}}$ input. In the second method, the data register of the port being used is read (in this case, MOVE PORTA,⟨ea⟩). b_7 is otherwise unaffected by any other system operation. The Q output from b_7 is fed to an AND gate that receives its gating signal from b_0's flip-flop. A one in b_0 passes the interrupt request on. A zero in b_0 stops it, masking the interrupt request from the $\overline{\text{IRQA}}$ line. The output of the AND gate is ORed with a similar line from CA2. The OR gate is inverted and sent out of the PIA as $\overline{\text{IRQA}}$.

To program port A of PIAs 2 and 3 to be input ports that also can pass a low to high transition on CA1 as an interrupt request to the 68000, start by configuring the ports to be input interfaces as we did earlier:

```
CLR CONHI    * Clears b2 of control register A
CLR KEYHI    * Selects port A as all inputs
```

To do the rest, you must supply the necessary data to the control register. Set "don't care" (unused) bits to zero so that b_3–b_7 will be low. The remaining control bits are set as follows. b_2 of the control register is set to a one (high) to allow the user to read the data register. For a low to high transition to be recognized as an interrupt request, b_1 must be set high. Last, to pass the interrupt request to the 68000, b_0 also must be set high (unmasked). The data to be put into the control register are 0000 0111, or $07:

```
MOVE #$0707,CONHI    * Data for the control register
```

Do not forget that b_6 and b_7 are unaffected by both the CLR instruction and by the data now loaded into the control register.

The system is not ready to handle the interrupt request yet, but we are at least able to pass it from the I/O device via CA1 to $\overline{\text{IRQA}}$ and to the $\overline{\text{IPL}}$ inputs of the 68000 via the encoder. CA1 and CB1 are always used for interrupt request data from the I/O device. CA2 and CB2 can be used either as additional interrupt request input leads or as handshake output leads sending control information back to the I/O device.

CA2 and CB2 Input Mode

b5 of the control register selects CA2 as either an input or output control lead. A zero in b_5 makes CA1 an input lead, while a one in b5 selects one of the output configurations. If CA2 is selected as an input lead, b_4 performs the transition choice, while b_3 acts as the mask bit, performing the same tasks as b_1 and b_0 did for CA1.

New control register data are necessary to configure CA2 to pass an interrupt request in response to a high to low transition along with CA1. b_6 and b_7 are again set low (something must be put into those bit positions even though they have no effect on the control register bits). b_5 is set low to select CA2 as an input control lead. b_4 is made low to recognize a high to low transition as an interrupt request. b_3 is high to pass the interrupt request to the $\overline{IRQA}$ line. b_0, b_1, and b_2 are high, as before. The control word is now 0000 1111, or $0F. Control register b6 is the interrupt request flag bit for CA2 and is set only by a correct transition on CA2 as an input lead. It is also reset by a $\overline{RESET}$ signal or a read of the port A data register.

CA2 and CB2 Steady State Output Mode

The simplest of the output modes for CA2 is the *steady state* or *bit following* mode. Refer to the control register (Table 5.2) and note that to select CA2 as an output lead, b_5 is set high. Once b_5 is set high, use the information in the table for b_3 and b_4 below the long double line. For CA2 to be in the steady state mode, b_4 also must be high. Selecting b_4 high leaves b_3 with the job of setting CA2 always low or always high. Since CA2 mirrors b_3 in this mode, it is at times called the bit following mode.

5.16 HANDSHAKE MODES

With b_5 high and b_4 low, CA2 and CB2 now take on different meanings as does the PIA in general. When operating in the pulse or handshake mode (b_4 low), port A must be configured as an input port, while port B is an output port. The responses of CA2 and CB2 are predicted on this requirement. CA2 and CB2 are put in the pulse mode by setting control register b_5 high; b_4, low; and b_3, high. When this is done, a pulse is generated on these lines under select conditions. To understand fully what occurs, let us observe an interaction between the PIA and the I/O device.

Pulse Mode Handshake

The I/O device begins by sending data to port A and signaling an interrupt request on CA1. The 68000, in response to the interrupt request, reads port A (MOVE PORTA,⟨ea⟩). A pulse is generated on CA2 and sent to the I/O device. In response to the pulse, the I/O device sends the next data word. Again, the 68000 reads the data and a pulse is sent out on the CA2 line. The process is repeated until all data are read by the 68000. Note that the I/O device signals the PIA only once to start the action. From that point on, the control process is a one-way affair from the PIA (in response to the MOVE PORTA,⟨ea⟩ instructions) to the I/O device. Data transfer ceases once the microprocessor is done. The pulse automatically is generated in this mode. Applications for the pulse mode involve high-speed transfer devices such as floppy disks. While the pulse mode has the advantage of speed, it has the disadvantage of the one-way control.

On the output side, using port B again, an I/O device signals the start of the sequence with an interrupt request on CB1, after which the 68000 stores data into port B. Each time a MOVE ⟨ea⟩,PORTB instruction is executed, a pulse is sent out on CB2 to the I/O device. The device responds to this pulse by accepting the data sent from the PIA. This too is one-way control system. The microprocessor continually sends data and PIA pulses until the microprocessor is finished. The I/O device has no say in the matter. The 68000 also must make a "dummy read" of port B to clear control register b6, the interrupt request flag, which was set by the output device's initial interrupt request.

Figure 5.8 is the timing diagram for the pulse mode of operation. The instructions to transfer data between the PIA ports and the microprocessor are represented as pulses in the diagram for the purpose of relative timing position and are not to be taken literally.

In each case, the pulse developed on the control line is a one-way handshake informing the peripheral device that data have been transferred in one direction or the other. Fast devices, such as disk drives, would use this quick handshake type of control.

Full Handshake

For slower devices such as keyboards or printers, a two-way, asynchronous handshake control is preferred. The peripheral device signals the computer that data either are ready to be input or have been received (depending on direction). Likewise, the computer signals the I/O device that data transfer is ready or has been completed, again dependent on the data direction. The last operational mode, called the full handshake mode, provides for this control. As in the pulse mode, port A must be an input port and port B must be an output port. b4 and b5 of the control register

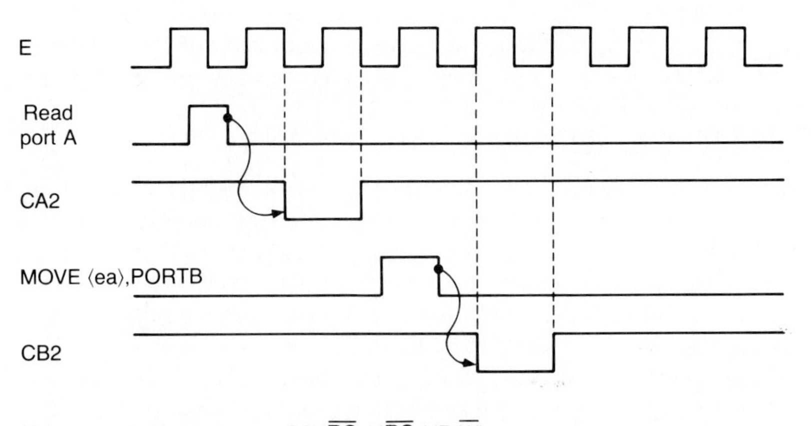

Notes: 1. Read port A = $(CS)(\overline{RS_1})(\overline{RS_0})(R/\overline{W})$

2. MOVE⟨ea⟩,PORTB = $(CS)(\overline{RS_1})(\overline{RS_0})(\overline{R/\overline{W}})$

Figure 5.8 CA2/CB2 pulse timing

are again set low (output mode function), but this time b3 also is set low, placing the PIA into the handshake mode of operation.

Figure 5.9 is a schematic diagram for the full handshake mode of operation, and Figure 5.10 (page 170) shows the timing diagram for this mode. As before, the pulses called read PORTA and MOVE ⟨ea⟩,PORTB represent the total control signals necessary to load or read the data from the ports indicated.

To illustrate the full handshake, start by assuming that the I/O device initiates the action. While this is generally the case, it is not an absolute. The I/O device connected to port A is a keyboard and the device connected to port B is a printer. Again, this is not the only arrangement that would use the handshake mode, but it is used for the purposes of this discussion. To begin operation, an operator presses a key on the keyboard, which sends a pulse onto CA1. This pulse can be negative or positive, depending on how the control register on side A is configured to recognize an interrupt on CA1. Upon receipt of the pulse, b7 of control register A is set; and an interrupt request is sent to the 68000. In addition, line CA2 is made low. These actions all occur as a result of logic within the PIA in response to the CA1 pulse. What is each action saying to the device involved? First, the pulse on CA1 tells the interface, from the keyboard, to pay attention because data will be sent. The data on the peripheral data lines to port A are sent to the data register of port A at the same time that the CA1 line is pulsed. The PIA acknowledges the receipt of these data and the pulse on CA1 by making CA2 high.

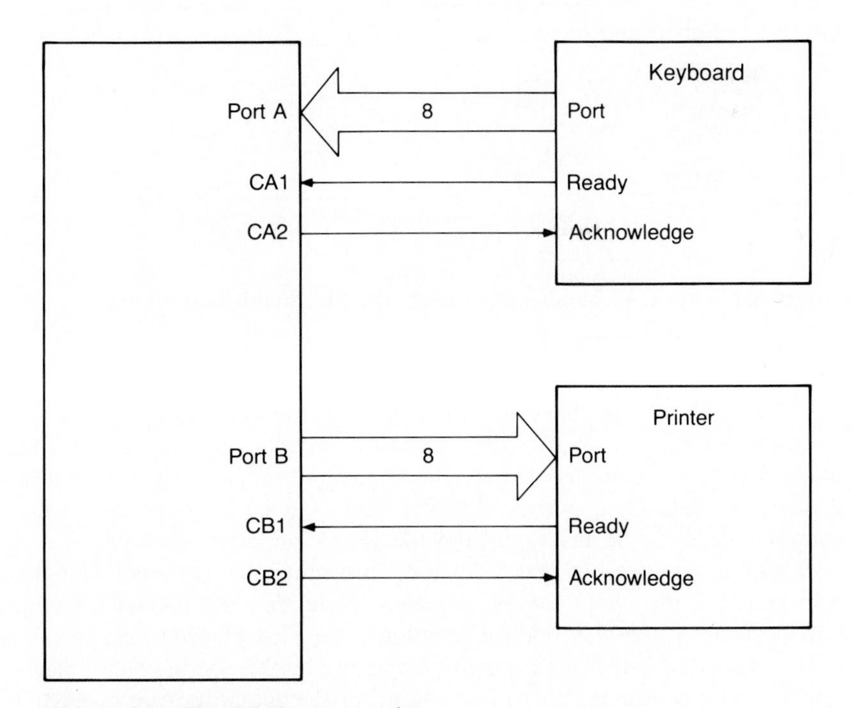

Figure 5.9 Handshake connections

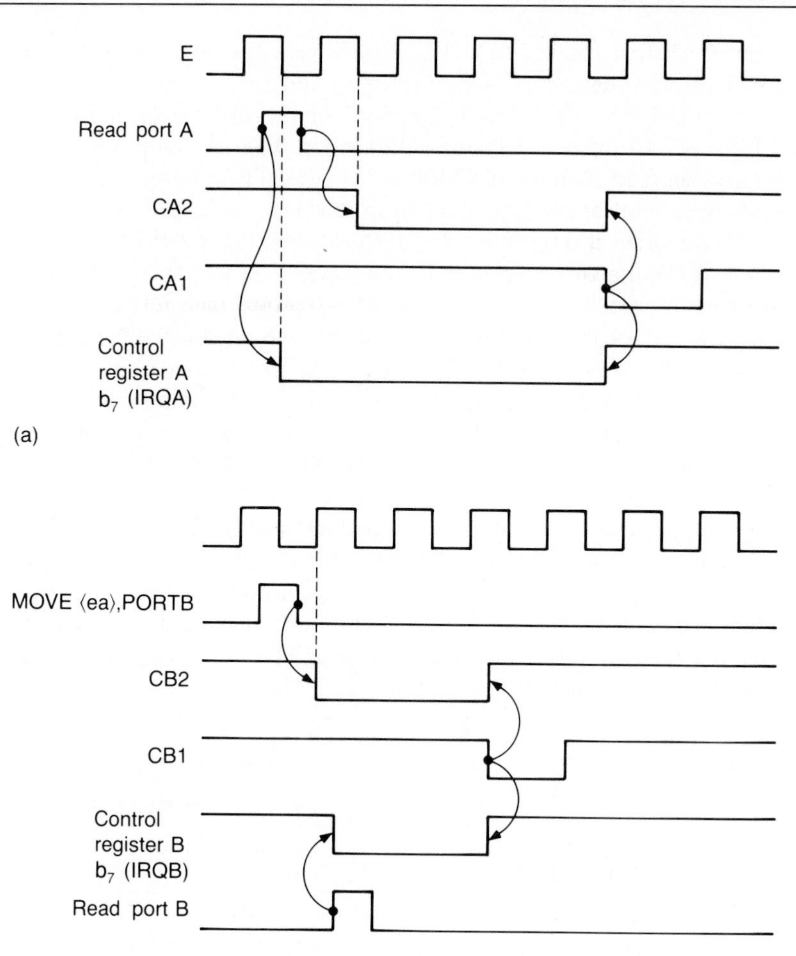

Figure 5.10 (a) CA2 handshake timing. (b) CB2 handshake timing.

The PIA informs the microprocessor that the data are in port A by sending an interrupt request via $\overline{\text{IRQA}}$ to the interrupt encoder. Nothing further happens until the 68000 responds to and services the interrupt request. When the 68000 does this, it reads the data from port A (MOVE PORTA,⟨ea⟩). This action causes b7 of the control register to be reset, making the $\overline{\text{IRQA}}$ go inactive (high). In addition, the read of the data register sets CA2 low, informing the keyboard that the data were read and that the cycle can be repeated. Note that the handshake control signals now perform a two-way control function. The time elapsed between data sent and read (acknowledged) is determined completely by the action of the interrupt routine. This is an asynchronous control situation, for the actual transfer of any data can take place within microseconds or (improbably) hours. Once the keyboard receives an acknowledge that its last data have been read, the next data are sent at the leisure of the operator.

The output transfer acts similarly. First, the printer tells the computer that it is ready to receive data by sending a pulse on the CB2 line. Once again, the type of pulse is determined by the setting of control register b_1. Upon receipt of the pulse, b_7 of control register B is set and an interrupt request is sent along $\overline{IRQB}$ to the interrupt encoder. CB2 is made high by receipt of the pulse on CB1, informing the printer that its request was received; but no data have been sent to port B yet. In response to the interrupt request, the 68000 stores data into port B. CB2 goes low, telling the printer that data are available in port B. The printer reads the data and awaits the next data word. Before the next data can be sent, b_7 of the control register must be reset and CB2 must be made high again. The sequence of these actions is important to maintain correct data control information. b_6 and b_7 of the control register can be reset only by a reset signal or by reading the data register associated with that port. In this case, a system reset is not expected; therefore, a dummy read of port B must follow the MOVE ⟨ea⟩,PORTB to cause b7 to be reset. CB2 is made high by the receipt of a pulse on CB1 from the printer.

Handshake Signal Summary

The following is a summary of handshake signals:

CA1 pulse from the keyboard says data are available at port A.

CA2 high acknowledges to the keyboard that data are at port A.

$\overline{IRQA}$ (b_7 of control register A) tells the 68000 that data are available at input port A.

CA2 low (in response to a read of port A) tells the keyboard that data were read.

CB1 pulse says that the printer is available to receive data from port B.

CB2 high acknowledges to the printer that it is ready.

$\overline{IRQB}$ (b_7 of control register B) says that port B is ready to receive data from the microprocessor.

CB2 low indicates that data are available at port B for the printer to take.

Configuring the Basic System PIAs

To use the basic system in the handshake mode, it is first necessary to configure the PIAs. To make the arrangement easier, the interrupt request from either device is set to be recognized by a high to low transition on the appropriate control lien (CA1 or CB1). Both CA2 and CB2 must be set to the handshake mode. Refer to the control register diagram in Table 5.2 and note that for the conditions stated, the data sent to the control registers are the same:

b_0 = 1 to allow interrupts (unmask them)

b_1 = 0 for a high to low transition

b_2 = 1 to access the data registers

b_3 = 0 for handshake mode

b_4 = 0 for mode control rather than steady state

b_5 = 1 for output control on CA2 and CB2

b_6 and b_7 are don't cares since they are interrupt flags

Once b_6 and b_7 of the data word to be transferred are made low to complete eight bits of information, the data word sent to the control registers to configure the PIA for handshake under these conditions is $25. This is the final control register data and follows direction selection as illustrated before. Using the auxiliary PIAs 3 and 4 of our system, the complete set of instructions to configure these PIAs is shown in Program 5.8.

```
CLR ACONHI               * Set b2 low
CLR BCONHI
CLR PORTAHI              * Port A: Input port
MOVE #$FFFF,PORTBHI      * Port B: Output port
MOVE #$2525,ACONHI       * Selects handshake mode
MOVE #$2525,BCONHI
```

Program 5.8

It should be apparent that two separate routines must be run to effect correct data transfers—an input routine and an output routine. The 68000 has several interrupt levels selectable by the encoder connected to the $\overline{IPL}$ lines. $\overline{IRQA}$ and $\overline{IRQB}$ are wired to the encoder. It is now up to the **interrupt programs** at the selected vector locations associated with the encoded interrupt to service the keyboard or the printer, depending on which sent the interrupt request.

5.17 THE 68230 PARALLEL INTERFACE TIMER

One of Motorola's direct support devices for the 68000 family of microprocessors is the parallel interface timer (PIT) (see Figure 5.11). It contains an 8-bit data bus for interfacing with the 68000 data bus. Two PITs can be used in similar fashion as the PIAs to use the full 16-bit width of the 68000 data bus.

The PIT is a dual-ported device that can be programmed to transfer data to and from these ports individually or together on a 16-bit peripheral bus. Keep in mind that there is only a single 8-bit data bus connecting a PIT to the microprocessor. Data transferred between a 16-bit port and the microprocessor must be handled one byte at a time through a single PIT. In this mode, the 16 bits from the I/O device can be read or output at one time, but the actual data between the PIT registers and the 68000 must travel a byte at a time along the PIT's 8-bit data bus. In this respect, the PIT is not a true 16-bit device.

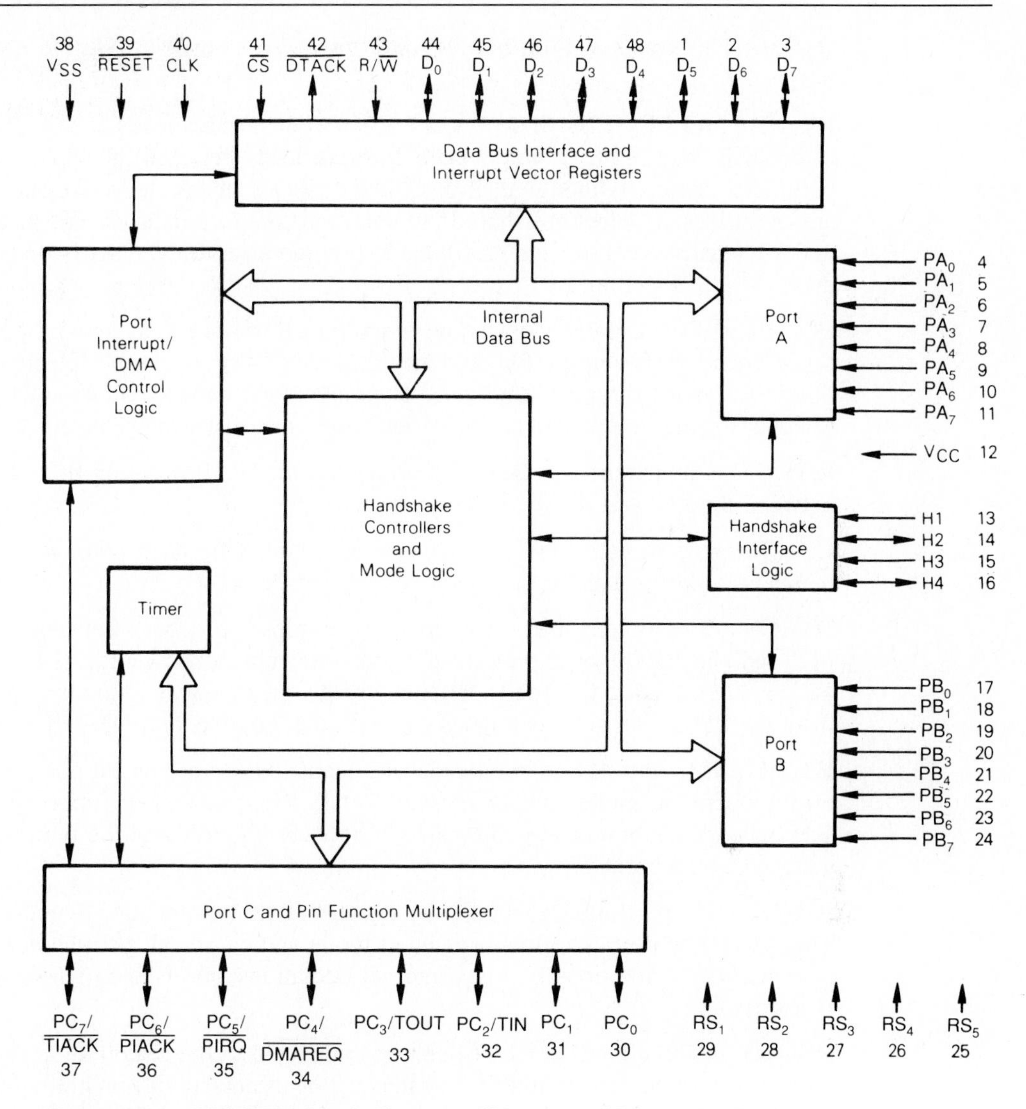

Figure 5.11 MC68230 block diagram (Courtesy of Motorola, Inc.)

In addition to the two data ports (A and B), there is a 24-bit timer and an optional third port (port C). This third port can be programmed as a general-purpose data port or, alternatively, to supply functions related to the timer and interrupt control.

PIT Pin Descriptions

Besides the bidirectional data bus (D_0–D_7) and peripheral data leads (PA_0–PA_7 and PB_0–PB_7), other leads are made available to handle the signals necessary to connect the PIT directly to the 68000 and to interface the ports to an I/O device.

D_0–D_7 This standard 8-bit bidirectional data bus is connected to the 68000's data bus.

PA_0–PA_7 and PB_0–PB_7 Port A's and B's peripheral data leads can be programmed to be individual input or output data carrying leads between the PIT ports and the peripheral device. Although each individual bit can be selected independently, there are overall mode selections that allow data transfers on selected leads to be handled under handshake control inputs from the peripheral devices. This is discussed later under Mode Selection.

PC_0–PC_7 Port C data leads can be programmed to operate as general-purpose data transfer leads between the PIT and an I/O device, or they can be selected to handle timer and interrupt control functions. Their actual use depends on the operating mode selected for the timer. Timer control selection is described later in this chapter.

$R/\overline{W}$ This input signal indicates the direction of data flow along the 8-bit data bus between the PIT and the microprocessor.

$\overline{CS}$ The chip select allows the chip to be enabled from an address decoder in a manner similar to that for the memory and PIA devices.

$\overline{DTACK}$ This output lead is asserted in response to a data transfer between the PIT and the 68000 or in response to an interrupt acknowledge (FC_2, FC_1, and FC_0 = 111). During the interrupt acknowledge cycle, an exception number is placed from the PIT's interrupt vector register onto the data bus and $\overline{DTACK}$ is asserted.

H1, H2, H3, and H4 Four handshake leads serve various purposes, depending on the operating modes of port A and port B. Generally, they supply strobing and acknowledgement and status information between the PIT and the peripheral device attached to the ports.

$\overline{RESET}$ The reset input to the PIT clears all PIT control and data direction registers. This places the ports into the 8-bit input mode and causes all peripheral leads to act as input leads. Additionally, most internal operations are disabled whenever $\overline{RESET}$ is asserted.

RS_1–RS_5 The register select inputs are connected to the 68000 address bus. They allow each of the internal PIT registers to be accessed by an individual address from the microprocessor. Table 5.6 presents the register complement of the PIT with respect to the states of the register select inputs.

Operating Modes

There are four basic operating modes for data ports A and B, selected by programming the desired levels of b_6 and b_7 of the port general-control register (see Table 5.6). These modes, in combination with submode selection (b_6 and b_7 of the individual port A or port B control register), select the primary data flow direction between the ports and the peripheral devices. The actual data direction flow on each bit is selected by loading data into the corresponding port direction register. A low in a bit position of the direction register selects the corresponding peripheral data lead of the port as an input lead. A high in a direction register bit position selects

Table 5.6 Register List (Courtesy of Motorola, Inc.)

Register Select Bits 5	4	3	2	1	REGISTER MODEL 7	6	5	4	3	2	1	0	
0	0	0	0	0	Port Mode Control		H34 Enable	H12 Enable	H4 Sense	H3 Sense	H2 Sense	H1 Sense	Port General Control Register
0	0	0	0	1	*	SVCRQ Select		Interrupt PFS		Port Interrupt Priority Control			Port Service Request Register
0	0	0	1	0	Bit 7	Bit 6	Bit 5	Bit 4	Bit 3	Bit 2	Bit 1	Bit 0	Port A Data Direction Register
0	0	0	1	1	Bit 7	Bit 6	Bit 5	Bit 4	Bit 3	Bit 2	Bit 1	Bit 0	Port B Data Direction Register
0	0	1	0	0	Bit 7	Bit 6	Bit 5	Bit 4	Bit 3	Bit 2	Bit 1	Bit 0	Port C Data Direction Register
0	0	1	0	1	Interrupt Vector Number						*	*	Port Interrupt Vector Register
0	0	1	1	0	Port A Submode		H2 Control			H2 Int Enable	H1 SVCRQ Enable	H1 Stat Ctrl.	Port A Control Register
0	0	1	1	1	Port B Submode		H4 Control			H4 Int Enable	H3 SVCRQ Enable	H3 Stat Ctrl.	Port B Control Register
0	1	0	0	0	Bit 7	Bit 6	Bit 5	Bit 4	Bit 3	Bit 2	Bit 1	Bit 0	Port A Data Register
0	1	0	0	1	Bit 7	Bit 6	Bit 5	Bit 4	Bit 3	Bit 2	Bit 1	Bit 0	Port B Data Register
0	1	0	1	0	Bit 7	Bit 6	Bit 5	Bit 4	Bit 3	Bit 2	Bit 1	Bit 0	Port A Alternate Register
0	1	0	1	1	Bit 7	Bit 6	Bit 5	Bit 4	Bit 3	Bit 2	Bit 1	Bit 0	Port B Alternate Register
0	1	1	0	0	Bit 7	Bit 6	Bit 5	Bit 4	Bit 3	Bit 2	Bit 1	Bit 0	Port C Data Register
0	1	1	0	1	H4 Level	H3 Level	H2 Level	H1 Level	H4S	H3S	H2S	H1S	Port Status Register
0	1	1	1	0	*	*	*	*	*	*	*	*	(null)
0	1	1	1	1	*	*	*	*	*	*	*	*	(null)
1	0	0	0	0	TOUT/$\overline{TIACK}$ Control			Z D Ctrl.	*	Clock Control		Timer Enable	Timer Control Register
1	0	0	0	1	Bit 7	Bit 6	Bit 5	Bit 4	Bit 3	Bit 2	Bit 1	Bit 0	Timer Interrupt Vector Register
1	0	0	1	0	*	*	*	*	*	*	*	*	(null)
1	0	0	1	1	Bit 23	Bit 22	Bit 21	Bit 20	Bit 19	Bit 18	Bit 17	Bit 16	Counter Preload Register (High)
1	0	1	0	0	Bit 15	Bit 14	Bit 13	Bit 12	Bit 11	Bit 10	Bit 9	Bit 8	(Mid)
1	0	1	0	1	Bit 7	Bit 6	Bit 5	Bit 4	Bit 3	Bit 2	Bit 1	Bit 0	(Low)
1	0	1	1	0	*	*	*	*	*	*	*	*	(null)
1	0	1	1	1	Bit 23	Bit 22	Bit 21	Bit 20	Bit 19	Bit 18	Bit 17	Bit 16	Count Register (High)
1	1	0	0	0	Bit 15	Bit 14	Bit 13	Bit 12	Bit 11	Bit 10	Bit 9	Bit 8	(Mid)
1	1	0	0	1	Bit 7	Bit 6	Bit 5	Bit 4	Bit 3	Bit 2	Bit 1	Bit 0	(Low)
1	1	0	1	0	*	*	*	*	*	*	*	ZDS	Timer Status Register
1	1	0	1	1	*	*	*	*	*	*	*	*	(null)
1	1	1	0	0	*	*	*	*	*	*	*	*	(null)
1	1	1	0	1	*	*	*	*	*	*	*	*	(null)
1	1	1	1	0	*	*	*	*	*	*	*	*	(null)
1	1	1	1	1	*	*	*	*	*	*	*	*	(null)

*Unused, read as zero.

that peripheral data lead as an output. The primary data direction responds to control from the handshake signals, while those leads selected in opposition to the primary direction do not.

8-Bit Input and Output Modes Mode 0, submode 00, places ports A and B into the 8-bit, double-buffered, input mode. This mode is selected when the control registers are cleared by an active $\overline{\text{RESET}}$ signal. Handshake signals H1 and H3 act as data strobe signals from the peripheral device by causing the data on the peripheral input lines to be strobed into the PIT ports. Any data leads that have been selected to be outputs in this mode ignore the data placed onto them. These output leads are loaded with data sent to them by the microprocessor.

Mode 0, submode 01, puts ports A and B into the 8-bit, double-buffered, output mode with H1 and H3 acting as data acknowledge inputs from the PIT. These input handshake signals inform the PIT that data were received from it on the port output leads. Any port data leads programmed as inputs in this mode are nonlatching; that is, they are continually updated and their value can be read any time by the microprocessor.

16-Bit Port Operation Mode 1 performs the same type of function as mode 0, except that both ports are considered as one 16-bit port. In the input submode 00, H3 serves the purpose of strobing in the full 16 bits into both ports. Port A is considered the upper byte and port B the lower byte of the 16-bit word. Again, any leads programmed as outputs in this mode ignore the H3 signal. Submode 01 in mode 1 places the two ports into the 16-bit output mode and, like the 8-bit mode, any input leads are nonlatching and can be read at any time. In this case, H3 serves as the acknowledge handshake as it did for port B in mode 0.

Bidirectional Mode Mode 2 selects the 8-bit bidirectional mode for port B. The direction of data flow for port B is controlled by the application of the handshake signals. H1 supplies the acknowledge signal for the output direction, and H3 supplies the input data strobe. Port A in this mode is strictly a bit I/O port, with each data direction selected by the state of the bits in the direction register. Input data are not latched, and output data are supplied by the microprocessor.

Mode 3 is the 16-bit bidirectional mode. In this mode, both ports are placed in the bidirectional mode and data transfers are totally under control of the handshake signals. The direction registers in this mode are ignored. H1 again supplies output data control, and H3 supplies input strobe control from the peripheral device.

A set of alternate port registers provides a second address for reading input port data. These registers are nonlatching and allow the user to read the condition of the input lines at any time, while the base port data registers supply data that were present at the time that the handshake control inputs were asserted. Data transfers from the microprocessor to these alternate registers are ignored, but the transfers do generate an active $\overline{\text{DTACK}}$ so that the 68000 does not receive a false bus error condition.

Timer Registers and Operation The 24-bit timer operation is selected by programming the timer control register (see Table 5.7). Timer outputs and inputs are directed through the alternate functions of port C.

Table 5.7 Port General Control Register

7	6	5	4	3	2 1	0
0 = PC7 data	0 = PC3 data 1 = PC3 TOUT	X	0 = Reload counter from preload	0	00 = PC2 data (CLK and pre-scaler used)	0 = Disable timer
1 = PC3 TOUT	0 = PC7 TIACK	0 = Timer interrupt disabled	1 = Roll over counter on 0		01 = PC2 TIN (CLK and pre-scaler used)	1 = Enable timer
	1 = PC7 data	1 = Timer interrupt enabled			10 = PC2 TIN (pre-scaler used)	
					11 = PC2 TIN (pre-scaler not used)	

The functions of port C pins 3 and 7 are selected by programming b_6 and b_7 of the timer control register. These functions perform either as pure data handling leads or as the timer output (TOUT) and timer interrupt acknowledge input (TIACK). When used as TOUT, the port C pin produces an output depending on the operating mode of the timer. This output can be a squarewave produced as the timer decrements through its count or can show up as a change of level generated when the timer count is through (reaches $000000). If the timer interrupt is enabled, TOUT serves as the timer interrupt request.

In response to an interrupt acknowledge from the 68000 (FC_2, FC_1, and FC_0 = 111) that has been decoded into an active TIACK (and if the timer interrupt is enabled), the vector number associated with the timer is placed onto the data bus from the PIT and $\overline{\text{DTACK}}$ is asserted. Details of interrupt and exception handling are discussed in Chapter 8.

Timer control b_5 enables or disables the timer interrupt function. If the interrupt is disabled, the PIT ignores the TIACK input.

Zero detect control is selected using timer control b_4. The counter either can be reloaded from the preload register and hence restarted, or rolled over to $FFFFFF when it reaches 0 ($000000). The counter then resumes counting as long as it remains enabled. The preload registers are loaded with a starting count value before the timer is enabled initially.

b_3 is unused and usually is set to 0. b_1 and b_2 select the function of port C bit 2. This function depends on whether the timer is externally triggered or synchronized to the CLK input and prescaler. The alternate function of PC_2 is the timer input (TIN). When used, the timer runs when it is enabled and TIN is high. When the CLK input and prescaler are used, the prescaler is decremented on the falling edge of the CLK input and the timer is decremented or loaded from the preload register when the prescaler rolls over from $00 to $1F. When only the prescaler is used, it

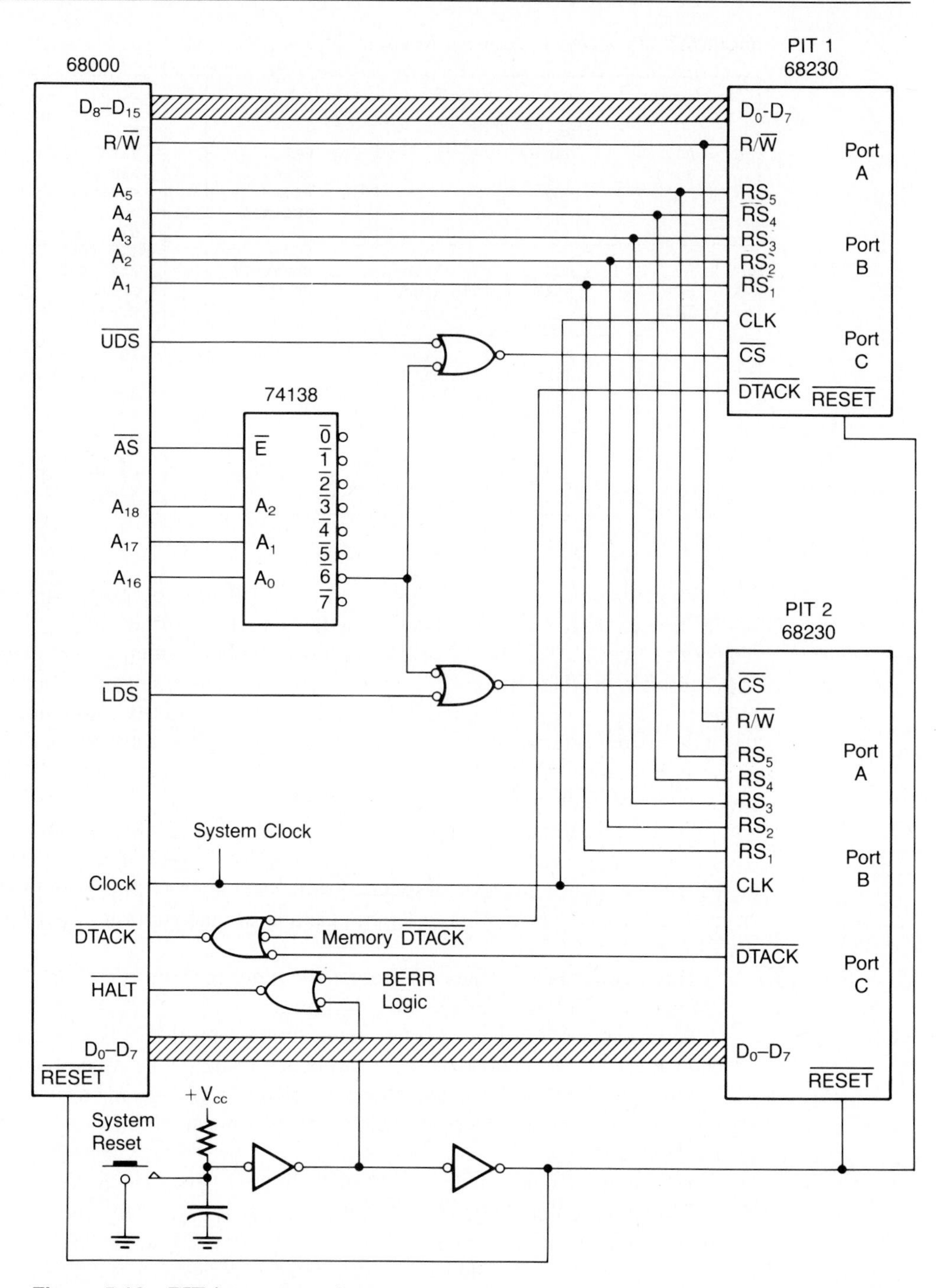

Figure 5.12 PIT interconnections

is decremented on the rising edge of TIN and the timer is decremented or reloaded from the preload register as before. If neither the CLK or the prescaler is used (timer control register b_2 and b_1 are both high), the timer is decremented or reloaded on the rising edge of TIN.

Finally, the timer is enabled or disabled using b_0 of the timer control register. Enabling causes the selected timer operation to commence, while disabling the timer places it into the halt state.

The starting count for the timer is programmed into three preload registers. These are each one byte long and are labeled high, middle, and low preload. The current count value can be read at any time by accessing the three count registers: high, middle, and low count. The three registers are required for each function to hold the 24-bit value of the timer.

Interconnecting the PIT to the 68000

Figure 5.12 diagrams the interconnections between the PIT and the 68000. Remember that inputs like $\overline{\text{DTACK}}$ come from several sources and may be required to be ORed with additional inputs. Two PITS are connected to illustrate the use of the full 16-bit data bus of the 68000. However, the PIT connected to the upper data bus (PIT 1) either must use the autovector function when generating interrupt requests or must dismiss using the interrupt function entirely. This situation occurs because the 68000 must read the vector number on the lower half of the data bus in response to an interrupt acknowledge. Details of the interrupt process are discussed in Chapter 8.

Two additional port C pins carry the alternate functions of a port interrupt request (PIRQ) and a port interrupt acknowledge (PIACK). When used, they respond to the handshake inputs associated with the ports. PIRQ must be encoded into an interrupt level supplied to the IPL inputs of the 68000. Port C connections are not shown in Figure 5.12.

Reset logic allows a reset to be asserted to the PIT either from the system reset switch or from the reset output of the 68000 in response to a RESET instruction. The $\overline{\text{HALT}}$ signal is driven from an OR gate whose inputs are from the system reset or from bus error logic. The 68000 $\overline{\text{RESET}}$ is buffered a second time to prevent the reset output function of the 68000 (RESET instruction) from asserting $\overline{\text{HALT}}$. The RC (resistor-capacitor) network connected to the system reset switch generates an active reset when power is first applied to the system. This RC network also holds the reset signal active for a short time after the system reset switch is released.

The register select lines are connected to the lower address lines to allow register accessing through memory transfer instructions. The 74138 decoder is used as it had been for the PIAs. This diagram is drawn to allow the PITs to replace the PIAs. Table 5.8 on page 180 is the replacement line for the system memory map developed in this and previous chapters.

PROBLEMS

5.1 Write a program segment to configure port A of PIAs 3 and 4 of our basic system to interface with two output devices. Each device has eight data leads and two interrupt

Table 5.8 PIT Memory Map

23 --- 19	18	17	16	15 --- 6	5	4	3	2	1	0	Device
	1	1	0		r5	r4	r3	r2	r1	U	PIT 1
	1	1	0		r5	r4	r3	r2	r1	L	PIT 2

request lines. The PIA is to allow all four interrupt requests to pass to the 68000. An interrupt request is recognized as a low to high transition on each of the control leads.

5.2 Write the additional instructions to send $3D to PIA 3 port A and $Å2 to PIA 4 port A of Problem 5.1.

5.3 How is b_7 of both port A PIA control registers set and reset in Problem 5.1?

5.4 Make a logic diagram that illustrates the process which CA2 goes through to become an interrupt request line. Do not include the logic for the output conditions of CA2.

5.5 In your own words, define what CA1, CA2, CB1, and CB2 signify when the PIA is in the pulse mode.

5.6 Write a polling routine to decipher which one of four control leads creates an interrupt. The leads are CA1, CA2, CB1, and CB2 of PIA 3.

5.7 Define each bit of control register A and tell what each signifies or selects with respect to the PIA.

5.8 List both an input and an output device that would use the handshake mode (other than a keyboard and printer).

5.9 Write a program to output a squarewave (approximately 50% duty cycle) on PIA 3's CA2 line.

6

SERIAL DATA INTERFACING WITH THE 68000

CHAPTER OBJECTIVES

The objective of this chapter is to describe serial I/O devices and their uses with the 68000 microprocessor. An introduction to data communications is provided to explain the need for these devices.

6.1 GLOSSARY

alphanumeric characters Letters, numbers, and punctuation.
answer station Responding station in a data communications link.
baud rate Rate of serial data transfer.
break A stop during serial data transfer.
data communications Transfer of digital information between two or more stations.
data link control characters Characters that define the message and type of data transfer channel.
digital communications Transfer of information between two or more stations by means of data in digital form.
duplex Type of serial data transfer channel.
frame Start and end of a character or message.
framing bits Start and stop bits.
frequency shift keying (FSK) Form of data modulation in which audio tones represent 1's and 0's.
full duplex Two-way simultaneous serial data transfer.
graphics characters Characters that determine how a message actually appears on the screen or paper.
half duplex Two-way serial data transfer, one way at a time.
mark Logic 1.

modem Device that converts between audio tones and digital data.
originate station Station that establishes the data communications link channel.
parity Method of error checking based on the number of 1's in a data character.
simplex One-way data transfer.
space Logic 0.

6.2 INTRODUCTION TO SERIAL DATA TRANSFER

The PIA is a parallel interface device that transfers information between an I/O device and the microprocessor, eight bits at a time, through eight parallel peripheral lines. Regardless of the direction of data transfer on those lines, all eight bits of a port are accessed using the PIA. The purpose of this type of interface is to carry information for short distances—from a keyboard to a computer, for instance. For longer distances, say from Phoenix, Arizona, to Levittown, New York, maintaining eight separate data lines would be impractical. Too much wiring and maintenance would be involved. Instead, it is preferable to convert the parallel data to a serial stream of information that can be sent on a single data line.

The actual data transfer between the two cities is covered under the broad heading of **data communications.** A subtopic of that heading is communications using the existing telephone network. Serial transmission of digital data using telephone lines is done by first converting the parallel data from the computer into a serial stream using a *universal asynchronous receiver/transmitter* (UART). Then this data stream is converted into analog signals using audio tones in place of the two digital levels; that is, the DC levels for a 1 and a 0 are converted into two different tones that are transmittable using telephone lines. The device that performs this conversion is the modulator-demodulator or **modem.**

6.3 DIGITAL MODULATION TYPES

Modems differ in the manner in which the digital tones are used to modulate some form of AC carrier, which is in turn transmitted across a medium from one station to another. One type of modulation is *amplitude modulation,* where the amplitude of a carrier is changed between two levels, one representing a digital 1 and the other representing a digital 0. In data communications this is more commonly called *amplitude shift keying* (ASK). *Shift keying* is a Morse code term to describe using a key to shift between two signals, one for off and one for on.

A second method of altering an AC signal is to change its phase from a reference. A change of one phase from another indicates a 1, and change to a second phase represents a 0. This method is a form of *phase shift keying* (PSK). More sophisticated devices combine several data bits and use the various bit combinations to generate different phases for each combination. This method allows higher rates of data transmission by sending out a single signal called a *symbol* to represent several bits.

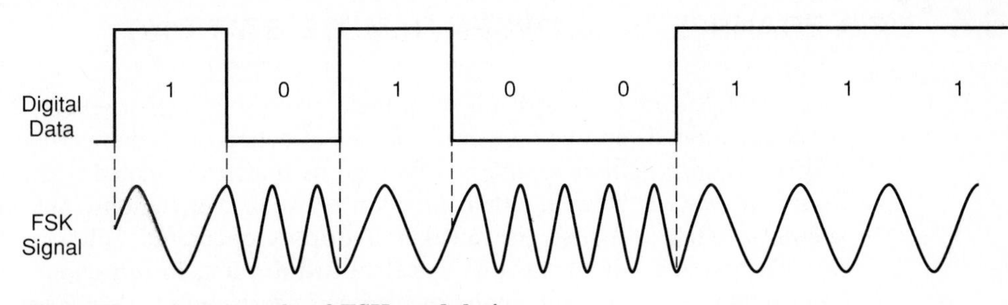

Figure 6.1 An example of FSK modulation

Combining PSK and ASK into a form of modulation called *quadrature amplitude modulation* (QAM) permits still higher rates of data transmission. With this method of modulation, bit combinations cause differences in phase and amplitude between each signal transmitted. Larger numbers of bits can be used to cause the generation of each unique signal pattern.

Another method of digital modulation, **frequency shift keying** (FSK), will be used in the discussion of the serial interface in this chapter. With this method, an audio tone is generated for a digital 1 and a different tone for a digital 0, as shown in Figure 6.1.

6.4 FSK MODEMS

An FSK modem is a device that receives digital data in serial format and converts the two DC levels (1's and 0's) into two separate AC signals that vary by frequency. When digital signals are converted to AC signals, telephone lines can be used to transfer the data over long distances. The frequency of these signals, or tones, falls between 300 Hz and 3 kHz—the practical bandwidth of the telephone network. The higher of the two frequencies represents a logic 1 and is called the **mark** tone. The lower **space** frequency represents a logic 0.

FSK modems contain a modulating circuit built around a *voltage-controlled oscillator* (VCO). The VCO's output frequency changes as the voltage applied to it is changed. By applying a DC level to the VCO representing a 1, a signal is generated at a set frequency. Applying a second DC level representing a data 0 produces a second set frequency.

The receive portion of the modem accepts the signals generated by a similar modem's modulator and demodulates them back into DC signals. The circuit generally used to perform the demodulation is a *phase lock loop* (PLL), which generates an error voltage whenever an incoming signal differs in phase or frequency from a reference signal. This error voltage is used to establish the digital data levels of 1 and 0 equivalent to the incoming tone patterns. In the case of an FSK modem, the reference signal is the center or carrier frequency. The value of this frequency lies between the frequency values of the mark and space tones.

6.5 DATA TRANSFERS BETWEEN REMOTE STATIONS

One way to create a data path is to provide a single pair of lines (for signal and return) and allow data to be transmitted in one direction only; that is, one station sends and one remote station receives. This type of transfer is called a **simplex** transfer. Permitting data to flow in either direction between two stations, but with only one station sending at a time, is called **half duplex** operation. Allowing transmission in both directions simultaneously is called **full duplex.** In full duplex transmission, there might be two separate pairs of data paths to handle the simultaneous communications. This method, called *four-wire* transmission, is preferred for long distance communications, for signal deterioration is a concern.

Table 6.1 ASCII Codes

MSB (hex)	0	1	2	3	4	5	6	7
LSB (hex)								
0	NUL	DLE	SP	0	@	P	'	p
1	SOH	DC1	!	1	A	Q	a	q
2	STX	DC2	"	2	B	R	b	r
3	ETX	DC3	#	3	C	S	c	s
4	EOT	DC4	$	4	D	T	d	t
5	ENQ	NAK	%	5	E	U	e	u
6	ACK	SYN	&	6	F	V	f	v
7	BEL	ETB	'	7	G	W	g	w
8	BS	CAN	(	8	H	X	h	x
9	HT	EM	)	9	I	Y	i	y
A	LF	SUB	*	:	J	Z	j	z
B	VT	ESC	+	;	K	[	k	{
C	FF	FS	,	<	L	\	l	|
D	CR	GS	–	=	M	]	m	}
E	SO	RS	.	>	N	↑	n	~
F	SI	US	/	?	O	←	o	DEL

MSB = b6, b5, b4

LSB = b3, b2, b1, b0

A *two-wire* (one signal pair) path, which requires the use of two different sets of tones for the different directions of data flow, is more economical for **duplex** transfers. In general, one station originates a call to a remote answer station. To allow transmission between both stations at the same time, a separate pair of frequencies is assigned to the **originate station** and another pair to the **answer station.** The originate station's transmitting mark and space frequencies are the same as the answer station's receive frequencies. Similarly, the originate station's receive frequencies must match the answer station's transmitting frequencies.

6.6 DIGITAL FORMATTING CODES

Data in digital form are used to represent information to be sent between two stations. For the digital combinations to have meaning, the raw digital data must be arranged in a coded order. Several digital codes are used to represent letters and numbers (alphanumeric), graphics, and data link control characters. **Alphanumeric characters** also include punctuation, the percent sign, and other characters. **Graphics characters** are used to format a printed page or a screen message on a terminal. They include carriage return, line feed, and tabs. **Data link control characters** define the boundaries of a message and any special characteristics of that message or a portion thereof. They include such characters as start of text (STX) and end of transmission (EOT).

The two most popular character codes in use today are EBCDIC and ASCII. EBCDIC (Extended Binary-Coded Decimal Interchange Code) is an 8-bit character code developed and chiefly used by International Business Machines Corporation (IBM). Each character is represented by eight binary bits in this code. ASCII (American Standard Code for Information Interchange) uses seven digital bits to represent each character. The ASCII code chart is shown in Table 6.1.

EXAMPLE 6.1

What are the ASCII codes for the letter A and the number 7?

Solution:

From the ASCII chart in Table 6.1, we see that

A = 100 0001, or $41

7 = 011 0111, or $37

Before we assemble an ASCII coded character for transmission, we must explain the concept of *parity.* **Parity** is a form of error checking. In ASCII, the parity bit is added after the MSB of data. At the sending site the ones in a character word are summed, producing either an odd or an even count. There are two types of parity systems: odd parity and even parity. The condition of the parity bit is determined by the result of the sum and the type of the parity system used. If an odd parity system is used, the parity bit is set or reset to make the total count of the data ones and the parity bit an odd number, Similarly, even parity makes the total count even.

EXAMPLE 6.2

What are the total character codes for the letter A and the number 7 using both parity systems?

Solution:

For A (A = 100 0001), the number of ones is two, an even number. Using an even parity system causes the parity bit to be reset (zero) to maintain the even count. This bit is b7 of the character word, making the complete ASCII code 0100 0001, or $41, for A. An odd parity system produces a high parity bit, making the ASCII code for A now 1100 0001, or $C1.

For 7 (7 = 011 0111), the number of ones is five, an odd number. For even parity, the parity bit is now set to make the sum an even number, or six. The ASCII 7 with even parity is 1011 0111, or $B7. For an odd parity system, the ASCII 7 becomes 0011 0111, or $37.

The character is sent and demodulated at the remote site. The receiver then takes the seven character bits and again generates a parity bit. The parity bit sent with the character and the new one generated by the receiver are compared. If they are the same, then no error is presumed. A mismatch indicates that something is wrong with the received data.

6.7 TRANSMITTING SERIAL DATA

Data are transmitted at a rate characterized by the number of bits sent per second (bps). Actual information transfer, which is the rate at which coded information is sent, is called the **baud.** Baud includes bit rate information plus character size and character transfer rate data. Baud values can be confusing because they can be given in either bps or symbols per minute. On a conceptual level, the difference between baud and bps representations of the data stream is not significant. For the following discussion, consider bps to be synonymous with baud.

Data can be sent synchronously (timed with a clocking pulse) or asynchronously (lacking any clocking). Asynchronous data require the addition of **framing bits** to identify the beginning (start) and ending (stop) of a character. The advantage of the synchronous system is the lack of additional bits, allowing for higher baud rates (smaller digital word sizes for each character result in higher bit rates). A serious disadvantage of synchronous data is the need to recover the timing pulses necessary for synchronization. The discussion in this chapter is limited to asynchronous data types.

6.8 SERIAL DATA STREAM

To put the total asynchronous character word together to see how it looks when it is transmitted, one must determine the level of the start bit. When a communications link is established, a mark tone is sent on the line to verify that a link exists. To signify when a character begins, the line must be changed from this mark condition. This change forces the start bit at the beginning of a character word to be a space, or zero logic level. What follow next are the seven character-code bits starting with the LSB of the code. The LSB is always sent first. After the seven character-code bits are sent, the parity bit, if used, comes next. Once the character is sent, the line must be returned to a mark, or idle line one condition. The stop bit or bits must be ones to generate the mark frequency. If two stop bits and even parity are used, the ASCII characters A and 7 are transmitted as shown in Table 6.2.

6.9 RS232C INTERFACE

To interconnect any one of the many modems on the market to any one of the computer systems made today, a standard interface connector was devised by the Electronics Institute of America (EIA). This standard, the RS232C, designates for the manufacturers the connector type to be used, the levels of the signals involved, pin designations of the connector, and other various electronic parameters. Note that RS stands for "recommended standard." A manufacturer is not obliged to use it, although it is certainly advantageous to do so.

The most significant designations of the interface deal with data lines and communications link signals. The serial data lines are transmit data (TxDATA) and receive data (RxDATA). (The actual interface standard uses other lettered designations, but for purposes of discussion, the more common and direct abbreviations will be used.)

The RS232C data specifications require that the voltage level for a mark data (logic one) lie between −5 V and −15 V from the transmitter and be at least −3 V into the receiver of the modem. Space data (logic zero) voltages are specified as between +5 V and +15 V from the transmitter and at least +3 V to the receiver. Thus, the data lines are using negative logic since the logic one level is more negative than the logic zero level.

For the control lines, the voltage levels have the same value, but positive logic is used. The on or logic one condition of a control lead uses positive voltages, and

Table 6.2 Asynchronous ASCII Code

Bit	10	9	8	7	6	5	4	3	2	1	0
Character	Stop		Par.	MSB-----Code------LSB							Start
A	1	1	0	1	0	0	0	0	0	1	0
7	1	1	1	0	1	1	0	1	1	1	0

the off or logic zero levels are negative. Two important signals are request to send (RTS), sent by the computer to the modem, and clear to send (CTS), sent in reply from the modem to the computer. RTS begins the process by which a data link is established. The modem, after a short delay, responds by sending CTS. These two signals act to establish the local data link between the computer and the modem. After these signals are set active, a mark tone is generated on the telephone lines.

When a receiver detects the mark tone sent as a result of the distant station coming on line (RTS and CTS made active), it responds with a carrier detect (CD) signal to its computer. This tells the computer that a data link is completed between it and the distant station.

Data terminal ready and data set ready (DTR and DSR) are interlocked-type control signals that either tell the modem or data set that the computer or terminal is connected to it (DTR is active) or tell the computer that the modem is connected to it (DSR is active).

Other lines of the RS232C interface include grounds and common signal returns as well as secondary channel control and data lines. Within the specification, the maximum voltage that can be applied to any lead without causing damage is limited to ±25 V. Other limitations imposed by the standard include cable length between the computer terminal and the modem (50 ft), cable impedance (300 Ω minimum for drivers and 7000 Ω maximum for loads), maximum capacitance per foot (50 pf for a maximum total of 2500 pf), and pin designations within a specified 25-pin connector.

6.10 UNIVERSAL ASYNCHRONOUS RECEIVER/TRANSMITTER—THE UART

The job of converting parallel data from the terminal to serial data used by the modem and ultimately by the telephone lines in order to transfer the message to a remote site lies within a device called a UART (universal asynchronous receiver/transmitter). Parallel data are loaded into the UART from the computer's parallel data bus. In the UART, the data are moved to a shift register where the start, stop, and parity bits are first inserted in their correct locations. The completed data character is then shifted out along a single serial data line. Upon completion of the shift process, the next word is transferred to the shift register and the process is repeated.

On the receive side of the UART, the serial data enter when the start bit is detected. They are shifted into the shift register on the receive side and the parity bit is checked. If parity is all right, the character bits are separated from the data stream and stored in a receive data buffer. The computer then reads the parallel data output of the buffer register as the next received word is shifted into the shift register.

Figure 6.2 illustrates, in block form, the entire communications link for two stations using FSK modems and telephone lines.

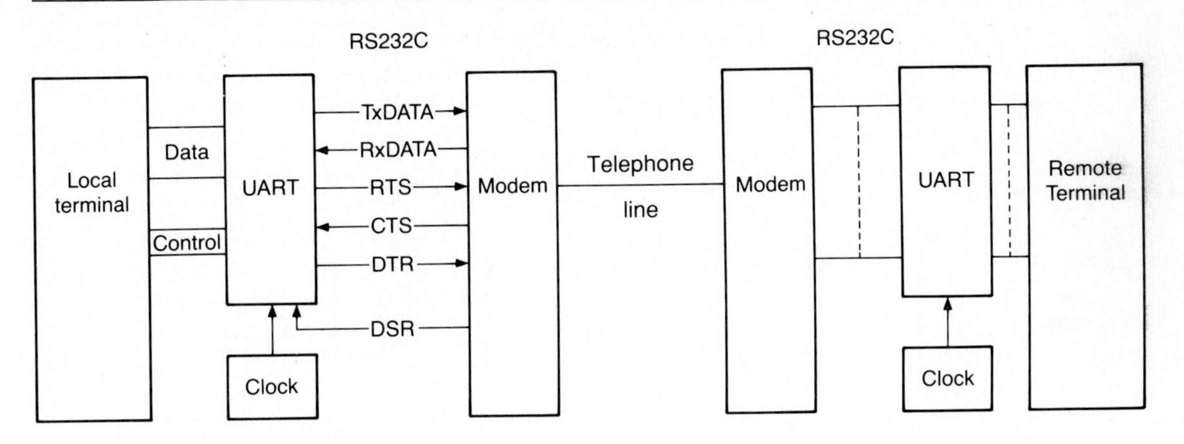

TXDATA RXDATA Data Link

Figure 6.2 Data link

6.11 THE ACIA (ASYNCHRONOUS COMMUNICATIONS INTERFACE ADAPTER)

Motorola's version of the UART is the ACIA (asynchronous communications interface adapter), pictured in Figure 6.3 on page 190. It performs all the functions of a general-purpose UART including serial-to-parallel and parallel-to-serial conversions, insertion of framing and parity bits, and error detection. Discussion of this chip is prefaced by a brief description of the signal pins and their functions.

ACIA Pin Descriptions

D_0–D_7 Pins D_0–D_7 are the standard 6800/68000-family data bus that carries data between the microprocessor and the ACIA.

CS_0, $\overline{CS_1}$, CS_2 These are the ACIA's chip selects used for address decoding.

RS (Register Select) and R/$\overline{W}$ (Read/Write Control) These two control signals combine to select which of four registers is accessed when the ACIA is addressed. Table 6.3 (page 190) is a truth table for these control signals. They are normally connected to the lower address bus and/or $\overline{UDS}$ or $\overline{LDS}$.

The control and status registers share the same address (RS is low for both), while the two data registers are accessed when RS is high. The registers are separated by the type of operation being performed. The control and transmit data registers are write-only registers (R/$\overline{W}$ low) while the status and receive data registers are read-only registers.

E (Enable) This enable input is used by the ACIA for internal timing. As with the PIA, it is connected to the 68000 E output clock.

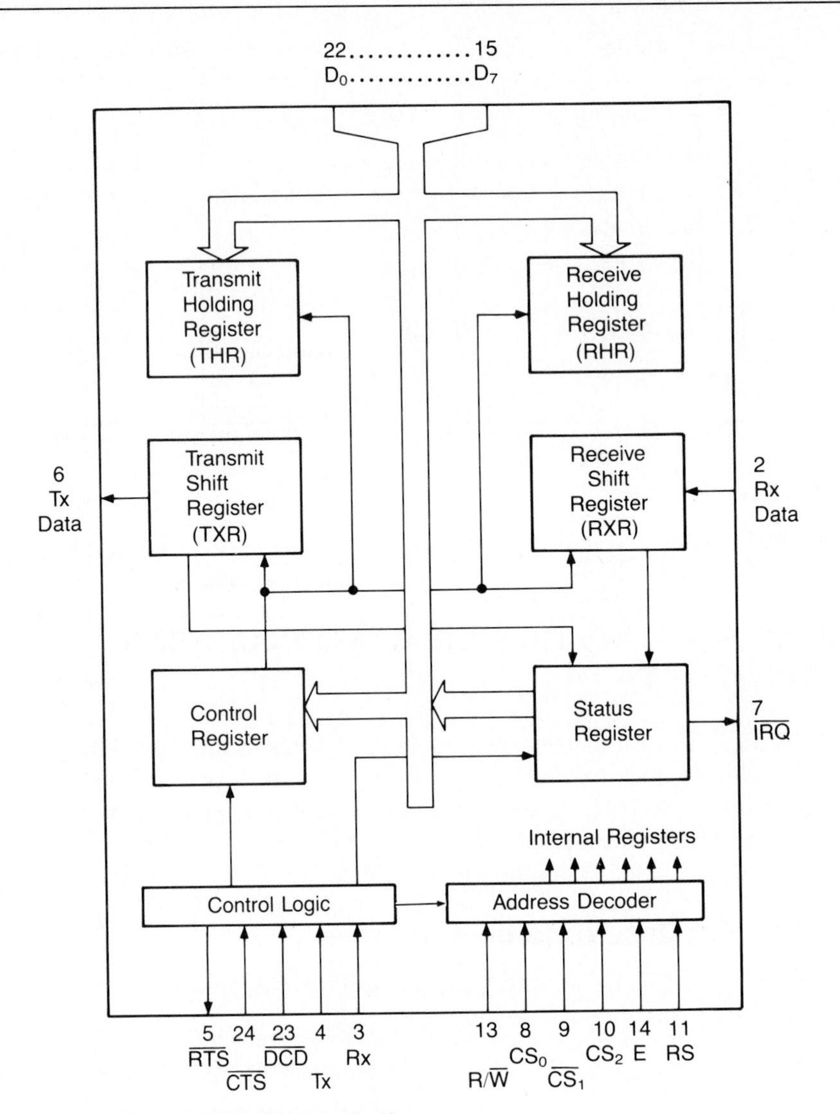

Figure 6.3 **6850 ACIA block diagram**

Table 6.3 Register Select Truth Table

RS	$R/\overline{W}$	Register
0	0	Control
0	1	Status
1	0	Transmit
1	1	Receive

$\overline{\text{IRQ}}$ (Interrupt Request) $\overline{\text{IRQ}}$ sends an interrupt request to the 68000 interrupt priority logic when ACIA interrupts are enabled.

TxDATA and RxDATA These are the actual RS232C-designated serial data lines that handle data transfers between the ACIA and the modem. Note that even though the ACIA uses RS232C designations, the voltage and current levels are those developed by NMOS (n-channel metal oxide semiconductor) technology, which uses 0 V and 5 V logic levels and does *not* conform to the standard. Users are thereby advised that RS232C interface drivers—such as the National Semiconductor's transmission line driver (DS1488) and receiver (DS1489)—which convert between TTL levels (0 and 5 V) and RS232C levels are required between these lines and their equivalent modem connections.

TxCLOCK This input sets up the bit rate of the transmitted data. The data are sent at a bit rate dependent on the clock frequency and a selected divide counter factor. This factor can be set to divide the clock frequency by 1, 16, or 64.

RxCLOCK This input determines at what rate the receive data are shifted into and sampled by the ACIA. The rate depends on the frequency of the receive clock and the same divide counter factor selected for the transmit side. Note that while the factor is set the same for both, each side (transmit and receive) could have different clock frequencies. The rate at which data are transmitted and received then will be different. However, this is not normally the case. Data are more commonly sent and received at the same bit rates.

Data Link Control Leads The remaining leads are control leads based on the RS232C standard. The caution note applies here as it did with TxDATA and RxDATA.

$\overline{\text{RTS}}$ (Request to Send) and $\overline{\text{CTS}}$ (Clear to Send) $\overline{\text{RTS}}$ is generated by the computer and sent to the modem, asking the modem if it is ready to send and receive data. In response to $\overline{\text{RTS}}$ and after a short delay, the modem returns a $\overline{\text{CTS}}$, informing the computer that data can now be sent to it.

$\overline{\text{DCD}}$ (Data Carrier Detect) Once communications are established, a tone is sent on the lines from one station to the other and detected by the distant modem. In turn, the modem generates a carrier detect signal to the computer, informing it that the communications link is fully established and data transfer may begin. This $\overline{\text{DCD}}$ signal is the last in a sequence of handshaking to establish the data communications link over telephone lines. For a more detailed discussion of data and **digital communications** links, refer to any of the data communications books listed in the bibliography at the end of the text.

6.12 USING THE ACIA

To allow the ACIAs to be used in any system, begin by assigning addresses to the two ACIAs: Connect ACIA 1 to the upper half (D_8–D_{15}) of the data bus and ACIA 2 to the lower half (D_0–D_7). A_{18} and A_{17} are made high to isolate the ACIA from

Table 6.4 ACIA Memory Map

Address Lines																								
23	22	21	20	19	18	17	16	15	14	13	12	11	10	9	8	7	6	5	4	3	2	1	0	Device
					1	1	1															R	U	ACIA 1
					1	1	1															R	L	ACIA 2

R = Register Select U = $\overline{UDS}$ L = $\overline{LDS}$

all memory chips. Next, A_{16} is made active high for each ACIA to avoid contention between the ACIAs and PIAs. As with the PIA chips, the address decoder is used by connecting the appropriate output (O_7) to one of the ACIA's chip selects. The RS input of the ACIA is connected to A_1 to provide register selection via the ACIA addresses. Finally, $\overline{UDS}$ is wired to ACIA 1 and $\overline{LDS}$ to ACIA 2. To make the memory map complete, add the line shown in Table 6.4 (which summarizes the ACIA connections) to Table 6.5.

There are four addresses related to the ACIAs that are formulated by applying the different possible logic combinations to the ACIA. Those addresses and their labels are summarized in Table 6.5.

Figure 6.4 illustrates the interconnections between a pair of ACIAs and the 68000 basic system described in preceding chapters.

The address decoder (74138) added to handle the decoding for the PIAs is extended to supply a chip select for the ACIAs from output $\overline{O_7}$. The inverter is necessary to convert the active low output from the decoder to the active high chip select of the ACIAs. Notice again that the decoder is enabled by $\overline{VMA}$ instead of $\overline{AS}$ since 6800 peripheral devices are being used. The E signal and R/$\overline{W}$ lines are connected directly to both chips to provide the necessary timing and control functions. A separate active low chip select on each chip is connected to the data strobes.

On the peripheral side of each ACIA, a Motorola 6860 modem is connected directly. The modems are, in turn, connected through a local switching network to the telephone lines. Data from the upper half of the data bus are routed through ACIA 1 to the telephone line and, from the lower data bus, through ACIA 2 to the telephone line.

Table 6.5 ACIA Address Labels

Address	Label	ACIA Register
07 0000	CONST1	ACIA 1 Control/Status
07 0001	CONST2	ACIA 2 Control/Status
07 0002	SDATA1	ACIA 1 Data
07 0003	SDATA2	ACIA 2 Data

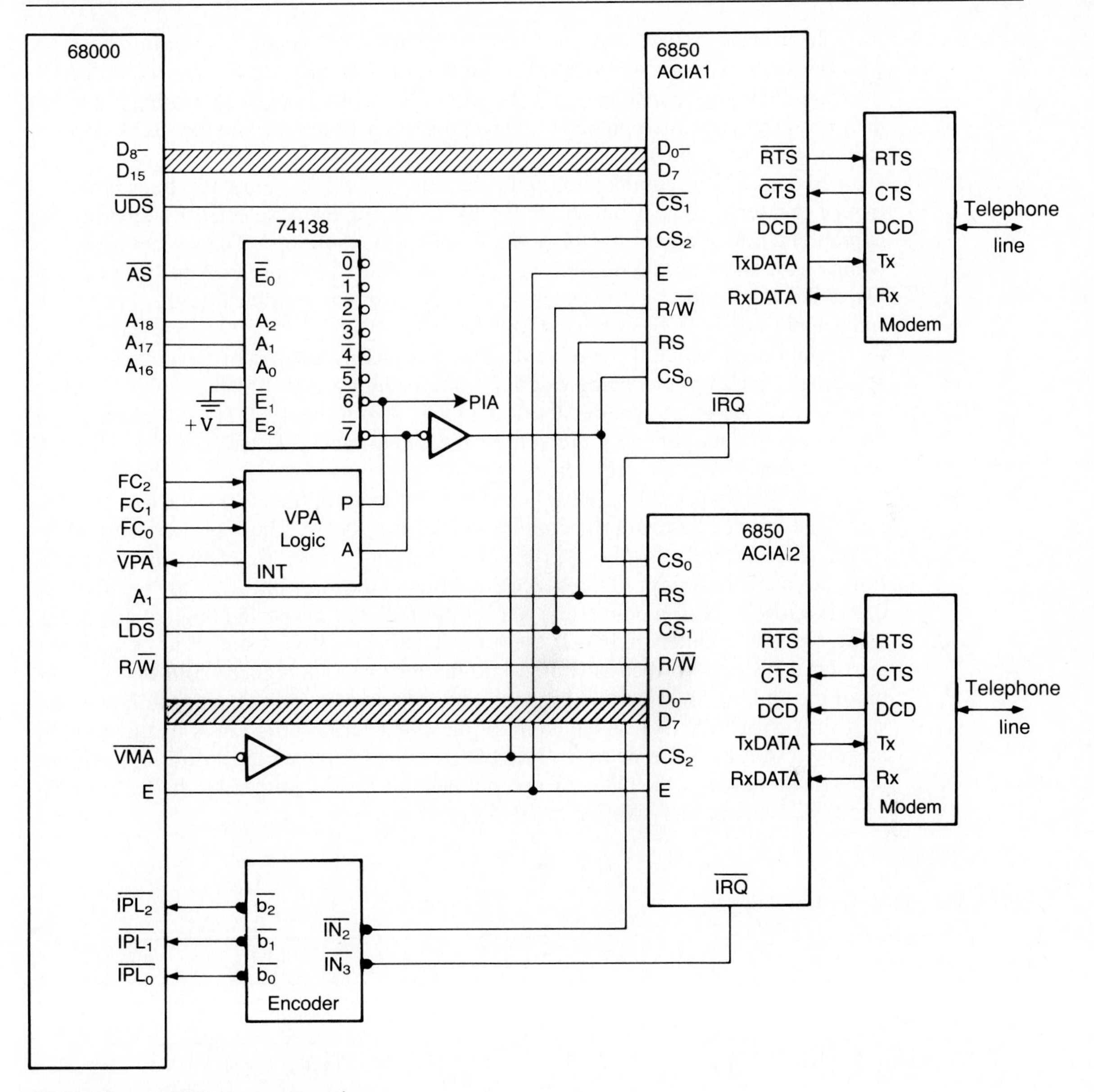

Figure 6.4 **ACIA interconnections**

Additional logic circuitry is required to generate the $\overline{VPA}$ signal because of its alternate use as an autovector indicator in response to an interrupt acknowledge (IACK). Also, the interrupt requests ($\overline{IRQ}$) from the ACIAs, along with those from the PIAs, are directed through a priority encoder chip to the $\overline{IPL}$ inputs of the 68000. Details on interrupt handling are presented in Chapter 8.

To use the ACIA, one must first configure it to handle a specific data type. Choices such as character code size, number of stop bits, odd, even, or no parity, interrupt enabling, and so on, must be made. Selection is done by storing a specific data word from the microprocessor, via a program instruction, into the ACIA's control register (see Table 6.6).

b1 and b0, in combination with the transmit clock, select the transmitted bit rate of the serial data. Counter divide by 1, 16, or 64 is selected by entering the appropriate one and zero combination as shown in Table 6.6. The selection process begins with knowing the desired bit rate of the data. Let's use 300 bps as an example. Selecting divide by 16, which is a fairly common relationship, the TxCLOCK frequency must be 300 × 16, or 4800 Hz.

The rate at which the receive data are sampled is also set by the counter divide selection. Using the preceding example, the incoming data stream at 300 bps using a counter divide of 16 requires the RxCLOCK also to be 4800 Hz. Since incoming and outgoing data generally are sent at the same rate, TxCLOCK and RxCLOCK can be connected to the same clock signal.

Why do we use this relationship instead of making the clocks equal to the data rate? The answer lies with the receive side of the ACIA. The input Rx data line is monitored constantly for the start bit of a character word. To ensure that a change from the mark condition to the space condition is the start bit, after the receive side detects a low level, it monitors the start bit for half its time period (eight clock pulses using a divide counter of 16). If the level is still low, it is assumed to be a correct start bit. The start bit then is shifted into the receive data register, followed by each bit of the data word, which is shifted in on every sixteenth clock pulse following the start shift pulse. That is, each data bit of the character is shifted in at its midpoint by creating a shift pulse for every 16 RxCLOCK pulses. Figure 6.5 illustrates the timing relationship between the data stream and the shift pulses created by the RxCLOCK. The ASCII character A is used for this illustration.

Table 6.6 ACIA Control Register

7	6	5	4	3	2	1	0
Receive Interrupt	RTS	Transmit Interrupt	Word Size	Stop and Parity		Counter Divide	
0 = Disable	0 = On	0 = Disable 1 = Enable	0 = 7 bit	0 = 2 stop 1 = 1	0 = Even 1 = Odd	0 0 = ÷1 0 1 = ÷16 1 0 = ÷64 1 1 = Reset	
1 = Enable	1 = Transmit Interrupt Enable	0 = RTS off 1 = RTS on Transmit Break	1 = 8 bit	0 = No parity	0 = 2 stop 1 = 1 stop		
				1 = 1 stop	0 = Even 1 = Odd		

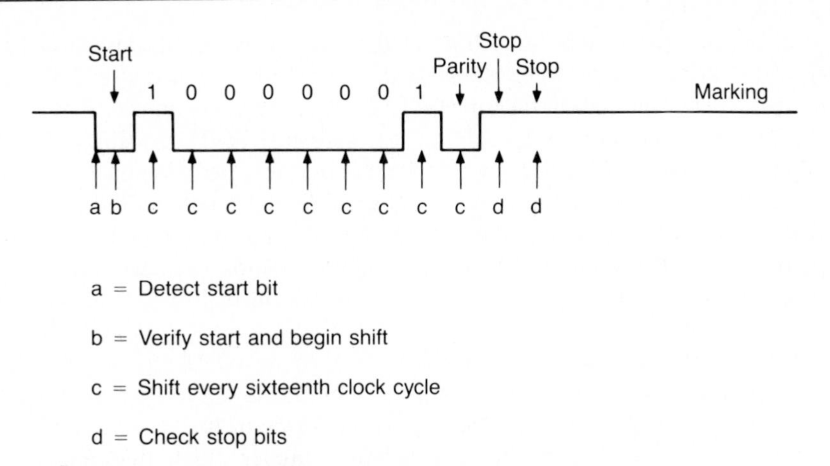

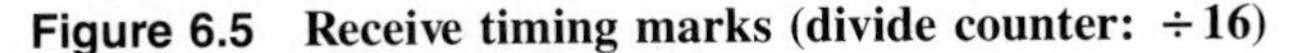

Figure 6.5 Receive timing marks (divide counter: ÷16)

By shifting in each bit at its center, the receive clock does not need to have precisely the same frequency as the transmit clock that generated the original data. It would be fairly difficult for these clocks, one at the originating station and one at the remote site, to have exactly the same frequency. They are as close as possible, but rarely *exactly* the same. The receive clock can be significantly different from the transmit clock without loss of any data because the shift process is reinitialized with every start bit. As long as the clock frequency allows the shifting in of the entire character word, it has done its job. The following example shows how far off a receive clock can be and still successfully shift in the data. The example uses a 300-bps data rate to transmit a full ASCII message (each character has 7 bits, a parity bit, 1 start bit, and 2 stop bits—for a total of 11 bits per character). The ideal receive clock using a counter divide of 16 is 4800 Hz.

At 300 bps, the time of a digital bit is 3.333 ms. The total time from the start shift pulse to the last stop pulse is 10 times the time of a digital data bit period (the first half of the start and last half of the second stop bit fall beyond the shift pulses) or 33.33 ms, as illustrated in Figure 6.6. To ensure that the completed data are shifted in, the last stop bit also must be included. This is successfully done as

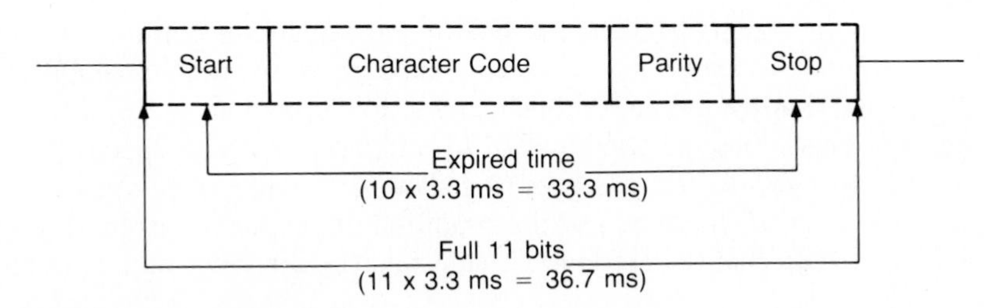

Figure 6.6 ASCII time measurement at 300 bps

long as the last shift pulse occurs at the same time as the last stop bit. In Figure 6.6, the time limits between the start shift pulse and the last stop shift pulse are also shown. Since one half a bit time is 1.667 ms, the total time is either 31.667 ms at the shortest (beginning edge of the last stop bit) or 35 ms at the longest (trailing edge of the last stop bit). There are 16 clock pulses between each shift pulse, so for 10 shift pulses, there is a total of 160 receive clock pulses. Dividing the two time periods by the number of pulses results in the period of each clock pulse. These periods are 198 μs and 218.8 μs. Taking their reciprocals, we find the range for the receive clock to shift ASCII character information in full is from 4570 Hz to 5050 Hz at 300 bps.

Dividing the clock frequency by 64 instead of 16 divides the sampling rate into smaller increments. This allows for more accuracy in detecting and shifting the data bits. This also provides for larger variations in the receive and transmit clock frequencies. Note that in this case a much higher clock frequency is required to generate a given bit rate. For 300 bps, the transmit and receive clocks must be 19.2 kHz. This is not necessarily an advantage or a disadvantage.

Interestingly enough, the originate station can use a counter divide of 16 to send data and sample the received data, while the remote site can use a counter divide of 64. The originate station would have to use a 4800-Hz clock frequency and the remote site a 19.2-kHz clock frequency to communicate successfully using this scheme at 300 bps.

The divide by one feature allows communications at considerably higher bit rates using synchronous data. The use of synchronous data eliminates the need to detect a start bit. Synchronization is done by generating a clock from the data stream itself. Typical bit rates that use synchronous data are 4800, 9600, and faster.

The last combination of ones and zeros for control registers b_1 and b_0 is 11. This generates a master reset of the ACIA. Initially, the ACIA is reset internally when power is applied to the chip. This causes the $\overline{IRQ}$ and $\overline{RTS}$ signals to be made inactive (high), preventing an interrupt from being sent to the 68000 or a request to send ($\overline{RTS}$) from being sent to the modem. These signals are made high by setting their corresponding bits in the status register high as a result of a power on reset. The flags of the status register (see Table 6.7) all are made inactive by master reset except $\overline{CTS}$. The $\overline{CTS}$ flag responds to the CTS signal from the modem. Since $\overline{RTS}$ is made inactive by the reset, $\overline{CTS}$ also becomes inactive. It does this because the modem, in response to an inactive RTS, makes CTS inactive. The status flag reflects this condition.

To help you understand the use of the remaining control bits, let us use an actual example to set up the ACIA to handle an asynchronous ASCII character with even parity, transmitted at 300 bps using a divide counter of 16. The divide counter selection determines the condition of b_1 and b_0: 0 and 1, respectively.

b_2, b_3, and b_4 are assigned together to provide selection of character code length, number of stop bits, and the condition of the parity system. For this example, there are 7 data bits (b_4 set low), 2 stop bits (b_3 set low), and even parity (b_2 also set low).

b_5 and b_6 of the control register turn $\overline{RTS}$ on or off and enable or disable the transmit interrupt request. The interrupt request is generated by a transmit buffer

Table 6.7 ACIA Status Register

b7	b6	b5	b4	b3	b2	b1	b0
$\overline{IRQ}$	PE	OVRN	FE	$\overline{CTS}$	$\overline{DCD}$	TDRE	RDRF

RDRF = Receive Data Register Full
TDRE = Transmit Data Register Empty
DCD = Data Carrier Detect
CTS = Clear to Send
FE = Framing Error
OVRN = Overrun
PE = Parity Error
IRQ = Interrupt Request

empty signal. This flag in the status register is set after a character is loaded into the transmit buffer register, transferred to the transmit data register, and shifted out on the TxDATA line. The transmit data register empty (TDRE) goes active after the last bit is shifted out and generates an interrupt request on the $\overline{IRQ}$ line if the transmit interrupt is enabled. For this example, to make $\overline{RTS}$ active and to disable the transmit interrupt, both b_5 and b_6 are set low. A special condition, called a **break** condition, exists if b_5 and b_6 both are set high; it forces the TxDATA line to transmit a steady stream of zeros and is used when a break in transmission is required before the normal end of transmission. It allows the data link to remain active by keeping a tone on the line and indicates that a change or correction is forthcoming.

The last bit of the control register, b_7, is the receive interrupt request enable bit. A high in this bit position enables interrupt requests generated by a loss of audio tones on the line ($\overline{DCD}$—data carrier detect inactive), the occurrence of a receive overrun, or filling the receive buffer. The receive buffer or holding register becomes full after a character is shifted into the receive data register from the Rx data line and transferred to the holding register. An overrun occurs if the 68000 does not read the data in the holding register before the next data word begins to be shifted into the receive data register. In effect, the unread data word is written over by the new word and becomes lost. The receive interrupt is enabled by setting b_7 of the control register high.

The data sent to the control register for this example and the meaning of each selected bit are summarized below:

$b_7 = 1$ Enable receive interrupt requests
$b_6 = 0$ $\overline{RTS}$ active
$b_5 = 0$ Disable transmit interrupt requests
$b_4 = 0$ Character word size is 7 bits
$b_3 = 0$ Selects 2 stop bits
$b_2 = 0$ Selects even parity
$b_1 = 0$ Divide counter is set to divide
$b_0 = 1$ Divide by 16

The instruction needed to store this information into the control register of ACIA 2 in the basic system is

```
MOVE.B #$81,CONST2
```

Before the control register is set to handle data, it should initially be used to reset the ACIA. Realize that power on resets the ACIA, so the reason for resetting it a second time is to allow the ACIA to be reset anytime the program is run after the system has been powered up. It is possible that execution of the program has been completed and the program has been rerun without powering down the system. To reset the system, b_1 and b_0 of the control register are set to ones. b_6 also is set to turn off the $\overline{RTS}$ line, and b_5 and b_7 are made low to disable interrupts. The data sent to the control register becomes 0100 0011 to reset the ACIA. The first two lines of an ACIA program using ACIA 2 of our basic system are

```
MOVE.B #$43,CONST2     * Resets the ACIA
MOVE.B #$81,CONST2     * Configures ACIA 2
```

6.13 DATA LINK USING THE ACIA

Now that the control register is set to configure the ACIA and it is sending an $\overline{RTS}$ to the modem, we can illustrate the sequence of operations for the transfer of data between two stations. Assume that each station has an ACIA configured as described in the preceding paragraphs. Both modems are on and have returned a $\overline{CTS}$ to each station. A telephone call is made by the originating station to the remote site. The answer station comes on line and sends a mark tone back to the originate station. The originate modem detects the mark tone and sets its $\overline{DCD}$ line low. It also returns a mark tone to the remote station, causing its $\overline{DCD}$ line to be made active, which completes the handshaking that establishes the data link between the two stations.

At the originate station, the program is running, causing the microprocessor to read the ACIA status register (MOVE.B CONST2,D1). b3 ($\overline{CTS}$) of the status register is checked first to ensure that the modem is on. b2 ($\overline{DCD}$) is checked next to assure the originate station that the remote station is on the line. Following this, b1 (TDRE—transmit data register empty) is examined; if it is high, a data word is sent to the ACIA using a MOVE.B ⟨data⟩,SDATA2 command. The data are either supplied immediately or more likely retrieved from memory through previous programming. TDRE goes low while the data are loaded into the transmit buffer register, transferred to the transmit data register, and shifted out onto the TxDATA line. The ACIA performs these functions automatically in response to the data loaded into the transmit buffer register ($RS = 1$ and $R/\overline{W} = 0$).

After the last bit is sent, the TDRE returns high. Once it sends the data, the microprocessor keeps reading the status register, waiting for the TDRE bit to return high. When it does, the microprocessor sends the next word and repeats the sequence until all data are sent. The digital serial data on the TxDATA line are sent to the modem, converted to the mark and space tones, and put onto the telephone lines. At the remote site, the tones are received and converted back to digital data. The data are sent on the RxDATA line to the remote station's ACIA.

On the receive side, a steady mark (idle line ones) initially sets and holds the $\overline{DCD}$ line. As long as a tone is present on the line, the $\overline{DCD}$ line remains active. The status register reflects the status of the receive side as well as the transmit side. b_0, the receive data register full bit, initially is low, indicating that no data have been received. b_2, the $\overline{DCD}$ line, indicates by a low condition that a mark tone is on the line and a link is established. b_4, b_5, and b_6 are error status flags. After a word has been received, they detect if there was a framing error (the first stop bit was not detected); a parity error (the received parity did not match the transmitted parity); or a receive overrun (the microprocessor failed to read a data word before the next word is fully received and shifted in). Before data are received, these bits all are low. b_7 of the status register indicates if an interrupt request has been generated by the ACIA. b_7 is made inactive by reading the receive data register (for receive interrupt requests) or loading the transmit data register (transmit interrupt requests).

When the start bit is detected by the receive side of an ACIA, it starts the receive process. The character word is shifted into the receive data register and transferred to the receive buffer register. b_0 of the status register (RDRF) is set and if the receive interrupt is enabled (as in this example), b_7 ($\overline{IRQ}$) is set low and an interrupt request is sent to the microprocessor. The parity bit is checked for correct parity and the parity error flag is set if there is an error. If the first stop bit was detected at the proper time, the framing error is reset. In response to the interrupt, the microprocessor first reads the status register. The goal is to determine what caused the interrupt. Recall that an overrun, a loss of carrier, or receive data register full as well as a full receive data buffer register could cause the interrupt. The microprocessor determines which occurred by checking the various bits of the status register. If no error is present, the microprocessor then reads the received data (MOVE.B SDATA2,D1). This reading causes the $\overline{IRQ}$ b_7 of the status register to be made high, resets the receive data register full b_0, and prevents the overrun bit, b_5, from being set. The microprocessor then stores the data and awaits the next interrupt request indicating the arrival of the next data word. Keep in mind that data are received as a serial stream at a rate of 300 bps. There is ample time for the microprocessor to execute many instructions between the arrivals of each data word. Naturally, if an error occurs, the microprocessor must execute a different program from the one that would read in the data. That program may do no more than send a break to the sending station followed by a message requesting a retransmission of the data. The sequence of operations to send data from the remote site to the originating site is the same, except that the call must already have been made by the originate station. The remote station can send data back to the originate station as long as the communications link remains established.

6.14 READING THE DATA SYSTEM

An oscilloscope can be used to verify that transfer of serial data is correct, but it is difficult to keep track of the data as they are sent out one at a time. A special type of

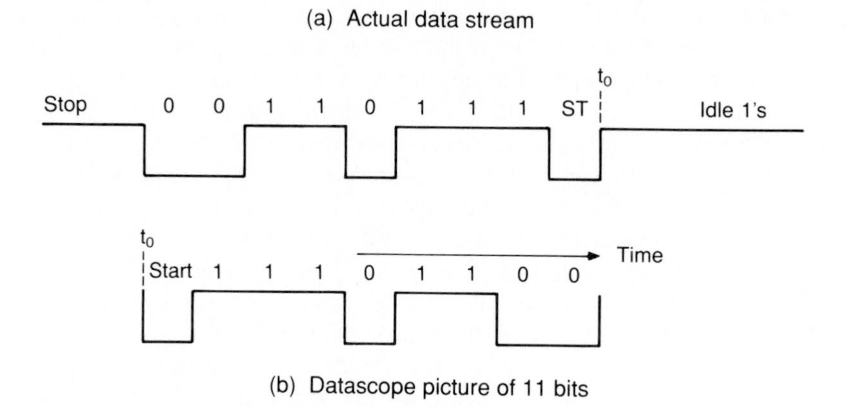

Figure 6.7 Data stream comparisons

test equipment called a *datascope* is available to facilitate the reading of serial data. The datascope displays a serial stream of data as it is applied to the vertical input of the datascope. The first bit sent to it is the start bit. This is followed by the LSB of the character word and the remaining bits. The data stream then is displayed in the reverse order from what was transmitted. The start bit is printed at the leftmost edge of the display screen because it was the first bit read by the scope. The LSB follows, along with the remaining bits of the character word. Figure 6.7 is a comparison of the data stream as it leaves the TxDATA line of the ACIA with the image displayed on the datascope.

6.15 THE 68681 DUART (DUAL UNIVERSAL ASYNCHRONOUS RECEIVER/TRANSMITTER)

Motorola has developed a dual universal asynchronous receiver/transmitter (DUART), the MC68681, as a support chip for the 68000. As shown in Figure 6.8, the data bus size to the DUART is eight bits, allowing two support chips to be connected to the 16-bit bus of the 68000. Each DUART contains independent full duplex data channels (A and B) capable of transferring serial data at any one of 18 fixed bauds from 50 to 38,400. Like most UARTs, data size (from five to eight bits), parity, and the number of stop bits (1, 1.5, or 2) are programmable. There is a 6-bit general-purpose input port, an 8-bit general-purpose output port, and a counter/timer. Additionally, the input and output ports can be programmed for special functions. Internally, each section has command and status registers in addition to its interface registers. The interrupt control area includes a vector register as well.

Pin Descriptions

X1/CLK and X2 There is an internal clock circuit in the DUART that requires a crystal to be connected to X1 and X2. To achieve standard data baud rates from the

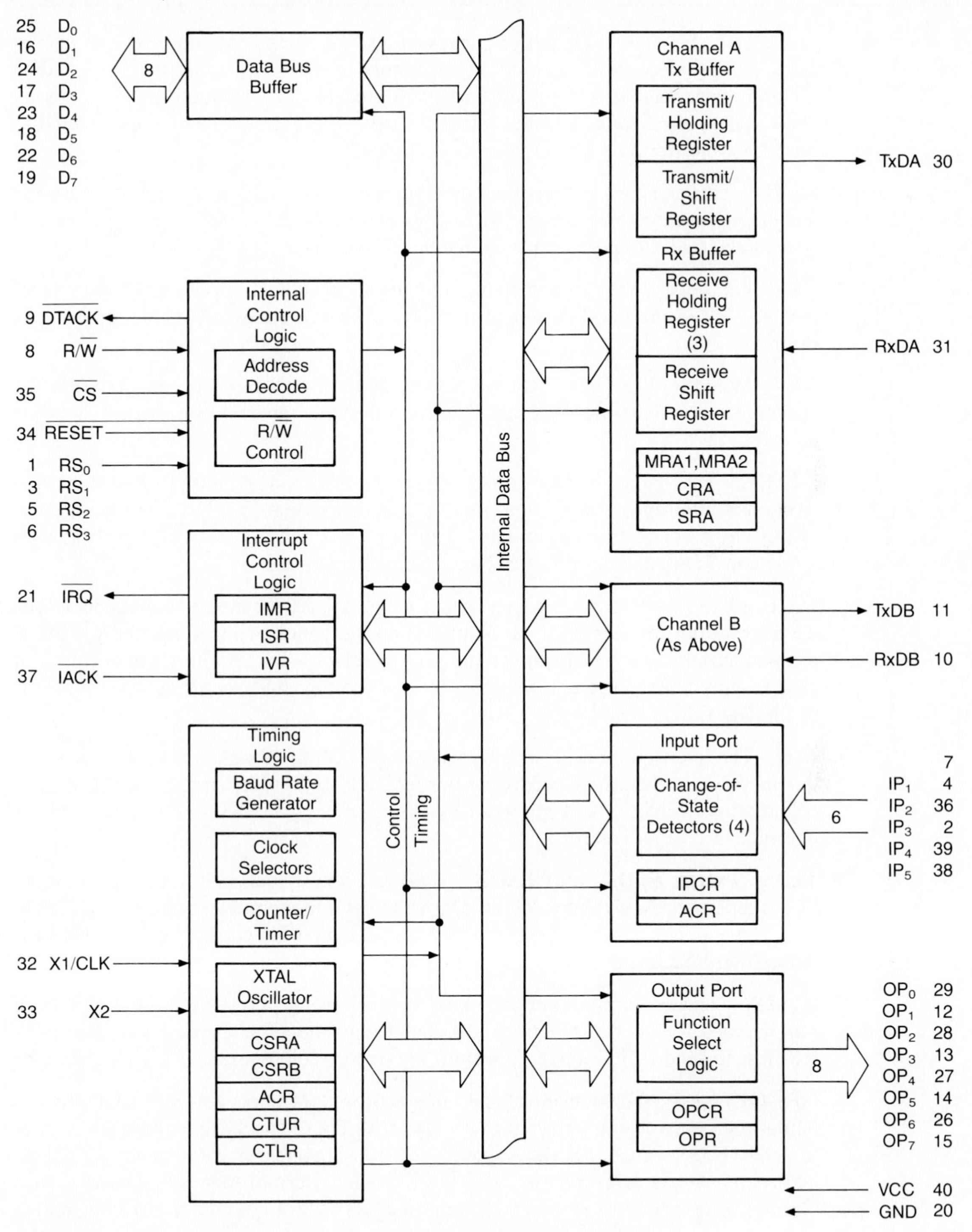

Figure 6.8 **DUART blocks and pin designations (Courtesy of Motorola, Inc.)**

two UART channels, a crystal frequency of 3.6864 MHz is required. If the internal clock generator is used, decoupling capacitors of 10–15 pf must be connected to X1 and capacitors of 0–5 pf must be connected to X2. Alternatively, an external clock can be used instead of the internal clock. The external clock is connected to X1/CLK, and the X2 input is returned to ground.

D_0–D_7 This standard bidirectional data bus carries all parallel information between the microprocessor and the DUART. This information includes data, control and status data, and interrupt vector numbers.

$\overline{CS}$ Chip select lets you select the chip from an address decoder. Outputs from the DUART to the 68000 are held in high-impedance states when this input is not asserted.

R/$\overline{W}$ When R/$\overline{W}$ is high, data are placed onto the data bus and transferred to the microprocessor. When R/$\overline{W}$ is low, data from the microprocessor are written into the DUART.

$\overline{DTACK}$ This signal is returned to the microprocessor in response to a normal read, write, and interrupt acknowledge (IACK) cycle involving the DUART chip. Because of this direct connection, additional circuitry to generate the $\overline{DTACK}$ signal (as with RAM and ROM discussed in Chapter 2) is not required.

$\overline{IACK}$ The interrupt acknowledge input to the DUART causes the contents of the interrupt vector register (IVR) to be placed on the data bus and generates a $\overline{DTACK}$ to the microprocessor. This allows the user to select a specific interrupt vector (and hence, a particular interrupt program) by first loading that vector number into IVR of the DUART in response to $\overline{IACK}$.

$\overline{IRQ}$ The interrupt request signal is asserted as a result of eight maskable conditions that include receive errors, receive buffer full, transmit buffer empty, data break, and counter/timer ready. These allow all functions of the DUART to be handled under interrupt or polling control.

TxDA, TxDB, RxDA, and RxDB These are the actual serial data lines output (TxD) and input (RxD) between the DUART and a modem connected to each channel. The modem can be connected directly if it is TTL compatible or connected through an RS232C buffer.

IP_0–IP_5 This 6-bit input port can function as a general-purpose input port or can be programmed for special functions associated with the serial ports. These functions are clear to send (CTS) inputs, transmit, and receive clock inputs for both channels.

OP_0–OP_7 Like the 6-bit input port, this 8-bit output port can serve as a general-output data port or else carry special functions. These special functions are request to send (RTS) signals for the modems of each channel, transmit ready (TxRDY) and receive ready or buffer full (RxRDY/FFULL) interrupt request signals for each channel, transmit or receive clock output signals, and a counter ready/timer output signal.

$\overline{\text{RESET}}$ The reset input to the DUART initializes the device specifically by the following:

1. Clearing the status registers (SRA and SRB) for both serial channels.
2. Clearing the interrupt mask (IMR) and status (ISR) registers. This clears current pending interrupts and masks future interrupts.
3. Clearing the output port (OPR) and output port configuration (OPCR) registers, removing any selected special functions.
4. Loading the data $0F into the interrupt vector register (IVR).
5. Setting OP_0–OP_3 high.
6. Placing the counter/timer into the timer mode.
7. Putting channels A and B into inactive states and generating a high level for idle line 1's on both serial transmit data lines (TxDA and TxDB).

RS_0–RS_3 These register select inputs allow the programmer to access the internal registers of the DUART via the address bus of the 68000. Table 6.8 lists the registers and the corresponding states of RS_0–RS_3.

Operation of the DUART

Before the DUART can handle data transfers, it must be programmed. Some functions are identical to the ACIA in that the nature of the data character, data sampling rates (X_1 or X_{16}), and interrupt status must be selected. Additionally, choices concerning

Table 6.8 Register Addressing and Address Triggered Demands (Courtesy of Motorola, Inc.)

RS_3	RS_2	RS_1	RS_0	Read ($R/\overline{W}=1$)	Write ($R/\overline{W}=0$)
0	0	0	0	Mode Register A (MR1A, MR2A)	Mode Register A (MR1A, MR2A)
0	0	0	1	Status Register A (SRA)	Clock-Select Register A (CSRA)
0	0	1	0	Do Not Access*	Command Register A (CRA)
0	0	1	1	Receiver Buffer A (RBA)	Transmitter Buffer A (TBA)
0	1	0	0	Input Port Change Register (IPCR)	Auxiliary Control Register (ACR)
0	1	0	1	Interrupt Status Register (ISR)	Interrupt Mask Register (IMR)
0	1	1	0	Counter Mode: Current MSB of Counter (CUR)	Counter/Timer Upper Register (CTUR)
0	1	1	1	Counter Mode: Current LSB of Counter (CLR)	Counter/Timer Lower Register (CTLR)
1	0	0	0	Mode Register B (MR1B, MR2B)	Mode Register B (MR1B, MR2B)
1	0	0	1	Status Register B (SRB)	Clock-Select Register B (CSRB)
1	0	1	0	Do Not Access*	Command Register B (CRB)
1	0	1	1	Receiver Buffer B (RBB)	Transmitter Buffer B (TBB)
1	1	0	0	Interrupt-Vector Register (IVR)	Interrupt-Vector Register (IVR)
1	1	0	1	Input Port (Unlatched)	Output Port Configuration Register (OPCR)
1	1	1	0	Start-Counter Command**	Output Port Register (OPR): Bit Set Command**
1	1	1	1	Stop-Counter Command**	Output Port Register (OPR): Bit Reset Command**

*This address location is used for factory testing of the DUART and should not be read. Reading this location will result in undesired effects and possible incorrect transmission or reception of characters. Register contents may also be changed.

** Address triggered commands.

port configuration and interrupt vector numbers must be made. Numerous registers in the chip allow these selections to be made and provide a way to monitor the status of the device and pass data to and from the microprocessor.

Mode Selection Two mode registers (MR_1 and MR_2) are available for each channel, as illustrated in Table 6.9. They select the operating mode and type of data character to be handled by each channel. The number of bits per character is selected by setting b_1 and b_0 of MR_1 to the appropriate states. Parity selection is made by setting the conditions of b_2–b_4 to one or zero. Making b_3 and b_4 low includes the parity bit with the data character. b_2 then allows selection of odd or even parity. The force parity mode ($b_4 = 0$ and $b_3 = 1$) causes the parity bit to be always either high ($b_2 = 1$) or low ($b_2 = 0$). Setting b_4 to a one and b_3 low selects no parity, causing b_2 to become

Table 6.9 Mode Registers (Courtesy of Motorola, Inc.)

CHANNEL A MODE REGISTER 1 (MR1A) AND CHANNEL B MODE REGISTER 1 (MR1B)

Rx RTS Control	Rx IRQ Select	Error Mode	Parity Mode	Parity Type	Bits-per-Character
Bit 7	Bit 6	Bit 5	Bit 4, Bit 3	Bit 2	Bit 1, Bit 0
0 = Disabled 1 = Enabled	0 = RxRDY 1 = FFULL	0 = Char 1 = Block	0 0 = With Parity 0 1 = Force Parity 1 0 = No Parity 1 1 = Multidrop Mode*	With Parity 0 = Even 1 = Odd Force Parity 0 = Low 1 = High Multidrop Mode 0 = Data 1 = Address	0 0 = 5 0 1 = 6 1 0 = 7 1 1 = 8

*The parity bit is used as the address/data bit in multidrop mode.

CHANNEL A MODE REGISTER 2 (MR2A) AND CHANNEL B MODE REGISTER 2 (MR2B)

Channel Mode	Tx RTS Control	CTS Enable Transmitter	Stop Bit Length
Bit 7, Bit 6	Bit 5	Bit 4	Bit 3, Bit 2, Bit 1, Bit 0
0 0 = Normal 0 1 = Automatic Echo 1 0 = Local Loopback 1 1 = Remote Loopback	0 = Disabled 1 = Enabled	0 = Disabled 1 = Enabled	see below

Bits 3–0	6-8 Bits/Character	5-Bits/Character
(0) 0 0 0 0 =	0.563	1.063
(1) 0 0 0 1 =	0.625	1.125
(2) 0 0 1 0 =	0.688	1.188
(3) 0 0 1 1 =	0.750	1.250
(4) 0 1 0 0 =	0.813	1.313
(5) 0 1 0 1 =	0.875	1.375
(6) 0 1 1 0 =	0.938	1.438
(7) 0 1 1 1 =	1.000	1.500
(8) 1 0 0 0 =	1.563	1.563
(9) 1 0 0 1 =	1.625	1.625
(A) 1 0 1 0 =	1.688	1.688
(B) 1 0 1 1 =	1.750	1.750
(C) 1 1 0 0 =	1.813	1.813
(D) 1 1 0 1 =	1.875	1.875
(E) 1 1 1 0 =	1.938	1.938
(F) 1 1 1 1 =	2.000	2.000

NOTE:
If an external 1X clock is used for the transmitter, MR2 bit 3 = 0 selects one stop bit and MR2 bit 3 = 1 selects two stop bits to be transmitted.

a "don't care" bit. If both b_4 and b_3 are made high, a special mode called *multidrop* is selected, in which case the parity bit becomes an address or data bit as selected by b_2.

Error mode selection (b_5) provides a choice of error detection by character (check each parity) or by block. In error detection by block, parity error is ignored and the last character in the message is checked as an error character. This character is called a *block check character* (BCC) or *frame check sequence* (FCS). Enabling or disabling the receive interrupt request and receive RTS signal is handled using b_6 and b_7, respectively.

The length of the stop bit is selected using b_1 and b_0 of MR_2. Standard lengths of 1, 1.5, and 2 are among the values available. Transmit RTS and enable on CTS control are enabled or disabled by b_5 and b_4, respectively. The RTS signals begin the handshaking between the DUART and the modem connected to the channel. Additionally, the transmit function can be enabled by the receipt of a CTS signal from the modem, completing the local handshake.

Channel mode selection is done using the last two MSBs of MR_2. The normal mode (both bits low) allows full duplex operation of the DUART channel. Parallel data are fed from the microprocessor to the DUART and transmitted out through the Txd line associated with that channel. Receive data are sensed on the Rxd line, shifted in, checked for errors, and sent to the microprocessor on the data bus. The actual parallel transfers are done using normal 68000 instructions. The microprocessor is informed when to perform the transfer by an interrupt generated by the DUART or by polling the status register associated with each channel.

Clock Selection and Command Functions The baud rate for the data transmission is selected by the byte sent to the clock select registers (CSRA and CSRB) for each channel (see Table 6.10, page 206). The standard baud rates from 50 to 38,400 can be selected along with a number of nonstandard rates. The standard rates are generated if a crystal frequency of 3.6864 MHz is in use. Other frequency crystals cause different sets of rates to be generated. The last two combinations for the transmit and receive clock selection allow data rates to be generated by external clock inputs supplied through the input port. These clock inputs are divided by 1 or 16, as selected, to create the desired baud rate.

Individual reset and enabling functions are performed using the command registers of the DUART (see Table 6.11, page 207). The individual reset function initializes that portion of the DUART without affecting the remainder of the device. For a complete system reset, the hardware $\overline{\text{RESET}}$ input must be asserted, causing the chip to be initialized as described earlier.

The command to reset the receiver (b_6, b_5, b_4 = 010) immediately disassembles the receiver and clears the receiver ready (RxRDY) and FIFO full (FFULL) status bits in the corresponding channel's status register. The receive FIFO (RxFIFO) is reinitialized and no other registers are affected by the command. Similarly, the reset transmitter command disables the transmitter and clears the TxRDY and transmit buffer empty (TxEMT) flags in the status register.

Error status bits in the character mode are cleared by the reset error command. These include the received break (RB), parity error (PE), framing error (FE), and

Table 6.10 Clock Select Registers (Courtesy of Motorola, Inc.)

CLOCK-SELECT REGISTER A (CSRA)

Receiver-Clock Select Bit 7 Bit 6 Bit 5 Bit 4	Baud Rate Set 1 ACR Bit 7=0	Baud Rate Set 2 ACR Bit 7=1	Transmitter-Clock Select Bit 3 Bit 2 Bit 1 Bit 0	Baud Rate Set 1 ACR Bit 7=0	Baud Rate Set 2 ACR Bit 7=1
0 0 0 0	50	75	0 0 0 0	50	75
0 0 0 1	110	110	0 0 0 1	110	110
0 0 1 0	134.5	134.5	0 0 1 0	134.5	134.5
0 0 1 1	200	150	0 0 1 1	200	150
0 1 0 0	300	300	0 1 0 0	300	300
0 1 0 1	600	600	0 1 0 1	600	600
0 1 1 0	1200	1200	0 1 1 0	1200	1200
0 1 1 1	1050	2000	0 1 1 1	1050	2000
1 0 0 0	2400	2400	1 0 0 0	2400	2400
1 0 0 1	4800	4800	1 0 0 1	4800	4800
1 0 1 0	7200	1800	1 0 1 0	7200	1800
1 0 1 1	9600	9600	1 0 1 1	9600	9600
1 1 0 0	38.4k	19.2k	1 1 0 0	38.4k	19.2k
1 1 0 1	Timer	Timer	1 1 0 1	Timer	Timer
1 1 1 0	IP4-16X	IP4-16X	1 1 1 0	IP3-16X	IP3-16X
1 1 1 1	IP4-1X	IP4-1X	1 1 1 1	IP3-1X	IP3-1X

NOTE: Receiver clock is always a 16X clock except when CSRA bits seven through four equal 1111.

NOTE: Transmitter clock is always a 16X clock except when CSRA bits three through zero equal 1111.

CLOCK-SELECT REGISTER B (CSRB)

Receiver-Clock Select Bit 7 Bit 6 Bit 5 Bit 4	Baud Rate Set 1 ACR Bit 7=0	Baud Rate Set 2 ACR Bit 7=1	Transmitter-Clock Select Bit 3 Bit 2 Bit 1 Bit 0	Baud Rate Set 1 ACR Bit 7=0	Baud Rate Set 2 ACR Bit 7=1
0 0 0 0	50	75	0 0 0 0	50	75
0 0 0 1	110	110	0 0 0 1	110	110
0 0 1 0	134.5	134.5	0 0 1 0	134.5	134.5
0 0 1 1	200	150	0 0 1 1	200	150
0 1 0 0	300	300	0 1 0 0	300	300
0 1 0 1	600	600	0 1 0 1	600	600
0 1 1 0	1200	1200	0 1 1 0	1200	1200
0 1 1 1	1050	2000	0 1 1 1	1050	2000
1 0 0 0	2400	2400	1 0 0 0	2400	2400
1 0 0 1	4800	4800	1 0 0 1	4800	4800
1 0 1 0	7200	1800	1 0 1 0	7200	1800
1 0 1 1	9600	9600	1 0 1 1	9600	9600
1 1 0 0	38.4k	19.2k	1 1 0 0	38.4k	19.2k
1 1 0 1	Timer	Timer	1 1 0 1	Timer	Timer
1 1 1 0	IP2-16X	IP2-16X	1 1 1 0	IP5-16X	IP5-16X
1 1 1 1	IP2-1X	IP2-1X	1 1 1 1	IP5-1X	IP5-1X

NOTE: Receiver clock is always a 16X clock except when CSRB bits seven through four equal 1111.

NOTE: Transmitter clock is always a 16X clock except when CSRB bits three through zero equal 1111.

Table 6.11 Command Register (Courtesy of Motorola, Inc.)

CHANNEL A COMMAND REGISTER (CRA) AND CHANNEL B COMMAND REGISTER (CRB)

Not Used*	Miscellaneous Commands	Transmitter Commands	Receiver Commands
Bit 7	Bit 6 Bit 5 Bit 4	Bit 3 Bit 2	Bit 1 Bit 0
X	0 0 0 No Command	0 0 No Action, Stays in Present Mode	0 0 No Action, Stays in Present Mode
	0 0 1 Reset MR Pointer to MR1	0 1 Transmitter Enabled	0 1 Receiver Enabled
	0 1 0 Reset Receiver	1 0 Transmitter Disabled	1 0 Receiver Disabled
	0 1 1 Reset Transmitter	1 1 Don't Use, Indeterminate	1 1 Don't Use, Indeterminate
	1 0 0 Reset Error Status		
	1 0 1 Reset Channel's Break-Change Interrupt		
	1 1 0 Start Break		
	1 1 1 Stop Break		

* Bit seven is not used and may be set to either zero or one.

overrun error (OE). In the block mode, all error status bits are cleared after a block of data has been received. The remaining miscellaneous commands deal with break conditions. The first (b_6, b_5, b_4 = 101) resets break change interrupt and clears the channel break change detect interrupt status flag in the status register. The start and stop break commands generate zeros (start) or idle-line ones (stop) the TxDATA line. The low condition is started after two bit times if the transmitter is idle, at the completion of the transmission of a current word if the transmit buffer register is empty, or after the transmission of the character from the transmit buffer register if both the transmit shift and buffer registers contain data. The line returns high after two bit times when the break stop command is issued by the microprocessor.

Three transmit commands (b_3 and b_2) and three receive commands (b_2 and b_1) are available for each channel. These commands are

1. No status change command, which retains the receiver or transmitter in its current state.
2. An enable command.
3. A disable command. Disabling the transmitter occurs after the current transmit character is completed and causes the TxRDY and TxEMT flags to be reset. The receiver, on the other hand, is disabled immediately and no status flags are affected by the disable receive command.

Status Registers The status register (see Table 6.12) contains the receive status flags for receive break, receive ready, framing error, parity error, and overrun error. The error flags are set by the same conditions described for ACIA. Additionally, a FIFO full status bit is included that indicates when the receive FIFO (first in, first out) register bank is full. The microprocessor is required to read a received word from the FIFO before another word is received and entered into the FIFO. If the microprocessor fails to do this, the overrun error flag is set and will generate a receive interrupt request if the interrupts are enabled.

Table 6.12 Status Register (Courtesy of Motorola, Inc.)

CHANNEL A STATUS REGISTER (SRA) AND CHANNEL B STATUS REGISTER (SRB)

Received Break	Framing Error	Parity Error	Overrun Error	TxEMT	TxRDY	FFULL	RxRDY
Bit 7*	Bit 6*	Bit 5*	Bit 4	Bit 3	Bit 2	Bit 1	Bit 0
0 = No 1 = Yes	0 = No 1 = Yes	0 = No 1 = Yes	0 = No 1 = Yes	0 = No 1 = Yes	0 = No 1 = Yes	0 = No 1 = Yes	0 = No 1 = Yes

*These status bits are appended to the corresponding data character in the receive FIFO and are valid only when the RxRDY bit is set. A read of the status register provides these bits (seven through five) from the top of the FIFO together with bits four through zero. These bits are cleared by a reset error status command. In character mode, they are discarded when the corresponding data character is read from the FIFO.

Transmit status bits include TxRDY and TxEMT. The TxRDY indicates that the transmitter is enabled and operational, and the TxEMT flag indicates that the transmit buffer is available to receive the next word from the microprocessor. The TxEMT flag is set as soon as the data in the buffer are transferred to the transmit shift register and the transmission of that character has begun. The flag is cleared by a transmit disable command, by a hardware reset, or when data have been transferred to the transmit buffer register by the microprocessor and the previous word is still being transmitted.

Counter/Timer Functions

The DUART is supplied with a 16-bit counter/timer (CT) whose value is contained in two registers, the CT upper byte register (CTUR) and the CT lower byte register (CTLR). The values that can be preloaded into these registers range from $0002 to $FFFF, and they can be changed at any time. The CT mode and source selection is made by programming b_4, b_5, and b_6 of the auxiliary control register (ACR). A table for these bits is shown in Table 6.13.

b6 selects the counter ($b_6 = 0$) or the timer ($b_6 = 1$) mode. In either mode, an external input to the CT can be supplied from IP_2 of the input register if b5 and b6 of the ACR are set low. The remaining timer and counter sources shown in the table are self-explanatory.

Counter Mode The microprocessor starts the count sequence by issuing a start counter command. The counter counts down from the preloaded value until it reaches $0000 (terminal count). The terminal count (CT) status bit in the interrupt status register is set, and an interrupt is generated on the $\overline{IRQ}$ line or output port pin OP_3. The choice of interrupt request line depends on how the device was programmed. The counter value wraps around to $FFFF and continues counting. The microprocessor can stop the counter by issuing a stop count command. This causes the CT interrupt status to be cleared and stops the count process. The current count value can then be read from CTUR and CTLR. A new value can be placed into these registers at any time, but it will not be recognized by the CT until it receives a new start counter command.

Table 6.13 Auxiliary Control Register (ACR) (Courtesy of Motorola, Inc.)

BRG SET Select*	Counter/Timer Mode and Source**			Delta*** IP3 IRQ	Delta*** IP2 IRQ	Delta*** IP1 IRQ	Delta*** IP0 IRQ
Bit 7	Bit 6	Bit 5	Bit 4	Bit 3	Bit 2	Bit 1	Bit 0
0 = Set 1 1 = Set 2	Bits 6 5 4	Mode	Clock Source	0 = Disabled 1 = Enabled	0 = Disabled 1 = Enabled	0 = Disabled 1 = Enabled	0 = Disabled 1 = Enabled
	0 0 0	Counter	External (IP2)****				
	0 0 1	Counter	TxCA – 1X Clock of Channel A Transmitter				
	0 1 0	Counter	TxCB – 1X Clock of Channel B Transmitter				
	0 1 1	Counter	Crystal or External Clock (X1/CLK) Divided by 16				
	1 0 0	Timer	External (IP2)****				
	1 0 1	Timer	External (IP2) Divided by 16****				
	1 1 0	Timer	Crystal or External Clock (X1/CLK)				
	1 1 1	Timer	Crystal or External Clock (X1/CLK) Divided by 16				

*Should only be changed after both channels have been reset and are disabled.
**Should only be altered while the counter/timer is not in use (i.e., stopped if in counter mode, output and/or interrupt masked if in timer mode).
***Delta is equivalent to change-of-state.
****In these modes, because IP2 is used for the counter/timer clock input, it is not available for use as the channel B receiver-clock input.

Timer Mode In the timer mode, a squarewave that can be output on OP_3 is generated by the CT circuit. This squarewave has a time period that is twice the value loaded into CTUR and CTLR times the time period of the clock source. The timer runs continuously and cannot be stopped by a command from the microprocessor. It can be reinitialized by performing a read using the start counter command address. The timer ceases its current count, inverts its output, and reinitializes itself from the value from CTUR and CTLR. The level of the squarewave is inverted on each terminal count ($0000), at which time the timer reinitializes itself from the value registers and resumes counting. A new value can be preloaded into these registers, but the timer will not recognize it until the timer reaches a terminal count.

DUART Operating Modes

When configured to operate in the normal mode, each DUART channel performs like a standard UART or ACIA. At the selected baud rate, it transmits the character generated by combining the data with start, stop, and parity bits. On the receive side, the incoming data stream is monitored for a start bit (change from idle line ones to a low condition) and the character is shifted in. Framing and parity errors are checked and the data are transferred to the microprocessor by a read of the receive FIFO. The

FIFO itself contains three buffer registers, allowing as many as three characters to be input to the receiver before one is read by the 68000. When all three of the registers are full, FFULL status is set and an interrupt is generated to the microprocessor (assuming this interrupt function is not masked). When the microprocessor fails to read a word from the full FIFO, the overrun error occurs.

Looping Modes Looping modes are provided for troubleshooting purposes. In the local loopback mode, the transmitted output is directed back through the receiver to allow the local operator to check the performance of a specific channel. Data transfers between the microprocessor and both the transmitter and receiver portions of a channel are functional. In the automatic echo mode, received data are transmitted back on a bit-by-bit basis. Receiver-to-microprocessor operations are enabled, while microprocessor-to-transmitter transfers are disabled. In this mode, the sending station receives a copy of the message just transmitted. In the remote loopback mode, the local station is taken out of the loop. Received data are retransmitted to the remote site, which can monitor the data returned. Both receiver and transmitter data transfers with the microprocessor, at the local site, are disabled.

Multidrop Mode In a multidrop system, several remote or secondary stations are connected to a single data channel controlled by a single primary station. The primary station controls the data link, determining with which secondary it will communicate by sending that station's address to the desired secondary station in a polling routine. The secondary station that recognizes its address comes on-line to complete the data channel. In the multidrop mode of the DUART, the receiver continually monitors the incoming data whether or not the receiver has been enabled. The normal parity bit position (A/D) is checked to determine if the character on the line is a data (A/D = 0) or an address (A/D = 1) character. If it is a data character, it is ignored and the receiver continues to monitor the line. If the A/D bit is set, the receiver compares that character with its own address to determine if the primary station is selecting that channel. If that channel is selected, the receiver is enabled and data transfers commence. If the address does not match the channel's address, the receiver resumes monitoring the data stream on the RxDATA line. Since the parity bit is now being used as an address/data flag, error detection is no longer performed by parity. Error detection now requires software to perform an additional function. Either an unused bit assumes the parity role or an additional character is added to the data stream to perform error detection. These block check characters can be created by using any number of error-detecting methods, such as checksum, Hamming codes, or cyclic redundancy checking (CRC). Consult the data communications sources in the bibliography for detailed information on these error codes.

Interrupt Selection Using the DUART

An interrupt request signal to the 68000 from the DUART is generated by a number of functions within the chip. These requests are, in turn, serviced or denied such service (masked). Individual bits in the control registers already discussed enable or disable some of the interrupts. Additional request control is found in the remaining bits of the auxiliary control register (ACR) and the interrupt mask register (IMR). An interrupt status register (ISR) can be used to monitor the interrupt status of any of the requests whether or not they have been masked. This provides a way to detect why an interrupt was generated or to poll the various conditions if interrupts have been masked. The former method allows hardware control (through an interrupt) of the various DUART functions, while the latter lets the programmer maintain software control of the DUART functions.

Interrupt Mask and Status Registers The interrupt mask register (IMR) (see Table 6.14) allows an interrupt request to be generated on the $\overline{IRQ}$ line if a bit in a given position is set. Each individual request is masked by programming a zero into its corresponding bit position in the IMR. IRQs are generated by asserted transmit or receive ready or FIFO full signals. A counter/timer ready signal also can be made to generate an active $\overline{IRQ}$. If a change of state occurs on input port pins 0, 1, 2, or 3, and the input port bit in the IMR is set, an $\overline{IRQ}$ is asserted. Selecting which of these bits causes the interrupt request is done by programming a high into its corresponding bit in the auxiliary control register. For example, to generate an $\overline{IRQ}$ when input port pins 1 and 2 change state, ACR b_1 and b_2 must be set, while b_0 and b_3 must be cleared. Additionally, IMR b_7 also must be set to pass the interrupt requests to the $\overline{IRQ}$ line. The remaining two mask bits in the IMR control interrupt requests generated by a change in the break status for either channel A or B as a result of a start or stop break command.

The interrupt status register (ISR) contains interrupt status flags matching each of the interrupt mask bits in the IMR. These status bits are set as a result of the corresponding condition regardless of the state of its IMR bit. For instance, a change in the status of channel A's break generated by a start break command causes b_2 of the ISR to be set. If, in addition, b_2 of the IMR was previously set to a one, $\overline{IRQ}$ is asserted. As long as the interrupt status bit remains set, an $\overline{IRQ}$ can be generated any time its IMR mask bit is set. The ISR is completely reset by the assertion of $\overline{RESET}$. While each individual status bit is set by the appropriate condition, resetting them individually is different for each bit.

The input port status bit is cleared by reading the input port change register. The change in break status bit is reset by issuing a reset break change interrupt command from the applicable channel command register. The receiver ready/FIFO full bit is set when a received data word is completely shifted in (RxRDY) or when all three FIFO buffers are full (FFULL). It is reset by reading the receive register. The CT ready status flag is set whenever the counter or timer reaches terminal count and

Table 6.14 Interrupt Mask Register (IMR) (Courtesy of Motorola, Inc.)

Input Port Change $\overline{IRQ}$	Delta Break B $\overline{IRQ}$	RxRDYB/ FFULLB $\overline{IRQ}$	TxRDYB $\overline{IRQ}$	Counter/ Timer Ready $\overline{IRQ}$	Delta Break A $\overline{IRQ}$	RxRDYA/ FFULLA $\overline{IRQ}$	TxRDYA $\overline{IRQ}$
Bit 7	Bit 6	Bit 5	Bit 4	Bit 3	Bit 2	Bit 1	Bit 0
0 = Masked 1 = Pass	0 = Masked 1 = Pass	0 = Masked 1 = Pass	0 = Masked 1 = Pass	0 = Masked 1 = Pass	0 = Masked 1 = Pass	0 = Masked 1 = Pass	0 = Masked 1 = Pass

is reset by issuing a stop counter command. This command also stops the counter but does not otherwise affect timer operation. Finally, the TxRDY flags are set when the transmit word is transferred from the transmit buffer register to the transmit shift register. It is reset when the microprocessor sends a data word to the transmit buffer.

Interrupt Vector Register (IVR) In response to an interrupt request, the 68000 supplies an interrupt acknowledge (IACK) by setting all function code lines high (FC_0, FC_1, and FC_2 = 111). These must be decoded into active low $\overline{IACK}$ inputs to the DUART. In response to these signals, the DUART places the contents of the IVR onto the data bus and asserts $\overline{DTACK}$. The 68000, in turn, reads this data and computes an interrupt vector. This vector is placed on the address bus and the contents of that location are placed in the program counter. From this location, the interrupt program is begun. This register is initialized to $$0F by the assertion of $\overline{RESET}$.

Applications of the devices described in this chapter are discussed in Chapter 9. A firm understanding of the interrupt process as handled by the 68000 is needed to apply the functions of these chips.

PROBLEMS

6.1 EBCDIC, IBM's character code, uses eight character code bits, one stop bit, and no parity. What is the control register word that will configure an ACIA to handle EBCDIC using a counter divide of 64? Both transmit and receive interrupts are to be enabled.

6.2 A parity error is detected on a received word. What are the contents of the status register read as the first part of a receive interrupt program? Assume that no other error occurred.

6.3 What word is sent to the control register to send a break using an asynchronous ASCII code with even parity and two stop bits? The receive interrupt is disabled and the counter divide is 16.

6.4 Two tones used for an FSK mode are 2025 Hz and 2225 Hz. Which is the marking frequency and which is the space?

6.5 Using odd parity and two stop bits, what is the bit pattern (LSB to the right) of the ASCII character e?

6.6 Using a counter divide of 64, what are the transmit and receive clock frequencies for a 1200-bps data rate?

6.7 Using the ASCII-type data described in Problem 6.3, what are the lowest and highest frequencies the receive clock can use in Problem 6.6 without losing any bit information?

6.8 An ASCII letter A, using even parity and one stop bit, is sent and received at 1200 baud. The receive clock has drifted and become slightly higher in frequency than the maximum allowable frequency. Which error flag in the status register, if any, is set and why? If none is set, why not?

6.9 Write a receive program to read in a character from the receive buffer register of ACIA 2. If no errors occur, store the character at location SAVE. Use appropriate labels to indicate the error routines that would be run if errors are detected.

6.10 Write a program to configure ACIA 1 and ACIA 2 of our basic system to meet the following requirements:

Specification	ACIA 1	ACIA 2
Divide-by rate	16	64
Number of data bits	7	8
Number of stop bits	2	1
Parity used	Odd	None
Interrupts	Receive only	Transmit only

6.11 The transmit and receive clocks of Problem 6.10 are all 4800 Hz. What are the data rates for both ACIAs?

6.12 The message HELP! is sent using asynchronous ASCII with no parity and one stop bit. A data storage oscilloscope is used to read the transmit line. What is the bit pattern that appears on the oscilloscope's CRT?

6.13 Create a line in the basic system memory map and draw the schematic diagram to use the DUART in place of the two ACIAs described in the chapter.

6.14 Which registers and what contents are required to set the DUART to transfer an ASCII message using two stop bits and odd parity on both channels of the DUART? Select appropriate label names to access the registers. The chip uses a crystal at a frequency of 3.6864 MHz. Normal mode at a baud rate of 2400 bps is selected for this transmission. The transmit rate and receive sampling times are divided by 16 from the transmit and receive clocks. Input and output ports are configured as general-purpose ports. Enable RTS and CTS controls, and select the error mode to be character. Place the counter/timer into the timer mode and start the timer using a count of $02BC. Interrupts are allowed for all conditions except input port changes. The interrupt vector is $12. The receiver generates an interrupt request when the FIFO is full.

7

TUTOR 1.3 AND EXORmacs ASSEMBLERS

CHAPTER OBJECTIVES

In this chapter Motorola's development and training aids for the 68000 are described. A brief discussion of the MEX68KECB MC68000 Educational Computer Board is followed by in-depth coverage of the TUTOR 1.3 line assembler used with the system. The EXORmacs system, a more complete assembler which runs with the VERSADOS disk operating software, is then explored. Program applications using the assemblers conclude the chapter. MEX68KECB, TUTOR 1.3, VERSADOS, and EXORmacs are copyrighted trademarks of Motorola, Inc.

7.1 MEX68KECB EDUCATIONAL COMPUTER BOARD

The MEX68KECB Educational Computer Board (ECB) was developed by Motorola as an inexpensive training aid to allow students and users to familiarize themselves with a 68000-based system and its instruction set. Figure 7.1 on page 216 is the functional block diagram for the system, which includes the 68000 microprocessor, 48K of total memory (with user RAM space at addresses $00 0800–$00 7FFF), a parallel interface (MC68230, discussed in Chapter 5), and two MC6850 ACIAs (described in Chapter 6). The system is incorporated into an existing keyboard and CRT terminal. To operate the system, the terminal is powered up and a reset switch on the board is pressed. This initializes the system and displays TUTOR 1.3, followed by the command prompt (>) and a blinking cursor in the upper left corner of the CRT screen.

7.2 TUTOR 1.3 LINE ASSEMBLER

The TUTOR 1.3 line resident assembler, stored in ROM, is a line-by-line assembler. The source program is not retained in memory, but is assembled into object code as

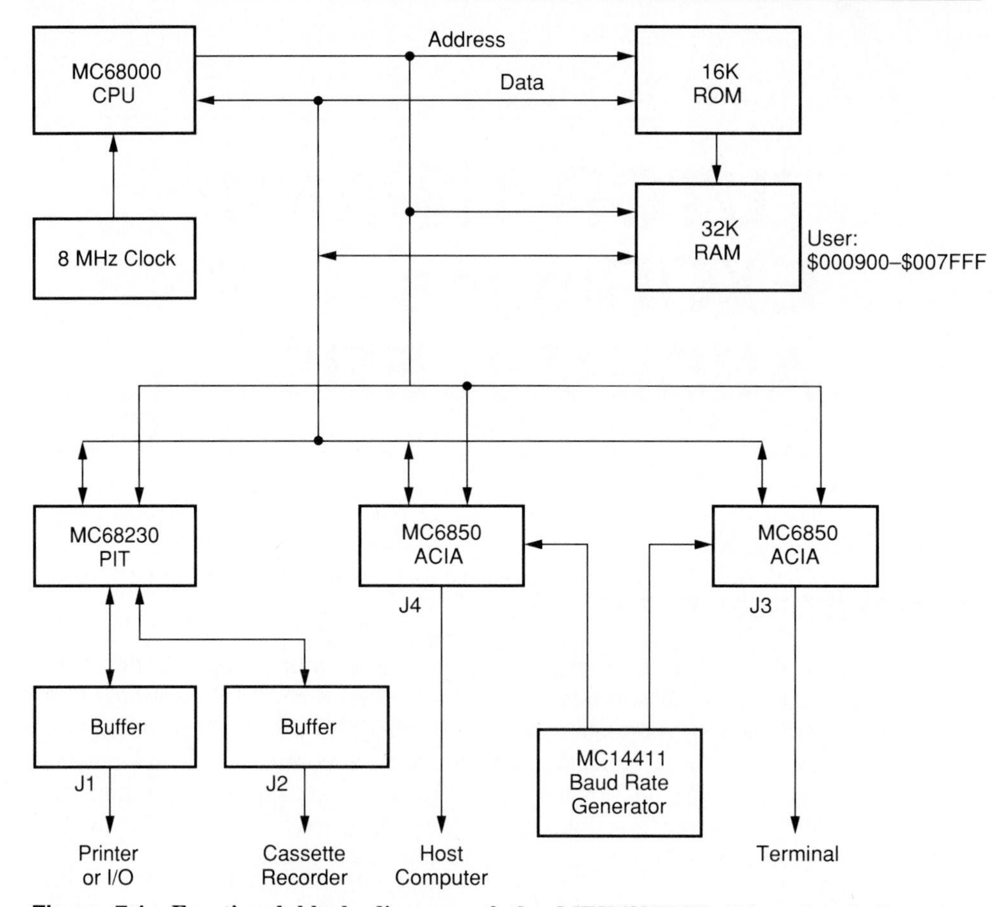

Figure 7.1 Functional block diagram of the MEX68KECB Educational Computer Board (Courtesy of Motorola, Inc.)

each instruction is written. Writing and assembling a program is initiated by typing the modify memory command after the > prompt:

```
TUTOR 1.3 > MM 1000;DI    <RETURN>
```

MM is the mnemonic for the modify memory command; and DI is the abbreviation for disassemble, which allows the user first to view the current disassembled contents of memory and then to assemble a new instruction into that same memory. 1000 is used here as a starting address for the assemble process. Commands in TUTOR require neither a $ nor leading 0's for address or data values. The assembler always assumes that commands use hexadecimal values. If nothing has yet been placed in memory following reset, the screen shows the following information after the RETURN key is pressed (the [designates the blinking cursor):

```
001000 FFFF DC.W $FFFF?[
```

The starting address (00 1000) and its contents (FFFF) are followed by the pseudo op DC.W. The disassemble process has detected the all-high condition of memory location $00 1000 and interpreted this information as a constant value since $FFFF is not an opword for an instruction. As such, the disassemble process shows this data to be a result of a declare constant (DC) pseudo op of word data size. The second $FFFF is the source operand for the DC.W pseudo op. The ? following this operand is the assemble prompt. It must be followed by a space and the new instruction to be entered at location $00 1000. If something was previously assembled at location $00 1000, it would be displayed in place of the DC.W pseudo op.

EXAMPLE 7.1

How would the instruction MOVEA.L #$2000,A0 at address $00 4000 appear on the CRT screen before it has been assembled? Assume nothing had been previously stored in memory.

Solution:

The preassembled instruction is typed following a space after the ? prompt:

```
004000 FFFF DC.W $FFFF? MOVEA.L #$2000,A0
```

When you press the RETURN key after typing the instruction, it is assembled and stored in memory in place of the previous data (the DC.W line in this example).

EXAMPLE 7.2

How does the instruction in Example 7.1 appear after it has been assembled?

Solution:

The new instruction is assembled and a full line of assembly code replaces the old DC.W line:

```
004000 207C00002000      MOVE.L #$2000,A0
004006 FFFF              DC.W $FFFF?[
```

A line assembler assembles each line as it is entered. If a syntax error occurs, instead of assembling the instruction, the assembler prints an X?, in place of the new line. The user must retype the instruction correctly and reenter it. The X? appears under the incorrect instruction so that the user can still view it.

EXAMPLE 7.3

What are the results of attempting to enter the instruction MOVEA.B A1,A0 into the program at location $00 4006?

Solution:

Because byte size data cannot be used with an address register, this instruction would cause the X? to appear. The assembler does not inform the user of the nature of the error. It only informs the user that an error has occurred. The results appear as

```
004006      MOVEA.B A1,A0
                    X?[
```

Typing the correct instruction replaces X? with that instruction. In this example, typing MOVEA.W A1,A0 followed by a RETURN key produces this composite result:

```
004006          MOVEA.B A1,A0
004006 3049     MOVEA.W A1,A0
004008 FFFF     DC.W $FFFF?[
```

Labels and relative addresses are not allowed with the TUTOR 1.3 assembler. Instead, all address references are absolute values. The assembler performs the task of converting a branch absolute address to the correct relative operand for the instruction. For the programmer, this restriction imposes some difficulty when writing branch instructions. Branches that cause program execution to branch back—that is, to repeat a portion of a program—present no problem since the branch return address is known as some previous location within the program. Forward branches *do* present a problem since the actual address being branched to is not readily known by the programmer. In this case, the programmer must enter an arbitrary dummy address when the instruction is first written. This dummy address is then replaced by the actual address once it is known. To understand the problem and its solution more clearly, let us use an example.

EXAMPLE 7.4

What are the steps required to enter and assemble the following program into memory, starting at location $00 1000, using the TUTOR 1.3 assembler?

```
        MOVE $2004,D0
        BMI.S STOP
        MOVE D0,$2000
DONE    BRA.S DONE
STOP    MOVE D0,$2002
        BRA.S DONE
```

Solution:

First, type in the assemble command

```
MM 1000;DI
```

Then press the RETURN key. If nothing was entered into memory previously, the DC.W line appears on the screen. Following the ? prompt, type and enter the first instruction. The results appear on the screen as

```
001000 30382004     MOVE.W $2004,D0
001004 FFFF         DC.W $FFFF?
```

The short form of the BMI instruction is limited to a signed byte size offset only. This restricts the branching amount to +127 and −128 locations. At this point in the assembly process, the user has no idea where the actual address for the label STOP is. Since the assembler does not allow labels, the label for the BMI operand cannot be entered. The actual address also cannot be entered since it is unknown. Instead, the programmer must enter a dummy value that cannot exceed the byte size offset limitation once it is assembled. As a result, the safest value to use at this point is the address of the branch instruction itself. This address creates an offset within the correct bounds once the instruction is assembled. Writing the branch operand in this manner also absorbs the correct number of program locations to hold the BMI instruction. Replacing the dummy address with the actual address later simply causes the new BMI instruction to replace the old one. The screen shows these first two instructions as

```
001000 30382004     MOVE.W $2004,D0
001004 6BFE         BMI.S $1004
001006 FFFF         DC.W $FFFF?
```

$FE is the actual operand for the relative address of the BMI instruction. The next MOVE instruction is entered as before and occupies two word locations, $001006–$001009, in the program list. The BRA is an endless loop entered in the same manner as the BMI instruction, but the branching address of the BRA is known. In this case, it is the address of the BRA opword since it branches on itself. The address used in this instance is the actual one and it will not be changed later. The BRA instruction occupies memory location $00 100A and is followed by the last of the MOVE instructions at location $00 100C (the value of STOP). The last instruction, BRA.S DONE, branches back to DONE, whose value ($00 100A) is also known. In this case, the actual value of the branch instruction is used and it need not be changed. This last instruction assembles as

```
00100A 60FB     BRA.S $100A
```

The program is complete except for the BMI instruction, which still has the dummy address. To change that instruction, first exit the assemble mode by pressing the BREAK key. Reenter the assemble mode using an MM command at the BMI location:

```
TUTOR 1.3 > MM 1004;DI     ⟨RETURN⟩
```

After the ? prompt, retype the BMI instruction with the corrected actual address. To this point, the screen shows the old instruction and the new retyped instruction:

```
001004 6BFE BMI.S $1004? BMI.S $100C
```

Following the RETURN, the new instruction, with the correct relative offset, replaces the old one:

```
001004 6B08         BMI.S $100C
001006 3E002000     MOVE.W D0,$2000?
```

When you press the RETURN key without typing anything after the ? prompt, the current instruction is unchanged and the next instruction is shown. In this manner, you can examine the program by repeatedly striking the RETURN key. The completely assembled program is

```
001000    30382004    MOVE.W $2004,D0
001004    6B08        BMI.S $100C
001006    3E002000    MOVE.W D0,$2000
00100A    60FE        BRA.S $100A
00100C    3E002002    MOVE.W D0,$2002
001010    60F8        BRA.S $100A
```

7.3 EXECUTING THE PROGRAM IN TUTOR

Before the program in Example 7.4 can be run, data must be entered into location \$00 2004 so that they can be fetched by the first instruction. Memory is modified by using a modify memory command that is similar to the assembly command:

```
MM (address) (size)    ⟨RETURN⟩
```

The size refers to the data size to be entered into memory and allows selection of the three major data sizes of the 68000: byte, word, and long word. The default (typing nothing for size) is byte size. After you press the RETURN key, the screen shows the address, its current contents, and a cursor prompt:

```
MM 2004

002004 XX_
```

To change the data, simply type in a byte of information and press the RETURN key. The next location and its contents appear. Repeat the process until you fill in as many locations as are required for your program. Exit the MM mode, as you would any command mode, by pressing the BREAK key.

Another command used to enter data into memory is the memory set (MS) command. This allows the user to enter data—in groups of eight bytes each, separated by spaces—into memory starting at a specified address by simply typing in a continuous stream of data.

EXAMPLE 7.5

Enter the data \$123456789ABCDEF0 starting at location \$00 4000.

Solution:

Type the command

```
MS 4000 12345678 9ABCDEF0     (RETURN)
```

The contents of as many as 16 memory locations can be displayed by using a memory display (MD) command, followed by the first location you want to examine and the number of locations to be displayed. The default is 16 locations. To the right of the hexadecimal data is a listing of their equivalent ASCII codes.

EXAMPLE 7.6

Using the program list of Example 7.4, what is displayed as a result of typing the command MD 1000?

Solution:

The first 16 program memory locations are displayed along with their ASCII equivalents to the right:

```
001000  30 38 20 04 6B 08 3E 00 20 00 60 FE 3E 00 20 02 08 .k.>...>...
```

The program listing is displayed by using the MD command followed by ;DI. The count value must include every byte of the program.

EXAMPLE 7.7

What is the command to list the program in Example 7.4?

Solution:

The program was assembled starting at location \$00 1000 and contains 17 bytes. The command to list that program is

```
MD 1000 11;DI
```

This causes the assembled program to display in a standard assembly format listing.

EXAMPLE 7.8

What is displayed by the command MD 4000 8?

Solution:

The eight digits of hexadecimal data entered using the MS command in Example 7.5 are displayed as follows:

```
004000     12 34 56 78 9A BC DE F0
```

After the program has been correctly assembled and data have been entered as needed, the program can be run successfully using the GO command:

```
GO 1000     ⟨RETURN⟩
```

The program executes each instruction until it encounters the BRA.S DONE instruction. At this point, it keeps executing the loopback on itself. To exit this condition, press the ABORT switch on the ECB. As a result another routine is run that displays the current contents of the 68000 registers as they would appear at the end of your program. This same set of registers, which includes the data and address registers, the program counter, user and supervisor stack pointers, and the flag register, can be displayed anytime using the display format (DF) command. Individual registers can be displayed using a dot command, which is a dot (period) followed by the register name. Each register is individually modified by following the dot command with a new data value.

EXAMPLE 7.9

Write the command to change register D2's contents to $014C 2130.

Solution:

The dot command is typed as

```
.D2 014C2130
```

To verify the contents, type .D2 to display the value of D2.

7.4 ADDITIONAL TUTOR FUNCTIONS

Additional TUTOR assembler functions facilitate program debugging. Chief among these are the single-step or trace (T) command and the breakpoint commands.

TRACE Mode

After assembling and running a long program, you discover that the results are incorrect. One method to find the problem in the program is to execute each instruction one at a time and observe the results of each instruction. This is commonly called single-step execution or trace mode execution. The TUTOR assembler allows a user to enter the trace mode by typing a T command. You can continue to execute each instruction by pressing the RETURN key after each instruction has been executed. The instruction executes and the 68000 registers display, showing the contents after the instruction is finished. The source code for the next instruction is also displayed.

The trace mode can be started from any memory location by preceding the T command with a dot program counter command followed by the address of the first instruction to be executed. As long as no other command is typed between each instruction execution, the trace mode execution can be continued simply by pressing the RETURN key after each instruction. If another command is used following the execution of an instruction, the trace mode must first be reentered by using a T command to continue single-stepping through the program.

EXAMPLE 7.10

What are the steps to single-step through the program in Example 7.3?

Solution:

After the program has run initially, the program counter contains the value of the location of the BRA.S DONE instruction. To single-step through the program, the program counter must be reinitialized with the first address of the program. This is done using .PC 1000 ⟨RETURN⟩. Following this with a T command causes the first instruction to be executed and the contents of the 68000 registers to be displayed. As you repeatedly press the RETURN key, each instruction in turn is executed and the updated register contents are displayed.

Using Breakpoints

Breakpoints are another way to find programming errors. They allow the program to be executed up to a designated address (the breakpoint), after which the register complement of the 68000 is displayed as with the trace mode. The breakpoint address is set by using the breakpoint set (BR ⟨address⟩) command. The program is then begun using the GO (starting address) command, which causes the program to run until the breakpoint address is encountered. Breakpoints provide a way to execute a loop one pass at a time.

EXAMPLE 7.11

A problem exists in the loop portion of the following program. How is the breakpoint used to allow the loop to be executed and checked one pass at a time?

```
002000        MOVE #$30,D0
002002        MOVEA #$3000,A0
002006        MOVEA #$4000,A1
00200A   LOOP MOVE A0,A1
00200C        SUBQ #1,D0
00200E        BMI.S LOOP
002010   DONE BRA.S DONE
```

Solution:

To check the results of the loop, the execution of the program must be suspended at the BMI instruction (location $00 200E). To do this, type the following breakpoint set command:

```
TUTOR 1.3 > BR 200E    ⟨RETURN⟩
```

Using the GO command begins execution of the program.

```
TUTOR 1.3 > GO 2000    ⟨RETURN⟩
```

The program runs to the breakpoint address and then the registers of the 68000 are displayed. The loop executes a second time when you press the RETURN key. At the end of each loop pass, the updated contents of the registers are displayed. This particular program example will stay in the loop until D0 reaches −1. The programmer's intention may have been to use the BNE instead of BMI instruction so that the loop is repeated until D0 is zero. In reality, you would discover this by running the program normally and observing the displayed registers after pressing the ABORT switch on the ECB. However, the program serves here as a simple example of breakpoint usage.

Breakpoints are cleared by typing the NOBR command. A temporary breakpoint can be used by applying the go to breakpoint (GT ⟨breakpoint address⟩) command. To use this, the program counter is first initialized to the starting address of the program (as with the TRACE operation) and then the GT command is issued. The program runs to the breakpoint address and stops, displaying the contents of the registers at that point. The temporary breakpoint is reset once it has been encountered.

EXAMPLE 7.12

Use the GT command to stop the program of Example 7.3 before it enters the eternal loop instruction.

Solution:

This command can be used to stop your program before it locks up in the eternally looping BRA.S instruction. The program begins at location $00 1000 with the BRA.S instruction at location \$00 1010. The first step to use the temporary breakpoint to halt the program is to initialize the program counter:

```
.PC 1000
```

and then issue the GT command:

```
GT 1010
```

The program runs up to the BRA.S instruction and stops, displaying the contents of the registers at that point.

Block Memory Commands

A number of commands facilitate the handling of a block of memory and can be used to perform some special user tasks. One such command is the block test (BT) command, the purpose of which is to test the ability of the RAM to accept data. The primary purpose of the command is to leave a designated memory area cleared (all zeros), with a secondary purpose of clearing a block of memory. The syntax of the command is

```
BT ⟨address 1⟩ ⟨address 2⟩
```

A second block command, block fill (or BF), allows a designated data word to be entered into a block of memory. This command causes every word throughout the block of data to be the same and is written as

```
BF ⟨address 1⟩ ⟨address 2⟩ ⟨word⟩
```

The block search command, BS ⟨address 1⟩ ⟨address 2⟩ ⟨byte⟩, searches through memory to find a data match and then displays those addresses that contain that byte of data.

There are no editing commands, such as delete or insert line, used with the TUTOR 1.3. However, you can insert or delete a line by using a block move command. The command itself moves a designated block of memory to a new area of memory. Its syntax is

```
BM ⟨address 1⟩ ⟨address 2⟩ ⟨address 3⟩
```

The memory contents of the locations between the first two addresses are moved to memory starting at the location of address 3. How is this used to insert or delete lines in an assembled program? The trick is to move a portion of the program down (for insert) or up (for delete). Moving the program up causes the unwanted line to be written over by the move process. Moving down causes a line to be repeated twice. Care must be taken in moving a total number of locations necessary to account for the instruction size to be deleted or inserted. Using the MM ⟨address⟩;DI command following the block move allows the new instruction to be inserted into the program.

EXAMPLE 7.13

In the program for Example 7.11, delete the instruction MOVEA #$4000,A1.

Solution:

The objective is to move the balance of the program following the undesired instruction up four locations: two for the opword and two for the operand. The command is written

```
TUTOR 1.3 > BM 200A 2011 2006     ⟨RETURN⟩
```

The data at location $00 2006 are replaced with the data from location $00 200A; location $00 2007, from $00 200B; and so forth until the entire block of data is moved. The effect is to write over the unwanted instruction. Notice that it is necessary to include every byte of the program block that is to be moved to perform the deletion correctly.

EXAMPLE 7.14

In the original program of Example 7.11, insert the instruction MOVE (A0),D2 after the MOVEA #$4000,A1 instruction.

Solution:

The new instruction requires two memory locations to store the opword. The procedure is to move the portion of program following the MOVEA #$4000,A1 down two locations to make room for the new instruction. The command to do this is

```
TUTOR 1.3 > BM 200A 2011 200C
```

This causes the MOVE A0,A1 instruction to move from location $00 200A to location $00 200C; then the remaining instructions move two locations down. To insert the new instruction, MM 200A;DI is performed. The new instruction is typed following the ? prompt and replaces the data at location $00200A.

Data Conversion Command

The DC or data conversion command displays the decimal and hexadecimal equivalents of a designated number. The syntax of the command is

```
DC ⟨data⟩
```

After you press the RETURN key, the display always shows

```
$data = &data
```

The $ indicates the hexadecimal value, and the & indicates its decimal equivalent. In the data field of the command, several data forms can be used, each with its own designator. Typing a number by itself or preceding it with a $ designates the original data as hexadecimal and allows a maximum of eight digits to be entered into the data field. Preceding the data in the command with the & designates the original data as decimal and allows a maximum of 10 digits. A % preceding the data indicates that it is a binary number with a maximum of 32 binary bits allowable. Finally, an @ designates octal values as long as 10 digits in the data field.

EXAMPLE 7.15

What is displayed following this command?

```
DC @ 247    ⟨RETURN⟩
```

Solution:

The equivalent decimal and hexadecimal values are displayed:

```
$A7 = &167
```

7.5 EXORmacs

The TUTOR 1.3 assembler and the ECB together make up an inexpensive training aid for the 68000. Incorporated into a basic terminal, the TUTOR 1.3 is a local interactive assembler. The limitations imposed by the exclusion of labels and comment lines as well as a true edit function prevent it from being a good development tool. This, of course, was not the intent of Motorola in producing that assembler and system. A more complete assembler is the EXORmacs system, which runs with the VERSADOS disk operating software. This section is concerned with the basic use of the EXORmacs system. Keep in mind that it is quite extensive. One would require a separate volume, by itself, to describe all functions of the EXORmacs adequately.

Signing on to EXORmacs

With both TUTOR and EXORmacs in place, turn on the terminal to access the TUTOR mode. The EXORmacs is a timesharing system when it is used in a multiple-terminal environment, requiring each user to sign onto the system with an individual user number.

After the TUTOR prompt appears, use the command called the transparent mode (TM) to switch to EXORmacs. Press the RETURN key after this command to place the terminal under VERSADOS control. Then press the BREAK key to start the EXORmacs software. An initial "cover page" appears on the screen telling you that you are in Motorola's development software, followed by this message:

```
ENTER USER NO.=
```

The equal sign (=) is the command prompt for EXORmacs. Any number from 1 to 256 followed by a RETURN is entered following the = prompt. A status list is generated containing an area designated for the programmer's unique user code. At this point, the code reflects only the user number just entered. The programmer must now enter a unique user code so that files can be stored later on in an area of disk memory associated with that user code. The commands to enter this code are

```
= USE ⟨user number⟩.⟨name⟩     ⟨RETURN⟩
```

The user number is the same one initially used, and the name can be any combination of letters and numbers, up to 6 characters in length, so long as it begins with an alphabetic character. Once this is done, the programmer is now signed onto the system.

Entering a Program

All programs are written and edited using the edit mode of the EXORmacs. This mode is entered by typing the command

```
= E ⟨filename⟩.SA;A        ⟨RETURN⟩
```

The SA delimiter designates a source file to be created at the end of the editing process. The ;A sets the tabs for the control fields within the edit frame. The fields are standard assembly format fields:

LABEL   OPWORD   OPERAND   COMMENT

EXORmacs requires at least one space between each field. Fields can be tabbed over to by using the →| key on the left of the keyboard. Arrow (→) keys are used for cursor movement. At the bottom of the screen are several soft-key functions that correspond to a row of F keys at the top of the keyboard. Soft-keys are those that allow an operation to be performed using a single key instead of multiple key presses. These functions allow the user to scroll the screen up or down a line or a page at a time. Another function key, F_1, switches between the edit command mode and the actual edit mode. This moves the edit cursor (<) between the edit field and the F_1 key itself. Entering or changing information within the edit field is done in a straightforward manner. Labels, which must begin with alphabetic characters, can be entered into the label field. The opwords are typed into the opword field, followed by at least one space and the operand value. Short comments can be added under the comment field. If longer comments are required between lines of code, an entire line can be typed as long as it is preceded with an asterisk (*). Note that the actual functions of the F keys are software dependent; that is, they may perform different functions depending on the software being used.

Once the program is entered and/or edited, the edit mode is exited first by pressing F1 to place the edit into the command mode. Then type the word QUIT and press the RETURN key. This returns you to the EXORmacs command mode.

Assembling the Program

Exiting the edit mode creates the SA source field. To create the object code file which will be eventually executed by the 68000, the source program must be assembled. This is done using the assemble command

```
= ASM ⟨filename⟩.SA,,,;R     ⟨RETURN⟩
```

The assembly is performed to the best of the system's ability. However, if a syntax error or other problem is encountered, a particular line will not be assembled, but will be flagged instead. At the end of the assemble process, the total number of errors and warnings is displayed. These errors and warnings must be corrected and eliminated before any further processes can be performed. Using the list command causes error and warning messages associated with the problem code lines to be displayed, along with a full program listing.

```
= COPY ⟨filename⟩.LS,#
```

Other delimiters can be used, but LS lists the source and object code along with error and warning messages. The screen scrolls through the program. A CONTROL-S (pressing and holding the CONTROL and S keys simultaneously) will freeze the scrolling so that a section of the program can be viewed. Pressing any other key resumes the scroll. The listing process stops on its own at the end of the listing or can be terminated at any time by pressing the BREAK key.

An alternate way to view the program with its error and warning messages is to use a form of the edit command.

```
E ⟨filename⟩.LS     ⟨RETURN⟩
```

The result of this command is a list of the first screen page of your program, including both error and warning information under any bad line of code. Soft-keys are available to allow you to move the program up or down by a line or a page. Additionally, you can enter the line-editing mode and modify the listing. However, this editing does not alter the SA file. You still must change that file and reassemble the program. This edit function is used solely as an alternative method to view the program. The advantage is to allow you control over the area of the program you want to examine. The COPY command scrolls through the program once, allowing no way to back up and examine the beginning of the program once it has scrolled through.

Errors are either syntax or operand errors. Warnings are indications that, although the instruction is correct, there may be a better way to do it or the execution of it may cause a problem. For example, using a long branch when a short branch would suffice would generate a warning. The short branch uses one less word of program space and executes faster. The system interprets a long branch as one that might cause the program to branch beyond the available RAM. Hence, a warning is generated that says that a short branch is preferred. Error and warning messages occur at the line that is in error and always appear with an error or warning number. The error and warning numbers are defined in the EXORmacs manual. All errors and warnings must be cleared before the next step can be performed. They are corrected by re-entering the edit mode and making the necessary changes.

Linking the File to TUTOR

Running the program in EXORmacs causes the system to be busy for that one terminal while the program is executed, locking out other terminals on the system. To reduce the time needed to execute a program, execution is done in TUTOR only; that is, at the local terminal site rather than through the timeshared EXORmacs system. To provide the TUTOR with the object code, it must be linked and downloaded from EXORmacs. The linking process is done with a link command:

```
= LINK ⟨filename⟩.RO,,#;Q    ⟨RETURN⟩
```

After linking the program, the TUTOR mode is entered by pressing CONTROL-A. This brings up the TUTOR 1.3 > message, whereupon the download command is typed as follows:

```
> LO;X = COPY ⟨filename⟩.MX,#    ⟨RETURN⟩
```

Memory can be modified to hold program data, and the program itself can be executed as described in the TUTOR sections of this chapter.

Linking with a Different User A special form of the LINK command allows the current user to pull a copy of a file from another user's directory and append it to a file in the current user's directory. The command requires knowledge of the other user's USE number and has the syntax

```
= LINK ⟨filename⟩/⟨USE number⟩.⟨filename⟩,,#;Q
```

The first filename is the file to be drawn from the other user's directory. It is followed by a slash (/) and that user's USE number. The second filename following the period is the file, under your user number, to which the first file will be appended.

Making a Hard Copy and Deleting Files

Once the program is verified as good and a hard copy (printout) is desired, the EXORmacs system is returned to using the TM command. This time, only an = prompt appears since the user is still signed onto the system. The command to make a hard copy requires sending the program listing to the printer through a spooler port.

The spooler stores the program in a buffer and drives the printer at the port from this buffer. The command to send the program out and initiate the printer process is

```
= SPOOL P ⟨filename⟩.LS,#PR2     ⟨RETURN⟩
```

where PR2 designates the port to which the printer is connected. (This designation may be different for other systems.) Any error in typing the SPOOL command causes the timesharing system to lock up as it searches for the incorrect spooler. In such a case, to unlock the system so that others can use it, the spool command must be aborted. This is done using the following sequence:

```
= SPOOL     ⟨RETURN⟩
> CANCEL     ⟨RETURN⟩
OK TO CANCEL ALL YOUR FILES (Y/N)? Y
>
```

After a hard copy is made and you are satisfied that your program is working, remove the created files from the disk to prevent it from becoming full. This is important in a multiterminal environment. The process has created many files associated with one program. They are designated by ⟨filename⟩.⟨del⟩ where ⟨del⟩ can be .SA, .LS, .RO, or .MX. Each one must be deleted separately using the delete command:

```
= DEL ⟨filename⟩.⟨del⟩     ⟨RETURN⟩
```

After all files are deleted (or at any time), the amount of files and their names can be checked by typing DIR (for directory). A listing of all files still existing under your user code is displayed.

Signing Off EXORmacs

Since EXORmacs is a timesharing system, it is as important to sign off the system as to sign on. Within the software is a timer that checks the activity on any terminal within the system. If it senses that the terminal is inactive but is still signed on for an allotted time period, it locks up the card interconnecting that terminal with the system. This prevents that terminal from being in the service loop when it is not being used. The problem with this is that several terminals usually are serviced by a single interface card. If any of the terminals on a particular card is inactive and on-line, all terminals connected to that card are locked out. The only way to free up the system is to power down and start over again. That is, everyone on the system must terminate and save their programs so that the system can be shut off and brought back up.

To prevent this, when you finish using the system, sign off by using one of two commands: OFF or BYE. OFF returns a sign-off message and BYE does not. To ensure that you are off the system, follow either of these commands with a CONTROL-A to return you to the TUTOR mode and then shut off the terminal.

7.6 PSEUDO OPS AND PROGRAM COMMANDS

The EXORmacs assembler supports a full set of pseudo operations as well as some special commands included within the file itself and executed in response to other EXORmacs commands. One pseudo op previously discussed is the declare constant (DC) operation. The complete syntax for this pseudo op is

⟨label⟩ DC.⟨size⟩ data1,data2,data3, ...

The size field designates any of the three 68000 data sizes: byte, word, or long word.

The ORG pseudo op tells the assembler where in memory to place the program during the ASM command. Its syntax is

⟨label⟩ ORG ⟨address⟩

The label field is optional with this command. A label can be equated directly to a value using the EQU pseudo op. This value later can be used throughout a program as data or to represent an address. The syntax for this pseudo op is

⟨label⟩ EQU ⟨data⟩

The SET pseudo op is identical to the EQU pseudo op, except that a label can be used in the data field as well as raw data. This allows a programmer to use two different labels to define a single bit of data.

The REG pseudo op equates a label to a register list to be used by a MOVEM instruction. Once the list is established, any MOVEM instruction can use it as its operand. For example, you can use the REG command to designate a list as follows:

```
KEEP REG D0-D7/A0-A6
```

This can be followed by the instruction MOVEM.L KEEP,−(A7) to save the environment during an exception operation and by MOVEM.L (A7)+,KEEP at the end of the exception routine to restore that environment.

A block of memory can be set aside for future data storage by using the DS pseudo op. The format of that operation allows selection of the data size and the number of data values that will be stored. It appears as

⟨label⟩ DS.⟨size⟩ ⟨data count⟩

One of the program commands causes the title of a program to be printed at the top of each page of a program printout. This command must appear as the first line of the program (preceding the ORG statement) and has the form

TTL ⟨program title⟩

There is no label used with this command. A second command allows you to insert a file from your directory into the current program being edited. This INCLUDE command is placed at the point in which the file is to be inserted into the current program. Again, a label is not used with this command. Its syntax is

INCLUDE ⟨filename⟩

File insertion is done when the current program is assembled using the ASM command.

EXAMPLE 7.16

A program entitled DISPLAY MESSAGE is to be written using the subroutine SHOW to display the data MESSAGE at a fixed location determined by the SHOW subroutine. Each word of the message is translated by the SHOW subroutine and displayed one character at a time. A delay program is stored under filename DELAY and is to be used between the displaying of each character. How does this program appear when editing is completed?

Solution:

The program begins with a TTL line that causes the title "DISPLAY MESSAGE" to appear at the top of every page of the listing when it is printed. This is followed by an ORG statement that tells the assembler where to store the program and a DC pseudo op that defines the location and values of the data. Within the body of the program, an INCLUDE statement tells the assembler to fetch and insert file DELAY. The edited listing is

```
          TTL       DISPLAY MESSAGE
          ORG       $2000
SEGMENTS  DC.B      'THE MESSAGE'
REDO      MOVE.B    #$A,D1
          MOVE.L    #SEGMENT,A0
SCAN      MOVE.B    (A0)+,D0
          BSR       SHOW
          INCLUDE   DELAY
          DBF       D1,SCAN
          BRA.S     REDO
```

The apostrophes (') surrounding the data (THE MESSAGE) store the ASCII codes for each character into memory, beginning at location \$00 2000. There are 11 characters (including the space), resulting in 11 bytes of data being placed in memory. The DBF instruction maintains the loop until D1 reaches -1. Since the value originally placed into D1 is 10 (\$A), the total number of loops is 11, accounting for each character of the message.

PROBLEMS

7.1 What are the TUTOR commands to initialize the program counter to \$01 BB00 and the user stack pointer (US) to \$CC 2000?

7.2 What are the TUTOR commands that can be used to store the following data into memory, starting at location \$00 4000?

```
$FEEDB00B1CC4DEED006B22BC98050EAC
```

7.3 What TUTOR command is used to display the data stored in Problem 7.2, and how do those data appear on the screen?

7.4 A program starting at location \$00 3040 appears to have an error between locations \$00 30A2 and \$00 30B0. What are the steps to execute the program to location \$00 30A2 and single-step through to location \$00 30B0?

7.5 Write the TUTOR command that causes locations \$10 C000 through \$10 FFFE to be cleared (set to zero).

7.6 What is the TUTOR command that displays the first 24 bytes of a program starting at location \$00 6B00?

7.7 After EXORmacs has assembled a file called TESTDATA, a message is displayed informing you that there are two errors and three warnings. Write the EXORmacs commands that can be used to display the program listing with the error and warning messages.

7.8 What is the EXORmacs command that allows the user whose USE number is 5.CEDE to append a file called HEREME under USE number 18.NOTME to a current file, THISTIME?

7.9 Write the pseudo op that stores the following word size data at a location labeled HOME:

```
$AB12 BB55 78DE CC10
```

7.10 What is the complete pseudo op to equate the label HELPUS with the label AIDHERE?

7.11 What EXORmacs command is used within the program to insert file ADDBYTES to the current file following the instruction MOVE.B DATABANK,D5?

7.12 What are the first two lines of a file that cause the program titled AVERAGE POINTS to be printed on the listing pages and the program to be assembled starting at location \$02 3C00?

7.13 Write the EXORmacs command to assemble the file BEGONE.

7.14 Write the TUTOR command that downloads the EXORmacs linked file FORME.

7.15 What are the commands or keystrokes that do the following:
(a) Exit the EXORmacs edit mode
(b) Return to TUTOR 1.3 from EXORmacs
(c) Change from the EXORmacs line editor to edit command mode
(d) Stop the screen from scrolling as a result of the COPY command
(e) Sign off of EXORmacs
(f) Go to EXORmacs from TUTOR 1.3
(g) Display all registers' contents in TUTOR
(h) Display the hexadecimal and decimal equivalents of the binary number 10011001101101011110

7.16 What is displayed as a result of the command in Problem 7.15(h)?

8

EXCEPTION PROCESSING

CHAPTER OBJECTIVE

The objective of this chapter is to describe the exception processing performed by the 68000. Exception programs include all processes except normal user programs. Some of these programs, such as interrupts, are initiated by an external signal, and others, such as address errors, respond to an internal condition or program instruction (e.g., TRAPs).

8.1 GLOSSARY

environment Present contents of the program model registers of a microprocessor.

exception programs Any 68000-executable programs other than user programs.

interrupt acknowledge Microprocessor-produced signal or set of signals that acknowledges receipt of an interrupt request. This acknowledge informs the interrupting source that its request is being honored and serviced.

interrupt program Exception program run in response to a request from a peripheral device. The term *interrupt* is derived from the process that occurs in response to the request; that is, the current program is interrupted momentarily so that the interrupt program can run.

interrupt request Signal sent to the microprocessor by a peripheral device requesting to interrupt the current program so that an interrupt program may be run.

privilege instruction 68000 instruction that can be executed only in the supervisor mode.

timesharing Sharing computer access time among several users.

TRACE (single-step) mode Allows user programs to be executed one instruction at a time for debugging purposes, causing a TRACE exception routine to be called at the completion of each instruction.

TRAP Software interrupt. Placing this instruction in a user program causes a TRAP exception routine to be called and run.

8.2 TIMESHARING

Many would-be computer users, especially private companies and small businesses, cannot afford the initial capital expenditure for a computer system. These individuals can rent or lease computer access time on someone else's system instead. The term applied to multiple use of computer availability is **timesharing,** since the users literally share the available computer time. The main disadvantages of timesharing are the wait to get onto the system and the limited amount of access time available to each customer. The microprocessor feature that makes timesharing possible is called an **interrupt.**

Timesharing implies that different user programs are run at different times. Some of these programs are more important or have higher priorities than others. If this were not so, each user would simply await his turn and would be granted access as soon as the preceding user had finished. Priority is determined in many ways, for example, by economics or by degree of importance. In a priority hierarchy, processing of a less important program is temporarily stopped or interrupted while the higher priority program is run. Once that program is finished, the original program is resumed. The original program can be resumed at the point where it was interrupted instead of returning to the beginning to be rerun. Interrupt routines are just one type of **exception programming,** which includes any processing other than normal user programming. Where these programs begin is a function of the type of exception and its associated vector address.

8.3 VECTOR ADDRESSING

The locations of exception programs must be known so that they can be accessed and executed in response to an exception call. The method of identifying the starting addresses of the exception programs is called *vector addressing*. Stored in the lower addresses of ROM are the starting addresses for the exception routines. Each address occupies four memory locations to accommodate the 24-bit address within the word size structure of the 68000 system data bus. The additional upper byte of the 32-bit (two-word) in the four memory locations is set to zero. Table 8.1 contains a listing of the vector addresses associated with each exception in order of its exception number. The vector addresses are derived by multiplying the exception number by 4. The vector numbers range from 0–255 with corresponding addresses from $00 0000 (SSP information) to $00 03FF (user-defined vector). Note that the first vector number, associated with the RESET exception, is not a true vector in the sense of retrieving a starting address for a program. It is used to initialize the supervisor stack pointer. The RESET vector is vector one, which contains the starting address of the initialization program.

8.4 EXCEPTION PROCESSING

Interrupt programs are one type of 68000 exception processing. Exception processing also includes any processing that is different from user programming, such as

Table 8.1 Exception Vectors and Addresses

Exception Number	Exception Address Range	Exception Type
0 1	00 0000-00 0003 00 0004-00 0007	RESET--Supervisor Stack Pointer RESET--Program Counter
2 3	00 0008-00 000B 00 000C-00 000F	Bus error Address error
4	00 0010-00 0013	Illegal instruction
5	00 0014-00 0017	Divide by zero
6 7	00 0018-00 001B 00 001C-00 001F	CHK instruction TRAPV instruction
8	00 0020-00 0023	Privilege violation
9	00 0024-00 0027	TRACE
10 11	00 0028-00 002B 00 002C-00 002F	Opword high 1010 Opword high 1111
12-23	00 0030-00 005F	Reserved by Motorola
24 25 26 27 28 29 30 31	00 0060-00 0063 00 0064-00 0067 00 0068-00 006B 00 006C-00 006F 00 0070-00 0073 00 0074-00 0077 00 0078-00 007B 00 007C-00 007F	Spurious interrupt Level 1 interrupt Level 2 interrupt Level 3 interrupt Level 4 interrupt Level 5 interrupt Level 6 interrupt Level 7 interrupt
32-47	00 0080-00 00BF	TRAP vectors
48-63	00 00C0-00 00FF	Reserved by Motorola
64-255	00 0100-00 03FF	User-defined vectors

bus error, TRAP, and divide by zero error programs. All these programs are executed in the supervisor mode of operation, which is entered into automatically whenever the 68000 senses and services an exception routine request. Some exceptions are initiated internally in the microprocessor, whereas some are begun in response to an external signal.

8.5 INTERNALLY GENERATED EXCEPTIONS

Internally generated exceptions are executed in response to a software problem or direction. They include addressing errors, privilege instruction violations, and others. A brief description of each of these routines follows.

Address Error

Word and long word operations must be accessed on even addresses. Failure to do this causes the microprocessor to stop at the invalid instruction and execute the address error routine.

Privilege Violation

A number of instructions can be executed in the supervisor mode only. An attempt to use the instruction in the user mode causes a **privilege** violation exception to be called. These instructions include STOP, RESET, RTE, MOVE to SR, ANDI to SR, EORI to SR, and MOVE USP. Consult the instruction set in Appendix A for a description of these instructions.

TRAP and TRAPV

A **TRAP** is a software-generated, non-maskable interrupt. Whenever either of these two instructions appears in a program, the microprocessor vectors to a routine associated with it. TRAPV stands for "trap if overflow (V) bit is set," and it executes the exception if the flag is set in the condition code register portion of the status register.

CHK

CHK is another software instruction that checks the lower word of a data register against data at an effective address (ea); if that value is less than zero or greater than that at the effective address, a CHK exception is run.

Divide by Zero

When you use the DIVU or DIVS instruction, an attempt to divide by zero causes this routine to be called.

TRACE

Debugging a program by single-step execution can be done by using the 68000 **TRACE** function. If the trace (T) bit in the status register was set before the execution of a user instruction, the trace exception is vectored to when that instruction is completed. What occurs in the TRACE exception routine is determined by the system designer and is included in ROM along with the other internally generated exception routines.

Illegal and Unimplementable Instructions

Two bit patterns used as the upper nibble of an opword will cause an unimplementable exception to commence. They are 1010 or $A and 1111 or $F. Three additional opwords are reserved and cause an illegal instruction exception to be generated. They are $4AFA, $4AFB, and $4AFC.

8.6 EXCEPTION PROCESSING FOR INTERNALLY GENERATED ROUTINES

In response to an internally generated exception call, the 68000 first copies the current contents of the status register into a temporary holding register. Consequently, the current condition of the interrupt level masks, the supervisor status (S), and trace mode (T) flags are saved in this register. The S flag in the status register is now set and the T flag is reset. This action places the microprocessor into the supervisor mode and disables the trace function; the latter is necessary to prevent the exception routine from becoming a single-step program. Each exception has an exception number associated with it which is multiplied by four (shifted left twice) to form the vector address location. This vector address is placed into the address buffers and memory is accessed through a read cycle. The fetched data are stored into the data buffers. The data copied into the temporary register from the original status register are pushed onto the stack and the supervisor stack pointer is decremented twice (for word size data) for this push operation. Next, the current contents of the program counter are pushed onto the stack. The supervisor stack pointer is decremented four times as these data are saved on the stack. The data in the data buffers now are loaded into the program counter and the exception routine is begun.

8.7 RETURNING FROM AN EXCEPTION

At the end of some exception routines, you might want to return to the original program that was interrupted. In an address error exception, however, the problem must be corrected so that the user program can run properly. One must use a manual reset or restart to return the program to a user mode.

For the cases involving a software return at the completion of the exception routine, a return from exception (RTE) instruction is provided. This instruction causes the data pointed to by the supervisor stack pointer to be pulled or retrieved and stored into the program counter, restoring the opword address of the next instruction to be executed by the original program into the program counter. The stack pointer is incremented four times during this operation, and the data now pointed to are pulled from the stack and loaded into the status register, restoring the condition codes and status flags as they were before the exception occurred. The original program now can be resumed from the point where it was interrupted.

8.8 EXTERNAL EXCEPTIONS

An external exception requires a hardware signal input to the 68000 before it is called and run. The most common external exception is the interrupt-type operation. Two additional external exception routines are BUS ERROR and RESET.

Interrupt Requests

An **interrupt request** is a signal generated by an interface device that requests the microprocessor to interrupt its current program and divert to a different routine specified by the interrupting device. There are essentially two types of interrupt requests. The first in a non-maskable interrupt (NMI), which the microprocessor cannot ignore. It must respond and service the request. A second, maskable type is the more general form of interrupt request that allows the programmer the option of servicing it (responding to an interrupt request) or not servicing it (masking the interrupt request). For the 68000, the mask is developed by a level priority scheme involving the three interrupt mask bits of the status register. These bits, I_0, I_1, and I_2, are set to one of eight possible binary combinations or levels. For an interrupt request (placed on input control leads $\overline{IPL}_0$, $\overline{IPL}_1$, and $\overline{IPL}_3$) to be recognized and serviced, its binary level must be greater than the binary level of the interrupt mask bits. Be careful to recognize that the IPL inputs are active low whereas the interrupt mask bits are active high. A 000 on the interrupt mask bits of the status register is a level 0 mask, meaning that any level of interrupt above 0 is serviced. In contrast, 111 on the $\overline{IPL}$ leads is a level 0 and is interpreted by the 68000 as a noninterrupt request since all three leads are inactive. On the other hand, a level 7 (000) on the interrupt input lines is the non-maskable interrupt level that always is serviced regardless of the state of the interrupt mask bits in the status register.

In handling an interrupt request, the 68000 first compares the inputs from the IPL leads with the interrupt mask bits in the status register. If the level coming in is less than or equal to the mask, the request is ignored and processing continues normally. If the level is greater than the mask, an exception sequence is begun.

EXAMPLE 8.1

What are the states of the $\overline{IPL}$ inputs for serviceable interrupt levels when the interrupt mask bits are set to 101 (level 5)?

Solution:

Interrupt requests above level 5 are serviceable by the microprocessor under these conditions. A level 6 request occurs when the IPL leads are $\overline{IPL}_2 = 0$, $\overline{IPL}_1 = 0$, and $\overline{IPL}_0 = 1$, and a level 7 is recognized when all three IPL leads are held low.

Once the microprocessor recognizes and accepts a serviceable interrupt request, the current status register is saved in the temporary register as before, then the S flag is set and the T flag is reset. In addition, the interrupt mask bits now are updated to the level of the interrupt request, preventing lower priority interrupts from interrupting the program about to be run.

The function control outputs (FCs) from the microprocessor are set high, providing an **interrupt acknowledge** that the interrupting device can sense. Additionally, the lower three address lines (A_1, A_2, and A_3) are set to match the interrupt request inputs of the IPLs. This action verifies that the correct interrupt is being serviced. Also, if two interrupt requests try to enter the microprocessor at the same time, the interrupt being serviced can be determined by monitoring the three address lines.

In response to the interrupt acknowledge, the interrupting device can do one of two things. First, it can return an exception number via the data bus and a $\overline{\text{DTACK}}$ to the 68000. Second, it can return a $\overline{\text{VPA}}$ signal to the microprocessor. The latter action causes the 68000 to perform an auto-vector process that internally generates an exception number associated with the interrupt level (see Table 8.1). Regardless of how the exception number is determined, it is shifted twice to the left (binary multiplication by four) to form the vector address. The status register and program counter are pushed onto the stack, and the exception starting address is retrieved from the vector location as already discussed. Figure 8.1 is a timing diagram showing the relationship of the various signals during an interrupt request process, and the step-by-step exception handling process is flowcharted in Figure 8.2 on page 242.

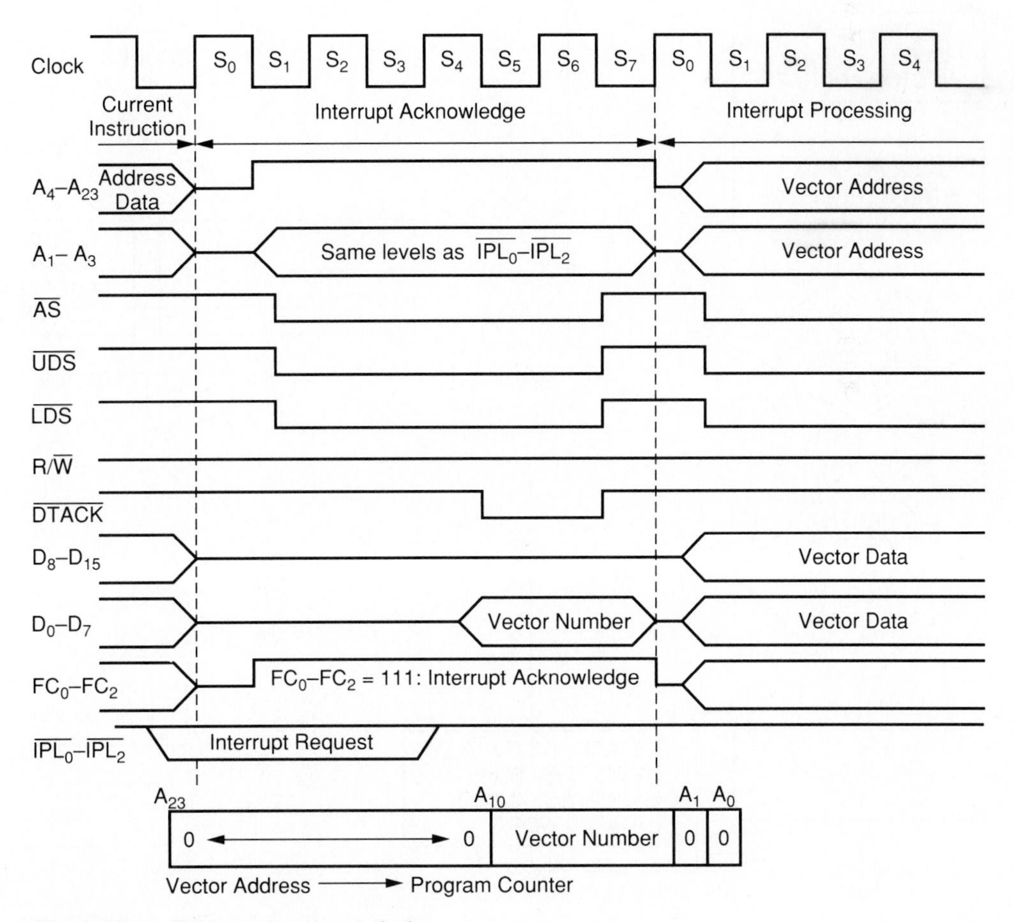

Figure 8.1 Interrupt request timing

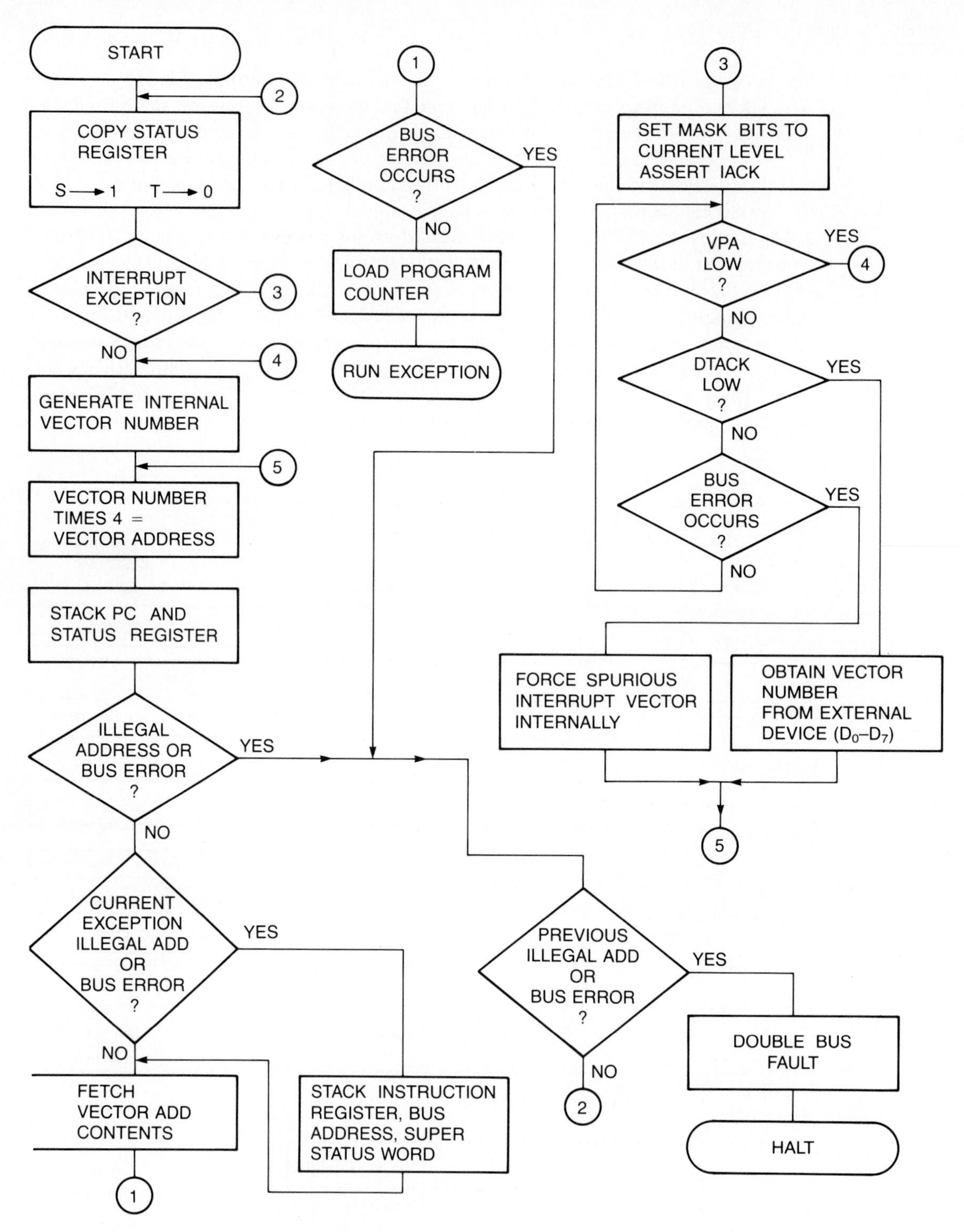

Figure 8.2 68000 exception flowchart

EXAMPLE 8.2

What is the vector address for a level 3 interrupt if the interrupter returns an active $\overline{VPA}$ signal in response to the interrupt acknowledge?

Solution:

The active $\overline{VPA}$ signals the microprocessor to perform an auto-vector process for this interrupt. From Table 8.1, you can discover that the exception number associated with interrupt level 3 is 27. This number is multiplied by 4. The resulting hexadecimal equivalent of the product (108) is the vector address, \$00 006C.

RESET Exception

The RESET exception is a unique routine because it initializes the system. Thus, it is not required to save either the status register or the program counter information of the current program. The reset process is begun by the simultaneous receipt of active $\overline{RESET}$ and $\overline{HALT}$ signals, which causes the S flag to be set and the T flag to be reset. In addition, the interrupt mask bits all are set to a level 7 mask. The contents of locations \$00 0000–\$00 0003 are retrieved and loaded into the supervisor stack pointer, after which the program counter is loaded with the contents of locations \$00 0004–\$00 0007. The address loaded into the program counter causes an initialization program to start. The system has no way to return, by an RTE instruction, to a current program interrupted by a $\overline{RESET}$ signal. In practice, to return to the original program after a RESET, you must rerun the program in the same manner that it was begun originally.

Bus Error Exception

The bus error exception is begun in response to the $\overline{BERR}$ signal returned in place of a $\overline{DTACK}$ during a bus read or write cycle. The signal indicates that something within the system caused the cycle to be aborted or not completed successfully. The 68000, in response to the $\overline{BERR}$ signal, begins, as it does with every exception, by saving the status register and setting the S flag. The T flag is reset, and the program counter information and the preserved status register data are saved on the stack, as with all exceptions. For bus error exceptions, the microprocessor also builds additional stack information containing the instruction register data, the current address of the failed cycle, and the cycle word. This information is necessary to determine the cause of the bus error. Table 8.2 on page 244 diagrams the bus error stack. A similar stack is generated for the internal address error exception. The exception number is converted to an exception address and that routine is begun. Usually, there is no return from this routine because some corrective action must be taken before the user program can be successfully rerun.

Table 8.2 Bus Error Stack

	b15 b0
SSP-14	Next available location
SSP-12	Access type
SSP-10	Current cycle address (high word)
SSP-8	Current cycle address (low word)
SSP-6	Instruction register
SSP-4	Status register
SSP-2	Program counter (high word)
SSP	Program counter (low word)

SSP: Supervisor Stack Pointer

Access Type

b15—Undefined—b5	b4	b3	b2	b1	b0
	R/W	I/E	FC_2	FC_1	FC_0

R/W: 0 = Write cycle 1 = Read cycle
I/E: 0 = Instruction 1 = Exception
FC: Function code lines

8.9 EXCEPTION PRIORITY

Besides being classified as internal or external, exceptions also are placed into priority groupings associated with their functions and types. Group 0 exceptions, consisting of RESET, address error, and bus error, have the highest priority; that is, they are serviced before any others and cannot be disrupted by them. Group 0 exceptions cause the current instruction to be aborted and the exception process to be started within two clock cycles of its initiation. The next level of priority is the group 1 exceptions, consisting of TRACE, interrupts, and illegal instruction and privilege violation. This group allows the current instruction to be completed and the current program to resume once the exception has been completed. Finally, Group 2 includes the instruction-generated exceptions: TRAP, TRAPV, CHK, and divide by zero.

8.10 DOUBLE BUS ERROR FAULT

The stacking operation during a bus error routine uses several additional bus cycles to move the data into memory. If a second bus error occurs during this operation, it is called a *double bus error* and the microprocessor enters a halt state. The system must be reset to exit this halt state. A double bus error condition also occurs when there is a bus error while the microprocessor is accessing the vector table to retrieve an exception vector. The microprocessor is halted to prevent the occurrence of erroneous processing and to allow time for the user to perform corrective action.

8.11 BUS ERROR RERUN

When a $\overline{\text{BERR}}$ is received in place of $\overline{\text{DTACK}}$, there is a failure to complete the current bus cycle. If a $\overline{\text{HALT}}$ is also returned with the $\overline{\text{BERR}}$, the 68000 attempts to rerun the failed bus cycle to see whether the error was caused by a spurious signal. Once the $\overline{\text{HALT}}$ and $\overline{\text{BERR}}$ signals have been asserted, the cycle is rerun after they have been removed. As long as these two signals are returned in place of $\overline{\text{DTACK}}$, the current cycle will be rerun. It is usually good practice to provide a means, through external logic, to limit the number of times the rerun attempt is made. Failure to do so can result in placing the system in a continuous rerun state if the error is real and not spurious.

8.12 MULTIPLE EXCEPTIONS

When dealing with any type of processing that deviates from the normal user mode, one must pay close attention to the way the system handles these processes. For instance, if two exception routines are requested simultaneously, the 68000 handles the one with the highest priority. At times, this means that the other exception request is lost.

EXAMPLE 8.3

What occurs if a RESET and an interrupt level 3 exception request arrive at the microprocessor simultaneously?

Solution:

The RESET, which has the highest priority, takes precedence. One result of receiving the RESET ($\overline{\text{RESET}}$ plus $\overline{\text{HALT}}$) is that the interrupt mask bits (I_0, I_1, I_2) in the status register are set to level 7. This prevents the level 3 interrupt from disturbing the RESET or initialization routine in process until that routine alters those status bits.

Keeping track of what occurs during exception routines can be complicated. Much depends on the exception in process and the new exception request. The easiest case of multiple exceptions involves different interrupt level requests. One effect of servicing an interrupt exception is that the I bits in the status register are updated to the level of the interrupt request being handled. Consequently, lower level interrupt requests are masked off until the current level interrupt is finished. However, if a higher level request arrives, it can cause a second interrupt exception to occur. No conflict arises because of the second request. The new interrupt is processed like any such request. The status register is saved into a temporary register, the S bit is set, and the T bit reset as before. Notice that since an exception was already in progress, these two bits of the status register actually are unchanged by this action. The interrupt mask bits are upgraded to the new interrupt level, and the program counter and previously saved status register are pushed onto the stack. The stack now contains data about the interrupted exception and the interrupted original program. Since the stack pointer is decremented as data are stored onto the stack, it is always pointing to the next empty location in the stack memory. The second exception eventually ends in an RTE instruction, causing the stack to be pulled (popped), restoring the status register and program counter information pertaining to the next instruction to be executed by the first exception routine. That first exception program then runs until it, too, reaches an RTE instruction. The original program's status register and program information then are recovered from the stack and that program is resumed.

EXAMPLE 8.4

If no additional register information is placed on the stack, how many locations are filled with data when these three interrupt requests occur one after the other: level 3, level 6, and level 5? Assume the original interrupt mask level is level 1.

Solution:

The status register occupies two memory locations for a word of data, and the program counter requires four locations to hold long word data for each exception. The level 3 interrupt is serviced, upgrading the interrupt mask to level 3. The level 6 request following closely to the level 3 also is served, causing the I bits to be changed to that level. When the level 5 also attempts to interfere, it is not serviced. Twelve stack locations are thus being used to hold the status register and program counter data of the original program and the level 3 exception program.

8.13 TRACE MODE MULTIPLE EXCEPTIONS

The TRACE mode is entered at the completion of an instruction if the T bit is set. If that instruction is not completed because a different exception occurred, then the TRACE mode is not operable. The explanation here is straightforward and reinforces the reason for resetting the T bit in the status register each time an exception is granted. The resetting of the T bit prevents the TRACE mode from becoming operable. If you are executing the TRACE exception and a higher priority exception occurs, that exception is serviced and executed. The only higher priority exceptions are RESET, bus error, and address error, which generally cause current programming to be aborted because some corrective or restart activity must occur before processing can resume. An interrupt occurring while the TRACE routine is in process is serviced at the completion of the TRACE program. Again, the interrupt routine will not be single-stepped because the T bit has been reset by the exception process dealing with the interrupt request. Never forget that the original status register is the first item saved (protected) during an exception process activity. It then is restored fully (most notably the T flag and interrupt mask bits) at the end of the exception routine. Returning from the TRACE exception causes the T bit to be restored to its active state by recovering the original status register data from the stack. As such, the next instruction in the program is executed and the TRACE exception is called once again at the completion of that instruction.

TRACE exception routines usually include instructions that allow the user to examine the contents of the registers as well as those of the user status register and program counter. This information lets the user know the results of the instruction executed before entering the TRACE routine. Thus, the TRACE function is a convenient tool for debugging programs one instruction at a time.

8.14 SAVING THE ENVIRONMENT

To ensure that an original program can be resumed after an exception is run, the contents of more than the program counter and status register need to be saved. This process is called saving the **enviroment.** Restoring that data is called retrieving the environment. The 6800 8-bit microprocessor automatically saved its registers on the stack as a result of an interrupt. This was relatively easy because an interrupt was the only exception besides a RESET, and the number of registers was relatively small (two accumulators, a condition code register, an index register, and a program counter). The amount of stack area was only eight locations for a single interrupt. The 68000 is entirely different. Besides the program counter and status register, there are 8 data and 7 address registers, all of which are 32 bits wide. Sixty-six locations must be set aside for all data to be stored during a single exception. But with multiple exceptions, an additional 66 locations are required per exception.

The actual process of saving or retrieving the environment for the 68000 is simple. A MOVEM instruction using the predecrement and postincrement functions supplies the method. The instructions placed at the beginning of an exception program to save and, at the end, to retrieve the environment are

```
MOVEM.L D0-D7/A0-A6,-(A7)     * Save
MOVEM.L (A7)+,D0-D7/A0-A6     * Retrieve
```

A_7 is not included in the register list because it is the supervisor stack pointer during an exception routine. In essence, the user stack pointer is not accessible during an exception, so its contents cannot be affected by those routines. The exception itself causes the program counter and status register to be saved, while the return from exception (RTE) instruction retrieves this data. It is technically possible to have large RAM space available in a 68000 system, but in reality much of this space may not be occupied by actual memory chips. Our basic system is an example of a limited memory-mapped system, which restricts the handling of multiple exceptions with restore ability.

8.15 SETTING THE INTERRUPT MASK LEVEL

The upper status register containing the interrupt mask bits can be modified in one of three ways: (1) by receiving $\overline{\text{RESET}}$ and $\overline{\text{HALT}}$ simultaneously; (2) by servicing an exception; and (3) by any number of instructions during an exception routine. These bits cannot be altered by a user program. The interrupt mask bits are set initially to a level 7 (111) by the RESET exception operation. The reset exception routine usually contains programming to set the interrupt level at some predetermined value. If one wants to change the interrupt mask level, one must first place the 68000 into the supervisor mode in order to have access to the status register data. This is usually done by using a TRAP instruction.

There are 16 different TRAP exception vectors. The opword of the TRAP instruction specifies the one being used. To be useful, the vector associated with the TRAP instruction to be used must vector to user RAM space. To begin, the programmer first stores a short program into this RAM space that modifies the interrupt bits. When the TRAP exception routine is run, this program is the one executed. The process is easier to understand through a practical example based on the basic system developed earlier. TRAP number 15 ($F) vectors to $00 8C00, and we want to set the interrupt mask to level 4. Starting at location $00 8C00, the program to effect the necessary change in the interrupt mask bits is

```
TRAPF   ORI #$0700,6(A7)
        ANDI #$FCFF,6(A7)
        RTE
```

The ORI instruction sets the interrupt mask bits all high but does not affect the remaining status bits of the original data saved on the stack when the TRAP routine was called. Since the 68000 is in supervisor mode, the reference to A_7 is the supervisor stack pointer. The offset six added to A_7 by the address indirect plus displacement acts as a pointer to the first stack location where the status register contents were saved. The ANDI instruction actually sets the level 4 interrupt mask by causing those interrupt mask bits to be set to 100, leaving the remaining status flag bits unchanged. When the RTE is executed, the original status data with the modified interrupt bits are returned to the status register. You must take care when you modify status register data. Privileged instructions that deal directly with the status register (such as ANDI #⟨data⟩, Sr, and other manipulations already described) affect the entire contents of the status register, including the condition code flags. Mishandling status register manipulations can result in unexpected consequences when the original program resumes.

For this example, the instruction in the user program that puts the 68000 into supervisor mode and allows the updating of the interrupt mask bits is TRAP #$F. ROM location $00 00BC, the fifteenth TRAP vector, must contain the data $0000; and location $00 00BE must contain the data $8C00, which is the starting address of the TRAP exception routine.

8.16 INTERRUPT PROGRAM

The most common exception routines are interrupts. They are useful for data transfers from peripheral devices at unspecified times while a current program is in progress. There are two methods to deal with such data transfers: polling and interrupts. *Polling* is a method whereby a data or a control register of a peripheral interface is read and checked for a predetermined indication that a data transfer is ready. In the case of the PIA, the control register is read using a MOVE instruction and the interrupt flags (b7 and b6) are tested to see if they have been set. Polling requires that the control register read be performed fairly regularly to keep a reasonably accurate check for data transfer ready. Reading the PIA control registers in a specific order establishes a form of priority; that is, the first control register examined becomes the highest priority peripheral interface. The system requires no additional programming or external signaling to use polling. The biggest disadvantage of polling lies in the time needed to achieve the register reads and the frequency with which they must be done to maintain a constant scan of the keyboard.

The alternative method of determining when a data transfer is set to take place is to use interrupts. Once the body of the original program is being processed, no control register reads are required. However, there are other requirements to handle the interrupt. First, an external signal called an interrupt request is presented to the microprocessor. In Chapter 6, we saw how a pulse sent from the peripheral device on CA1 or CB1 generated an active interrupt request on the corresponding $\overline{\text{IRQA}}$ or $\overline{\text{IRQB}}$ line from the PIA if the masks were not set. These in turn are encoded into various interrupt levels and are sent on the $\overline{\text{IPL}}$ lines to the 68000. The degree of the interrupt level determines the priority of each peripheral interface.

As the original program runs, the status of the PIAs associated with various peripherals is ignored until an interrupt exception occurs. At that point, the current processing ceases, the exception is serviced, and the interrupt program begins. The original program is resumed at the finish of the interrupt program and is completed unless it is interrupted again.

Interrupt programs can reside in either RAM or ROM area depending on their use and nature. The following discussion concerns interrupt programs that the programmer stores in RAM locations.

When interrupt programs are stored in RAM locations, the vector address associated with the interrupt exception number must point to a RAM address. As an example, we will use interrupt level 5, whose auto-vector is $00 0074. The value at that vector address is $00 8050, a RAM location. To make the example a bit more practical, we will have the program process a keyboard entry from an interrupt processed keyboard illustrated in Figure 8.3 on page 250. This circuit is incorporated into the basic system described in Chapters 5 and 6. Thus, there is now a method that allows the keyboard to be monitored for a key entry while normal program execution occurs.

The keyscan subroutine in Chapter 5 is called by the interrupt routine, and key data are stored in memory before returning from the interrupt. What the user wants the microprocessor to do with the actual key data depends on the programs

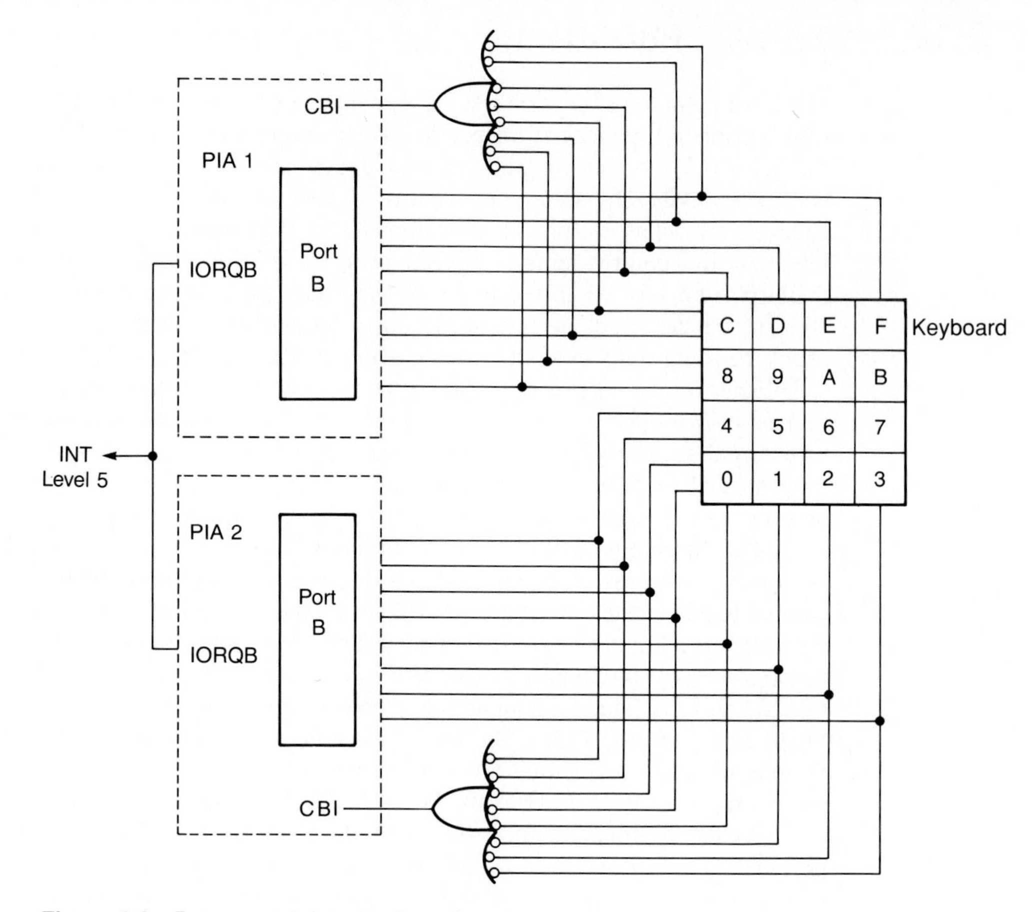

Figure 8.3 Interrupt driven keyboard

previously written. An interrupt-handling routine to scan the keyboard begins by saving the environment and freeing up the registers to be used by the subroutine within the exception program. The results of the subroutine are stored in memory, the environment is retrieved, and an RTE is executed. The original program is resumed and may or may not use the key information as required by the programmer.

Program Details

The interrupt program starts at location $00 8050 by saving the environment with the multiple move instruction

```
MOVEM.L D0-D7/A0-A6,-(A7)
```

A_7 is actually A_7' (the supervisor stack pointer [SSP]), which had been initialized by the reset process with data from $00 0000–$00 0003 and modified by the pushes of the interrupt exception action.

EXAMPLE 8.5

What are the contents of A_7' if \$00 0000 contains \$0001 and \$00 0002 contains \$5000 after executing the MOVEM.L D0–D7/A0–A6,−(A7) instruction for a level 5 interrupt?

Solution:

Initially, the SSP (A_7') is loaded with \$0001 5000 as a result of a system reset. When the interrupt first is serviced, the status register is pushed onto the stack and the SSP is decremented twice to \$0001 4FFE. Next, the program counter is saved on the stack as a continuation of the exception-handling operation and the SSP is decremented four times to \$0001 4FFA. As the MOVEM.L instruction is executed, the SSP is again decremented four times with each register transfer. There are 15 registers whose contents are pushed onto the stack by this instruction, so 60 bytes (15×4) are saved and A_7' is decremented 60 times. The final contents of the SSP are \$0001 4FBE.

After the environment is saved, actual processing can commence. In our interrupt program, a JSR KEYWORD followed by a MOVE D2, SAVEKEY causes two key presses to be stored at location SAVEKEY. To complete the interrupt exception routine, the environment is retrieved by using a MOVEM.L (A7)+,D0–D7/A0–A6. Besides restoring the original contents of the registers, this instruction also increments the SSP back to \$0001 4FFA (see Example 8.5). An RTE, ending the interrupt program, pulls the program counter and status register information from the stack; this instruction also restores the original data in the SSP through incrementing. The original program is resumed with all data restored that it had at the time of the interrupt plus a word of new data at location SAVEKEY.

Creating the Interrupt Program

The level 5 program that is being used can be entered manually into memory like any other program or it can be stored into memory as part of the original user program. The advantage of loading the interrupt program manually is that the assembler can do the actual translation from source to object code. The disadvantage is the need to reload the interrupt program independently every time you load in the user program. This disadvantage is overcome by having the user program transfer the object code into the necessary RAM locations. Then, each time the user program is loaded into memory, the interrupt program also is loaded in automatically. The problem with using this method is that the interrupt source code must first be translated into object code. This object code can then be loaded as data by using a series of MOVE instructions.

EXAMPLE 8.6

Write a program segment that stores the level 5 interrupt routine described in this section starting at location INT5 (\$00 8050). The location of the interrupt mask TRAP routine is labeled TRAPF.

Solution:

Step 1 is to assemble the object code for the interrupt program. To facilitate this example, we will designate KEYWORD as address $00 2040 and SAVEKEY as $00 1100. The machine code becomes

```
MOVEM.L D0-D7/A0-A6,-(A7)  = $48E7FE
JSR KEYWORD                = $4EB82040
MOVE D2,SAVEKEY            = $31C21100
MOVEM.L (A7)+,D0-D7/A0-A6  = $4CDF7F
RTE                        = 4E73
```

Also:

```
ORI #$0700,6(A7)  = $006F07000006   * Preset mask
ANDI #$FCFF,6(A7) = $026FFCFF0006   * Set interrupt mask
```

The second step is to write a series of MOVE instructions to place the data into memory:

```
MOVE.L #INT5,A0               * Initialize pointer
MOVE.L #$48E7FE4E,(A0)+       * Load interrupt 5 program
MOVE.L #$B8204031,(A0)+
MOVE.L #$C211004C,(A0)+
MOVE.L #$DF7F4E73,(A0)+
MOVE.L #TRAPF,A0              * Reset pointer
MOVE.L #$006F0700,(A0)+       * Load mask routine
MOVE.L #$FCFF0006,(A0)+
MOVE.W #$4E73,(A0)+
TRAP #15                      * Go to TRAP 15
```

Placing the segment of Example 8.6 at the beginning of the user program causes the interrupt program to be stored in memory and the interrupt mask bits to be modified to level 4. While this process is a little cumbersome, it is done only once and has the advantage of providing both the interrupt program and the interrupt mask update each time the user program is entered and executed. The keyboard itself requires modification (see Figure 8.3) to supply an interrupt request whenever a key is pressed.

8.17 SINGLE-STEP (TRACE) EXCEPTION ROUTINE

An assembler, when executing a program in a single-step mode, displays the contents of all registers onto a CRT so that the results of the execution of the last instruction can be examined. The basic system example contains a 7-segment LED display bank that requires a slightly different approach to the problem of displaying the register contents.

To use the single-step function of the 68000, the trace exception must be called at the end of each instruction. This is done by initially setting the T bit in the status register high by a process similar to setting the interrupt flag bits to a specific level.

EXAMPLE 8.7

Using TRAP 15, write a program that updates the interrupt mask bits to a level 4 and places the 68000 into the TRACE mode. All other status bits are to remain unchanged.

Solution:

The TRAPF program used earlier to set the interrupt mask bits to a level 4 can be used here with one minor change. The data for the ORI instruction is \$8700 instead of \$0700. This causes b15, the T flag, to be set high. This bit is not affected by the ANDI instruction as shown.

TRACE Exception Program

The TRACE program itself must perform a number of basic tasks. First, it must save the environment so that the next instruction in the user program can be executed successfully after the TRACE routine is completed. Second, it must create a method to allow the register contents to be identified and displayed on the 7-segment displays. After each of the registers is viewed, the TRACE routine restores the environment and returns to the user program to execute the next instruction.

The TRACE exception, like most other exceptions, pushes the status register and program counter data onto the stack. Using the same multiple MOVE instruction to save the environment as before but also including A_7 in the register list, it places the remaining data and address registers onto the stack. This data and the contents of the stack pointer A_7' are to be left untouched so that when one is ready to return to the user program, these data are available to be replaced into the applicable registers. Note that the contents of A_7' must *not* be altered while instructions in the TRACE routine are executed. Changing A_7' could cause the stack to become lost and, with it, the original data from the user program necessary to resume its execution.

Retrieving TRACE Program Data

A_7' now points to the next available location on the stack with all data sitting in address locations of larger value (recall that the stack pointer is decremented as data are pushed onto it). There is no need to create a new address pointer, but there is a need to allow that pointer to be changed as we use the available data. A_7', however, must not be changed. The paradox is resolved easily by using the index addressing mode. Let us select D_3 as an index register and D_4 as a counter to keep track of how much data have been accessed. There are 16 data and address registers as well as the program counter and status register to be displayed, for a total of 18 registers. The question now becomes, how much of a register's contents are visible at one time on the display? Our display is set normally to show an address and its 16-bit contents or, in effect, the address bus and the data bus. This yields 6 displays for the address area and 4 for data, making a total of 10 seven-segment LED displays.

Using A for address and d for data, each of the general-purpose registers can be identified using two units of the display to hold the appropriate letter and register number. Further, the same two units can show program counter (PC) and status register (Sr). This leaves eight units for the contents—just enough to illustrate all 32 bits of each register. Figure 8.4 illustrates the formation of data on the display.

To begin the TRACE program, after saving the environment, the index register is cleared and the counter register is set to $12. The keyboard is incorporated at this point. The computer displays the next register's contents whenever key 1 is pressed.

```
TRACE   MOVEM.L D0-D7/A0-A6,-(A7)   * Save the environment
        CLR.L D3                    * Clear index register
        MOVE #$12,D4                * Initialize counter
NEXT    JSR SHOW                    * Display data
        JSR KEYSCAN                 * Scan keypad
        CMPI #1,D2                  * Look for key 1
        BNE NEXT                    * Test loop
```

Program 8.1

You might want to set up an initial message before jumping to SHOW; otherwise, the display will show the last entries from the keyboard before the TRACE program is begun. This is left up to the programmer.

The next step is to get the data on the stack and cause them to be displayed. Keep in mind that these locations can be read as many times as one wants without destroying the data there.

```
MOVE.L 4(A7,D3.W),D5    * Get data
```

The location accessed is found by adding the value in A_7 to the value in D_3 ($0000 0000) to offset $4. This points to a location four addresses higher than where

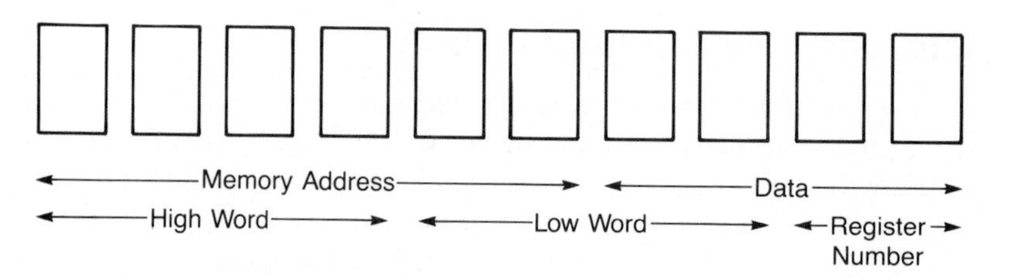

Figure 8.4 68000 system display

the stack pointer is pointing. This location is the high byte of the last data placed on the stack by the MOVEM instruction, which is the contents of D_0. This index mode instruction creates the correct pointer without causing any register to change except the destination register, D_5. After the data are retrieved, some form of coding routine, called CODE, is required to translate all 32 bits to their seven-segment codes and to store those codes in the appropriate memory locations for use by the SHOW subroutine. The CODE routine also is required to check the counter and to develop the codes for the register designations. Returning from the CODE routine, the index register is increased by 4 and the counter is decremented and tested.

```
JSR CODE                    * Create 7-segment code
ADDQ #4,D3                  * Move index pointer
DBF D4,NEXT                 * Loop back to SHOW
MOVEM.L (A7)+,D0-D7/A0-A6   * Restore environment
RTE
```

Program 8.2

Since the DBF instruction returns to SHOW, the register's contents are displayed until key 1 again is pressed. When the counter finally reaches -1, the loop falls through, the environment is retrieved, and the user program is resumed. The RTE, by returning the status data to the status register, sets the T flag high again, so that after each user instruction the TRACE routine is run again.

PROBLEMS

8.1 Which exceptions generally do *not* end in RTE?

8.2 Which exception has the highest priority? The lowest?

8.3 What is the vector address for exception 12?

8.4 Which exception tests the condition of a flag in the condition code register portion of the status register? Which flag is tested and what level must it be to allow the exception to be called?

8.5 Which status flags are affected immediately by calling a TRAP exception? What status do they become?

8.6 How is an interrupt acknowledge indicated by the 68000?

8.7 How does a peripheral device verify that its interrupt is the one being serviced?

8.8 Which code on the IPL lines is used as a non-maskable interrupt? As a level 3 interrupt request? To indicate no interrupt request?

8.9 What are two methods to develop an interrupt exception number?

8.10 The interrupt mask bits are set to a level 3. A level 6 interrupt request is sent to the 68000 on the IPL lines. List the activity that occurs from the moment the interrupt request is received and the first exception instruction is executed. Assume auto-vectoring is used.

8.11 An exception routine uses registers D0, D1, D2, A0, A3, and A5 only. Write the instruction(s) to save this limited environment.

8.12 How many times will the 68000 rerun the same bus cycle if a $\overline{\text{BERR}}$ and $\overline{\text{HALT}}$ are returned each time?

8.13 List the bus error stack. Use SSP as a label for the initial stack pointer contents.

8.14 The interrupt status flags are set to level 0. What is the maximum number of different consecutive interrupts that can be serviced assuming that they occur in the correct sequence and before an RTE is executed?

8.15 What prevents the TRACE exception routine from being executed in single-step if the T bit was previously set?

8.16 Write a TRAPF program to set interrupt level 2 and place the microprocessor into the TRACE mode.

8.17 What is the vector address of a TRAP 7 exception?

8.18 The supervisor stack pointer holds $0000 413C after the environment is saved during an interrupt exception. Register D3 holds $0018 0012. Write an index mode instruction to load 32 bits starting at location $00 4150 into register D0.

8.19 Using discrete logic gates, design a circuit to return an active $\overline{\text{VPA}}$ in response to interrupt acknowledge for interrupt level 6.

8.20 Design a circuit to deliver exception vector 70 on data lines D_0–D_7 in response to a level 3 interrupt acknowledge.

9

68000 APPLICATION PROGRAMS

CHAPTER OBJECTIVES

Now that the hardware for a basic system has been developed, we can use it to generate a number of application programs to extend our understanding of the 68000's instruction set and its uses. The programs in this chapter use the address map and labels developed in earlier chapters, which are illustrated in the basic system block diagram of Figure 9.1 on page 258.

9.1 THE BASIC SYSTEM

The 68000-based system, thus far developed, includes the microprocessor (the 68000 itself); ROM, which holds the initialization program and exception routine vector addresses along with an EXORmacs assembler; a page of user RAM in address spaces \$01 0000–\$01 FFFF; and PIAs interfacing the display and keyboard of the system. The external connections to the system are a 16-bit parallel port (PIAs 5 and 6) and two serial ports (ACIAs 1 and 2). The block diagram also shows modems connected to the ACIAs, but the serial interfaces are not limited to data communications applications through these modems. They can be disconnected and the serial ports used for additional applications. It also is assumed that the necessary hardware and software is in place to handle $\overline{\text{DTACK}}$, $\overline{\text{BERR}}$, interrupt encoding, and auto-vectoring, if applicable.

9.2 INITIALIZING THE SYSTEM

The initialization program resident in ROM has several tasks to perform. Included among these tasks is the configuring of PIAs 1–4 to drive a CRT display circuit and input data from a full-ASCII keyboard. Additionally, the interrupt mask level is selected to be level zero (all three I bits in the supervisor status register are set to zero), and the operating mode is returned to the user. This is done by clearing the first

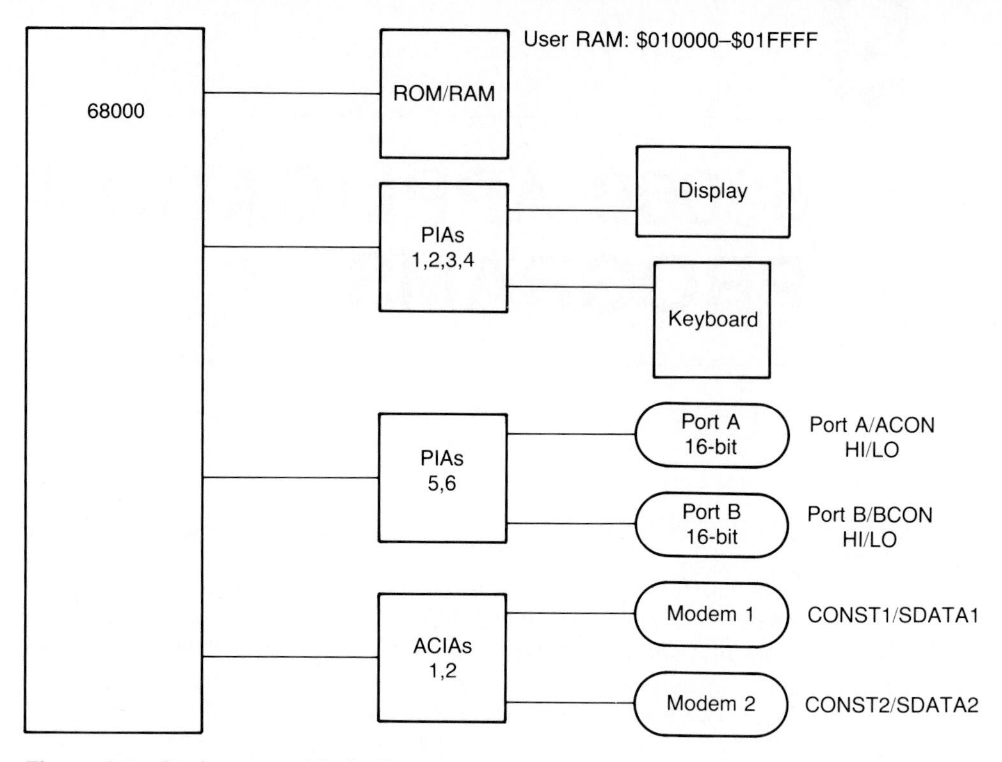

Figure 9.1 Basic system block diagram

stack location in memory, followed by storing a selected starting address into the next stack area. Executing an RTE instruction loads the status register and program counter from the stack, placing the system into the user mode (S flag is set to zero from zero data on the stack).

9.3 RELAY TEST STATION APPLICATION

Relays (see Figure 9.2) are electromechanical devices that isolate controlling signals from switched signals. A current is applied to a coil of wire wound about a ferric core. The current creates a magnetic field that causes a movable wiper arm to break contact with a station contact (normally closed contact) and move toward the core until it is stopped by a second stationary contact (normally open contact). If the current is removed from the coil, the magnetic field collapses, releasing the hold on the wiper. A spring causes the wiper to return to the normally closed stationary contact.

The quality control tests performed on this device include

1. **Coil Resistance** The resistance of the coil is selected to be within designated limits so that a voltage applied to the coil produces enough current to cause the wiper to move to the normally open contact.

Figure 9.2 Relay

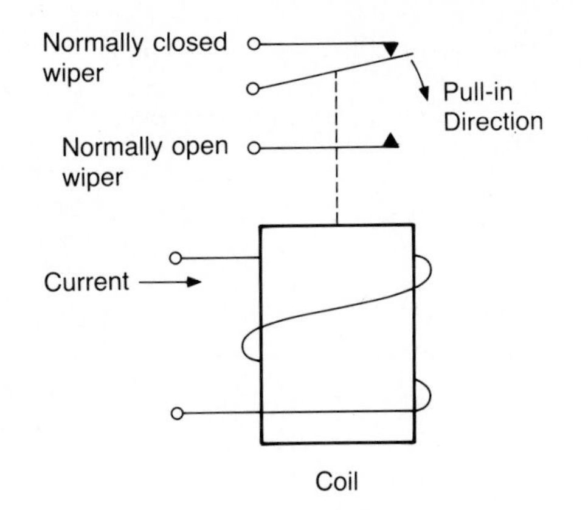

2. **Contact Resistance** The resistances between the wiper and the stationary contacts are measured to ensure that a solid closure is made when the wiper connects with either contact.
3. **Pull-in Time** The time it takes for the wiper to move from the normally open contact to the normally closed contact is another parameter of the relay.
4. **Drop-out Time** The amount of time it takes for the wiper to return to the normally closed contact once coil voltage is removed is as important as the pull-in time.
5. **Contact Bounce** After the wiper makes initial contact with either stationary contact, it tends to bounce off the contact, breaking connection, and then returns to make connection again. This bouncing occurs several times before the wiper settles onto the contact permanently. Contact bounce is the measure of time from the first closure until the last bounce closure.
6. **Pull-in Voltage** The minimum amount of voltage required to develop sufficient current to move the wiper to the normally open contact is measured.
7. **Drop-out Voltage** After the relay is pulled in, it takes less current to hold the contact against the normally open contact. The coil voltage is reduced to discover the minimum voltage necessary to hold the wiper in place. The point at which the wiper releases contact with the normally open contact is called the drop-out point.

Relay Test Station

The relay station is designed to be a totally automated test station. A relay is inserted in a test socket equipped with a set of contacts that connect to several test areas within the station. A photodiode senses when the relay is in place and sends a high logic level to port A pin PA_{14}. When the system detects this signal, it issues a stepper pulse signal from PA_{12}, causing the test socket to be moved into test position 1. This position performs the static resistance test measurements of the coil and the contacts illustrated in Figure 9.3 on page 260.

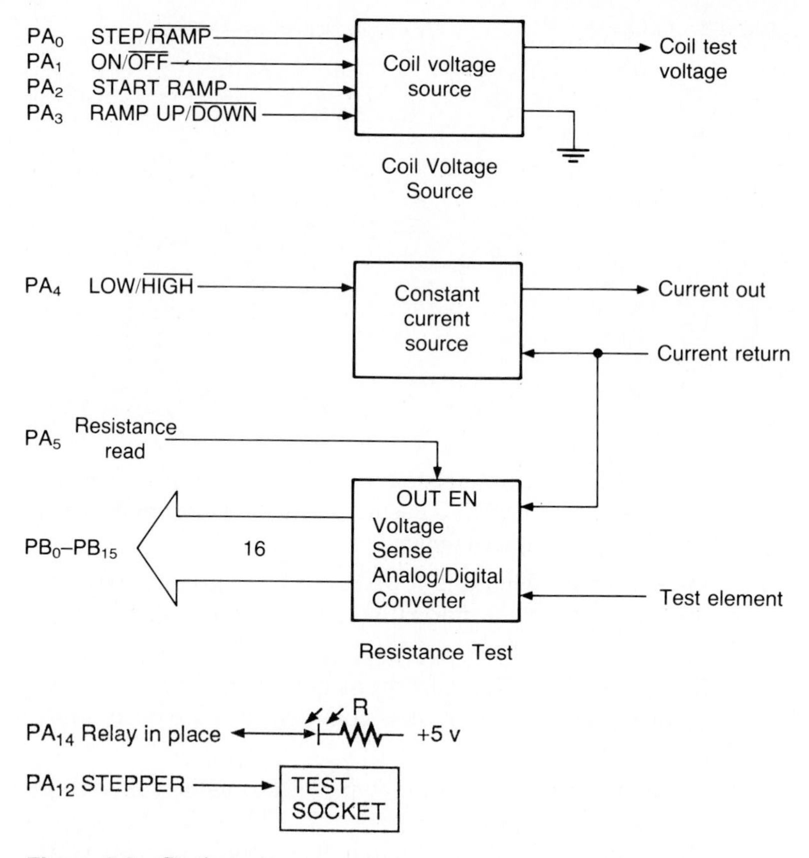

Figure 9.3 Static tests

Initially, the constant current source is selected to be in the high mode so that the larger coil resistance can be measured. The voltage sense circuit measures the voltage drop across the coil created by the current passing through it. This current is considerably lower than that necessary to activate the relay. The voltage read is then converted into a 16-bit digital quantity. After a short delay to account for the initial application of the current to the coil, the system sends out a resistance read signal on PA_5 and the digital output from the voltage sense circuit is placed on the 16-bit bus connected to port B (PB_0–PB_{15}). The computer can then read the results from port B.

Following time delay, the test socket is stepped to the next test area. Here, the constant current source and the voltage sense circuit are connected to the wiper and stationary contacts. The constant current source, which is now set low, is connected so that the current is applied to both stationary contacts simultaneously. The voltage sense circuit monitors the return from the wiper. At this point, the coil voltage source, connected across the coil, is set off so that the voltage sense circuit measures the contact resistance between the wiper and the normally closed contact. After this is read by the computer, it sets the coil voltage source to pull in the relay. This is done by setting PA_0 to STEP and PA_1 to ON. START RAMP and the RAMP UP/$\overline{\text{DOWN}}$ are ignored when the coil source is in the step mode. Once the relay pulls in, the contact resistance between the wiper and the normally open stationary contact can be measured.

Dynamic Tests

The dynamic tests involve activating and deactivating the relay and measuring time and voltage parameters. After the resistance tests are completed, the stepper pulse is issued once more to cause the test socket to be moved to the first dynamic test position, PIDO. PIDO is an abbreviation for pull-in and drop-out voltage tests. The coil source, connected as shown in Figure 9.4, is set to the ramp-up mode and turned on. Once a ramp start has issued from PA_2, PA_{10} is monitored for a change of

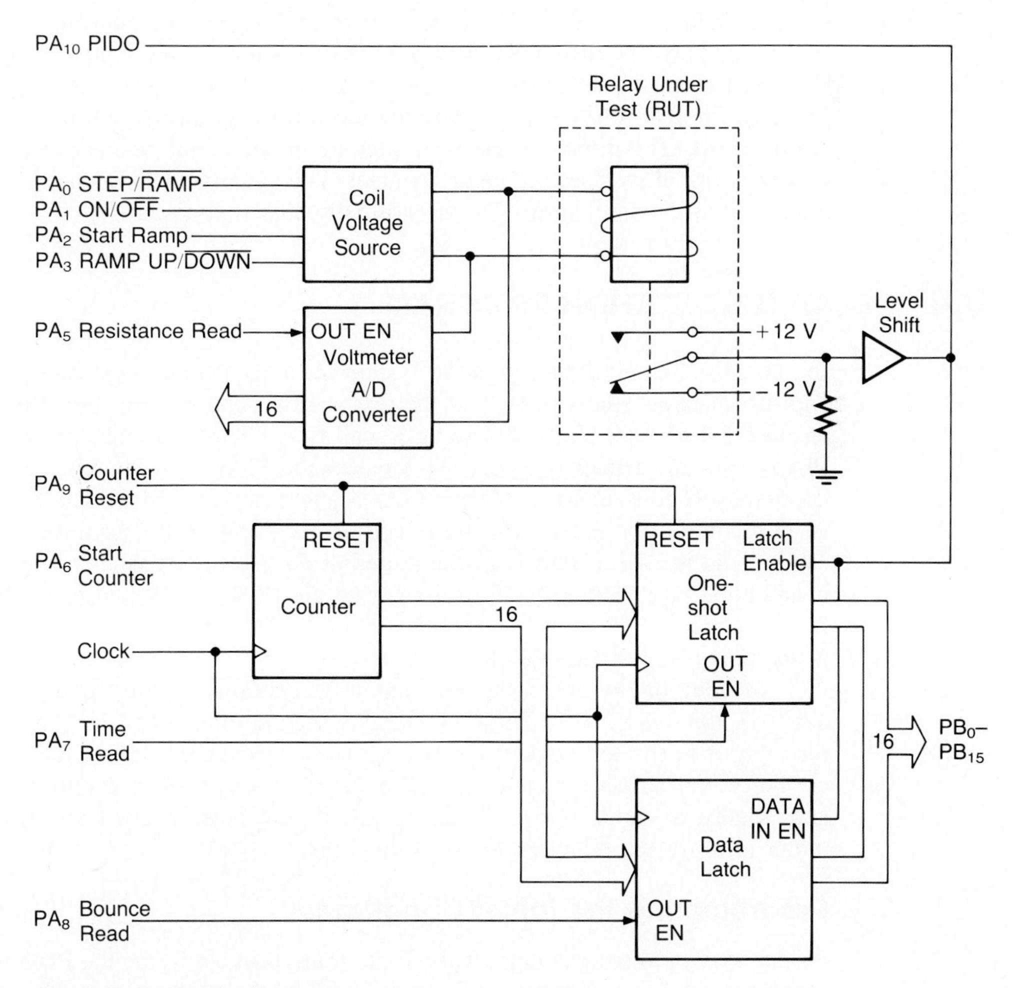

Figure 9.4 Dynamic tests

state, which occurs when the relay makes contact with the normally open wiper. The voltage read at port B is the pull-in voltage just necessary to activate the relay. The ramp continues to full coil voltage and remains there. To test for the drop-out level, PA_3 is set low to cause the coil source voltage to ramp down. A START RAMP is issued again, and PA_{10} is monitored for a change of state. The change now occurs when the relay wiper remakes contact with the normally closed stationary contact.

The remaining circuitry, illustrated in Figure 9.4, is used for time measurements. The single counter initially is reset. Then the coil source voltage is set to a step voltage again and is turned on at the same time the counter is started. When the relay switches, the signal from the wiper causes the one-shot latch to hold the count when the first contact was made with the normally open contact. This same data is loaded into the data latches; but because these are dynamic latches, the data is updated continually every time contact is broken and remade with the normally open contact. Eventually, the wiper settles onto the normally open contact and this count is held in the latch. The computer reads both counts, storing the first as the pull-in time and the difference between the two counts as the bounce time. Drop-out time measurements are done in the same manner, by first resetting the counter and the latch (high on PA_9) and then starting the counter at the same time the coil source is turned off. This time the one-shot latch holds the count when contact is resumed with the normally closed stationary contact. The dynamic data latches hold the count after the wiper comes to rest on the normally closed contact.

9.4 RELAY TEST STATION PROGRAM

Initially, the PIAs interfacing to the system must be configured to handle the signals and data between the system and the test station. The system reset function partly ensures that no false starts can occur. Recall from Chapter 5 that a reset signal to the PIAs clears all of their registers. This places the PIAs into the input mode, causing no signals to be issued from them. PIAs 5 and 6 are used in tandem as 16-bit ports so that control commands can be issued at once and 16-bit resolution on the data can be read in. All of port B is configured as an input port, whereas most of Port A is an output port. The exceptions for port A are PA_{14}, which senses that the relay is in its socket, and PA_{10}, used for PIDO control. Table 9.1 summarizes the allocation of the port pins for the relay test station.

In addition to the basic coil and test functions, station control signals are available through port A. These include the second input line, PA_{14}, to detect when the relay is in the test socket; a reject signal to eject relays into a reject box if they are defective; an accept eject, to cause the relay to be ejected into an accept bin; and, finally, a stepper pulse signal, to initiate the pulse needed to cause a stepper motor to move the relay test socket to the next test area.

Programming the Initial Conditions

For the system to operate correctly, the program must configure the PIAs to pass data in the correct direction between the computer and the test station. Refer to Chapter 5

Table 9.1 Relay Station Interface

Pin	Function	Group
PA_0	STEP/$\overline{\text{RAMP}}$	Coil Voltage Source
PA_1	ON/$\overline{\text{OFF}}$	
PA_2	START RAMP	
PA_3	RAMP UP/$\overline{\text{DOWN}}$	
PA_4	LOW/$\overline{\text{HIGH}}$	Constant Current
PA_5	Resistance Read	Static Tests
PA_6	Start Counter	Dynamic Tests
PA_7	Time Read	
PA_8	Bounce Read	
PA_9	Reset Counter/Latch	
PA_{10}	PIDO Detect	
PA_{11}	Reject	Station Control
PA_{12}	Stepper Pulse	
PA_{13}	Accept Eject	
PA_{14}	Relay in Place	

Port A: Output control functions
Port B (PB_0–PB_{15}): 16-bit input port to read in test results

to refresh your memory on how to configure a PIA. Table 5.2 is the control register of the PIA that is referenced throughout this chapter. To select the direction of the individual port peripheral data leads, b2 of the control registers first must be set low. This is done by clearing the control registers of the PIAs in the following manner:

```
CLR.W ACONHI
CLR.W BCONHI
```

By selecting the port A side control registers of both PIAs for the first instruction and both B side registers for the second instruction, all four control registers of PIAs 5 and 6 of the basic system are cleared. To make port B lines all inputs, its associated direction register bits are set to 0's. Port A's output lines are selected by setting their associated direction bits high, and port A's inputs are set by making their associated direction leads low. For the port A direction registers, the high side control register connected to the upper data bus requires b14 and b10 to be low and the rest to be high. All lower port side leads are outputs, making the direction register bits all high. The combined direction register data (setting b15 low) is 0011 1011 1111 1111, or \$3BFF. The next two instructions set the direction as required for the test station:

```
CLR.W PORTBHI
MOVE.W #$3BFF,PORTAHI
```

This application does not use interrupts because this is a dedicated system; that is, the system does not handle any other tasks other than testing the relay. The control registers now are set so that interrupts are masked, and b2 is made high to allow data to be moved to and from the data registers of the PIAs:

```
MOVE.W #$0404,ACONHI
MOVE.W #$0404,BCONHI
```

To complete the system's initial setup conditions, the program must send data to the output port to prevent a false start. The relay in place pin, PA_{14}, then is monitored to detect when the system is ready to be used. The relay is not to be rejected or ejected initially (PA_{13} and PA_{11} low), and no stepper pulse is to be issued (PA_{12} low). The first tests are the static tests that require the coil voltage to be a step input (PA_0 high); but, initially, that voltage is off (PA_1 low). The ramp control leads (PA_2 and PA_3) are not used in the step mode and are set initially low. The first test is a coil resistance measurement requiring the constant current source to be in the high mode (PA_4 high). The counter and read control function are initially off, making their lines low. Finally, a reset signal is sent to the counter and latch circuits, initializing them to 0. Putting these in order produces the data 0000 0000 0001 0001, or $0011. After these data are sent to the port, b14 is monitored for a high condition, which indicates that the test relay has been inserted into its socket:

```
         MOVE.W #$0011,PORTAHI
RELAYIN  MOVE.W PORTAHI,D0
         ANDI.W #$40,D0
         BEQ.S RELAYIN
```

9.5 THE TEST SEQUENCE

A short delay sequence, written as a subroutine, allows the relay under test to settle into its tested condition before a reading takes place. This subroutine, called DELAY, is similar to those discussed earlier.

After the relay has been detected as loaded into its socket, a stepper pulse is initiated to move the relay into the first step area. Here a coil resistance is measured and compared with test limits. If the relay does not meet specifications, it is rejected and the socket is stepped on through to the initial loading area. If the relay passes this first test, it is stepped to the next station where it is tested for contact resistance. Two dynamic test areas for voltage and time follow. A total of four test areas are used, eventually requiring the issuance of four step pulses. The first of these moved the relay into the coil resistance test area, leaving three additional stepper moves to be made later. A count of the step pulses is maintained for the reject program to allow the relay test socket to be returned to the loading position in case of failure.

Coil Resistance Test

The following program segment is used to test the coil resistance of the relay:

```
MOVE.B #$3,D7              * Stepper pulse count
MOVE.W #$2011,PORTAHI      * Issue first step pulse
MOVE.W #$0031,PORTAHI      * Make measurement
BSR DELAY                  * Settling delay
MOVE.W PORTBHI,D1          * Read voltmeter
CMPI.W HILIMIT,D1          * Check high limit
BHI.S REJECT
CMPI.W LOLIMIT,D1          * Check low limit
BCS.S REJECT
MOVE.W #$2001,PORTAHI      * Issue second step pulse
```

This segment first initializes the reject step counter to 3 and issues the first step command. The step command activates a one-shot circuit to generate a pulse, which drives a stepper motor to move the relay socket a specified distance to each test area. The next instruction turns on the output buffer of the voltmeter/analog-to-digital converter circuit and resets the stepper pulse command line. The delay allows enough time for the socket to move into position and for the test circuit to settle in and the coil resistance measurement to stabilize. After the program returns from the DELAY subroutine, it reads the actual data from port B and compares them with the upper and lower coil resistance limits. If one or the other limit fails, the program execution is diverted to the REJECT routine. If they both pass, a new stepper pulse command is issued. The data for this command are essentially the same as before, except that the constant current source is changed into the low mode. The next station and program segment are used to measure contact resistance, which is much lower than the coil resistance.

Contact Resistance Measurements

The following program segment is used to measure contact resistance:

```
SUBQ.B #$1,D7              * Decrement stepper count
MOVE.W #$0021,PORTAHI      * Make measurement
BSR DELAY                  * Settling delay
MOVE.W PORTBHI,D1          * Read value
CMPI.W #LIMIT,D1           * Exceeds maximum value?
BHI.S REJECT
MOVE.W #$0023,PORTAHI      * Activate relay and make reading
BSR DELAY
MOVE.W PORTBHI,D1          * Read value
CMPI.W #LIMIT,D1           * Exceeds maximum value?
BHI.S REJECT
MOVE.W #$2008              * Issue stepper command
```

The stepper count is decremented by 1 to account for the command issued to move the relay to the contact resistance test area. Because the minimum for contact resistance is 0, only one comparison is made, that is, for maximum amount. First, the normally closed contact is measured with respect to the wiper arm. Then the coil voltage is turned on and the resistance between the normally open contact and the wiper is measured and tested. If everything is within limits, another stepper command is issued. Again, the data are the same except that the next tests are dynamic, requiring a changing (ramp) coil voltage instead of a step voltage (PA_0 set low).

Pull-In and Drop-Out Voltages

The following program segment is used to test pull-in and drop-out voltages:

```
            SUBQ.B #$1,D7                * Decrement count
            BSR DELAY                    * Settling delay
            MOVE.W #$002E,PORTAHI        * Start ramp
PULLIN      MOVE.W PORTAHI,D1            * Read port A
            ANDI.W #$0400,D1             * Isolate PIDO
            BEQ.S PULLIN
            MOVE.W PORTBHI,D1            * Read pull-in voltage
            CMPI #MAX,D1                 * Maximum voltage?
            BHI.S REJECT
            BSR DELAY
            MOVE.W #$0026,PORTAHI        * Start down ramp
DROPOUT     MOVE.W PORTAHI,D1            * Read port A
            ANDI #$0400,D1               * Isolate PIDO
            BEQ.S DROPOUT
            MOVE.W PORTBHI,D1            * Read drop-out voltage
            CMPI #MIN,D1                 * Minimum voltage?
            BCS.S REJECT
            MOVE.W #$2281,PORTAHI        * Issue stepper command
```

This time, the delay is issued immediately after the stepper count has been decremented. The ramp-up coil voltage is applied to the coil of the relay and the normally open contact is monitored for a closure (indicated by PIDO going high). When the closure is detected, the coil voltage is read from port B. This pull-in voltage is compared with the maximum allowable voltage. A second delay allows the ramp to continue to the rated coil voltage before the drop-out test begins. At this point, a command is issued to cause the coil voltage to begin to ramp down from the rated value. PIDO is monitored to detect when the normally closed contact makes contact with the wiper arm. Again, the coil voltage is read and compared. This time a minimum drop-out value is checked. Following the successful completion of this test, the last stepper command is issued. The data are changed to set up the system to test for pull-in and drop-out times. The voltmeter/analog-to-digital converter output is disarmed (PA_5 low), and the time measurement buffers are activated (PA_7 high). The coil voltage is returned to a step mode (PA_0 high), and the counter/latch is reset (PA_9 high).

Time Measurements

The following program segment is used to determine time needed for pull-in and drop-out:

```
SUBQ.B #$1,D7               * Decrement count
BSR DELAY
MOVE.W #$00C3,PORTAHI       * Start counter
BSR DELAY
MOVE.W PORTBHI,D1           * Read pull-in time
CMPI.W #PIMAX,D1            * Maximum time?
BHI.S REJECT
MOVE.W #$0303,PORTAHI       * Set up for bounce
MOVE.W PORTBHI,D2           * Read bounce value
SUB.W D1,D2                 * Compute bounce
CMPI.W BNCMAX,D2            * Maximum time
BHI.S REJECT
MOVE.W #$00C1,PORTAHI       * Remove coil voltage
BSR DELAY
MOVE.W PORTBHI,D1           * Read drop-out time
CMPI.W #DOMAX,D1            * Maximum time
BHI.S REJECT
MOVE.W #$0301,PORTAHI       * Set up for bounce
MOVE.W PORTBHI,D2           * Read bounce
SUB.W D1,D2                 * Compute bounce
CMPI.W BNCMAX,D2            * Maximum time
BHI.S REJECT
MOVE.W #$2000,PORTAHI       * Issue accept/reject
BSR DELAY
MOVE.W #$1000,PORTAHI       * Issue last stepper command
BRA.S RELAYIN               * Return to start loop
```

The time tests rely on the hardware circuitry of the counter and the latching network. Pull-in and drop-out times are latched upon detection of the first connection by the wiper to the appropriate contact. The value entered by the last PIDO pulse is the bounce time. Clearing the counter and the latch does not change the contents of the bounce buffer. After the stepper count has been decremented one last time, a delay allows the relay test socket to settle into place. The counter is started and coil voltage is applied to the relay. A second DELAY subroutine is executed to allow time for pull-in and bounce values to be latched into their registers. The pull-in time then is read and the command to set up for a read of the bounce register is issued.

Notice that the coil voltage remains on when this command is issued. The bounce buffers are read, and the actual bounce value is calculated by taking the difference between the two times. Each has a maximum value to be met and is tested once the actual quantity is known.

The counter is reset at the same time that the bounce read is set up. After the pull-in readings, a drop-out command is issued by starting the counter and removing the coil voltage. Again, the DELAY subroutine is called to allow time for the measurements to be loaded into their proper registers. The drop-out time is read and tested. A bounce setup and counter reset command is issued once more, and the bounce time is read and computed as before. If these tests prove the relay is good, it is ejected into the accept bin and a final stepper command is issued to place the now empty socket into loading position.

Reject Processing

The following program segment is used to set up the REJECT subroutine:

```
REJECT    MOVE.W #$1000,PORTAHI      * Stepper command
          BSR DELAY
          SUBQ.B #$1,D7              * Decrement count
          BNE.S REJECT
          MOVE.W #$0800,PORTAHI      * Reject command
          BSR DELAY
          MOVE.W #$1000,PORTAHI      * Move to load area
          BRA.S RELAYIN              * Return to start
```

Upon a test failure, the REJECT routine is executed. The relay is moved into the last test area and ejected into the reject bin instead of the test bin. Finally, the empty socket is moved into the load area and the system waits until it detects that the next relay has been inserted into the test socket.

Final Comments

The program developed in this section is used for a GO/NOGO test arrangement. However, there are enough time and data available to accumulate actual test results for the relays sent through the system. These data easily can be stored into memory following the reading of port B for each test made and then be transferred to a printer after the acceptance or rejection of the relay. Additional programming can process accumulated information on all relays run. These data can be printed out or assimilated into an average for each test quantity, as required by the quality testing facility.

The system described allows the manual loading of each relay. An automatic loader can be added to load another relay after each one is accepted or rejected. Depending on the sophistication of the hardware and the additional software management required, a system can be made to load and test relays continually. Tests at one test area can be embedded into the time delays required at other test areas. One way to do this is by using interrupt control at each test area, whereupon a return from an interrupt from one test area causes the test process required by a different test area to resume.

9.6 ROBOTIC ARM APPLICATION

This application uses interrupts and assorted interface configurations to manipulate dangerous radioactive material at a remote site. The control commands are supplied by voice, and several testing conditions are used. The overall system is minimal and is included only to demonstrate the use of the 68000 and associated interfaces. The central item in the system is a robot arm (illustrated in Figure 9.5) capable of five different motions:

1. Clockwise body rotation
2. Up and down movement of the entire arm
3. Clockwise rotation of the arm
4. Up and down movement of the hand portion
5. Opening and closing the hand

In addition to voice control of the arm movements, there are several test monitoring conditions as well as an emergency operation to protect the users.

Robot Interface

The top portion of the interface diagram (see Figure 9.6, page 270) indicates the voice recognition input to port A. The commands are interpreted by software control. The

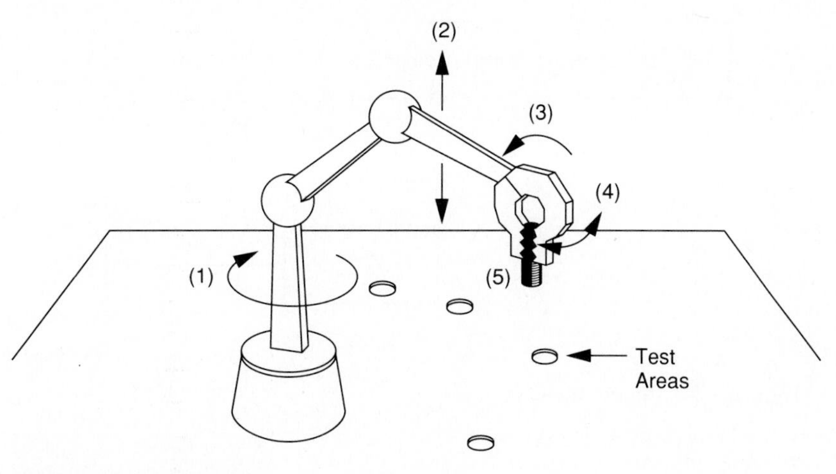

Figure 9.5 Robot arm

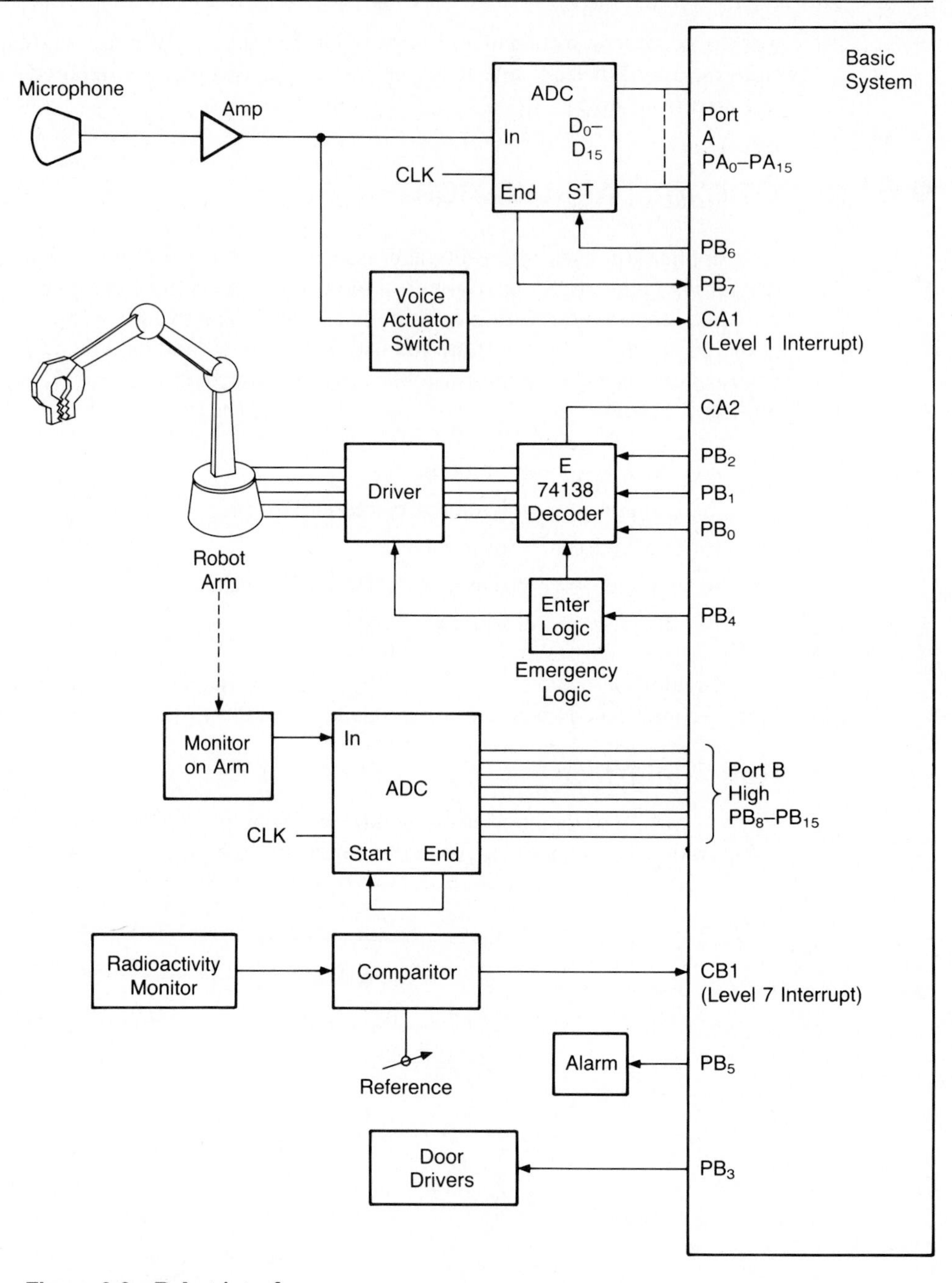

Figure 9.6 Robot interface

microphone and amplifier pick up the communication and deliver it to two places. The first is a full-wave rectifier used as a voice-actuated switch. When the audio level is high enough, the voice-actuated circuit puts a negative pulse on the CA1 line. This line, the interrupt request input to PIA 5, generates a level 1 interrupt to the 68000.

The second circuit driven by the voice amplifier is a 16-bit successive approximation analog-to-digital converter (ADC). The conversion begins when a positive pulse is placed on the start conversion (ST) input of the ADC. When the conversion is complete, the END output of the ADC goes high. The digital information is transferred to port A and is read by the microprocessor. The input is sampled many times, with the resulting digital information compared with a lookup table stored in memory. This table contains the digital data sequence for the following commands:

1. UP, to move the arm up
2. DOWN, to move the arm down
3. TURN, to cause the body to rotate
4. CLOSE, to close the hand
5. OPEN, to open the hand
6. HAND, to rotate the hand
7. ABOVE, to move the hand up
8. UNDER, to move the hand down
9. STOP, to stop the current action

Once the instruction is interpreted by the 68000, the correct code is sent to a 74138 decoder chip connected to port B pins 0, 1, and 2 (see Table 9.2). The 74138 in turn sets one of eight lines active, causing a corresponding driver to drive the motors necessary to carry out the command.

The peripheral handshake lead CA2 is used in the steady state output mode as an enabling/disabling signal to the 74138 driver decoder. This signal is set high by any of the commands except STOP. STOP causes this signal to be returned to a low state and disables the 74138. In the disabled condition, all 74138 outputs are driven to their inactive conditions regardless of the input to the device.

Table 9.2 Driver Commands

PB_2	PB_1	PB_0	Decoder	Robot Action
0	0	0	0	Open Hand
0	0	1	1	Close Hand
0	1	0	2	Arm Up
0	1	1	3	Arm Down
1	0	0	4	Rotate Body
1	0	1	5	Rotate Hand
1	1	0	6	Hand Up
1	1	1	7	Hand Down

Note: CA2 as output to enable/disable driver

An additional output from port B pin 4 (PB_4) is used as an emergency override. A set pattern is generated by the emergency logic to store the radioactive material into a safe container, should an emergency situation develop. A Geiger counter connected to a comparator monitors radioactivity levels for possible danger. The expected safe limit is set by a reference adjust input to the comparator. If the radioactivity exceeds the safe limit, a level 7 (non-maskable interrupt) is generated that starts a program to sound an alarm and initiate the emergency sequence.

Test monitoring uses a monitor connected to the robot arm. The data delivered by this monitor continually are converted to an 8-bit digital data word and fed to the port B high side (PB_8–PB_{15}). The continual analog-to-digital conversion is performed by connecting the END output to the ST input of the ADC. Whenever a conversion is completed, the ADC sends a pulse out on the END line. By connecting this lead to the ST input, the pulse needed to start the next conversion then is supplied by the END pulse of the last conversion.

The remaining port B output lead (PB_3) is a signal that closes the lead doors covering the glass enclosure either manually or as part of the emergency routine.

Configuring the System

The initial task of the program is to configure the PIAs to handle the hardware connected to the ports. Refer to Chapter 5 for port address labels (see Table 5.4) and the control register contents (see Table 5.2) during the following discussion. Initially, the control registers are cleared to make b2 low, allowing access to the direction registers of the PIAs. Next, the direction registers are loaded with the data needed to select port A as a 16-bit input port (all direction bits low) and to have port B handle the control functions, as illustrated in Figure 9.6 (bits PB_0–PB_6 are outputs and PB_7–PB_{15} are inputs). After the direction selection has been made, the control registers are configured to allow interrupt requests to be generated on $\overline{IORQA}$ and $\overline{IORQB}$ of PIA 5 in response to a low signal applied to CA1 and CB1, respectively.

```
START   CLR ACONHI              * Set b2 low
        CLR BCONHI              * Set b2 low
        CLR PORTAHI             * Port A-input
        MOVE #$007F,PORTBHI     * Configure port B
        MOVE #$3D04,ACONHI      * CA1 and CB1 interrupts
        MOVE #$0404,BCONHI      * Set b2 high
        CLR PORTBHI             * Set all outputs low
```

Data \$04 into a control register sets b2 high and disables interrupts initiated from that PIA's CA and CB lines. \$3D into port A's high-side control register enables the interrupt requests in response to a low transition on CA1 and CB1. Additionally, CA2 is selected as an output control lead and enables or disables the driver decoder. Making all output leads (including CA2) low effectively disables the system. The arm driver and alarm are shut off and the lead doors remain closed. During this condition, the canister in the test area is monitored continually for radioactivity, and the results are recorded as part of a life test of the sample.

```
LIFE    MOVE #255,D5           * Sampling count
        MOVEA.L #MON,A5        * Storage pointer
READ    MOVE PORTBHI,(A5)+     * Read data
        BSR DELAY              * Sampling delay
        DBF D5,READ            * Data loop
        BSR TRANSFER           * Move data to disk
        BRA.S LIFE
```

Two hundred fifty-six readings are made (inner loop) before the data are moved as a block to bulk storage area. This part of the system is not illustrated, but would consist of a DMA device that takes the data from the system RAM and loads it onto a hard disk storage medium. After the transfer is completed, the life test is resumed. The test is interrupted when a test technician activates the voice circuit with one of the commands.

Voice Exception Routine

A sufficiently loud voice command actuates the voice-controlled switch, sending a low pulse on the CA1 line. Exception level 1 is detected by the 68000. Current program counter and status register data are pushed onto the stack and the exception program begins. The first instruction of that program saves the contents of D_5 and A_5 on the stack as well, so that the life test can be resumed when the exception is completed. Then the system is set up to handle the command made by the technician.

```
VOICE   MOVEM.L D5/A5,-(A7)
        MOVEA.L # TABLE,A0     * Code table
        MOVE #SAMPLE,D3        * A/D sample count
```

A_0 contains the value for the pointer to the memory area that holds the incoming A/D values. D_3 holds the count of the number of information samples needed to determine the command. The greater the number of samples, the more accurate is the interpretation of the word spoken. The large size of the directly accessible memory of a 68000 system provides for a large number of samples to be taken, stored, and compared with an extensive lookup table.

```
CONV    MOVE.B #$48,PORTBLO    * Set ST high
        DSR DELAY 2
        CLR.B PORTBLO          * Set ST low
TEST    BTST.B #8,PORTBLO      * Check END
        BEQ.S TEST             * Is END low?
        MOVE PORTAHI, (A0)+    * Store voice data
        DBF D3,CONV            * Start next conversation
        BSR COMPARE            * Check data
        MOVE.B D0,PORTBLO      * Motion code
        MOVEM.L (A7)+,D5/A5    * Get D5 and A5
        RTE                    * Back to LIFE test
```

The ST line is pulsed and the END line monitored to perform the conversion from analog data created by the voice command into digital data. The digital data then are compared with a lookup table accessed in the COMPARE subroutine. Once the command is determined, the COMPARE subroutine returns the command code in D_0. This code is sent to the arm driver decoder and the motion is begun. While the arm is moving, the LIFE test resumes so that minimal data are lost. This condition remains until the next command is given, at which time the process is repeated. Once the canister holding the sample is in place, any number of tests can be performed. The software for these tests might include further manipulation of the canister by the arm or use of the arm monitor for data collection. These data are read from port B's high byte. The software chosen to perform these tasks depends on the test to be made.

Emergency Exception

The hardware for an emergency includes a monitor for radioactivity and a comparator manually set by the operator at a desired safety limit. Once this limit is exceeded, a low pulse is placed on CB1, causing a level 7 interrupt to be initiated. This exception sends out an alarm, closes the lead doors (opened as part of the data sent to port B in the VOICE exception routine), and initiates an emergency sequence to the arm drivers. This sequence begins by setting the emergency line (PB_4) high. The remaining action is under hardware control to reduce the time involved in carrying out the routine. The circuitry in the emergency area requires monitoring the current positions of the arm and test sample. The sample then would be moved to a lead safety area and sealed, to contain the radioactivity and any leakage. Once the sample is secured, the life test is resumed to reduce loss of data and to monitor the sample further.

```
EMERG   MOVEM.L D5/A5,-(A7)     * LIFE test register
        MOVE.B #$38,PORTBLO     * Emergency code
        MOVEM.L (A7)+,D5/A5
        RTE
```

Note that the life test is resumed, even as the emergency sequence is being executed, so that data regarding the cause and result of the emergency situation continue to be gathered. The operator reads the data and then determines the next move and issues the necessary voice command.

PROBLEMS

9.1 Using the basic system circuit (see Figure 9.1) and the labels and devices discussed in Chapters 5 and 6, write an interrupt program that is initiated by an incoming ring signal from the telephone lines. This 110-Vac, 20-Hz signal is rectified and applied to CA1 of PIA 5. In response to the ring signal, ACIA 1 is configured to send and receive asynchronous ASCII data with odd parity and one stop bit. Transmit control is maintained under interrupt control using a level 2 interrupt. Receive transfers are under a level 3 interrupt control generated by the ACIA. The ring interrupt is a level 1 interrupt generated by PIA 5. Actual transmitted data are sent out under direction of a SEND subroutine, with the number of characters to be sent held in location COUNT. The last receive character is followed by a break sequence (more than 10 consecutive 0 data bits).

9.2 Develop the hardware necessary to interface the basic system for an application selected by your instructor.

9.3 Write the software programs required for the application in Problem 9.2.

9.4 Create the hardware and software necessary for an application of your own choice following the guidelines set by your instructor.

9.5 Draw a schematic diagram of the voice-actuated circuit for the robot arm application in Section 9.6. Include level detection, rectification, and pulse generation in your circuit.

10

THE 68000 FAMILY OF MICROPROCESSORS

CHAPTER OBJECTIVES

The 68000 16-bit microprocessor is the principal device heading a full line of devices that include a byte-oriented version, the 68008; a virtual memory processor, the 68010/12; and a 32-bit CPU, the 68020. This chapter deals with the 68008 and the 68010/12. Among the new concepts introduced with these microprocessors is the idea of virtual memory, which allows a system designer to forgo the outlay of a large physical memory area. Instead, the microprocessor detects when access is made to a logical rather than a physical area of memory and then performs a memory swap. This chapter explains virtual memory in relation to the 68010 and 68012 microprocessors.

10.1 GLOSSARY

bus arbitration Process by which control of the system buses is designated in a direct memory access environment.

cache Local RAM included in a device or separate from it that allows quick data accesses.

CPU space Designated functional operations as a result of exception processing. Examples are the interrupt acknowledge function in response to interrupt exceptions and breakpoint processes in response to designated illegal instruction exceptions.

instruction continuation Allows an instruction to be continued at the bus cycle within the instruction cycle where the interrupt occurred.

memory management unit In a virtual memory environment, controls data swapping between physical and logic memory address spaces.

virtual memory System by which a small amount of physical memory appears to be much larger. Memory accesses are handled in physical memory, whose data content is changed as needed with external mass-storage contents.

10.2 THE 8-BIT SPIN-OFF: THE MC68008

While the 68000 16-bit microprocessor was making an important impact on the industry with its expanded data size (compared with the 8-bit predecessors) and increased processing power, it also created a serious concern for current microprocessor users. The 8-bit microprocessors were the hearts of many byte-oriented systems. Data transfers and memory maps were established upon fixed 8-bit word buses. However, the processing capabilities and 16 megabytes of direct memory addressing of the newer 16-bit microprocessors were attractive to these users. In response to the desire to fit the benefits of the 68000's processing power and memory accessibility into byte systems, Motorola created the MC68008, an 8-bit version of the 68000.

The 68008 is a scaled-down device consisting of 48 pins instead of the 68000's 64 pins. Part of that reduction obviously comes from fewer data pins (8 versus 16), but that does not account for the entire reduction. As a compromise, Motorola reduced the address bus from 24 lines to 20 lines, allowing for one megabyte of direct addressing. Because the 68008 is designed for an 8-bit bus, there is no need to distinguish between the upper and lower halves of a word size data arrangement. As a result, all address lines (A_0–A_{19}) leave the package through individual pins, unlike the 68000, which does not supply the A_0 line from the device. Additionally, the data strobe signals, $\overline{UDS}$ and $\overline{LDS}$, are replaced by a single data strobe line, $\overline{DS}$.

Figure 10.1 **68008 pin diagram (Courtesy of Motorola, Inc.)**

Signal	Pin	Pin	Signal
A_3	1	48	A_2
A_4	2	47	A_1
A_5	3	46	A_0
A_6	4	45	FC0
A_7	5	44	FC1
A_8	6	43	FC2
A_9	7	42	$\overline{IPL2/0}$
A_{10}	8	41	$\overline{IPL1}$
A_{11}	9	40	$\overline{BERR}$
A_{12}	10	39	$\overline{VPA}$
A_{13}	11	38	E
A_{14}	12	37	$\overline{RESET}$
V_{CC}	13	36	$\overline{HALT}$
A_{15}	14	35	GND
GND	15	34	CLK
A_{16}	16	33	$\overline{BR}$
A_{17}	17	32	$\overline{BG}$
A_{18}	18	31	$\overline{DTACK}$
A_{19}	19	30	$R/\overline{W}$
D_7	20	29	$\overline{DS}$
D_6	21	28	$\overline{AS}$
D_5	22	27	D_0
D_4	23	26	D_1
D_3	24	25	D_2

Table 10.1 Interrupt Priority Levels

$\overline{IPL}_1$	$\overline{IPL}_2/\overline{IPL}_0$	Interrupt Level
0	0	7 (NMI)
1	0	5
0	1	2
1	1	0 (No request)

The pin diagram for the 68008 is shown in Figure 10.1. Almost all of the signals that were present on the 68000 are included with the 68008.

The modifications discussed in the preceding paragraph are apparent in Figure 10.1. A less obvious change concerns the interrupt priority lines (IPLs). The 68000 has three ($\overline{IPL2}$, $\overline{IPL1}$, and $\overline{IPL0}$). They are all present in the 68008, except that $\overline{IPL2}$ and $\overline{IPL0}$ are connected internally to the same pin, designated here as $\overline{ILP2/0}$. This does not mean that this pin is multiplexed between these two inputs. Rather, it means that both IPL inputs are at the same level. This reduces the number of interrupt request levels that the 68008 can recognize. These levels are summarized in Table 10.1.

The remaining change in signal allotment for the 68008 is the omission of the valid memory address ($\overline{VMA}$) signal used to interface to 6800 peripheral devices. This control line must be created externally and is illustrated later in the chapter.

10.3 MEMORY DATA ORGANIZATION

Table 10.2 illustrates the effects of the data strobe control line ($\overline{DS}$) and the read/write line (R/$\overline{W}$). By combining these with the function control, address strobe, and address lines, memory chip selection is performed. The process is much like it was with the 68000, except that two separate data strobes are not required to distinguish between the upper data bus (b8–b15) and the lower data bus (b0–b7). Now the programmer must specifically keep memory accesses correct. Word and long word data, as well as opwords, still must be accessed starting at even address bounds. Figure 10.2 on page 280 illustrates how data are organized in memory for purposes of transferring the various data sizes using the 68008's limited data bus.

Table 10.2 Data Strobe Truth Table

$\overline{DS}$	R/$\overline{W}$	Data Bus (D_0–D_7)
1	X	No Valid Data
0	0	Write Data Valid

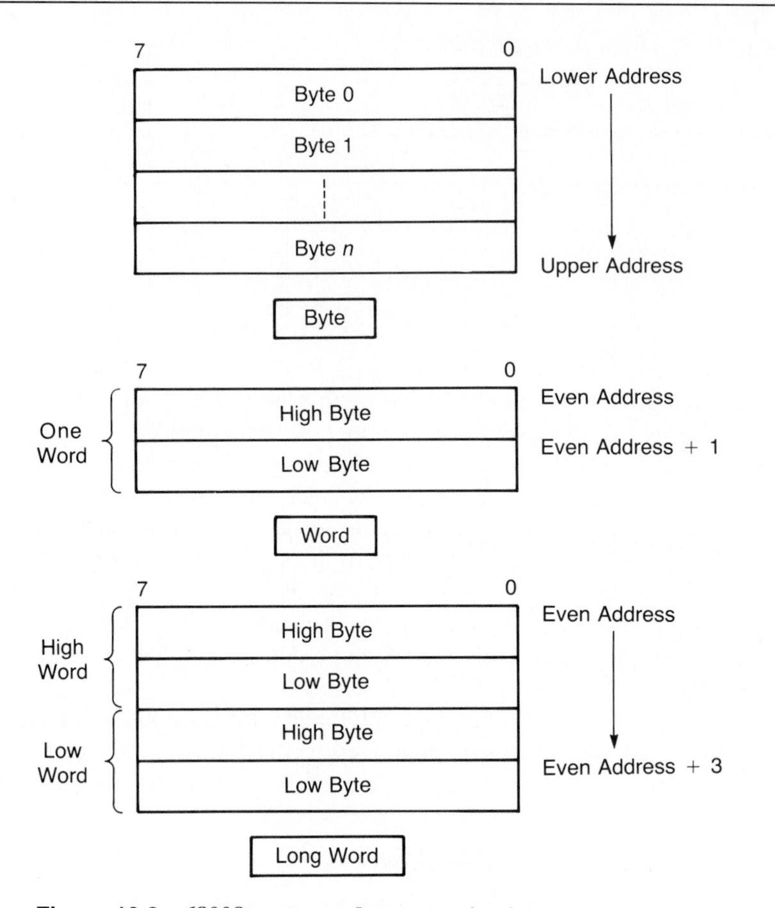

Figure 10.2 68008 memory data organization

Byte size data present no new problems to the user who is familiar with Motorola's previous microprocessor lines. Byte transfers are performed to any memory location, with multiple bytes transferred from lower memory addresses to higher addresses when done sequentially. Similarly, all sequential transfers of any size are moved to memory in ascending address order, with the exception of stack maneuvers.

Word size data are transferred along the 8-bit bus with the high byte located at the lower even address, followed by the lower byte at the next highest address location (even address plus 1). Long word data extend the concept through four bytes of information, with the highest byte transferred to memory first, as shown in Figure 10.2.

10.4 CAPABILITIES OF THE 68008

The 68008's programming model is identical to the 68000's, as is the instruction set. Because the registers within the 68008 are the same as in the parent device, data

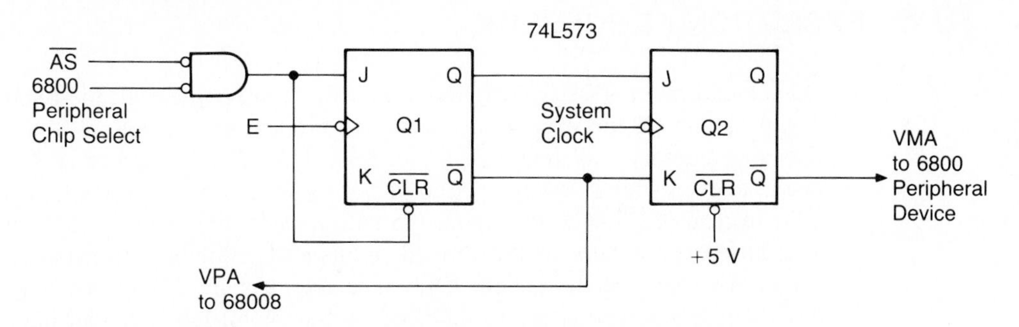

Figure 10.3 **$\overline{\text{VMA}}$ and $\overline{\text{VPA}}$ generation (Courtesy of Motorola, Inc.)**

sizes and functions are performed in the same manner as with the 68000. The major difference is in the memory transfer times because of the multiple read and write cycles that must be performed to move word and long word data between memory and the microprocessor. Once the opword and operands are on board, the processing mirrors the 68000.

A slightly different bus arbitration scheme handles direct memory accessing (DMA). The external device requesting use of the buses still asserts the bus request ($\overline{\text{BR}}$) line and receives a bus grant ($\overline{\text{BG}}$) in return when the 68008 is ready to release control of the buses. However, there is no $\overline{\text{BGACK}}$ input to the 68008. The requesting device must hold the $\overline{\text{BR}}$ active until it is finished with the DMA operation. $\overline{\text{BG}}$ usually is asserted following the completion of the bus cycle in which the $\overline{\text{BR}}$ was first received.

10.5 INTERFACING WITH 6800 PERIPHERAL DEVICES

Nothing new arises when connecting the 68008 to 6800 devices such as the PIA and ACIA. The only change is the lack of separate data strobes and $\overline{\text{VMA}}$ signal, as mentioned earlier. The system designer is required to create a $\overline{\text{VMA}}$ signal as part of the chip-enabling process when accessing these peripheral devices. Figure 10.3 is a diagram of one scheme that can generate the missing control signal.

When the decoding scheme detects a 6800 peripheral access, it causes flip-flop Q1 to be set in coincidence with the address strobe ($\overline{\text{AS}}$) and E clock signal. The output of the first flip-flop is sent back to the microprocessor as a $\overline{\text{VPA}}$ (indicating a valid peripheral operation) and as input information to the second flip-flop, Q2. A short time after that (keep in mind that the E signal is 1/10 the system clock frequency), the second flip-flop is set, making $\overline{\text{VMA}}$ active. When either the $\overline{\text{AS}}$ or peripheral decoding is no longer asserted, the first flip-flop is cleared, followed shortly by the resetting of Q2.[1]

[1]*MC68008 16-Bit Microprocessor with 8-Bit Data Bus,* August, 1983, by Motorola, Inc., p. 4-4.

10.6 EXCEPTION PROCESSING

Once again, the 68008 is designed to be fully compatible with the 68000. As such, it can handle all the exceptions of its big brother. Its exception table is accessed, like any memory transfer, in byte sections with all accesses being word size and begun at even address bounds. Auto-vectoring for interrupts is available by returning $\overline{VPA}$ instead of $\overline{DTACK}$ in response to an interrupt acknowledge (FC_0–FC_2 = 111). The difference occurs in the limited number of available interrupt priority levels created by marrying $\overline{IPL_0}$ with $\overline{IPL_2}$. The interrupt mask bits can be set to any of the seven levels, and space is allocated on the vector tables for all interrupt request levels. However, because of the physical limitation on the number of input request lines, only interrupt levels 0 (for no request), 2, 5, and 7 (non-maskable interrupt) are recognized within the 68008. Additional interrupt levels can be decoded externally to the microprocessor, but once the $\overline{IPL}$ inputs are applied, only those vectors associated with the interrupt levels just described are used. Simply stated, more than one external interrupt request can be serviced by one exception routine.

To retain compatibility with the 68000, in response to an accepted interrupt request, the function control leads all are set high and the interrupt level of the request is placed on A_1–A_3. Address lines A_0 and A_4–A_{19} are driven high. The interrupt vector number then is either read from the data bus if a $\overline{DTACK}$ is returned or derived internally in response to a $\overline{VPA}$. The actual order and stack process, as well as fetching the vector data, are identical to that of the 68000.

10.7 THE 68008 IN SUMMARY

The 68008 introduced no new concepts, but instead filled a need based on the current system technology; that is, it adapted byte size systems to the world of 16-bit processing. The alterations to the chip accommodate the smaller data bus and the lower performance needs of these well-established bus systems. A compromise is made in address size and interrupt request prioritizing. The higher level of processing internal to the device is retained along with the power realized from the instruction set and available addressing modes. The next 68000-family microprocessor, the 68010/12, also was created to meet a need. The need this time was not to adapt to present systems, but rather to incorporate the idea of **virtual memory.**

10.8 VIRTUAL MEMORY: WHAT IS IT AND WHY USE IT?

The 68000 can access 16 megabytes of memory directly through its 24-bit address bus. However, maintaining such a large physical memory is expensive. If a smaller physical memory is used, there must be a way to replace the information in portions of that memory as the need arises. These data can be replaced from external mass-storage devices, such as hard or floppy disks, associated with the system. A disk can hold data that do not need to be held in the computer's user RAM all the time. The problem, however, is how to transfer the data when needed without the user's having to call up the files each time. This requirement is the basis for virtual memory.

Virtual Memory Access

As long as a program accesses memory locations that exist in the physical system, memory tranfers are performed easily and efficiently by the 68000 family of microprocessors. However, when the microprocessor receives an order to access an address that is not within the physical system, it detects a $\overline{\text{BERR}}$ instead of a $\overline{\text{DTACK}}$ in response to the bus cycle request. This results in a bus error exception routine that detects the location of the problem so that it can be corrected.

In a virtual memory environment, this access is performed on purpose. The bus error routine first must decide whether the $\overline{\text{BERR}}$ was returned because of a legitimate error condition or because of a virtual memory access. In the first case, normal bus error processing is called and the system usually is locked up until the error is corrected. If a virtual memory access is detected, the exception process is directed to fetch a block of data from the disk and swap this block with the least used block of data within the system RAM. After the swap, the bus cycle is resumed, completing the data transfer. The microprocessor then goes on to the next step in the program. The new data remain in physical memory, whereas the replaced data are held on the disk until such time as they are needed again.

10.9 THE 68010 AND 68012 16-BIT MICROPROCESSORS

The programming model for the 68010 and 68012 microprocessors is shown in Figure 10.4 on page 284. This model is very similar to the 68000's but has a few new registers. The most important of these is the vector base register (VBR), which allows the exception vector table to be located anywhere in memory. The contents of the VBR are added to the vector address, which is computed by multiplying the vector number by four. The sum then becomes the vector location, as shown in Table 10.3 on page 284.

EXAMPLE 10.1

The vector base table contains $0012 2000. What location holds the starting address of the interrupt level 2 exception?

Solution:

The interrupt level 2 exception number is 26. Multiplying 26 by 4 gives a vector address of $0068. This address is added to the VBR to give the sum $0012 2068. Truncated to 24 bits, the vector location that contains the starting address of the level 2 exception is $12 2068.

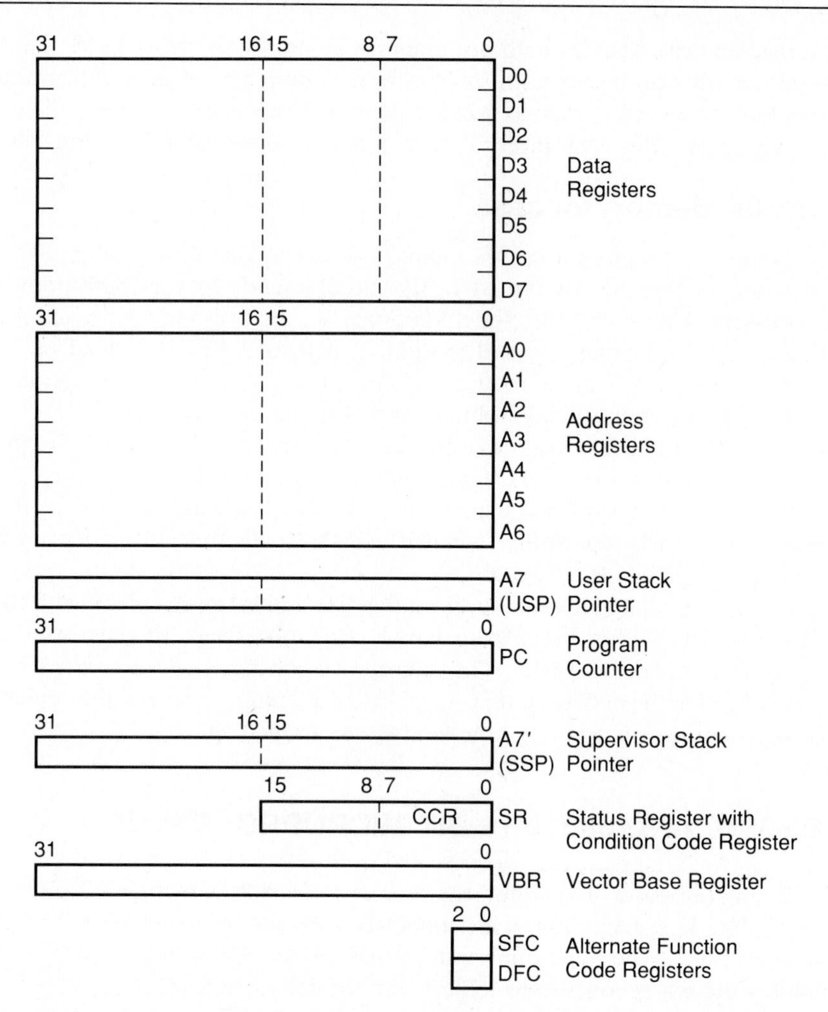

Figure 10.4 68010 programming model (Courtesy of Motorola, Inc.)

Table 10.3 68010/68012 Interrupt Vector

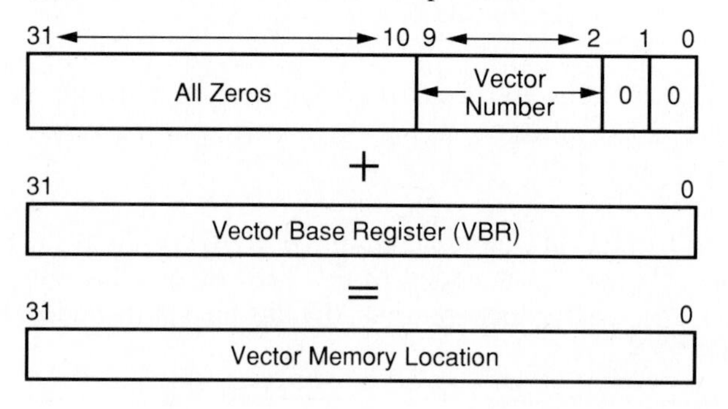

Two additional 3-bit alternate function code registers allow the operator in the supervisor mode to alter the data on the function control lines so that user memory space can be accessed in the supervisor mode.

The pin layout for the 68010 is identical to that of the 68000. Both devices are made in a grid array package. The grid array was developed to handle the increased number of pins required by the enhanced capabilities and larger bus sizes of the 68000 family. The grid array for the 68012 is very similar to that of the 68010, with the exception of the address bus size (Figure 10.5, page 286). The 68012 has address lines A_1–A_{29} and A_{31} available for direct memory accessing. A_0 still controls $\overline{UDS}$ and $\overline{LDS}$, whereas A_{31} differentiates between two noncontiguous memory sections. This configuration provides for one gigabyte of direct accessing. The other difference between the 68010 and the 68012 is the addition in the 68012 of a control pin, $\overline{RMC}$, which indicates when an indivisible read-modify-write cycle is in process.

10.10 EXCEPTION PROCESSING: CPU SPACE

When the 68000 receives an interrupt request, the function control lines are made high and the interrupting priority level is copied onto lines A_1, A_2, and A_3. This action still occurs in the 68010, but the all-high condition of the function control lines now includes **CPU space.** During an interrupt, the CPU space is the interrupt acknowledge. However, the CPU space also can indicate additional uses, such as breakpoint processing (to be discussed later). Distinction must be made electrically between the types of CPU space activity in process (interrupt or breakpoint). Address lines A_{16}–A_{19} code the CPU space activity. That is, whenever the function control lines are all high, A_{16}–A_{19} define the operation in process. For the 68010 and 68012, the only two codes used are 1111, for interrupt acknowledge, and 0000, for breakpoint acknowledge. Additionally, during a breakpoint acknowledge, the remaining address lines are set low. Interrupt acknowledge CPU space forces all address lines high except A_1–A_3, which hold the interrupt level number. Using 1111 as the code for the interrupt acknowledge maintains compatibility with the 68000 acknowledge condition (A_4–A_{23} all high). Use of the CPU space code area is expanded in the 32-bit microprocessor, the 68020 (see Chapter 11).

Exception Stack

Table 10.4 on page 287 shows the stack produced whenever an exception occurs. There are essentially two lengths of stack frames, one containing 4 words and one containing 29 words. The shorter frame is used for most of the exception processes, whereas the longer frame is used for bus error exceptions.

Both frame formats start out identically and in a similar manner to the 68000 by stacking the status register and program counter information. Both push a new data word onto the stack consisting of format information (to distinguish between a short and a long frame) and the vector offset (vector number multiplied by 4). This word is read first by the microprocessor in response to an RTE to determine

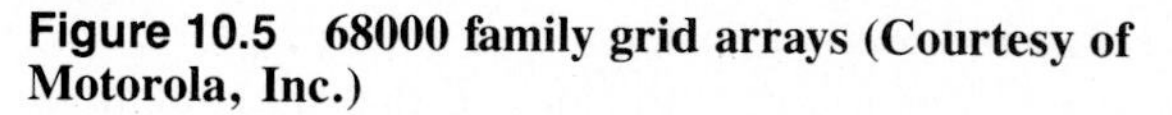

Figure 10.5 **68000 family grid arrays (Courtesy of Motorola, Inc.)**

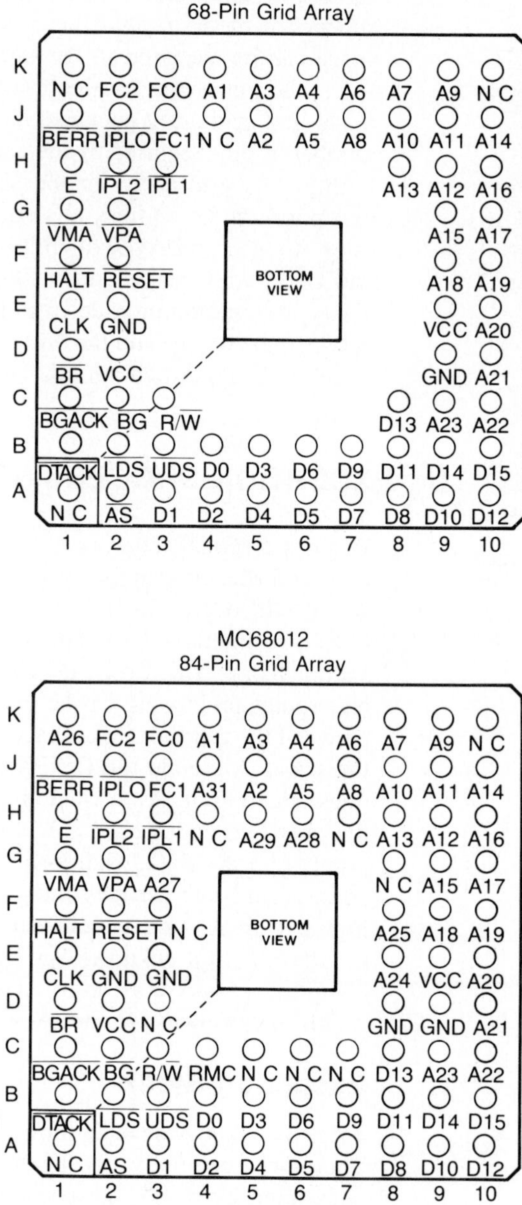

the stack size to be fetched back into the microprocessor. Stack size is particularly important during virtual memory operations. The two legal formats for the 68010 also are shown in Table 10.4.

The long frame, illustrated in Table 10.5, contains the same initial four words, but also holds information about the microprocessor at the time the bus error excep-

Table 10.4 68010 Stack Frame

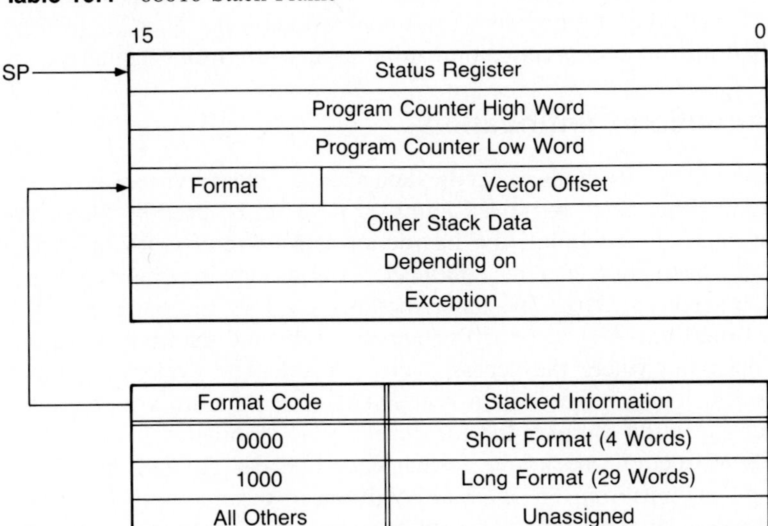

Format Code	Stacked Information
0000	Short Format (4 Words)
1000	Long Format (29 Words)
All Others	Unassigned

Table 10.5 Bus and Address Error Stack for the 68010

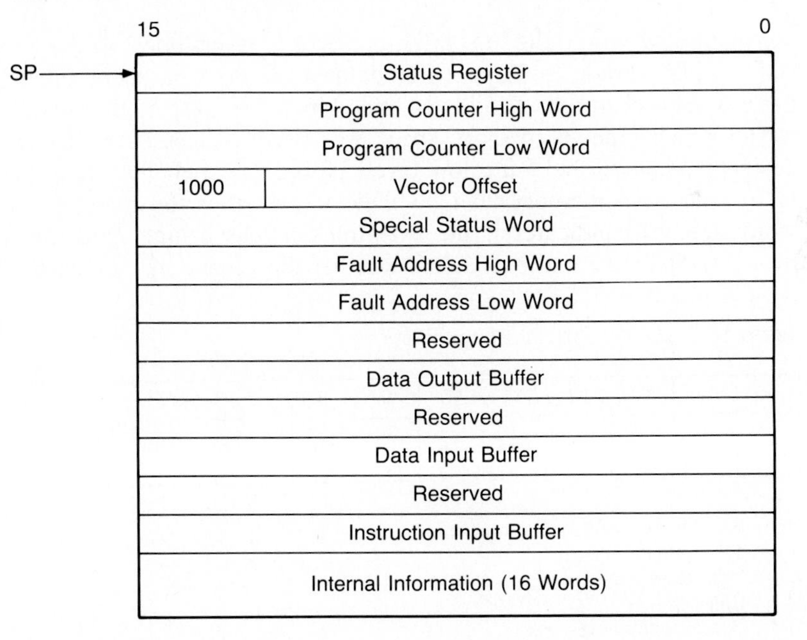

tion was taken. These data are used either to determine the bus error cause or to continue the interrupted bus cycle after the memory swap for virtual accesses.

The additional data contain the fault address and the contents of the data and instruction buffers. The block of internal information contains the contents of various registers within the 68010 at the time of the exception. The prefetch pipeline and any housekeeping registers are examples of the types of data saved in those locations. By having this information saved and restorable by an RTE instruction, a process called instruction continuation can be used with virtual memory management.

Instruction Continuation

There are two methods to handle data accesses during virtual memory swapping. The data in either case are moved into user RAM to be used by the instruction requiring that data. At this point, the instruction can be rerun with the data now available, or the instruction can be continued from where it stopped when trying to fetch the nonexistent data. The latter method, used by the 68010, is called **instruction continuation.** All the 68010's internal works and data are saved on the long stack at the point where the access cycle stopped. The cycle was stopped because the physical location was not in user RAM. The data are swapped as part of the bus error exception routine, and the exception is terminated in an RTE instruction. The RTE instruction checks the format code and detects and causes the long stack to be pulled, restoring the internal condition of the microprocessor at the time it was stopped. The cycle now is complete, accessing the data it could not find earlier.

Special Status Word

An additional word, the *special status word,* is included in the long stack. This status word, shown in Table 10.6, defines the type of operation in process when the bus error occurred. The lower three bits are a copy of the function codes, indicating which type of memory space was being accessed. b3–b7 and b14 are not designated and are held at a low level. b8 indicates whether the access was a write to or read from memory, and b9 indicates whether the transfer was byte size or word size. b10 indicates if the byte transfer was a high byte (D_8–D_{15}) or a low byte (D_0–D_7). b12 and b13 define whether the access was a data or an information

Table 10.6 68010 Special Status Word

15	14	13	12	11	10	9	8	7 6 5 4 3	2 1 0
RR	0	I	D	RM	H	B	$R/\overline{W}$	——0's——	FC_2–FC_0

RR: 0 = Microprocessor rerun (default)
 1 = Software rerun
I: Instruction fetch
D: Data fetch
RM: Read-modify-write cycle
H: High byte data transfer
B: 0 = Word transfer
 1 = Byte transfer
$R/\overline{W}$: 0 = Write cycle
 1 = Read cycle
FC: Function control signals

fetch operation. Finally, b15 indicates if the cycle is to be rerun under software control or as a function of the microprocessor sequence. Combining information about the special status word with the rest of the data on the stack defines the cause of the bus error exception.

Added as the lowest priority exception is the new format exception. During the RTE operation, the microprocessor first examines the format to determine the size of the stack to be recovered. If the format is correct, the stack frame size is determined and the remaining stack data are pulled and restored into the microprocessor. The microprocessor then continues processing from the point where it was interrupted or discontinued. If the format is not one of the two legal formats, 0000 or 1000, a format exception process is begun.

Interrupt Exception Process

The procedure followed by the 68010 during an interrupt exception cycle is similar to that of the 68000; however, because of the additional information generated by the stack and the establishment of CPU spaces, there are some significant differences. The process starts with the receipt of an interrupt request from an external device.

The incoming level of the interrupt request is compared with the level set by the interrupt mask bits in the status register. Once the current instruction is completed, if the incoming interrupt request level is higher than that of the interrupt mask, the interrupt process begins. As with the 68000, the status register is copied into a holding register and the S bit of the status register is set. The T bit is reset and the interrupt mask bits are upgraded to the level of the interrupt request.

The function control lines all are set high along with A_4–A_{23}. This forces the CPU space bits (A_{16}–A_{19}) to be high, indicating that this CPU space operation is an interrupt acknowledge. A_1–A_3 are set to the interrupt request level to complete the interrupt acknowledge process.

Control leads $\overline{AS}$, $\overline{UDS}$, and $\overline{LDS}$ are asserted and the microprocessor awaits a return of $\overline{DTACK}$ or $\overline{VPA}$. If $\overline{DTACK}$ is sent to the 68010, it also must be accompanied by a vector number placed on the lower data bus supplied by the interrupting device. An asserted $\overline{VPA}$ in place of the $\overline{DTACK}$ causes the 68010 to perform the auto-vector operation. In this case, a vector number associated with the interrupt request level is generated internally in the microprocessor. Once the 68010 has the vector number latched into a holding register, $\overline{AS}$, $\overline{UDS}$, and $\overline{LDS}$ are negated and the vector number is shifted left twice (multiplied by four).

The format register then is cleared (set to 0000), indicating a short stack frame. The vector offset developed from the shift of the vector number is sign extended and added to the VBR to form the actual vector address location.

The contents of the program counter and the temporary register holding the original status register data are pushed onto the stack, followed by the word length combination of the format (0000) and the vector offset (vector number $\times$ 4) information. The contents of the vector address are read into the microprocessor and transferred to the program counter. This value becomes the starting address of the first instruction of the interrupt program.

10.11 RETURNING FROM AN EXCEPTION

Upon executing an RTE instruction, the microprocessor first reads the format/offset word to determine if the stack is short or long. From this information, the number of stack locations to be retrieved and the destination of the incoming data are determined. After the stack is emptied and the contents of the internal registers of the 68010 are restored, the execution of the interrupted instruction is resumed. Note that for an interrupt, the last instruction is completed before the exception occurs. As a result, the instruction executed after the return process is complete is the next regular instruction in the original program.

For bus error exceptions, the result is the creation of a long stack. The information on the long stack includes the contents of the working registers within the microprocessor. This detail is important because a bus error exception occurs as a result of any bus cycle interruption and not necessarily at the end of an instruction cycle. The exception could occur as easily during the read cycle of a memory data fetch. In the case of virtual memory operation, this type of disruption does occur. The microprocessor begins a read cycle in an attempt to fetch data from a memory location it cannot find in the system memory. In place of a $\overline{\text{DTACK}}$, a $\overline{\text{BERR}}$ is returned and the bus error exception is taken.

As a part of the process of the bus error exception, the long stack is built and the starting address of the exception routine is fetched. The first job of the bus error routine is to determine whether a real bus error occurred or a memory swap is required to support virtual memory. In the latter case, the swap is performed under software control within the bus error exception routine. This routine ends in RTE as does any other exception routine. The format/offset word is read, and the microprocessor determines that a long stack is to be retrieved and begins the process. The working registers as well as the usual status register and program counter information are restored into the 68010. With all registers returned to their original contents, the interrupted bus cycle is resumed. This time, when the memory read is performed, there are data at the addressed location and $\overline{\text{DTACK}}$ is returned to complete the cycle.

10.12 BREAKPOINTS

Breakpoints assist the software designer in debugging a program. They allow the program to be interrupted under software control. An exception routine generated by the breakpoint is executed, enabling the designer to examine the registers by reading their data from a stack or other designated memory space created by a MOVEM instruction within the routine. Breakpoint exceptions are generated by the use of the illegal exception using $4848–$484F illegal opword codes. An additional breakpoint acknowledge cycle (see Figure 10.6) is inserted between the vector address fetch and the beginning of the stack operation.

Other illegal instructions do not generate the breakpoint acknowledge cycle, but they still cause the illegal instruction exception to be taken. By detecting a breakpoint acknowledge cycle or lack thereof, this exception determines whether the exception

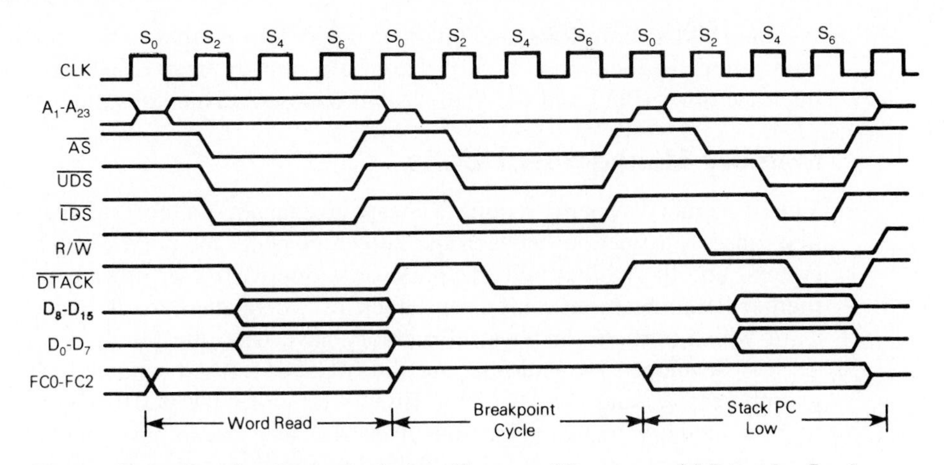

Figure 10.6 Breakpoint cycle timing diagram (Courtesy of Motorola, Inc.)

is being executed because of an illegal instruction or because of a breakpoint operation. During a breakpoint acknowledge cycle, the function control leads are set high to indicate a CPU space. The address lines, including A_{16}–A_{19}, are driven low, indicating that the operation is a breakpoint and not an interrupt acknowledge. While $\overline{AS}$, $\overline{UDS}$, and $\overline{LDS}$ are asserted during the breakpoint acknowledge cycle, no data are being transferred on the data bus. The breakpoint cycle is terminated by an active $\overline{DTACK}$, $\overline{VPA}$, or $\overline{BERR}$. However, the microprocessor continues the illegal exception processing by building the stack and executing the illegal instruction exception (vector number 4).

10.13 68010/12 IN SUMMARY

The 68010 and the 68012 are extensions of the original 68000 microprocessor. They add the concepts of virtual memory and the VBR to the basic 16-bit machine. Additionally, they supply alternate function control registers that are accessible by privilege instructions only. These registers allow supervisor mode programs to access user memory space in a function control defined system. Function control code 111 now designates CPU space, which for these devices is shared by the interrupt acknowledge and breakpoint operations. Additionally, the 68012 has an expanded address bus and an additional control lead to designate when a read-modify-write cycle is in progress.

10.14 16-BIT SUPPORT DEVICES

A number of devices have been developed as direct supports to the 68000 family of microprocessors. They include memory management units to assist virtual memory systems, DMA controllers, and interface devices for peripheral devices and data communications applications. Two of these have already been discussed: the peripheral interface/timer (PIT) and the dual asynchronous receiver transmitter (DUART).

Memory Management Units

Virtual memory systems require a means to manage memory allocation. One must first make a distinction between the known physical memory, located as part of the system, and the logical bulk memory, on a floppy disk or other external device. A **memory management unit** keeps track of where memory is and is not, as well as of which portions of physical memory have been the last least used. Two such devices developed by Motorola are the MC68851 (used with the MC68020 32-bit microprocessor) and the MC68461 (used with either the 68010/12 or the 68020).

A memory management unit uses basically one of two methods to perform memory swapping: segmented and paged memory management. In segmented operation, three variables programmed into the unit control how memory data are managed. These variables are the starting logical address, an offset to physical memory, and the length of usable physical addresses (Figure 10.7).

Paged addressing (see Figure 10.8) is similar to segmented management, except that the length of physical memory addresses is fixed to a page of specified length. The MC68851 is a paged memory management unit (PMMU).

Managing blocks of memory between logical and physical allocations is facilitated by an on-board cache in the 68851. A **cache** is a localized block of read/write memory that is internally accessible by the chip in which it resides. Data from memory are initially loaded into the cache along with related addresses called *tags*. When the CPU attempts to access memory, the cache first is checked for a match, called a *hit,* between the access address and the tag. If a hit occurs, then access is made directly from the cache rather than from memory. If a miss occurs—that is, no match—data are accessed from memory as usual. The data also are written into the cache, causing the oldest data to be erased.

In the PMMU, the cache is used to hold a table of tag addresses corresponding to a set of logical addresses associated with the bulk storage device connected to

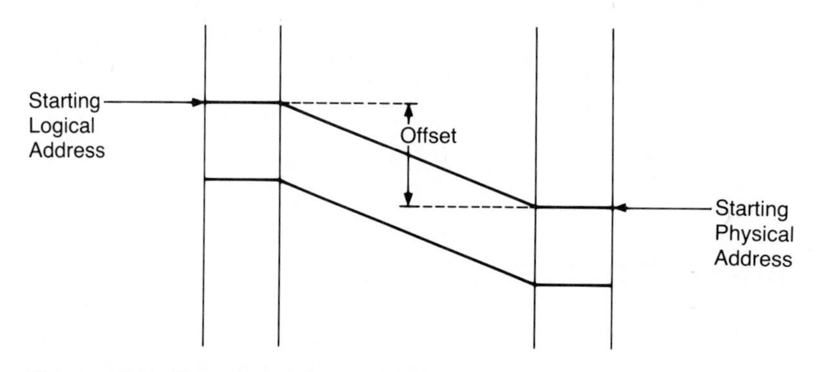

Figure 10.7 Segmented descriptor

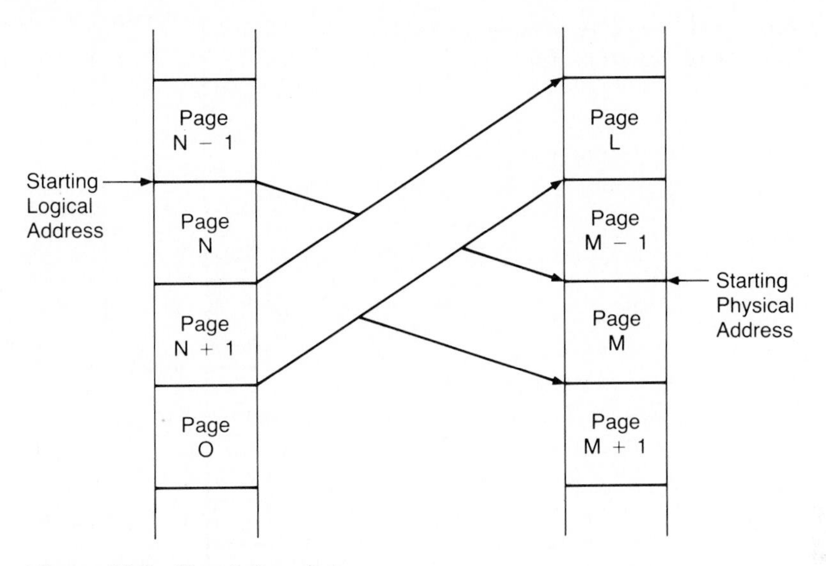

Figure 10.8 Paged descriptor

the system. If the data are located already in physical memory, the access is made directly between memory and the CPU. If the access is not to physical memory, the PMMU checks for a tag match (the hit). Once found, the data in the cache are used to address the external memory device (the disk drive, for instance). The data read from the disk are placed in physical memory corresponding to the logical address of the access. If a miss occurs (that is, no logical address is found, and thus no physical or logical address is accessed), a bus error ($\overline{BERR}$) is generated. The PMMU is used as a coprocessor in conjunction with the 68000 family of microprocessors. The PMMU is written to and read from with regular 68000 instructions. In turn, the processes that it performs as part of an exception routine use a short number of instructions called *coprocessor instructions*. These instructions are detailed in Chapter 12 in the discussion of coprocessor interfaces.

DMA Bus Arbitration

The MC68452 provides the bus request ($\overline{BR}$) and bus grant acknowledge ($\overline{BGACK}$) signals required for DMA operation in a 68000 environment. This device sends out the $\overline{BR}$ in response to a signal from the peripheral device for which it is providing the interface. Upon receipt of the $\overline{BG}$ from the 68000, the 68452 sends and holds $\overline{BGACK}$ until a signal is received from the peripheral device indicating that the transfer is complete. This device supports either cycle-by-cycle or block transfers. The data bus is buffered by this module so that data transfer is under its control.

A second DMA device, the MC68450 direct memory access controller (DMAC), is a full DMA device (Figure 10.9, page 294). That is, the 68450 has all necessary control functions needed for data transfers within a 68000-based system. When this device is granted control of the buses, it supplies the timing and control

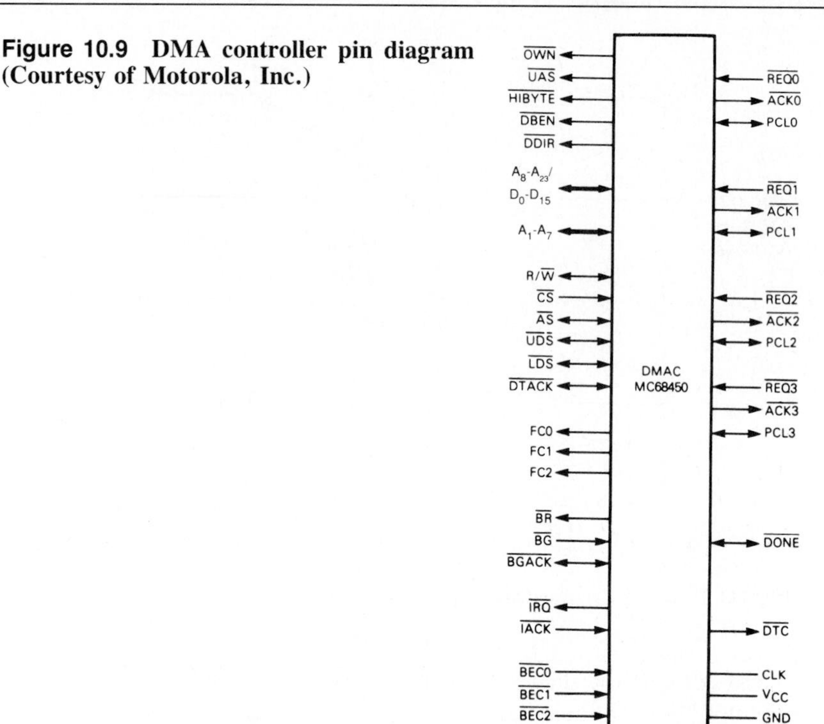

Figure 10.9 DMA controller pin diagram (Courtesy of Motorola, Inc.)

data for the transfer. On the peripheral interface side, control lines are provided to direct data transfers with one of four independent DMA channels. The control functions to and from the peripheral devices include a request line ($\overline{REQ}$) for each channel which allows that peripheral device to request line ($\overline{REQ}$) for each channel which allows that peripheral device to request access to the 68000 buses, an acknowledge line ($\overline{ACK}$) which tells the device that it has control, and a peripheral control line ($\overline{PCL}$) which allows incidental control of the peripheral device. The four channels are prioritized programmably. Transfer can take place from memory to memory or between memory and an I/O device at rates up to four megabytes per second.

Other Support Devices

Other devices developed for the 68000 family include an intelligent peripheral controller, an enhanced programmable communications interface for serial data interfacing through a modem, a DUART for asynchronous data transfers, a multiprotocol communications controller for recognizing various data communications protocols such as SDLC and Bisync, and other unique interfacing devices. Table 10.7 is an overview of the devices available from Motorola to support the 68000 family of microprocessors.

Table 10.7 MC68000 Family Fact Sheet (Courtesy of Motorola, Inc.)

	DEVICE #	DEVICE NAME	DESCRIPTION	SOURCES	SPEEDS	PACKAGE TYPE
PROCESSORS	MC68000	16/32 bit MPU	16 bit external/32 bit internal MPU. 17 general purpose 32 bit registers. 16 MB linear address space.	M, H, MK, R, S, T	8, 10, 12.5 MHz	64 lead L, LC, P 68 lead R, ZB FN (2Q85)
PROCESSORS	MC68008	8/32 bit MPU	8 bit external/32 bit internal MPU. 17 general purpose 32 bit registers. 1 megabyte linear address space (4 MB with FN package option).	M, MK, S, T	8, 10 MHz	48 lead L, P 52 lead FN (2Q85)
PROCESSORS	MC68010	Virtual Machine 16/32 bit MPU	16 bit external/32 bit internal MPU. 17 general purpose 32 bit registers. Virtual memory/machine. 16 megabyte linear address space.	M, S	8, 10, 12.5 MHz	64 lead L, LC, P 68 lead R, RC
PROCESSORS	MC68012	Extended Virtual 16/32 bit MPU	16 bit external/32 bit internal MPU. 17 general purpose 32 bit registers. Virtual memory/machine. 2 gigabyte linear address space.	M	8, 10, 12.5 MHz	84 lead RC
PROCESSORS	MC68020	32 bit MPU	Complete 32 bit microprocessor. 4 gigabyte linear address space. Coprocessor Interface. Instruction Cache. Dynamic Bus Sizing. 2-3 MIPS performance.	M	12.5, 16.67 (1Q85) MHz	114 lead RC
PROCESSORS	MC68881	Floating Point Coprocessor (FPCP)	Meets full IEEE spec for advanced floating point calculations. Single, double and extended precision.	M (1Q85)	12.5, 16.67 MHz	68 lead RC
MEMORY MANAGEMENT	MC68451	Memory Management Unit (MMU)	Ideal MMU for non-demand paged MC68000, 68010 systems.	M, MK	8, 10 MHz	64 lead L, LC 68 lead R, RC
MEMORY MANAGEMENT	MC68851	Paged MMU (PMMU)	32 bit demand virtual paged memory management unit for MC68020 based systems.	M (3Q85)	12.5, 16.67 MHz	124 lead RC
MEMORY MANAGEMENT	MC68461	Memory Management Cntlr (MMC)	Gate array implementation of PMMU functional subset. Can be used with 68020, 68010, or 68012. Bipolar.	M	N/A	149 lead RC or mezz board
DMA CONTROL	MC68440	Dual DMA (DDMA)	Dual channel, high speed Direct Memory Access controller. Capable of 5 MB/sec data transfer rates.	M	8, 10, 12.5 (3Q85) MHz	64 lead L, LC, P 68 lead R, RC FN(4Q85)
DMA CONTROL	MC68450	DMA Controller (DMAC)	Four channel Direct Memory Access Controller. Capable of very complex "chained" data transfers.	M, H	8, 10 MHz	64 lead L, LC 68 lead R, RC
DMA CONTROL	MC68442	Expanded DDMA	32 bit address version of DDMA. Support for 4 gigabyte range of MC68020. Pin compatible with 68440/68450.	M (1Q85)	8, 10, 12.5 (3Q85) MHz	68 lead R
EM INTERFACE	MC68153	Bus Interrupt Module (BIM)	Routes interrupts from 4 independent sources to any of 7 M68000 MPU interrupt levels. Bipolar.	M	200 ns access time (16 MHz clock)	40 lead L, P
EM INTERFACE	MC68452	Bus Arbitration Module (BAM)	Arbitrates access of an M68000 system bus between up to 8 local masters. Bipolar.	M	50 ns arbitration time	28 lead L, P
EM INTERFACE	MC68172	VMEbus Controller (E-BUSCON)	Performs VMEbus/local bus arbitration, VMEbus requests, bus transceiver control.	S (2Q85)	N/A	28 lead L, P
EM INTERFACE	MC68173	VMSbus Controller (S-BUSCON)	Handles interface of operations between high speed serial peripheral VMSbus and controlling VMEbus.	S (2Q85)	N/A	28 lead L, P
EM INTERFACE	MC68174	VMEbus Arbiter (E-BAM)	Performs round robin and 4 level priority arbitration for VMEbus based systems. Bipolar.	M (2Q85)	N/A	20 lead L, P

Table 10.7 (continued)

	DEVICE #	DEVICE NAME	DESCRIPTION	SOURCES	SPEEDS	PACKAGE TYPE
DATA COMMUNICATION	68561	Multi-Protocol Comm Cntlr 2 (MPCC-2)	Single channel. M68000 interface. Asynch, bisynch, SDLC.	R (1Q85)	4 Mb/s	48 lead L
	68562	Dual Universal Serial Comm Cntlr (DUSCC)	Dual channel. Asynch. Byte control (bisynch, DDCMP, X.21). Bit oriented (HDLC/ADCCP, SDLC X.25). DMA interface. Counter/timer.	S (1Q85)	4 Mb/s	48 lead L
	68564	Serial I/O (SI/O)	Dual channel. Asynch, bisynch, SDLC.	MK	1 Mb/s	48 lead L, P
	MC68652 MC2652	Multi-Protocol Comm Cntlr (MPCC)	Single channel. Byte control. Bit oriented. CRC (error correction) circuitry. (MC2652 has generic bus interface.)	M, S	2 Mb/s	40 lead L, P
	MC68653 MC2653	Polynomial Generator Checker (PGC)	Error correction, code generation/comparitor circuit. Excellent companion chip for 68652 MPCC or 68661 EPCI. (MC2653 has generic bus interface.)	M, S	4 Mb/s	16 lead L, P
	MC68661 MC2661	Enhanced Peripheral Comm I/F (EPCI)	Universal synch/asynch. Double buffered receiver/transmitter. Internal baud rate clock. (2661 generic bus I/F)	M, S	1 Mb/s	28 lead L, P
	MC68681 MC2681 MC2682	Dual UART (DUART)	Dual channel. Quad buffered receiver. Double buffered transmitter. Independent baud rate selection. 1 Mb/s. (2681 has generic bus interface; 2682 offers partial functionality in a smaller package.)	M, S	1 Mb/s	40 lead L, P (MC2682-28 lead)
LAN CTRL	68590	LAN Controller for Ethernet (LANCE)	Provides complete IEEE specified Ethernet communication control for M68000 based systems. Plus DMA controller.	MK	10 Mb/s	48 lead L
	68802	IEEE 802.3 LAN Controller (LAN)	Provides communication control between M68000 based systems and the IEEE 802.3 LAN protocol.	R	10 Mb/s	40 lead P
DISK CONTROL	68454	Intelligent Multiple Disk Controller (IMDC)	Controls up to four disks. Any combination of single/dual density, floppy or hard. 256 byte FIFO. 4 GB DMA ctrlr.	S	2-10 Mb/s	48 lead L, P
	68459	Disk Phase Lock Loop (DPLL)	Companion device to 68454 IMDC. Used for interfacing more than one drive to the IMDC.	S	N/A	24 lead L, P
	68465	Floppy Disk Cntrlr (FDC)	Interfaces 2 single or double density floppy disks to the M68000 bus.	R	N/A	48 lead L, P
GENERAL I/O	MC68120/ 68121	Intelligent Peripheral Controller (IPC)	Provides peripheral control for M68000 or M6800 systems.	M, S	1, 1.25 MHz	48 lead L
	MC68230	Parallel Interface/ Timer (PI/T)	Unidirectional/bidirectional, 8/16 bit, double buffered parallel interface. 24 bit timer with 5 bit prescaler. M68000 I/F.	M, MK S, T	8, 10 MHz	48 lead L, P
	MC68901	Multi Function Peripheral (MFP)	Single channel USART. 8 source interrupt controller. 8 parallel I/O lines. Four 8 bit timers.	M, MK	4 MHz 1 Mb/s USART	48 lead L, P
GRAPHICS	MC68486 (RMI) MC68487 (RMC)	Raster Memory System (RMS)	Provides functionality required by bit mapped or object-oriented graphics systems. Features include object definition & manipulation, collision detection, light pen input, x/y capture & interrupt, 32 of 4096 colors, visual/virtual screens.	M (2Q85)	N/A	48 lead L, P (both)

Sources—M = Motorola
H = Hitachi
MK = United Technologies/Mostek
R = Rockwell
S = North American Philips/Signetics
T = Thomson CSF

Packaging: L = Ceramic DIP
LC = Ceramic DIP, Gold Lead Finish
P = Plastic DIP
R = Pin Grid Array
RC = Pin Grid Array, Gold Lead Finish
ZB = Socketable LCC
ZC = Surface Mount LCC
FN = Plastic Quad Pack

PROBLEMS

10.1 List three significant physical differences between the 68000 and the 68008.

10.2 What interrupt request levels are available to the 68008? Which interrupt mask levels are available in the 68008?

10.3 Draw a logic circuit that allows all interrupt request levels to the 68008 to generate interrupts. Structure your logic so that the interrupt priority scheme of the 68000 family is not violated.

10.4 Compare the ways the 68008 and the 68000 handle DMA arbitration.

10.5 What does a virtual memory system provide that a regular memory system does not?

10.6 List three major differences between the 68010 and the 68000. What are two additional differences between the 68012 and both the 68000 and the 68010?

10.7 The VBR contains $0020 3C46. An interrupt level 5 is received and processed using auto-vectoring. What is the vector location that holds the starting address of the interrupt program?

10.8 How is an interrupt acknowledge identified using the 68010?

10.9 What CPU spaces are designated by the 68010 and how are they identified?

10.10 What are the lengths of the two stack frames used by the 68010? How does the microprocessor know the frame size upon execution of an RTE?

10.11 What new exceptions have been added in the 68010? Describe them in terms of priority.

10.12 Which exception type does the breakpoint use?

10.13 What are the opwords for breakpoint instructions?

10.14 What process allows the 68010 to perform instruction continuation as opposed to instruction rerun during a virtual memory operation?

10.15 Which exception is used for virtual memory operations?

10.16 With which memory area (user RAM or disk) are the PMMU tag addresses associated?

10.17 Under what conditions will a PMMU generate an actual $\overline{\text{BERR}}$?

10.18 When does a hit occur when using a cache?

10.19 Which DMA control signals are generated by the 68452 bus arbitration device?

10.20 To which DMA control signals does the 68452 respond?

11

MC68020 32-BIT MICROPROCESSOR

CHAPTER OBJECTIVES

The next step in microprocessor evolution is the 32-bit CPU. Motorola's 68020, which is upwardly compatible to the 68000/68010, internally performs 32-bit processing along a 32-bit internal bus, using long word registers, transfers, and operations. *Upwardly compatible* means that programs written for a 68000 or 68010 system can be used for a 68020 system. This chapter explores the 68020, explaining concepts such as cache memory and dynamic bus sizing. The function of coprocessing, using the 68020 along with a floating point coprocessor, is discussed in Chapter 12.

11.1 GLOSSARY

aligned data Data that begin in memory at specified address bounds.
bit field Designated portion of memory that holds up to 32 bits of data.
dynamic bus sizing Variable length data and/or memory port sizes for data transfers.
gigabyte 1,073,741,824 bytes.
misaligned data Data that begin at incorrect address boundaries.
packed BCD Binary coded decimal data organized as two digits per byte.
postindexed Memory indirect address mode that includes indexing after the memory pointer is established.
preindexed Memory indirect address mode that includes indexing in the computation of the data pointer itself.
unpacked BCD BCD data organized as one digit per byte.

11.2 THE 68020 MICROPROCESSOR ARCHITECTURE

The programming model for the 68020 is shown in Figure 11.1 on page 300. Notice that the data and address register complement is essentially the same as for the 68000; that is, there are eight 32-bit data registers and seven address registers.

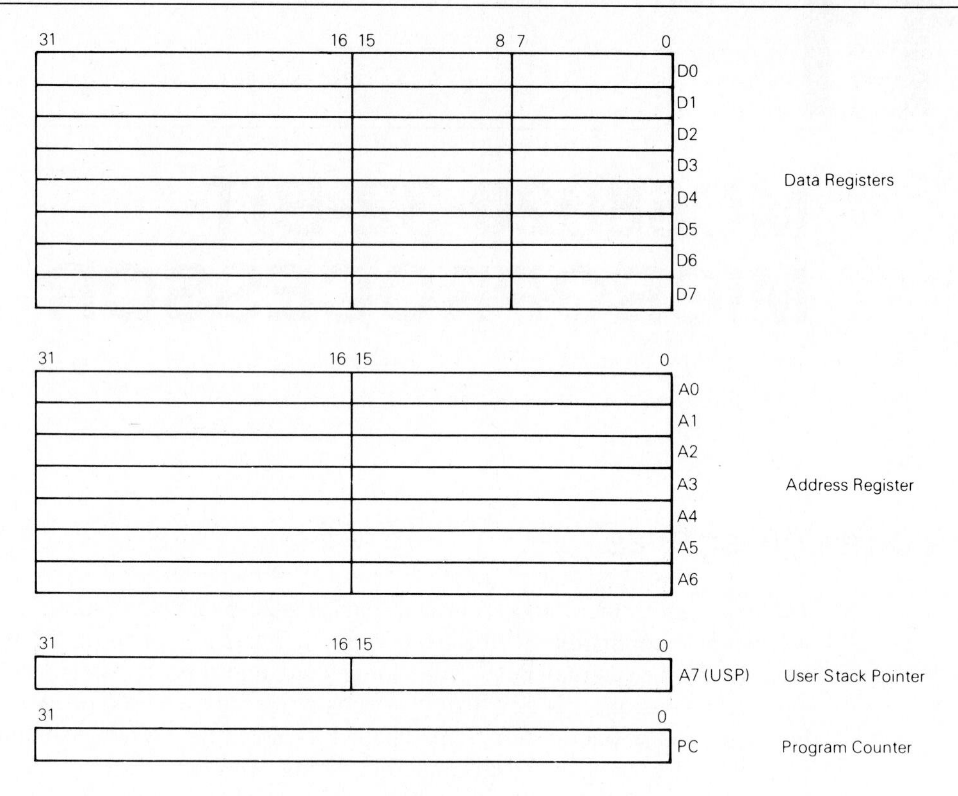

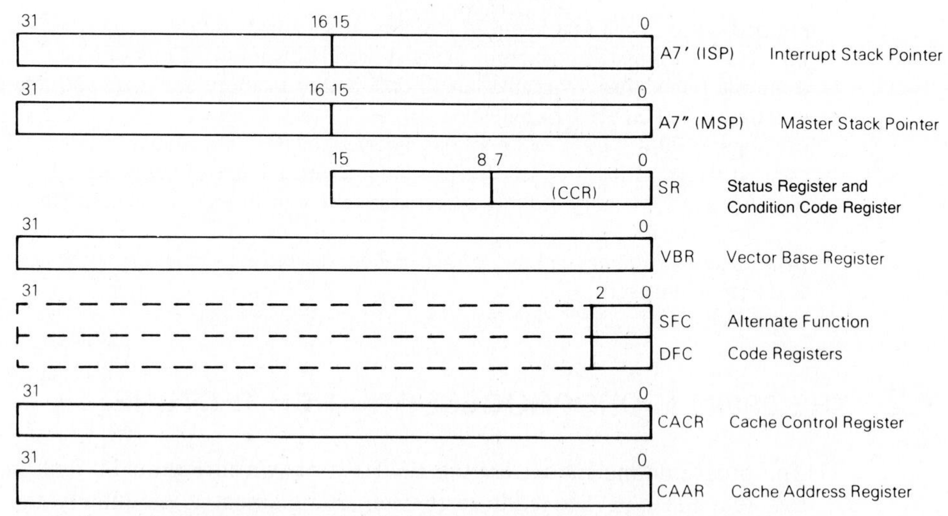

Figure 11.1 68020 programming model (Courtesy of Motorola, Inc.)

A_7 still serves as the user stack pointer in the user mode and as a supervisor stack pointer in the supervisor mode. However, in the supervisor mode, there is a choice between two separate stack pointers, the interrupt stack pointer (A_7') and the master stack pointer (A_7''). Alternate function code registers used in the 68010 also are present in the 68020, allowing the 68020 in the supervisor mode to access the user memory area as well as the supervisor memory area. As in the 68010, interrupt vector tables can be located anywhere in memory because of the VBR.

Two new registers, the cache control and cache address registers, allow manipulation of the on-board 256-byte instruction cache. A cache is sequential memory directly accessible by the microprocessor. In the case of the 68020, the cache is located internally in the microprocessor and holds a portion of the program being executed. This configuration reduces execution time by removing continual instruction fetches during a looping process. For example, let us suppose an earlier program used a loop for a time delay by loading a register with a value and decrementing it until it reached zero. Each time the decrement and branch test instructions were executed, they were fetched first from program memory space. With a cache, there is no need to prefetch the loop instructions because the looping instructions are retained in the on-board instruction cache until the looping process is complete.

11.3 THE 68020 CACHE

In a cache environment, a portion of the program is prefetched and loaded into the cache. As each instruction is emptied into an instruction pipeline from the cache, the next program data (opword or operand) is loaded into the cache from memory. The cache therefore remains full as the program executes. Fetch cycle times are buried within execute cycles. If a loop (such as a delay loop) is encountered, the microprocessor keeps returning to the cache for instructions instead of accessing them from memory. Again, no time is lost because of instruction fetches.

Cache Types

There are two basic types of caches: the fully associative (Figure 11.2, page 302) and the set associative (Figure 11.3, page 302). The cache access address contains a *tag,* which is the actual data location within the cache memory. This access tag is compared with all cache memory tags. If it matches one of them, a hit occurs. In the fully associative cache, each tag location has its own comparator so that all locations can be checked simultaneously against the access tag. The hit enables a logic switch that connects the data lines of the cache and the system data bus.

In a set associative cache, the cache memory is split into blocks or groups. Each group has a single comparator. The cache access address now contains an index and a block field as well as the tag. The index specifies which group is being accessed, whereas the tag identifies the access location within the group. The block field controls which cache block within a multiple-cache environment is being accessed. In the set associative arrangement, each location's tag within a group is checked sequentially against the access tag, using the block group's single comparator. If a

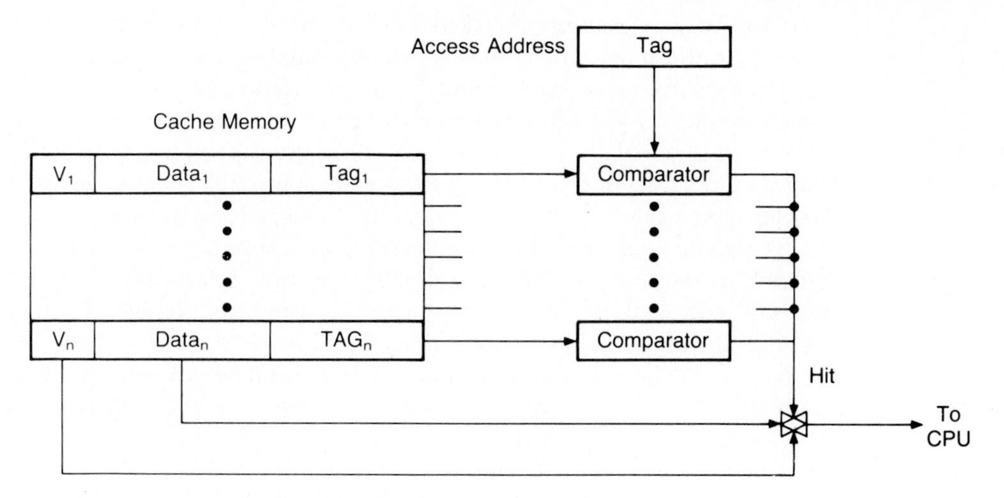

Figure 11.2 Fully associative cache

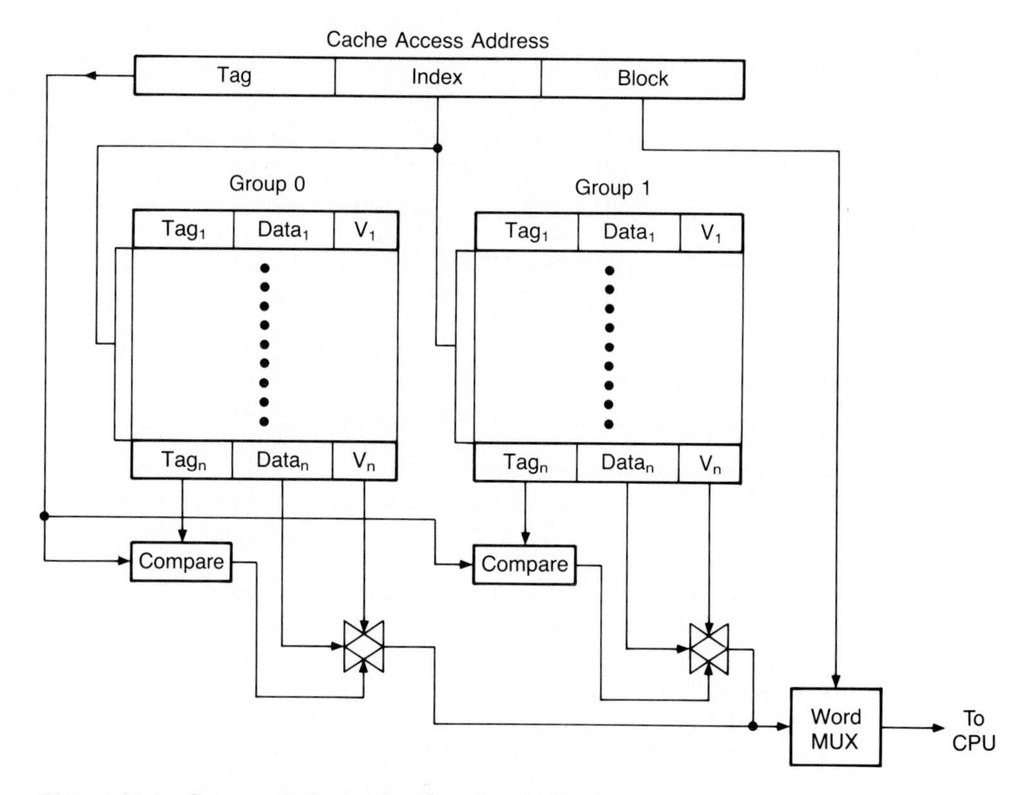

Figure 11.3 Set associative cache (Set size = 2)

match occurs, a hit is signaled and the data at that location are passed out to the system bus. In both types of caches, if no match is made to the access tag, a miss occurs, preventing any data from leaving the cache.

The cache can be loaded manually under software control or automatically by one of two methods. In the first method, the cache is filled with a new piece of data every time a miss occurs until a hit finally is made. This method is not practical because the hit data might be a long way down through memory. In the second method, the cache is filled each time a hit occurs and data are fetched directly from memory for any misses. In essence, this method creates a first in, first out (FIFO) type of local memory that can be used as an instruction cache. The cache is filled as it is emptied by each instruction accessed from the cache and loaded into the instruction pipeline. A memory access then is made from the program memory area to replace the vacated space in the cache.

68020 Instruction Cache

The 68020 on-board instruction cache is illustrated in Figure 11.4. The block at the top of the diagram shows the form of the cache access word. The function control line b2 is used with the tag field from the address bus to form the cache tag. The remaining bits of the access address form the index and block to the single set associative cache. The cache itself has one set and, therefore, one comparator to

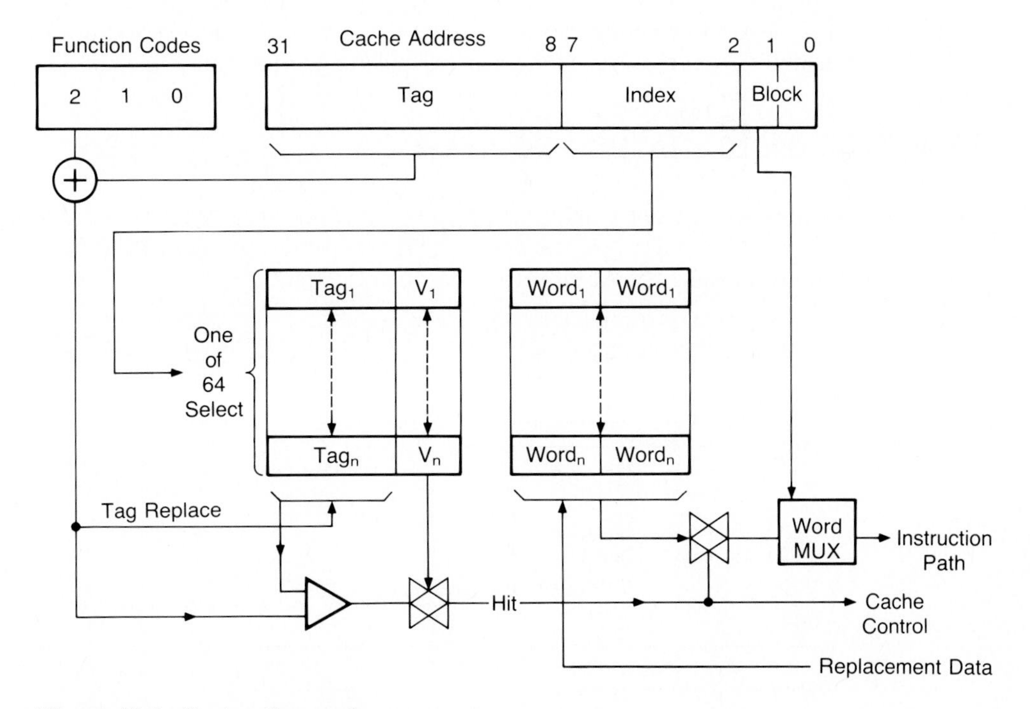

Figure 11.4 Instruction cache

search for a hit or a miss. Each time an instruction is fetched from the cache, it is replaced in the cache with the next instruction from the program memory area. In this manner, the cache is kept full while normal program execution is in progress. Data fetches still are made directly to user memory.

Cache Control and Address Registers

The cache registers that allow software control of the cache are listed in Table 11.1. The control register shows four control bits. Setting the clear cache (C) bit causes all cache entry valid bits to be set to zero or cleared. The clear entry (CE) bit causes the valid bit of the cache location specified in the cache address register to be cleared. The freeze cache (F) bit allows the cache to remain full and operative, but no entries are lost or added to the cache. Finally, the enable cache (E) bit enables the cache when set and disables the cache when cleared.

The cache address register is used when the CE bit is set in the control register. First, the address must be loaded into this register, then the CE bit is set in the control register. This combination causes the specified location within the cache to be cleared.

The instruction that allows these registers to be read or written to is MOVEC (move to/from control register). The coding of the opword specifies which register within the operation of the instruction is involved. Other registers accessed by this instruction are the alternate function control registers, the VBR, and the three stack pointers. Because of the types of registers involved with this instruction, it is a privileged instruction that can be used only in the supervisor mode. Use of this instruction in the user mode causes a privilege violation exception to occur.

11.4 THE 68020 STATUS REGISTER

The status register contains the flag bits that we are already familiar with: S, I bits, X, N, Z, V, and C. As shown in Table 11.2, there are additional flags in the 68020 status register. The M bit selects between the interrupt (M = 0) and master (M = 1)

Table 11.1 Cache Registers

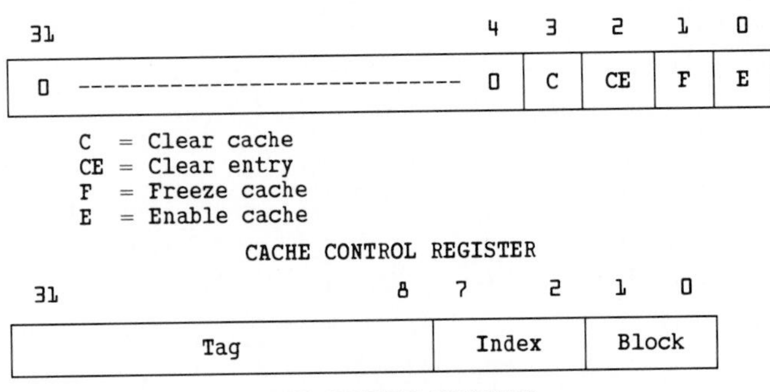

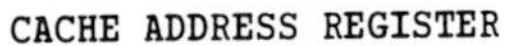

Table 11.2 Status Register of the 68020

15	14	13	12	11	10	9	8	7	6	5	4	3	2	1	0
T_1	T_0	S	M	0	I_2	I_1	I_0	0	0	0	X	N	Z	V	C

T_1	T_0	TRACE Function
0	0	No TRACE
0	1	TRACE on Change of Flow
1	0	TRACE on Instruction Execution
1	1	Undefined

M: 0 = Interrupt stack pointer
1 = Master stack pointer

stack pointers when in the supervisor mode. The trace mode is represented by two T flags instead of one. The truth table shown in Table 11.2 defines the meaning of these bits.

11.5 EXTERNAL DIFFERENCES BETWEEN THE 68020 AND THE 68000/68010

One obvious difference between the 68020 and the 68000/68010 is the package used for the 68020 (Figure 11.5, page 306). In Chapter 10 we discussed the need to go to a grid array package for the 68012 because of the increased number of pins for the device. The packaging shown in Figure 11.5 shows an increase in pins, due in large part to the 32-bit data and 32-bit address buses. This packaging allows four **gigabytes** of direct memory access. Table 11.3 on page 307 lists the 68020 pins and their functions. A description of the pin functions follows.

1. Transfer size (SIZ0 and SIZ1) indicates the number of bytes of an operand to be transferred during a bus cycle. The 68020 can write and read data using dynamic bus sizes. This capability also allows word and long word transfers using odd address bounds. (See Section 11.7 for a discussion of variable bus sizing.)
2. External start cycle ($\overline{ECS}$) output is asserted during the first half clock cycle of every bus cycle.
3. Operand cycle start ($\overline{OCS}$) output is asserted during the first half clock cycle of an operand transfer or instruction prefetch.
4. Data strobe ($\overline{DS}$) indicates (a) that there are valid data from the microprocessor during a write cycle and (b) when data are expected from an external device during a read cycle.

	INCHES		INCHES	
DIM	MIN	MAX	MIN	MAX
A	34.18	34.90	1.345	1.375
B	34.18	34.90	1.345	1.375
C	2.67	3.17	.100	.150
D0	.46	.51	.017	.019
G	2.54 BSC		.100 BSC	
K	4.32	4.82	.170	.190
V	1.74	2.28	.065	.095

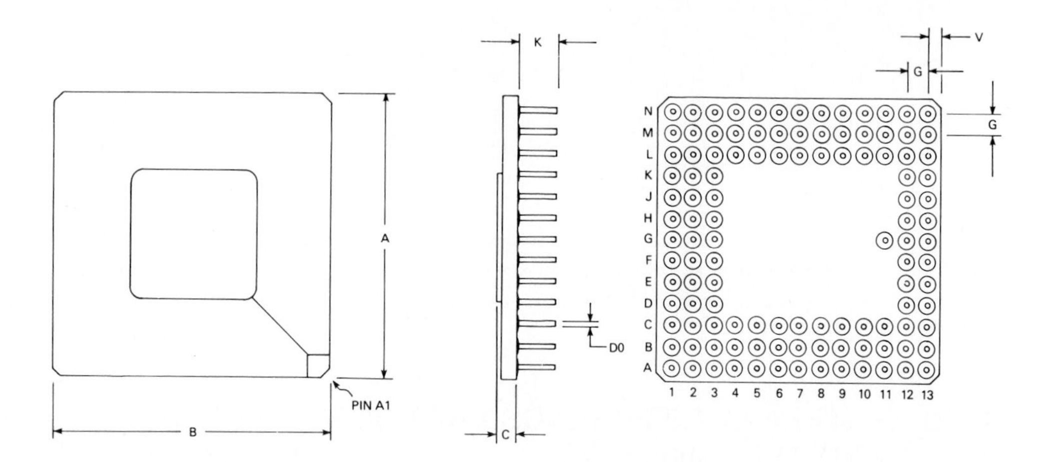

Figure 11.5 68020 package dimensions and pin assignments (Courtesy of Motorola, Inc.)

5. Data buffer enable ($\overline{\text{DBEN}}$) output controls external tri-state data buffers.
6. Data transfer and size acknowledge ($\overline{\text{DSACK0}}$ and $\overline{\text{DSACK1}}$) serve the same purpose as $\overline{\text{DTACK}}$ for the 68000, but they also include dynamic bus size data. (See Section 11.7 for a discussion of variable bus sizing.)
7. Cache disable ($\overline{\text{CDIS}}$) disables the on-board instruction cache.
8. Interrupt pending ($\overline{\text{IPEND}}$) indicates that the interrupt request on the IPL pins is higher than the current interrupt mask and that the 68020 is about to enter an interrupt service process.
9. Auto-vector ($\overline{\text{AVEC}}$) replaces $\overline{\text{VPA}}$ for the auto-vectoring function when you use interrupt exceptions.
10. Read-modify-write cycle ($\overline{\text{RMC}}$) indicates that an indivisible read-modify-write cycle is in progress.

11.6 ALTERED PIN FUNCTIONS

A number of 68020 signals are similar in nature to those of the 68000, but have been either enhanced or changed to meet the needs of the new microprocessor.

Table 11.3 Pin Designations for the 68020 (Courtesy of Motorola, Inc.)

Pin Number	Function
A1	$\overline{BGACK}$
A2	A_1
A3	A_{31}
A4	A_{28}
A5	A_{26}
A6	A_{23}
A7	A_{22}
A8	A_{19}
A9	V_{CC}
A10	GND
A11	A_{14}
A12	A_{11}
A13	A_8
B1	N.C.
B2	$\overline{BG}$
B3	$\overline{BR}$
B4	A_{30}
B5	A_{27}
B6	A_{24}
B7	A_{20}
B8	A_{18}
B9	GND
B10	A_{15}
B11	A_{13}
B12	A_{10}
B13	A_6
C1	$\overline{RESET}$
C2	CLOCK
C3	N.C.
C4	A_0
C5	A_{29}
C6	A_{25}
C7	A_{21}
C8	A_{17}
C9	A_{16}
C10	A_{12}
C11	A_9
C12	A_7
C13	A_5

Pin Number	Function
D1	V_{CC}
D2	V_{CC}
D3	N.C.
D4–D11	—
D12	A_4
D13	A_3
E1	FC0
E2	$\overline{RMC}$
E3	V_{CC}
E4–E11	—
E12	A_2
E13	$\overline{OCS}$
F1	SIZ0
F2	FC2
F3	FC1
F4–F11	—
F12	N.C.
F13	$\overline{IPEND}$
G1	$\overline{ECS}$
G2	SIZ1
G3	$\overline{DBEN}$
G4–G10	—
G11	V_{CC}
G12	GND
G13	V_{CC}
H1	$\overline{CDIS}$
H2	$\overline{AVEC}$
H3	$\overline{DSACK0}$
H4–H11	—
H12	IPL2
H13	GND
J1	$\overline{DSACK1}$
J2	$\overline{BERR}$
J3	GND
J4–J11	—
J12	$\overline{IPL0}$
J13	$\overline{IPL1}$

Pin Number	Function
K1	GND
K2	$\overline{HALT}$
K3	N.C.
K4–K11	—
K12	D_1
K13	D_0
L1	$\overline{AS}$
L2	$R/\overline{W}$
L3	D_{30}
L4	D_{27}
L5	D_{23}
L6	D_{19}
L7	GND
L8	D_{15}
L9	D_{11}
L10	D_7
L11	N.C.
L12	D_3
L13	D_2
M1	$\overline{DS}$
M2	D_{29}
M3	D_{26}
M4	D_{24}
M5	D_{21}
M6	D_{18}
M7	D_{16}
M8	V_{CC}
M9	D/3
M10	D_{10}
M11	D_6
M12	D_5
M13	D_4
N1	D_{31}
N2	D_{28}
N3	D_{25}
N4	D_{22}
N5	D_{20}
N6	D_{17}
N7	GND
N8	V_{CC}
N9	D_{14}
N10	D_{12}
N11	D_9
N12	D_8
N13	N.C.

The V_{CC} and GND pins are separated into three groups to provide individual power supply connections for the address bus buffers, data bus buffers, and all other output buffers and internal logic.

Group	V_{CC}	GND
Address Bus	A9	A10, B9
Data Bus	M8, N8	L7, N7
Logic	D1, D2, E3, G11, G13	G12, H13, J3, K1

$\overline{\text{RESET}}$

Asserted by itself, $\overline{\text{RESET}}$ acts as the master reset. It is no longer necessary to assert $\overline{\text{HALT}}$ with $\overline{\text{RESET}}$ to perform a system reset. As an output, $\overline{\text{RESET}}$ responds to the RESET instruction as it did with the 68000.

$\overline{\text{HALT}}$

Used as an individual input to halt internal processes of the 68020, $\overline{\text{HALT}}$ causes the following to occur: (1) all control signals are placed in their inactive state; (2) the R/$\overline{\text{W}}$, function control codes, size signals, and the address bus remain driven as they were by the halted bus cycle; and (3) the $\overline{\text{RMC}}$ is driven inactive and the data bus is placed in the tri-state condition.

As an output, $\overline{\text{HALT}}$ is driven active by a double bus fault in a manner similar to that of the 68000.

Function Control Codes (FC_2, FC_1, and FC_0)

The function control codes of the 68020 are the same as those of the 68010 and the 68000, except for code 111. This code indicates a CPU space instead of an interrupt acknowledge. Additional decoding of the address bus during a CPU space cycle is necessary to define the exact operation taking place. Examples of CPU space functions are interrupt acknowledge and breakpoints, as discussed in Chapter 10.

11.7 VARIABLE MEMORY DATA SIZING

With the 68000, word and long word accesses to or from memory require even bound addressing. That is, the high byte of the data always starts at an even address location and the remaining bytes follow sequentially through upper address locations. Memory to accommodate data transfers for the 68000 is designed through word organization. The data bus of the 68000 is connected directly to the memory data bus so that during word transfers, when the $\overline{\text{UDS}}$ is active, the upper data byte is sent or accessed from an even address. Similarly, an asserted $\overline{\text{LDS}}$ allows transfer between the 68000's lower data byte and an odd address location. Keep in mind that we are discussing word and long word and not byte transfers at this point.

Variable Memory Port

The 68020 can transfer data from the 68000 data bus to any multiple-byte size of memory starting on any address bound. A memory size, for purposes of discussion, is called a *memory port*. For instance, a memory organized as word size (16 bits) is called a 16-bit, or word, port. The 68020 can handle port sizes of byte, word, or long word. Data transferred to these memory ports may be aligned or misaligned.

Aligned Memory Transfer

An **aligned memory** transfer is illustrated in Figure 11.6. The data register line indicates the contents of one of the 68020's registers. The next line shows the data

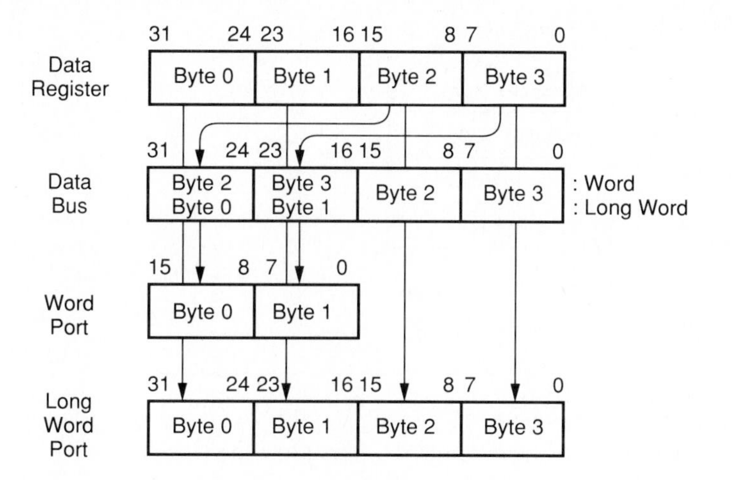

Figure 11.6 Aligned data transfer

bus and how word and long word ports are connected to the bus. An aligned word transfer to a word port sends bytes 2 and 3 of the 32-bit register to memory on an even address bound. The fourth line illustrates an aligned transfer of a long word to a long word port. To be fully aligned, long words begin at a long word boundary, designated as address $xxxx xxx0 or any multiple of four bytes thereafter ($xxxx xxx4, $xxxx xxx8, and $xxxx xxxC). For instance, $0120 03B8 is a long word bound, but $0120 03B6 is not.

The first two examples of an aligned transfer appear straightforward; and they are, as long as you are transferring word data to word ports and long word data to long word ports. However, when you transfer word data to a byte port or long word data to a byte or word port, take note of when the data are aligned and when they are not. When aligned and misaligned data are transferred, a method called **dynamic bus sizing** provides a means for managing data transfer sizes of 1, 2, 3, or 4 bytes using multiple bus cycles as needed to complete the entire transfer. The size of the current transfer is identified by the size signals SIZ1 and SIZ0.

SIZ1 and SIZ0

The truth table for the size pins is shown in Table 11.4 on page 310. At first, it looks fairly easy to deal with; but to understand this table, transfer size must be explained. With the 68000, a long word transfer is accomplished by doing two word transfers. After the upper word is transferred, there remains an additional word to be transferred—the lower word. The 68020 attempts to transfer all long words at once using the full 32-bit data bus. However, because of restrictions imposed by the size of the memory port in use or by the alignment of the transfer, not all bytes are always transferred during a single bus cycle. After determining the size of the memory port, the 68020 uses the size pins to keep track of the number of bytes to be transferred initially by a current bus cycle. On the next bus cycle, the size pins are updated, indicating how many additional bytes still must be moved.

Table 11.4 Data Transfer Sizes

SIZ1	SIZ0	Transfer Size
0	0	Long Word: 4 Bytes
0	1	Byte: 1 Byte
1	0	Word: 2 Bytes
1	1	3 Bytes

EXAMPLE 11.1

The memory port size is byte, and a long word is to be transferred to it from register D_3. How many bus cycles are required to complete the transfer? What is the condition of the size pins for each transfer?

Solution:

A byte size port can accept a single byte of data for each bus cycle. As a result, it will take four separate write cycles to transfer the 32-bit long word data from D_3. The first cycle sends the highest byte of data to memory, and the size pins indicate four bytes (both SIZ1 and SIZ0 are low) will be transferred. One byte is successfully written into memory, then the second write cycle is commenced. For this cycle, since one byte already has been sent, there are only three bytes left to be transferred. This is indicated by the 68020 making both size pins high. Similarly, the third cycle indicates a 2-byte transfer (SIZ1 = 1 and SIZ0 = 0); and, last, cycle 4 shows the last byte to be transferred by making SIZ1 = 0 and SIZ0 = 1.

For aligned transfers to word or long word ports, only word and long word transfers are required, using one or two write bus cycles for each transfer.

EXAMPLE 11.2

How many bus cycles are required to transfer a long word data to an aligned address within a word port? What are the states of SIZ1, SIZ0, A_1, and A_0 during the transfer?

Solution:

Figure 11.7 illustrates the solution to this example.

For the first cycle, the data from the register is transferred to the data buffers. A_1 and A_0 are low, indicating a long word address bound. The size pins are low, indicating the need to transfer four bytes. The word port accepts the upper word, as expected, and indicates to the 68020 that one word has been transferred.

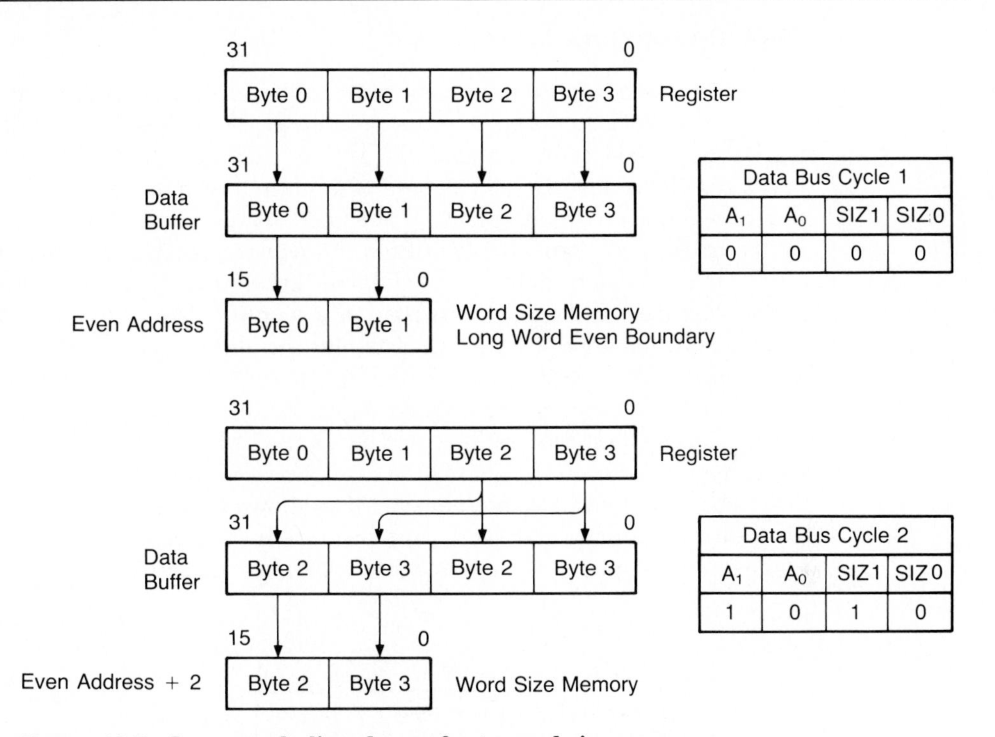

Figure 11.7 **Long word aligned transfer to word size memory**

The second cycle shows the lower word being transferred to the lower and upper words of the data buffers. The upper word of the data buffer again is transferred to memory. A_1 is high and A_0 is low, indicating an even address, but not a long word bound. The long word is aligned because it began on the long word bound at the start of the process and the second word was moved to the next sequential word address. SIZ1 is high and SIZ0 low, indicating a word (2-byte) transfer is required.

Why are data from the register copied into the data bus in the second cycle of Example 11.2? Consider the problem of transferring data from a 32-bit data bus into a memory that may be organized in one of three port sizes. The data port is fixed, beginning at the upper data lines of the 68020 bus. In other words, a long word port is connected so that data pin 31 of the 68020 data bus goes to b31 of the long port memory. For a word port, data pin 31 is connected to b15 of the memory port. Finally, for a byte port, data pin 31 is connected to b7 of the memory port. The remaining port data lines are connected to declining sequential data lines of the 68020. As a result, data always are transferred through the higher data lines of the 68020 to the memory port being used.

Misaligned Data Transfers

The ability to handle various data port sizes leads to an additional capability to transfer data to a misaligned memory address. Word and long word data can be transferred to odd address bounds as well as even ones, regardless of the memory port size. Figure 11.8 is one example of a misaligned transfer. A long word is being transferred to a word port beginning at an odd address. Part of the process performed by the 68020 is to move the data from the register into the data buffers so that the correct transfer can be done. The word port lines (b0–b15) are connected to b16–b31 of the data bus. To begin the transfer of a word of data to this port starting at an odd address, the upper byte of the data must be aligned into the data buffers so that it is transferred to the correct location. b0–b7 of the word port are accessed by odd addresses. These bits are physically connected to b16–b23 of the 68020 data bus. As a result, the first process the cycle performs is to move the upper byte (byte 0) to the correct position in the data buffers, as shown in the second line of Figure 11.8.

The other bytes also are transferred as shown to facilitate a misaligned transfer to a long word port. The word port memory only accepts the data written to the odd address location. On the next write cycle, bytes 1 and 2 are aligned correctly by

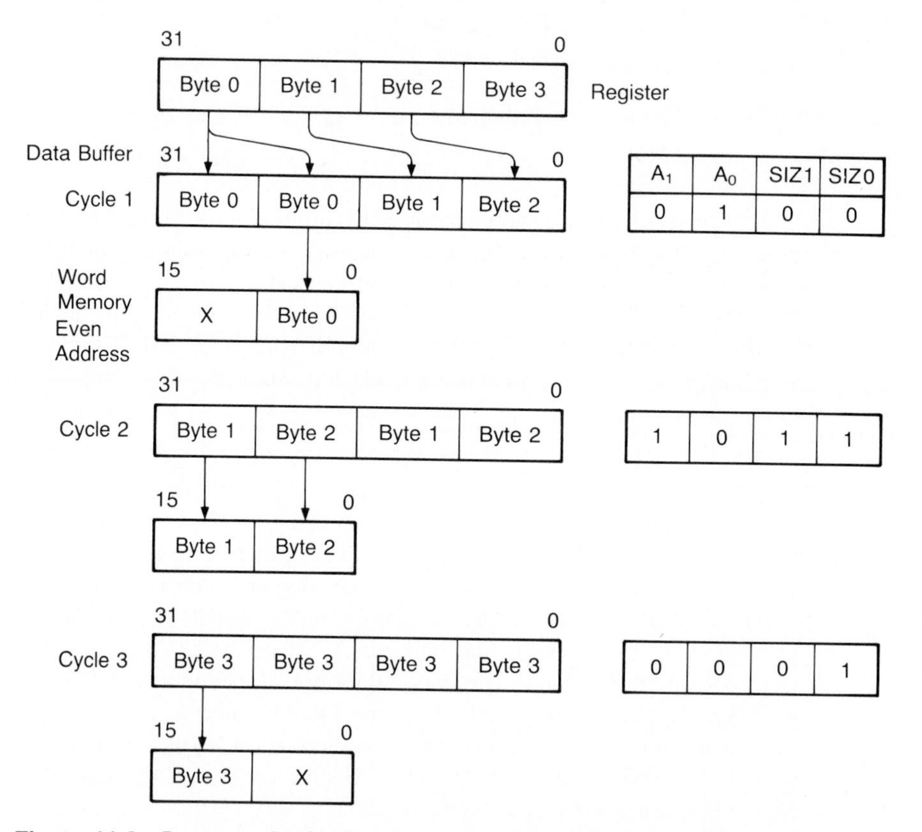

Figure 11.8 **Long word misaligned on odd address to word size memory**

the 68020 into the upper data buffer locations and then transferred as a word to the memory. The third cycle accepts the lowest byte of data to complete the transfer.

Table 11.5 details the alignment of data in the data buffer for all combinations of data transfers. A_1 and A_0 direct the type of alignment—even or odd, long word aligned or not aligned. Note that all even address accesses are word aligned, whereas all odd address accesses are word misaligned. Alignment is significant in determining the transfer requirements at the start of the memory cycle. Each individual write or read cycle following the initial cycle is dictated by the original alignment.

SIZ1 and SIZ0 indicate the number of bytes required to be transferred. The data in the data bus are arranged to permit successful cycle-by-cycle transfer of all sizes of data to/from any size of memory data port. Figure 11.9 on page 314 is an example of how to interpret Table 11.5.

In this case, a long word data is being transferred to a byte port. The port is connected to the upper byte of the data bus, so all data must be transferred through these lines. Four cycles are required to complete the transfer; and for each, the correct byte must be placed in the upper byte of the data buffers before the actual transfer. Since it is an aligned transfer, the first address causes A_1 and A_0 to be low. SIZ1 and SIZ0 are also low, informing the external memory port that four bytes must be written into memory. All four bytes appear on the data bus in correct sequence. However, only the upper byte of the data bus is connected to the memory port,

Table 11.5 Data Bus Transfers for Variable Sizing

						31			0
					Register	Byte 0	Byte 1	Byte 2	Byte 3
A1	A0	SIZ1	SIZ0	TB	Alignment	Data Bus			
						31---24	23---16	15---8	7---0
X	X	0	1	1	All	Byte 3	Byte 3	Byte 3	Byte 3
X	0	1	0	2	Even Add.	Byte 2	Byte 3	Byte 2	Byte 3
X	1	1	0	2	Odd Add.	Byte 2	Byte 2	Byte 3	Byte 2
0	1	1	1	3	Long Odd	Byte 1	Byte 1	Byte 2	Byte 3
1	0	1	1	3	Word Even	Byte 1	Byte 2	Byte 1	Byte 2
1	1	1	1	3	Word Odd	Byte 1	Byte 1	Byte 2	Byte 1
0	0	1	1	3	Long Even	Byte 1	Byte 2	Byte 3	Byte 0
0	0	0	0	4	Long Even	Byte 0	Byte 1	Byte 2	Byte 3
0	1	0	0	4	Long Odd	Byte 0	Byte 0	Byte 1	Byte 2
1	0	0	0	4	Word Even	Byte 0	Byte 1	Byte 0	Byte 1
1	1	0	0	4	Word Odd	Byte 0	Byte 0	Byte 1	Byte 0

TB = Total number of bytes desired to complete the transfer

Long word alignments are also word alignments.

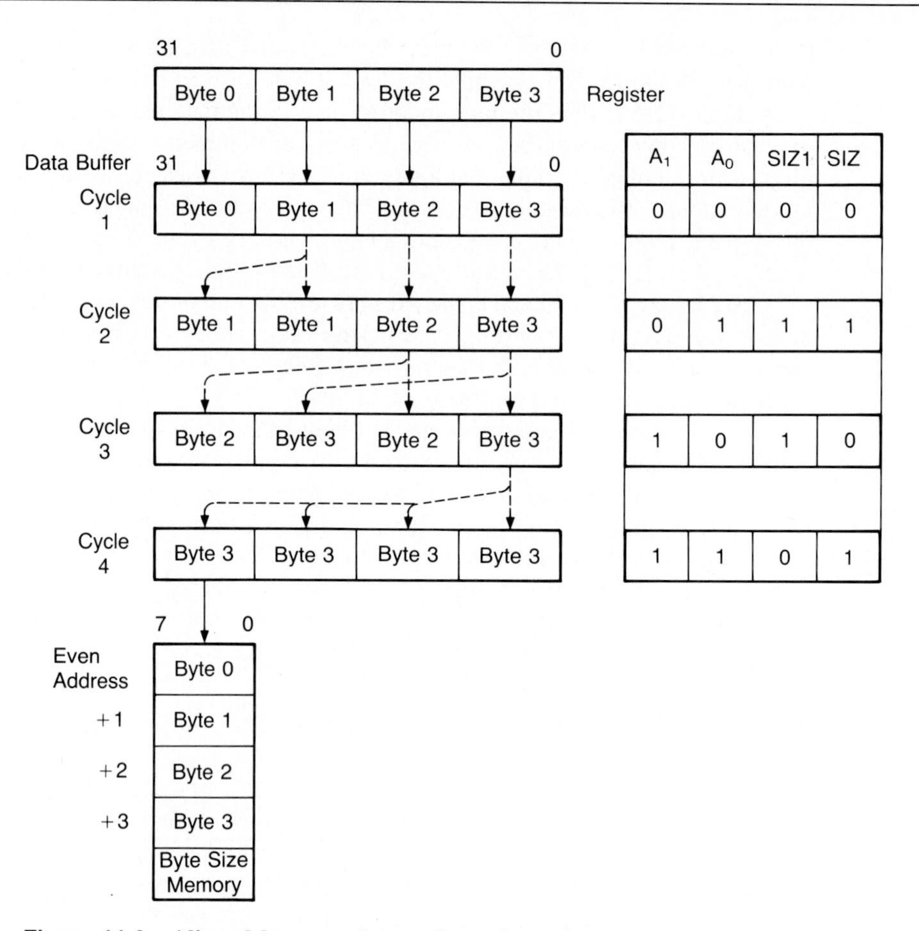

Figure 11.9 **Aligned long word transfer to byte size memory**

and the data on those leads are the only data successfully written into memory. The 68020 is informed that this is a byte port and starts the second write cycle, addressing the next highest address ($A_1 = 0$ and $A_0 = 1$). This condition can be found in two places in Table 11.5. Next, we need to determine how many bytes still must be transferred. One byte was successfully written to memory by the first cycle, leaving three more to be moved. SIZ1 and SIZ0 are both set high to indicate this fact. In Table 11.5, find the only line that has $A_1 = 0$, $A_1 = 1$, and both size pins high. The line that fits these needs (the fourth one down) shows the data buffer alignment used for the second write cycle. Again, for the byte port, only the uppermost byte is written into the memory and the process must go on for two more cycles.

EXAMPLE 11.3

Cycle 3 of Figure 11.9 is used for this byte port transfer. For what other transfer is it used?

Solution:

Cycle 3 of Figure 11.9 also is used for the lower word transfer of a long word aligned to a word port (bytes 2 and 3 moved to the word port).

One last example of misalignment is diagrammed in Figure 11.10, showing a long word transfer misaligned to a long word port. The first cycle writes b0 to memory, leaving three bytes to be transferred. The second write cycle transfers those remaining bytes to memory, completing the instruction cycle.

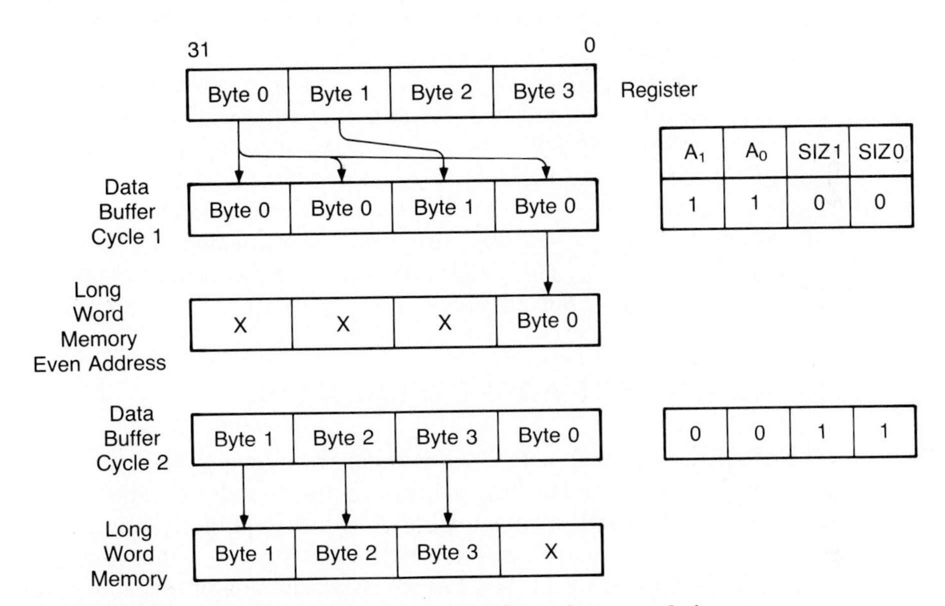

Figure 11.10 **Misaligned long word transfer to long word size memory**

11.8 DETERMINING PORT SIZE

The preceding discussion mentioned that the microprocessor is informed in some way of the memory port size in order to generate the correct SIZ1 and SIZ0 states. The input pins used for this process are $\overline{\text{DSACK1}}$ and $\overline{\text{DSACK0}}$, which stand for "data size acknowledge." They recognize the memory port size and acknowledge a successful data transfer in much the same manner as $\overline{\text{DTACK}}$ did for the 68000. Table 11.6 on page 316 is the truth table for the $\overline{\text{DSACK}}$ pins. For all data transfers, the 68020 awaits the return of the $\overline{\text{DSACK}}$ signals or $\overline{\text{BERR}}$. If these input pins remain inactive (all high), the 68020 inserts wait cycles to allow for the use of slow memories or peripheral devices with the 68020.

Internally, the 68020 examines the $\overline{\text{DSACK}}$ and $\overline{\text{SIZ}}$ information to determine how many bytes of the memory still must be transferred. If additional bus cycles are

Table 11.6 Data Size Acknowledge

$\overline{DSACK1}$	$\overline{DSACK0}$	Data Port Size
1	1	Insert Wait Cycles
1	0	Byte: 8 Bits
0	1	Word: 16 Bits
0	0	Long Word: 32 bits

needed, the correct byte data are supplied to the data buffers and the size pins and address lines are set accordingly.

With the availability of dynamic bus sizing and the acceptance of misaligned data accesses along with virtual memory, the $\overline{BERR}$ signal is not used so often as it was with the 68000. $\overline{BERR}$ is used if the access is to an area that is in neither physical nor logical memory. An address bus error is produced when opword accesses are made to an odd address. The 68020 maintains the requirement of accessing opwords on even bounds to allow 68000 software to be run using the 32-bit microprocessor. Beyond this restriction, additional accesses can be done aligned or misaligned without generating an address bus error.

11.9 MEMORY INDIRECT ADDRESSING MODE

The 68000 used several methods of indirect addressing with the address register as the pointer. The 68020 retains these addressing modes and uses a new mode called *memory indirect,* which allows one memory location to point to another memory location where the data are held. There are two forms of this address mode—**postindexed** and **preindexed** (Figure 11.11). During the discussion of this addressing mode, keep in mind that most portions of syntax can be omitted if they are not applicable to a specific instruction. We discuss each of them as a whole to explore the flexibility of this mode.

Postindexed Mode

In the postindexed form of the memory indirect mode, the value of the index register is added to the indirect memory address to formulate the effective address. The syntax for the postindexed mode is

([bd,An],Xn.SIZE*SCALE,od)

where bd is the sign-extended base displacement; An is any address register; Xn is any data or address register used as the index register; SIZE is the word or long word size selection of the index register; SCALE is an index register multiplier value of 1, 2, 4, or 8; and od signifies the outside displacement that is added to the form

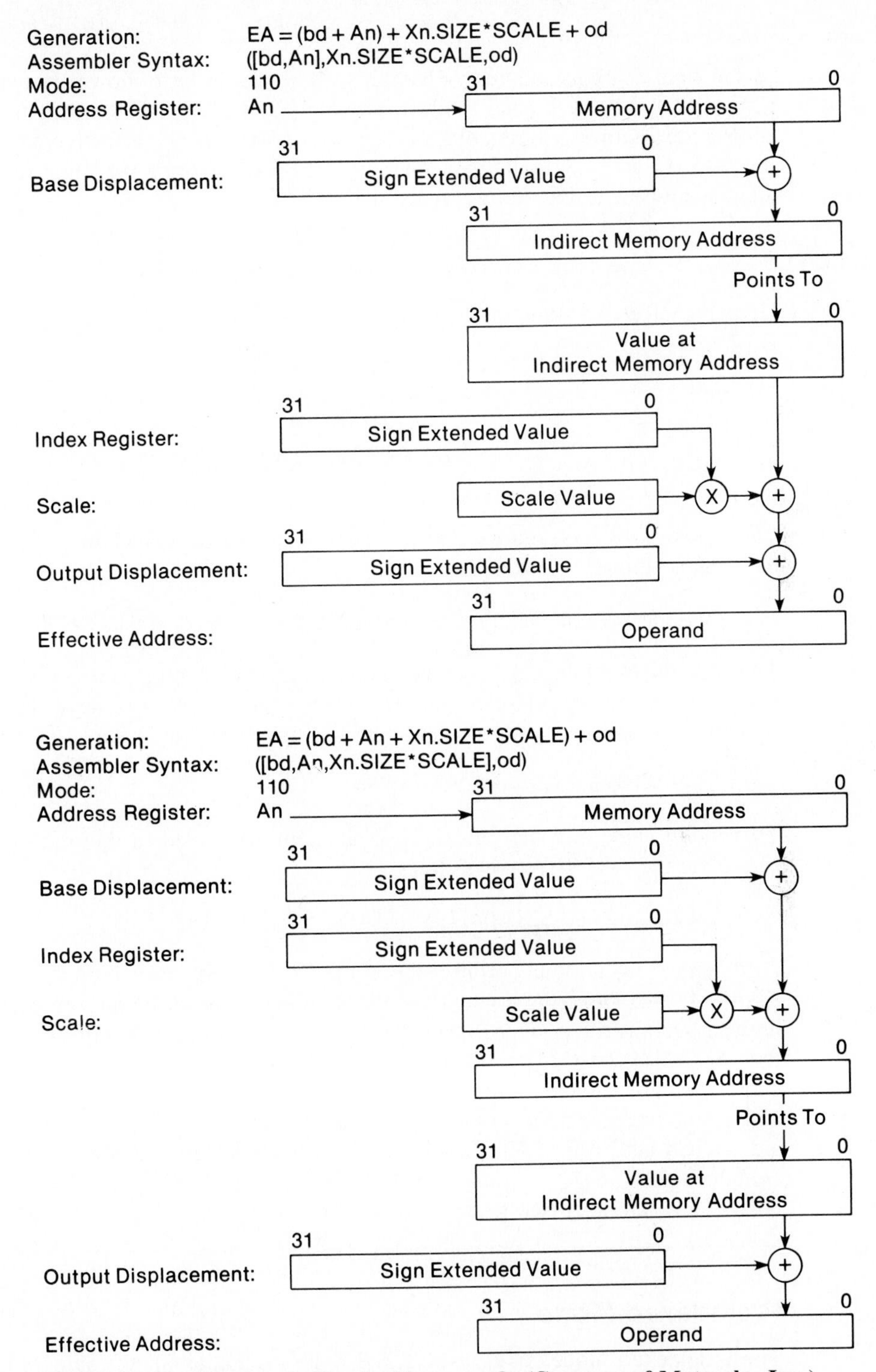

Figure 11.11 Memory indirect address mode (Courtesy of Motorola, Inc.)

of the final effective address. How this all works is best shown by example. Use Figure 11.8 as a reference while following Example 11.4. Keep in mind that the central idea is to use a memory location as a pointer to the effective address base of the data. In effect, this indirect memory location performs the same function as an address register in the address register indirect mode.

EXAMPLE 11.4

Give the effective address of an instruction using the following syntax: ([6,A2], D3.L*2,4).

A2 = $0012 CC2A
D3 = $1450 03C2
($0012 CC30) = $4010 2B34

Solution:

The contents of the address register are transferred to the address buffers and the base displacement is added to it:

Indirect memory address = $0012 CC2A + $6 = $0012 CC30

The contents of this location are fetched and transferred to the address buffers. (Thus, the indirect memory address has been used as a pointer to retrieve the actual data pointer.) The index register contents are multiplied by the scalar

$1450 03C2 × $2 = $28A0 0784

This index value and the outside displacement are added to the address buffers to formulate the final effective address:

$4010 2B34 + $28A0 0784 + $4 = $68B0 32BC

After removing all the options, the syntax of the memory indirect instruction is ([An]). In this case, the contents of the memory location pointed to by the address register become the pointer to the effective address of the data. For example, assume that A_1 = $0012 CC24 and location $0012 CC24 contains $1234 5678. Using the syntax as shown, the effective address of the data is $1234 5678. The contents of the address register are transferred to the address buffers, and the data at that location are fetched and in turn transferred to the address buffers. At this point, the buffers contain $1234 5678, the data at $0012 CC24. The data pointed to by the address buffers now are accessed to be used by the instruction.

Preindexed Mode

The preindexed mode, like postindexed mode, contains all the same variable (and optional elements. However, this time the index register becomes part of the initial address pointer to the memory pointer. That is, the index register computation is done before the address pointer is retrieved from memory. The syntax for the preindexed mode is

([bd,An,Xn.SIZE*SCALE],od)

Notice that the index register elements as well as the address register are contained within the square brackets. To find the effective address, add the contents of the address register (An), the base displacement (bd), and the index value. The index value is found, as before, by multiplying the contents of the index register by the SCALE factor. The sum derived is an address in memory. The contents of this memory location and the next three consecutive bytes added to the outside displacement (od) form the address of the data to be used by the instruction.

EXAMPLE 11.5

Describe where the data go in this instruction:

MOVE.L #$12345678,([$4C,A2,D4.L*4],$CC)
A2 = $2004 6BBA   D4 = $1001 201A
$6008 EC5E = $00DE $6008 EC61 = $FA02

Solution:

The first step is to compute the index value by multiplying the contents of D4 by 4:

$1001 201A × 4 = $4004 8058

To this value, add the contents of A_2 and the base displacement:

$4004 8058 + $2004 6BBA + $4C = $6008 EC5E

From this location, fetch the memory indirect pointer:

($6008 EC5E–$6008 EC61) = $00DE FA02

These data are added to the outside displacement to form the address of the data to be used by the instruction:

$00DE FA02 + $CC = $00DE FACE

As with the postindexed mode, if none of the options are used, the syntax takes the form ([An]). The address register points to a memory location that contains the data pointer. In this example, the syntax would be

MOVE.L #$12345678,([A2])

Location $2004 6BBA contains the memory indirect pointer information. For instance, if location $2004 6BBA contains $0000, location $2004 6BBC contains

$1234, location $0000 1234 holds $33CC, and $0000 1236 holds $DEA2, then the address to which the data are going is pointed to by the contents of the memory location pointed to by the address register. The contents of A_2 ($2004 6BBC) and the contents of the next address form the memory pointer $0000 1234. The contents of this location and the next one are used to direct the data of the move instruction. $12345678 is stored starting at location $33CC DEA2.

You must be careful when you write instructions in assembler form because of the similarity of the memory indirect instruction with no options,

MOVE ⟨ea⟩,([An])

and an address indirect instruction,

MOVE ⟨ea⟩,(An)

11.10 NEW INSTRUCTION TYPES

Besides byte, BCD, word, and long word data manipulations, other instructions manipulate bits in a bit field and alter the storage of BCD data. The latter category is called packing and unpacking BCD numbers. Newly added instructions, such as TRAPCC, expand upon the existing 68000 concepts and instructions.

Packing and Unpacking BCD

BCD data used by the 68000 are restricted to a two-digit byte size format. This format is called **packed BCD.** In contrast, **unpacked BCD** digits allow each byte to contain a single digit in the lower nibble. The upper nibble is made all low. For example, 0305 is considered unpacked, whereas 0035 is packed. Table 11.7 shows the memory or register contents of packed and unpacked BCD data.

Two instructions allow the programmer to maneuver BCD data in either direction—PACK and UNPK. For UNPK, two BCD digits within the source operand (byte size data) are unpacked into two separate bytes. In our previous example, $35

Table 11.7 BCD Data Types

15	14	13	12	11	10	9	8	7	6	5	4	3	2	1	0
0	0	0	0	BCD 0				0	0	0	0	BCD 1			

Unpacked BCD = 1 decimal digit per byte

15	14	13	12	11	10	9	8	7	6	5	4	3	2	1	0
BCD 0				BCD 1				BCD 2				BCD 3			

Packed BCD = 2 decimal digits per byte

would be converted to \$0305. An additional adjustment value is allowed, which is added hexadecimally to the unpacked result before the final value is stored into the destination. The syntax for the UNPK instruction takes one of two forms:

UNPK −(Ax),−(Ay),#⟨adj⟩

or

UNPK Dx,Dy,#⟨adj⟩

EXAMPLE 11.6

What are the contents of D_3 after this instruction, if D_1 contains \$1234 5678? D3 initially holds \$ABCD 12EF.

UNPK D1,D3,#\$2060

Solution:

The first step is to unpack the lower byte of D1 into 0708. The adjustment then is added to this result to form 2768. That result is stored into the lower word of D3 without affecting the upper word:

D3 = \$ABCD 2768

Packing BCD from two separate bytes into a single byte is performed by the PACK instruction. The adjustment is added to the lower word of the source; then it is packed and stored into the destination. The upper word and byte of the lower word of the destination remain unaffected by the instruction.

EXAMPLE 11.7

What are the contents of D3 after this instruction, if D1 contains \$8812 84A5? D3 holds \$ABCD EFCC initially.

PACK D1,D3,#\$8060

Solution:

The adjustment is added hexadecimally to the source to become

\$8812 84A5 + \$8060 = \$8812 0505

This lower word result is packed and sent to the destination without affecting the remaining register contents:

D3 = \$ABCD EF55

TRAPCC

The TRAP and TRAPV instructions have been expanded to allow traps on all condition code combinations used in the branch instruction. The vector is the same one used for TRAPV (vector number 7) of the 68000.

CAS

Compare and swap (CAS) takes the syntax form

CAS.⟨size⟩Dc,Du,⟨ea⟩

The contents of the Dc data register are compared with the contents at the effective address; if they match, the contents of the Du register are sent to the effective address. If they do not match, the contents of the effective address are sent to the Dc data register and the Du register has no effect on the outcome. This is one of the indivisible, read-modify-write cycle instructions that causes the $\overline{RMC}$ pin to be asserted during its execution. Effective addresses are limited to memory reference modes only.

EXAMPLE 11.8

D0 contains $0123 4567, D1 holds $3CC4 BF10, and A6 contains $2FC1 32B0. What are the contents of these registers as a result of the instruction CAS.L D0,D1,A6?

Solution:

D0 is compared with the contents of the location pointed to by A6. Because they do not match, the contents of this location are sent to D0. D0 and the location indicated by A6 hold $2FC1 32B0, and D1 retains $3CC4 BF10 at the end of the instruction.

EXAMPLE 11.9

What are the contents of D0, D1, and A6 if a second CAS.L D0,D1,A6 instruction is executed following the one in Example 11.8?

Solution:

Both D0 and the memory location pointed to by A6 contain $2FCC1 32B0. Because they match, the contents of D1 are copied into the location indicated by A6. D0 = $2FC1 32B0, and D1 and the location indicated by A6 contain $3CC4 BF10 when the second CAS instruction has been completed.

Other New Instructions

Additional instructions affect data moves to the alternate function control registers, and set breakpoints and an exception return that deallocates parameters. This return pulls the program counter from the stack and resets the stack pointer upon a return from an exception. Other parameters are not pulled from the stack.

Enhanced Instructions

Some instructions that appear in the 68000/68010 instruction set are enhanced with the 68020, making them more powerful or versatile. These include CMP2, branching, multiply, and divide instructions.

CMP2 Compare number 2 (CMP2) is identical to a check instruction except that an exception does not occur. Both instructions compare the data with an upper and lower bound. For the CHK2 instruction, an exception occurs if the data are out of bounds. For CMP2, no exception occurs, but some of the condition codes are set or reset as the results dictate.

Branch Instructions Branch instructions now can use long word displacements, increased from word size.

MULU and MULS MULU and MULS instructions can perform long word multiplies yielding quad word (64-bit) results. The syntax of these instructions can take one of three forms, depending on the desired results:

1. MULX.W ⟨ea⟩,Dn is for word operands capable of yielding up to long word results. X indicates either S for signed or U for unsigned.
2. MULX.L ⟨ea⟩,D1 is for long word operands expected to yield up to a long word result. If a quad word result occurs, the upper long word is lost and the overflow (V) bit is set.
3. MULX.L ⟨ea⟩,Dh:D1 is for long word operands that might result in quad word products. Dh holds the upper long word result, and D1 holds the lower long word.

DIVU and DIVS The two divide instructions also take advantage of quad word data values. The syntax for the divide instructions and their data sizes are detailed in Table 11.8 on page 324. For DIVX.L⟨ea⟩,Dq produces no remainder result. The remainder for that operation is discarded. If an overflow occurs with any of the forms (in the manner of the overflow example in Section 3.16), the overflow bit is set. The result of the division is retained in similar manner as that of the 68000—the lower half of the quotient is stored in the quotient register. Dividing by 0 causes the zero divide exception to be initiated.

11.11 BIT FIELDS

A **bit field** is any number of consecutive bits from 1 to 32 located in memory. The *base address* is defined as the location that contains b_0 of the field, and the *field offset number* is the last bit in the field. The number of bits in the field is called the *bit field width*. The offset number can be specified with an immediate value in the range of ±32. Another way to set the offset is to use a data register to hold the

Table 11.8 Divide Instruction Syntax

General Syntax: DIVX.⟨size⟩⟨source⟩,⟨destination⟩
Operation: Destination/Source ⟶ Destination
Specific Syntax Forms:

Syntax	Operation
DIVX.W ⟨ea⟩,Dn	Long/Word ⟶ WR : WQ
DIVX.L ⟨ea⟩,Dq	Long/Long ⟶ LQ
DIVX.L ⟨ea⟩,Dr:Dq	Quad/Long ⟶ LR : LQ
DIVXL.L ⟨ea⟩,Dr:Dq	Long/Long ⟶ LR : LQ

Dn = Any data register
Dr = Data register to hold remainder
Dq = Data register to hold quotient
WQ = Word size quotient
WR = Word size remainder
LQ = Long word quotient
LR = Long word remainder
X = U: unsigned S: signed

offset value. When a data register is used, the offset range can extend from 2^{31} to -2^{31} bits from reference b_0. The maximum width of the field is 32 bits. The general syntaxes of bit field instructions are

⟨opword mnemonic⟩ {⟨offset⟩:⟨width⟩}

and

⟨opword mnemonic⟩ {Dn:⟨width⟩}

Now we are ready to see how a bit is specified within a field. Using Table 11.9, locate b_0 at the base address. If we want to access the 5 bits shown, we would

Table 11.9 Bit Field Example

	Address	Bit Field Offsets							
	$3FFD	−24	−23	−22	−21	−20	−19	−18	−17
	$3FFE	−16	−15	−14	−13	[−12]	[−11]	[−10]	[−9]
	$3FFF	[−8]	−7	−6	−5	−4	−3	−2	−1
Base	$4000	0	1	2	3	4	5	6	7
	$4001	8	9	10	11	12	13	14	15
	$4002	16	17	18	19	20	21	22	23

Boxed numbers designate the field example used in the text.
Example offset: width = −12:5.

have to designate the field as offset number −12 and width 5. The last bit of the field is the leftmost bit, with the remaining 4 bits directly and consecutively to its right comprising the rest of the field. The instruction to clear those bits is BFCLR {−12 : 5}.

Bit Field Instructions

BFTST ⟨ea⟩ {⟨offset⟩:⟨width⟩} BFTST tests a bit field and sets the condition codes according to the result. The two significant condition code bits are the Z and N bits.

BFSET ⟨ea⟩ {⟨offset⟩:⟨width⟩} The bit field is tested, and the Z and N bits are set or reset accordingly. The bits in the field are all set high after the test.

BFCLR ⟨ea⟩ {⟨offset⟩:⟨width⟩} The bit field is tested, after which the bits in the field are set low.

BFCHG ⟨ea⟩ {⟨offset⟩:⟨width⟩} After the bit field is tested, all the bits are complemented.

BFEXTU ⟨ea⟩ {⟨offset⟩:⟨width⟩},Dn The bit field is extracted from memory and placed into the data register. This unsigned data is then zero extended to the full 32-bit complement of the data register.

BFEXTS ⟨ea⟩ {⟨offset⟩:⟨width⟩},Dn This instruction is the same as BFEXTU, except that the remaining bits in the data register are sign extended rather than zero extended.

BFFFO ⟨ea⟩ {⟨offset⟩:⟨width⟩},Dn Find the first high bit in the bit field. The bit offset number of the most significant high bit in the field is placed in the data register. Finding no high bits in the field results in the sum of the offset and width being placed into the data register.

BFINS Dn,⟨ea⟩ {⟨offset⟩:⟨width⟩} BFINS moves a bit field from the lower order of a data register to the specified effective address.

11.12 ADDRESS AND BUS ERROR EXCEPTIONS

As a result of the variable bus sizing and misaligned data transfers, there is essentially only one condition that causes an address error to occur—prefetching an instruction on an odd address bound. Motorola maintained the even address bound requirement for instruction fetches to allow other 68000 family programs to run using the 68020.

Bus Errors

A bus error exception is initiated whenever $\overline{BERR}$ is asserted, with $\overline{DSACK1}$ and $\overline{DSACK0}$ high (inactive) in response to a memory transfer. The exception process begins as others have by copying the status register into a temporary register and then setting the S bit to place the microprocessor into the supervisor mode. Both trace flags are reset to disable the trace function. The vector number associated with the bus error exception is multiplied by four (shifted left twice) to form the vector offset.

The contents of the program counter (address of the instruction being executed), followed by the status register copy, are sent out onto the stack. If the bus error occurred at the beginning of an instruction execution, 1010 is set as the format code. Code 1011 indicates a bus error that occurred somewhere in the middle of the instruction execution. These two formats indicate the formation of a short (1010) or a long (1011) stack (see Table 11.10). A combination word formed from the format code and the vector offset is sent onto the stack following the status register information.

After an internal register (not specifically defined by Motorola) is saved on the stack, a special status word is pushed out. This word (see Table 11.11, page 327) provides additional information about the bus error. FB and FC indicate that the fault occurred during pipelining activity. The instruction pipeline is the mechanism that allows the prefetch and cache to work together. DF indicates that the bus error fault occurred as part of a data transfer. The DF flag also doubles as an indicator to the microprocessor that the cycle should be rerun after the execution of an RTE instruction. The RTE would be the last instruction in the bus error exception routine. RB and RC allow the operator to select whether the microprocessor should rerun the instruction in the pipe after the RTE has been executed. The rerun is possible because the original data in the pipe as well as other internal registers have been saved on the stack and are retrieved when the RTE is used.

After the stack is complete, the vector offset is added to the contents of the VBR. This sum is placed in the address buffers, and the data retrieved from that location are placed in the program counter as the address of the bus error's first

Table 11.10 Bus Error Stack (Courtesy of Motorola, Inc.)

BUS ERROR STACK, SHORT

SP →	STATUS REGISTER
+2	PROGRAM COUNTER HIGH
	PROGRAM COUNTER LOW
+6	1 0 1 0 ¦ VECTOR OFFSET
	INTERNAL REGISTER
+10	SPECIAL STATUS WORD
+12	INSTRUCTION PIPE STAGE C
+14	INSTRUCTION PIPE STAGE B
+16	FAULT ADDRESS HIGH
	FAULT ADDRESS LOW
	INTERNAL REGISTER
	INTERNAL REGISTER
+24	DATA OUTPUT BUFFER HIGH
	DATA OUTPUT BUFFER LOW
	INTERNAL REGISTER
+30	INTERNAL REGISTER

BUS ERROR STACK, LONG

SP →	STATUS REGISTER
+2	PROGRAM COUNTER HIGH
	PROGRAM COUNTER LOW
+6	1 0 1 1 VECTOR OFFSET
	INTERNAL REGISTER
+10	SPECIAL STATUS WORD
+12	INSTRUCTION PIPE STAGE C
+14	INSTRUCTION PIPE STAGE B
+16	FAULT ADDRESS HIGH
	FAULT ADDRESS LOW
	INTERNAL REGISTERS (2 WORDS)
+24	DATA OUTPUT BUFFER HIGH
	DATA OUTPUT BUFFER LOW
+28	INTERNAL REGISTERS (4 WORDS)
+36	STAGE B ADDRESS HIGH
	STAGE B ADDRESS LOW
+40	INTERNAL REGISTERS (2 WORDS)
+44	DATA INPUT BUFFER HIGH
	DATA INPUT BUFFER LOW
+48	INTERNAL REGISTERS (22 WORDS)

Table 11.11 Special Status Word

15	14	13	12	11	10	9	8	7	6	5 4	3	2 1 0
FC	FB	RC	RB	0	0	0	DF	RM	RW	Size	0	FC_2-FC_0

FC = Instruction pipe stage C fault
FB = Instruction pipe stage B fault
RC = Stage C rerun flag
RB = Stage B rerun flag
DF = Data fault rerun flag
RM = Read-modify-write on data cycle
RW = Read/write for data cycle
Size = Size code for data cycle

FC_2 – FC_0: Address space code (function codes) for data cycle

instruction. If a bus error occurs during the bus error routine or during stacking operations, a double bus fault is signaled and the microprocessor is placed into a halt state, which asserts the $\overline{\text{HALT}}$ line. The microprocessor must be reset to release it from the halt condition.

Returning from the Bus Error Exception

Reading the format/vector offset word at the beginning of an RTE execution allows the microprocessor to determine the size of the stack to be recovered. If the 68020 reads a format that it does not recognize at this time, a format error exception is initiated and must be resolved before resuming the original return process. The stack is pulled (popped) and the internal register data are returned. Following this operation, the special status word is polled to discover the cause of the fault and whether the instruction of data fetch should be rerun (RB, RC, or DF asserted). Polling determines the type of cycle and action that the microprocessor must perform.

11.13 TRACE MODES

The 68020 expands the single-step or tracing function to allow selection between normal, every-instruction tracing and tracing on flow modifiers only. Flow-modifying instructions include the branch, jump, and call instructions. Refer to Table 11.2 for the TRACE flag truth table. Resetting both flags results in disabling the TRACE function. When T_0 is set and T_1 is reset, most instructions execute normally. After execution of a flow-altering instruction (such as BRA), the TRACE exception is initiated. Setting T_1 and resetting T_0 allow the microprocessor to enter the full TRACE mode with the TRACE exception initiated at the conclusion of every instruction execution. Keep in mind that both TRACE flags are reset as a result of the exception process. This disables the TRACE function while the TRACE exception executes.

11.14 BREAKPOINTS

The 68010 introduced breakpoints that operate essentially the same in the 68020. Breakpoints use illegal instruction codes \$4848 to \$484F. These cause an illegal instruction exception to occur when they are experienced in the program flow. If the $\overline{BERR}$ line is asserted after the fetch of an illegal instruction, the illegal instruction exception is taken. If $\overline{DSACK1}$ and $\overline{DSACK0}$ are both asserted, the data on the data lines replace the breakpoint instruction in the internal instruction pipe and this new instruction is executed. The fetched data are accessed by a CPU space address (function control lines all high) with the space code \$0 and the breakpoint number found on address lines A_2–A_4.

11.15 RESET

The assertion of a $\overline{RESET}$ causes the VBR and the cache control register to be cleared. The S flag in the status register is set, placing the microprocessor into the supervisor mode. The M flag is reset, selecting access to the interrupt stack pointer. Both trace flags are reset, disabling the TRACE function. The vector number and the vector offset for reset are \$0. When added to the contents of the VBR, the vector offset forms address \$0000 0000. The data at the first four locations are loaded into the interrupt stack pointer. The next four locations supply program counter information of the first instruction of the reset program.

PROBLEMS

11.1 Which cache type, fully associative or set associative, is more efficient from a time standpoint? Explain your answer.

11.2 What is the difference between the external start cycle pin and the operand start cycle pin? Are they ever asserted at the same time?

11.3 What are the different memory port sizes that the 68020 can recognize? How does the CPU know which one is being used?

11.4 What is the difference between a misaligned word and a misaligned long word?

11.5 What is the byte number sequence of the contents of the data buffers during the second write cycle of a long word transfer to word memory port address \$0120 443D?

11.6 What exactly do SIZ0 and SIZ1 indicate?

11.7 What memory location is written into with this instruction?

```
MOVE.B D3,([3,A1],D1.W*2,6)

D1 = $BB23 05C2     $0011 C210  = $2BA1
D3 = $1235 6BC1     $0011 C212  = $1CC0
A1 = $0011 C211     $0011 C214  = $AA00
                    $0011 C216  = $1234
                    $0011 CD98  = $ABCD
                    $0011 CD9A  = $CD00
                    $0011 CD9C  = $DA34
```

11.8 What memory location is written into with this instruction?

MOVE.B D3,([3,A1,D1.W*2],6)

Use the data in Problem 11.7.

11.9 What are the contents of D2 after this instruction? Initially, D1 = $3875 2318 and D2 = $8702 1546.

UNPK D1,D2,#$3050

11.10 What are the contents of D2 after this instruction? Initially, D1 = $2496 7382 and D2 = $3698 3345.

PACK D1,D2,#$9080

11.11 What are the contents of D1, D2, and D3 at the end of the following instructions? Use the contents in the address registers and memory locations designated in problem 11.7. Initial data for both instructions are

D1 = $1234 5678
D2 = $ABCD EF89
D3 = $1234 5678

(a) CAS D1,D2,(A1) (b) CAS D1,D3,(A1)

11.12 In the bit field table (Table 11.12), place an X in the bit boxes of the field designated in this instruction:

BFTST $2340 {7:5}

11.13 In Table 11.12, put circles in the bit boxes of the field designated by this instruction:

BFSET $2342 {−2:8}

11.14 Write an instruction to clear the last 10 bits in Table 11.12, using $0000 233D as the base address.

Table 11.12 Bit Field Table

Address	b7	b6	b5	b4	b3	b2	b1	b0
0000 233E								
0000 233F								
0000 2340								
0000 2341								
0000 2342								
0000 2343								

12

68881 FLOATING POINT COPROCESSOR AND 68020 MODULE SUPPORT

CHAPTER OBJECTIVES

In this chapter, the coprocessor capability of the 68020 is discussed in detail, followed by a description of the floating point coprocessor as used with the 68020. Finally, additional coprocessors in the 68000 family are briefly described.

12.1 GLOSSARY

access level Scheme by which access to various modules is controlled.
argument Data parameter.
coprocessor A dedicated processor that supplements a system.
descriptor Memory stack that describes a module.
floating point Decimal point used for handling large and fractional numbers.
microcode An instruction or program subset, usually resident in a device, that directs the internal operations of the device.
module A subroutine-type program that allows parameters to pass into and out of it.
not a number (NAN) A set of symbols that does not represent a number. 3B26 is a decimal NAN, because B is not a decimal number.
parameter Data to be used by a module.
RESTORE Instruction that performs a stack pull or pop for a coprocessor.
SAVE Instruction that performs the stack push operation for a coprocessor.

12.2 COPROCESSORS IN GENERAL

Most microprocessors can be programmed to perform a multitude of tasks, provided the appropriate software has been designed and the processing time is available to carry out the functions desired. If the system needs elaborate program coding or experiences delays in the execution time, a dedicated **coprocessor** can be added to the system. A coprocessor is designed with its own **microcode** to perform a specific task. An example of one such task is memory management in a virtual memory environment. There is little doubt that, with sufficient coding and time, the microprocessors in the 68000 family can handle memory swapping to accommodate virtual memory. With a memory management unit acting as a coprocessor, the time and code needed to handle the task are greatly reduced.

In a system using a coprocessor, the functions of the coprocessor appear as an extension to the main processor, from a software standpoint. That is, the operation of the coprocessor is transparent to the user. The programmer treats the instructions for the coprocessor as if they were included in the instruction set of the main processor.

12.3 ACCESSING A COPROCESSOR

The 68020 supplies several instructions that allow the coprocessor to do its processing. Note that this is not a direct memory access (DMA) procedure. For a DMA

Table 12.1 F-Line Instruction

15	14	13	12	11	10	9	8	7	6	5	4	3	2	1	0	Bit
1	1	1	1	Coprocessor ID			Type			Type Dependent						

Coprocessor ID	Coprocessor Identification Code
000	MC68851 Memory Management Unit (MMU)
001	MC68881 Floating Point Processor (FP)
010 to 101	Reserved
110 to 111	User Defined

Operation Type Code

Code	Operation Type
000	General
001	Conditional
010	Branch -- Word
011	Branch -- Long
100	SAVE
101	RESTORE
110	Undefined
111	Undefined

Table 12.2 General Instruction Format

<table>
<tr><td>15</td><td>14</td><td>13</td><td>12</td><td>11</td><td>10</td><td>9</td><td>8</td><td>7</td><td>6</td><td>5</td><td>4</td><td>3</td><td>2</td><td>1</td><td>0</td></tr>
<tr><td>1</td><td>1</td><td>1</td><td>1</td><td colspan="3">Coprocessor ID</td><td>0</td><td>0</td><td>0</td><td colspan="6">⟨ea⟩</td></tr>
<tr><td colspan="16">Coprocessor Command</td></tr>
</table>

function, bus mastership is released to the DMA device while the main processor is taken off-line. In a coprocessor action, the main processor still functions and usually interacts with the coprocessor. The instructions that tell the coprocessor to perform its job contain ones (1111 or $F) in the upper nibble of the opword and are called *F-Line* instructions. The format of these instructions is shown in Table 12.1.

In an F-Line instruction the first four bits are all high, identifying the instruction as a coprocessor type. The next three (b11, b10, and b9) contain the coprocessor code identifying which one of several different coprocessors connected to a single system is being accessed. b8, b7, and b6 identify what type of instruction is being used, and the last six bits relate specific information about the instruction itself.

General Instruction Type

The general instruction type used for data manipulation by the coprocessor has a type code of 000 in b8–b6. b5–b0 contain an effective address code defining the address mode and register number for the instruction. This F-Line instruction is followed by at least one word of extension containing the actual coprocessor command. Additional operands may be needed to complete the instruction (such as an absolute address, and so on). The syntax for a general instruction type is

cpgen ⟨parameters as defined by the coprocessor⟩

The machine code of its opword is shown in Table 12.2.

Branch Instruction Type

Based on its status bits, the coprocessor tests a condition in response to a cpbcc (coprocessor conditional branch) instruction. It returns a true or false condition to the main processor. If the condition is false, the next instruction in the program flow is executed. A true condition causes the branch to be taken. The format of the branch instruction is shown in Table 12.3. Again, note that this is an F-Line instruction.

Table 12.3 Branch Instruction Format

<table>
<tr><td>15</td><td>14</td><td>13</td><td>12</td><td>11</td><td>10</td><td>9</td><td>8</td><td>7</td><td>6</td><td>5</td><td>4</td><td>3</td><td>2</td><td>1</td><td>0</td></tr>
<tr><td>1</td><td>1</td><td>1</td><td>1</td><td colspan="3">Coprocessor ID</td><td>0</td><td>1</td><td>S</td><td colspan="6">Condition</td></tr>
<tr><td colspan="16">Word Displacement (S = 0; S = 1)</td></tr>
<tr><td colspan="16">Long Word Displacement (S = 1)</td></tr>
</table>

b8 is low and b7 is high, identifying the branch type instruction. b6 allows the programmer to select between a word size relative address offset (b6 low) or a long word size displacement. Finally, the lower six bits identify the condition to be tested by the coprocessor (see Table 12.4).

Decrement and Branch Type

This decrement and branch instruction is identical to the DBcc instruction of the 68000 family, except that it tests a coprocessor condition. The syntax of the instruction is

cpDBcc Dn,⟨label⟩

The format is shown in Table 12.5.

The register specified is the counter register for the decrement and branch function, and the condition is the coprocessor status being tested. A one-word displacement operand holds the relative address of the instruction.

Table 12.4 68881 Condition Tests

Mnemonic	Definition
Indicates true or false using the null function.	
F	False
EQ	Equal
OGT	Ordered Greater Than
OGE	Ordered Greater Than or Equal To
OLT	Ordered Less Than
OLE	Ordered Less Than or Equal To
OGL	Ordered Greater Than or Less Than
OR	Ordered
UN	Unordered
UEQ	Unordered or Equal To
ULT	Unordered or Less Than
ULE	Unordered or Less Than or Equal To
NE	Not Equal
T	True
Preinstruction operation with BSUN set; otherwise, uses the null function as above.	
SF	Signaling False
SEQ	Signaling Equal To
GT	Greater Than
GE	Greater Than or Equal To
LT	Less Than
LE	Less Than or Equal To
GL	Greater Than or Less Than
GLE	Greater Than or Less Than or Equal To
NGLE	Not Greater Than or Less Than or Equal To
NGL	Not Greater Than or Less Than
NLE	Not Less Than or Equal To
NLT	Not Less Than
NGE	Not Greater Than or Equal To
NGT	Not Greater Than
SNE	Signaling Not Equal To
ST	Signaling True

Table 12.5 Decrement and Branch

15	14	13	12	11	10	9	8	7	6	5	4	3	2	1	0
1	1	1	1	Coprocessor ID			0	0	1	0	0	1	68020 Dn		
Reserved										Condition					
Word Displacement															

The SAVE and RESTORE Functions

The **SAVE** instruction is used to store the internal state of the coprocessor into memory. It is similar to a stacking operation for the main processor. The syntax for this instruction is

cpSAVE ⟨ea⟩

where the effective address points to the memory area where the data will be stored. The general format of the internal state frame for the coprocessor is detailed in Table 12.6.

The **RESTORE** function is the reverse of the SAVE function for the coprocessor. The format is checked first; if it is correct, the information is retrieved from memory and returned to the registers of the coprocessor. Effective addresses for the

Table 12.6 Internal State Frame

15	14	13	12	11	10	9	8	7	6	5	4	3	2	1	0
Format								Length							
Reserved															
Coprocessor Dependent															
Information															

Format	Length	Definition
$00	xxx	Empty, Reset
$01	xxx	Not Ready, Come Again
$02	xxx	Invalid, Format Error
$03--$0F	xxx	Reserved, Format Error
$10--$FF	Length	Coprocessor Defined

SAVE instruction are the control modes (those that access a memory location) and the predecrement mode. For RESTORE, the effective addresses are limited to the control and postincrement modes.

SET and TRAP Instructions

SET and TRAP instructions test a coprocessor condition. If the condition is true, the operation of the instruction is performed. If the condition is false, the operation is not performed and the next program instruction is executed. The SET instruction causes a specified byte at an effective address to be set. The TRAP instruction forces a TRAP exception to be taken if the conditions are met. The syntaxes for these two instructions are

cpScc ⟨ea⟩ and cpTRAPcc #⟨data⟩

where the data are defined by the user and are used in the TRAP routine.

The user, then, needs to be familiar with the coprocessor instructions to complete the cp instruction types, just as the user needs to be familiar with the instruction set of any processor before attempting to use it.

12.4 COPROCESSOR INTERFACE

Many coprocessors require a certain amount of initializing. Also, in order to retrieve the results and to analyze them correctly, a number of coprocessor registers must be read. Table 12.7 shows the relationship between the coprocessor register set and the lower byte of the address bus used to access them. These registers are accessed using CPU space addressing (function control leads all high). The address bus takes the format of Table 12.8 during these accesses. b19–b16 identify the CPU space

Table 12.7 Coprocessor Register Set

Address	31 Register 16	15 Bits 0
$00	Response (RO)	Control (WO)
$04	Save (RO)	Restore (R/W)
$08	* Operation (WO)	Command (RO)
$0C	(Reserved)	Condition (WO)
$10	Operand (R/W)	
$14	* Register Select (RO)	(Reserved)
$18	* F-Line Instruction Address (R/W)	
$1C	* Operand Address (R/W)	

RO = Read Only; WO = Write Only; R/W = Read/Write
* = Optional register use

Table 12.8 CPU Space Encoding for Coprocessors

31 --- 20	19	18	17	16	15	14	13	12 -- 5	4 -- 0
0 0	0	0	1	0	Cop ID			0 . . 0	Cop Reg

Cop ID = Coprocessor ID Code
Cop Reg = Coprocessor Register Address

function as a coprocessor access, and b15–b13 define the coprocessor itself. Two such codes are 000 for the 68851 Memory Management Unit and 001 for the 68881 Floating Point Coprocessor. b4–b0 code the specific register being accessed; these bits are the address portions shown in Table 12.7.

When the 68020 fetches and recognizes an F-Line instruction, it generates a coprocessor CPU space access. The coprocessor is accessed as if a write to it is being performed; that is, the CPU space address is decoded to access the coprocessor and the F-Line instruction is passed along the data bus. The coprocessor determines how to react to the instruction—execute it, return information to the 68020, or request an operand or an exception. This information is conveyed by placing the contents of the response register onto the data bus and returning DSACKs. The 68020 will respond to the coprocessor response by filling its need, initiating an exception or continuing on to the next instruction.

Coprocessor Response Register

The coprocessor response register defines how the coprocessor is to react to the F-Line instruction. The response register is shown in Table 12.9. b15, called the come again flag, informs the main processor to recheck the coprocessor response. b14 tells the main processor to move the program counter contents to the F-Line instruction address register.

The function bits code the function type being used by the coprocessor. These functions include processor synchronization, instruction manipulation, exception handling, and operand and register transfers. The parameter field is used for register pointers, operand size definition, vector numbers, and condition evaluations.[1]

Other Coprocessor Registers

Remaining registers used for coprocessor functions include save and restore, operation and command, condition, operand, and F-Line registers. The save and restore

Table 12.9 Response Register

15	14	13	12	11	10	9	8	7	6	5	4	3	2	1	0
C	P	Function						Parameter							

C = Come Again; P = Pass Program Counter

[1]From *M68000 32-Bit Microprocessor Course Notes* (1984) by Motorola, Inc.

register contains pointer information about the location of the coprocessor state frame in memory. The operation and command register deals with the type of coprocessor operation being performed in response to an F-Line instruction. The condition register holds the status flags used by the coprocessor conditional instructions. The operand of the coprocessor command is held in the operand register, followed by an optional register select register. Finally, the F-Line address (see Table 12.8) is copied into the F-Line instruction address register, followed by the address of the command operand.

12.5 MC68881 FLOATING POINT COPROCESSOR

The MC68881 **floating point coprocessor,** designed specifically as a 68000 family coprocessor, has the necessary buses and control leads to interface directly into a 68020-based system. The MC68881 implements the IEEE standard for binary floating point arithmetic (ANSI/IEEE STD 754-1985).

Floating Point Arithmetic

Basic arithmetic performed on early computer systems used a fixed decimal point system, which restricted the maximum and minimum values of the data that could be manipulated. As an example, a fixed 8-bit binary system has a maximum positive integer value of 127 decimals (all bits are high except the sign bit) and a minimum positive value of 0. The negative values range from -1 ($FF) to -128 ($80). Unsigned values range from 0 ($00) to 255 ($FF). Fractional values were emulated in software, as were many mathematical operations. Processor arithmetic instructions were limited to addition and subtraction. Even with the 68000, this capability was extended to include only multiply and divide instructions. Operations using trigonometric and logarithmic functions had to be performed using program routines. The floating point math coprocessor resolved these problems by including real data types and an extended set of mathematical instructions. An additional benefit of the math coprocessor is the increased speed that executing floating point instructions in the processor brings compared to software emulation of these instructions.

MC68881 Data Formats

The MC68881 floating point (FP) coprocessor deals with numerous data formats, which can be divided into two basic groups:

1. *Integers,* or whole numbers. The integer group comprises the standard 68000 data types: byte, word, and long word (8-, 16-, and 32-bit precision, respectively).
2. *Real,* or scientific notation, numbers. This group, listed in Table 12.10, includes single, double, and extended precision values, each with a fixed exponent and mantissa size. An additional real type, packed BCD format, is shown in Table 12.11.

Table 12.10 Real Data Group Formats

Single precision real:

31	30 23	22 0
S	8-bit exp	two's complement 23-bit fraction

S = sign of the fraction

Double precision real:

63	62 52	51 0
S	11-bit exp	two's complement 52-bit fraction

S = sign of the fraction

Extended precision real:

95	94 80	79 64	63	62 0
S	15-bit exp	Zeros	I	two's complement 63-bit fraction

← 64-bit mantissa → (bits 63–0)

S = sign of the mantissa; I = integer; exp = exponent

Table 12.11 Packed BCD Real

95	94	93	92	91 80	79 68	67	0
S1	S2	I1	I2	3-digit exp	Zeros	I	16-digit fraction

I1, I2: ± infinity or NAN
S2: Sign of the exponent
S1: Sign of the mantissa
I: Integer
← 17-digit mantissa → (bits 67–0)

For single, double, and extended precision types, the sign bit is the sign of the mantissa and occupies the most significant bit position of each number. The following field is the **biased exponent** field. This exponent value is formed by adding the actual exponent to a bias value, which is equivalent to the maximum positive value that can occupy the exponent field.

EXAMPLE 12.1

A binary number has an exponent of −23. What is the value of the biased exponent stored in single precision format?

Solution:

The bias value for single precision is $7F, or 127. The actual exponent (−23) is added to the bias value to produce a biased exponent of 104 ($68).

Bias exponents are used so that the value in the exponent field is always positive, thus eliminating the need for a separate exponent sign bit. Comparing real numbers by size is facilitated by the lack of signed numbers. One exponent will be determined as greater than, less than, or the same as another without reference to whether it is more positive or more negative.

Binary integers in scientific format lie between 1.0000 . . . and 1.1111 . . . since the integer portion can only have a value of 1 (integer values of 0 are not normal scientific notation form). As such, the integer portion of the mantissa is implicitly understood to be a 1 and is not included in the actual storage of single or double precision numbers. This forces single and double precision values to always be **normalized.** The mantissa portion of these types is the mantissa's fractional part (numbers to the right of the decimal point). Since the integer part is not stored, the precision of the fraction is held in all the bits of the single and double precision mantissa. The IEEE standard designates a fraction mantissa as a **significand.** Positive values (sign bit 0) appear in the significand as true values, whereas negative values (sign bit 1) are in their 2's complement form.

EXAMPLE 12.2

Evaluate the single precision values $3EE8 0000 and $BEE8 0000.

Solution:

The most significant bit of $3EE8 0000 is a 0, making the number positive. The next eight bits (0111 1101) translate to a bias exponent value of $7D. Subtracting the bias value ($7F) produces an exponent of -2. The remaining digits form the significand. Including the implicit integer of 1, this becomes a mantissa of 1.1101, with the remaining bits low. Putting it all together yields a binary real value of $1.1101 \times 2^{-2} = 0.453125$.

The only difference with the $BEE8 0000 is that the sign bit is 1, indicating that the mantissa is negative. The integer value of 1 is *not* included when taking the 2's complement of the number. Only the significand is complemented, so that 1101 becomes 0011. Completing the number gives its value as -1.0011×2^{-2}, or -0.296875.

Extended precision format includes the same fields as single and double precision numbers except that the integer portion of the mantissa is explicitly included in the mantissa field. All internal calculations are performed using extended precision data. This requires conversion from the desired data type into extended precision real and allows for mixed-number type operations. The extended precision type is used for the calculations because it is the most precise, or accurate, data type to use. The format in Table 12.10 shows how extended precision real values are stored in memory. The sign and exponent fields are followed by a field of zeros. This field is included to fill out the data size to 96 bits (three long words) so that long word transfers can be used to move the data between memory and the coprocessor.

The explicit mantissa integer is provided to allow for the maximum possible precision in the extended precision data. Normalized values have already been introduced as having an integer value of 1. An additional restriction is that the biased exponent be any value except 0 or maximum. These values have special meaning that will be discussed shortly. **Denormalized** numbers are those that have an integer value of 0 and a non-zero fraction value. The biased exponent value is zero, the minimum exponent value (−16383). Use of denormalized numbers allows values that are less than the minimum normalized value. To increase accuracy of intermediate calculations, a third form, called **unnormal,** is used with extended precision real type numbers. This form is similar to the denormalized value except that the biased exponent is any non-zero value. This allows for an increased number of bits in the precision of a number by adjusting its position in the extended data form to fill the maximum number of bits. The exponent is then adjusted to compensate for the position change.

An exponent value of 0 and a mantissa of 0 is *zero*. The number is not evaluated any further but is simply accepted as 0. Zeros can be positive or negative according to the state of the sign bit. Similarly, *infinity* can be positive or negative, and it also requires no further evaluation beyond recognizing its form, which is a maximum exponent value and zero mantissa. A third special data form, called **not a number,** or **NAN,** comes in two forms: NANs and **signaling NANs (SNAN).** A regular NAN is created by an operation that does not produce a valid number result (such as dividing infinity by infinity). NANs are unordered numbers; that is, they cannot be evaluated as being larger or smaller than any other number (integer, real, or NAN). Regular NANs are recognized by a maximum exponent and some non-zero mantissa with a leading bit of 1. Signaling NANs have the same form as NANs except that the leading bit is a 0. SNANs are created by the user as a signaling mechanism. There is no FP mathematical operation that creates a SNAN. Table 12.12 summarizes the special number types described in the last two paragraphs.

An additional packed BCD real data form is shown in Table 12.11. It contains additional information not included in the other binary forms. The implicit decimal

Table 12.12 Special Number Types

Number Type	Biased Exponent	Integer	Significand
Normal	Not 0 Not maximum	1	Any value
Denormal	0	0	Not 0
Unnormal	Not 0 Not maximum	0	Not 0
Zero	0	0 or 1	0
NAN	Maximum	1	Not 0
Infinite	Maximum	1	0

point location conforms to standard scientific notation by following the first digit of the mantissa. The most significant bits of the data string specify the mantissa sign allowing the value to be positive or negative. The exponent of packed BCD is *not* biased, so an exponent sign bit is needed following the mantissa sign bit. The three-digit BCD exponent and sign bit provide for an exponent range of 1 to 10 ± 999. Two bits separating the sign bits and the exponent field are used for ± infinity and NAN indications; these two bits are otherwise held low. NANs for packed BCD result when a hexadecimal value appears in the number. A field of zeros, as with the extended precision type, fills the number to 96 bits. The mantissa contains 17 BCD digits (68 bits), including an explicit integer digit.

EXAMPLE 12.3

How are the numbers 3,235.78 and .000735 represented in the packed BCD real format?

Solution:

Both numbers are positive, making the mantissa sign bit low. Scientific notation form for the first number is 3.23578×10^3. The sign of the exponent is positive and the number is neither + or − infinity nor a NAN, so the next three bits are low. The exponent value is 003 and the mantissa is 323578. The packed BCD real form for the first number is

0003 0003 2357 8000 0000 0000

The scientific notation for the second number is 7.35×10^{-4}, yielding a negative exponent (−4) and a mantissa of 735. Again, the number is neither + or − infinity nor a NAN. Putting the parts together, this number as a packed BCD real is

4004 0007 3500 0000 0000 0000

The operand notations used by 68000 assemblers in context with this coprocessor are

B—Byte Integer
W—Word Integer
L—Long Word Integer
S—Single Precision Real
D—Double Precision Real
X—Extended Precision Real
P—Packed Decimal String Real

Notice that B, W, and L conform to standard 68000 family notations for byte, word, and long word data sizes.

Table 12.13 68881 Programming Model

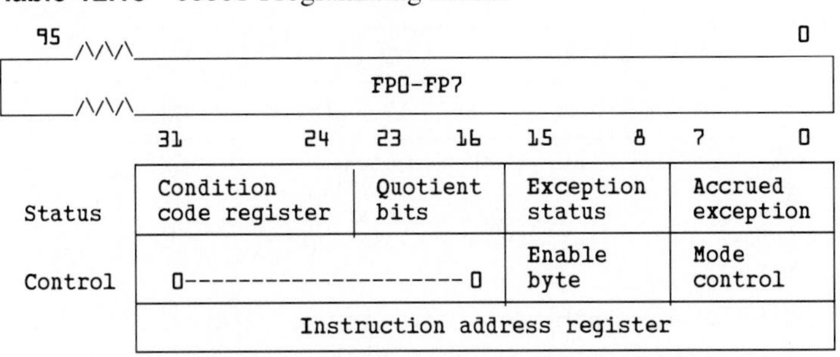

95 … 0				
	FP0–FP7			

	31 – 24	23 – 16	15 – 8	7 – 0
Status	Condition code register	Quotient bits	Exception status	Accrued exception
Control	0--------------------	-0	Enable byte	Mode control
	Instruction address register			

The 68881 Architecture

Inside the floating point coprocessor, there are eight 96-bit general purpose floating point data registers (FP_0–FP_7). As with any basic general purpose data register, any instruction may use any register. Additional register complement includes 32-bit control, status, and instruction address registers. The programmer may consider the programming model for the 68881 (Table 12.13) as an extension of the 68020 processor when it is included in a 68020 system.

Status Register

The status register is composed of four fields: condition codes, quotient, exception status, and accrued exception. The condition code byte, shown in Table 12.14, includes two flags—N for negative and Z for zero. The I bit indicates if positive or negative infinity resulted from the operation. NAN, as with the packed BCD format, indicates a result that is not a number.

The quotient byte contains the lower seven bits of the result from division and the sign of the entire quotient. The quotient value remains unchanged until cleared by the user or until a remainder modulo (FMOD) or remainder (FREM) instruction is executed.

The exception status byte contains flags that indicate which function during the last arithmetic operation caused the exception to occur. The exception with the highest priority is BSUN. The priority of the remaining exceptions depends on their bit positions in the status and enable words. Table 12.15 on page 344 diagrams this part of the status register. These flags are cleared at the beginning of most instructions so that a set flag indicates the cause of the current exception associated with the current operation.

If not a number (NAN) is set during one of the IEEE non-aware branch tests (Table 12.16, page 344), the BSUN flag is set. An exception is initiated if BSUN is

Table 12.14 Condition Code Byte

31	30	29	28	27	26	25	24
0	0	0	0	N	Z	I	NAN

Table 12.15 Exception Status Byte

15	14	13	12	11	10	9	8
BSUN	SNAN	OPERR	OVFL	UNFL	DZ	INEX2	INEX1

BSUN = Branch/Set on Unordered
SNAN = Signaling Not a Number
OPERR = Operand Error
OVFL = Overflow
UNFL = Underflow
DZ = Divide by Zero
INEX2 = Inexact Operation
INEX1 = Inexact Input

enabled in the exception enable byte of the control register. The SNAN is a user-defined indicator of a non-IEEE data type. Operand errors are caused by conflicts between the operation and the data type in use. Table 12.17 lists possible operand errors.

The overflow flag retains its basic meaning; it is set when a result is too large to be contained within the data size in use. *Underflow* is an overflow using negative exponents; that is, the number is too small to be represented by the data size in use. Divide by zero (DZ) flag includes using zero operands in FDIV, FLOG2, FLOG10, FLOGN, and FSGLDIV instructions. The DZ flag will also be set when using an odd multiple of $\pm\pi$ with an FTAN operation.

Finally, the inexact flags are set when the resulting number has too many digits for the mantissa to be represented in the current rounding precision. The number is rounded using the selected rounding mode, producing a result that is less accurate (inexact) than the original value. INEX1 is used for packed BCD real data, and INEX2 is used for non-decimal data forms.

The accrued exception byte (Table 12.18) contains a summary of all the exceptions that occurred since the last time the user cleared this byte as well as of those in progress. The accrued exception byte is cleared by a reset or a null state size restore operation.

Control Register

The control register contains two fields, the exception enable byte and the control mode byte. The programmer uses each field to configure the floating point copro-

Table 12.16 IEEE Non-Aware Condition Tests

Symbol	Definition
GT	Greater Than
NGT	Not Greater Than
GE	Greater Than or Equal To
NGE	Not Greater Than or Equal To
LT	Less Than
NLT	Not Less Than
LE	Less Than or Equal To
NLE	Not Less Than or Equal To
SF	Set False
ST	Set True
GL	Greater Than or Less Than (Not Equal To)
NGL	Not Greater Than or Less Than (Equal To)
SEQ	Set Equal To
SNE	Set Not Equal To

Table 12.17 Operand Errors

Instruction	Condition Causing Operand Error
FACOS	Source is ±infinity, $> +1$, or < -1
FADD	(+infinity)+(−infinity)
FASIN	Source is ±infinity, $> +1$, or < -1
FATANH	Source is $\geq +1$ or ≤ -1
FCOS	Source is ±infinity
FDIV	0/0 or infinity/infinity
FGETEXP	Source is ±infinity
FGETMAN	Source is ±infinity
FLOG10	Source < 0
FLOG2	Source < 0
FLOGN	Source < 0
FLOGNP1	Source ≤ -1
FMOVE to b, W, L	Integer overflow/underflow; source is NAN or ±infinity
FMOD FREM	FP data register is ±infinity; source is 0; other operand is NAN
FMOVE to P	Source exponent >999 or k-factor $> +17$
FMUL	One operand is 0; other is ±infinity
FSCALE	Source is ±infinity
FSGLDIV	0/0 or infinity/infinity
FSGLMUL	One operand is 0; other is ±infinity
FSIN	Source is ±infinity
FSINCOS	Source is ±infinity
FSQRT	Source is < 0
FSUB	Source or FP data register are +infinity or −infinity
FTAN	Source is ±infinity

Table 12.18 Accrued Exception Byte

7	6	5	4	3	2	1	0
IOP	OVFL	UNFL	DZ	INEX	0	0	0

IOP = Invalid Operation
OVFL = Overflow
UNFL = Underflow
DZ = Divide by Zero
INEX = Inexact Operation

cessor. The exception enable byte allows selection of conditions that will cause an exception to be taken. It has the same format as the exception status byte illustrated in Table 12.15. The control mode byte (Table 12.19, page 346) allows selection of a number's precision. The rounding mode is used to determine how precisely results should be rounded. Rounding precision selects at what bounds the rounding should occur. Rounding to the *nearest* number rounds a value to the closest numerical value. In cases where the distances to the nearest higher number or nearest lower number are the same, the even number is used.

EXAMPLE 12.4

What values are the following numbers rounded to using fourth place nearest rounding?

(a) 34.677 (b) 123.7568 (c) 1111.11

Table 12.19 Control Mode Byte

7	6	5	4	3	2	1	0
Rounding Precision		Rounding Mode		0	0	0	0

b7 b6	Rounding
0 0	Extended
0 1	Single
1 0	Double
1 1	(Reserved)

b5 b4	Mode
0 0	To nearest number
0 1	Toward zero (truncate)
1 0	To −infinity (down)
1 1	To +infinity (up)

Solution:

For example (a), 34.677 is rounded up to 34.68, the nearest fourth place value. In example (b), the determining value is 5, which is the midway value in a base 10 system. The number will therefore be rounded to the nearest even value. The answer for (b) is 123.8. In example (c), the result is rounded down to 1111.

EXAMPLE 12.5

If the values in Example 12.2 are all hexadecimal values, what are their nearest rounded values?

Solution:

For example (a), 7 is the midway value in a hex number system. Thus, the number is rounded to 34.68, its nearest even value. Examples (b) and (c) are rounded down to 123.7 and 1111, respectively.

Rounding to zero truncates the value. For the values in Example 12.4, rounding to zero produces 34.67 and 1111. Rounding to positive infinity is simply rounding up, and rounding to negative infinity is rounding down. Using the same values in Example 12.4, rounding to positive infinity yields 34.68 and 1112, and rounding to negative infinity produces 34.67 and 1111. The rounding precision selects the boundary of the mantissa where the rounding is to occur. Single precision is rounded to a 24-bit boundary, double precision to a 53-bit boundary, and extended precision to a 64-bit boundary.

12.6 THE 68881 COPROCESSOR SYSTEM

The pin layout diagram of the 68-pin grid package of the 68881 is shown in Figure 12.1. A number of the pins have the same designations as the 68000 family of processors, but their functions are the complements of those devices. For example, the two pins $\overline{\text{DSACK0}}$ and $\overline{\text{DSACK1}}$ supply the data transfer acknowledge and memory port size information to the 68020. (See the discussion of dynamic memory sizing in Chapter 11.)

Pin D_1: $\overline{\text{RESET}}$

The reset condition of the 68881 results from an active low $\overline{\text{RESET}}$ applied to this pin. Floating point registers are initialized to non-signaling not-a-numbers (NANs), and the control, status, and instruction address registers are cleared.

Pin B_4: $\overline{\text{SENSE}}$

The $\overline{\text{SENSE}}$ pin is used as an "interlock" pin that allows other devices in the system to "sense" the presence of the 68881. This pin is internally connected within the 68881 to ground. As long as the chip is in the system, this line is low. If the chip is removed, the low signal is removed.

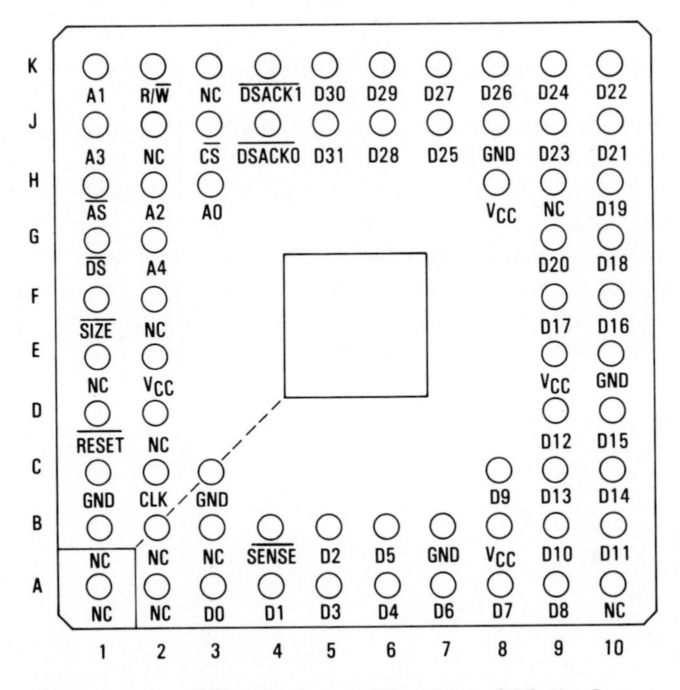

Figure 12.1 **68881 pin layout (Courtesy of Motorola, Inc.)**

Pin Columns 5–10 and A_4: D_0–D_{31}

D_0–D_{31} are the standard 68000 32-bit bidirectional data bus.

Pins K_1, H_2, J_1, G_2, and J_3: A_1–A_4 and $\overline{CS}$

These four input address lines combined with the $\overline{CS}$ allow the internal registers of the 68881 to be accessed by the 68020 using CPU space address decoding to generate the $\overline{CS}$. See Tables 12.7 and 12.8 for register select decoding and CPU space addressing.

Pins K_2, G_1, and H_1: $R/\overline{W}$, $\overline{DS}$, and $\overline{AS}$

$R/\overline{W}$, $\overline{DS}$, and $\overline{AS}$ are inputs from the 68020. $R/\overline{W}$ selects the direction of data flow between the two processors. $\overline{DS}$ and $\overline{AS}$ are active during a valid data transfer. $\overline{DS}$ indicates valid data on the data bus (or expected valid data in the case of a read cycle), and $\overline{AS}$ indicates a valid address on the address bus.

Pins F_1 and H_3: $\overline{SIZE}$ and A_0

$\overline{SIZE}$ and A_0 inform the 68881 which data bus size is used to interface to a particular microprocessor's data bus. Table 12.20 details the size selection.

Interconnections Between the 68881 and 68020

An example of an interconnection between the 68020 microprocessor and the 68881 floating point coprocessor is shown in Figure 12.2. Because this coprocessor was designed for this system, note the direct and fairly uncomplicated interconnection. The coprocessor and the system clocks do not need to be the same because transfers between the devices are handled asynchronously. From the programmer's standpoint, this interconnection extends the 68020 programming model, as seen in Table 12.21 on page 350. All standard programs for the 68020 will run on this system as well as those which access the 68881 coprocessor.

Table 12.20 Data Bus Sizing

A0	Size	Data Bus
--	0	Byte
0	1	Word
1	1	Long Word

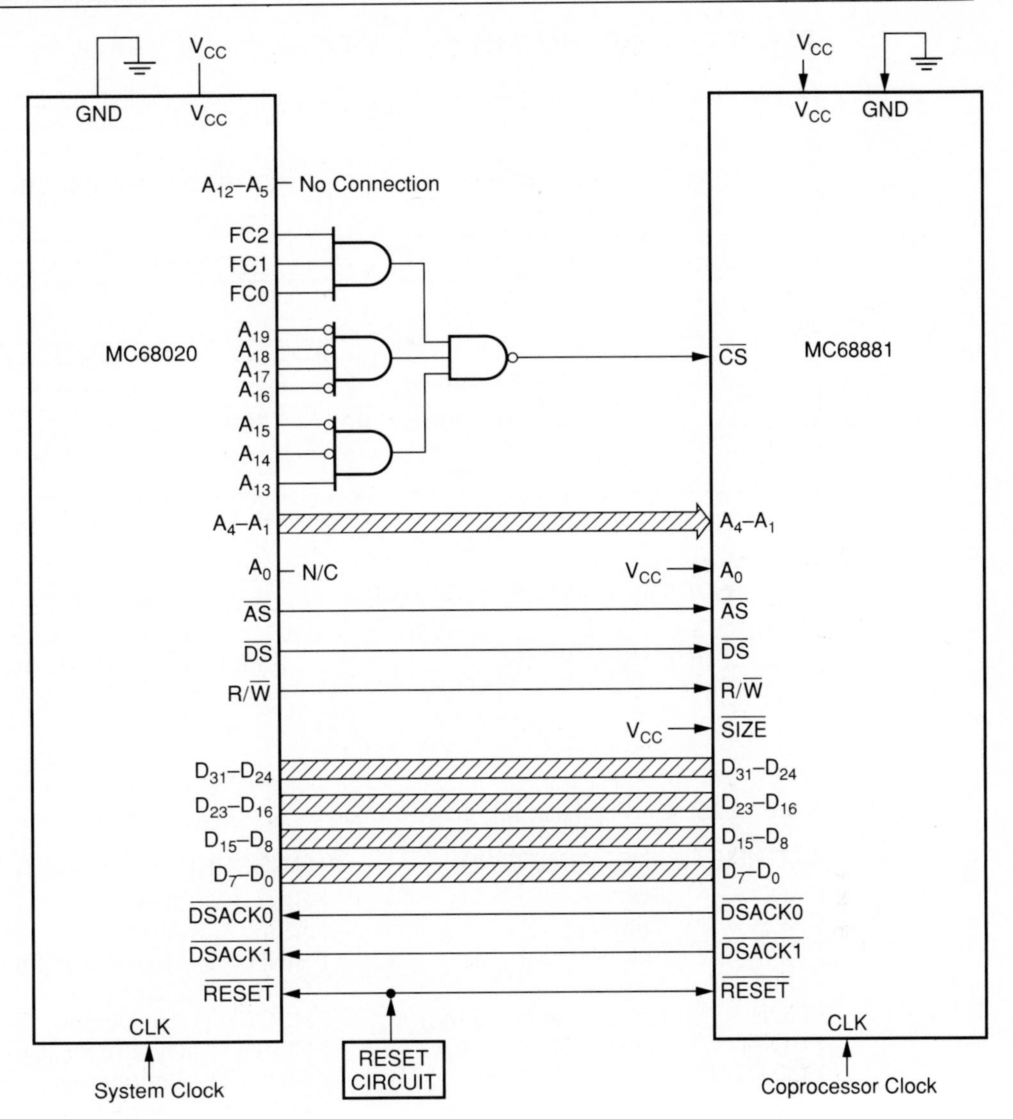

Figure 12.2 **Coprocessor interconnections (Courtesy of Motorola, Inc.)**

12.7 THE 68881 INSTRUCTION SET

Besides extending the programming model of the 68020, the 68881 also adds numerous instructions to the system instruction set (Table 12.22, page 350). These instructions, when used in the system assembler, form one of the F-Line formats discussed earlier in this chapter. Once entered into the 68881, the instructions are further decoded to perform their indicated function.

Table 12.21 68020/68881 Programming Model (Courtesy of Motorola, Inc.)

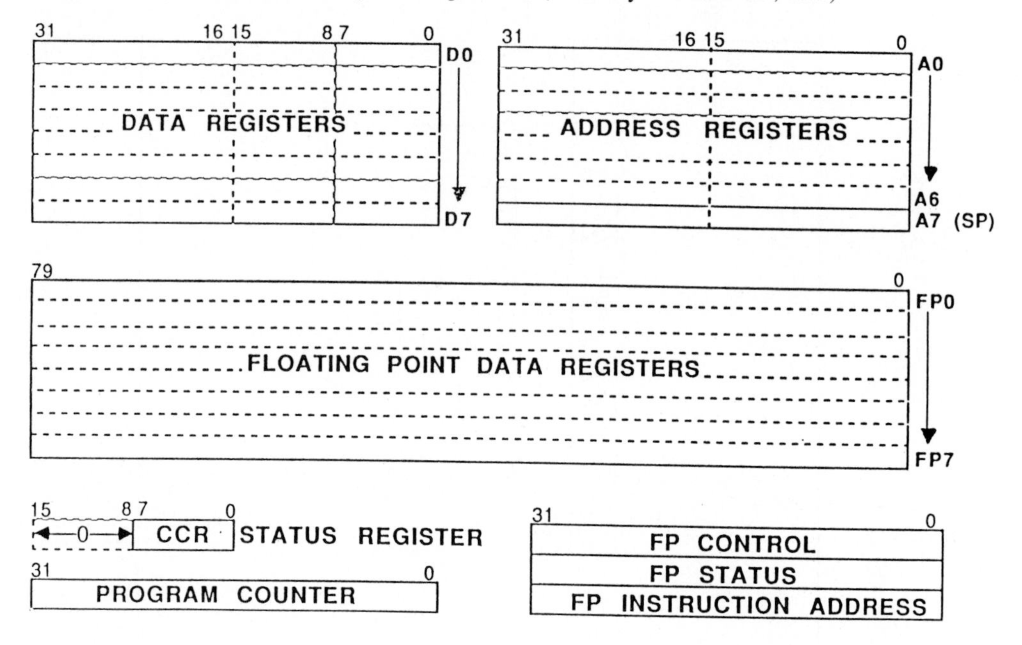

Table 12.22 Floating Point Instructions (Courtesy of Motorola, Inc.)

ARITHMETIC INSTRUCTIONS:

FADD	Add
FSUB	Subtract
FINT	Take Integer part
FNEG	Negate
FNOP	No Operation (Sync)
FSQRT	Square Root
FGETMAN	Return Mantissa
FTST	Test the Operand
FGETEXP	Return Exponent
FCMP	Compare
FMUL	Multiply
FDIV	Divide
FMOD	Modulo
FREM	Remainder
FSCALE	Scale Exponent
FSGLMUL	Single Prec. Multiply
FSGLDIV	Single Prec. Divide

TRANSCENDENTAL INSTRUCTIONS

FSIN	Sine
FASIN	Arc Sin
FSINH	Hyperbolic Sine
FCOS	Cosine
FACOS	Arc Cosine
FCOSH	Hyperbolic Cosine
FSINCOS	Simultaneous Sine/Cosine
FTAN	Tangent
FTANH	Hyperbolic Tangent
FATAN	Arc Tangent
FATANH	Hyperbolic Arc Tangent
FETOX	e to the XPower
FETOXM1	e to the (X-1) Power
FTENTOX	10 to the X Power
FTWOTOX	2 to the X Power
FLOG10	Logarithm base 10
FLOG2	Logarithm base 2
FLOGN	Logarithm base e
FLOGNP1	Logarithm base e of (X+1)

Many of these instructions appear as high-level commands rather than as machine-level instructions. For instance, FSIN, which produces the sine of a radian angle, is more complete and complex than a simple add or subtract instruction. From the programmer's standpoint, the FSIN instruction is written into the program like any 68000 family instruction. The microcode within the 68881 handles the necessary operations to complete the task of the instruction.

12.8 FLOATING POINT COPROCESSOR APPLICATION

The basic format for most 68881 arithmetic instructions is

F⟨function⟩.⟨data type⟩ ⟨ea⟩,FPRN

where FPRN is any one of the floating point general registers. The FMOVE type instruction uses the same basic format except that the ⟨ea⟩ and FPRN can be reversed depending on the direction of data movement. Program 12.1 is a simple direct program example using floating point coprocessor instructions. This subroutine, called MULT, multiplies the double precision data at location $0001 2000 by 5. The data is transferred into the floating point processor in integer form and converted to double precision real within the processor. The double precision multiplication is performed and the results returned to memory in integer form. That is, an integer value for the exponent and integer values for the mantissa are stored in memory starting at location $0001 2000. Notice that the first instruction (initializing A_0) is *not* an F-Line coprocessor instruction, but is instead a regular 68000 family instruction. This is also true for the RTS instruction terminating the subroutine. The program itself appears to take standard assembly form, except that the data size delimiter is a D for double precision rather than the more familiar B, W, or L. The assembler recognizes these F-Line instructions as floating point coprocessor types and assembles them accordingly.

```
MULT    MOVEA.L #$12000,A0      * Initialize pointer
        FMOVE.D (A0)+,FP0       * Moves data into FP0
        FMUL.D #5,FP0           * Multiply by 5
        FMOVE.D FP0,-(A0)       * Replace data
        RTS
```

Program 12.1

12.9 FLOATING POINT APPLICATION PROGRAM: SERIES LRC CIRCUIT

To illustrate an application of the floating point coprocessor, a series circuit (Figure 12.3, page 352) will be analyzed for the following parameters: component reactance (X_C and X_L) at the given frequency of the source generator (GEN); total

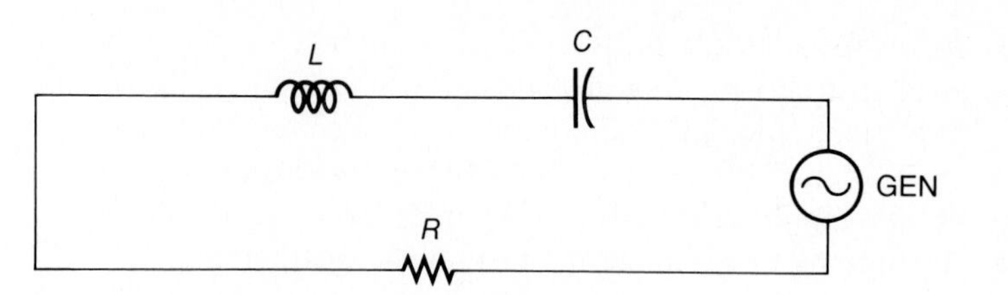

Figure 12.3 Series *LRC* circuit

circuit impedance (Z) and current (I); individual voltage drops (V_R, V_C, and V_L); and reactance (X_R), quality (Q) of the circuit, and its bandwidth (BW) at the natural circuit resonance. The formulas used to perform the analysis are at the generator frequency (F) and voltage (V):

1. Capacitor reactance $(X_C) = 1/(2\pi FC)$ (12.1)
2. Inductive reactance $(X_L) = 2\pi FL$ (12.2)
3. Total circuit reactance $(X_T) = X_L - X_C$ (12.3)
4. Total circuit impedance $(Z) = \sqrt{R^2 + X_T^2}$ (12.4)
5. Impedance phase angle $(\theta) = \arctan(X_T/R)$ (12.5)
6. Circuit current $(I) = V/Z$ (12.6)
7. Resistor voltage drop $(V_R) = I * R$ (12.7)
8. Capacitor voltage drop $(V_C) = I * X_C$ (12.8)
9. Inductive voltage drop $(V_L) = I * X_L$ (12.9)

The asterisks in Equations 12.7–12.9 signify multiplication.

Formulas for the calculations at the resonant circuit frequency (F_r) are as follows:

10. Resonant frequency $(F_r) = 1/(2\pi\sqrt{LC})$ (12.10)
11. Resonant reactance $(X_r) = 2\pi F_r L$ (12.11)
12. Circuit quality $(Q) = X_r/R$ (12.12)
13. Circuit bandwidth $(\text{BW}) = F_r/Q$ (12.13)

All of the calculations use decimal (BCD) data with original component values retrieved from a block of memory and results returned to memory. The floating point coprocessor supports packed BCD values which occupy 96 bits of memory (three long words) each (see Table 12.11). When entering data into memory, one must exercise care to put it into its proper format. The two uppermost bits of a packed BCD data format are the signs of the mantissa and the exponent of the BCD data in scientific notation. The two I bits which follow should be set to zero to indicate that the number is not infinity or a NAN. The three-digit exponent occupies b80–b91 followed by three nibbles of zeros (or don't cares). Finally, a 17-digit mantissa fills out the rest of the 96-bit data value. This mantissa is always expressed in scientific notation, with the decimal point following the most significant digit.

Circuit Analysis Data Bank

The data bank for this problem begins at a location labeled RESIS, which contains the three long word values for R in the circuit of Figure 12.3. The remaining given values follow in sequence: the capacitor value (CAP), the inductance (INDUCT), the generator voltage (GEN) and frequency (FREQ), and a constant (CONST) for the value of 2π (6.283185). [*Note:* Because italic type and subscripts cannot be "understood" by the assembler, the notation used in Equations 12.1–12.13 will not be followed *exactly* in the discussion of the program.]

Calculated results are entered into the memory bank following the CONST value. They start with capacitive (XC) and inductive (XL) reactances and circuit impedance (ZIMP). The phase angle of the circuit impedance (ANGLE) in radians is stored next, followed by circuit current (AMP). Individual voltage drops across the resistance (VR), capacitor (VC), and inductor (VI) are next in line, completing the fixed frequency analysis of the circuit. The resonance quantities are placed in the bank in sequence following the fixed frequency values: resonant frequency (RFREQ), reactance (RXL), quality of the circuit (QUAL), and circuit bandwidth (BW). The total number of quantities of the given value and results is 18. Each result is three long words in length, for a total of 54 long words. Since each long word occupies 4 bytes, the total amount of memory required for the data bank is 216 bytes. This is a minimal quantity considering that the 68020 can address 4 gigabytes.

The LRC Program

The program itself is a straightforward process of gathering the data and computing the quantities needed. The floating point instructions make the program appear to be a higher level program rather than one at an assembly level. The data acted upon are in decimal form, and the results are returned in decimal form.

Initial Conditions The first section of the program (Program 12.2) initializes some needed constants and retrieves data from the memory bank. The only nonfloating point instruction is the first one, which loads the address register A_0 with pointer data. Floating point registers FP_0–FP_5 receive the component values from the data bank. Register FP_7 is set to a constant value of one, which will be used when it is necessary to divide a quantity into one [as in the formula $1/(2\pi FC)$].

```
MOVE.L #RESIS,A0        * Initialize pointer register
FMOVE.P (A0)+ FP0       * FP0 = R
FMOVE.P (A0)+ FP1       * FP1 = C
FMOVE.P (A0)+ FP2       * FP2 = L
FMOVE.P (A0)+ FP3       * FP3 = Vs
FMOVE.P (A0)+ FP4       * FP4 = F
FMOVE.P (A0)+ FP5       * FP5 = 2π
FMOVE.P #1,FP7          * FP7 = 1
```

Program 12.2

Computing the Fixed Frequency Values Floating point multiply, divide, subtract, and add instructions are similar to 68020 instructions of the same type. The difference lies with the data type and size in use. .P in the size area of each instruction syntax stands for packed BCD and allows instruction execution involving 96-bit BCD numbers, as described earlier.

The capacitive and inductive reactances are calculated first (Program 12.3). After the individual reactances are found, total circuit reactance is computed and used to find the circuit impedance. The magnitude and phase angle of the impedance are calculated using a rectangular-to-polar conversion. This conversion combines reactance (X) and resistance (R) values into the magnitude of the impedance (Z) using the following relationship:

$$Z = \sqrt{R^2 + X^2} \tag{12.4}$$

The phase is calculated by finding its tangent angle:

$$\theta = \text{ARCTAN}(X/R) \tag{12.5}$$

```
FMUL.P FP5,FP4          * FP4 = 2πF
FMUL.P FP4,FP1          * FP1 = 2πFC
FDIV.P FP1,FP7          * FP7 = XC = 1/(2πFC)
FMOVE.P FP7,(A0)+       * Save XC
FMUL.P FP4,FP2          * FP2 = XL = 2πFL
FMOVE.P FP2,(A0)+       * Save XL
FSUB.P FP7,FP2          * FP2 = X = XL - XC
FMOVE.P FP2,FP1         * Copy X into FP1
FMUL.P FP2,FP2          * FP2 = X2
FMOVE.P FP0,FP6         * Copy R into FP6
FMUL.P FP6,FP6          * FP6 = R2
FADD.P FP2,FP6          * FP6 = Z2
FSQRT.P FP6             * FP6 = Z
FMOVE.P FP6,(A0)+       * Save Z
FDIV.P FP0,FP1          * FP1 = X/R
FTAN.P FP1              * FP1 = θ in radians
FMOVE.P FP1,(A0)+       * Save θ
```

Program 12.3

Circuit Voltages Total current is found by dividing the source voltage in FP_3 by the magnitude of the impedance now held in register FP6. The individual voltage drops across each component are then calculated by multiplying the circuit current, which is the same throughout a series circuit, by the resistance or reactance of each component.

It is important to remember that each voltage drop has a corresponding phase angle (θ), which is calculated by adding the current phase angle of the circuit (the negative, or reciprocal, of the impedance angle) with each individual component's phase angle (0° for the resistor, +90° for the inductor, and −90° for the capacitor). These phase angle calculations have not been included in Program 12.4.

```
FDIV.P FP6,FP3          * FP3 = I = V/Z
FMOVE.P FP3             * Save I
FMUL.P FP3,FP0          * FP0 = VR = IR
FMOVE.P FP0,(A0)+       * Save VR
FMUL.P FP3,FP7          * FP7 = VC = IXC
FMOVE.P FP7,(A0)+       * Save VC
FMUL.P -72(A0),FP3      * FP3 = VL = IXL
FMOVE.P FP3,(A0)+       * Save VL
```

Program 12.4

Resonant Quantities Registers FP1 and FP2 are reloaded with the original capacitor and inductor values from the data bank so that these values can be used in the resonant frequency computations. Each series LRC circuit has a natural frequency at which it resonates; that is, the circuit sustains oscillations at a natural circuit frequency. Natural frequency, the point at which the capacitive and inductive reactances are identical ($X_C = X_L$), can be found by equating the two reactance formulas:

$$2\pi F_r L = 1/(2\pi F_r C)$$

With some algebraic manipulation and solving for F_r, the equation for the resonant frequency (F_r) becomes

$$F_r = 1/(2\pi\sqrt{LC}) \tag{12.10}$$

After the original capacitor and inductor values are gathered, the resonant frequency is computed. The inductive reactance is then calculated using the resonant frequency. The value of the capacitive reactance is equal to the inductive reactance, by definition.

The bandwidth of the resonant circuit is determined by adjusting the frequency of the source generator and recording the voltage across the resistor. For a series LRC circuit at resonance, the total reactance is zero ($X_L - X_C = 0$), leaving the total circuit impedance equal to the resistance value. Circuit impedance is lowest at the resonant frequency. Changing the frequency of the generator causes more reactance (inductive above the resonant frequency and capacitive below) and increases the circuit impedance. When the impedance is at its lowest value (at the resonant frequency), the circuit current (V/Z) is at its highest. This current, in turn, causes the largest voltage drop across the resistor at resonance. As the impedance increases, the current decreases and the voltage across the resistor drops. The frequencies at which the voltage is 0.707 times the peak voltage at resonance define the limits of the bandwidth of the circuit. Mathematically, the difference between these two frequencies (the total bandwidth) is calculated by dividing the resonant frequency by the quality (Q) of the circuit ($Q = X_L/R$). These calculations are the last group of instructions in this program, as shown in Program 12.5.

```
          FMOVE.P CAP,FP1           * FP1 = C
          FMOVE.P INDUCT,FP2        * FP2 = L
          FMUL.P FP2,FP1            * FP1 = LC
          FSQRT.P FP1               * FP1 = √LC
          FMUL.P FP5,FP1            * FP1 = 2π√LC
          FMOVE.P #1,FP7            * FP7 = 1
          FDIV.P FP1,FP7            * FP7 = Fr = 1/(2π√LC)
          FMOVE.P FP7,(A0)+         * Save Fr
          FMUL.P FP7,FP5            * FP5 = 2π Fr
          FMUL.P FP5,FP2            * FP2 = Xr = XL = XC = 2π FrL
          FMOVE.P FP2,(A0)+         * Save Xr
          FDIV.P FP0/FP2            * FP2 = Q = Xr/R
          FMOVE.P FP2,(A0)+         * Save Q
          FDIV.P FP2,FP7            * FP7 = BW = Fr/Q
          FMOVE.P FP7,(A0)+         * Save BW
DONE      BRA.S DONE                * End of program
```

Program 12.5

There are no tests and no looping instructions performed by the LRC circuit program. This example shows an application of the calculating power of the floating point processor. Imagine the number of instructions that might be necessary to manipulate the size of data and perform the functions used in this program if the floating point were not used!

12.10 MODULE SUPPORT

Modular programming, though not new in concept, has become the foremost method in programming. Program sections can be written independently and then married into one final program. The beauty of this method, besides breaking the task of programming into smaller segments, is that the various program **modules** can be used for future programs as well.

Entry into each section or module is accomplished chiefly through the use of subroutine calls—either JUMP TO subroutine, CALL subroutine, or GOSUB commands. Data developed in the main program are used by these modules, and resulting data are made available to the main program after the execution of each module. Higher-level languages have adopted and facilitated this modular concept by providing a means to pass different types of data forms, called **arguments** or **parameters,** into the subroutines. Pascal, for example, achieves the passing of data variables by creating data types at the top of the program. Any established variables can be passed into a module, called a PROCEDURE, by being referred to within that module. Additional local parameters can be established within the module as necessary. The established or global parameters as modified by the procedure can then be passed out of the module at its completion.

Two 68020 instructions allow for modular formatting at the assembly level: CALLM (call module) and RTM (return from module). These instructions are similar to subroutine call and return instructions, but they are used to pass to the module a list of arguments for the module. The format for the call module instruction is

CALLM #⟨argument count⟩,⟨ea⟩

The argument count is the number of argument bytes to be passed into the module. The effective address points to the memory location where the module **descriptor** frame stack resides (Table 12.23).

The descriptor stack, which is organized as long word data, starts with a description of the module type. The remaining long words follow in ascending address locations. Two option (OPT) codes are presently recognized by the 68020: 000 and 100. All others will cause a format exception to be initiated. Option 000 specifies that an argument list follows the module stack frame (see Table 12.24, page 358). The 100 option directs the called module to access the arguments indirectly, from locations pointed to by a pointer in the stack frame (module data area pointer). When option 100 is used, any arguments following the stack frame are not accessed unless, for some reason, the data area pointer contains the address of the arguments following the frame.

Two type codes presently in use are \$00 and \$01. Again, any other combination initiates a format error exception. Type \$00 indicates no change in access level

Table 12.23 Module Descriptor

31 ----- 29	28 ----- 24	23 ----- 16	15 ----- 8	7 ----- 0
OPT	TYPE	Access Level	00000000	00000000
Module entry word pointer				
Module data area pointer				
Module stack pointer (optional)				
User-defined information				

Table 12.24 Module Stack Frame

15 14 13	12 11 10 9 8	7 6 5 4 3 2 1 0
OPT	TYPE	Saved access level
0 0 0 0 0 0 0 0		Condition code byte
0 0 0 0 0 0 0 0		Argument count
(Reserved)		
Module descriptor pointer (high word)		
Module descriptor pointer (low word)		
Saved program counter data (high word)		
Saved program counter data (low byte)		
Saved module data area pointer (high word)		
Saved module data area pointer (low word)		
Arguments (optional)		

and causes the stack built by the called module to be copied on top of the stack built by the calling module. Type $01 indicates a module that may change access level and/or cause its stack to be created elsewhere in memory. Type $01 passes the access level information to external hardware to cause the present access level to be changed to a higher or lower access level. The hierarchical arrangement of access levels will be discussed later in this chapter.

The module entry word pointer indicates where the module begins by pointing to the first word of the module (Table 12.25). The next word is the first opword of the module program. The register specified in the module entry word is the descriptor data area address that is saved on the stack frame and used with option 100. Additional descriptor information contains optional stack pointer information for type $01 modules and any further user-defined information.

Module Stack Frame

The stack built when a CALLM is executed (Table 12.24) starts by saving the module description (option, type, and present access level). The current condition

Table 12.25 Module Entry Word

15	14 13 12	11 ------------ 0
$\overline{D}$/A	Register	0 ------------ 0

code register flag information follows next, with the argument count extracted from the CALLM instruction. The argument count lets the CALLM and RTM instructions know how many arguments are included on the stack.

Significant address pointers are included next on the stack: the descriptor pointer, the program counter contents of the instruction following the CALLM, the module data area pointer for option 100, and the current stack pointer information for type $01 operation. The argument list follows the stack frame when option 000 is used, but may also be included (though not accessed) when option 100 is used.

Access Level

Access levels and meanings are user defined, but, once established, can serve numerous purposes. Basically, limited access modules are allowed to call a module with greater access rights. External hardware verifies the legality of the access and performs the access change, or detects attempts to access levels that a particular operation is not entitled to. In this manner, the access structure becomes more than simply a priority device. Prioritizing usually imposes limits in an upward fashion; that is, a higher priority access overrides a lower one, while the lower one can be shut out by the higher one. In the access level scheme, an access change is requested and passed to the external equipment. Logic at that level determines if the access change is permissible. If it is, the change is performed and the CALL operation is continued. If the access change is not allowable, a format error exception is initiated.

During a type $01 module CALL an access change is indicated, and the request is initiated by a CPU space address (Table 12.26) with b19–b16 set to 0001. The lower seven bits of the address indicate access level directions as listed in Table 12.27 on page 360.

The current access level (CAL) contains the current access level rights and results in no level change. IAL and DAL indicate requests to increase (IAL) or decrease (DAL) the current access levels. The contents of these three locations are used by the external hardware to determine the legality of the access change or operation. The hardware returns the status of the results of the request, and the processor can access this information by using the STATUS location and interpreting the status codes as shown in Table 12.26. The remaining locations hold specific descriptor information related to each of the function control memory area codes.

Return from Module (RTM)

In response to a CALLM, the system reads the first descriptor word to determine if an access level change is required. If so, it is performed first; then, the CALLM is begun by building the stack frame. Once the critical information is saved, the module entry word is accessed from the descriptor stack to determine the data area

Table 12.26 CPU Space Access Level Address

31 ------ 20	19 18 17 16	15 ------ 7	6 ------------- 0
0 ------ 0	0 0 0 1	0 ------ 0	Access level register

Table 12.27 Access Level Memory Map

Address	31 -------------------- 24	23 -------------------- 0
$00	CAL	
$04	Status	
$08	IAL	
$0C	DAL	
$40	Function code descriptor 0	
$44	Function code descriptor 1: User data	
$48	Function code descriptor 2: User program	
$4C	Function code descriptor 3	
$50	Function code descriptor 4: Supervisor data	
$54	Function code descriptor 5: Supervisor program	
$58	Function code descriptor 6	
$5C	Function code descriptor 7: CPU space	

STATUS CODES:

Value	Validity	Processor Action
00	Invalid	Format error
01	Valid	No change in access rights
02–03	Valid	Change access rights; no change in stack pointer
04–07	Valid	Change access rights and stack pointer
Other	Undefined	Undefined; format error exception

pointer register, if needed. The module begins to execute its program, pulling data from the argument list as needed and writing new data into the list as it is developed. Upon an RTM instruction at the end of the module program, the reverse process is begun. The format of the return instruction is RTM Rn, where Rn is any one of the data or address registers and specifies the module data area pointer information. The processor checks the argument count to determine how many locations are used for the list. The stack frame is restored into the processor to allow the next instruction of the main program to be executed. Data can be accessed from the argument list as needed by the main program (or other modules). Access is referenced from the data area pointer information in the RTM-specified register or from the module descriptor stack ($08 locations from the base address of the module descriptor).

The CALLM instruction, from a programmer's standpoint, is inserted into the program in a similar manner to a jump to subroutine instruction (JSR). The nature of the RTM instruction, however, is similar to that of the return from subroutine (RTS) instruction. CALLM differs from JSR and RTS in the information saved on the stack. The process itself and the time required to perform the operations also differ, but these differences are of relatively little concern to the programmer. A programmer must know what modules are available and where the descriptor and data area pointers are in order to use the module instructions and process correctly.

The flowcharts for the CALLM instruction shown in Figures 12.4 and 12.5 (page 362) summarize the step-by-step process of executing a module call. Type 0 is fairly direct because it does not require an access level change. Contrast this with type 1, where additional events must occur before the module is finally called and allowed to run.

Figure 12.4 Type 0 flowchart

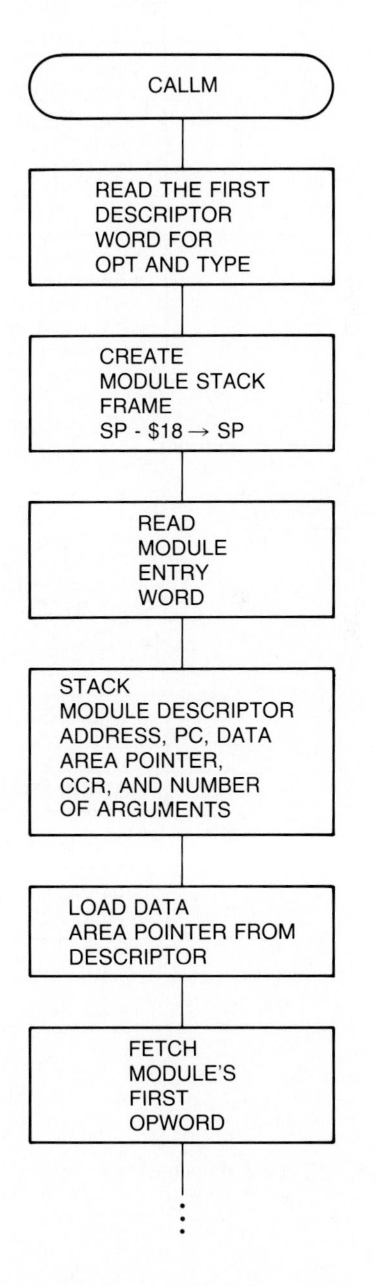

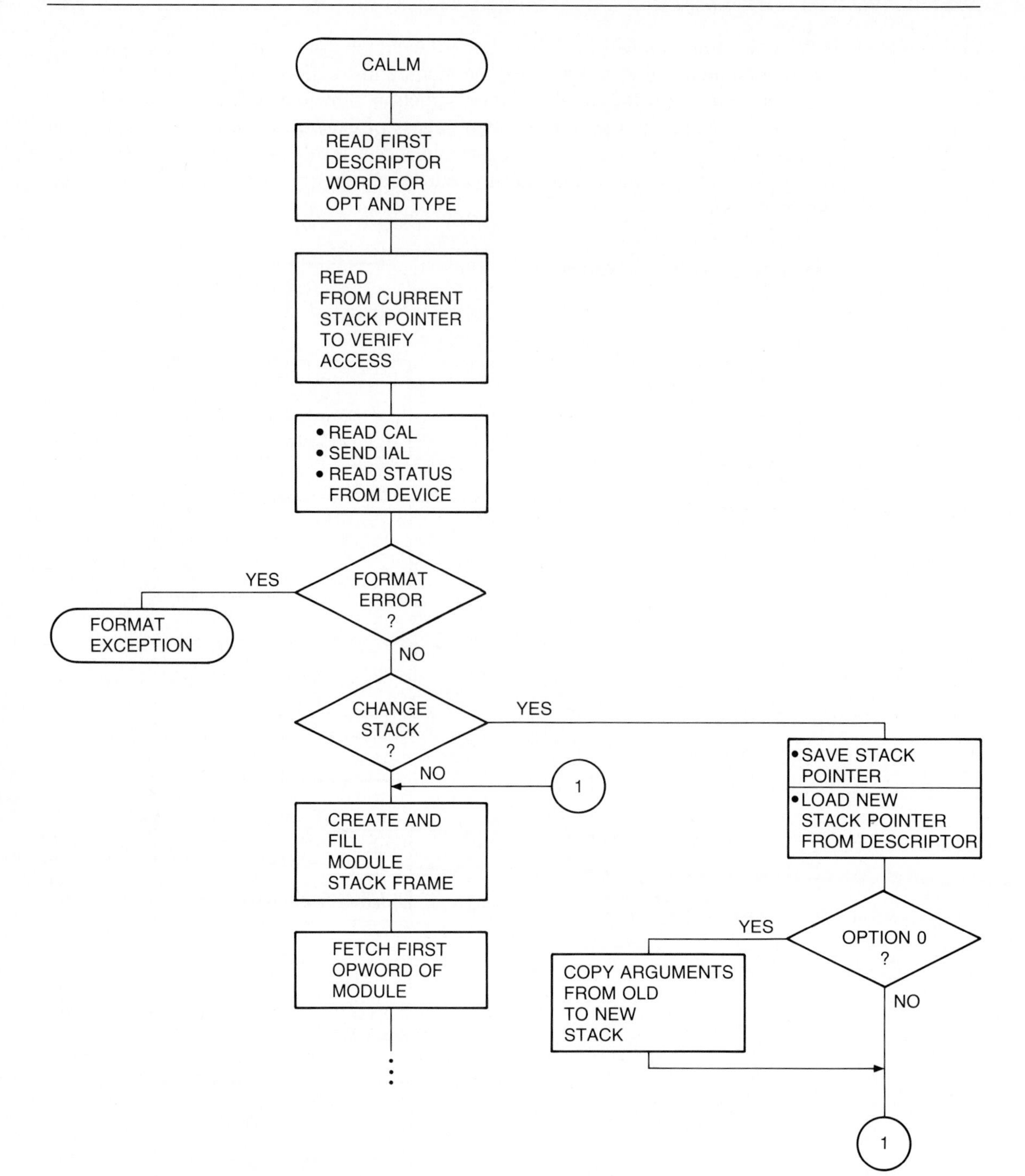

Figure 12.5 Type 1 flowchart

PROBLEMS

12.1 Describe two main differences between a coprocessor and a DMA device.

12.2 How does the 68020 recognize a coprocessor instruction after it has been fetched from memory?

12.3 In the following program, which branch is executed and why?

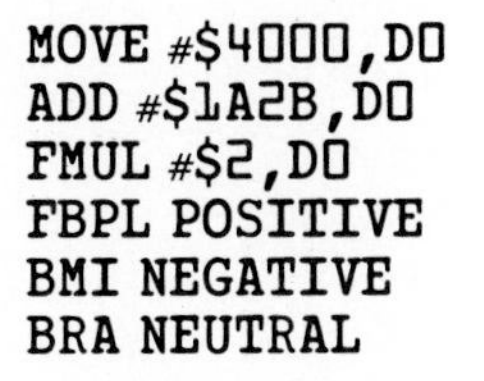

```
MOVE #$4000,D0
ADD #$1A2B,D0
FMUL #$2,D0
FBPL POSITIVE
BMI NEGATIVE
BRA NEUTRAL
```

12.4 Draw a circuit diagram of the logic necessary to select a floating point coprocessor in a 68020-based system. Pay particular attention to address and CPU space decoding. Refer to Figure 12.2.

12.5 The 68020 reads \$A5 from the 68881 RESPONSE register. What action should the 68020 take next?

12.6 What is the value of this packed BCD real number?

8002 0003 5213 9000 0000 0000

12.7 Describe the number that produced this 68881 status register value: \$ 08B51248.

12.8 What kind of rounding is to be performed with the control data \$0000 6C80?

12.9 Which 68881 exceptions are enabled using the data in Problem 12.7?

12.10 What is the value returned when the following numbers are rounded to fifth place NEAREST?
(a) 18.67699 (b) 771.2254 (c) 82828.212
(d) \$3A.7F77 (e) \$B.17ABC2 (f) \$111.6F4C

12.11 What are the values returned in Problem 12.10 if rounding is to the fourth place and rounded to zero? to positive infinity? to negative infinity?

12.12 In the LRC circuit analysis program, the inductive reactance quantity was retrieved from memory to be used in the multiply computation for V_L. Assign the value \$0100 2C00 to the label RESIS, and show that the offset used in the effective address of the FMUL.P −72(A0),FP3 is correct.

12.13 Using a 68020 system with a 68881 coprocessor, write a program that converts a set of polar coordinates (magnitude and angle) into their real and imaginary components. The integer magnitude is stored in the upper word of D_0, and the radian angle is stored in the lower word of the same register. The two values returned as the real and imaginary rectangular values are to be double precision real numbers stored with the real component starting at location ANSWER.

12.14 The first long word on a module descriptor stack is \$8102 0000. Describe the module.

12.15 Which register contains the data area pointer if the first word of the module program is \$3000?

12.16 The third word of the module stack frame is $001A. What information is contained in this word?

12.17 The module stack frame starting at location $0000 3000 is full, including the saved stack pointer value. The data pointer locations hold the value $0100 4600. Where does the argument list begin for option 000 and option 100 modules?

12.18 What type of access level change is being requested by address $0001 0008?

12.19 What CPU address is sent to maintain the current access level when accessing a type $01 module?

13

68030 32-BIT MICROPROCESSOR

CHAPTER OBJECTIVES

This chapter explores the next step in the Motorola 16/32-bit microprocessor generation, the 68030. The memory management functions that were available as a coprocessor with the 68020 are now included on the same chip as the main processor in the 68030. Therefore, details of the operation of memory management are also discussed in this chapter. Additional comparisons to the previous processors in the 68000 line are made to illustrate the performance enhancement of the 68030.

13.1 GLOSSARY

ATC (**address translation cache**) A device within the memory management unit used for translating logical addresses to physical addresses.
BIU (**bus interface unit**) The section of the 68030 that interfaces with external buses.
burst fill Cache entry fill of one line.
CISC Complex instruction set computer.
EU (**execution unit**) Portion of the 68030 which performs instruction execution.
line 4 long word cache entry.
root pointer Register which points to the first address of the translation table in memory.
translation table Section in memory which holds pointers within a translation tree used in the translation of logical to physical addresses by the memory management unit.
write-through Process by which the same data are written into memory and into a data cache simultaneously.

13.2 INTRODUCTION TO THE 68030

The major conceptual advance contributed by the 68020 32-bit microprocessor was coprocessing capabilities. The "020" was designed to operate with both memory management and floating point coprocessors, as described in Chapters 11 and 12.

An additional functional advance was achieved in the 020 by the addition of an instruction cache which led to improved performance through allowing program looping instructions to be fetched from the internal instruction cache instead of from memory. By taking both of these concepts a step further, the 68030 increases processing performance. The memory management unit (MMU) functions derived from the paged memory management unit (PMMU) in the 020 are included on the same silicon chip with the processor in the 030. This combination reduces the time needed for operation of memory management functions because external bus cycles are not required for the processor to interface with the MMU. Also, besides the instruction cache, there is now a data cache to reduce still further the need for external bus cycles during the execution of a program.

13.3 THE 68030 IN GENERAL

Figure 13.1 is a functional pin diagram of the 68030 microprocessor. Many of the control functions and buses included in the 020 are present in the 030. A full 32-bit address bus is accessible, permitting four gigabytes of direct addressing. The 32-bit data bus can be connected to any one of three memory port buses since the 030 supports dynamic bus sizing just like the 020. As with the 020, data can be written and read to any address location. Usually address bus errors occur only when there is an attempt to access an opword or instruction operand from an odd address. No other access restrictions limit the transfer of data to or from the 030.

The three function codes are used as before, to define memory areas and CPU space activity. SIZ0, SIZ1, $\overline{\text{DSACK0}}$, and $\overline{\text{DSACK1}}$ pin functions perform the asynchronous handshake between the processor and memory access circuitry as they did with the 020. This operation facilitates dynamic bus sizing and enhances the ability to access data starting at any address. R/$\overline{\text{W}}$, $\overline{\text{AS}}$, $\overline{\text{DS}}$, and $\overline{\text{DBEN}}$ perform their functions of supplying the necessary control functions for asynchronous data transfers.

Operand cycle start ($\overline{\text{OCS}}$) and external cycle start ($\overline{\text{ECS}}$) are outputs which are brought low during the first clock period of an external bus cycle. $\overline{\text{ECS}}$ is made active during any bus cycle while $\overline{\text{OCS}}$ indicates when an operand fetch is in process.

Read-modify-write cycle ($\overline{\text{RMC}}$) indicates when an indivisible read/write cycle is taking place. One example of this signal that has already been discussed is the execution of the compare and set (CAS) instruction in Section 11.10. Another example of its use is in the execution of a test and set (TAS) instruction. The TAS first reads an operand's data, tests a bit in that data (setting the Z flag according to the results), sets the bit, and then writes the new data back to the operand address. This operation is deemed indivisible because the processor does not recognize exceptions that can disrupt instructions midway until the read-modify-write cycle is completed.

Interrupt pending level inputs ($\overline{\text{IPL0}}$–$\overline{\text{IPL2}}$) still provide seven layers of interrupts including level 0 which is actually a "no-interrupt request" indication. The $\overline{\text{IPEND}}$ output pin, introduced in the 020, indicates than an interrupt request has been received but not yet serviced. In both the 020 and the 030, $\overline{\text{AVEC}}$ provides

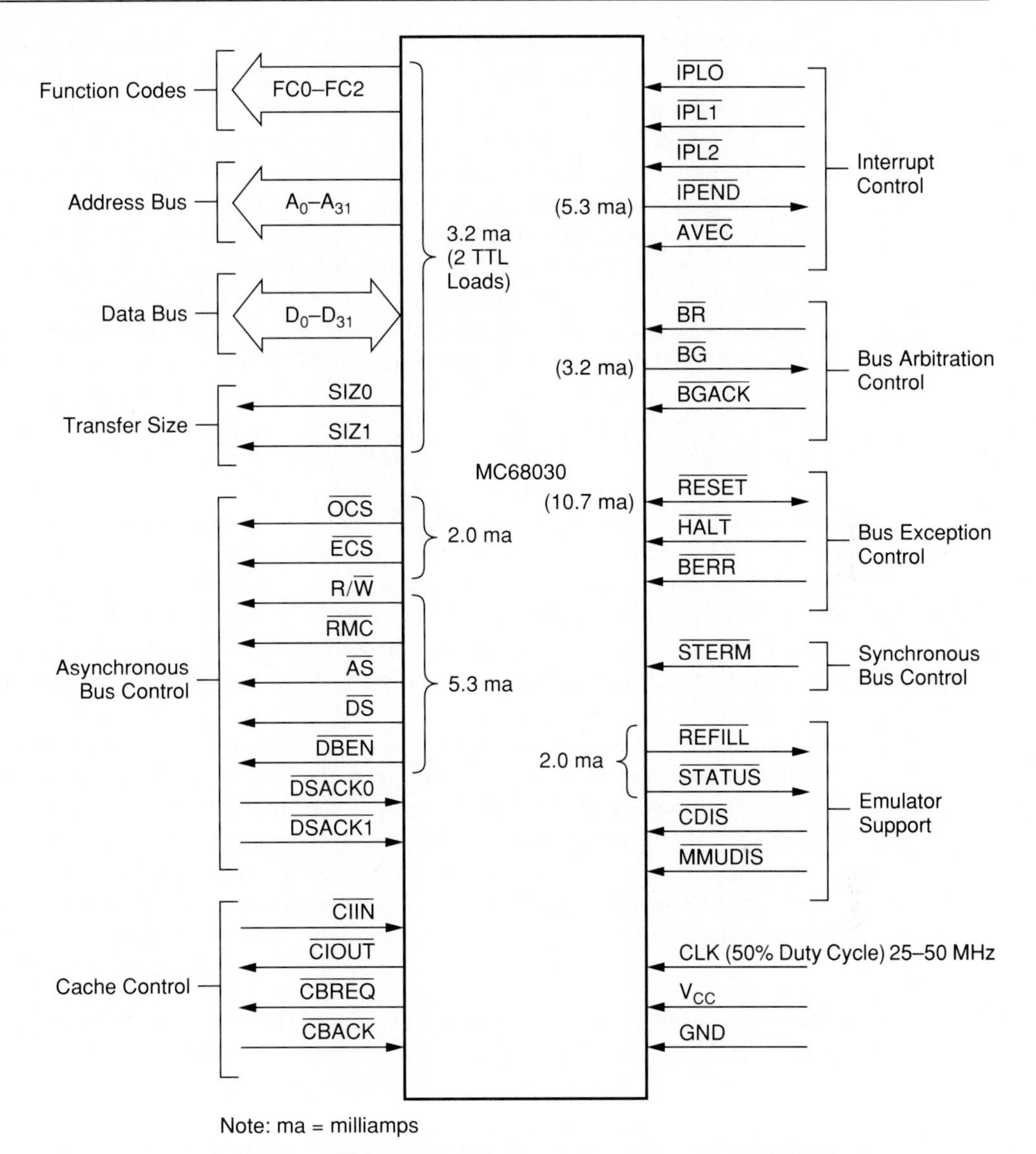

Figure 13.1 MC68030 functional signal groups (Courtesy of Motorola, Inc.)

the auto-vectoring option for interrupt exceptions previously performed by the $\overline{\text{VPA}}$ function in the 68000.

Bus arbitration for dynamic memory accessing (DMA) utilizes three control pins, as before. The DMA device sends a $\overline{\text{BR}}$ (bus request) for requesting mastership control of the buses to the processor. The 030 returns a bus grant ($\overline{\text{BG}}$) and in turn receives a bus grant acknowledge ($\overline{\text{BGACK}}$) from the DMA device which holds that pin active until it completes its transfers. This is the sequence of bus arbitration for a system with a single DMA device. Systems with multiple DMAs must be able to monitor the BGACK line to see if the 030 has granted control of the buses to another device. In this case, the 030 can still receive a second bus request and issue a bus grant. However, the second device will not gain control of the buses until the original DMA device releases the bus grant acknowledge. Once that device has released the acknowledge, the second device will take over the buses by issuing its own bus grant acknowledge. During this change from one DMA device to another, the 030 remains "off-line" and will continue to do so until its BGACK pin is negated.

The bi-directional RESET pin accepts a hardware reset when it acts as an input and asserts a reset to other devices when executing a reset instruction. The 030 will hold the RESET line low in response to a RESET instruction for 512 clock periods instead of the previous 124 clocks of the 68000. The faster clock speeds used with the 030 require the 030 to hold the RESET line low for more clock periods than the 68000 to achieve a reset time that is sufficiently long enough for external devices.

In previous members of the 68000 family, the HALT line was bi-directional like the RESET line. In the 030, the HALT functions are separated into two pins, HALT and STATUS. The bus error (BERR) pin, used to initiate bus error expectations, serves that same purpose in the 030. Table 13.1 summarizes the functions of the RESET, HALT, and BERR pins for the 68000 family up through the 030. This table resembles previous tables describing these functions. Major differences involve the manner in which the processors are reset and the indication for a double bus error condition.

Bus error exceptions and cycle rerun are initiated as they were with the 68000 by the correct assertion of the BERR and HALT signals. Reset is achieved in the 68000/010 by asserting RESET and HALT and in the 020/030 by asserting RESET by itself. All of the 68000 family processors can be placed in a halt condition suspending processor activity by asserting HALT by itself.

Double bus error faults caused by catastrophic errors during external bus accesses place the 68000 family processors into a halt condition. In the 68000, 010

Table 13.1 68000 Family RESET, HALT, and BERR Logic

Inputs	RESET	HALT	BERR	Function
Inputs	1	1	1	Normal operation
	1	1	0	Bus Error Exception
	1	0	1	CPU HALT
	1	0	0	Bus Cycle Rerun
	0	1	1	68020/030 RESET
	0	1	0	68020/030 RESET
	0	0	1	68000/010 RESET
	0	0	0	68000/010 RESET
Outputs	0	1	x	RESET Instruction
*	1	0	x	Double Bus Error HALT

x — Input Only

* Note: Status pin 68030 indicates HALT ouput condition

and 020, this condition is detectable by an active $\overline{\text{HALT}}$ asserted by the processor. For the 030, this function is now provided on the $\overline{\text{STATUS}}$ output control pin.

In actuality, the $\overline{\text{STATUS}}$ signal is a synchronous control function used to indicate several different conditions. The information it conveys is based on the number of clock cycles the pin is held active by the 030. If it is asserted for only one clock cycle, then $\overline{\text{STATUS}}$ indicates that the 030 is at an instruction boundary. That is, the 030 is beginning an instruction execution. Held for two clock cycles, the $\overline{\text{STATUS}}$ pin indicates that a TRACE or interrupt exception process is beginning. Asserting the $\overline{\text{STATUS}}$ pin for three clock cycles indicates that an address error, bus error, reset, or spurious interrupt exception is underway. If the STATUS pin remains active for more than three clock cycles, a double bus error fault condition has occurred. Since this condition remains until the 030 is reset, the $\overline{\text{STATUS}}$ line will remain asserted until the fault condition is cleared.

The remaining pins are involved with memory management and caching functions and will be discussed in detail later in the chapter. A quick reference guide, Table 13.2 (page 370), summarizes all the pins with a very brief note about the function of each one.

13.4 68030 FUNCTIONAL BLOCKS

Performance is also improved in the 68030 because a number of activities occur simultaneously within the processor. Figure 13.2 on page 371 is a block diagram of the functional areas of the 030. External accesses to memory are controlled within the **bus interface unit (BIU)**, which includes the data and address pads (buffers), size multiplexer, memory management unit (MMU), and bus controller functions. An address required to fetch an instruction, operand, or data is treated internally by the 030 as a logical address and presented to the MMU. Within the MMU is an **address translation cache (ATC)** which determines the process required to translate a logical address into a physical address. The physical address is eventually derived and used to complete the transfer of data. Details on the translation process as well as its involvement with virtual memory operations is discussed later in the chapter.

Instructions executed by the 030 are fed to the four-stage pipeline from the instruction cache in similar manner to the 020. At the beginning, or stage A, of the pipeline is a cache holding register which accepts the opwords and operands from the instruction cache. Stage B is responsible for validating the instruction entering the pipeline. Stage C performs most of the opword decoding, while stage D starts the execution process. Essentially, three different opword/operand words of a program are being simultaneously operated on in the pipeline. The results of stage D guide the microsequencer and control blocks to select the microcode of the instruction being executed.

The **execution unit (EU)** contains the logic, arithmetic logic units (ALU), and assorted registers and control circuitry that actually execute the instruction as directed by the microcode from the microsequencer. Any data accesses required by the instruction are sent to the data cache and MMU simultaneously. If a hit occurs in

Table 13.2 68030 Pin Summary (Courtesy of Motorola, Inc.)

Signal Name	Mnemonic	Function
Function Codes	FC0–FC2	3-bit function code used to identify the address space of each bus cycle.
Address Bus	A_0–A_{31}	32-bit address bus used to address any of 4,294,967,296 bytes.
Data Bus	D_0–D_{31}	32-bit data bus used to transfer 8, 16, 24, or 32 bits of data per bus cycle.
Size	SIZ0/SIZ1	Indicates the number of bytes remaining to be transferred for this cycle. These signals, together with A0 and A1, define the active sections of the data bus.
Operand Cycle Start	$\overline{\text{OCS}}$	Identical operation to that of $\overline{\text{ECS}}$ except that $\overline{\text{OCS}}$ is asserted only during the first bus cycle of an operand transfer.
External Cycle Start	$\overline{\text{ECS}}$	Provides an indication that a bus cycle is beginning.
Read/Write	R/$\overline{\text{W}}$	Defines the bus transfer as an MPU read or write.
Read-Modify-Write Cycle	$\overline{\text{RMC}}$	Provides an indicator that the current bus cycle is part of an indivisible read-modify-write operation.
Address Strobe	$\overline{\text{AS}}$	Indicates that a valid address is on the bus.
Data Strobe	$\overline{\text{DS}}$	Indicates that valid data is to be placed on the data bus by an external device or has been placed on the data bus by the MC68030.
Data Buffer Enable	$\overline{\text{DBEN}}$	Provides an enable signal for external data buffers.
Data Transfer and Size Acknowledge	$\overline{\text{DSACK0/DSACK1}}$	Bus response signals that indicate the requested data transfer operation is completed. In addition, these two lines indicate the size of the external bus port on a cycle-by-cycle basis.
Cache Inhibit In	$\overline{\text{CIIN}}$	Prevents data from being loaded into the MC68030 instruction and data caches.
Cache Inhibit Out	$\overline{\text{CIOUT}}$	Reflects the CI bit in ATC entries or a transparent translation register; indicates that external caches should ignore these accesses.
Cache Burst Request	$\overline{\text{CBREQ}}$	Indicates a miss in either the instruction or data cache.
Cache Burst Acknowledge	$\overline{\text{CBACK}}$	Indicates that accessed device can operate in burst mode.
Interrupt Priority level	$\overline{\text{IPL0-IPL2}}$	Provides an encoded interrupt level to the processor.
Interrupt Pending	$\overline{\text{IPEND}}$	Indicates that an interrupt is pending.
Autovector	$\overline{\text{AVEC}}$	Requests an autovector during an interrupt acknowledge cycle.
Bus Request	$\overline{\text{BR}}$	Indicates that an external device requires bus mastership.
Bus Grant	$\overline{\text{BG}}$	Indicates that an external device may assume bus mastership.
Bus Grant Acknowledge	$\overline{\text{BGACK}}$	Indicates that an external device has assumed bus mastership.
Reset	$\overline{\text{RESET}}$	System reset.
Halt	$\overline{\text{HALT}}$	Indicates that the processor should suspend bus activity.
Bus Error	$\overline{\text{BERR}}$	Indicates an invalid or illegal bus operation is being attempted.
Synchronous Termination	$\overline{\text{STERM}}$	Bus response signal that indicates a port size of 32 bits and that data may be latched on the next falling clock edge.
Cache Disable	$\overline{\text{CDIS}}$	Dynamically disables the on-chip cache to assist emulator support.
MMU Disable	$\overline{\text{MMUDIS}}$	Dynamically disables the translation mechanism of the MMU.
Microsequencer Status	$\overline{\text{STATUS}}$	Status indications for debug purposes.
Pipe Refill	$\overline{\text{REFILL}}$	Indicates when the instruction pipe is beginning to refill.
Clock	CLK	Clock input to processor.
Power Supply	V_{CC}	−5 volt = 5% power supply.
Ground	GND	Ground connection.

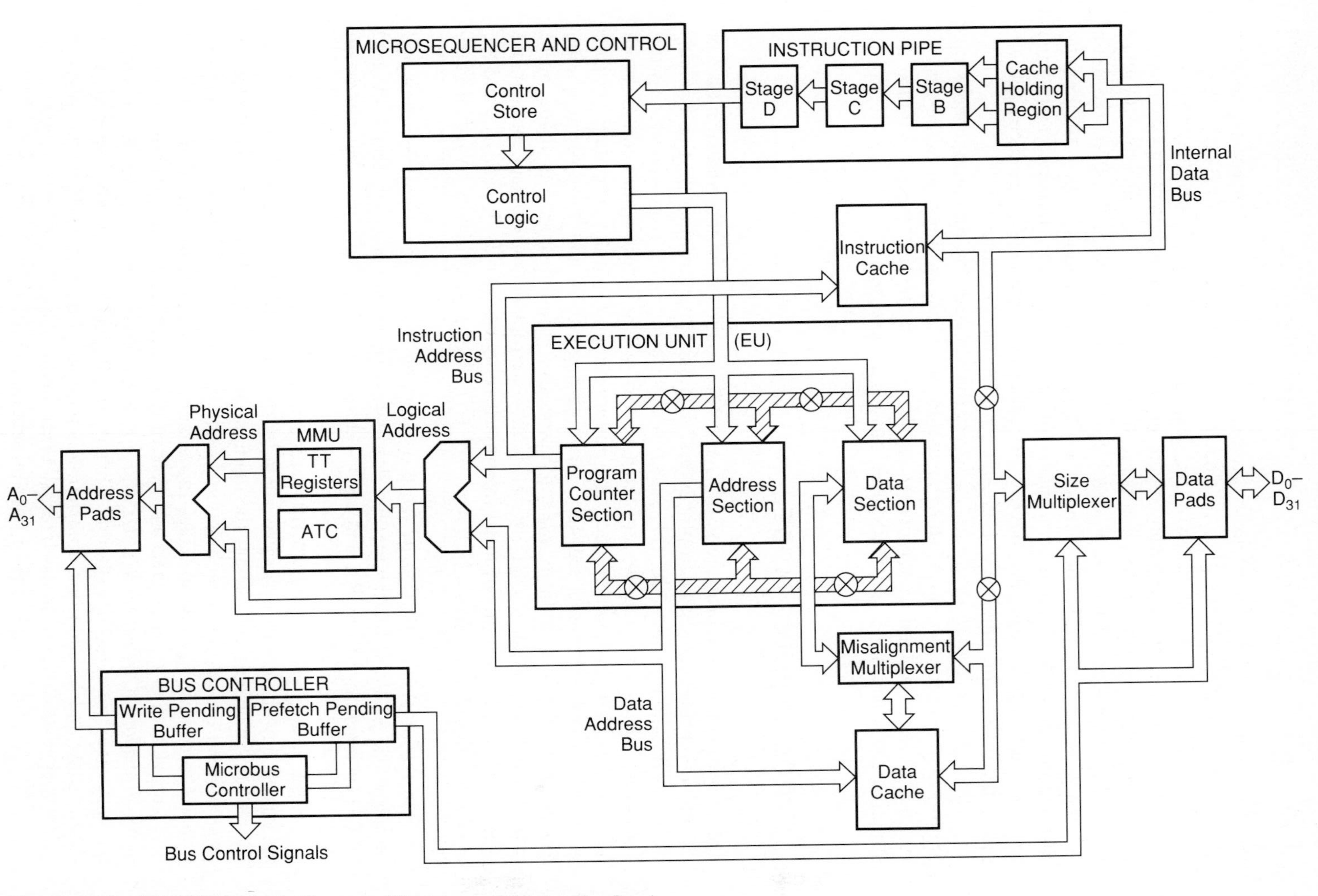

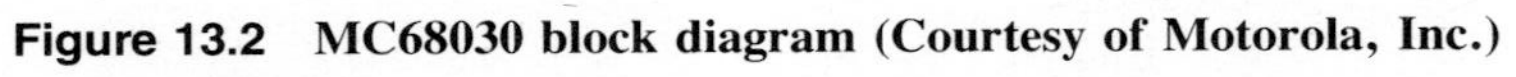

Figure 13.2 MC68030 block diagram (Courtesy of Motorola, Inc.)

the data cache, the potential memory access from the MMU is aborted and the data are obtained from the cache. If a miss occurs in the data cache, the MMU supplies the physical address, associated with the desired data access, to memory.

Instructions from the program stream in memory are prefetched and sent to the instruction cache as each instruction from the cache is sent to the pipeline. In other words, the instruction cache is updated as it is used. As with the 020, the main advantage in caching program instructions is for program loops. Once the looping instructions are cached, there is no need to return to external memory to re-fetch those instructions. Instead, the looping instructions are retained in the cache and are fed to the instruction pipeline continually until the loop is exited.

In summary, 68030 performance is greatly enhanced by the simultaneous operation of the following: 1) MMU translation of logical to physical addresses; 2) instruction decoding and execution; 3) internal data cache accesses; 4) internal instruction cache accesses; and 5) external bus cycle accesses.

13.5 68030 PROGRAMMING MODEL

The programming model for the 030 (Figure 13.3) builds from the general model of all previous 68000 family microprocessors. The user programming portion contains eight 32-bit data registers (D0–D7) which operate using byte, word, and long word data sizes, seven 32-bit address registers (A0–A6) which handle word and long word data sizes, a 32-bit program counter, and 32-bit user stack pointer. The status register is identical to the 020 status register containing flag bits T_1 and T_0 to define tracing modes; an S flag bit to indicate when the processor is in the supervisor mode; an M bit that selects between master ($M = 1$) and interrupt ($M = 0$) stack pointer use when in the supervisor mode; three interrupt level mask bits (I_2, I_1, and I_0); and the condition code flags, extend (X), negative (N), zero (Z), overflow (V), and carry (C).

In the supervisor portion of the programming model, the 32-bit interrupt and master stack pointers, the source and destination function code registers (SFC and DFC), cache control (CACR), and cache address (CAAR) registers—all introduced in the 020—are included.

The vector base register (VBR) added in the 68010 to allow vector tables to be placed anywhere in memory is included in the 030. New registers to provide for memory management use are the main additions in the 030 to the 68000 family programming model. They include CPU and supervisor **root pointer** registers used to point to address translation tables in memory. A *translation control (TC)* register sets virtual memory page sizes and determines which translation tables are to be used. *Transparent translation (TT0 and TT1)* registers allow the user to set aside memory areas for direct translation. That is, the logical and the physical addresses are the same and no additional translation is required. Lastly, a *MMU status register (MMUSR)* is provided to assist internal MMU sequencing. How these registers and the MMU function are described later in Sections 13.12 and 13.13.

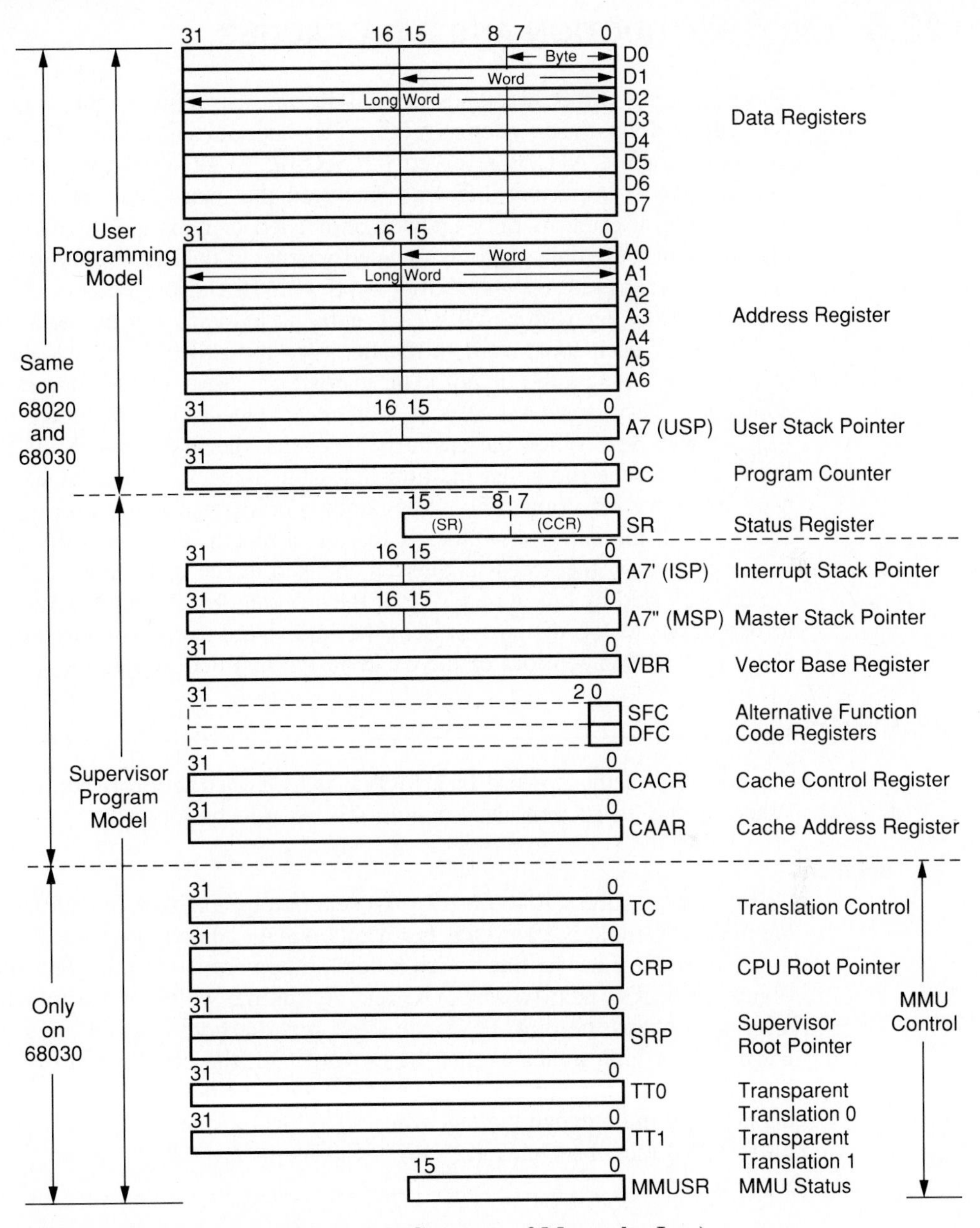

Figure 13.3 **Programming model (Courtesy of Motorola, Inc.)**

13.6 68030 INSTRUCTION AND DATA CACHES

There are three different caches in the 030, an instruction cache, a data cache, and an **address translation cache (ATC)**. This section will discuss the first two while discussion of the ATC is reserved for Section 13.12 on the memory management unit. The most important difference between the instruction and data cache is that the instruction cache is only updated as needed to maintain instruction flow to the pipeline while the data cache is updated whenever data is written to memory.

Both caches are set associative type with 16 entries. Each entry, called a **line**, contains four long words. With each entry is associated a tag address. Four valid bits, one for each long word in the line, are used to indicate whether each of the four long words is valid or not. The instruction cache access address (Figure 13.4) contains a 24-bit tag (A_8–A_{31}), a 4-bit index (A_4–A_7), and three word select bits (A_1, A_2 and A_3). When the cache is accessed, the index bits point to a particular entry in the cache (note that the four index bits have 16 unique binary combinations used for entry selection). The comparator compares the tag portion of the access address with the entry tag value. If there is a match, then the valid bit for the long word selected by the access address is checked to see if the entry is valid. The long word is selected by bits A_2 and A_3. If the selected long word is valid, bit A_1 is used to determine which of the words in the long word (upper or lower) is to be used. This word is then sent out of the cache to the instruction pipeline.

EXAMPLE 13.1

The location of the opword of a MOVE D0,D2 instruction is address $0400 6522. Where in the instruction cache would this opword be located?

Solution:

Using access address $0400 6522, first separate the tag from the remaining bits. The tag in this example is $0400 65. Index bits A_4–A_7 contain the value $2, which points to cache entry #2. The line which holds the opword is the third line from the top in Figure 13.4. The next two bits (A_3 and A_2) define which of the four long words in the line contains the opword. Since these bits are both low, the long word is LW0. Bit A_1 being high indicates that the opword is in the upper word of long word 0 of line 2 in the instruction cache. For the opword to be released from the cache, the tag at line 2 must match the tag word ($0400 65) of the access address and the valid bit for LW0 must be high indicating the contents are valid.

The combination of a matching tag address and a good valid bit causes a hit to occur. If the tag does not match or the valid is not set, a miss occurs and the cache entry must be updated through an external bus cycle access to memory. Cache entries are updated with an aligned long word access regardless of which word caused the miss to occur. That is, if a miss occurs on, say, the upper word of LW0 of line 2, both the upper and lower words will be updated by the fetch from memory, even if the lower word was already valid.

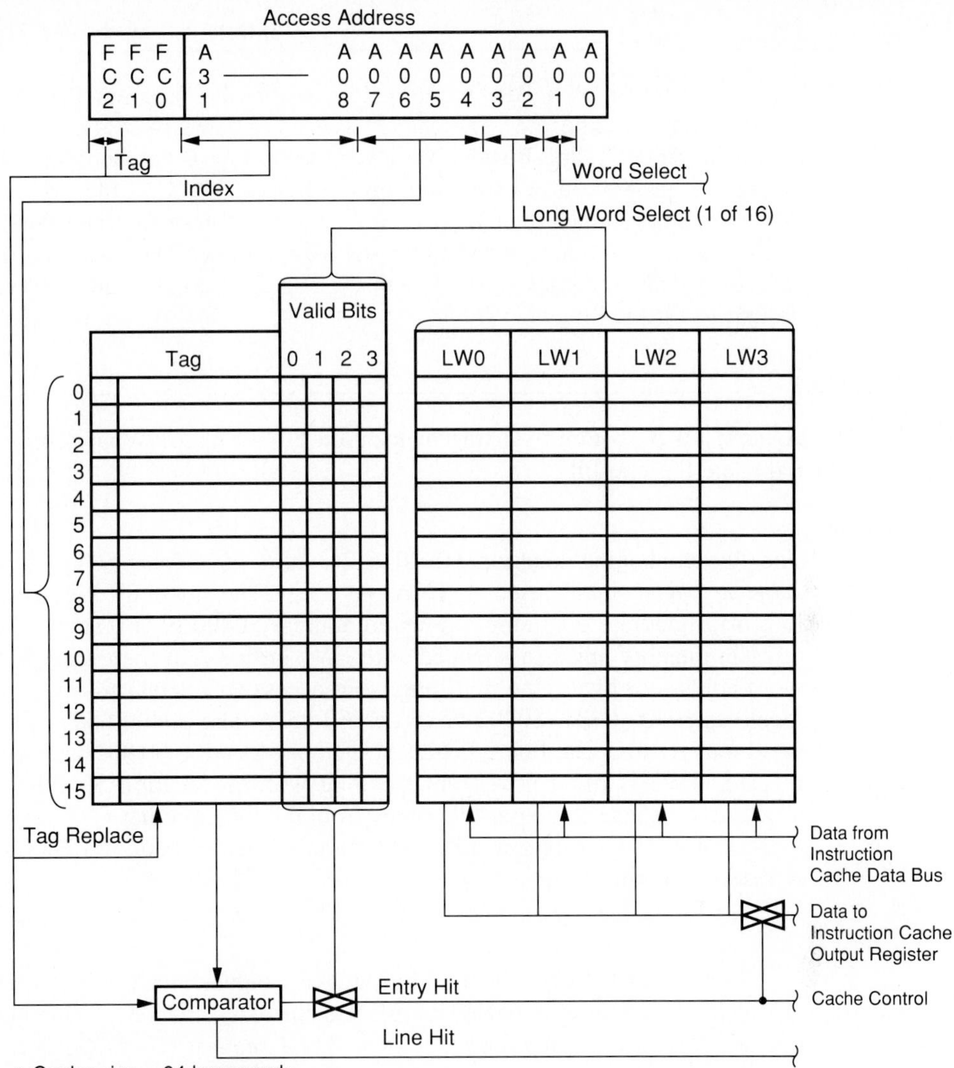

- Cache size = 64 long words
- Line size = 4 long words
- Set size = 1 (direct mapped)
- For an entry hit to occur
  - The access address tag field (A8–A31 and FC2) must match the tag field selected by the index field(A4–A7)
  - The selected long word entry (A2–A3) must be valid
  - The cache must be enabled in the cache control register
- Instructions are always prefetched from long word aligned bus cycles. Both words are updated in cache regardless of which word caused the miss.
- If a prefetch ends in a bus error the entry selected by A7–A2 of the access address is invalidated.
- Only prefetched instructions are cached.
- In-circuit emulators can assert CDIS to disable cache and force all bus cycles external.

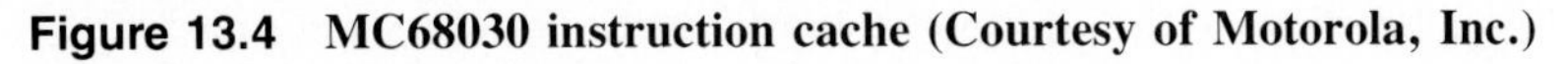

Figure 13.4 MC68030 instruction cache (Courtesy of Motorola, Inc.)

In the case of a BRANCH instruction in the program listing, a change of program flow is expected. The cache must be updated from the new program area. A special mode, called **burst fill,** makes the filling of the cache in this (and similar changes of program flow) situation more efficient. It allows the 030 to fill the cache with a full line rather than limit it to a long word transfer. The burst fill can start from any one of the long word addresses within a line. The operation then will make four long word accesses starting with the accessed address and, if necessary, wrap around to fill the line.

EXAMPLE 13.2

A burst fill is required starting at location \$0400 6526. What sequence is used to complete the line fill?

Solution:

The line containing location \$0400 6526 starts at address \$0400 6520 and is the same as cache line 2 used in Example 13.1. The burst fill begins at the aligned long word address containing accessed address \$0400 6526. This is location \$0400 6524 in memory and is associated with LW1 in line 2 of the cache. Since each long word holds four bytes, the first long word access will actual fill LW1 with data from locations \$0400 6524, \$0400 6525, \$0400 6526, and \$0400 6527. The second long word transfer to cache line 2 begins at memory location \$0400 6528 and moves data to LW2. LW3 is filled next from the long word at location \$0400 652C. The last access to complete the burst fill begins from the location that is "wrapped around" to LW0 of line 2. This long word access starts at memory address \$0400 6520, which is associated with LW0.

When using the burst fill option, a 2-bit counter must be included in the memory addressing logic connected to address lines A_2 and A_3. While the burst fill is operating, a control signal called *cache burst request* ($\overline{CBREQ}$) is asserted and is used to preset the counter to accommodate the first long word access. The logic circuitry used with the burst fill must return a *cache burst acknowledge* ($\overline{CBACK}$) to indicate to the 030 that the external system is able and ready to transfer four long words to fulfill the burst fill requirement. As each long word access is made, the counter is incremented so the resultant change in address lines A_2 and A_3 points to the correct aligned long word location. The other address lines remain stationary through the entire process. In this way, the correct addressing sequence to fill a single cache line is maintained throughout the burst fill activity. Note that even though time is required to perform four long word accesses to complete a burst fill, in the long run time is saved. After any change in program flow, the cache entries need to be updated with new instructions from the program listing. Without a burst fill, a single long word update occurs before the instruction can enter the pipeline. The next instruction is then missing from the cache, so a second miss causes another fetch from memory, and so forth. With a burst fill, four long words are loaded into the cache; that is, the first eight words of the program are now in the cache and can be fed to the pipeline. Initially, four memory accesses were performed to load the cache; this avoided time that would have been lost through the cache miss operations had the burst fill not been used.

While the cache is filling as a result of a change of program flow, the 030 asserts a control pin called $\overline{\text{REFILL}}$ to indicate that the instruction pipeline as well as the instruction cache are being refilled. This output is used to assist system designers in debugging hardware problems that may be related to program execution.

The data cache (Figure 13.5, page 378) is configured identically to the instruction cache except that data are accessed differently. An operand directs data to be fetched from a particular location. If a hit occurs in the cache, access is made directly from the cache; during a data read operation, data are read from the cache and no updating occurs. If a miss occurs during a read cycle, the cache will be updated with an aligned long word access; the data cache can also be updated using the burst fill method to move a line of data into the cache. During a memory write operation, a hit causes a **write-through** to occur. Both the cache entry and memory are written into so that the cache contains the same data as the corresponding memory location. A cache miss during a write cycle causes only memory to be written to since the cache does not contain a corresponding entry location.

The *cache control register (CACR)* for the 030 contains the same functions as the CACR for the 020 (Figure 13.6, page 379). One set of control bits is used for the instruction cache and another for the data cache. Other bits are used to enable (or disable) the use of the burst fill for the *instruction cache* (*IBE*) and the *data cache* (*DBE*). Normally, the data cache is not updated if a miss occurs during a data write access. However, the *write allocation (WA)* bit in the CACR can be set to cause the data cache to be updated by these misses. With both the 020 and 030 caches, the *cache address register (CAAR)* is used to point to the cache entry to be invalidated by the clear cache entry function of either cache.

Besides the software control of cache operations via the CACR, there are hardware control signals to allow operation of the 68030 without the caches. These signals are used in debugging situations when it is desirable to observe the sequence of 030 operations. When asserted, *cache disable* ($\overline{CDIS}$) disables caching functions. Instructions and data are fetched directly from memory through normal external bus cycles. The caches are not flushed (entries are not invalidated) so that when the signal is negated, normal caching sequences can resume.

The *cache inhibit input* ($\overline{CIIN}$) is a synchronous input signal which inhibits the updating of instruction and data caches with new data from memory. Other caching operations within the 030 still take place. The $\overline{\text{CIIN}}$ pin is ignored during all write cycles so that a write to memory will still allow the data cache to be updated in the write-through mode. The *cache inhibit output* ($\overline{CIOUT}$) pin is associated with the address translation cache and will be discussed in Section 13.12.

13.7 68030 ON-BOARD MEMORY MANAGEMENT UNIT

The process of translating an accessed address into a memory address is performed in the on-board memory management unit (MMU) of the 030. All requested addresses

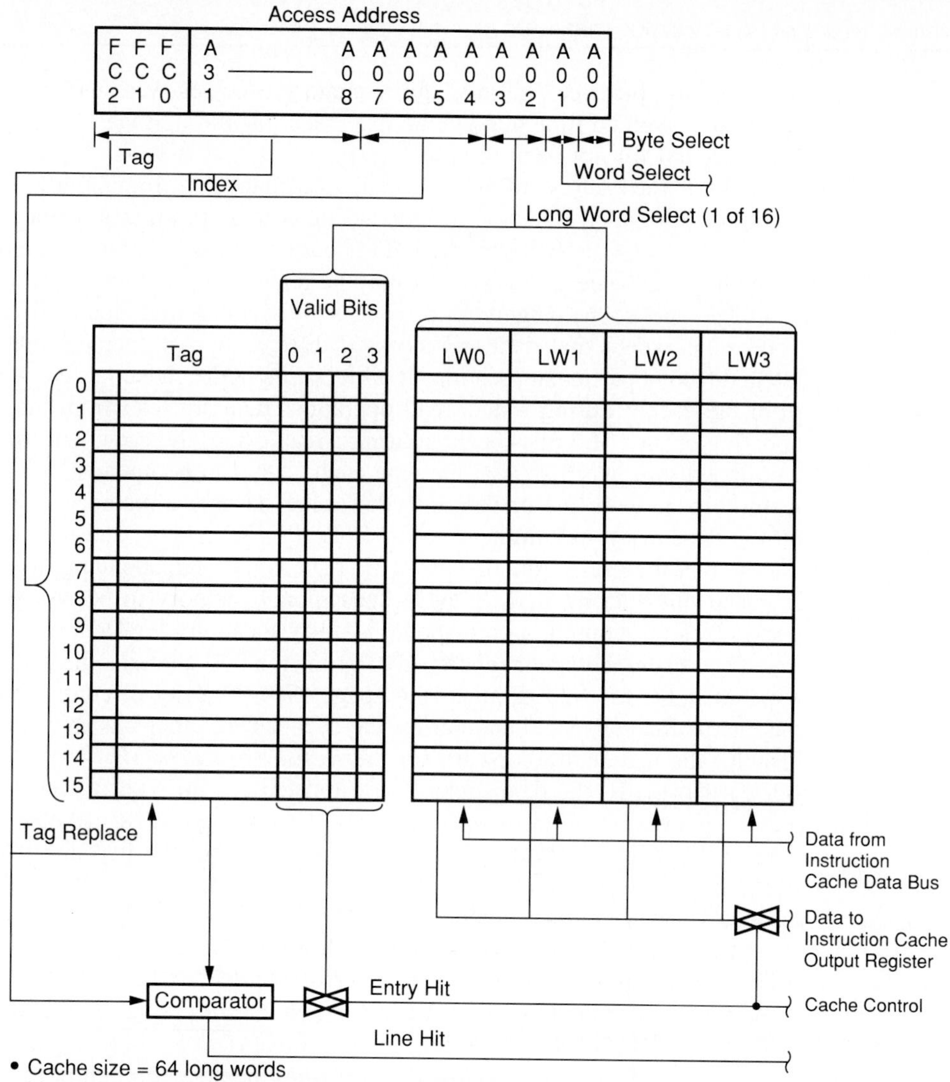

- Cache size = 64 long words
- Line size = 4 long words
- Set size = 1 (direct mapped)
- For an entry hit to occur
  - The access address tag field (A8–A31, FC0–FC2) must match the tag field selected by the index field (A4–A7)
  - The selected longword entry (A2–A3) must be valid
  - The cache must be enabled in the cache control register

- When a hit occurs on a write cycle, the data are written both to the cache and to external memory, regardless of the operand size, and even if the cache is frozen (write-through policy).

- The data cache is the only one that can be written into by user. Write-through writes to cache and memory so cache and memory data match.

- When a miss occurs on a write cycle, data are written only to memory.

Figure 13.5 **MC68030 data cache (Courtesy of Motorola, Inc.)**

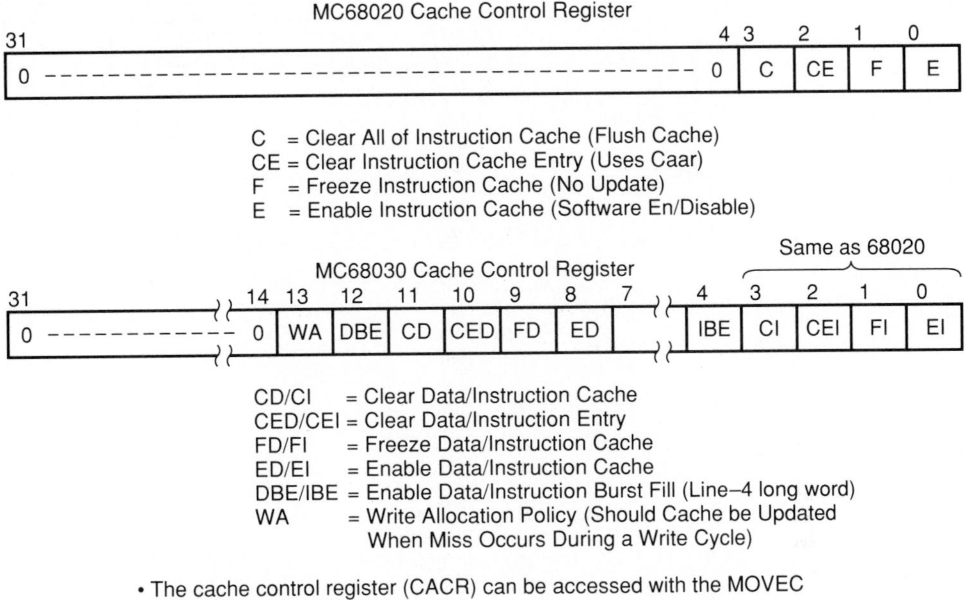

Figure 13.6 **Cache control registers for the MC68020 and MC68030 (Courtesy of Motorola, Inc.)**

are considered logical addresses by the 030. Some of them will be translated into a different physical address to access a piece of information in memory. Some of them will be directly translated into a physical address with the same value as the logical address and sent to access a memory location. Still other logical addresses will produce a hit in a cache and not be sent to memory at all. The memory management unit (MMU) inside the 030 controls what happens to an access address and how the translation is performed.

13.8 MEMORY ADDRESS TRANSLATION TABLES

Several registers briefly described in Section 13.5 are used to establish a set of **translation tables** in memory; these establish the translation rules that create a physical address from a given logical address. First the program uses the *translation control (TC)* register (Figure 13.7, page 380) to establish how many levels of translation tables are to be used in the translation tree. The most significant bit of that register enables or disables MMU functions. If the MMU is disabled, then all access addresses are used to access the instruction cache, data cache, and memory directly. When enabled, the MMU directs address translation, including the possibility of virtual memory sequencing. Bit 25 is the *supervisor root pointer enable (SRE)* flag which selects whether the CPU root pointer (CRP) or the supervisor root pointer (SRP) will be used to point to the top of the translation tree located in physical memory. The translation

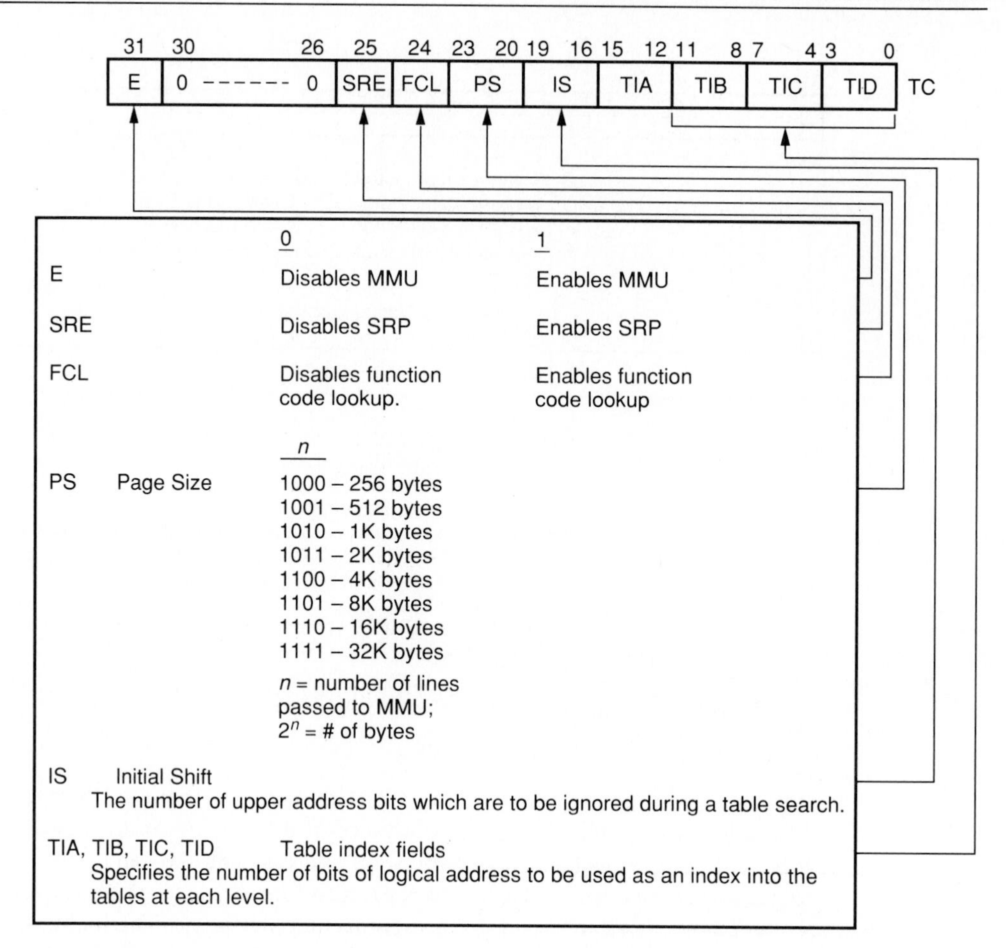

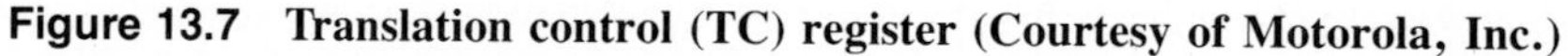
Figure 13.7 Translation control (TC) register (Courtesy of Motorola, Inc.)

tables in the tree contain the translation rules or sequence used to change a logical address to a physical address.

Selection of the use of function codes to separate user and supervisor program and data translation tables is made by setting the *function code lookup (FCL)* pin in the TC. The page field allows the page size for page accesses and virtual memory swaps to be selected from 256 to 32K bytes wide. As the binary value of these four bits increases from 0000 to 1111, the page size doubles, starting at 256 bytes for 0000 and finishing with 32K bytes for 1111.

The *initial shift (IS)* bits are used to choose how large each translation table is to be. This is effected by selecting how many logical address bits between 17 and 32 are used for comparison by each table. The lower the number of compared bits, the smaller the number of entries in a translation table. Each table index size is selectable by the remaining four fields (TIA through TID) in the TC. Each field is four bits and therefore allows the selection of an index value into the table of 0–15. By setting the index value to zero, the program effectively removes this table from the tree.

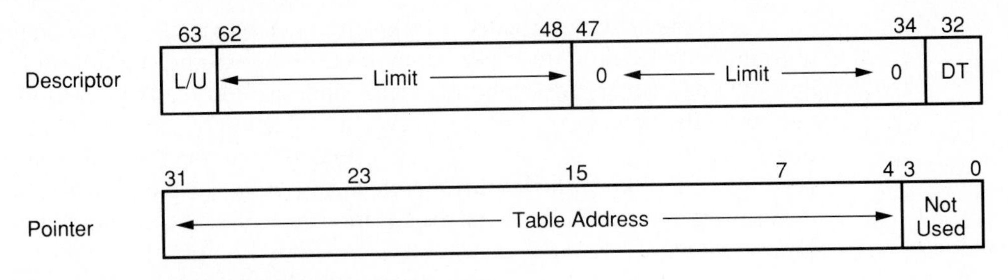

L/U – Lower or upper page range for limit
D/T – Descriptor type

0 0	Invalid
0 1	Page Discription
1 0	Short (4 byte) Format
1 1	Long (8 byte) Format

Limit – Index size (table size) limit

Figure 13.8 Root pointer format

Once it has been determined that an external memory bus cycle will be made, the selected root pointer value is placed on the address bus. This pointer points to the top of the first translation table in memory. The root pointer descriptor is made up of two long words (Figure 13.8). The first long word contains the actual descriptor. The second long word holds the pointer to physical memory where the first translation table (TIA) is to be found. The most significant bit of the descriptor long word determines if the limit value in bits 48–62 are an upper limit or lower limit value. This limit sets the maximum index size for table TIA. If the limit is an upper one, then the total maximum number of allowable entries for TIA is one more than the limit. If the limit is a lower one, then the total number of entries equals the difference between $7FFF and the limit value plus one.

EXAMPLE 13.3

How many entries can TIA table contain if an upper limit value is $15? A lower value of $7100?

Solution:

For an upper limit of $15, the total number of entries is 21 plus 1, or 22 (essentially, $0–$15). To find the lower limit, first subtract the value from $7FFF. $7FFF − $7100 = $EFF. Then add 1 to equal the total number of entries of $F00, or 3840. In essence, the entries are $7100–$7FFF.

The least significant bits of the descriptor for the root pointer are called the *descriptor type (DT)* and define the type of descriptor used for table TIA. Choices of DT as shown in Figure 13.8 are page, short, long, and invalid.

Sections of the original logical address act as pointers to the translation tables used in the translation tree. The value of these indexes must conform to the rules set up in the descriptors for each table. The value held at the table location is used as a pointer to the next translation table. Again, a portion of the logical address forms an index into this table, and the sequence is repeated into the third and fourth tables. The contents of the fourth table plus the physical portion of the logical address combine to form the physical address of the location in memory associated with the original logical address.

The format of the indexes within the logical address follow the descriptor descriptions and are illustrated in the translation tree examples. Figure 13.9 contains the short and long descriptor forms used for translation tables. The *short descriptor* contains the pointer value address to the next table and four status bits. The status bits include the *update (U)* and *write protect (WP)* flags. The U bit is set whenever this table entry is used as part of a translation. The write protect bit says that areas in memory resulting from this pointer are write protected. That is, those locations can be read but not written into. WP protects program area accesses against accidental deletions or changes. DT is used to describe the type of descriptor for the next table in the translation tree in the same manner as the DT bits in the root pointer descriptor.

For the *long descriptor* format, the descriptor is separated into two long words much like the root pointer descriptor. The first long word contains the same data as

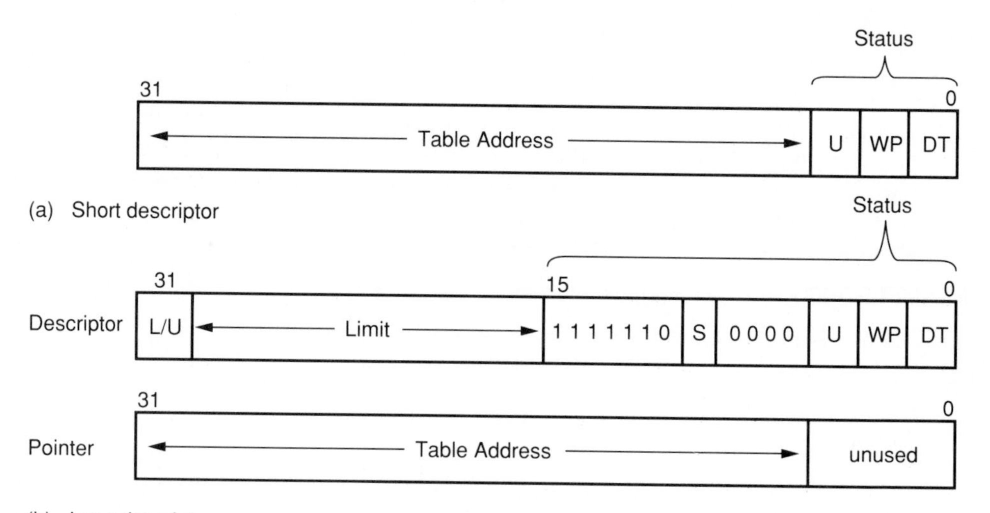

Figure 13.9 Short and long descriptors

the root pointer descriptor as well as the status information of the short descriptor. In addition, an S flag is added to indicate that this table should be accessed only when the 030 is in the supervisor mode. The second long word of the long descriptor contains the pointer to the next table.

Valid *page descriptors* (Figure 13.10) are similar to the short and long descriptors just described. Instead of pointing to the next table, the address portion of the descriptors points to the top of the page in memory that contains the accessed data location. In brief, the page descriptors are used as entries in the last translation table. Two new status bits are included with these descriptors. The *modified page* or M bit is set following a write to the page pointed to by the address in the descriptor. The cache inhibit (CI) bit is made high if the data accessed are not to be cached in either the data or the instruction cache. When this bit is set and this entry is used, the 030 will assert the cache inhibit out (CIOUT) control pin to indicate that the caches are inhibited. Inhibiting the caches only prevents them from being updated. They are not disabled; a hit can still occur if a valid entry exists in the cache.

Two special types of descriptors, shown in Figure 13.11 on page 384, are used to handle special situations. The *early termination page descriptor*, which looks much like the long descriptor, is used when it is desirable to reduce the time required for the table search process. This descriptor replaces a short or long descriptor that would normally point to another table in the tree. The table exists because the translation control register contains a value for the index size for that table. If the

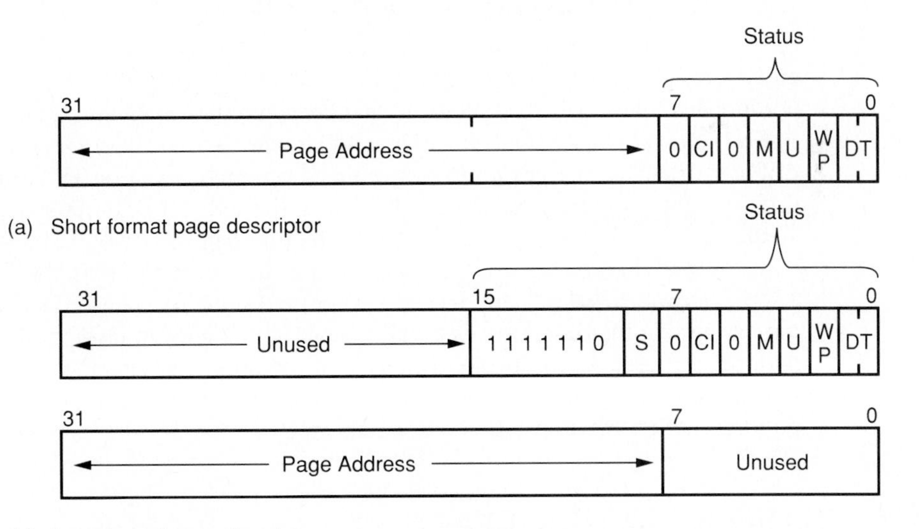

Figure 13.10 Valid page descriptors

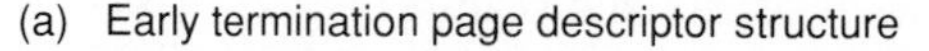

(a) Early termination page descriptor structure

- Short
  Same as the short normal page descriptor
- Long

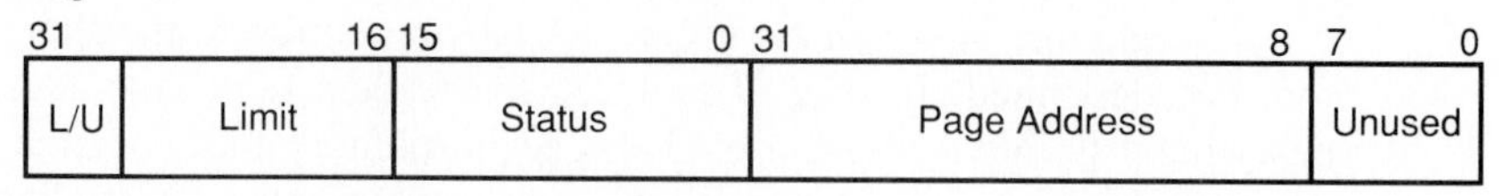

(b) Indirect descriptor structure

- Short

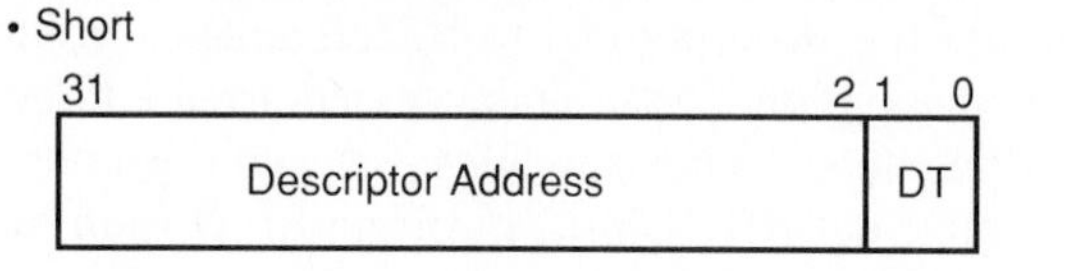

- Long

31 2 1 0 31 2 1 0

Unused	DT	Descriptor Address	UN

- DT
  - Only 10 or 11 is allowed.
  - 10 = the size of the descriptor being pointed to is 4 bytes.
  - 11 = the size of the descriptor being pointed to is 8 bytes.

Figure 13.11 Special descriptors (Courtesy of Motorola, Inc.)

early termination descriptor is used as a table entry, then the process will skip any additional tables and go directly to memory pointed to by the page address. The *indirect descriptor* is used when it is desirable to share memory pages from two different translation trees or to allow the same page location to be accessed from different entries within the same table. Since the indirect descriptor does not contain status bits, access can be made without affecting page descriptor status information.

Invalid descriptors use the same form as the short and long descriptors (Figure 13.10). However, all bits are unused except the DT fields. These fields contain zeros, indicating that the descriptor is invalid. That is, the current portion of the translation table is not used or not located in physical memory. Instead, the section of the tree may be on a virtual memory device such as a disk and a memory swap will be required before the translation can be completed.

13.9 TRANSLATION TREE EXAMPLES

To become familiar with how the process works, use Figure 13.12, which is a partial system involving a two table (or three level) translation tree. In this example, the CPU root pointer points to an address at the top of table TIA located in memory. The index into table TIA from the logical address points to an entry in that table. In turn, the contents of that location point to the top of table TIB. A second index

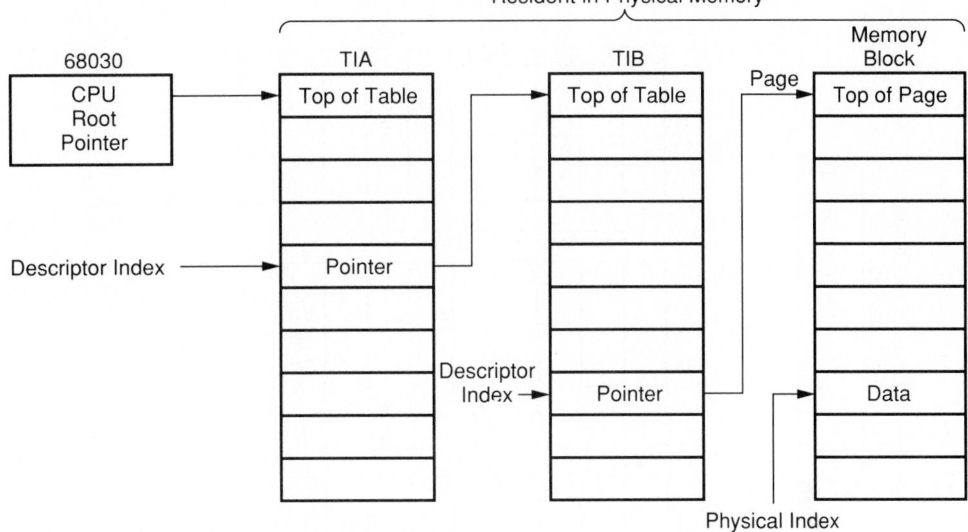

Figure 13.12 Two table translation tree

from the logical address selects a location in that table. These entries were short or long format descriptors. The entry in table TIB is a page descriptor pointing to the top of the memory page that contains the location sought after. A third index from the logical address contains the lower portion of the physical address which selects the location within the page pointed to by table TIB. The data at this location are cached into the instruction or data cache, whichever is appropriate, and the access cycle is rerun. This time a hit in the cache occurs and the access is completed.

Figure 13.13 on page 386 is a more detailed example of the use of the 030 translation tree. At the top of the illustration is a sample logical address that has been created from a program instruction like MOVE $15486C,D0. When that instruction is first decoded, the data cache in the 030 is checked to see if this data is present in the cache at the logical address of the operand. When a miss occurs, external memory must be accessed to retrieve the data. The memory management unit (MMU) next checks the address translation cache (ATC) to see if there is a direct translation already cached. If there is, then the contents of that ATC entry are used to access memory directly. In other words, the ATC does the translation from the logical to a physical address.

A miss in the ATC begins the translation tree search. The translation control register (TC) is checked for the size of the tree (how many tables are used) and the maximum size of the index (i.e., the entries) for each table. These are compared to the index values in the logical address. If any of the index values exceed those designated by the TC, then a bus error exception is taken. Next, if TC SRE is reset, the DT bits of CPU root pointer descriptor are checked to see if table TIA is used. The 030 makes this determination by the type of descriptor expected for the next level in the table. A short or long descriptor indicates that a table is next. A

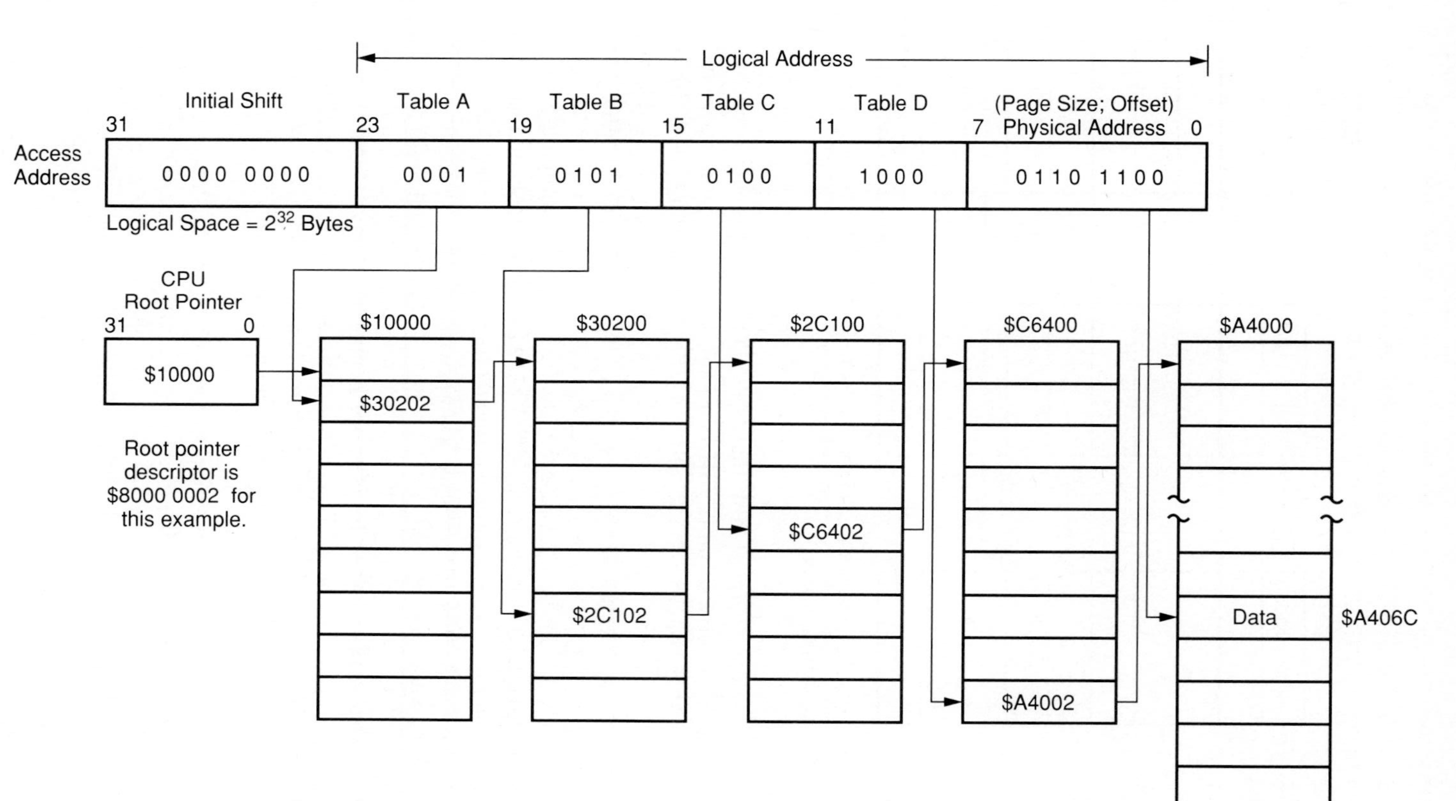

Note: Lower nibble of each table entry is the set of translation rules (status) for the next table; $2 = 0010, U = 0, WP = 0, DT = 10
Access has not been made previously (U = 0), and is not write protected (WP = 0).
Descriptors are short (DT = 10).

Figure 13.13 **Translating logical address $0015 486C to physical address $000A 406C**

page descriptor indicates that the page holding the data is next. Finally, an invalid descriptor indicates that the table is not available or that the next level is in virtual memory and a memory swap is required. A bus error exception must be initiated (recall that virtual memory, introduced with the 68010, used the bus error exception to facilitate memory swaps).

In the example of Figure 13.13, the root pointer descriptor is $8000 0002. This descriptor says that the limit ($0000) is a lower limit (L/U = 1) setting the table size to $8000 entries ($0–$7FFF). The DT field identifies the memory area pointed to by the root pointer as a table (short descriptor). The first table in the tree will be table TIA unless the TIA field in the TC is $0. In that case, TIA would be skipped and the next table used.

Bits 20–23 of the logical address contain the index into table TIA, which is $0001 in this example. The first location in the table is designated as entry $0, so $1 is the second entry. This entry is a short descriptor as denoted by the descriptor field of the root pointer. An entry value of $30202 says that the next pointer value is $30200 and at that area in memory is another table (DT bits = 10). Bits 16–19 of the logical address index into table TIB. The value in our example is $5, the sixth location in the table. The entry value at that location is $2C102, which is again interpreted as pointing to a table (DT = 10) that begins at address $2C100. Bits 12–15 provide the index into table TIC. The process is repeated, picking up a pointer value of $C6400 to table TID. Bits 8–11 index into the ninth location in table TID. The entry in this table turns out to be a page descriptor (DT = 01). The pointer value becomes the upper part of the physical address accessed by the logical access. The lower portion of that physical address is contained in logical address bits 0–7. In our example, these two values form the physical address $000A 406C. The data at that location are read and cached into the data cache. The bus cycle which initiated the table search is rerun. This time a hit occurs in the data cache and the data is moved to register D0 to complete the instruction that requested the original access.

13.10 TRANSPARENT OR DIRECT TRANSLATION

Two transparent translation registers—TT0 and TT1 (Figure 13.14, page 388)—are provided to allow areas in memory to be accessed directly instead of through the translation tree. In essence, the logical address is the same as the physical address. The upper eight bits of either the TT0 or TT1 register contain the upper bits of the logical address used for direct translation. The next eight bits are used to select the size of the memory area for direct translation. These bits allow the user to select which of the logical address bits between bit 16 and 23 should be included in the logical base and which should be treated as don't care bits. If a bit in this field is set, then logical addresses will be directly translated regardless of the state of that bit. A 0 in a logical address mask bit forces that bit in the logical address to be a 1 in order for the logical address to be directly translated to a physical address.

The enable (E) bit, which is cleared by a reset, is used to enable or disable the transparent translation function. The CI (cache inhibit) bit is used to select whether the data fetched by this access is to be cached (in either the data or instruction

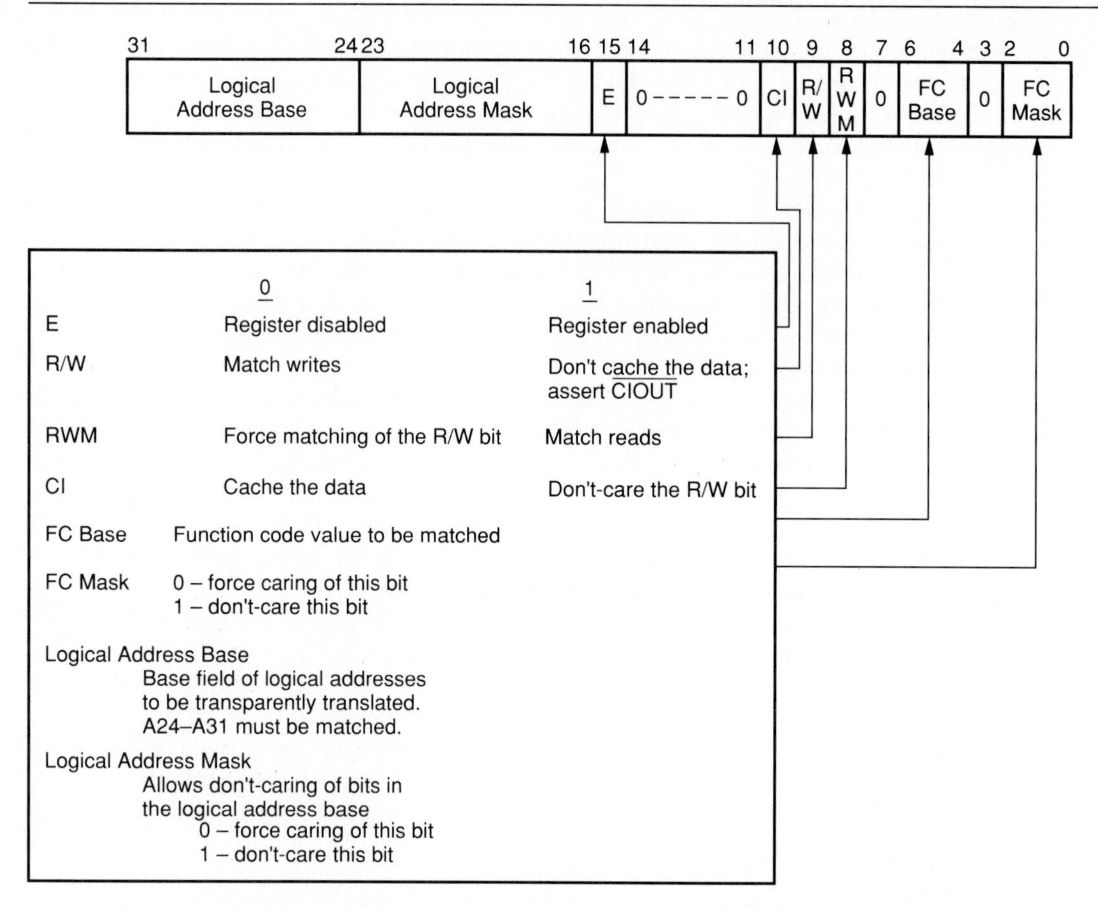

Figure 13.14 **Transparent translation registers (TT0 and TT1) (Courtesy of Motorola, Inc.)**

cache, whichever is appropriate). The read/write (R/W) bit determines if the transparent translation is to occur for a read or a write operation. However, if the RWM (read/write mode) bit is set, both read and write operations can use the transparent translation facility. The function codes can be used to confine transparent accesses to a particular area in memory depending on the type of access (user or supervisor data or program area). This confining function is enabled by setting the *FC base* to the type of access permitted. The *FC mask* allows the user to broaden the limitation created by the function code. By masking an FC bit and treating it as a don't care, the limitations can be changed from user program to just user area, for example. Likewise, they can be changed from user program to any program space by masking the corresponding FC bit.

When the transparent translation registers are in use (E bit set) and a logical access occurs that causes a miss in the address translation cache, the logical address bits in TT0 and TT1 are compared with the logical address. If either register matches the logical address and the remaining control bits correspond to the access being attempted, then the logical address is put onto the address bus and that memory location is accessed directly, ignoring the address translation tree.

EXAMPLE 13.4

What are the contents of the transparent translation register that allow user data to be read from memory locations \$C000 0000–\$DFFF FFFF and cached without requiring an address translation?

Solution:

The logical address base bits are set to \$C0. To allow the range to extend from \$C000 0000 to \$DFFF FFFF, only the three most significant bits must be monitored. \$C = 1100 and \$D = 1101. Only the three upper bits are fixed at 110. Therefore the logical address mask is set to \$1F so that only the upper three bits must be matched to allow a direct translation. The register is enabled (E = 1) and the data will be cached (CI = 0). Since the access is to be a read type only, the R/W bit is set to a 1 and the RWM is set low. The FC base bits are set to 001, the user data function code. All FC bits are monitored so the FC mask is set to 000. Figure 13.15 shows the transparent translation register with the bits set according to the needs of the present example.

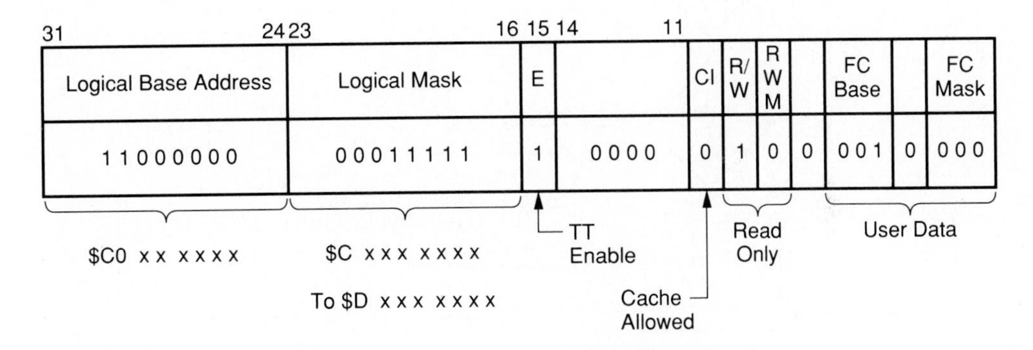

Figure 13.15 **Transparent translation register example**

A flow chart that summarizes the sequence the memory management unit takes for an instruction operand data access is shown in Figure 13.16 on page 390. The address of the operand of the instruction is treated as a logical address. It is first compared to the address tags in the data cache and if a hit occurs, the data is accessed from the cache and the instruction is completed. Since this is an operand access there will not be a hit in the instruction cache. If a miss occurs in the data cache, the logical address is next checked for a hit in the address translation cache of the memory management unit. If a hit occurs, the contents of that cache entry are sent out on

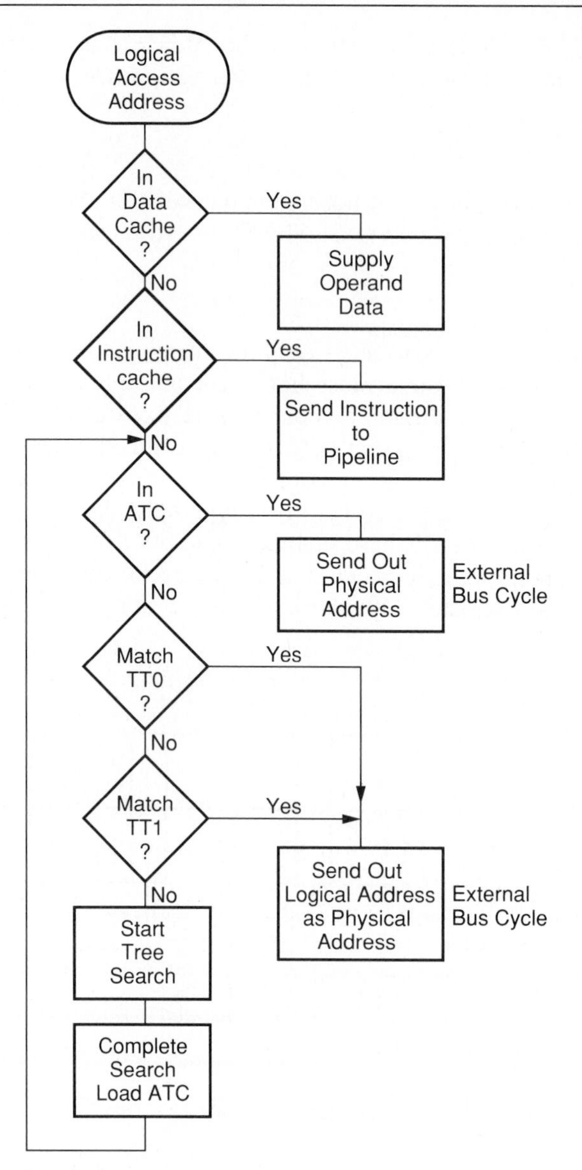

Figure 13.16 Valid memory management

the address bus as a physical address. The data is accessed from memory and cached into the data cache and the access cycle is rerun. If there is a miss in the ATC, the transparent translation registers are compared with the logical address. If a match occurs, the logical address is placed onto the address bus as a physical address and the data is again accessed and cached. Finally, if the logical address does not match either transparent translation register, then a table search is initiated. At the completion of the search, the physical address derived from the table is cached into the ATC and the access cycle is rerun.

13.11 ADDRESS TRANSLATION CACHE

The 68030 memory management unit address translation cache (ATC) is a 22-entry fully associative cache. When an access is made in the 030 that requires the fetching of information from memory, the logical address is first compared with the entries in the ATC. If a hit occurs, the contents of that entry are used as the physical address to memory that is associated with the logical address. If a miss results, the ATC aborts the access cycle and initiates a translation tree search. Upon completion of the search, the resultant physical address is cached in the ATC and the access cycle is rerun. The format for an ATC entry is shown in Figure 13.17. The 28-bit tag contains the logical address entry for an ATC line. The most significant bit is the valid bit followed by three function code bits. Again, the function code bits can be used to limit access to specific areas of memory. The remaining tag bits hold the upper 24 bits of a logical address page.

The physical translation contents of the ATC entry hold the upper 24 bits of the physical address associated with the logical address tag. Additionally, there are four status bits familiar to the reader. Bit 27 is the bus error (B) bit, indicating if a bus error occurred during the access process. Next is the cache inhibit (CI) bit which inhibits the caching of information to the instruction and data caches and also asserts $\overline{\text{CIOUT}}$. The write protect (WP) bit and the modify page (M) bit follow. WP denotes that physical page as a read only and the M bit indicates that this page has been previously written to. Obviously, a write protected page will not be modified and the M bit in that case will always remain reset.

13.12 MEMORY MANAGEMENT INSTRUCTIONS

Four instructions—PMOVE, PFLUSH, PLOAD, and PTEST—affect the memory management unit directly. They are similar to the MMU coprocessor instructions for the MC68851 (see Section 10.14) and contain similar bit coding, hence the use of the P preceding each mnemonic. P refers to *paged MMU*, indicating that the memory management functions operate on memory page sections.

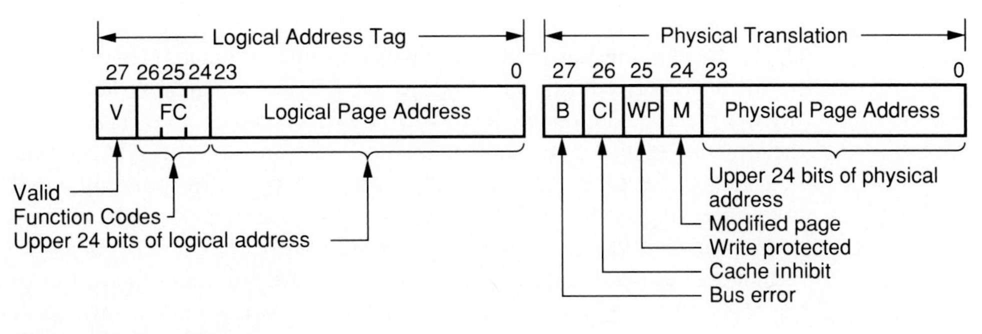

Figure 13.17 ATC entry

An ATC entry is loaded with a new physical translation using the PLOAD instruction. This causes descriptor used (U) and modified (M) bits to be set, indicating that an entry or table descriptor has been updated. The syntax for a PLOAD follows:

PLOADx ⟨fc⟩,⟨ea⟩

where x is either R (for simulated reads affecting U bits in the translation table descriptor) or W (for simulated writes causing U and M bits to be set as if a write cycle occurred). The fc field contains a function code value for memory area restriction. This value can be (a) an immediate value equivalent to a function code value, (b) a data register whose lowest three bits contain the function code value, or (c) either the source or destination function code (SFC or DFC) register contents. The ea field holds the effective address whose translation is to be loaded. When the PLOAD instruction is executed, the ATC is checked for the effective address. Once the address is located in the ATC, the translation table is searched for the descriptor associated with the effective address. After the descriptor is located, the ATC is loaded with a new entry so that the physical content of the logical effective address tag holds the new translation. If the effective address is not resident in the cache, then the cache replaces an existing entry tag with the effective address tag and the instruction is completed. Invalid entries are used first for replacement. If all entries are valid, then the 030 uses a pseudo least recently used sequence to select an entry to be replaced. This sequence examines the history of the cache entries used to determine which one has been used the least number of times. In case of several entries having the same history, the sequence selects the least-used entry that was used most recently. The M bit in the entry line is used to determine whether the entry has been used (modified) or not.

The PMOVE instruction is used to move data between an effective address and one of the MMU registers (TC, TT0, TT1, SRP, or CRP). The syntax for the instruction takes one of three forms:

PMOVE.⟨size⟩ MRn,⟨ea⟩	moves data from a MMU register.
PMOVE.⟨size⟩ ⟨ea⟩, MRn	loads a MMU register.
PMOVEFD.⟨size⟩ ⟨ea⟩, MRn	prevents the ATC from being flushed as a result of the move instruction.

Data size can be selected from word, long, or quad (16, 32, or 64 bits). Use of the PMOVE causes the ATC to be flushed when a write to a MMU register (MRn) occurs. Using the PMOVEFD syntax causes the *flush disable (FD)* bit in the opword of the instruction to be set. As a result, the ATC is unaffected by the instruction. Sending incorrect information to the corresponding MMU register initiates a *configuration error exception*. An example of such an error is writing a descriptor to a root pointer (SRP or CRP) with both descriptor type (DT) bits low indicating an invalid descriptor. Another cause for a configuration error is loading the translation control register (TC) with less than a long word of information.

PFLUSH is used to flush an ATC entry or the entire cache. In order to flush the entire ATC, a letter A is added to the PFLUSH mnemonic (PFLUSHA). The syntax to flush an entry is the following:

PFLUSH ⟨fc⟩, #⟨mask⟩,⟨ea⟩

Specific entries are flushed based on the page descriptor located at the effective address as long as the page is located in the boundary set by the function code bits. Omitting the ⟨ea⟩ field causes all entries associated with the function code area to be flushed.

PTEST causes a search of the ATC or the translation tables up to a specified level and returns status information corresponding to the condition of the last level tested. The memory management unit status register (MMUSR) shown in Figure 13.18 illustrates the type of information returned by the PTEST instruction.

The reader is already familiar with the B, S, and M bits in the MMUSR. The L bit indicates if the index value of the descriptor exceeds the limit set for that table. Invalid descriptors or entries are reflected by the I bit. An effective address that exceeds the range of a transparent translation register is shown by the T bit. The N bits indicate the number of translation tables up to and including the level of the PTEST search.

The PTEST instruction in a variety of formats is summarized by the following:

PTESTx ⟨fc⟩,⟨ea⟩, #⟨level⟩,An

The function codes, fc, and ea are used as they have been by earlier MMU instructions. The level value indicates the level to be tested by the instruction. Level 0 causes a search of the ATC only for the effective address to be tested. Levels 1–7 relate to different levels within the translation tree to be searched for the descriptor associated with the effective address. Register An is an option used if it is desir-

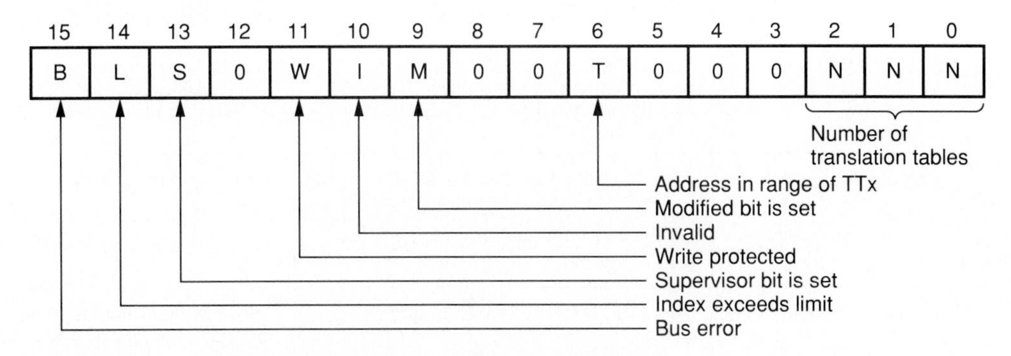

Figure 13.18 MMU status register (PTEST instruction)

able to store the physical translation value of the descriptor at the effective address being tested. The instruction terminates early in its search toward the level of the instruction if one of the following three conditions occur:

1. An invalid descriptor is detected along the way.
2. A limit violation prevents continuation through the tables.
3. $\overline{\text{BERR}}$ is asserted.

Table 13.3 summarizes the MMUSR results for searches of the ATC (level 0) and the translation tables (levels 1–7).

Table 13.3 MMUSR Bit Definitions

MMUSR Bit	PTEST, Level 0	PTEST, Level 1–7
Bus Error (B)	This bit is set if the bus error bit is set in the ATC entry for the specified logical address.	This bit is set if a bus error is encountered during the table search for the PTEST instruction.
Limit (L)	This bit is cleared.	This bit is set if an index exceeds a limit during the table search.
Supervisor Violation (S)	This bit is cleared.	This bit is set if the S bit of a long (S) format table descriptor or long format page descriptor encountered during the search is set, and the FC2 bit of the function code specified by the PTEST instruction is not equal to one. The S bit is undefined if the I bit is set.
Write Protected (W)	This bit is set if the WP bit of the ATC entry is set. It is undefined if the I bit is set.	This bit is set if a descriptor or page descriptor is encountered with the WP bit set during the table search. The W bit is undefined if the I bit is set.
Invalid (I)	This bit indicates an invalid translation. The I bit is set if the translation for the specified logical address is not resident in the ATC or if the B bit of the corresponding ATC entry is set.	This bit indicates an invalid translation. The I bit is set if the DT field of a table or a page descriptor encountered during the search is set to invalid or if either the B or L bits of the MMUSR are set during the table search.
Modified (M)	This bit is set if the ATC entry corresponding to the specified address has the modified bit set. It is undefined if the I bit is set.	This bit is set if the page descriptor for the specified address has the modified bit set. It is undefined if I is set.
Transparent (T)	This bit is set if a match occurred in either (or both) of the transparent translation registers (TT0 or TT1). If the T bit is set, all remaining MMUSR bits are undefined.	This bit is set to zero.
Number of Levels (N)	This 3-bit field is cleared to zero.	This 3-bit field contains the actual number of tables accessed during the search.

13.13 MMU ASSOCIATED EXCEPTIONS

A number of exceptions are initiated by MMU operations, usually because the 030 experienced a problem. These exceptions, summarized in Table 13.4, include bus error, F-Line exception, privilege violation, and configuration error. During the following discussion, remember that some of these exceptions have additional causes and that this discussion concentrates only on their initiation by a MMU function.

One of four events starts a bus error exception. The first is the assertion of a bus error signal ($\overline{BERR}$). $\overline{BERR}$ is asserted usually because there is nothing at the address of the current access whether that be an effective address or a translation table address. This type of bus error requires the task to be aborted and some corrective action to be taken.

Two more reasons for bus error are the existence of an invalid descriptor or a valid descriptor which exceeds the limit set for a particular table. In both these instances insufficient space is allocated for this section of the translation tree. The problem is corrected by loading in an additional page of memory and validating the descriptor or increasing the limit.

A bus error is also generated if a user attempts to write to a supervisor or write protected area in memory. In effect, these areas are not available to the user and the

Table 13.4 MMU Exceptions

Exception Type	Cause	Response
Bus Error a) Invalid descriptor	Page not in memory	Load page to memory (virtual swap) and rerun cycle
b) Limit violation	Page not in memory or range violation	Load page to memory and rerun cycle
c) $\overline{BERR}$ asserted	No device at access location	Abort task
d) User write to supervisor or write protected area	Bad task	Abort task
F-Line Exception F opcode with cpID = 0 not equal to PFLUSH, PMOVE, PTEST, or PLOAD	MC68851 instruction	Emulate MC68851 or rewrite code
Privilege Violation User attempt to execute P. . . . instruction	Bad task	Abort task
Configuration Error a) Load invalid descriptor (DT = 0) into root pointer b) PMOVE to translation control register with PS + IS + TIx ≠ 32	System program error	System crash

bus error results. The task is terminated and the code should be adjusted so that the user program only accesses user areas in memory.

The F-line exception occurs if a coprocessor instruction that does not exist in the 030 microcode is decoded. This usually occurs when a program written for the MC68020 (or 020) is imported into a 030 system. While the code for the main processor is fully upward compatible, the code for the MC68851 MMU coprocessor used with the 020 is not. Since the MMU functions are on chip with the 030, a reduced number of MMU instructions are used. Attempting to execute other MMU coprocessor instructions results in an F-Line exception. The problem can be averted either by rewriting that portion of the 020 code to reflect the change in MMU instructions or by emulating the MC68851 Coprocessor in the F-Line exception routine.

Privilege violations result from the inclusion of a MMU instruction (Pxxxx) in a user program. These instructions must only appear in supervisor programs and must be deleted from user programs.

The final exception associated with memory management is the configuration exception, which occurs if an incorrect attempt is made to load data into one of the memory management registers through the PMOVE instruction. Loading either root pointer with an invalid descriptor (DT = 0) says that the first table pointed to by the root pointer is invalid. Instead of locking up the system with an impossible situation, the processor initiates the configuration exception. Corrective action is required to avoid a system crash. A second situation that initiates the configuration exception is an attempt to load the translation control register with a word of data instead of a long word. This action would leave the upper half of the translation control register with undefined bits and functions. Again, the system would be locked up. The configuration exception diverts the problem to an exception routine informing the user to take some corrective action.

13.14 68030 SUMMARY

Performance of the 68000 line of processors is greatly enhanced by the MC68030. An on board data cache and memory management unit reduce the time for external bus cycles and logical to physical address translation. Increasing the concurrence of operation allows many internal and external operations to occur simultaneously, further enhancing performance. Many of the features developed through growth of the 68000 family culminated in the MC68030. A desire for further increases in performance led Motorola engineers to develop the MC68040 microprocessor. It adds floating point operations onto the same chip as well as separate instruction and data cache memory management units.

PROBLEMS

13.1 Compare the list of registers added to the user's portion of the 68030 microprocessor with that of the 68020.

13.2 Compare the list of the registers added to the supervisor portion of the 68030 with that of the 68020. Include the memory management unit registers.

13.3 Describe the caches of the 68030. What is the purpose of each one?

13.4 What functional control pins have been added to the 68030?

13.5 Which control pin(s) have changed function? How have they been changed?

13.6 What are the three main functional sections of the 68030? What tasks is each responsible for?

13.7 What is the purpose of the root pointer register?

13.8 What is a line and what process uses it?

13.9 A burst fill is set to update the instruction cache initially. What is the address sequence for the long words that fill the line of the cache that includes location $6CC0 346A?

13.10 Which signal does the 68030 assert to initiate a burst fill? Which signal must be returned by address decoding logic to acknowledge that the system is configured to manage the addressing needs for the burst fill?

13.11 What does write-through mean? With which cache is it associated?

13.12 What does an active $\overline{\text{CIOUT}}$ indicate? What is it used for?

13.13 What are the contents of the CPU root pointer descriptor that signifies that table TIA has an upper limit of $40000?

13.14 What are the contents of a logical address that contains the following information:
(a) user data access
(b) TIA index of $18
(c) TIB index of $4
(d) TIC index of $C
(e) TID index of $16
(f) physical index of $E6
(g) uses full access space

13.15 The 68030 instruction prefetch mechanism begins its process. What sequence is followed to fetch the instruction and load it into the instruction pipe if the instruction and its physical translation address are not yet fetched? Start the sequence from the initial presentation of the logical address of the instruction to the system.

13.16 What types of restrictions to memory accessing are placed by using the FC BASE bits? How are those restrictions modified using the FC MASK?

13.17 What is the difference between a short and long table descriptor? between a table and page descriptor?

13.18 What are early termination and indirect descriptors used for?

13.19 What are the contents of TT0 that allow the addresses $4000 0000 to $7FFF FFFF to be directly translatable for write protected supervisor program fetches? These fetches are to be cached as they are accessed.

13.20 Describe the contents of a line in the address translation cache.

13.21 Write an instruction to load the translation control register with the following data:
(a) enable the MMU
(b) select the cpu root pointer
(c) page size of 8K bytes
(d) initial shift of $2

(e) TIA and TIB are 10 entries long
(f) TIC and TID are 8 entries long

13.22 What are the contents of the MMU status register following a PTEST instruction to level 0 with an effective address to a user data RAM location? This location was recently written to before the PTEST instruction was executed. No faults were experienced during the execution of the instruction.

13.23 List four ways that an MMU associated action can generate a bus error exception.

13.24 Write a PMOVE instruction from D0 to the translation control register that would cause a configuration exception to occur.

14

68040 ADVANCED 32-BIT MICROPROCESSOR

CHAPTER OBJECTIVE

This chapter deals with the top of the line processor in the 68000 family, the MC68040, or, simply, the 040. The performance of the 040 far exceeds that of all previous members of the family. The 040 no longer includes a coprocessing interface. Instead, memory management functions that had been added to the 030 are included and expanded. Additionally, a floating point (FP) math unit is included into the chip.

14.1 GLOSSARY

bus snooping Monitoring buses for possible intervention.
denormalized number Scientific value with an integer value of zero, a mantissa fraction that is non-zero and a minimum exponent.
dirty bit Flag that indicates that the data cache has more current data than a related memory location.
not a number (NAN) Value returned that is not a finite number.
normalized number A scientific number with a binary integer of 1.
significand Fractional part of the mantissa.
unnormalized number A scientific value with an integer of zero, a non-zero fraction and an exponent that is greater than minimum value.

14.2 INTRODUCTION

The performance of the 68040, or, simply, the 040, exceeds that of the rest of the 68000 family; improvement has been achieved by expanding the number of functional units within the processor so that many activities can occur simultaneously. For example, the memory management area is divided so that separate MMUs manage program and data fetches. Internally, the instruction/operand operations operate on a separate bus from the data movement operations. This means that instruction decoding and operand computations as well as data manipulation actions occur concurrently.

The floating point (FP) unit included in the 68040 executes a subset of the MC68881/2 Floating Point Coprocessor instructions. The most commonly used instructions are incorporated, leaving the remaining instructions to be emulated, as needed, in software.

The 040 can operate directly in a multiprocessing environment through the use of a function called **bus snooping**. This function determines who has mastership of the buses and if the 040 should intervene into the current operation.

The burst fill process of the 030 has been extended through the introduction of a new instruction called *MOVE16* which allows a line of data to be moved between two memory areas. In effect, this instruction adds a new integer data type to the list of allowable data sizes and types in the 68000 family.

14.3 PROGRAMMING MODEL OF THE 040

Figure 14.1 summarizes the development of the 68000 family programming model from the 68000/8 to the 040. The family began with eight 32-bit data registers (D0–D7) and seven 32-bit address registers (A0–A6). Address register A7 was dedicated as a stack pointer for stacking operations. For the 68000, the 008 and the 010, there were two stack pointers. The user stack pointer (or A7) remained intact throughout the family evolution. The second or supervisor stack pointer (A7′) was used during execution of exception programs. The 020, 030, and 040 have a third stack pointer which required a name change for the supervisor stack pointer to the interrupt stack pointer. The new stack pointer (A7″) became known as the master stack pointer. The master stack pointer is accessible in the supervisor mode only.

While the size of the status register remained unchanged throughout the family evolution, the number of operational flags increased. The original status register contained two bytes—the condition code register (CCR) byte, which held flags to indicate results of instruction operations, and the supervisor status register byte. The CCR remained unchanged during the evolution of the family and the description of those flags given in Chapter 2 still applies. From the 68000 level to the 040 level, the supervisor status portion increased by two flags. The trace flag (T) was expanded to 2 bits (T_0 and T_1) to allow for different forms of tracing and the M flag was added to provide selection between the interrupt and master stack pointer in the supervisor mode.

The 010 programming model included three new registers, the vector base register (VBR) and the source and destination function code registers (SFC and DFC). The addition of the instruction cache into the 020 meant the addition of two registers, the cache control (CACR) and cache address (CAAR) registers. When memory management functions were brought on board the 030, MMU registers were added, including two root pointers, the CPU and Supervisor root pointer (CRP and SRP), a translation control register (TACR), and two transparent translation registers (TTO and TT1). Lastly, a MMU status register (MMUSR) rounded out the additions to the 030 programming model.

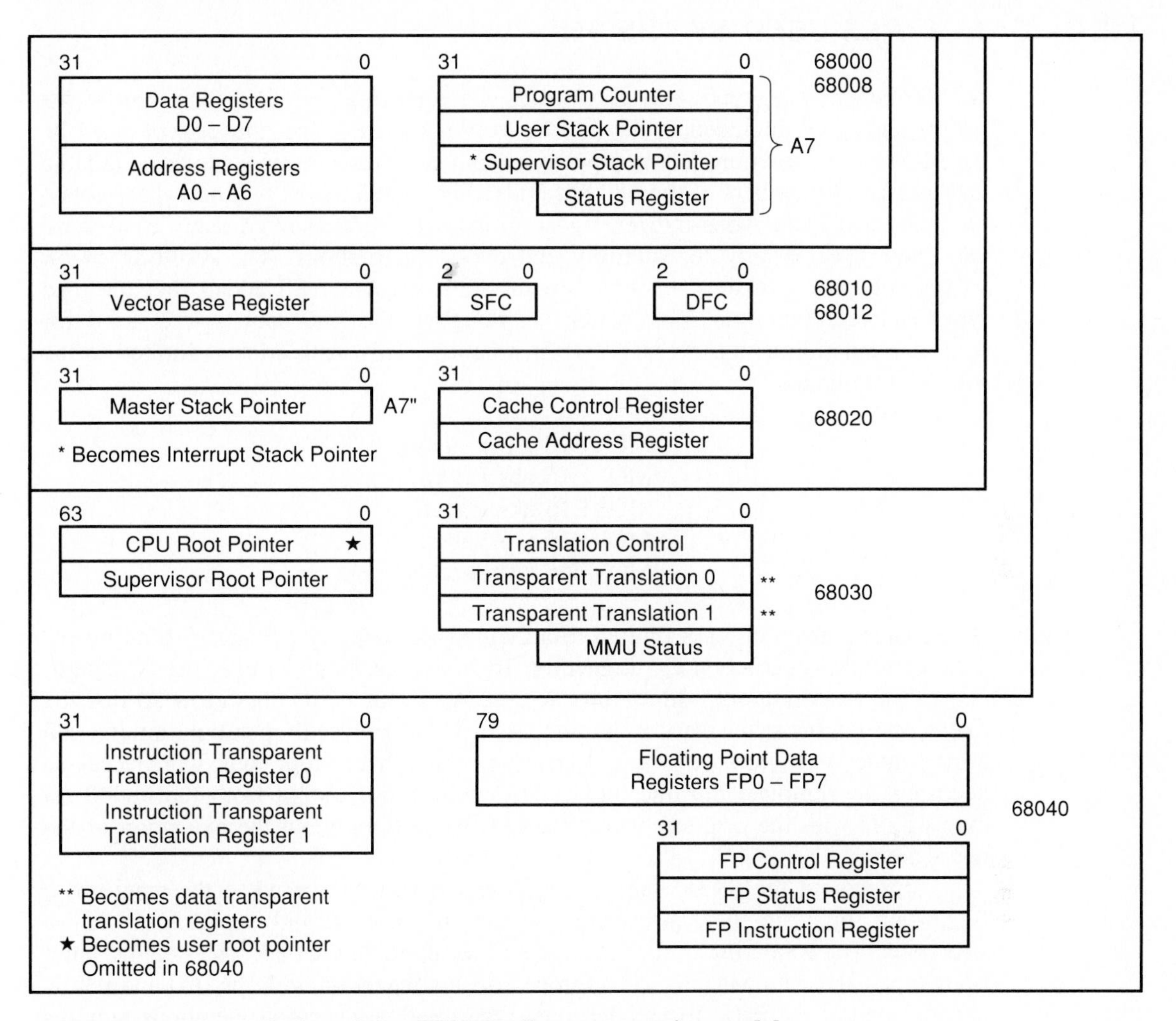

Figure 14.1 The evolution of the 68000 family programming model

The 040 includes everything so far mentioned except the CAAR, which has been omitted. The addition of a floating point unit onto the chip expands both the user portion of the programming model and the supervisor portion. In the user part, there are now eight 80-bit floating point data registers (FP0–FP7). In the supervisor section, registers associated with floating point operations have been added. These are FP control register, FP status register, and an FP instruction register.

Splitting memory management functions into two separate MMUs results in a second set of transparent translation registers. The original TT0 and TT1 are used with the data MMU and are now called data transparent translation registers (DTT0 and DTT1) and the new ones, associated with the instruction MMU, are designated as instruction transparent translation registers (ITT0 and ITT1).

14.4 DATA TYPES USED BY THE 040

All the integer data types—byte, word, and long word—previously usable by the 68000 family are included in the 040 as well as the new line integer type used by the MOVE16 instruction. Bit manipulations and BCD data forms were also available on the earliest members of the 68000 family. The 68020 added two new data forms, bit fields and quad word integer. To accommodate 32-bit source data, quad word data were used strictly for multiply and divide instructions. The resulting 64-bit values occupied two data registers. In the 030, the burst fill added the line data type, equivalent to four long words. In the 040, the line data type is used for data manipulation using the MOVE16 instruction. This instruction is limited to the following syntaxes:

```
MOVE16 (Ax)+,(Ay)+
MOVE16 abs.L,(An)
MOVE16 abs.L,(AN)+
MOVE16 (An),abs.L
MOVE16 (An)+,abs.L
```

Note that this instruction is limited to a memory to memory operation. Line moves directed by this instruction are aligned on 16-byte (line) boundaries. That is, regardless of the address used, address bits A_2 and A_3 are treated as both low so that the line access begins at the start of the line. As with the burst fill operation, an external 2-bit counter is required to track A_2 and A_3 through the four long word accesses necessary to complete the line move. The address bus output from the 68040 remains fixed with the original operand access value throughout the execution of the instruction.

Throughout the 68000 family, BCD data types are limited to the single byte BCD data for basic BCD arithmetic instructions. Packed and unpacked BCD instructions, first appearing in the 020, are still available in the 040. The floating point BCD type used with MC68881/2 coprocessor in connection with the 020 and 030 is *not* included in the 040. To use that data type for floating point operations requires software emulation.

With the incorporation of the floating point unit into the 040, a set of real data types previously associated with the MC68881/2 floating point coprocessor becomes part of the 040 data family. Only floating point instructions can make use of the real data types. Real data types used by the 040 are summarized in Figure 14.2. Each contains a sign indication in the most significant bit. A low level in this bit indicates that the mantissa of the number is positive, and the high signifies a negative mantissa. The remaining bits in the data formats are divided between exponent and mantissa quantities. The exponent used is a *biased exponent*. The actual value of the exponent is added to a bias value corresponding to the particular data type. The result is a biased exponent value ranging between 0 (the largest negative exponent value) and the maximum value held by the exponent size. This eliminates the need for a sign bit for the exponent and facilitates comparison of results between two exponent

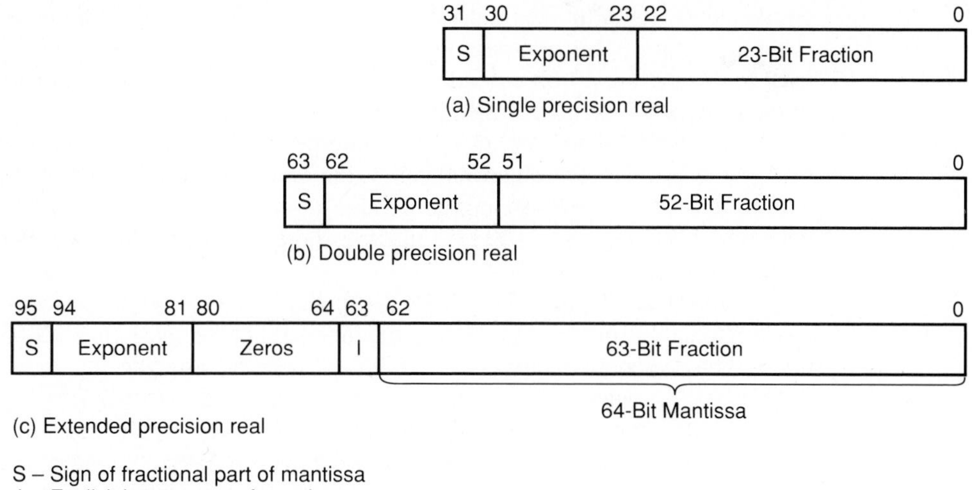

Figure 14.2 Floating point real data types

values. For instance, the bias value for the single precision real type is 127. The resulting range of biased exponent values is 1 (actual exponent value of −126) to 254 (maximum positive exponent value of 127). Exponent values of 0 and 255 have special meanings that are discussed later. The biasing of the exponent complies with American National Standards Institute/Institute of Electrical and Electronic Engineers Standard for Binary Floating Point Arithmetic (ANSI/IEEE STD 745-1985).

In the ANSI/IEEE standard, the mantissa has another aspect in addition to its normal mathematical definition. The reason for this is that the integer value for *binary* floating point real values is always 1 when using scientific notation (integer, mantissa, and exponent). That is, the mantissa portion of the number lies between 1 (1.0000..) and 2 (1.1111111....). As such, the integer portion of the mantissa is assumed to be 1 and is not included specifically when storing binary single or double precision numbers. Thus the mantissa can be more precise since all the bits represent the fractional portion of the mantissa. A binary number in this form is called a **normalized number** and the mantissa is specified as either a *fraction* or a **significand** to distinguish it from a full mantissa, which includes the integer and fractional parts. Binary numbers which include a binary integer of 0 and a non-zero fractional part are referred to as **denormalized numbers**. Since the exponent portion of this number is zero, denormalized numbers are smaller than normalized real values. A third category called **unnormalized** allows for non-zero exponent and non-zero fractional values with an integer of 0. Unnormal numbers are only used with the extended precision real data type. They provide for more exact values for intermediate calculations.

To illustrate the differences in normality of real numbers, take three possible values held in memory in extended precision form. A normalized value of $4012 0000 9C00 0000 is interpreted as follows. $4012, the biased exponent value, is equivalent to 16402 in decimal value. The bias value of 16383 for extended precision is subtracted from this number to yield an exponent value of 19. The next 16 zeros are used to expand the 80-bit extended real to a 96-bit (three long word) form for storage in memory. $9C when expanded into binary real is 1.001 11 and is equal to 1.109375 in decimal value. The number is fully interpreted as $1.109375 \times 2^{19} = 581632$.

An example of a denormalized value could be $0000 0000 1C00 0000. The exponent is 0, the smallest possible value (-16383); and the mantissa of $1C (binary 0.001 11) equals 0.109375, making the equivalent value less than the lowest normalized value.

Unnormal numbers contain an exponent value and are used to create the most precise number available. An example might be the value $0003 0000 00C0 0000. The exponent value is $3 - 16383 = -16380$ and the mantissa computes to 0.005859375, making the combined value of $0.005859375 \times 2^{-16380}$.

The differences between the three real data types are the sizes of the exponent and fractional parts. Single precision real have an 8-bit exponent size and a bias value of 127. Double precision real contain an 11-bit exponent and a bias value of 1023. Lastly, extended precision real numbers have a 15-bit exponent and a bias value of 16,383. The precision of the fractional mantissas for each type is 23 bits for single, 52 bits for double, and 64 bits for extended precision real data types. The integer portion of the mantissa is included explicitly for the extended precision type, to allow this data type to contain values less than one. A number of unused bits are held to 0; they are there to extend the 80-bit extended real to a full 96 bits for storage in three long word memory locations.

All floating point calculations are performed using the extended real data type to maximize the precision of the operation. Other data types are converted to extended real and the results of the operation are placed in one of the FP registers. When data is sent to memory, it is converted using selected rounding and precision to the data type specified in the move instruction. This function allows arithmetic operations to be performed using mixed data types. The microcode performing the operation is unaware of the forms of the original data since the data for the operation is always presented in extended precision form.

There are two special numbers which take the exponent value of 0 and of maximum. A maximum value of exponent signifies the number as *infinity*. The sign bit is still applicable allowing for representation of negative and positive infinities. When the exponent is zero, the number can take one of three forms, zero, **not a number (NAN)**, and *signaling not a number (SNAN)*. A zero number contains zeros in every bit of the exponent and mantissa. The sign bit again is valid allowing for negative and positive zeros. NANs have a maximum exponent value and a mantissa of some non-zero value. NANs result from operations that do not have an actual numerical answer, such as infinity divided by infinity. When used in additional arithmetic operations, NANs always return a NAN as a result. NANs are identified by a 1 in the most significant part of the mantissa fraction. They are also referred to as *non-signaling NANs*. SNANs, recognized by a 0 in the most significant fraction bit, are purposely created by the user as a test mechanism for conditional loops involving user-defined, non-standard IEEE data types.

14.5 68040 INSTRUCTION SET

The instruction set included in Appendix A is for the 68000 and 010 and does not include instructions added with the 020, 030, and 040. Figure 14.3 shows the evolution of the 68000 family instruction set. The 020 made the large addition of bit field and coprocessor interface instructions. The 030 did not add much, just the four MMU-related instructions discussed in Chapter 13. With the inclusion into the 040 of the floating point arithmetic unit, the instruction set greatly expanded to include the FP instructions. Note that this group of FP instructions is a subset of and does not include all of the MC68881/2 FP coprocessor instructions. Instead, these are the most commonly used FP instructions. If one wishes to use the full set of

MC68000/008

ABCD	Bcc	DIVS	MOVE	OR	SBCD
ADD	BCHG	DIVU	MOVE to CCR	ORI	Scc
ADDA	BCLR	EOR	MOVE SR	ORI to CCR	STOP
ADDI	BRA	EORI	MOVE USP	ORI to SR	SUB
ADDQ	BSET	EORI to CCR	MOVEA	PEA	SUBA
ADDX	BSR	EORI to SR	MOVEM	RESET	SUBI
AND	BTST	EXG	MOVEP	ROL	SUBQ
ANDI	CHK	EXT	MOVEQ	ROR	SUBX
ANDI to CCR	CLR	JMP	MULU	ROXL	SWAP
ANDI to SR	CMP	JSR	MULS	ROXR	TAS
ASL	CMPA	LEA	NEG	RTE	TRAP
ASR	CMPI	LINK	NEGX	RTR	TRAPV
	CMPM	LSL	NOP	RTS	TST
	DBcc	LSR	NOT		UNLK

MC68010/012

***BKPT	MOVE fm CCR	MOVEC	MOVES	RTD	ILLEGAL

MC68020

BFCHG	BFFFO	*CALLM	CHK2	**cpRESTORE	DIVUL
BFCLR	BFINS	CAS	CMP2	**cpSAVE	EXTB
BFEXTS	BFSET	CAS2	**cpBcc	**cpScc	PACK
BFEXTU	BFTST		**cpDBcc	**cpTRAPcc	*RTM
			**cpGEN	DIVSL	TRAPcc
					UNPK

MC68030

*PMOVE	**PLOAD	***PFLUSH	***PTEST

MC68040

MOVE16	CINV	CPUSH	FScc	FSADD	FDADD
FABS	FDBcc	FMUL	FSQRT	FSDIV	FDDIV
FADD	FDIV	FNEG	FSUB	FSMUL	FDMUL
FBcc	FMOVE	FRESTORE	FTST	FSSUB	FDSUB
FCMP	FMOVEM	FSAVE	FTRAPcc	FSABS	FDABS
				FSMOVE	FDMOVE
				FSNEG	FDNEG
				FSQRT	FDSQRT

* MC68030 & MC68040 do not have these instructions
** MC68040 does not have these instructions
*** These instructions' functionality differs among M68000 processors.

Note: BKPT instruction now synchronizes the pipeline (allows all instructions to complete) the runs acknowledge cycle.

Figure 14.3 Upward compatible instruction sets (Courtesy of Motorola, Inc.)

FP instructions, the remaining instructions must be emulated in software accessed through the F-Line exception.

All the sets of addressing modes used in previous processors of the 68000 family are used with the 040 with no new additions. However, with the inclusion of the FP registers, modes like data register direct (Dn) are expanded to include the FP registers (FP0–FP7).

14.6 68040 PIN FUNCTIONS

The pin-out of the 040 (Figure 14.4) is significantly different from the preceding members of the 68000 family. The 040 data access assumes that the memory port size is always 32 bits. The use of other size memory ports is not precluded. But accesses to memory are always 32 bits wide. Smaller size data ports can be used, but the user must ensure proper orientation of data to conform to the location of data on the 32-bit data bus and the actual addresses associated with them. Unused data lines not connected to smaller memory ports are ignored. Data may still be accessed from any address, but the access will be part of a long word access. For instance, data at location $0040 1231 will be fetched through the long word starting at location $0040 1230. If the original data requested is a long word, a second long word access starting at address $0040 1234 will be required to fetch the fourth byte. Internal logic in the 040 selects the desired data as it is presented along the data bus. Thus, the meaning of the SIZx pins is changed. Previously, the SIZx pins indicated the number of bytes desired to be transferred during the current cycle. Now they define the number of bytes that *are* being transferred during the current cycle, as indicated in Table 14.1. Note that the last line in the table indicates a transfer size of LINE (16 bytes). This indicates that LINE transfers are indivisible transfers. That is, once the transfer begins the full line is transferred in four separate bus cycles of one aligned long word each. As far as the 040 is concerned internally, this is an indivisible transfer of the four long words, hence both SIZx pins will remain set until the transfer is completed.

EXAMPLE 14.1

Compare the information conveyed by the SIZx pins for the 020 and 040 for a long word move to location $0040 1231.

Solution:

During the first cycle of the 020 move, the SIZx pins inform the decode logic that the 020 would like to transfer four bytes to memory starting at location $0040 1231. In contrast, the 040 sets the SIZx pins on the first bus cycle to indicate that the transfer will be a byte (location $0040 1231).

Having received DSACK indications on the second cycle that the transfer was made to a long word port, the 020 knows that three bytes were written to memory (locations $0040 1231, 1232, and 1233). As such, on the second cycle it sets the

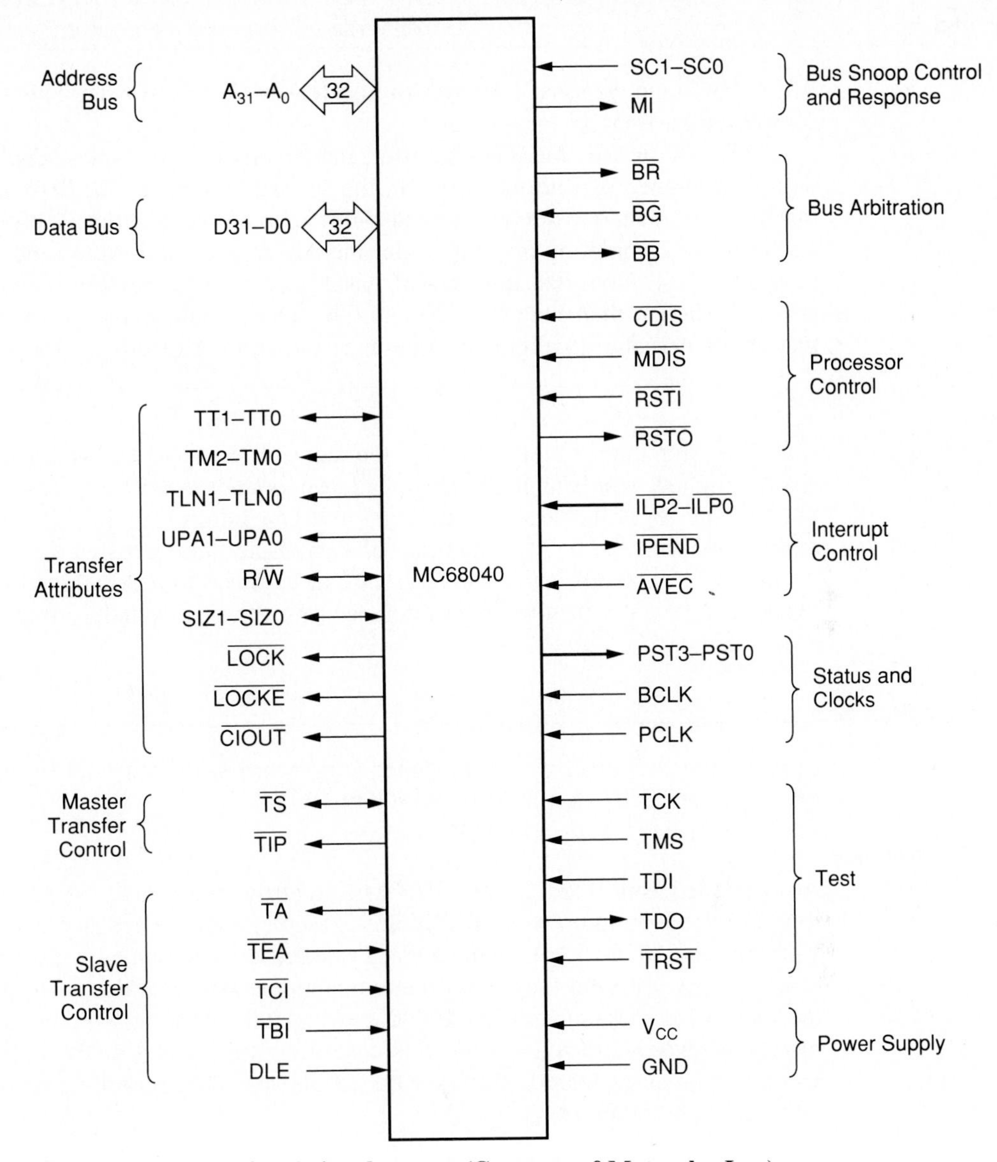

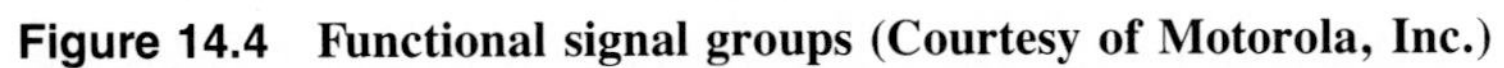

Figure 14.4 **Functional signal groups (Courtesy of Motorola, Inc.)**

Table 14.1 Transfer Size Encoding

SIZ1	SIZ0	Transfer Size
0	0	Long Word
0	1	Byte
1	0	Word
1	1	Line

SIZx pins to indicate that it desires to transfer an additional byte. This second cycle completes the transfer for the 020.

The 040 on the other hand transfers data in aligned bus cycles. The first transfer writes an aligned byte to memory. On the second bus cycle, the SIZx pins indicate that the 040 will transfer an aligned word of data (to locations $0040 1232 and 1233). After completing this bus cycle, the 040 requires a third to send the last byte to $0040 1234. Note that transfers of misaligned data to memory in an 040 system are less efficient than with the 020 and 030. One should make the effort to ensure that data is transferred aligned to minimize the number of bus cycles required.

The alignment of data follows the same rules as it did with the 020, bytes always aligned, words aligned on even address bounds and long words aligned when both A_1 and A_0 of the access address are low. One additional alignment specification that occurs with the 040 is for a line of data. Line data accesses are aligned when A_0 through A_3 are all low. Line accesses can be made to a misaligned location, but the data retrieved will always be referenced on the line boundary that contains the access address.

EXAMPLE 14.2

Contrast the data read by a misaligned long word with that read by a misaligned line. Both accesses start at location $0040 1231.

Solution:

A misaligned long word gathers data starting at the access address and includes the four sequential memory locations. In this example, data fetched is at locations $0040 1231 through $0040 1234. A misaligned line retrieves four long words in sequential order starting with the long word that contains the access address. In this example, the long word starts at location $0040 1230. After that long word is fetched, three more long words are fetched. In this case they are located at the next three long word bounds since $0040 1230 is a line boundary. Those are long words at $0040 1234 through $0040 123C.

EXAMPLE 14.3

What locations are read by a line access to location $0040 1237?

Solution:

As before, the access begins at the long word boundary which contains the access address. This is address $0040 1234. After this long word is fetched, the next long word starting at location $0040 1238 is read. This is followed by the long word at location $0040 123C. The fourth long word fetched begins at the address where A_2 and A_3 are returned to 00, or location $0040 1230. Notice that this is a "wrapped around" value. This example illustrates the point that line accesses begin with the long word that contains the access address but always include the four long words referenced to the line boundary that contained the long word which has the access address.

Throughout the line transfers, the SIZx pins retained the line code and the access address on the address bus remained unchanged. Changes in the state of A_2 and A_3 during the transfer must be implemented with external logic or else must be included in the accessed memory devices.

Two *transfer type pins* (TT0 and TT1) are used by the processor to supply coding information to address decoding circuitry that describes the type of bus cycle in progress. Table 14.2 summarizes this coding. Included in these codes are one for normal access and one for a MOVE16 (line) access to tell memory decoders and devices when to handle A_2 and A_3 decoding for the line access. The TTx pins also inform external circuitry when interrupt or breakpoint acknowledge cycles are occurring. These special read bus cycles contain information which the processor treats differently compared with a regular data fetch. Lastly, the TTx pins indicate when a bus access is being made that is diverted by the contents of one of the function code registers. Recall that the source and destination function code registers (SFC and DFC) introduced in the 010 were designed to allow a supervisor program to access any portion of memory by modifying the function code states. When TT1 = 1 and TT0 = 0 during an access, this modification of the function codes is occurring during that current memory access.

The function code pins themselves have been replaced in the 040 by *transfer mode (TM)* pins (Table 14.3, page 410) which contain the same codes as the function code pins with the following differences:

1. A code has been added to indicate a cache push access.
2. Two MMU table search access codes are included, one for the data cache and one for the instruction (code) cache.
3. CPU space code 111 has been deleted and is reserved by Motorola.

During acknowledgment cycles (TT1 and TT0 = 11), the TM pins contain the interrupt level for interrupt acknowledge and are held low for breakpoint acknowledge cycles.

Table 14.2 Transfer Type Encoding

TT1	TT0	Transfer Type
0	0	Normal Access
0	1	MOVE16 Access
1	0	SFC/DFC Modify
1	1	Acknowledge

Table 14.3 Transfer Modifier Encoding

TT0,TT1 = 11: Acknowledge access

Acknowledge Type	TM2	TM1	TM0	Address Bus
Breakpoint	0	0	0	$0000 0000
Interrupt	Interrupt Level			$FFFF FFFF

TT0,TT1 = 00 or 01 Normal and MOVE16 Access

TM2	TM1	TM0	Transfer Modifier
0	0	0	Data Cache Push Access
0	0	1	User Data Access
0	1	0	User Code Access
0	1	1	MMU Data Table Access
1	0	0	MMU Code Table Access
1	0	1	Supervisor Data Access
1	1	0	Supervisor Code Access
1	1	1	Reserved

The RMC pin of the 020 and 030 is replaced by the $\overline{LOCK}$ pin to indicate that the current cycle is part of an indivisible read-modify-write cycle. The LOCK signal is designed to lock out an alternate bus master from gaining access to the buses while the 040 is completing a read-modify-write cycle. In conjunction with the $\overline{\text{LOCK}}$ is the $\overline{LOCKE}$ or *lock end* signal which informs alternate bus masters that the 040 is in the last bus cycle of the read-modify-write operation and that the alternate bus masters can prepare for access to the buses upon completion of the current bus cycle.

The external cycle start ($\overline{\text{ECS}}$) pin, which indicated the beginning of an external bus cycle in previous 68000 family processors, is replaced by the *transfer start* ($\overline{TS}$) pin. In addition a *transfer in progress* ($\overline{TIP}$) pin indicates that a bus cycle is in progress. This pin is negated during idle bus cycles like wait states.

Since the only memory port size recognized by the 040 is long word, there no longer is a need for two DSACK pins. Instead, data transfers are acknowledged by input pin $\overline{TA}$ for *transfer acknowledge*. This pin must be asserted to indicate that a bus transfer has been acknowledged. Wait states are inserted as long as this pin remains negated following a transfer start sequence.

The functions of the $\overline{\text{BERR}}$ pin are replaced by a *transfer error acknowledge* ($\overline{TEA}$). When it is asserted the $\overline{\text{TEA}}$ initiates a bus error sequence. Asserting the $\overline{\text{TEA}}$ and $\overline{\text{TA}}$ at the same time causes the current bus cycle to be rerun; these pins perform the same function as $\overline{\text{BERR}}$ and $\overline{\text{HALT}}$ in earlier 68000 family processors. Table 14.4 lists the results of asserting $\overline{\text{TA}}$ and $\overline{\text{TEA}}$—normal, bus error, and cycle rerun.

Table 14.4 $\overline{TA}$ and $\overline{TEA}$ Truth Table

$\overline{TA}$	$\overline{TEA}$	Result
0	0	Rerun Bus Cycle
0	1	Normal Cycle Termination
1	0	Bus Error Exception
1	1	Insert Wait States

Transfer cache inhibit ($\overline{TCI}$) and *transfer burst inhibit* ($\overline{TBI}$) are used to inhibit the caching of data into the instruction and data cache and to inhibit the burst fill function. Inhibiting the caches only prevents them from being updated. Accesses still check the cache for a hit before they attempt an external bus access. To disable the instruction and data cache the *cache disable* ($\overline{CDIS}$) pin needs to be asserted. Disabling the caches disables their use but does not flush them. Assertion of the $\overline{TBI}$ pin indicates to the 040 that (a) the memory area accessed does not have bursting capabilities with which to load the cache and (b) the transfer will require four separate and individual long word transfers.

Compared with earlier 68000 processors, bus arbitration in the 040 is performed in the reverse manner in order to support multiple bus masters. Instead of a direct memory device (DMA) or other processor requesting the buses from the 040, the 040 makes the requests to gain access to the buses. *Bus request* ($\overline{BR}$) is asserted by the 040 to a bus arbiter to request use of the bus for at least one transfer cycle. The reasoning is that many of the operations of the 040 occur internally once the caches are filled and execution processing has begun. Thus the buses are idle and available a good deal of the time for other processors to use. When the 040 needs to regain use of the buses, it asserts $\overline{BR}$ again. The arbiter responds with a *bus grant* ($\overline{BG}$) if the buses are not in use by another processor and are available to the 040. A third arbitration signal, *bus busy* ($\overline{BB}$) as an input, informs the 040 that the buses are indeed busy. Once the 040 asserts $\overline{BR}$, it monitors the $\overline{BG}$ and $\overline{BB}$ signals. The 040 takes command of the buses once $\overline{BG}$ is asserted and $\overline{BB}$ is negated. The arbiter can assert $\overline{BG}$ in response to the $\overline{BR}$ of the 040 and not be concerned that the 040 will begin operations prematurely so long as the $\overline{BB}$ signal remains asserted by the device currently using the bus. Once the 040 detects an active $\overline{BG}$ and an inactive $\overline{BB}$, it asserts $\overline{BB}$ as an output and begins its required external bus cycles.

The bi-directional function of the reset line is separated into two pins, *RESET IN* ($\overline{RSTI}$) and *RESET OUT* ($\overline{RSTO}$). The 040 enters a reset state by asserting RSTI and then performing the sequence illustrated in Figure 14.5 on page 412. In the status register the S bit is set, placing the 040 in the supervisor mode; T bits and M bits are reset to prevent tracing of the initialization program and to select the interrupt stack pointer. The interrupt mask (I) bits are set to level 7, the highest interrupt level mask. The vector base register is cleared so that the reset exception vector is always accessed from the same locations in memory, \$0000 0000 through \$0000 0007. The cache control register (CACR) is cleared to inhibit the instruction and data caches.

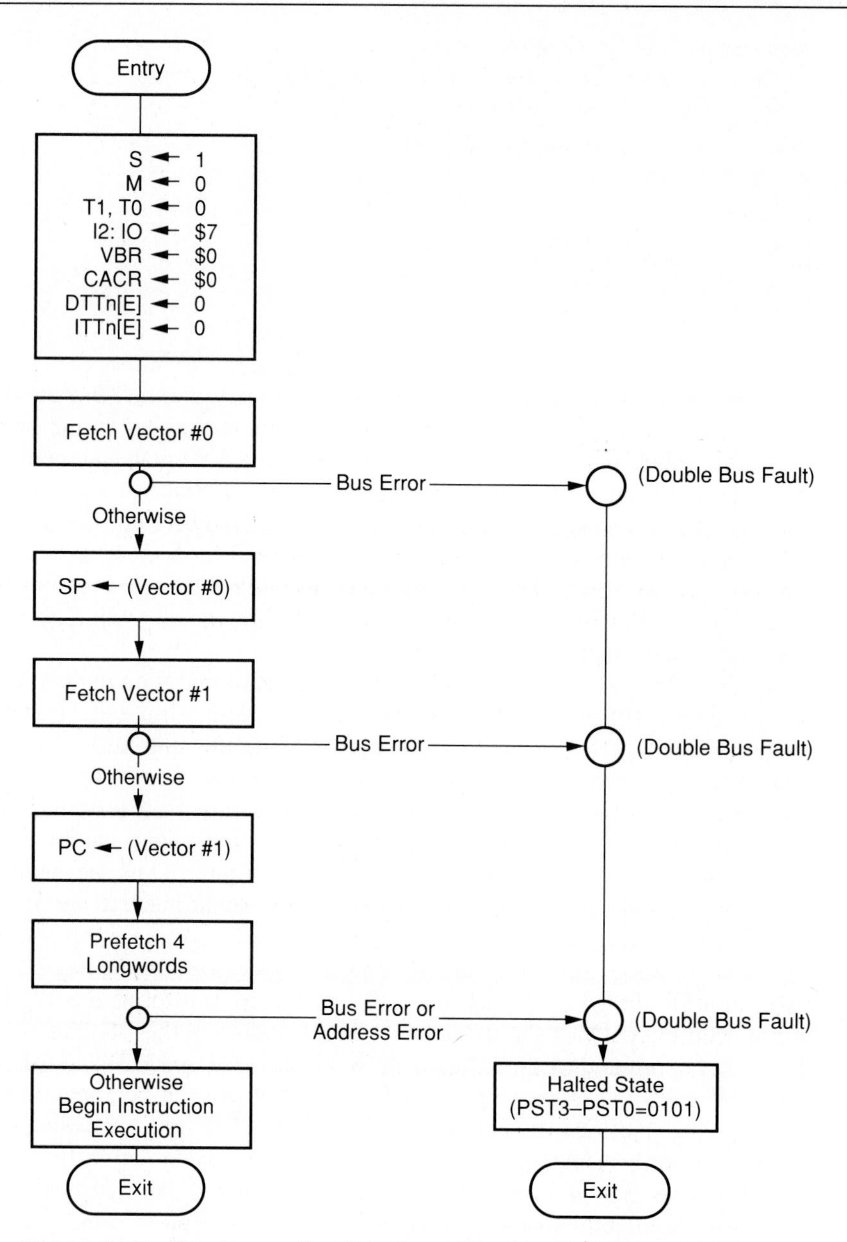

Figure 14.5 Reset operation flowchart (Courtesy of Motorola, Inc.)

Lastly, the translation control registers are cleared to allow direct translation for the reset exception table accesses. A direct long word access is made to location $0000 0000 to fill the interrupt stack pointer. A bus error occurring at this point would be a catastrophic failure and send the 040 into a double bus error condition, halting the process. Once the stack pointer data is fetched and loaded into the interrupt stack pointer, location $0000 0004 is accessed to fetch information for the program counter. This information is the starting address of the initialization routine. A bus error occurring during this fetch also would place the 040 into a double bus error fault condition. Finally, four long words are prefetched to be placed into the instruction pipeline to begin execution of the initialization program. Again, a bus error at this point would place the 040 into a double bus error condition.

A RESET instruction asserts $\overline{RSTO}$ for 512 clock cycles to reset devices external to the 040. Internal registers and operations of the 040 are not affected by the assertion of $\overline{RSTO}$.

Interrupt pending level input pins ($\overline{IPL0}$–$\overline{IPL2}$) are still used to detect an interrupt request. Additionally, in the 040 they are used during a processor reset as inputs to select output driver characteristics for three groups of signals (Table 14.5). A high on the $\overline{IPL}$ lead sets buffers to small (low current) and a low sets them for high current capability. The selected state on these pins is sampled immediately following the negation of RSTI and must be returned to normal operating levels (highs) before the completion of the internal reset time frame of 128 clock cycles.

As they did in the 020, interrupt pending and auto-vectoring ($\overline{IPEND}$ and $\overline{AVEC}$) indicate that an interrupt exceeding the interrupt mask has been received ($\overline{IPEND}$) and that internal vectors are to be used instead of expecting one on the data bus ($\overline{AVEC}$). The $\overline{TA}$ signal must accompany the $\overline{AVEC}$ signal when requesting auto-vectoring for interrupt exceptions.

The *processor status* ($\overline{PST0}$*–*$\overline{PST3}$) pins indicate the current state of the processor with respect to the action it is performing. Table 14.6 on page 414 summarizes the 16 internal states defined by these pins. They include instruction starts, change of flow due to a branch instruction, tables searches, exception stacking, a return from exception (RTE), stopped and halted (double bus error) conditions. By telling the emulator software what activity the 040 was attempting to perform, these signals can be used in diagnosing 68040 based systems.

Two clocks are used by the 040, which performs all of its actions in a synchronous manner. The clocks are *processor clock (PCLK)* and *bus clock (BCLK)*. The BCLK has exactly one half of the frequency of the PCLK. All internal timing is synchronized by the PCLK and all bus signal timing uses the BCLK. Timing references for external signals are made to the leading edge of the first BCLK cycle.

The remaining pins are used for specific functions described in later sections of the chapter. Table 14.7 on pages 415–416 briefly summarizes the pins of the 040

Table 14.5 $\overline{IPL}$ Buffer Select During Reset

Signal	Control Pin Buffering
$\overline{IPL2}$	Data Bus (D0–D31)
$\overline{IPL1}$	Address Bus (A0–A31); SIZ0; SIZ1; $\overline{LOCK}$; $\overline{LOCKE}$; R/$\overline{W}$; TLN1; TLN0; TT1; TT0; TM0–TM2; UPA0; UPA1
$\overline{IPL0}$	$\overline{BR}$; $\overline{BB}$; $\overline{IPEND}$; $\overline{MI}$; PST0-PST3; $\overline{TA}$; $\overline{TIP}$; $\overline{TS}$; RSTO

Note: High on $\overline{IPL}$ selects low current buffer. Low on $\overline{IPL}$ selects high current buffer.

Table 14.6 MC68040 Internal Status (Courtesy of Motorola, Inc.)

PST Pins				
3	2	1	0	**Internal Status Indication**
0	0	0	0	User start/continue current instruction
0	0	0	1	User end current instruction
0	0	1	0	User branch not taken and end current instruction
0	0	1	1	User branch taken and end current instruction
0	1	0	0	User table search
0	1	0	1	Halted state due to double bus fault
0	1	1	0	(Reserved)
0	1	1	1	(Reserved)
1	0	0	0	Supervisor start/continue current instruction
1	0	0	1	Supervisor end current instruction
1	0	1	0	Supervisor branch not taken and end current instruction
1	0	1	1	Supervisor branch taken and end current instruction
1	1	0	0	Supervisor table search
1	1	0	1	Stopped state due to STOP instruction
1	1	1	0	RTE instruction executed
1	1	1	1	Exception stacking

When execution unit has not completed an instruction it indicates:
- Start/continue current instruction

When execution unit completes an instruction, one of three indications is given:
- End current instruction
- Branch taken and end current instruction
- Branch not taken and end current instruction

(c.f. Figure 14.4). It includes their names, abbreviations, a brief functional description, and whether the indicated pin is an input (IN), output (OUT), or bi-directional (IN/OUT).

14.7 DATA CACHE

The two internal caches, instruction and data, are enlarged in the 040 and each has its own memory management facilities. They are identical to each other in all respects except for the addition of dirty bits in the data cache. This will be described shortly. The instruction cache is used for read accesses only since its main function is to supply opwords and operands to the instruction pipeline. The data cache, on the other hand, is involved with both read and write operations. Both caches are four way set associative with 64 sets of four line entries.

Table 14.7 040 Pin Summary (Courtesy of Motorola, Inc.)

Signal Name	Mnemonic	Function	IN/OUT
Address Bus	A_{31}–A_0	32-bit address bus used to address any of 4 Gbytes.	IN/OUT
Data bus	D_{31}–D_0	32-bit data bus used to transfer up to 32 bits of data per bus transfer.	IN/OUT
Transfer Type	TT1,TT0	Indicates the general transfer type: normal, MOVE16, alternate logical function code, and acknowledge.	IN/OUT
Transfer Modifier	TM2,TM0	Indicates supplemental information about the access.	OUT
Transfer Line Number	TLN1,TLN0	Indicates which cache line in a set is being pushed or loaded by the current line transfer.	OUT
User Programmable Attributes	UPA1,UPA0	User-defined signals, controlled by the corresponding user attribute bits from the address translation entry.	OUT
Read/Write	R/$\overline{W}$	Identifies the transfer as a read or write.	IN/OUT
Transfer Size	SIZ1/SIZ0	Indicates the data transfer size. These signals, together with A0 and A1, define the active sections of the data bus.	IN/OUT
Bus Lock	$\overline{LOCK}$	Indicates a bus transfer is part of a read-modify-write operation, and that the sequence of transfers should not be interrupted.	OUT
Bus Lock End	$\overline{LOCKE}$	Indicates the current transfer is the last in a locked sequence of transfers.	OUT
Cache Inhibit Out	$\overline{CIOUT}$	Indicates the processor will not cache the current bus transfer.	OUT
Transfer Start	$\overline{TS}$	Indicates the beginning of a bus transfer.	IN/OUT
Transfer in Progress	$\overline{TIP}$	Asserted for the duration of a bus transfer.	OUT
Transfer Acknowledge	$\overline{TA}$	Asserted to acknowledge a bus transfer.	IN/OUT
Transfer Error Acknowledge	$\overline{TEA}$	Indicates an error condition exists for a bus transfer.	IN
Transfer Cache Inhibit	$\overline{TCI}$	Indicates the current bus transfer should not be cached.	IN
Transfer Burst Inhibit	$\overline{TBI}$	Indicates the slave cannot handle a line burst access.	IN
Data Latch Enable	DLE	Alternate clock input used to latch input data when the processor is operating in DLE mode.	IN
Snoop Control	SC1,SC0	Indicates the snooping operation required during an alternate master access.	IN
Memory Inhibit	$\overline{MI}$	Inhibits memory devices from responding to an alternate master access during snooping operations.	OUT
Bus Request	$\overline{BR}$	Asserted by the processor to request bus mastership.	OUT
Bus Grant	$\overline{BG}$	Asserted by an arbiter to grant bus mastership to the processor.	IN
Bus Busy	$\overline{BB}$	Asserted by the current bus master to indicate it has assumed ownership of the bus.	IN/OUT
Cache Disable	$\overline{CDIS}$	Dynamically disables the internal caches to assist emulator support.	IN
MMU Disable	$\overline{MDIS}$	Disables the translation mechanism of the MMUs.	IN
Reset In	$\overline{RSTI}$	Processor reset	IN
Reset Out	$\overline{RSTO}$	Asserted during execution of a RESET instruction to reset external devices.	OUT

Table 14.7 *(continued)*

Signal Name	Mnemonic	Function	IN/OUT
Interrupt Priority Level	$\overline{IPL2-IPL0}$	Provides an encoded interrupt level to the processor.	IN
Interrupt Pending	$\overline{IPEND}$	Indicates an interrupt is pending.	OUT
Autovector	$\overline{AVEC}$	Used during an interrupt acknowledge transfer to request internal generation of the vector number.	IN
Processor Status	PST3–PST0	Indicates internal processor status.	OUT
Bus Clock	BCLK	Clock input used to derive all bus signal timing.	IN
Processor Clock	PCLK	Clock input used for internal logic timing. The PCLK frequency is exactly 2X the BCLK frequency.	IN
Test Clock	TCK	Clock signal for the IEEE P1149.1 Test Access Port (TAP).	IN
Test Mode Select	TMS	Selects the principle operations of the test-support circuitry.	IN
Test Data Input	TDI	Serial data input for the TAP.	IN
Test Data Output	TDO	Serial data output for the TAP.	OUT
Test Reset	$\overline{TRST}$	Provides an asynchronous reset of the TAP controller.	IN
Power Supply	V_{CC}	Power supply.	—
Ground	GND	Ground connection.	—

Figure 14.6 shows the data cache. The logical address for the access is divided into several fields. The first field contains (a) the supervisor (S) bit, which can restrict access to the cache to the supervisor mode and (b) the page frame area. Logical address bits A_{12}–A_{31} (labeled LA_{12}–LA_{31} for clarity) are compared with the tag addresses in the address translation cache (ATC) in the same way as the memory management unit (MMU) of the 030 did. The ATC translates the page frame from a logical value to a physical value (PA_{12}–PA_{31}). These physical values are combined with bits A_{10} and A_{11} of the logical address to form a data cache physical tag address.

Logical address bits A_4–A_9 comprise the index or page offset field which is used to select one of the four line entries of the data cache. The tags of the four lines are simultaneously compared with the physical tag address from the ATC. If a hit occurs, logical address bits A_2 and A_3 select which of the four long words in the matched line are to be used for the desired access.

EXAMPLE 14.4

Which long word of the data cache is accessed by the instruction MOVE.L D0,$2204?

Solution:

The logical address, $0000 2204, is presented to the address translation cache of the data cache. The upper word, $0000 2 is compared with the ATC tags. Let us

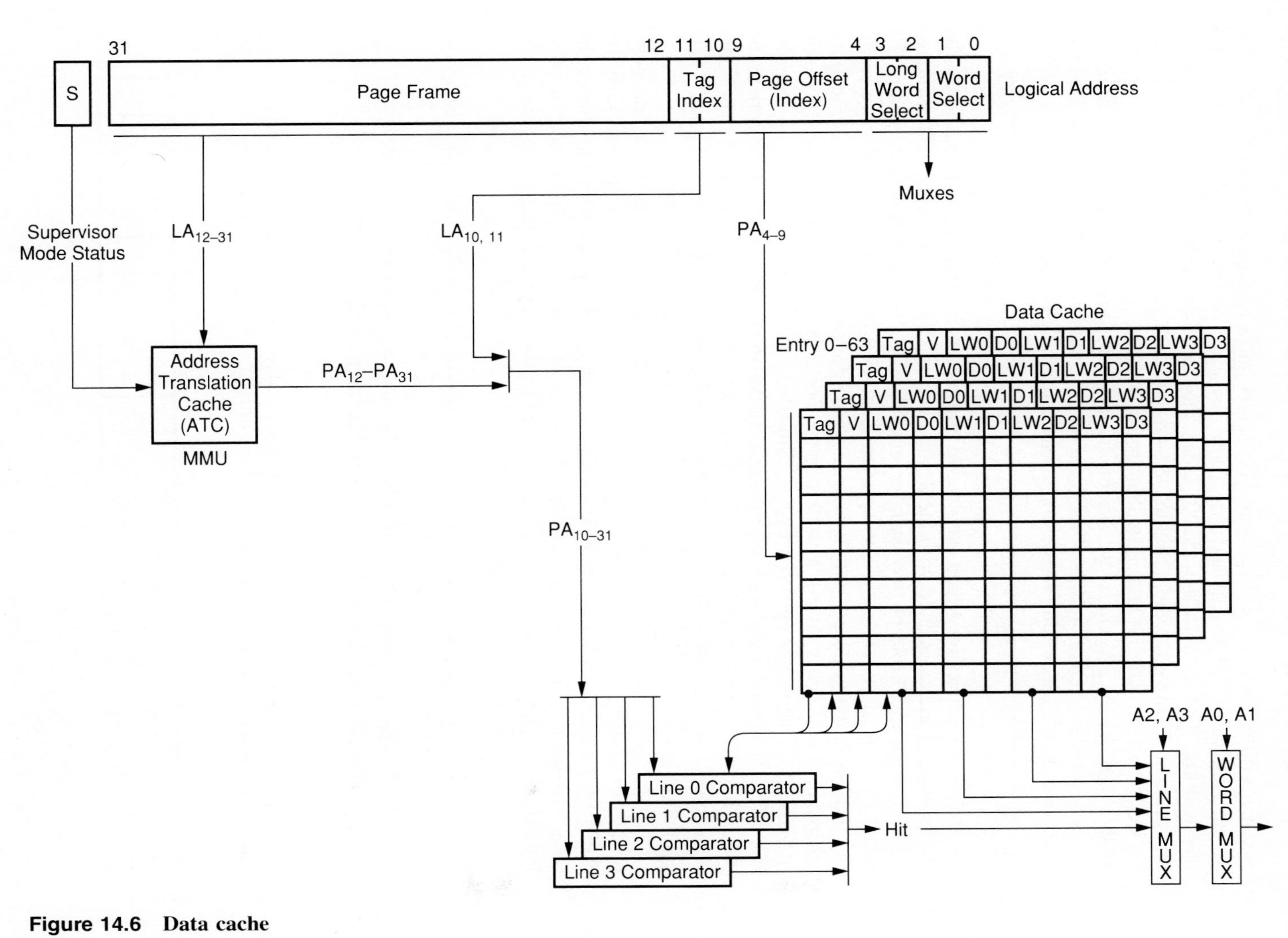

Figure 14.6 Data cache

assume a hit occurs. As a result the ATC sends out a replacement physical address to the data cache. Bits A_{10} and A_{11} (both low) are appended to form the 22-bit physical tag. The next 16 bits ($220) index into the data cache. Note that these 16 bits are contained in the three nibbles shown and the full hex representation is used for convenience. The binary equivalent is 10 0010 00 and comes from the logical address bits A_4–A_9. Again, if we assume a hit occurs, the line whose tag caused the hit is accessed. Logical address bits A_3 and A_2 (0 and 1) select long word 1 of the four long words of the line at the entry where the hit occurred.

The procedures by which data cache entries are updated are influenced by the *cache mode (CM)* associated with the desired access. Each entry in the ATC contains CM bits that select one of two write access functions for each four-line data cache entry. The functions are *write-through* (as discussed with the 030) and *copy back*. In the write-through mode, data is written to memory and its associated cache entry whenever a hit occurs during a write access. For copy back, time is saved by only updating the cache during write cycles. Since memory is not updated, no external bus cycle is performed. With copy back, however, a problem arises with cache coherency. The data in the cache entry is now more recent than the "dirty" data in memory. To manage this apparent discrepancy, data cache entries (Figure 14.7) contain an additional status bit for each long word called a **dirty bit**.

The entry begins with a 22-bit physical address tag followed by a line valid bit (V) and four long word data fields (LWn). Each data field has its own dirty bit (Dn). In the copy back mode when a write access results in a hit in the cache, the data is written into the associated long word and the dirty bit associated with that long word is set. The memory location related to the dirty cache entry is written to under one of three circumstances:

1. The line containing the dirty data indication is replaced due to a miss in the data cache;
2. A non-cacheable access matches the line containing the dirty data;
3. The line is pushed using the CPUSH instruction.

In all three cases, the data in the line is expected to be changed. Since the data in memory is not current due to the earlier write, it is first written to memory before one of the three mentioned actions takes place.

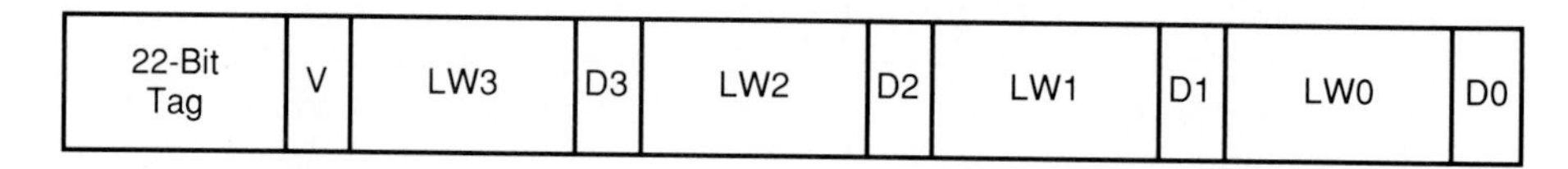

V – Line Valid Bit
LWn – Long Word Data
Dn – Dirty Bit

Figure 14.7 Data cache entry

Cache line entries are replaced in the cache by misses during read and write accesses as long as the cache is enabled and not inhibited. Most generally, a miss causes a cache line to be replaced with a burst fill operation. The selection of which line to replace begins by seeking invalid entries. If there are no invalid entries then a pseudo random replacement is used to select the line to be replaced. When the replaced line contains a set dirty bit, the line is first written to memory before it is replaced with the new line. Tags in the ATC and data cache are updated with the logical and physical values of the new line so that future accesses to it will cause a hit to occur in both caches.

EXAMPLE 14.5

How is the MOVE.L D0, $2204 instruction access completed if there is a miss in the cache using copy back mode?

Solution:

First, a new line in the cache is fetched from memory. This line will start at the line boundary containing the access address ($0000 2200). This line will replace any invalid line in the cache. If no invalid lines are available, then a valid line is randomly selected. If that valid line contains any dirty data, it is written to memory first and then replaced by the new line. The access is completed by the data from D0 being written to LW1 of the new line. The dirty bit for LW1 is set to indicate that it contains more current data than physical memory location $0000 2204.

14.8 BUS SNOOPING

The 040 is designed to exist in multiple bus master processor environments. The reversal of the bus arbitration sequence described in the pin descriptions is one result of this design. Another is called **bus snooping**. Many of the control pins involving external bus transfers are bi-directional, as indicated in Table 14.6 in Section 14.6. When the 040 is not the bus master, it monitors bus activity by snooping inputs on these lines to determine if it needs to intervene in an external bus cycle initiated by the current bus master. Intervention is required if the data cache of the 040 contains a dirty bit condition and the bus master attempts to access memory associated with that cache entry. The data in memory is not current and the 040, acting as a subordinate or slave device, must supply the correct data.

Snooping begins when the 040 detects an active transfer start ($\overline{TS}$) signal. The state of the *snoop control (SC)* input determines whether the 040 is to snoop or not. Table 14.8 on page 420 lists the types of snoop control available. Setting both control lines low or both high will inhibit snooping. For read accesses, the snooping 040 can supply the dirty data from its cache but still not update memory (SC1 = 0; SC0 = 1). In this case, the current data is sent to the master processor, but it is not written to memory. Future read accesses of the same data will require the slave 040 to supply

Table 14.8 Snoop Control Functions

SC 1	SC 0	Read Access	Write Access
0	0	Inhibit snooping.	Inhibit snooping.
0	1	Supply dirty data. Leave dirty data bit set.	Write data to data cache. Set dirty bit.*
1	0	Supply dirty data. Invalidate line.	Invalidate line.
1	1	Reserved: Inhibit snooping.	Reserved: Inhibit snooping.

* A snooped write that hits a *valid* line causes the cache line to be invalidated.

the data again. The same SC input tells the slave 040 to write the data into its cache if the cached line contains dirty data. After the write is completed, the dirty bit remains set since the data in the cache is again more recent than the data in memory. A future read of that line requires the slave 040 to supply the data.

When SC1 and SC0 are set to 1 and 0, respectively, the accessed data cache line in the slave 040 is invalidated. In the case of a read cycle, the dirty data is first read and then invalidated. Write accesses are invalidated without any other effect on the cache.

The slave 040 monitors the address bus, SIZx, and transfer type pins to decide if intervention is required. The address and SIZx determine where and how large an access is being made. The transfer type pins will restrict snooping to normal data and MOVE16 type transfers. Instruction accesses are not snooped since the instruction cache does not contain any dirty data conditions. The R/W pin is examined to see if the external access is a read or write cycle. If snooping is permitted, the 040 asserts *memory inhibit* ($\overline{MI}$) while checking the data cache for matching lines. If there is neither a matching line nor any dirty data in the line, the 040 negates $\overline{\mathrm{MI}}$ and the access is allowed to continue normally.

Once dirty data is detected at a matching address, the 040 reverts to a slave state and the access is completed using the cache location in place of the memory location the bus master was seeking. The slave 040 then monitors the $\overline{\mathrm{TA}}$, $\overline{\mathrm{TEA}}$, and $\overline{\mathrm{TBI}}$ lines to determine how to complete the access. External logic still generates the appropriate signals, $\overline{\mathrm{TA}}$ for successful access, $\overline{\mathrm{TEA}}$ for bus error, and $\overline{\mathrm{TBI}}$ if burst fill is inhibited. Additionally, simultaneous assertion of $\overline{\mathrm{TA}}$ and $\overline{\mathrm{TEA}}$ causes the current bus cycle to be rerun.

As a bus master, the 040 can request snooping by a slave 040 through interconnection of the slave's SC lines to the master's *user programmable attribute (UPA)* lines. The states of these lines are programmed on a page basis to match the requested snooping states of the SC lines. The selected functions are maintained in the ATC of the data cache so that snooping can be selected for each page that is cached.

Performance of snooping does cause some loss of time. However, it ensures cache coherency (correct data correlation between data cache, memory, and access request). The time saved using copy back to reduce external data cycles is significantly more than that lost by snooping. While the 040 has bus mastership, writing new data to the same location within a single program is a common occurrence (in a loop for instance). When another bus master accesses that same location, its only interest is in the most current data, so the location is usually only snooped once.

14.9 EXCEPTION PROCESSING

Many of the exceptions used in earlier versions of the 68000 family are used in the 040. The uses of some are modified or expanded while others are identical to their original application. Table 14.9 lists the exceptions by priority and supplies a

Table 14.9 MC68040 Exception Priority Groups (Courtesy of Motorola, Inc.)

Group/ Priority	Exception and Relative Priority	Characteristics
0	Reset	Aborts all processing (instruction or exception) and does not save old context.
1	Data Access Error (ATC Fault or Bus Error)	Aborts current instructions—can have pending trace, FP post instruction, or unimplemented FP instruction exceptions.
2	Floating-Point Pre-Instruction	Exception processing begins before current floating-point instruction is executed. Instruction is restarted on return from exception.
3	BKPT #n, CHK, CHK2, Divide by Zero, FTRAPcc, RTE, TRAP #n, TRAPV	Exception processing is part of instruction execution.
	Illegal Instruction, Unimplemented Line A and Line F, Privilege Violation	Exception processing begins before instruction is executed.
	Unimplemented Floating-Point Instruction	Exception processing begins after memory operands are fetched and before instruction is executed.
4	Floating-Point Post-Instruction	Only reported for FMOVE to memory. Exception processing begins when FMOVE instruction and previous exception processing are completed.
5	Address Error	Reported after all previous instructions and associated exceptions complete.
6	Trace	Exception processing begins when current instruction or previous exception processing is completed.
7	Instruction Access Error (ATC Fault or Bus Error)	Reported after all previous instructions and associated exceptions complete.
8	Interrupt	Exception processing begins when current instruction or previous exception processing is completed.

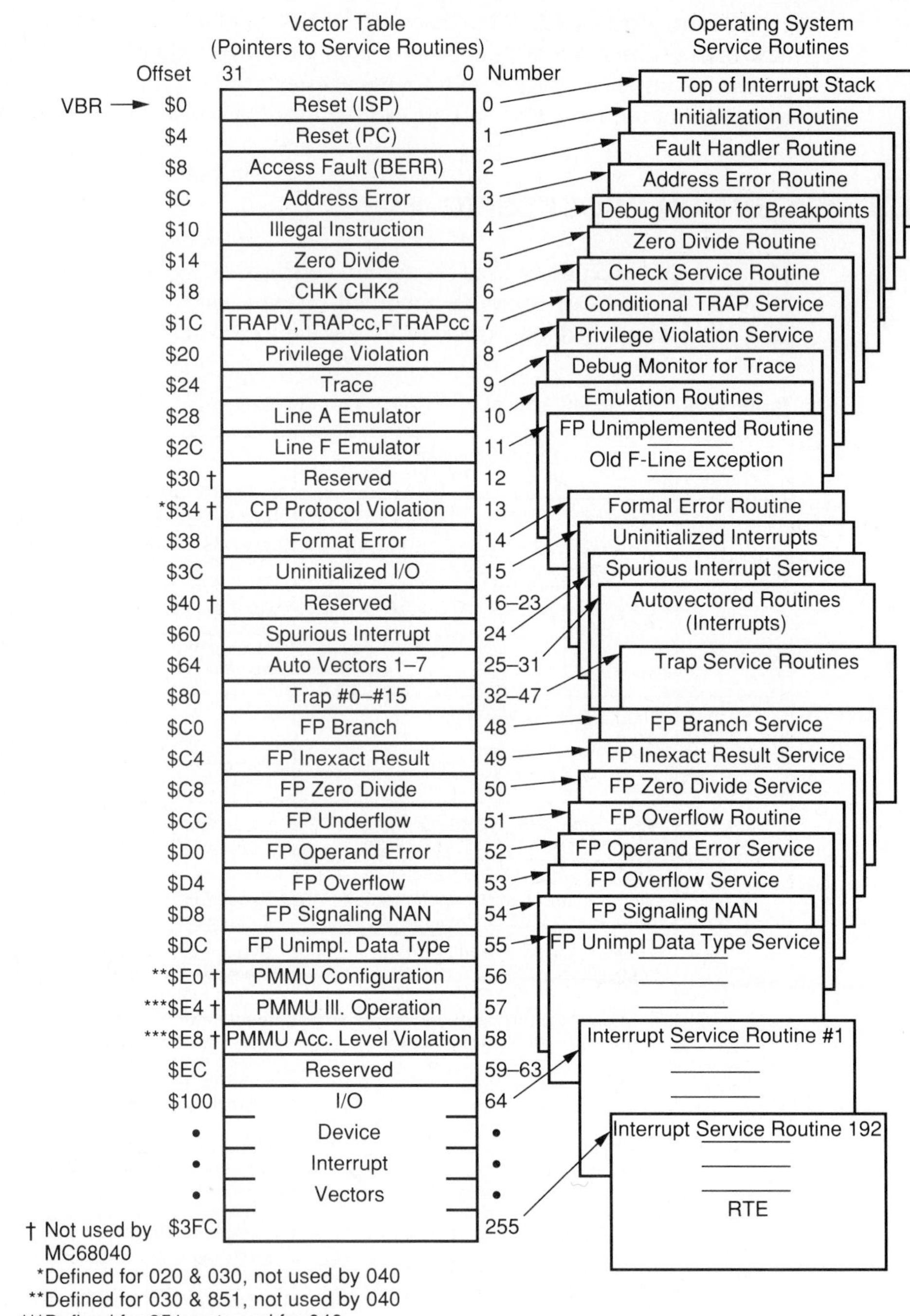

Figure 14.8 **Exception vector table (Courtesy of Motorola, Inc.)**

brief description of each one. Figure 14.8 is a diagram of the exception vector table including offset values associated with each exception number. Recall that the vector offset is created by multiplying a vector number by four. The actual table address of the exception vector is then developed by using this vector offset as the lower word of the address and the vector base register as the upper word.

The reset and bus error exceptions retain the highest priority, as in previous 68000 family processors. Part of the reset exception process is to clear the vector base register (VBR) to $0 so that the vector table address for the reset vector will always be $0000 0000 through $0000 0007. The first four locations hold the initializing contents for the interrupt stack pointer (ISP) and the next four hold the starting address of the initialization exception program which is loaded into the program counter (PC).

14.10 ACCESS ERROR EXCEPTION

Access error exception is the name applied to a group of problems including bus errors and MMU address translation faults. Several internal and external conditions produce this exception. Assertion of the transfer error acknowledge ($\overline{TEA}$) pin in response to an external bus access initiates the access error exception in the same manner as the $\overline{BERR}$ pin in earlier processors did for the bus error exception. Detection of the need for a virtual memory page swap is handled in the access error exception routine as it was in the bus error routine.

When one of the MMUs cannot complete an address translation, the access error exception is taken. Possible causes of failure to complete the translation include detection of a write protect during a write access or a supervisor only during a user program access. A translation also fails if the translation itself is not resident, that is, it does not exist. An invalid descriptor during the table search or the return of a $\overline{TEA}$ when trying to access a translation table will also generate an access error exception.

The first step that an access error exception takes is to copy the status register into an internal temporary register (TEMP). The supervisor bit in the status register is set to place the 040 into the supervisor mode, and the trace bits (T0 and T1) are reset to inhibit tracing. Exception number 2, the access fault exception number, is shifted left twice (multiplied by four) to form the vector address for the access error exception routine. Exception numbers are located in the 040 and are used in the same manner described in Section 8.3. Next, the access error stack (Figure 14.9, page 424) is placed onto the stack using the current supervisor stack pointer, either interrupt (ISP) or master (MSP) stack pointer.

The first four words for all exception stacks are standard and contain the same data supplied for 020 stacks—status register value from the TEMP register, program counter contents of either the next instruction or the current instruction depending on exception type, and the vector offset and format word. The format for an access error exception is $7 and the vector offset is $8.

The effective address words contain values that are specific to the type of access error that occurred. If the access error occurred during the execution of a

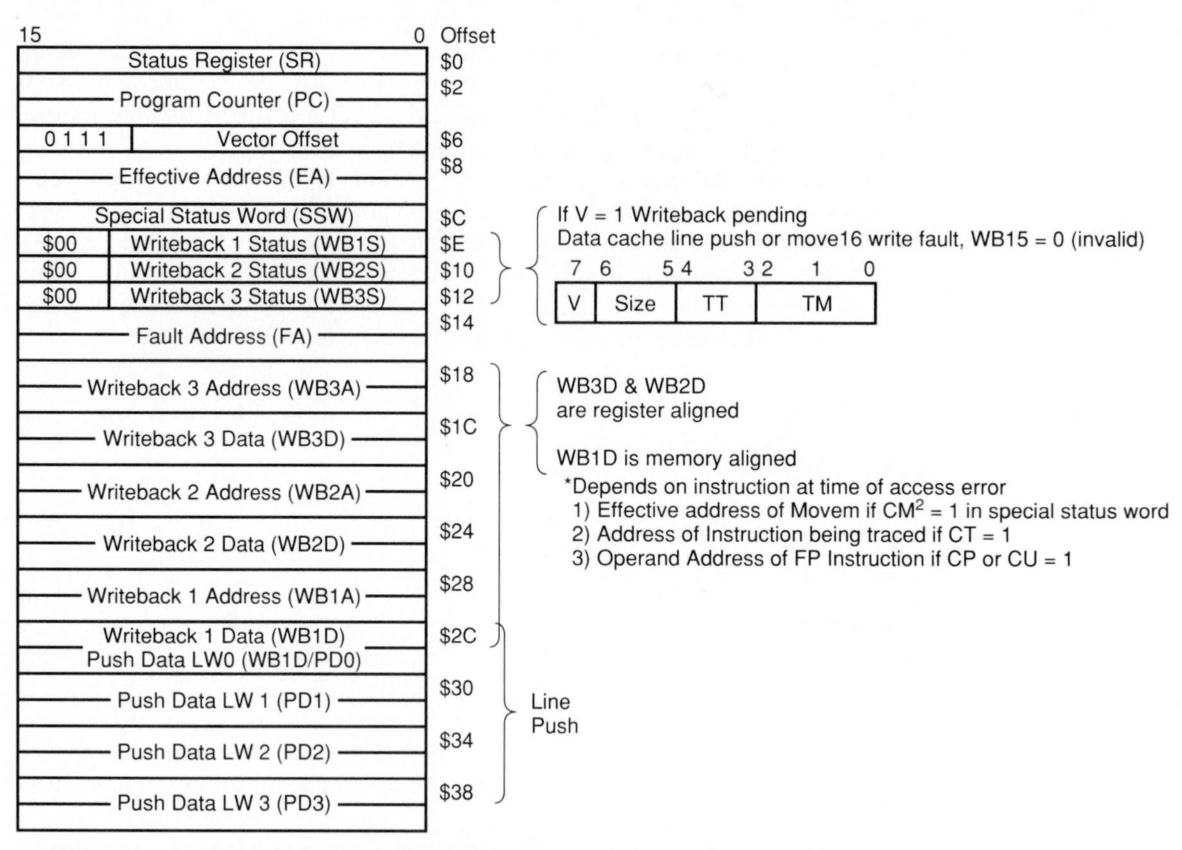

Figure 14.9 **Access error stack frame (Courtesy of Motorola, Inc.)**

MOVEM instruction, then the effective address values placed onto the stack are those of the calculated effective address for the MOVEM instruction. If the instruction is to be continued upon returning from the MOVEM access error, this calculated effective address is restored and the instruction is restarted from the point normally following the effective address calculation. This is called *continue MOVEM (CM)* and is indicated by setting a bit in a special status word found on the access error exception stack. For an access error occurring during a trace (single step) routine the effective address contains the address of the instruction being traced. Finally, for unimplementable and floating point unimplementable instructions, the effective address is the operand address of the unimplementable instruction. The status of the bits in the *special status word (SSW)* that is the next word on the stack determines what the 040 is to do with this effective address and other information.

The special status word (Figure 14.10) contains four continuation bits that al-

Exclusive Continuation Bits

15	14	13	12	11	10	9	8	7	6 5	4 3	2 1 0
CP	CU	CT	CM	MA	ATC	LK	RW	X	SIZE	TT	TM

The first five fields listed below correspond to the TMn, TTn, SIZn, R/$\overline{W}$, and $\overline{LOCK}$ signals for the faulted access.

TM – Transfer Modifier
TT – Tranfer Type
Size – Transfer Size
  Corresponds to Original Access Size. If Data Cache Line results in a Read Miss and line is Bus Error, Size Field Indicates Size of the Original Read
X – Undefined
RW – Read/Write
LK – Locked Transfer
ATC – ATC Fault
  Set for ATC Fault due to:
  – Non-resident entry (Invalid Descriptor or BERR during Tablewalk)
  – Privilege Violation (Write Protected or Supervisor-only)

  Cleared for Bus Error
  – Instruction Access
  – Data Access
  – Cache Line Push Access

MA – Misaligned Access
  Set if an ATC Fault Occurs for the Second Page for an Access Which Spans Two Pages in Memory
Continuation bits – what is pending when access error occurred
  CM – Movem Instruction
  CT – Trace
  CU – Unimplementable FP Instruction
  CP – FP Exception (Post Instruction)

Figure 14.10 MC68040 special status word

low an operation to continue once a return from the access error exception routine is executed. The continue MOVEM has already been described in the previous paragraphs. The *continue trace (CT)* directs the 040 to continue the trace operation of the instruction located at the effective address value on the stack. Continuing unimplemented (CU) and floating point unimplemented (CP) instructions support emulation of these instructions in software. In contrast, initial unimplementable instructions, which are not directly executed by the 040, cause an unimplementable exception to occur. The 040 floating point unit contains only a subset of the 688881/2 floating point coprocessor instruction set. The 040 considers as unimplementable those instructions which are not included and thus produces the F-line exception. These two exceptions are used to emulate CU and CP instructions. Should an access error occur during this emulation the special status word is used to indicate that the emulation should be resumed once the access error has been satisfied and its exception returned from. An example of the use of the access error continuation of an emulated instruction is the procedure involving a virtual memory swap during a floating point unimplementable instruction emulation.

The remaining bits in the special status word detail the status of the system during the time of the access error. The transfer type (TT) and transfer mode (TM) are coded identically to the definition of the corresponding control pins. The SIZE field defines the size of the original transfer when the access error occurred and matches the state of the SIZx pins. The RW designates whether the error access was a read or write cycle. LK specifies whether the transfer was bus locked ($\overline{\text{LOCK}}$ control pin = 0) as part of an indivisible read-write-modify cycle. ATC is set if the fault was due to an invalid descriptor or bus error during a translation table search, write protect, or supervisor only violation during a translation operation. Lastly, the *misaligned access (MA)* bit is set if an ATC fault occurs on the second page of an access which spans two pages of memory. For example, access may be made to a memory location beginning at the end of a page of memory and requiring access to the next page to complete the access. If a fault occurs on the second page access, the MA bit is set.

Following the special status word on the access error stack (Figure 14.9) are three writeback status bytes. An access error can occur while a writeback process to any one of the caches is required. Since many operations are concurrent, up to three writebacks could be in progress when the fault occurs. These three bytes reflect the status of the writeback mechanisms and include the transfer type and transfer mode (TT and TM) for each writeback and whether the writeback was pending or valid (V = 1). For data cache line pushes and MOVE16 instruction faults, writeback status V bit is always cleared, invalidating writeback 1. Writeback addresses and data are stacked for each valid writeback. Invalid writebacks have invalid data in these locations. The one exception occurs during the line push or MOVE16 fault that uses the location for writeback 1 data to hold long word 0 (LW0) related to the line of data being used. Then the last three long words hold LW1–LW3 for line push or MOVE16 faults.

Between the writeback status and address/data entries is a fault address entry, that is, the initial address for the access where the fault occurred. Access errors, which occur during the execution of access error, address error, or reset exceptions or while popping the stack during the execution of an RTE instruction, produce a double bus error condition. The processor enters a halt state, which is indicated by processor status bits (PST3–PST0) set to 0111. The double bus error halt condition can be escaped only by resetting the processor ($\overline{\text{RSTI}}$ brought low).

14.11 EXCEPTION STACKS

The remaining exception stacks are illustrated in Figure 14.11. They begin with the basic stack whose format code is $0. This four word stack contains the first four words described for the access error stack (c.f. Figure 14.9). These four words begin each exception stack and are copies of the status register, program counter, and offset/format code data. Exceptions which require no additional 040 internal information to be saved use this stack. A list of these exceptions is included next to the stack itself. In parentheses is an indication of the information held in the program

Stack Frame	Exception Type	What Stacked Program Counter Value Points to
15 ... 0 SP → Status Register + $2 Program Counter Low + $4 Program Counter High + $6 Format Code | Vector Offset a) Four word stack Format code $0 : Basic $1 : Throw away	Interrupt	Next Instruction
	Format Error	RTE or Restore Instruction
	TRAPn	Next Instruction
	Illegal Instruction	Illegal Instruction
	A-Line Instruction	A-Line Instruction
	F-Line Instruction	F-Line Instruction
	Privilege Violation	First word of instruction causing violation
	Floating Point Pre-instruction	Floating point instruction that returned exception
15 ... 0 SP → Status Register + $2 Program Counter Low + $4 Program Counter High + $6 0010 | Vector Offset + $8 *Address + $A b) Six word stack Format code $2	CHK, CHK2, TRAPcc, FTRAPcc, TRAPV, trace, divide by zero	Next Instruction *Address is the address of the instruction that caused the exception
	Address Error	Instruction that caused the exception *Address is the address being accessed (odd value)
	Unimplemented Floating Point Instruction	Next Instruction *Address – is effective address of floating point operand
c) Six word stack Format code $3	Floating Point Post-instruction	Next Instruction *Address is effective address of floating point operand

Figure 14.11 Exception stacks

counter long word on the stack. Most of these exceptions have been discussed in earlier sections of this chapter or in earlier chapters. They include interrupt, format error, TRAPn, illegal, A-line, F-line, privilege violation, and pre-instruction floating point exceptions. Of that group, we will discuss only the interrupt and the pre-instruction floating point exceptions.

The pre-instruction exception is detected by the floating point unit before it executes the current floating point instruction. Essentially, the floating point unit decides during the decoding of the floating point instruction that it cannot execute the instruction for one reason or another and instead initiates the pre-instruction exception. Upon returning from this exception, the floating point instruction is retried.

In all 68000 family processors, interrupts are begun by applying an interrupt level onto the $\overline{\text{IPLx}}$ pins. The 040 monitors the $\overline{\text{IPLx}}$ on the leading edge of BCLK. If the same interrupt level is detected on two successive BCLK cycles, as shown in Figure 14.12, the 040 accepts that level as a valid interrupt request. Next, the $\overline{\text{IPLx}}$ level is compared with the interrupt mask bits (I2–I0); if the $\overline{\text{IPLx}}$ level is higher than I2–I0. $\overline{\text{IPEND}}$ is asserted on the next BCLK positive transition. $\overline{\text{IPEND}}$ is used to inform external devices that an interrupt is pending and will be serviced at the completion of the current instruction.

From this point on the interrupt exception sequence follows that of earlier members of the 68000 family with two minor changes. An interrupt acknowledge is indicated by high states on both transfer type (TT1 and TT0) pins (Figure 14.13) instead of the function code (FC) pins. Also, the address bus (A_0–A_{31}) contains all ones and the interrupt level appears on the transfer mode (TM0–TM2) pins instead of the lower address lines, A_1–A_3. In response to the interrupt acknowledge, the interrupting device can now remove the interrupt request from the $\overline{\text{IPLx}}$ pins. That device also must either return an interrupt vector number onto the data bus or assert $\overline{\text{AVEC}}$, the auto-vector input signal. Both of these must be accompanied by an active TA. The 040 then uses the vector number from the data bus or the internal vector number associated with the interrupt level (in the case of auto-vectoring) to compute the vector address. The exception sequence continues by building the stack and fetching the starting address for the exception routine from the vector table. A summary of responses to the interrupt acknowledge is listed in Table 14.10 on page 430. Besides the usual $\overline{\text{TA}}$ and $\overline{\text{AVEC}}$ assertions, there is a third signal to which the 040 responds

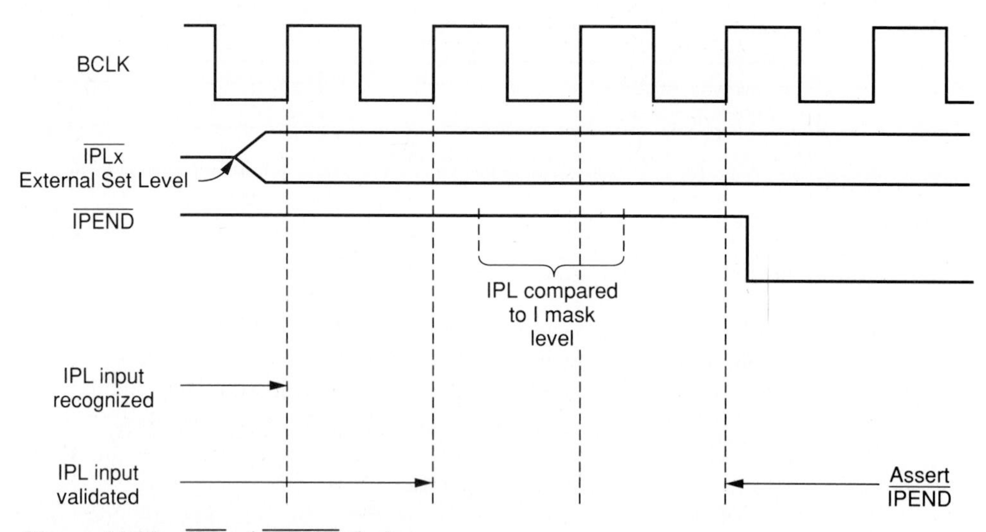

Figure 14.12 $\overline{\text{IPLx}}$ / $\overline{\text{IPEND}}$ **timing**

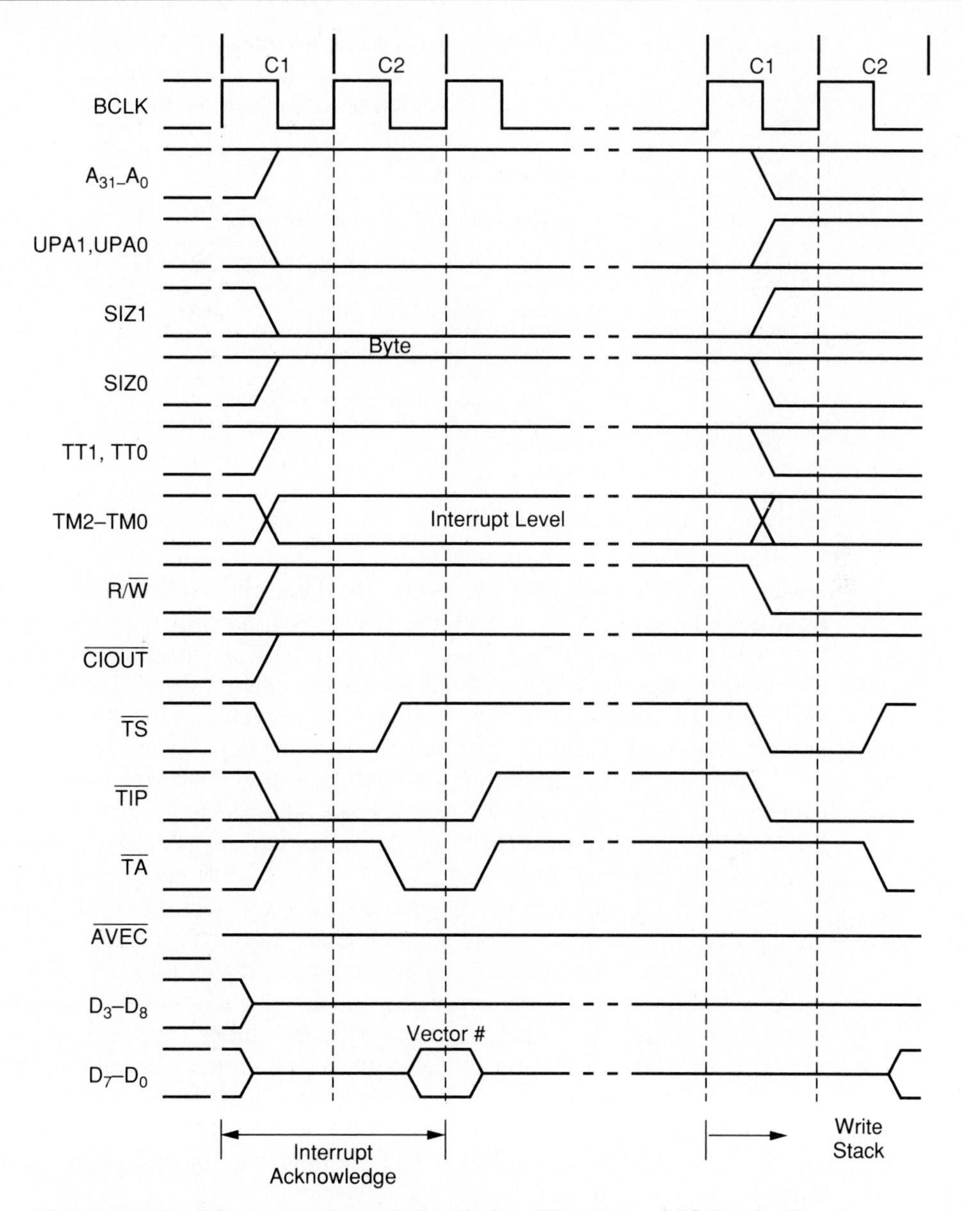

Figure 14.13 **Interrupt acknowledge timing (Courtesy of Motorola, Inc.)**

Table 14.10 Interrupt Ackowledge Termination (Courtesy of Motorola, Inc.)

$\overline{TA}$	$\overline{TEA}$	$\overline{AVEC}$	Termination Condition, or Response to Interrupt Acknowledge Cycle
1	1	X	Insert Waits
1	0	X	Take Spurious Interrupt Exception
0	1	1	Latch Vector Number on D7-D0 and Take Interrupt Exception
0	1	0	Take Autovectored Interrupt Exception
0	0	1	Retry Interrupt Acknowledge Cycle
0	0	0	Take Spurious Interrupt Exception

X = Don't Care
TT1/TT0 = 11 to indicate an acknowledge bus cycle.
A31–A0 = $FFFFFFFF
TM2, TM1, TM0 = interrupt request level

during an interrupt acknowledge cycle, the $\overline{TEA}$ pin. Assertion of $\overline{TEA}$ by itself or with both $\overline{TA}$ and $\overline{AVEC}$ causes the **spurious interrupt** exception to be initiated in place of the interrupt exception. The assertion of $\overline{TEA}$ in these cases says that the interrupt request was not actually intended. Assertion of $\overline{TA}$ and $\overline{TEA}$ together without $\overline{AVEC}$ restarts the interrupt acknowledge cycle, giving the interrupting device another chance to place a vector number onto the data bus and return $\overline{TA}$ alone.

There are two possible four word stacks that the 040 may build for an interrupt exception (c.f. Figure 14.11). Stack format $0 is created when the interrupt stack pointer (ISP) is being used (M bit in the status register is reset). Throw away stack, format $1, is made when the master stack (M = 1) has been selected. Figure 14.14 is a flowchart for the sequence using a throw away stack frame. It begins normally enough. A copy is made of the status register to the TEMP register and the S bit is set and T bits are reset in the status register. At this point, S = 1 and M = 1 indicate that the 040 is in the supervisor mode using the master stack pointer. A four word stack, format 0 is made; and the 040, recognizing that the master stack is in use, returns to the beginning of the sequence to create a second stack. This time, after the status register contents have been copied, the M bit is cleared to select the ISP and the second stack is made with a format code of $1. Now, the status register has S = 1 and M = 0, indicating that the 040 is in the supervisor mode using the ISP.

Upon execution of the RTE instruction at the end of the interrupt exception routine, the 040 examines the format code on the stack. Detecting a code of $1, the 040 increments the interrupt stack pointer by eight and restores the status register contents. The rest of the stacked information is discarded or "thrown away". The M bit in the status register is returned high, thus selecting the master stack pointer. RTE is repeated and this time a format code of $0 is detected. The master stack is popped, the appropriate data are returned to the 040, and the original program is resumed.

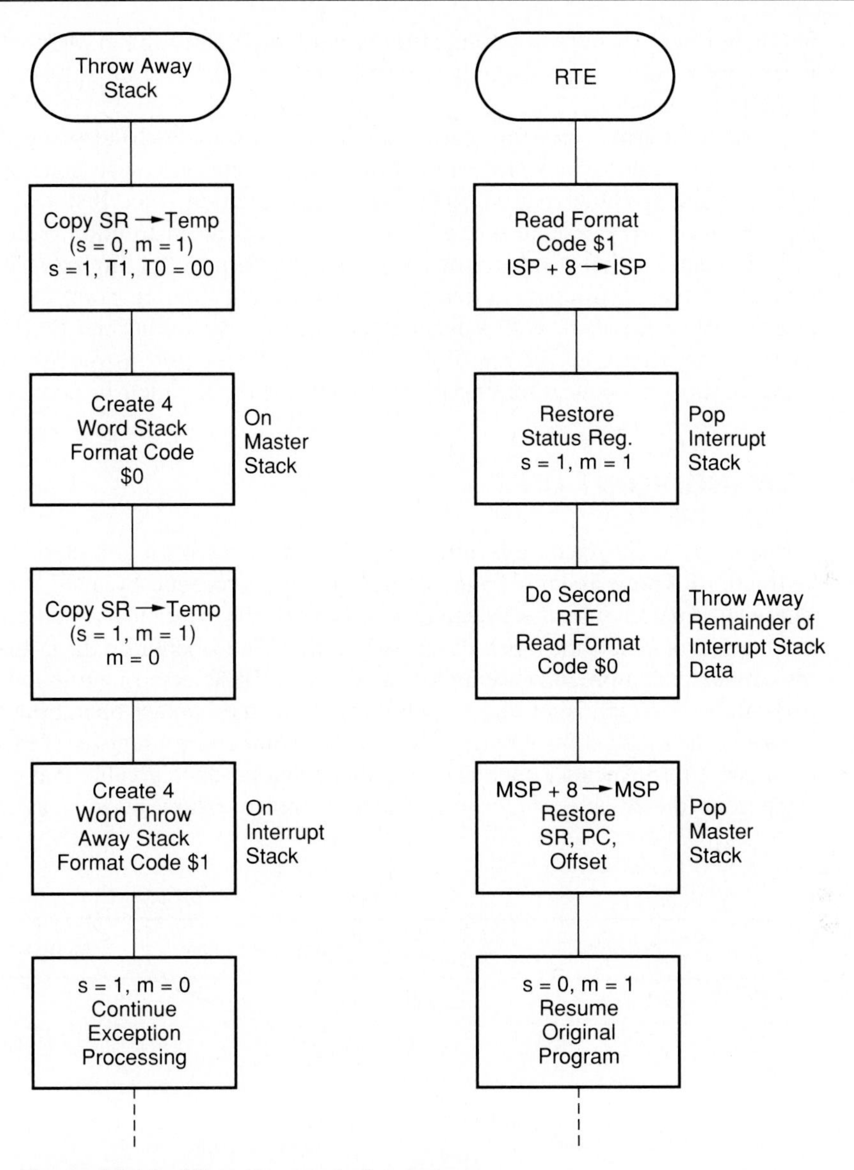

Figure 14.14 Throw away stack sequence

Instruction traps, unimplemented floating point instructions and address errors produce a six-word stack with a format code of $2. Besides the four words normally stacked, an additional long word is saved containing an ADDRESS field. The contents of this field are specified in Figure 14.11. All of the instructions shown (or slight variations of each) have been discussed in earlier chapters and will not be repeated here. The unimplemented floating point exception allows emulation of floating point coprocessor instructions that are not included in the 040 floating point instruction set. This assures compatibility with the 020 using the MC68881/2 floating point coprocessor. These unimplementable instructions must be emulated in software using this exception.

Address errors are only caused by an attempt to fetch an opword or operand from an odd address in a program listing. This exception is also generated if an odd relative address value is used with a branch instruction since that would also force the 040 to try to access an odd address location for an instruction prefetch.

Exceptions that occur following the execution of a floating point instruction as a result of that instruction initiate a post-instruction floating point exception which builds a six-word stack with a format code of \$3. The additional field on this stack contains the operand effective address of the floating point instruction which caused the exception to occur. Floating point exceptions are detailed in Section 14.12.

14.12 FLOATING POINT UNIT

The floating point execution unit of the 040 operates on a subset of floating point instructions whose assembly mnemonics are each preceded by a 'F' such as FADD. These instructions use the selected data type of the operand, which can be any of the integer or real data types described earlier. The operand data is first converted into the extended precision real format and then the selected instruction operation is performed. The precision and rounding mode of the inexact operations are selected by use of the mode control byte in the floating point control register (FPCR) as shown in Figure 14.15. Inexact operations are those that produce a value that contains more digits than the selected precision. These do not overflow as long as the exponent

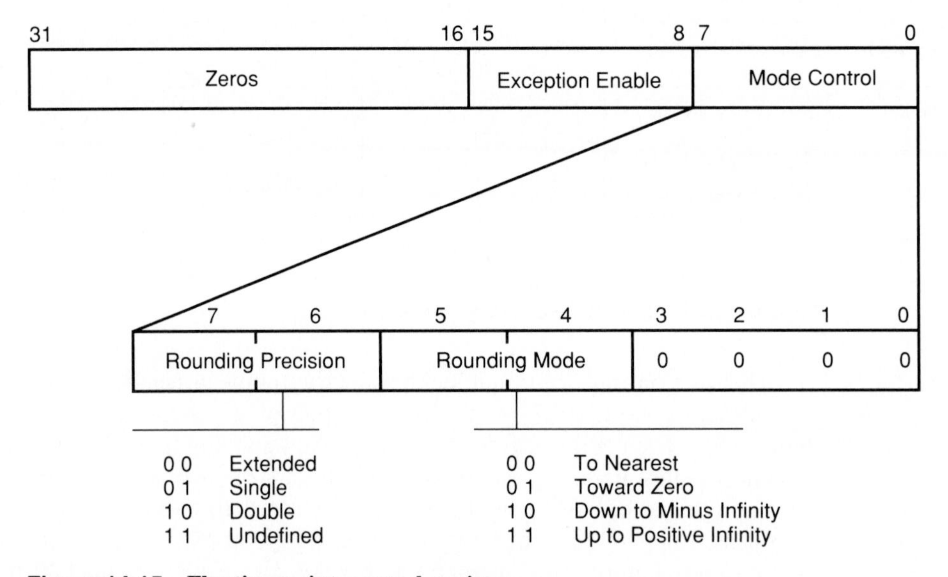

Figure 14.15 Floating point control register

value remains within the range of the data type. The number will be rounded to produce an inexact (rounded) value. As an example, let us say the results of some operation have produced the number 1.01101101×2^4 and the exponent value, 4, is within the precision range of the data type. If a precision limited to four places is selected, the number will be rounded to reduce the mantissa fraction to four bits to 1.0111. The resultant rounded number is inexact.

In the mode control byte, the rounding precision is selected from one of the three real number forms, single, double, or extended precision. Additionally, the method of rounding is programmable as one of four rounding modes. *Rounding to the nearest* rounds a number to its closest decimal value. For instance, a value of 1.63786 rounded to four places is 1.6379, while a value of 1.63782 is rounded to 1.6378. A "tie" value rounds to the nearest even number. Both 1.63775 and 1.63785 round to 1.6378. *Rounding toward zero* truncates the number to the size of the rounding precision. Rounding 1.637864567 toward zero with a four place precision is 1.6378 regardless of the numbers that follow. *Rounding down toward minus infinity* simply rounds down to the lower value regardless of the value of the next number. Both 1.63781 and 1.63789 round down to 1.6378. *Rounding up toward positive infinity* always rounds up to the next value. Both 1.63781 and 1.63789 round up to 1.6379. Notice that in all of these cases an inexact value results since the rounded value has less precision and accuracy than the original value.

Also included in the FPCR is an exception enable byte (Figure 14.16). This byte and a corresponding exception status byte in the floating point status register (FPSR), Figure 14.17 (page 434), deal with exceptions related to floating point operations. Some of these exceptions occur before the floating point instruction is executed and

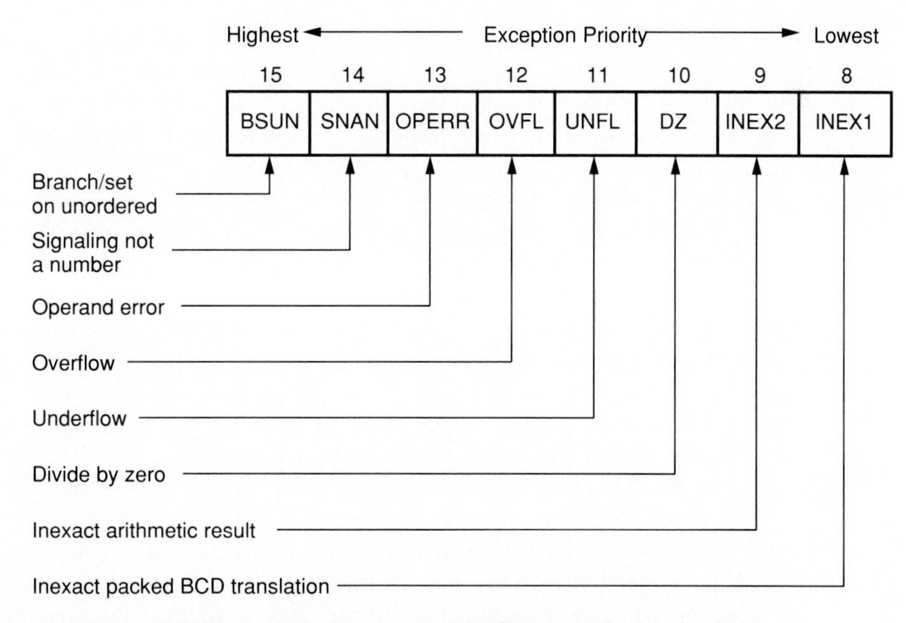

Figure 14.16 FPCR exception enable and FPSR exception status byte

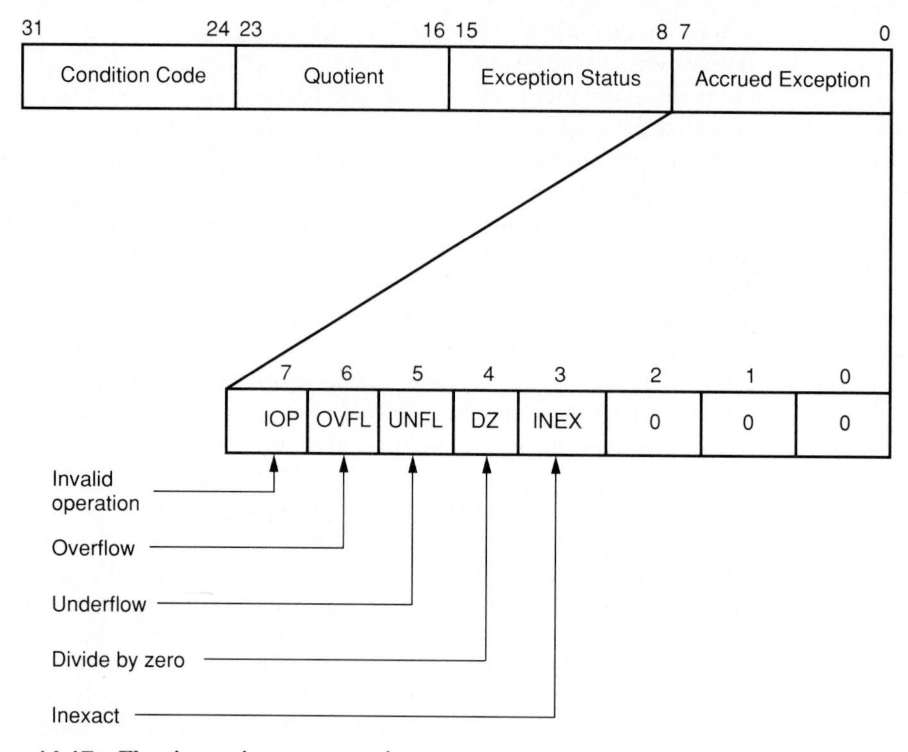

Figure 14.17 Floating point status register

are called *pre-instruction exceptions* while others result from a floating point operation (*post-instruction exception*). For an exception to be taken and processed, the floating point unit must detect the cause of the exception and set the respective status bit. If the corresponding exception enable bit is also set, then the exception is initiated. Under certain circumstances, some of the exception types are non-maskable and will initiate the exception regardless of the status of its exception enable bit.

The branch/set on unordered (BSUN) is a pre-instruction exception condition that occurs if a conditional instruction is attempted using data that is unordered (not a number, or NAN). A comparison between a NAN and any other value cannot produce a distinct result since NANs are unordered. That is, a value which is not an actual number cannot be larger or smaller than any other number of any type. When the floating point (FP) unit receives a conditional instruction, such as FBcc or FScc, it checks the status register to see if the number the condition is based on is a NAN. If it is, the BSUN bit is set and the exception taken.

NANs are created as a result of a FP arithmetic instruction, but signaling NANs (SNAN) are not generated by the 040 at all. The user creates them by setting the SNAN bit in the FPSR with a FMOVE instruction. An example of a use for SNANs is the termination of a floating point program sequence through a SNAN exception. Moving SNAN values to memory initiates a post-instruction exception while an attempt to move a SNAN to a FP data register results in a pre-instruction FP exception.

Operand errors (OPERR) result from the absence of any mathematical interpretation for a given operand. They produce an FP pre-instruction exception. Overflow (OVFL) and underflow (UNFL) exceptions are post-instruction types that reflect a result that is too large or too small for the selected precision. Divide by zero (DZ) exceptions occur when an attempt to divide by zero is made either for the FDIV instruction or as an intermediate operation in other FP instructions.

The two inexact bits (INEX2 and INEX1) indicate when an inexact result is produced. INEX2 is set when a number is rounded to the selected precision by the chosen rounding method and an inexact value results. INEX1 is not set by the 040 itself since it used, with packed BCD numbers, a data type not used with the 040; but it is used with the MC68881/2 floating point coprocessors. When the BCD data type is used under emulation and conversion to extended precision would produce an inexact result, the emulation software must set the INEX1 bit in the FPSR.

Besides the exception status byte in the FPSR (Figure 14.17), there are also condition code, quotient, and accrued exception bytes. The accrued exception byte contains a history of exceptions taken since the last time the byte was cleared. The bits in the byte are set according to the following logical combinations:

1. IOP (invalid operation) is logic OR of itself and the SNAN and OPERR status bits.
2. OVFL is logic OR of itself and the OVFL status bit.
3. UNFL is logic OR of itself and the UNFL and INEX2 status bits.
4. DZ is logic OR of itself and the DZ status bit.
5. INEX is logic OR of itself and INEX1, INEX2, and OVFL status bits.

The quotient byte contains the lower seven bits of a FDIV quotient plus the sign of the quotient. It is used when the full quotient value is not needed to make a mathematical determination.

The condition code byte of the FPSR (Figure 14.18, page 436) contains the status flags used by the conditional FP instructions for decision making. The four active bits are **negative (N), zero (Z), infinity (I),** and **not a number (NAN)**. The IEEE floating point standard utilizes four basic tests:

1. Equal to (EQ), indicated by a high state of the Z bit
2. Greater than (GT), resulting from the N, NAN, and Z bits all being low
3. Less than (LT), caused by a set N bit and NAN low and Z low
4. Unordered (UN), indicated by a not a number (NAN = 1)

Other tests that are included by Motorola in the 040 can be made, some IEEE aware and others nonaware. These tests are listed in Table 14.11 on page 436 and are all self-explanatory.

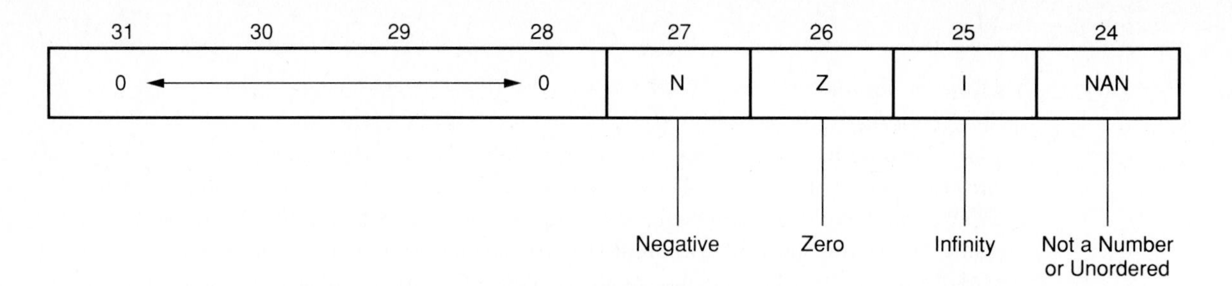

Condition Code versus Results Data Type

N	Z	I	NAN	Results Data Type
0	0	0	0	+ Normalized or Denormalized
1	0	0	0	− Normalized or Denormalized
0	1	0	0	+ 0
1	1	0	0	− 0
0	0	1	0	+ Infinity
1	0	1	0	− Infinity
0	0	0	1	+ NAN
1	0	0	1	− NAN

- Resultant data type determines setting of four condition code bits.
- FPU generates only 8 combinations shown; loading FPCC with other values may cause unexpected branch condition.

Figure 14.18 Condition code byte (FPCC) (Courtesy of Motorola, Inc.)

Table 14.11 IEEE Aware and Nonaware Tests

Test Mnemonic		Definition	Boolean True
Aware	Nonaware		
EQ	EQ	Equal	Z
NE	NE	Not Equal	$\overline{Z}$
OGT	GT	Greater Than	$\overline{NAN} \cdot \overline{Z} \cdot \overline{N}$
ULE	NGT	Less Than or Equal	$NAN + Z + N$
OGE	GE	Greater Than or Equal	$Z + \overline{NAN} \cdot \overline{N}$
ULT	NGE	Not Greater or Equal	$NAN + N \cdot \overline{Z}$
OLT	LT	Less Than	$N \cdot \overline{NAN} \cdot \overline{Z}$
UGE	NLT	Not Less Than	$NAN + Z + \overline{N}$
OLE	LE	Less Than or Equal	$Z + N \cdot \overline{NAN}$
UGT	NLE	Not Less Than or Equal	$NAN + \overline{N} \cdot \overline{Z}$
OGL	GL	Greater or Less Than	$\overline{NAN} \cdot \overline{Z}$
UEQ	NGL	Unordered or Equal	$NAN + Z$
OR	GLE	Ordered	$\overline{NAN}$
UN	NGLE	Unordered	NAN

O — ordered U — unordered (not a number) G — greater L — less
E — equal N — not

PROBLEMS

14.1 What additions have been made to the USER portion of the 68000 programming model with the 040?

14.2 What additions have been made to the SUPERVISOR portion of the 68000 family programming model with the 040?

14.3 What deletions have been made to the 68000 family programming model with the 040?

14.4 How has memory management been changed in the 040?

14.5 What new integer data type has been added with the 040? Which instruction uses this data type?

14.6 What is a biased exponent? What are the advantages of using it?

14.7 Explain the differences between single, double, and extended precision data formats.

14.8 What is the difference between normal, denormal, and unnormal numbers?

14.9 What is the decimal equivalent of the extended real precision value of $BFFE 0000 D600 0000 stored in memory?

14.10 How are zero, infinity, and NAN values recognized in real number formats?

14.11 What information do the following pins convey?
(a) SIZx (b) TTx (c) TMx (d) $\overline{\text{LOCK}}$ (e) $\overline{\text{LOCKE}}$
(f) $\overline{\text{TS}}$ (g) $\overline{\text{TIP}}$ (h) $\overline{\text{BR}}$ (i) $\overline{\text{RSTO}}$ (j) $\overline{\text{IPEND}}$

14.12 What do the following input signals signify?
(a) $\overline{\text{TA}}$ (b) $\overline{\text{TEA}}$ (c) $\overline{\text{IPLx}}$ (d) $\overline{\text{RSTI}}$ (e) $\overline{\text{AVEC}}$
(f) $\overline{\text{TCI}}$ (g) $\overline{\text{TBI}}$ (h) $\overline{\text{CDIS}}$ (i) $\overline{\text{BG}}$

14.13 What is a dirty bit? What does it signify?

14.14 Why is bus snooping used?

14.15 What conditions cause an access error exception to be initiated?

14.16 What is the purpose of the continue bits in the special status word of the access error stack?

14.17 What is the difference between stack frame $0 and $1, the basic and throw away stacks? When is each stack built?

14.18 Explain under what conditions an RTE detects a throw away stack. What is thrown away?

14.19 What are the indications of an interrupt acknowledge?

14.20 How does the response to an interrupt acknowledge differ between the 040 and the 68000?

14.21 What causes the initiation of an address error in the 040?

14.22 Which exception is begun when the 040 detects an unimplementable floating point instruction?

14.23 What is an inexact number?

14.24 Explain the differences between rounding to the nearest, toward zero, toward negative infinity, and toward positive infinity.

14.25 List the floating point exceptions and give a brief description of each.

14.26 What do INEX1 and INEX2 each indicate?

14.27 Define the fields in the floating point status register and explain what information each supplies.

14.28 What are the IEEE standard floating point test conditions? Write a Boolean expression for each one.

APPENDIX A THE 68000 INSTRUCTION SET

EFFECTIVE ADDRESS CODING SUMMARY

Codes Used as Part of the Opword Coding

Effective Address	Mnemonic	ea	Rn
Data register direct	Dn	000	Register number
Address register direct	An	001	Register number
Address register indirect	(An)	010	Register number
Postincrement	(An)+	011	Register number
Predecrement	−(An)	100	Register number
Indirect with displacement	d(An)	101	Register number
Indirect with index	d(An,Xi)	110	Register number
Absolute short	Abs.W	111	000
Absolute long	Abs.L	111	001
Program counter with displacement	d(PC)	111	010
Program counter with index	d(PC,Xi)	111	011
Immediate	Imm	111	100

68000 INSTRUCTION SET

ABCD: Add Decimal

Operation: Add the decimal data in the source to the decimal data in the destination. Add the X flag to the sum and store the results in the destination.

Data Size: Byte

Syntax: ABCD Dy,Dx
ABCD −(Ay),−(Ax)

Effective Addresses: Effective addresses are limited to data register direct and address register indirect with predecrement as illustrated by the syntax.

Condition Codes: N: Undefined
Z: Cleared if result is non-zero; unchanged, if result is zero
V: Undefined
C: Set for a decimal carry. Reset for no carry.
X: Follows the C bit

Opword Coding:

15	14	13	12	11 10 9	8	7	6	5	4	3	2 1 0
1	1	0	0	— Rx —	1	0	0	0	0	R	— Ry —

R: 0 = data register; 1 = address register

ADD: Add Binary (Hexadecimal)

Operation: Add the contents of a data register to the contents of an effective address. Store the results at the destination.

Data Size: Byte, word, long word

Syntax: ADD.size ⟨ea⟩,Dn
ADD.size Dn,⟨ea⟩

Effective Addresses:

Address Type	Source	Destination
Dn	Yes	No
An	Yes*	No
(An)	Yes	Yes
(An)+	Yes	Yes
-(An)	Yes	Yes
d(An)	Yes	Yes
d(An,Xi)	Yes	Yes
Absolute	Yes	Yes
Immediate	Yes	No
d(PC)	Yes	No
d(PC,Xi)	Yes	No

* Word and long word data only.

Condition Codes: All set or reset according to the results of the addition.

Opword Coding:

15	14	13	12	11 10 9	8 7 6	5 4 3	2 1 0
1	1	0	1	— Dn —	- Mode -	— ea —	— Rn —

Mode:

Byte	Word	Long	Destination
000	001	010	Dn
100	101	110	⟨ea⟩

ADDA: Add Binary to an Address Register

Operation: Add the contents of an effective address to an address register and store the results in the address register.

Data Size: Word, long word. Word data is sign extended and added to the full destination register.

Syntax: ADDA.size ⟨ea⟩,An

Effective Addresses: All effective addresses are usable for this instruction.

Condition Codes: No condition codes are affected by this instruction.

Opword Coding: 15 14 13 12 11 10 9 8 7 6 5 4 3 2 1 0

1 1 0 1	— An —	— Size —	— ea —	— Rn —

Size: 011 = word; 111 = long word

ADDI: Add Immediate

Operation: Add the data in the instruction operand to the destination.

Data Size: Byte, word, long word

Syntax: ADDI.size #data, ⟨ea⟩

Effective Addresses:

Address	Destination
Dn	Yes
An	No
(An)	Yes
(An)+	Yes
-(An)	Yes
d(An)	Yes
d(An,Xi)	Yes
Absolute	Yes
Immediate	No
d(PC)	No
d(PC,Xi)	No

Condition Codes: All set or reset based on the result of the addition.

Opword Coding: 15 14 13 12 11 10 9 8 7 6 5 4 3 2 1 0

0 0 0 0 0 1 1 0	Size	— ea —	— Rn —

Size: 00 = byte; 01 = word; 10 = long

Operand Format: Byte size data: b_0–b_7*
Word size data: b_0–b_{15}
Long word data: Upper word followed by lower word

* For byte size data, data bits 8–15 are don't cares but must be supplied so that word data is always fetched during program read read cycles.

ADDQ: Add Quick

Operation: Add the immediate data between 1 and 8 to the destination register.

Data Size: Byte, word, long word. When using address register direct as a destination, only word and long word data sizes can be used. The source data is sign extended from a word to long word and added to the address register.

Syntax: ADDQ.size #data,⟨ea⟩

Effective Addresses (Destination Only): All effective addresses are usable for this instruction except immediate, d(PC), and d(PC,Xi).

Condition Codes: N: Set if the results are negative; reset if positive.
Z: Set if the results are zero; reset if results are not zero.
V: Set if overflow occurs; reset if not.
C and X: Set if a carry occurs; reset if not.

Opword Coding:

15	14	13	12	11	10	9	8	7	6	5	4	3	2	1	0
0	1	0	1	– Data –			0	Size		— ea —			— Rn —		

Size: 00 = byte; 01 = word; 10 = long
Data: 001–111 = 1–7; 000 = 8

ADDX: Add Binary with Extended

Operation: Add the contents of the source and the extend flag to the contents of the destination.

Data Size: Byte, word, long word

Syntax: ADDX.size Dy,Dx
ADDX.size −(Ay),−(Ax)

Effective Addresses: This instruction uses only data register direct and address register indirect with predecrement as shown in the syntax.

Condition Codes: N: Set if the results are negative; reset if positive.
Z: Reset if results are not zero; unchanged if results are zero.
V: Set if overflow occurs; reset if not.
C and X: Set if a carry occurs; reset if not.

Opword Coding:

15	14	13	12	11	10	9	8	7	6	5	4	3	2	1	0
1	1	0	1	— Rx —			1	Size		0	0	R	— Ry —		

Size: 00 = byte; 01 = word; 10 = long
R: 0 = data register; 1 = address register

AND: Logical AND

Operation: Logical AND each bit in the source with each bit in the destination. Place results in the destination.

Data Size: Byte, word, long word

Syntax: AND.size ⟨ea⟩,Dn
AND.size Dn,⟨ea⟩

Effective Addresses:

Address	Source	Destination
Dn	Yes	No
An	No	No
(An)	Yes	Yes
(An)+	Yes	Yes
−(An)	Yes	Yes
d(An)	Yes	Yes
d(An,Xi)	Yes	Yes
Absolute	Yes	Yes
Immediate	Yes	No
d(PC)	Yes	No
d(PC,Xi)	Yes	No

Condition Codes: N: Set if results are negative; reset if positive.
Z: Set if results zero; reset if not.
V: Reset
C: Reset
X: Unaffected

Opword Coding:

15	14	13	12	11	10	9	8	7	6	5	4	3	2	1	0
1	1	0	0	— Dn —			Opmode			— ea —			— An —		

Opmode:

Byte	Word	Long	Destination
000	001	010	Dn
100	101	110	⟨ea⟩

ANDI: AND Immediate

Operation: Logically AND each bit of the destination with the data in the instruction operand.

Data Size: Byte, word, long word

Syntax: ANDI.size #data,⟨ea⟩

Effective Addresses (Destination Only): Use all except An, Immediate, d(PC), and d(PC,Xi).

Condition Codes: N: Set if results negative; reset if positive.
Z: Set if results zero; reset if not.
V: Reset
C: Reset
X: Unaffected

Opword Coding:

15	14	13	12	11	10	9	8	7	6	5	4	3	2	1	0
0	0	0	0	0	0	1	0	Size		— ea —			— Rn —		

Size: 00 = byte; 01 = word; 10 = long

Operand Words: Byte size data: b_0–b_7*
Word size data: b_0–b_{15}
Long word data: Upper word followed by lower word

* Upper bytes (b_8–b_{15}) are don't cares but must be supplied to maintain word size program accesses.

ANDI to CCR: AND Immediate to Condition Codes Register

Operation: Logically AND each bit of the CCR (lower byte of the status register) with the data in the instruction operand.

Data Size: Byte

Syntax: ANDI #data,CCR

Effective Address: Specified in the syntax

Condition Codes: Reset if the corresponding bits of data are low; unaffected if they are high.

Data bit:	4	3	2	1	0
CCR flag:	X	N	Z	V	C

Opword Coding:

15	14	13	12	11	10	9	8	7	6	5	4	3	2	1	0
0	0	0	0	0	0	1	0	0	0	1	1	1	1	0	0

Operand Word: b_8–b_{15} all low; b_0–b_7 hold the data

ANDI to SR: AND Immediate to Status Register

This is a privileged instruction.

Operation: Logically AND each bit of the status register with the data of the operand.

Data Size: Word

Syntax: ANDI #data,SR

Effective Address: Specified in the syntax

Condition Codes: Reset if the corresponding bit of data is low; unaffected if the data bit is high.

Data bit:	15	13	10	11	9	4	3	2	1	0
Status flag:	T	S	I2	I1	I0	X	N	Z	V	C

Opword Coding:

15	14	13	12	11	10	9	8	7	6	5	4	3	2	1	0
0	0	0	0	0	0	1	0	0	1	1	1	1	1	0	0

Operand: Word size data follows the opword.

ASL; ASR: Arithmetic Shift Left and Right

Operation: The bits in the destination are shifted in the direction specified and the form shown in Figures A.1 and A.2. The number of shifts is specified in one of three ways:

1. Immediate data
2. Contents of a data register
3. Unspecified defaults to a quantity of one

Data Size: Byte, word, long word *

*⟨ea⟩ memory shifts are word size only

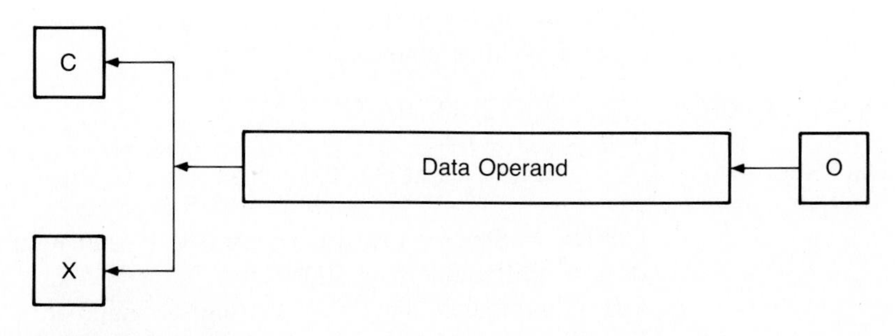

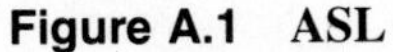
Figure A.1 ASL

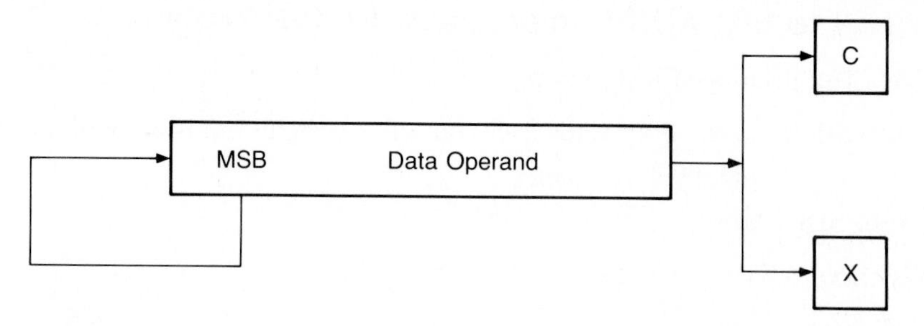

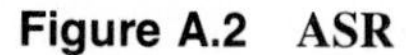
Figure A.2 **ASR**

Syntax: ASd.size #data,Dy
ASd.size Dx,Dy
ASd.size ⟨ea⟩

Effective Address (Destination Only):

Address	Destination
Dn	No
An	No
(An)	Yes
(An)+	Yes
−(An)	Yes
d(An)	Yes
d(An,Xi)	Yes
Absolute	Yes
Immediate	No
d(PC)	No
d(PC,Xi)	No

Condition Codes:
N: Set if results negative; reset if positive.
Z: Set if results are zero; reset if not.
V: Set if the MSB changes at any time during the shift process; reset if it doesn't.
C: Set according to the last bit shifted by the process; reset if shift count is zero.
X: Set according to the last bit shifted by the process; unaffected by zero shift count.

Opword Coding:

	15	14	13	12	11 10 9	8	7 6	5	4	3	2 1 0
Data Register:	1	1	1	0	· Cnt/Dx ·	d	Size	i	0	0	— Dy —

Cnt/Dx = Specifies shift count or Dx register number
d: 0 = shift right; 1 = shift left
i: 0 = immediate data; 1 = Dx register number
Size: 00 = byte; 01 = word; 10 = long word

⟨ea⟩ Shift:

15	14	13	12	11	10	9	8	7	6	5 4 3	2 1 0
1	1	1	0	0	0	0	d	1	1	— ea —	— An —

d: 0 = shift right; 1 = shift left

Bcc: Branch upon Condition

Operation: A condition, based on the state of the CCR flags, is tested to see if it is met. If it is not, program execution continues to the next instruction. If it is met, the next instruction is executed at an address that is developed by adding the operand of the branch instruction to the current contents of the program counter (relative zero).

Data Size: The relative displacement size is byte or word.

Syntax: Bcc LABEL
Bcc disp

Effective Address: LABEL or relative displacement specified in the syntax

Condition Codes: All unaffected

Opword Coding:

15 14 13 12	11 10 9 8	7 6 5 4 3 2 1 0
0 1 1 0	—— cc ——	—— 8-bit displacement ——

Operand: Word displacement is used if 8-bit displacement is 0.

Condition Code Tests:

cc	Name	Test	cc Bits
CC	Carry clear	C = 0	0100
CS	Carry set	C = 1	0101
EQ	Equal (to 0)	Z = 1	0111
GE	Greater or equal to	N ⊕ V = 0	1100
GT	Greater than	N ⊕ V + Z = 0	1110
HI	Higher than	C + Z = 0	0010
LE	Less than or equal to	N ⊕ V + Z = 1	1111
LS	Lower than or the same as	C + Z = 1	0011
LT	Less than	N ⊕ V = 1	1101
MI	Minus	N = 1	1011
NE	Not equal (to zero)	Z = 0	0110
PL	Plus	N = 0	1010
VC	No overflow	V = 0	1000
VS	Overflow	V = 1	1001

BCHG: Test a Bit and Change it

Operation: The condition of a selected bit of the destination is mirrored by the Z flag. The bit's state is then complemented. Any one of a data register's

32 bits can be selected. For a memory destination, a byte is read from the location, the bit tested and changed, and the new byte written back to memory. The number of the bit to be tested is specified in the source as the contents of a data register or as immediate data.

Data Size: Byte, long word

Syntax: BCHG Dn,⟨ea⟩
BCHG #data,⟨ea⟩

Effective Addresses (Destination Only):

Address	Destination
Dn*	Yes
An	No
(An)	Yes
(An)+	Yes
−(An)	Yes
d(An)	Yes
d(An,Xi)	Yes
Absolute	Yes
Immediate	No
d(PC)	No
d(PC,Xi)	No

*Long word only.
All others are byte size only.

Condition Codes: All are unaffected except the Z flag, which is set or reset according to the condition of the bit tested.

Opword Coding:

15	14	13	12	11	10	9	8	7	6	5	4	3	2	1	0
0	0	0	0	— Dn —			1	0	1	— ea —			— Rn —		

or

15	14	13	12	11	10	9	8	7	6	5	4	3	2	1	0
0	0	0	0	1	0	0	0	0	1	— ea —			— Rn —		

Followed by a word operand whose upper byte is zero and lower byte is the bit number.

BCLR: Test a Bit and Clear It

Operation: The specified bit is copied by the Z bit of the CCR. The bit is then set to zero. Any one of the 32 bits of a destination data register can be specified as the test bit. Memory destinations are limited to byte size data which is read, the bit tested and cleared, and the new byte written back to memory.

Data Size: Byte, long word

Effective Addresses (Destination Only):

Address	Destination
Dn*	Yes
An	No
(An)	Yes
(An)+	Yes
-(An)	Yes
d(An)	Yes
d(An,Xi)	Yes
Absolute	Yes
Immediate	No
d(PC)	No
d(PC,Xi)	No

*Long word only.
All others are byte size only.

Condition Codes: All unaffected except the Z bit, which is set or reset according to the bit being tested.

Opword Coding:

15	14	13	12	11	10	9	8	7	6	5	4	3	2	1	0
0	0	0	0	— Dn —			1	1	0	— ea —			— Rn —		

or

15	14	13	12	11	10	9	8	7	6	5	4	3	2	1	0
0	0	0	0	1	0	0	0	1	0	— ea —			— Rn —		

Followed by a word operand whose upper byte is zero and lower byte is the bit number.

BRA: Unconditional Branch

Operation: The program execution resumes at the specified relative address plus the current contents of the program counter. The relative address is either the displacement value of the opword or a word size operand following the BRA opword with a displacement of zero.

Data Size: Byte, word

Syntax: BRA LABEL
BRA displacement

Effective Address: The actual effective address is a function of instruction and is in effect d(PC).

Condition Codes: Not affected

Opword Coding:

15	14	13	12	11	10	9	8	7	6	5	4	3	2	1	0
0	1	1	0	0	0	0	0	—— Byte displacement ——							

Followed by a word displacement if the byte displacement is zero.

BSET: Test a Bit and Then Set It

Operation: A specified bit in the destination is copied by the Z flag of the CCR and the bit then is set. Any bit of a 32-bit data register can be tested. When a memory location is specified, a byte is read and the bit is copied and set. The byte is written back to memory.

Data Size: Byte, long word

Syntax: BSET Dn,⟨ea⟩
BSET #data,⟨ea⟩

Effective Addresses (Destination Only):

Address	Destination
Dn*	Yes
An	No
(An)	Yes
(An)+	Yes
−(An)	Yes
d(An)	Yes
d(An,Xi)	Yes
Absolute	Yes
Immediate	No
d(PC)	No
d(PC,Xi)	No

*Long word only.
All others are byte size only.

Condition Codes: All unaffected except the Z flag, which is set or reset according to the bit tested.

Opword Coding:

15	14	13	12	11	10	9	8	7	6	5	4	3	2	1	0
0	0	0	0	— Dn —			1	1	1	— ea —			— Rn —		

or

15	14	13	12	11	10	9	8	7	6	5	4	3	2	1	0
0	0	0	0	1	0	0	0	1	1	— ea —			— Rn —		

Followed by a word operand whose upper byte is zero and lower byte is the bit number.

BSR: Unconditional Branch to Subroutine

Operation: The long word address of the instruction following the BSR is pushed onto the stack. The relative address operand is added to the program counter so that the execution of the next instruction is performed at the new subroutine location.

Data Size: Byte, word

Syntax: BSR LABEL
BSR displacement

Effective Address: Relative addressing is used and is represented by d(PC).

Condition Codes: Unaffected

Opword Coding:

15	14	13	12	11	10	9	8	7	6	5	4	3	2	1	0
0	1	1	0	0	0	0	1	— Byte displacement —							

One word of displacement follows as an operand if the byte displacement is zero.

BTST: Test a Bit

Operation: A specified bit in the destination is copied by the Z flag of the condition codes.

Data Size: Byte, long word

Syntax: BTST Dn,⟨ea⟩
BTST #data,⟨ea⟩

Effective Address: All address modes are usable, with the exception of address register direct (An), with the following limitations:

1. Data direct (Dn) as a destination uses long word data only.
2. All others use byte data only; a byte is read from memory and the bit is tested.
3. Immediate mode is not usable if the source is immediate.

Condition Codes: All unaffected except the Z bit, which is set or reset according to the condition of the bit being tested.

Opword Coding:

15	14	13	12	11	10	9	8	7	6	5	4	3	2	1	0
0	0	0	0	— Dn —			1	0	0	— ea —			— Rn —		

or

15	14	13	12	11	10	9	8	7	6	5	4	3	2	1	0
0	0	0	0	1	0	0	0	0	0	— ea —			— Rn —		

Followed by a word operand whose upper byte is zero and lower byte is the bit number.

CHK: Check Register Against Bounds

Operation: The low word of a data register is checked against an upper limit. If the word is greater than that limit or less than zero, CHECK exception processing is initiated.

Data Size: Word

Syntax: CHK ⟨ea⟩,Dn

Effective Addresses (Source Only): All address modes can be used except address register direct (An).

Condition Codes: N: Set if the word is less than zero; reset if the word is greater than the upper limit. Undefined otherwise.
X: Unaffected
All others: Undefined

Opword Coding:

15	14	13	12	11	10	9	8	7	6	5	4	3	2	1	0
0	1	0	0	— Dn —			1	1	0	— ea —			— Rn —		

CLR: Clear an Operand

Operation: The destination operand is cleared to zero.

Data Size: Byte, word, long word

Syntax: CLR ⟨ea⟩

Effective Addresses:

Address	Operand
Dn	Yes
An	No
(An)	Yes
(An)+	Yes
−(An)	Yes
d(An)	Yes
d(An,Xi)	Yes
Absolute	Yes
Immediate	No
d(PC)	No
d(PC,Xi)	No

Condition Codes: N: Reset
Z: Set
V: Reset
C: Reset
X: Unaffected

Opword Coding:

15	14	13	12	11	10	9	8	7 6	5 4 3	2 1 0
0	1	0	0	0	0	1	0	Size	— ea —	— Rn —

Size: 00 = byte; 01 = word; 10 = long word

CMP: Compare Two Data Values

Operation: Compare the data at the effective address with the contents of a data register by subtraction. The original data are unaffected by the operation.

Data Size: Byte, word, long word

Syntax: CMP.size ⟨ea⟩,Dn

Effective Address (Source Only): All address modes are usable.

Condition Codes: N: Set if the signed data at the effective address is higher than the data in Dn; reset otherwise.
Z: Set if both numbers are the same; reset otherwise.
V: Set if an overflow occurred; reset otherwise.
C: Set if the unsigned data at the effective address is larger than the data in Dn; reset otherwise.
X: Unaffected

Opword Coding:

15	14	13	12	11 10 9	8 7 6	5 4 3	2 1 0
1	0	1	1	— Dn —	Opmode	— ea —	— Rn —

Opmode: 000 = byte; 001 = word; 010 = long

CMPA: Compare to the Contents of an Address Register

Operation: Compare the data at an effective address with the data in an address register by subtraction. The original data is unaffected by the operation.

Data Size: Word, long word

Syntax: CMPA.size ⟨ea⟩,An

Effective Address (Source Only): All effective addresses are usable.

Condition Codes: N: Set if the signed data at the effective address is higher than the data in the address register; reset otherwise.
Z: Set if both numbers are equal; reset otherwise.
V: Set if an overflow occurred; reset otherwise.
C: Set if the unsigned data at the effective address is larger than at the address register; reset otherwise.
X: Unaffected

Opword Coding:

15	14	13	12	11	10	9	8	7	6	5	4	3	2	1	0
1	0	1	1	— An —			Opmode			— ea —			— Rn —		

Opmode: 011 = sign extended word; 111 = long word

CMPI: Compare Immediate

Operation: Compare the contents of an effective address with the operand data by subtraction. The data at the effective address is unaffected by the operation.

Data Size: Byte, word, long word

Syntax: CMPI.size #data,⟨ea⟩

Effective Address (Destination Only):

Address	Destination
Dn	Yes
An	No
(An)	Yes
(An)+	Yes
−(An)	Yes
d(An)	Yes
d(An,Xi)	Yes
Absolute	Yes
Immediate	No
d(PC)	No
d(PC,Xi)	No

Condition Codes: N: Set if the signed immediate data is higher than the reference data; reset otherwise.
Z: Set if the data is the same; reset otherwise.
V: Set if an overflow occurred; reset otherwise.
C: Set if the immediate unsigned data is larger; reset otherwise.

Opword Coding:

15	14	13	12	11	10	9	8	7	6	5	4	3	2	1	0
0	0	0	0	1	1	0	0	Size		— ea —			— Rn —		

Size: 00 = byte; 01 = word; 10 = long word
Followed by the immediate data operand(s).

CMPM: Compare Memory Contents

Operation: Compare the data at the location pointed to by Ay with the data pointed to by Ax by subtraction. The original data is not affected by the operation.

Data Size: Byte, word, long word

Syntax: CMPM.size (Ay)+,(Ax)+

Effective Address: Specified in the syntax as address register indirect with postincrement.

Condition Codes: N: Set if the signed data at (Ay) is higher than (Ax); reset otherwise.
Z: Set if the data is the same; reset otherwise.
V: Set if an overflow occurred; reset otherwise.
C: Set if the unsigned data at (Ay) is larger than (Ax); reset otherwise.
X: Unaffected

Opword Coding:

15	14	13	12	11	10	9	8	7	6	5	4	3	2	1	0
1	0	1	1	— Ax —			1	Size		0	0	1	— Ay —		

Size: 00 = byte; 01 = word; 10 = long word

DBcc: Conditional Test Flags, Decrement Counter and Branch

Operation: The specified condition code test is made. If it is true, the branch is not taken and the next instruction in the program list is executed. If the test is false, the lower word in the specified data register is decremented. If the results are not equal to −1 ($FFFF), then the branch is performed. The next instruction to be executed is at the location specified by the relative address operand plus the current contents of the program counter. If the data register reaches −1, the branch is not taken and the next listed instruction is executed.

Data Size: Word

Syntax: DBcc Dn, LABEL
DBcc Dn, Displacement

Effective Address: The relative address represented by d(PC) is the effective address of this instruction.

Condition Codes: Unaffected

Opword Coding:

15	14	13	12	11	10	9	8	7	6	5	4	3	2	1	0
0	1	0	1	—— cc ——				1	1	0	0	1	— Dn —		

Followed by a word of displacement.

Condition Tests (cc):

cc	Name	Test	Bit Pattern
CC	No carry	C = 0	0100
CS	Carry	C = 1	0101
EQ	Equal to zero	Z = 1	0111
F	Always false	None	0001
GE	Greater than or equal to	$N \oplus V = 0$	1100
GT	Greater than	$N \oplus V + Z = 0$	1110
HI	Higher than	C + Z = 0	0010
LE	Less than or equal to	$N \oplus V + Z = 1$	1111
LS	Lower than or the same as	C + Z = 1	0011
LT	Less than	$N \oplus V = 1$	1101
MI	Minus	N = 1	1011
NE	Not equal to zero	Z = 0	0110
PL	Plus	N = 0	1010
T	Always true	None	0000
VC	No overflow	V = 0	1000
VS	Overflow	V = 1	1001

Optional Form (check assembler): DBRA = DBF

DIVS: Division of Signed Data

Operation: Divide the long word destination by the lower word of the source. Data is signed and the results are found in the destination as follows:

1. The quotient is the lower word.
2. The remainder is the upper word. The sign of the remainder is the same as the dividend unless the remainder is zero (always positive).
3. Division by zero causes an exception process to be initiated.
4. In case of an overflow (quotient larger than a word), the contents of the registers remain unchanged and the V flag of the condition codes is set.

Data Size: Specified in the operation

Syntax: DIVS ⟨ea⟩,Dn

Effective Address: All address modes except address register direct (An) are usable.

Condition Codes: N: Set if quotient is negative; reset otherwise. Undefined if overflow occurs.
Z: Set if quotient zero; reset otherwise. Undefined if overflow occurs.
V: Set if quotient larger than a word is detected; reset otherwise.
C: Reset
X: Unaffected

Opword Coding:

15	14	13	12	11 10 9	8	7	6	5 4 3	2 1 0
1	0	0	0	— Dn —	1	1	1	— ea —	— Rn —

DIVU: Divide Unsigned

Operation: Divide the long word unsigned data in the destination register by the unsigned word data at the source effective address. The results of the division are stored in the destination register as follows:

1. The upper word holds the remainder of the division.
2. The lower word contains the quotient. In the case of an overflow condition, the lower word of the quotient is stored in the lower word of the destination register, and the overflow flag is set.
3. Division by zero causes an exception routine to be initiated.
4. In case of an overflow (quotient larger than a word), the contents of the registers remain unchanged and the V flag of the condition codes is set.

Data Size: Word source and long word destination

Syntax: DIVU ⟨ea⟩,Dn

Effective Address: All address modes except address register direct (An) are usable with this instruction.

Condition Codes: N: Copies the MSB of the quotient. Undefined if an overflow occurs.
Z: Set if the quotient is zero; reset otherwise.
V: Set if the quotient is larger than word size; reset otherwise.
C: Reset
X: Unaffected

Opword Coding:

15	14	13	12	11 10 9	8	7	6	5 4 3	2 1 0
1	0	0	0	— Dn —	0	1	1	— ea —	— Rn —

EOR: Logical Exclusive OR

Operation: Exclusive OR the data in the destination with the data from the source register.

Data Size: Byte, word, long word

Syntax: EOR.size Dn,⟨ea⟩

Effective Address (Destination Only):

Address	Destination
Dn	Yes
An	No
(An)	Yes
(An)+	Yes
-(An)	Yes
d(An)	Yes
d(An,Xi)	Yes
Absolute	Yes
Immediate	No
d(PC)	No
d(PC,Xi)	No

Condition Codes: N: Set if the results are negative; reset otherwise.
Z: Set if the results are zero; reset if not.
V: Reset
C: Reset
X: Unaffected

Opword Coding:

15	14	13	12	11 10 9	8 7 6	5 4 3	2 1 0
1	0	1	1	— Dn —	Opmode	— ea —	— Rn —

Opmode: 100 = byte; 101 = word; 110 = long word

EORI: Exclusive OR with Immediate Data

Operation: Exclusive OR the data at the effective address with the data in the operand of the instruction.

Data Size: Byte, word, long word

Syntax: EORI.size #data,⟨ea⟩

Effective Address (Destination Only):

Address	Destination
Dn	Yes
An	No
(An)	Yes
(An)+	Yes
-(An)	Yes
d(An)	Yes
d(An,Xi)	Yes
Absolute	Yes
Immediate	No
d(PC)	No
d(PC,Xi)	No

Condition Codes: N: Set if the results are negative; reset otherwise.
Z: Set if the results are zero; reset if not.

V: Reset
C: Reset
X: Unaffected

Opword Coding:

15	14	13	12	11	10	9	8	7	6	5	4	3	2	1	0
0	0	0	0	1	0	1	0	Size		— ea —			— Rn —		

Followed by data operand(s).
Size: 00 = byte; 01 = word; 10 = long word

EORI to CCR: Exclusive OR CCR with Immediate Data

Operation: Exclusive OR the data in the condition code register (lower byte of the status register) with the data in instruction operand.

Data Size: Byte

Syntax: EORI #data,CCR

Effective Address: Specified by the syntax

Condition Codes: Set or reset by the exclusive OR operation

Opword Coding:

15	14	13	12	11	10	9	8	7	6	5	4	3	2	1	0
0	0	0	0	1	0	1	0	0	0	1	1	1	1	0	0

Followed by a word operand whose upper byte is zero and lower byte is the data.

EORI to SR: Exclusive OR Immediate Data with the SR

This is a privileged instruction.

Operation: Exclusive OR the contents of the status register with the data in the instruction operand.

Data Size: Word

Syntax: EORI #data,SR

Effective Address: Specified in the syntax

Condition Codes: Set or reset as a result of the exclusive OR operation.

Opword Coding:

15	14	13	12	11	10	9	8	7	6	5	4	3	2	1	0
0	0	0	0	1	0	1	0	0	1	1	1	1	1	0	0

Followed by a word operand.

EXG: Exchange Registers

Operation: Exchange the long word contents of two registers.

Data Size: Long word

Syntax: EXG Rx,Ry

Effective Address: Data register and address register direct only

Condition Codes: Unaffected

Opword Coding:

15	14	13	12	11	10	9	8	7	6	5	4	3	2	1	0
1	1	0	0	— Rx —			1	— Opmode —					— Ry —		

Opmode: 01000 = Rx and Ry data registers
01001 = Rx and Ry address registers
10001 = Rx data and Ry address registers

EXT: Sign Extend

Operation: Sign extend a byte to word (word size) or word to long word (long word size) in the specified data register.

Data Size: Word, long word

Syntax: EXT.size Dn

Effective Address: Data register direct only

Condition Codes: N: Set if the results are negative; reset otherwise.
Z: Set if the results are zero; reset otherwise.
V: Reset
C: Reset
X: Unaffected

Opword Coding:

15	14	13	12	11	10	9	8	7	6	5	4	3	2	1	0
0	1	0	0	1	0	0	Opmode			0	0	0	— Dn —		

Opmode: 010 = word; 011 = long word

ILLEGAL: Illegal Instruction

Operation: This instruction causes an illegal exception to be initiated.

Data Size: Not applicable

Syntax: None

Effective Address: Implied

Condition Codes: Unaffected

Opword Coding:

15	14	13	12	11	10	9	8	7	6	5	4	3	2	1	0
0	1	0	0	1	0	1	0	1	1	1	1	1	1	0	0

JMP: Jump to New Program Location

Operation: Program execution continues at the location specified by the effective address.

Data Size: Not applicable

Syntax: JMP ⟨ea⟩

Effective Address:

Address	Source
Dn	No
An	No
(An)	Yes
(An)+	No
−(An)	No
d(An)	Yes
d(An,Xi)	Yes
Absolute	Yes
Immediate	No
d(PC)	Yes
d(PC,Xi)	Yes

Condition Codes: Unaffected

Opword Coding:

15	14	13	12	11	10	9	8	7	6	5	4	3	2	1	0
0	1	0	0	1	1	1	0	1	1	— ea —			— Rn —		

JSR: Jump to Subroutine

Operation: The program counter contents are saved on the stack; then, the program execution is resumed at the address specified in the effective address.

Data Size: Not applicable

Syntax: JSR ⟨ea⟩

Effective Address:

Address	Source
Dn	No
An	No
(An)	Yes
(An)+	No
−(An)	No
d(An)	Yes
d(An,Xi)	Yes
Absolute	Yes
Immediate	No
d(PC)	Yes
d(PC,Xi)	Yes

Condition Codes: Unaffected

Opword Coding:

15	14	13	12	11	10	9	8	7	6	5	4	3	2	1	0
0	1	0	0	1	1	1	0	1	0	— ea —			— Rn —		

LEA: Load Effective Address

Operation: The long word data specified by the source effective address is loaded into the specified address register.

Data Size: Long word

Syntax: LEA ⟨ea⟩,An

Effective Address (Source Only):

Address	Source
Dn	No
An	No
(An)	Yes
(An)+	No
−(An)	No
d(An)	Yes
d(An,Xi)	Yes
Absolute	Yes
Immediate	No
d(PC)	Yes
d(PC,Xi)	Yes

Condition Codes: Unaffected

Opword Coding:

15	14	13	12	11	10	9	8	7	6	5	4	3	2	1	0
0	1	0	0	— An —			1	1	1	— ea —			— Rn —		

LINK: Link and Allocate

Operation: The long word contents of the specified address register are saved on the stack. The address register is loaded from the stack using the updated stack pointer as an effective address. (This is *not* a pull operation.) The displacement specified in the operand of the instruction is sign extended and added to the stack pointer.

Data Size: Not applicable

Syntax: LINK An,#displacement

Effective Address: Specified by the syntax

Condition Codes: Unaffected

Opword Coding:

15	14	13	12	11	10	9	8	7	6	5	4	3	2	1	0
0	1	0	0	1	1	1	0	0	1	0	1	0	— An —		

Followed by a word operand specifying the displacement.

LSL and LSR: Logic Shift Left and Logic Shift Right

Operation: The bits specified in the effective address or destination register (Dy) are shifted in the direction indicated and illustrated in Figures A.3 and A.4. The source register or immediate operand specifies the number of times the data is to be shifted. Immediate data is restricted to the quantities 1 through 8.

Operation Diagrams:

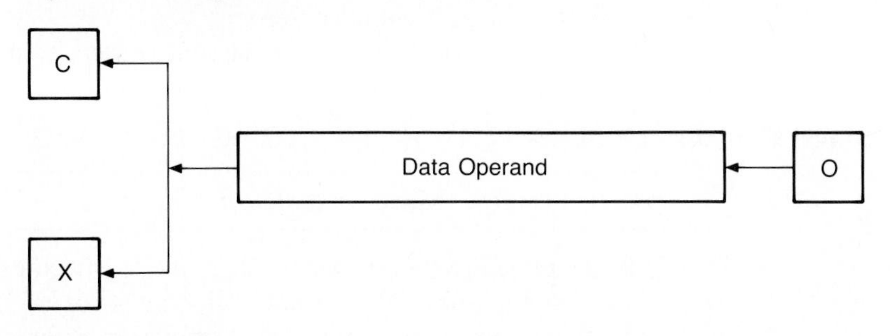

Figure A.3 LSL

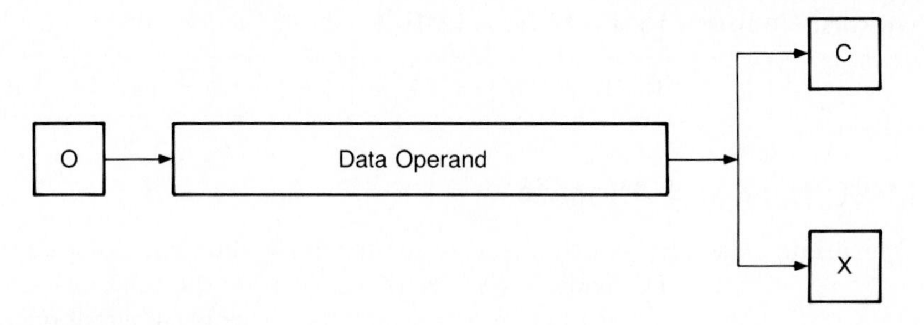

Figure A.4 LSR

Data Size: Byte, word, long word

Syntax: LSd Dx,Dy [d specifies direction (left or right)]
LSd #data,Dy
LSd ⟨ea⟩

Effective Address:

Address	Operand
Dn	No
An	No
(An)	Yes
(An)+	Yes
−(An)	Yes
d(An)	Yes
d(An,Xi)	Yes
Absolute	Yes
Immediate	No
d(PC)	No
d(PC,Xi)	No

Condition Codes: N: Set if the results are negative; reset otherwise.
Z: Set if the results are zero; reset if not.
V: Reset
C: Set or reset according to the last bit shifted from the data; reset for a shift count of zero.
X: Set or reset according to the last bit shifted from the data; unaffected for a shift count of zero.

Opword Coding:

15	14	13	12	11 10 9	8	7 6	5	4	3	2 1 0
1	1	1	0	− Mode −	d	Size	i	0	1	— Rn —

i: 0 = immediate shift count; then mode specifies the count with 000 = 8 and 001–111 = 1–7
i: 1 = register holds count; then mode specifies Dx's number.
d: 0 = shift right; 1 = shift left
Size: 00 = byte; 01 = word; 10 = long word

MOVE and MOVEA: Move Data from Source to Destination

Operation: The contents of the source register are moved to the destination register. MOVEA is a second format used when the destination register is an address register. When using an address register as a destination, data sizes are restricted to word and long word. Additionally, word size transfers to address registers are sign extended to a full 32 bits.

Data Size: Byte, word, long word

Syntax: MOVE.size ⟨ea⟩,⟨ea⟩
MOVEA.size ⟨ea⟩

Effective Address:

Address	Source	Destination
Dn	Yes	Yes
An	Yes	Yes*
(An)	Yes	Yes
(An)+	Yes	Yes
−(An)	Yes	Yes
d(An)	Yes	Yes
d(An,Xi)	Yes	Yes
Absolute	Yes	Yes
Immediate	Yes	No
d(PC)	Yes	No
d(PC,Xi)	Yes	No

*Word and long word only

Condition Codes: N: Set if results are negative; reset otherwise.
Z: Set if results are zero; reset if not.
V: Reset
C: Reset
X: Unaffected

Opword Coding:

15	14	13 12	11 10 9	8 7 6	5 4 3	2 1 0
0	0	Size	— Rd —	— Dea —	— Sea —	— Rs —

Size: 01 = byte; 11 = word; 10 = long word
Rd = destination register
Dea = destination effective address
Sea = source effective address
Rs = source register

MOVE from CCR: Move Data from the Condition Code Register

Operation: Move the contents of the CCR to the destination. A word move is performed; the upper byte is zero and the lower byte is the CCR data.

Data Size: Word

Syntax: MOVE CCR,⟨ea⟩

Effective Address:

Address	Destination
Dn	Yes
An	No
(An)	Yes
(An)+	Yes
-(An)	Yes
d(An)	Yes
d(An,Xi)	Yes
Absolute	Yes
Immediate	No
d(PC)	No
d(PC,Xi)	No

Condition Codes: Unaffected

Opword Coding:

15	14	13	12	11	10	9	8	7	6	5	4	3	2	1	0
0	1	0	0	0	0	1	0	1	1	— ea —			— Rn —		

MOVE from SR: Move Data from the Status Register

Operation: Move a word of data from the status register to a destination.

Data Size: Word

Syntax: MOVE SR,⟨ea⟩

Effective Address:

Address	Destination
Dn	Yes
An	No
(An)	Yes
(An)+	Yes
-(An)	Yes
d(An)	Yes
d(An,Xi)	Yes
Absolute	Yes
Immediate	No
d(PC)	No
d(PC,Xi)	No

Condition Codes: Unaffected

Opword Coding:

15	14	13	12	11	10	9	8	7	6	5	4	3	2	1	0
0	1	0	0	0	0	0	0	1	1	— ea —			— Rn —		

MOVE to CCR: Move Data to the Condition Code Register

Operation: A word of data is moved to the CCR from a destination. The upper byte is ignored, and the lower byte becomes the new CCR data.

Data Size: Word

Syntax: MOVE ⟨ea⟩,CCR

Effective Address: All effective addresses except address register direct (An) are usable.

Condition Codes: N: Matches b_3 of the data
Z: Matches b_2 of the data
V: Matches b_1 of the data
C: Matches b_0 of the data
X: Matches b_4 of the data

Opword Coding:

15	14	13	12	11	10	9	8	7	6	5	4	3	2	1	0
0	1	0	0	0	1	0	0	1	1	— ea —			— Rn —		

MOVE to SR: Move Data to Status Register

This is a privileged instruction.

Operation: A word of data is moved to the status register, affecting every bit of the SR.

Data Size: Word

Syntax: MOVE ⟨ea⟩,SR

Effective Address: All effective addresses are usable except address register direct (An).

Condition Codes: N: Matches b_3 of the data
Z: Matches b_2 of the data
V: Matches b_1 of the data
C: Matches b_0 of the data
X: Matches b_4 of the data

Status Flags: T: Matches b_{15} of the data
S: Matches b_{13} of the data
I_2: Matches b_{10} of the data

I_1: Matches b_9 of the data
I_0: Matches b_8 of the data

Opword Coding:

15	14	13	12	11	10	9	8	7	6	5	4	3	2	1	0
0	1	0	0	0	1	1	0	1	1	— ea —			— Rn —		

MOVE USP: Move User Stack Pointer

This is a privileged instruction.

Operation: The contents of the user stack pointer are moved to or from a specified address register.

Data Size: Long word

Syntax: MOVE USP,An
MOVE An,USP

Effective Address: Specified in the syntax as address register direct (An) only.

Condition Codes: Unaffected

Opword Coding:

15	14	13	12	11	10	9	8	7	6	5	4	3	2	1	0
0	1	0	0	1	1	1	0	0	1	1	0	d	— An —		

d: 0 = An to USP; 1 = USP to An

MOVEM: Multiple Register Move

Operation: The contents of one or more registers are moved to or from memory as specified in one of three ways. The transfer may be word or long word, but word data transfers to registers (both data and address) are sign extended to a full 32 bits. The order of transfer is designated by the transfer type.

Data Size: Word, long word

Syntax: MOVEM.⟨size⟩ ⟨register list⟩, ⟨ea⟩
MOVEM.⟨size⟩ ⟨ea⟩, ⟨register list⟩

Register List Syntax: Rn/Rn/Rn in any order, Rn–Rn, or a combination of both. Any register may be omitted from the list.

Effective Address:

1. *Control Mode Transfer:* The order of transfer is from D0–D7, then A0–A7. Any register may be omitted, and its place in the transfer list will be discarded. Transfer to or from memory begins at the specified address up through higher addresses. The effective addresses for the control mode are as follows:

Address	Source	Destination
Dn	No	No
An	No	No
(An)	Yes	Yes
(An)+*	Yes	No
-(An)**	No	Yes
d(An)	Yes	Yes
d(An,Xi)	Yes	Yes
Absolute	Yes	Yes
Immediate	No	No
d(PC)	Yes	No
d(PC,Xi)	Yes	No

*See pop/pull mode
**See push mode

2. *Push or Predecrement Mode:* This is a stack push type operation. The data are transferred from the registers to memory in the following order: A7–A0, then D7–D0. They are stored in memory from the designated address to lower addresses. The pointer address register is decremented by the pre-decrement function. Again, any register may be omitted from the register list.
3. *Pop/Pull or Postincrement Mode:* This is similar to a stack pop or pull operation. The order of transfer is the same as the control mode. The pointer register is incremented by the postincrement function.

Condition Codes: Unaffected

Opword Coding:

15	14	13	12	11	10	9	8	7	6	5	4	3	2	1	0
0	1	0	0	1	d	0	0	1	S	— ea —			— Rn —		

d: 0 = register to memory transfer
1 = memory to register transfer
S: 0 = word; 1 = long word

Followed by an operand containing the register mask field. The low order bit designates the first register transferred.

Control and postincrement modes:

15	14	13	12	11	10	9	8	7	6	5	4	3	2	1	0
A7 ——— A0								D7 ——— D0							

Predecrement mode:

15	14	13	12	11	10	9	8	7	6	5	4	3	2	1	0
D0 ——— D7								A0 ——— A7							

MOVEP: Move Data to/from Peripheral

Operation: Data is moved between a data register and alternate bytes of memory starting at the designated address and incrementing by two. The high order byte is transferred first.

Data Size: Word, long word

Syntax: MOVEP.⟨size⟩ Dn,d(An)
MOVEP.⟨size⟩ d(An),Dn

Effective Address: Specified by the syntax as data register direct and address register indirect with displacement.

Condition Codes: Unaffected

Opword Coding:

15	14	13	12	11	10	9	8	7	6	5	4	3	2	1	0
0	0	0	0	— Dn —			Opmode			0	0	1	— An —		

Opmode: 100 = word to data register
101 = long word to data register
110 = word to memory
111 = long word to memory

Opword is followed by one word of displacement.

MOVEQ: Move Immediate Data Quick

Operation: Move sign extended (to 32 bits) byte of immediate data to a data register.

Data Size: Immediate data is a byte. Data transfer is long word.

Syntax: MOVEQ #data,Dn

Effective Address: As specified in the syntax

Condition Codes: N: Set if results are negative; reset otherwise.
Z: Set if results are zero; reset otherwise.
V: Reset
C: Reset
X: Unaffected

Opword Coding:

15	14	13	12	11	10	9	8	7	6	5	4	3	2	1	0
0	1	1	1	— Dn —			0	——— Data ———							

MULS: Multiply Signed Data

Operation: Multiply a word of signed data by another word of signed data. The resulting long word product is stored in the destination register.

Data Size: Word

Syntax: MULS ⟨ea⟩,Dn

Effective Address: The destination address is always data register direct. All effective addresses except address register direct are usable for the source effective address.

Condition Codes: N: Set if results are negative; reset otherwise.
Z: Set if results are zero; reset otherwise.
V: Reset
C: Reset
X: Unaffected

Opword Coding:

15	14	13	12	11	10	9	8	7	6	5	4	3	2	1	0
1	1	0	0	— Dn —			1	1	1	— ea —			— Rn —		

MULU: Multiply Unsigned Data

Operation: Two unsigned words are multiplied together. The long word product is stored in the destination register.

Data Size: Word

Syntax: MULU ⟨ea⟩,Dn

Effective Address: Destination address is always data register direct. All effective addresses except address register direct are usable as a source address.

Condition Codes: N: Set if results are negative; reset otherwise.
Z: Set if results are zero; reset otherwise.
V: Reset
C: Reset
X: Unaffected

Opword Coding:

15	14	13	12	11	10	9	8	7	6	5	4	3	2	1	0
1	1	0	0	— Dn —			0	1	1	— ea —			— Rn —		

NBCD: 9's and 10's Complement Data

Operation: The binary coded decimal (BCD) data and extend bit are subtracted from zero, producing the 10's complement or negated value of the original data if the extend bit is set and the 9's complement if the extend bit is low.

Data Size: Byte

Syntax: NBCD ⟨ea⟩

Effective Address:

Address	Destination
Dn	Yes
An	No
(An)	Yes
(An)+	Yes
−(An)	Yes
d(An)	Yes
d(An,Xi)	Yes
Absolute	Yes
Immediate	No
d(PC)	No
d(PC,Xi)	No

Condition Codes: N: Undefined
Z: Reset if results not zero. Unaffected otherwise.
V: Undefined
C: Set if a borrow was generated; reset otherwise.
X: Copies the C bit

Opword Coding:

15	14	13	12	11	10	9	8	7	6	5	4	3	2	1	0
0	1	0	0	1	0	0	0	0	0	— ea —			— Rn —		

NEG: Negate Data

Operation: The specified data is subtracted from zero and saved at the destination.

Data Size: Byte, word, long word

Syntax: NEG.size ⟨ea⟩

Effective Address:

Address	Destination
Dn	Yes
An	No
(An)	Yes
(An)+	Yes
−(An)	Yes
d(An)	Yes
d(An,Xi)	Yes
Absolute	Yes
Immediate	No
d(PC)	No
d(PC,Xi)	No

Condition Codes: N: Set if results are negative; reset otherwise.
Z: Set if results are zero; reset otherwise.
V: Set if overflow is generated; reset otherwise.
C: Reset if results are zero; set otherwise.
X: Copies the C bit

Opword Coding:

15	14	13	12	11	10	9	8	7 6	5 4 3	2 1 0
0	1	0	0	0	1	0	0	Size	— ea —	— Rn —

Size: 00 = byte; 01 = word; 10 = long word

NEGX: Negate Data with Extend Bit

Operation: The specified data and the extend (X) bit are subtracted from zero, with the results saved in the destination.

Data Size: Byte, word, long word

Syntax: NEGX.size ⟨ea⟩

Effective Address:

Address	Destination
Dn	Yes
An	No
(An)	Yes
(An)+	Yes
−(An)	Yes
d(An)	Yes
d(An,Xi)	Yes
Absolute	Yes
Immediate	No
d(PC)	No
d(PC,Xi)	No

Condition Codes: N: Set if results are negative; reset otherwise.
Z: Set if results are zero; reset otherwise.
V: Set if overflow occurs; reset otherwise.
C: Set if a borrow occurs; reset otherwise.
X: Copies the C bit

Opword Coding:

15	14	13	12	11	10	9	8	7 6	5 4 3	2 1 0
0	1	0	0	0	0	0	0	Size	— ea —	— Rn —

Size: 00 = byte; 01 = word; 10 = long word

NOP: No Operation

Operation: There is no execute cycle for this instruction.

Data Size: Not applicable

Syntax: NOP

Effective Address: Not applicable

Condition Codes: Unaffected

Opword Coding:

15	14	13	12	11	10	9	8	7	6	5	4	3	2	1	0
0	1	0	0	1	1	1	0	0	1	1	1	0	0	0	1

NOT: 1's Complement

Operation: The data is 1's complemented and saved at the destination.

Data Size: Byte, word, long word

Syntax: NOT.size ⟨ea⟩

Effective Address:

Address	Destination
Dn	Yes
An	No
(An)	Yes
(An)+	Yes
−(An)	Yes
d(An)	Yes
d(An,Xi)	Yes
Absolute	Yes
Immediate	No
d(PC)	No
d(PC,Xi)	No

Condition Codes: N: Set if results are negative; reset otherwise.
Z: Set if results are zero; reset otherwise.
V: Reset
C: Reset
X: Unaffected

Opword Coding:

15	14	13	12	11	10	9	8	7	6	5	4	3	2	1	0
0	1	0	0	0	1	1	0	Size		— ea —			— Rn —		

Size: 00 = byte; 01 = word; 10 = long word

OR: Logical OR

Operation: Each bit of the source is logically ORed with its corresponding destination bit. The results are stored in the destination.

Data Size: Byte, word, long word

Syntax: OR.size ⟨ea⟩,Dn
OR.size Dn,⟨ea⟩

Effective Address:

Address	Source (Dn = Destination)	Destination (Dn = Source)
Dn	Yes	No
An	No	No
(An)	Yes	Yes
(An)+	Yes	Yes
−(An)	Yes	Yes
d(An)	Yes	Yes
d(An,Xi)	Yes	Yes
Absolute	Yes	Yes
Immediate	Yes	No
d(PC)	Yes	No
d(PC,Xi)	Yes	No

Condition Codes: N: Set if results are negative; reset otherwise.
Z: Set if results are zero; reset otherwise.
V: Reset
C: Reset
X: Unaffected

Opword Coding:

15	14	13	12	11 10 9	8 7 6	5 4 3	2 1 0
1	0	0	0	— Dn —	Opmode	— ea —	— Rn —

Opmode:

Direction	Byte	Word	Long Word
To Dn	000	001	010
From Dn	100	101	110

ORI: Logic OR Immediate Data

Operation: Logically OR each bit of the destination with the corresponding bit of immediate data. Store results at destination.

Data Size: Byte, word, long word

Syntax: ORI.size #data,⟨ea⟩

Effective Address:

Address	Destination
Dn	Yes
An	No
(An)	Yes
(An)+	Yes
−(An)	Yes
d(An)	Yes
d(An,Xi)	Yes
Absolute	Yes
Immediate	No
d(PC)	No
d(PC,Xi)	No

Condition Codes: N: Set if results are negative; reset otherwise.
Z: Set if results are zero; reset otherwise.
V: Reset
C: Reset
X: Unaffected

Opword Coding:

15	14	13	12	11	10	9	8	7	6	5	4	3	2	1	0
0	0	0	0	0	0	0	0	Size		— ea —			— Rn —		

Size: 00 = byte; 01 = word; 10 = long word
Followed by one or two words of operand.

ORI to CCR: Logic OR CCR with Immediate Data

Operation: Logically OR each bit of the condition code register with the corresponding bits of immediate data. Results are stored in the CCR.

Data Size: Byte

Syntax: ORI #data,CCR

Effective Address: Specified in the syntax as immediate

Condition Codes: Set or reset according to the results of the operation.
N: b_3
Z: b_2
V: b_1
C: b_0
X: b_4

Opword Coding:

15	14	13	12	11	10	9	8	7	6	5	4	3	2	1	0
0	0	0	0	0	0	0	0	0	0	1	1	1	1	0	0

Followed by a word operand whose upper byte is zero and lower byte is the data.

ORI to SR: Logic OR Immediate Data with the Status Register

This is a privileged instruction.

Operation: Logically OR each bit of the status register with the corresponding bits of the status register. Store the results in the SR.

Data Size: Word

Syntax: ORI #data, SR

Effective Address: Immediate, as specified in the syntax

Condition Codes: Set or reset as a result of the operation.
N: b_3
Z: b_2
V: b_1
C: b_0
X: b_4

Opword Coding:

15	14	13	12	11	10	9	8	7	6	5	4	3	2	1	0
0	0	0	0	0	0	0	0	0	1	1	1	1	1	0	0

Followed by one word of data operand.

PEA: Push Effective Address

Operation: The long word effective address is computed and stored onto the stack.

Data Size: Long

Syntax: PEA ⟨ea⟩

Effective Address:

Address	Source
Dn	No
An	No
(An)	Yes
(An)+	No
−(An)	No
d(An)	Yes
d(An,Xi)	Yes
Absolute	Yes
Immediate	No
d(PC)	Yes
d(PC,Xi)	Yes

Condition Codes: Unaffected

Opword Coding:

15	14	13	12	11	10	9	8	7	6	5	4	3	2	1	0
0	1	0	0	1	0	0	0	0	1	— ea —			— Rn —		

RESET: Reset External Devices

Operation: Asserts (makes active) the RESET line for 124 clock cycles.

Data Size: Not applicable

Syntax: RESET

Effective Address: Not applicable

Condition Codes: Unaffected

Opword Coding:

15	14	13	12	11	10	9	8	7	6	5	4	3	2	1	0
0	1	0	0	1	1	1	0	0	1	1	1	0	0	0	0

ROL and ROR: Rotate Left or Right

Operation: Rotate the data a specified number of times in the specified direction. The final condition of the most significant bit (ROL) or the least significant bit (ROR) is copied into the C bit of the CCR.

Data Size: Byte, word, long word

Syntax: ROd.WORD ⟨ea⟩: A single rotate is done.
ROd.size Dx,Dy: The number of rotations is specified in Dx.
ROd.size #count,Dy: The number of rotations is specified in the immediate count.

Effective Address:

Address	Destination
Dn	No
An	No
(An)	Yes
(An)+	Yes
-(An)	Yes
d(An)	Yes
d(An,Xi)	Yes
Absolute	Yes
Immediate	No
d(PC)	No
d(PC,Xi)	No

Condition Codes: N: Set if results are negative; reset otherwise.
Z: Set if results are zero; reset otherwise.

V: Reset
C: Copies the most/least significant bit
X: Unaffected

Opword Coding: Register Rotate:

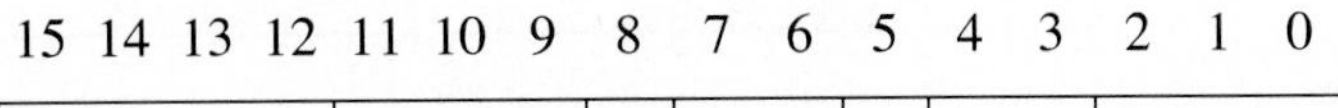

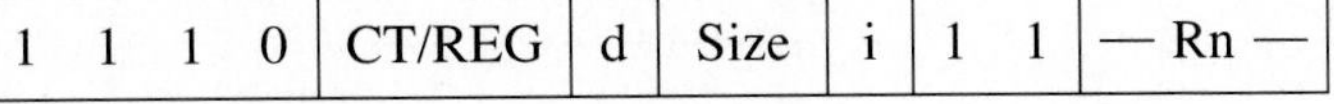

15	14	13	12	11 10 9	8	7 6	5	4	3	2 1 0
1	1	1	0	CT/REG	d	Size	i	1	1	— Rn —

CT/REG and i:
i = 0: CT/REG is the rotate count
000 = 8; 001–111 = 1–7
i = 1: CT/REG is the data register number that holds the count
d: 0 = right; 1 = left
Size: 00 = byte; 01 = word; 10 = long word

Memory Rotate:

15	14	13	12	11	10	9	8	7	6	5 4 3	2 1 0
1	1	1	0	0	1	1	d	1	1	— ea —	— Rn —

d: 0 = right; 1 = left

ROXL and ROXR: Rotate Left or Right with Extend Bit

Operation: Rotate the data at the specified destination a selected number of times. The rotation involves the data and the extend bit as diagrammed in Figures A.5 and A.6. The carry bit copies the extend bit.

Data Size: Byte, word, long word

Syntax: ROXd.WORD ⟨ea⟩
ROXd.size Dx,Dy
ROXd.size #data,Dy

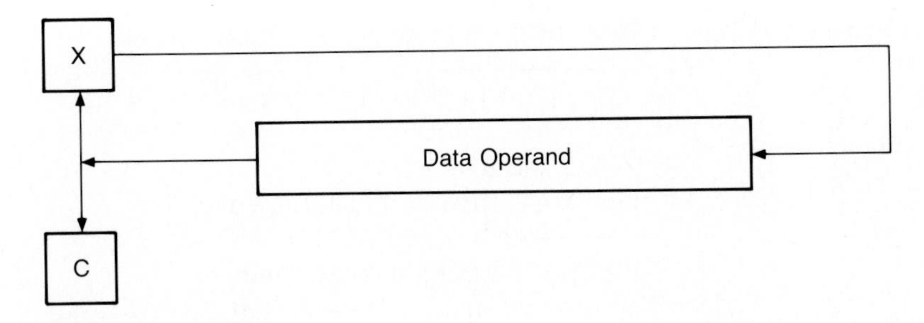

Figure A.5 ROXL

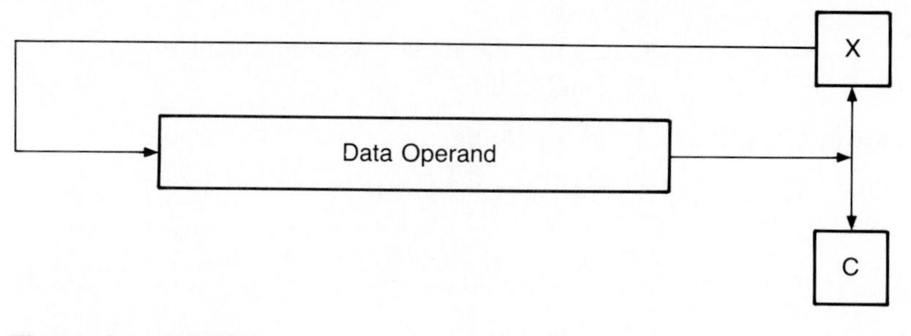

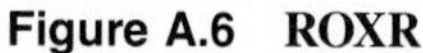
Figure A.6 ROXR

Effective Address: The data register direct mode specifies the rotate count as the data in Dx, while the actual data rotated are held in Dy. The immediate mode count is the immediate data, while the data rotated are held in Dy. The remaining address modes involve a single rotate at the effective address.

Address	Destination
Dn	No
An	No
(An)	Yes
(An)+	Yes
-(An)	Yes
d(An)	Yes
d(An,Xi)	Yes
Absolute	Yes
Immediate	No
d(PC)	No
d(PC,Xi)	No

Condition Codes: N: Set if results are negative; reset otherwise.
Z: Set if results are zero; reset otherwise.
V: Reset
C: Copies X bit
X: Copies last bit rotated out. Unaffected for a count of zero.

Opword Coding:

15	14	13	12	11 10 9	8	7 6	5	4	3	2 1 0
1	1	1	0	CT/REG	d	Size	i	1	0	— Dy —

CT/REG and i:
i = 0: CT/REG is immediate count
000 = 8; 001–111 = 1–7
i = 1: CT/REG holds Dx number
d: 0 = rotate right; 1 = rotate left
Size: 00 = byte; 01 = word; 10 = long word

or

15	14	13	12	11	10	9	8	7	6	5	4	3	2	1	0
1	1	1	0	0	1	0	d	1	1	— ea —			— Rn —		

d: 0 = rotate right; 1 = rotate left

RTE: Return from Exception

Operation: The status register and program counter are pulled from the stack and the original program is resumed.

Data Size: Not applicable

Syntax: RTE

Effective Address: Not applicable

Condition Codes: Set or reset according to data pulled from the stack.

Opword Coding:

15	14	13	12	11	10	9	8	7	6	5	4	3	2	1	0
0	1	0	0	1	1	1	0	0	1	1	1	0	0	1	1

RTR: Return and Restore Condition Codes

Operation: The condition code register portion of the status register and the program counter are pulled from the stack. The status codes are unaffected.

Data Size: Not applicable

Syntax: RTR

Effective Address: Not applicable

Condition Codes: Set or reset according to the data pulled from the stack.

Opword Coding:

15	14	13	12	11	10	9	8	7	6	5	4	3	2	1	0
0	1	0	0	1	1	1	0	0	1	1	1	0	1	1	1

RTS: Return from Subroutine

Operation: The program counter is pulled from the stack.

Data Size: Not applicable

Syntax: RTS

Effective Address: Not applicable

Condition Codes: Unaffected

Opword Coding:

15	14	13	12	11	10	9	8	7	6	5	4	3	2	1	0
0	1	0	0	1	1	1	0	0	1	1	1	0	1	0	1

SBCD: Subtract BCD Data with Extend Bit

Operation: The source BCD data and the X flag value are subtracted from the destination BCD data. The results are stored at the destination.

Data Size: Byte

Syntax: SBCD Dy,Dx
SBCD −(Ay),−(Ax)

Effective Address: Specified in the syntax as either data register direct or address register indirect with predecrement

Condition Codes: N: Undefined
Z: Reset if results are not zero. Unaffected otherwise.
V: Undefined
C: Set if borrow occurs; reset otherwise.
X: Copies C bit

Opword Coding:

15	14	13	12	11	10	9	8	7	6	5	4	3	2	1	0
1	0	0	0	— Rx —			1	0	0	0	0	M	— Ry —		

M: 0 = data direct; 1 = address register indirect
Rx = Dx or Ax Ry = Dy or Ay

Scc: Set Byte According to Condition Code

Operation: Set the specified byte to $FF if the condition code is set. Set the byte to $00 if the condition code is reset.

Data Size: Byte

Syntax: Scc ⟨ea⟩

Effective Address:

Address	Destination
Dn	Yes
An	No
(An)	Yes
(An)+	Yes
-(An)	Yes
d(An)	Yes
d(An,Xi)	Yes
Absolute	Yes
Immediate	No
d(PC)	No
d(PC,Xi)	No

Condition Codes: Unaffected

Opword Coding:

15	14	13	12	11	10	9	8	7	6	5	4	3	2	1	0
0	1	0	1	—	cc	—		1	1	—	ea	—	—	Rn	—

cc	Condition	Test	cc Code
CC	Carry bit reset	C = 0	0100
CS	Carry bit set	C = 1	0101
EQ	Zero bit set	Z = 1	0111
F	False	Always 0	0001
GE	Greater than or equal to	$N \oplus V = 0$	1100
GT	Greater than	$N \oplus V + Z = 0$	1110
HI	Higher than	$C + Z = 0$	0010
LE	Less than or equal to	$N \oplus V + Z = 1$	1111
LS	Lower than or the same as	$C + Z = 1$	0011
LT	Less than	$N \oplus V = 1$	1101
MI	Minus	N = 1	1011
NE	Zero bit reset	Z = 0	0110
PL	Positive	N = 0	1010
T	True	Always 1	0000
VC	Overflow bit reset	V = 0	1000
VS	Overflow set	V = 1	1001

STOP: Load Status Register and Stop

This is a privileged instruction.

Operation: Move the immediate data into the status register, advance the program counter to the next instruction, and stop executing instructions. Execution resumes when a TRACE, interrupt, or RESET exception occurs.
1. TRACE occurs if T = 1 when STOP is executed.
2. Interrupt exception occurs if request is at high enough level.
If the S bit is reset by the move, a privilege violation exception occurs.

Data Size: Word

Syntax: STOP #data

Effective Address: Specified by syntax as immediate

Condition Codes: Set or reset according to the data moved into the CCR portion of the status register.

Opword Coding:

15	14	13	12	11	10	9	8	7	6	5	4	3	2	1	0
0	1	0	0	1	1	1	0	0	1	1	1	0	0	1	0

Followed by a word of data.

SUB: Subtract

Operation: Subtract the source data from the destination. The results are stored in the destination.

Data Size: Byte, word, long word

Syntax: SUB.size ⟨ea⟩,Dn
SUB.size Dn,⟨ea⟩

Effective Address:

Address	Source (Dn = Destination)	Destination (Dn = Source)
Dn	Yes	No
An	Yes	No
(An)	Yes	Yes
(An)+	Yes	Yes
−(An)	Yes	Yes
d(An)	Yes	Yes
d(An,Xi)	Yes	Yes
Absolute	Yes	Yes
Immediate	Yes	No
d(PC)	Yes	No
d(PC,Xi)	Yes	No

Condition Codes: N: Set if results are negative; reset otherwise.
Z: Set if results are zero; reset otherwise.
V: Set if overflow occurs; reset otherwise.
C: Set if borrow occurs; reset otherwise.
X: Copies C bit

Opword Coding:

15	14	13	12	11 10 9	8 7 6	5 4 3	2 1 0
1	0	0	1	— Dn —	Opmode	— ea —	— Rn —

Opmode:

Dn	Byte	Word	Long Word
Destination	000	001	010
Source	100	101	110

SUBA: Subtract from Address Register

Operation: Subtract the source data from the destination address register and place results in the destination. Word size data is sign extended to 32 bits before the subtraction is done.

Data Size: Word, long word

Syntax: SUBA.size ⟨ea⟩,An

Effective Address: Destination specified in the syntax as address register direct (An). All address modes are usable for a source effective address.

Condition Codes: Unaffected

Opword Coding:

15	14	13	12	11 10 9	8 7 6	5 4 3	2 1 0
1	0	0	1	— An —	Opmode	— ea —	— Rn —

Opmode: 011 = word; 111 = long word

SUBI: Subtract Immediate Data

Operation: Subtract the immediate data from the destination, and place the results in the destination.

Data Size: Byte, word, long word

Syntax: SUBI.size #data,⟨ea⟩

Effective Address:

Address	Destination
Dn	Yes
An	No
(An)	Yes
(An)+	Yes
−(An)	Yes
d(An)	Yes
d(An,Xi)	Yes
Absolute	Yes
Immediate	No
d(PC)	No
d(PC,Xi)	No

Condition Codes: N: Set if results are negative; reset otherwise.
Z: Set if results are zero; reset otherwise.
V: Set if overflow occurs; reset otherwise.
C: Set if borrow occurs; reset otherwise.
X: Copies C bit

Opword Coding:

15	14	13	12	11	10	9	8	7 6	5 4 3	2 1 0
0	0	0	0	0	1	0	0	Size	— ea —	— Rn —

SUBQ: Subtract Immediate Data Quick

Operation: Subtract the immediate data from the destination register. The data range is 1–8.

Data Size: Byte, word, long word

Syntax: SUBQ.size #data,⟨ea⟩

Effective Address: Source mode is always immediate.

Address	Destination
Dn	Yes
An	Yes*
(An)	Yes
(An)+	Yes
−(An)	Yes
d(An)	Yes
d(An,Xi)	Yes
Absolute	Yes
Immediate	No
d(PC)	No
d(PC,Xi)	No

*Word and long word size only. Word is sign extended before the operation.

Condition Codes: For ⟨ea⟩ = An, the condition codes are unaffected, For all others,
N: Set if the results are negative; reset otherwise.
Z: Set if the results are zero; reset otherwise.
V: Set if overflow occurs; reset otherwise.
C: Set if a borrow occurs; reset otherwise.
X: Copies C bit

Opword Coding:

15	14	13	12	11 10 9	8	7 6	5 4 3	2 1 0
0	1	0	1	– Data –	1	Size	— ea —	— Rn —

Data: 000 = 8; 001–111 = 1–7
Size: 00 = byte; 01 = word; 10 = long word

SUBX: Subtract with Extend Bit

Operation: Subtract the source data and the extend bit from the destination. The results are stored in the destination.

Data Size: Byte, word, long word

Syntax: SUBX.size Dy,Dx
SUBX.size −(Ay),−(Ax)

Effective Address: Specified in the syntax as either data register direct or address register indirect with predecrement.

Condition Codes: N: Set if results are negative; reset otherwise.
Z: Set if the results are zero; reset otherwise.
V: Set if overflow occurs; reset otherwise.
C: Set if a borrow occurs; reset otherwise.
X: Copies C bit

Opword Coding:

15	14	13	12	11	10	9	8	7	6	5	4	3	2	1	0
1	0	0	1	— Rx —			1	Size		0	0	M	— Ry —		

Size: 00 = byte; 01 = word; 10 = long word
M: 0 = data register direct; 1 = address register indirect

SWAP: Swap Register Halves

Operation: The upper and lower words of a data register are swapped so that the former high word is the new low word, and vice versa.

Data Size: Word

Syntax: SWAP Dn

Effective Address: Data register direct

Condition Codes: N: Set if results are negative; reset otherwise.
Z: Set if the results are zero; reset otherwise.
V: Reset
C: Reset
X: Unaffected

Opword Coding:

15	14	13	12	11	10	9	8	7	6	5	4	3	2	1	0
0	1	0	0	1	0	0	0	0	1	0	0	0	— Dn —		

TAS: Test and Set an Operand

Operation: Test the destination byte for zero and negative. Set b7 of the operand data.

Data Size: Byte

Syntax: TAS ⟨ea⟩

Effective Address:

Address	Destination
Dn	Yes
An	No
(An)	Yes
(An)+	Yes
-(An)	Yes
d(An)	Yes
d(An,Xi)	Yes
Absolute	Yes
Immediate	No
d(PC)	No
d(PC,Xi)	No

Condition Codes: N: Set if results are negative; reset otherwise.
Z: Set if data is zero; reset otherwise.
V: Reset
C: Reset
X: Unaffected

Opword Coding:

15	14	13	12	11	10	9	8	7	6	5	4	3	2	1	0
0	1	0	0	1	0	1	0	1	1	— ea —			— Rn —		

TRAP: Trap Exception

Operation: Initiate a TRAP exception routine.

Data Size: Not applicable

Syntax: TRAP #vector

Effective Address: Specified in syntax as immediate mode.

Condition Codes: Unaffected

Opword Coding:

15	14	13	12	11	10	9	8	7	6	5	4	3	2	1	0
0	1	0	0	1	1	1	0	0	1	0	0	— Vector —			

Vector: TRAP exception vector number 0–15

TRAPV: TRAP Exception on Overflow

Operation: The TRAPV exception routine is initiated if the overflow bit is set.

Data Size: Not applicable

Syntax: TRAPV

Effective Address: Not applicable

Condition Codes: Unaffected

Opword Coding:

15	14	13	12	11	10	9	8	7	6	5	4	3	2	1	0
0	1	0	0	1	1	1	0	0	1	1	1	0	1	1	0

TST: Test an Operand

Operation: Test an operand and cause the condition code bits to reflect the status of the data. The data are unaffected by the instruction.

Data Size: Byte, word, long word

Syntax: TST.size ⟨ea⟩

Effective Address:

Address	Destination
Dn	Yes
An	No
(An)	Yes
(An)+	Yes
−(An)	Yes
d(An)	Yes
d(An,Xi)	Yes
Absolute	Yes
Immediate	No
d(PC)	No
d(PC,Xi)	No

Condition Codes: N: Set data is negative; reset otherwise.
Z: Set if data is zero; reset otherwise.
V: Reset
C: Reset
X: Unaffected

Opword Coding:

15	14	13	12	11	10	9	8	7	6	5	4	3	2	1	0
0	1	0	0	1	0	1	0	Size		— ea —			— Rn —		

Size: 00 = byte; 01 = word; 10 = long word

UNLK: Unlink

Operation: The stack pointer is loaded from the specified address register. The address register is then loaded with the long word pulled from the stack.

Data Size: Not applicable

Syntax: UNLK An

Effective Address: Specified by syntax as address register direct

Condition Codes: Unaffected

Opword Coding:

15	14	13	12	11	10	9	8	7	6	5	4	3	2	1	0
0	1	0	0	1	1	1	0	0	1	0	1	1	— An —		

APPENDIX B 68000 TIMING DIAGRAM DETAILS

Table B.1 AC Electrical Specifications: Clock Timing (Courtesy of Motorola, Inc.)

Characteristic	Symbol	8 MHz		10 MHz		12.5 MHz		Unit
		Min	Max	Min	Max	Min	Max	
Frequency of Operation	f	4.0	8.0	4.0	10.0	4.0	12.5	MHz
Cycle Time	t_{cyc}	125	250	100	250	80	250	ns
Clock Pulse Width	t_{CL} t_{CH}	55 55	125 125	45 45	125 125	35 35	125 125	ns
Rise and Fall Times	t_{Cr} t_{Cf}	– –	10 10	– –	10 10	– –	5 5	ns

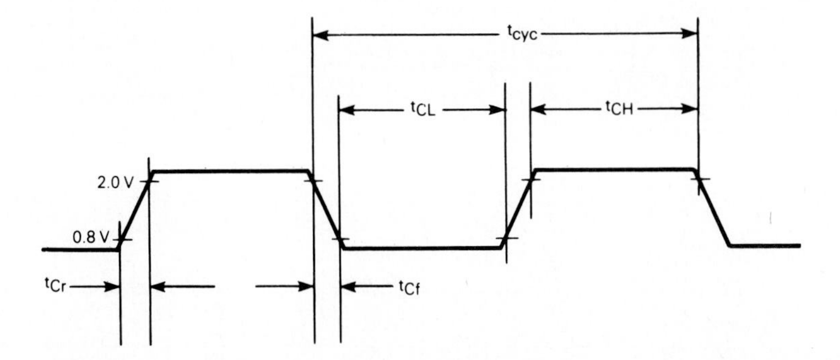

Figure B.1 **Clock input timing diagram (Courtesy of Motorola, Inc.)**

Table B.2 AC Electrical Specifications: Read and Write Cycles (Courtesy of Motorola, Inc.)

Num.	Characteristic	Symbol	8 MHz		10 MHz		12.5 MHz		Unit
			Min	Max	Min	Max	Min	Max	
1	Clock Period	t_{cyc}	125	250	100	250	80	250	ns
2	Clock Width Low	t_{CL}	55	125	45	125	35	125	ns
3	Clock Width High	t_{CH}	55	125	45	125	35	125	ns
4	Clock Fall Time	t_{Cf}	–	10	–	10	–	5	ns
5	Clock Rise Time	t_{Cr}	–	10	–	10	–	5	ns
6	Clock Low to Address Valid	t_{CLAV}	–	70	–	60	–	55	ns
6A	Clock High to FC Valid	t_{CHFCV}	–	70	–	60	–	55	ns
7	Clock High to Address, Data Bus High Impedance (Maximum)	t_{CHADZ}	–	80	–	70	–	60	ns
8	Clock High to Address, FC Invalid (Minimum)	t_{CHAFI}	0	–	0	–	0	–	ns
9[1]	Clock High to $\overline{AS}$, $\overline{DS}$ Low	t_{CHSL}	0	60	0	55	0	55	ns
11[2]	Address Valid to $\overline{AS}$, $\overline{DS}$ Low (Read)/ $\overline{AS}$ Low (Write)	t_{AVSL}	30	–	20	–	0	–	ns
11A[2,7]	FC Valid to $\overline{AS}$, $\overline{DS}$ Low (Read)/ $\overline{AS}$ Low (Write)	t_{FCVSL}	60	–	50	–	40	–	ns
12[1]	Clock Low to $\overline{AS}$, $\overline{DS}$ High	t_{CLSH}	–	70	–	55	–	50	ns
13[2]	$\overline{AS}$, $\overline{DS}$ High to Address/FC Invalid	t_{SHAFI}	30	–	20	–	10	–	ns
14[2,5]	$\overline{AS}$, $\overline{DS}$ Width Low (Read)/$\overline{AS}$ Low (Write)	t_{SL}	240	–	195	–	160	–	ns
14A[2]	$\overline{DS}$ Width Low (Write)	t_{DSL}	115	–	95	–	80	–	ns
15[2]	$\overline{AS}$, $\overline{DS}$ Width High	t_{SH}	150	–	105	–	65	–	ns
16	Clock High to Control Bus High Impedance	t_{CHCZ}	–	80	–	70	–	60	ns
17[2]	$\overline{AS}$, $\overline{DS}$ High to R/$\overline{W}$ High (Read)	t_{SHRH}	40	–	20	–	10	–	ns
18[1]	Clock High to R/$\overline{W}$ High	t_{CHRH}	0	70	0	60	0	60	ns
20[1]	Clock High to R/$\overline{W}$ Low (Write)	t_{CHRL}	–	70	–	60	–	60	ns
20A[8]	$\overline{AS}$ Low to R/$\overline{W}$ Valid (Write)	t_{ASRV}	–	20	–	20	–	20	ns
21[2]	Address Valid to R/$\overline{W}$ Low (Write)	t_{AVRL}	20	–	0	–	0	–	ns
21A[2,7]	FC Valid to R/$\overline{W}$ Low (Write)	t_{FCVRL}	60	–	50	–	30	–	ns
22[2]	R/$\overline{W}$ Low to $\overline{DS}$ Low (Write)	t_{RLSL}	80	–	50	–	30	–	ns
23	Clock Low to Data Out Valid (Write)	t_{CLDO}	–	70	–	55	–	55	ns
25[2]	$\overline{AS}$, $\overline{DS}$ High to Data Out Invalid (Write)	t_{SHDOI}	30	–	20	–	15	–	ns
26[2]	Data Out Valid to $\overline{DS}$ Low (Write)	t_{DOSL}	30	–	20	–	15	–	ns
27[6]	Data In to Clock Low (Setup Time on Read)	t_{DICL}	15	–	10	–	10	–	ns
28[2,5]	$\overline{AS}$, $\overline{DS}$ High to $\overline{DTACK}$ High	t_{SHDAH}	0	245	0	190	0	150	ns
29	$\overline{AS}$, $\overline{DS}$ High to Data In Invalid (Hold Time on Read)	t_{SHDII}	0	–	0	–	0	–	ns
30	$\overline{AS}$, $\overline{DS}$ High to $\overline{BERR}$ High	t_{SHBEH}	0	–	0	–	0	–	ns
31[2,6]	$\overline{DTACK}$ Low to Data In (Setup Time)	t_{DALDI}	–	90	–	65	–	50	ns
32	$\overline{HALT}$ and $\overline{RESET}$ Input Transition Time	$t_{RHr,f}$	0	200	0	200	0	200	ns
33	Clock High to $\overline{BG}$ Low	t_{CHGL}	–	70	–	60	–	50	ns
34	Clock High to $\overline{BG}$ High	t_{CHGH}	–	70	–	60	–	50	ns
35	$\overline{BR}$ Low to $\overline{BG}$ Low	t_{BRLGL}	1.5	90 ns +3.5	1.5	80 ns +3.5	1.5	70 ns +3.5	Clk.Per.

Table B.2 *(continued)*

Num.	Characteristic	Symbol	8 MHz		10 MHz		12.5 MHz		Unit
			Min	Max	Min	Max	Min	Max	
36[9]	$\overline{\text{BR}}$ High to $\overline{\text{BG}}$ High	t_{BRHGH}	1.5	90 ns + 3.5	1.5	80 ns + 3.5	1.5	70 ns + 3.5	Clk.Per.
37	$\overline{\text{BGACK}}$ Low to $\overline{\text{BG}}$ Low	t_{GALGH}	1.5	90 ns + 3.5	1.5	80 ns + 3.5	1.5	70 ns + 3.5	Clk.Per.
37A[10]	$\overline{\text{BGACK}}$ Low to $\overline{\text{BR}}$ High	t_{GALBRH}	20	1.5 Clocks	20	1.5 Clocks	20	1.5 Clocks	ns
38	$\overline{\text{BG}}$ Low to Control, Address, Data Bus High Impedance ($\overline{\text{AS}}$ High)	t_{GLZ}	—	80	—	70	—	60	ns
39	$\overline{\text{BG}}$ Width High	t_{GH}	1.5	—	1.5	—	1.5	—	Clk.Per.
40	Clock Low to $\overline{\text{VMA}}$ Low	t_{CLVML}	—	70	—	70	—	70	ns
41	Clock Low to E Transition	t_{CLET}	—	70	—	55	—	45	ns
42	E Output Rise and Fall Time	$t_{Er,f}$	—	25	—	25	—	25	ns
43	$\overline{\text{VMA}}$ Low to E High	t_{VMLEH}	200	—	150	—	90	—	ns
44	$\overline{\text{AS}}$, $\overline{\text{DS}}$ High to $\overline{\text{VPA}}$ High	t_{SHVPH}	0	120	0	90	0	70	ns
45	E Low to Control, Address Bus Invalid (Address Hold Time)	t_{ELCAI}	30	—	10	—	10	—	ns
46	$\overline{\text{BGACK}}$ Width Low	t_{GAL}	1.5	—	1.5	—	1.5	—	Clk.Per.
47[6]	Asynchronous Input Setup Time	t_{ASI}	20	—	20	—	20	—	ns
48[3]	$\overline{\text{BERR}}$ Low to $\overline{\text{DTACK}}$ Low	t_{BELDAL}	20	—	20	—	20	—	ns
49[11]	$\overline{\text{AS}}$, $\overline{\text{DS}}$ High to E Low	t_{SHEL}	−70	70	−55	55	−45	45	ns
50	E Width High	t_{EH}	450	—	350	—	280	—	ns
51	E Width Low	t_{EL}	700	—	550	—	440	—	ns
53	Clock High to Data Out Invalid	t_{CHDOI}	0	—	0	—	0	—	ns
54	E Low to Data Out Invalid	t_{ELDOI}	30	—	20	—	15	—	ns
55	R/$\overline{\text{W}}$ to Data Bus Driven	t_{RLDBD}	30	—	20	—	10	—	ns
56[4]	$\overline{\text{HALT}}$/$\overline{\text{RESET}}$ Pulse Width	t_{HRPW}	10	—	10	—	10	—	Clk. Per.
57	$\overline{\text{BGACK}}$ High to Control Bus Driven	t_{GABD}	1.5	—	1.5	—	1.5	—	Clk.Per.
58[9]	$\overline{\text{BG}}$ High to Control Bus Driven	t_{GHBD}	1.5	—	1.5	—	1.5	—	Clk.Per.

NOTES:

1. For a loading capacitance of less than or equal to 50 picofarads, substract 5 nanoseconds from the value given in the maximum columns.
2. Actual value depends on clock period.
3. If #47 is satisfied for both $\overline{\text{DTACK}}$ and $\overline{\text{BERR}}$, #48 may be 0 nanoseconds.
4. For power up, the MPU must be held in $\overline{\text{RESET}}$ state for 100 ms to allow stabilization of on-chip circuitry. After the system is powered up, #56 refers to the minimum pulse width required to reset the system.
5. #14, #14A, and #28 are one clock period less than the given number for T6E, BF4, and R9M mask sets.
6. If the asynchronous setup time (#47) requirements are satisfied, the $\overline{\text{DTACK}}$ low-to-data setup time (#31) requirement can be ignored. The data must only satisfy the date-in clock-low setup time (#27) for the following cycle.
7. For T6E, BF4, and R9M mask set #11A timing equals #11, and #21A equals #21. #20A may be 0 for T6E, BF4, and R9M mask sets.
8. When $\overline{\text{AS}}$ and R/$\overline{\text{W}}$ are equally loaded (±20%), subtract 10 nanoseconds from the values given in these columns.
9. The processor will negate $\overline{\text{BG}}$ and begin driving the bus again if external arbitration logic negates $\overline{\text{BR}}$ before asserting $\overline{\text{BGACK}}$.
10. The minimum value must be met to guarantee proper operation. If the maximum value is exceeded, $\overline{\text{BG}}$ may be reasserted.
11. The falling edge of S6 triggers both the negation of the strobes ($\overline{\text{AS}}$ and $\overline{\text{xDS}}$) and the falling edge of E. Either of these events can occur first, depending upon the loading on each signal. Specification #49 indicates the absolute maximum skew that will occur between the rising edge of the strobes and the falling edge of the E clock.

These waveforms should only be referenced in regard to the edge-to-edge measurement of the timing specifications. They are not intended as a functional description of the input and output signals. Refer to other functional descriptions and their related diagrams for device operation.

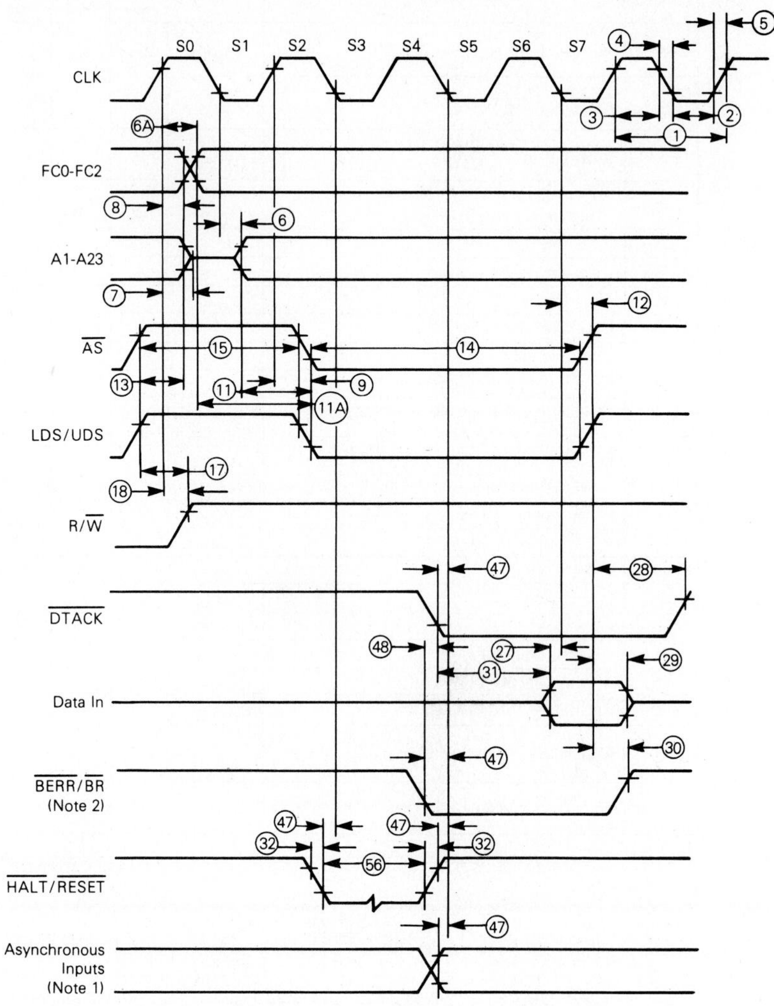

NOTES:
1. Setup time for the synchronous inputs $\overline{BGACK}$, $\overline{IPL0\text{-}2}$, and $\overline{VPA}$ guarantees their recognition at the next falling edge of the clock.
2. $\overline{BR}$ need fall at this time only in order to insure being recognized at the end of this bus cycle.
3. Timing measurements are referenced to and from a low voltage of 0.8 volt and a high voltage 2.0 volts, unless otherwise noted. The voltage swing through this range should start outside and pass through the range such that the rise or fall will be linear between 0.8 volt and 2.0 volts.

Figure B.2 Read cycle timing diagram (Courtesy of Motorola, Inc.)

These waveforms should only be referenced in regard to the edge-to-edge measurement of the timing specifications. They are not intended as a functional description of the input and output signals. Refer to other functional descriptions and their related diagrams for device operation.

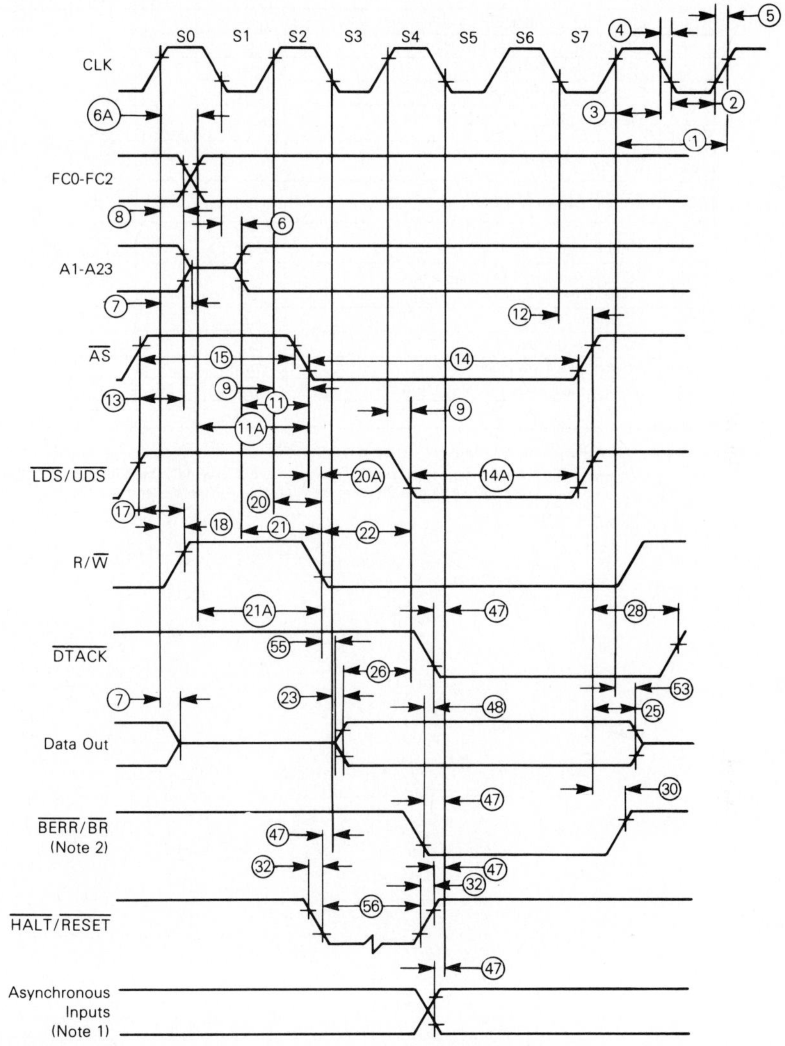

NOTES:

1. Timing measurements are referenced to and from a low voltage of 0.8 volt and a high voltage of 2.0 volts, unless otherwise noted. The voltage swing through this range should start outside and pass through the range such that the rise or fall will be linear between 0.8 volt and 2.0 volts.
2. Because of loading variations, R/$\overline{W}$ may be valid after $\overline{AS}$ even though both are initiated by the rising edge of S2 (Specification 20A).

Figure B.3 Write cycle timing diagram (Courtesy of Motorola, Inc.)

Table B.3 AC Electrical Specifications: Bus Arbitration (Courtesy of Motorola, Inc.)

Num.	Characteristic	Symbol	8 MHz		10 MHz		12.5 MHz		Unit
			Min	Max	Min	Max	Min	Max	
7	Clock High to Address, Data Bus High Impedance	t_{CHADZ}	–	80	–	70	–	60	ns
16	Clock High to Control Bus High Impedence	t_{CHCZ}	–	80	–	70	–	60	ns
33	Clock High to $\overline{BG}$ Low	t_{CHGL}	–	70	–	60	–	50	ns
34	Clock High to $\overline{BG}$ High	t_{CHGH}	–	70	–	60	–	50	ns
35	$\overline{BR}$ Low to $\overline{BG}$ Low	t_{BRLGL}	1.5	90 ns +3.5	1.5	80 ns +3.5	1.5	70 ns +3.5	Clk. Per.
36[1]	$\overline{BR}$ High to $\overline{BG}$ High	t_{BKHGH}	1.5	90 ns +3.5	1.5	80 ns +3.5	1.5	70 ns +3.5	Clk. Per.
37	$\overline{BGACK}$ Low to $\overline{BG}$ High	t_{GALGH}	1.5	90 ns +3.5	1.5	80 ns +3.5	1.5	70 ns +3.5	Clk. Per.
37A[2]	$\overline{BGACK}$ Low to $\overline{BR}$ High	t_{GALBRH}	20	1.5 Clocks	20	1.5 Clocks	20	1.5 Clocks	ns
38	$\overline{BG}$ Low to Control, Address, Data Bus High Impedance ($\overline{AS}$ High)	t_{GLZ}	–	80	–	70	–	60	ns
39	$\overline{BG}$ Width High	t_{GH}	1.5	–	1.5	–	1.5	–	Clk. Per.
46	$\overline{BGACK}$ Width Low	t_{GAL}	1.5	–	1.5	–	1.5	–	Clk. Per.
47	Asynchronous Input Setup Time	t_{ASI}	20	–	20	–	20	–	ns
57	$\overline{BGACK}$ High to Control Bus Driven	t_{GABD}	1.5	–	1.5	–	1.5	–	Clk. Per.
58[1]	$\overline{BG}$ High to Control Bus Driven	t_{GHBD}	1.5	–	1.5	–	1.5	–	Clk. Per.

NOTES:

1. The processor will negate $\overline{BG}$ and begin driving the bus again if external arbitration logic negates $\overline{BR}$ before asserting $\overline{BGACK}$.
2. The minimum value must be met to guarantee proper operation. If the maximum value is exceeded, $\overline{BG}$ may be reasserted.

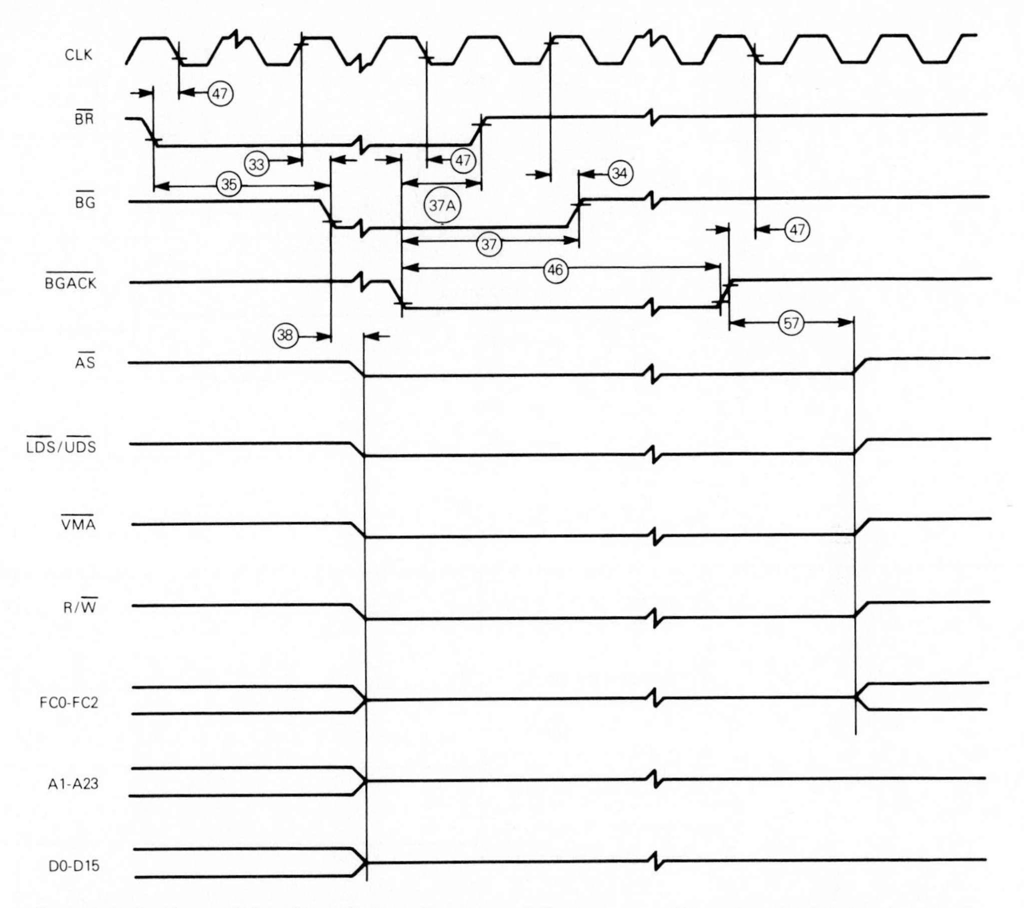

Figure B.4 Bus arbitration timing diagram: Idle bus case (Courtesy of Motorola, Inc.)

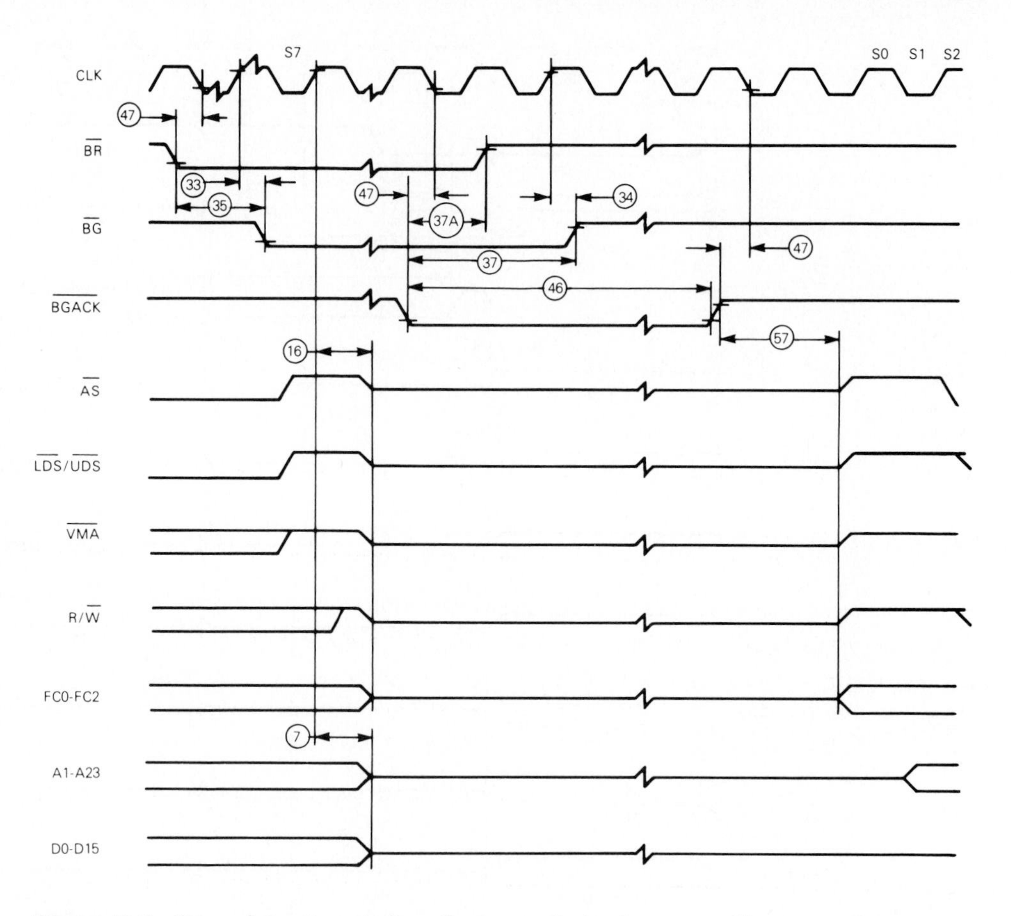

Figure B.5 **Bus arbitration timing diagram: Active bus case (Courtesy of Motorola, Inc.)**

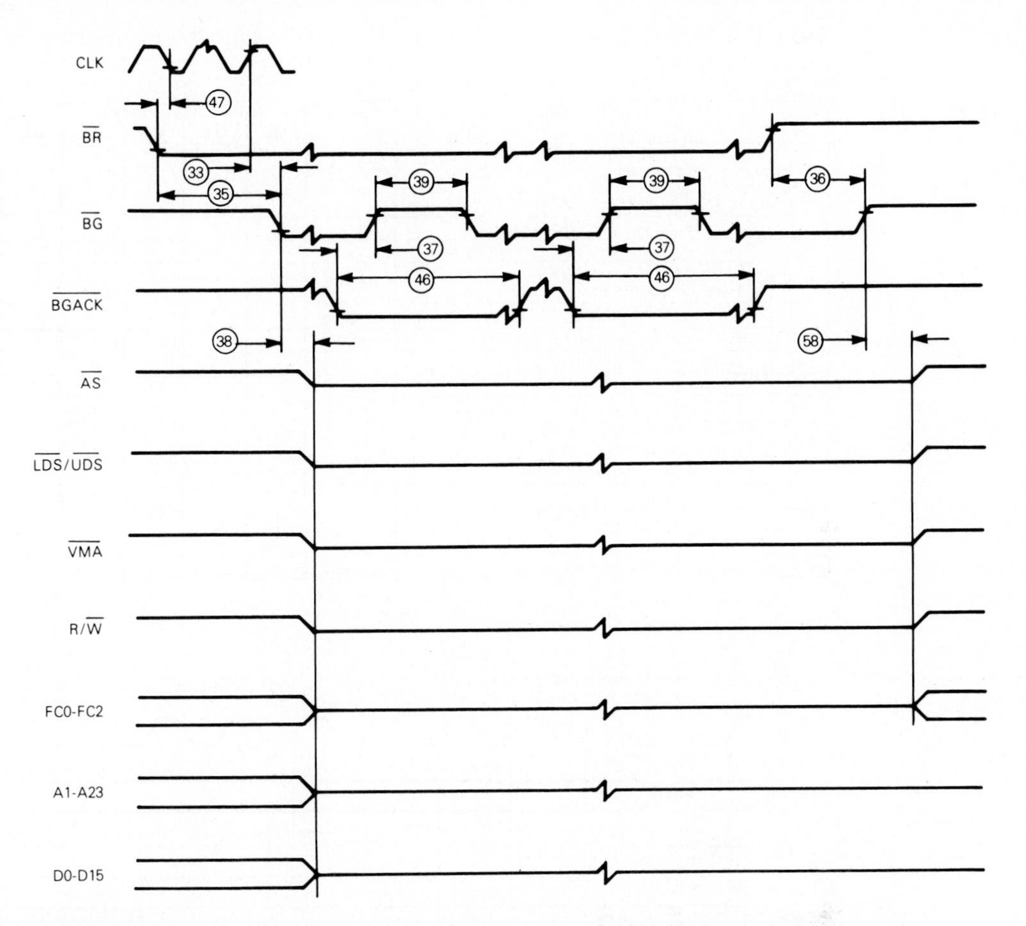

Figure B.6 Bus arbitration timing diagram: Multiple bus requests (Courtesy of Motorola, Inc.)

Table B.4 AC Electrical Specifications: MC68000 to MC6800 Peripheral (Courtesy of Motorola, Inc.)

Num.	Characteristic	Symbol	8 MHz		10 MHz		12.5 MHz		Unit
			Min	Max	Min	Max	Min	Max	
12	Clock Low to $\overline{AS}$, $\overline{DS}$ High	t_{CLSH}	–	70	–	55	–	50	ns
18	Clock High to R/$\overline{W}$ High	t_{CHRH}	0	70	0	60	0	60	ns
20	Clock High to R/$\overline{W}$ Low (Write)	t_{CHRL}	–	70	–	60	–	60	ns
23	Clock Low to Data Out Valid (Write)	t_{CLDO}	–	70	–	55	–	55	ns
27	Data In to Clock Low (Setup Time on Read)	t_{CLDO}	15	–	10	–	10	–	ns
29	$\overline{AS}$, $\overline{DS}$ High to Data In Invalid (Hold Time on Read)	t_{SHDII}	0	–	0	–	0	–	ns
40	Clock Low to $\overline{VMA}$ Low	t_{CLVML}	–	70	–	70	–	70	ns
41	Clock Low to E Transition	t_{CLET}	–	70	–	55	–	45	ns
42	E Output Rise and Fall Time	$t_{Er,f}$	–	25	–	25	–	25	ns
43	$\overline{VMA}$ Low to E High	t_{VMLEH}	200	–	150	–	90	–	ns
44	$\overline{AS}$, $\overline{DS}$ High to $\overline{VPA}$ High	t_{SHVPH}	0	120	0	90	0	70	ns
45	E Low to Control, Address Bus Invalid (Address Hold Time)	t_{ELCAI}	30	–	10	–	10	–	ns
47	Asynchronous Input Setup Time	t_{ASI}	20	–	20	–	20	–	ns
49[1]	$\overline{AS}$, $\overline{DS}$ High to E Low	t_{SHEL}	–70	70	–55	55	–45	45	ns
50	E Width High	t_{EH}	450	–	350	–	280	–	ns
51	E Width Low	t_{EL}	700	–	550	–	440	–	ns
54	E Low to Data Out Invalid	t_{ELDOi}	30	–	20	–	15	–	ns

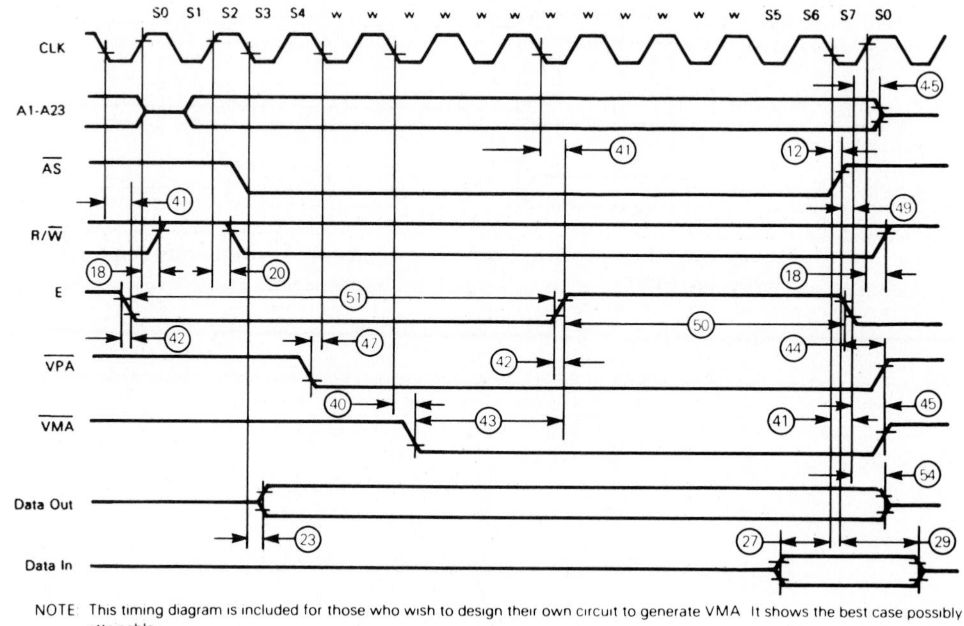

Figure B.7 **MC 68000 to M6800 peripheral timing diagram: Best case (Courtesy of Motorola, Inc.)**

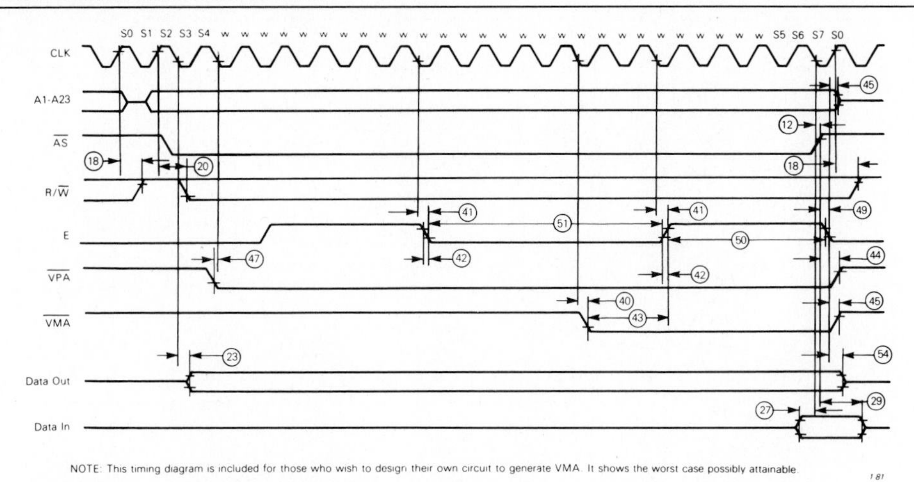

Figure B.8 MC68000 to M6800 peripheral timing diagram: Worst case (Courtesy of Motorola, Inc.)

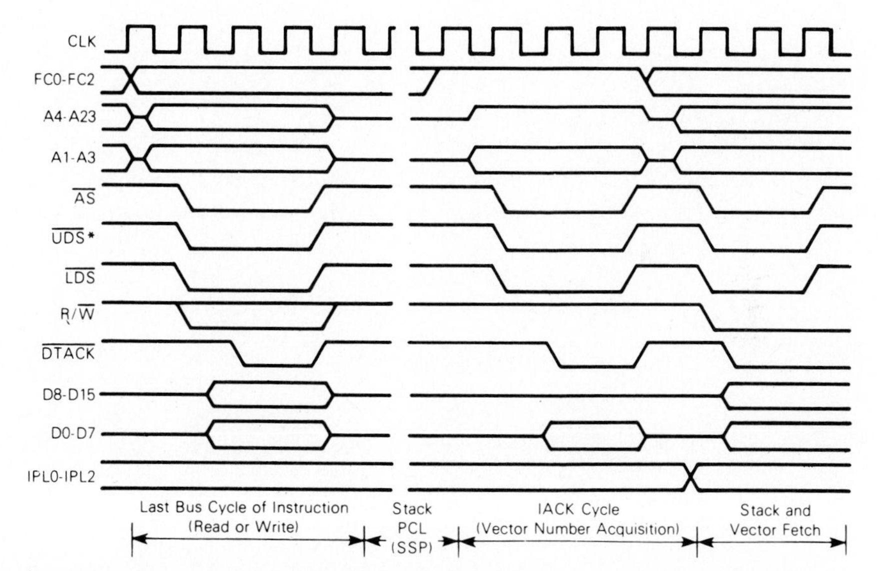

Figure B.9 Interrupt acknowledge cycle timing diagram (Courtesy of Motorola, Inc.)

APPENDIX C OVERALL SYSTEM SCHEMATIC DIAGRAM

The following schematic diagram shows the system as developed throughout the text. The system is created to illustrate various concepts in relationship to the 68000 microprocessor-based computer. This system is presently being breadboarded by students in the author's classes. Changes to the schematic design will occur as problems arise. Difficulty with timing considerations is expected to be responsible for most required changes. An updated version of this system will replace this initial effort once the system is operational. At that time, it is also anticipated that 68000 support devices in place of the 6800 ones in use now will be included.

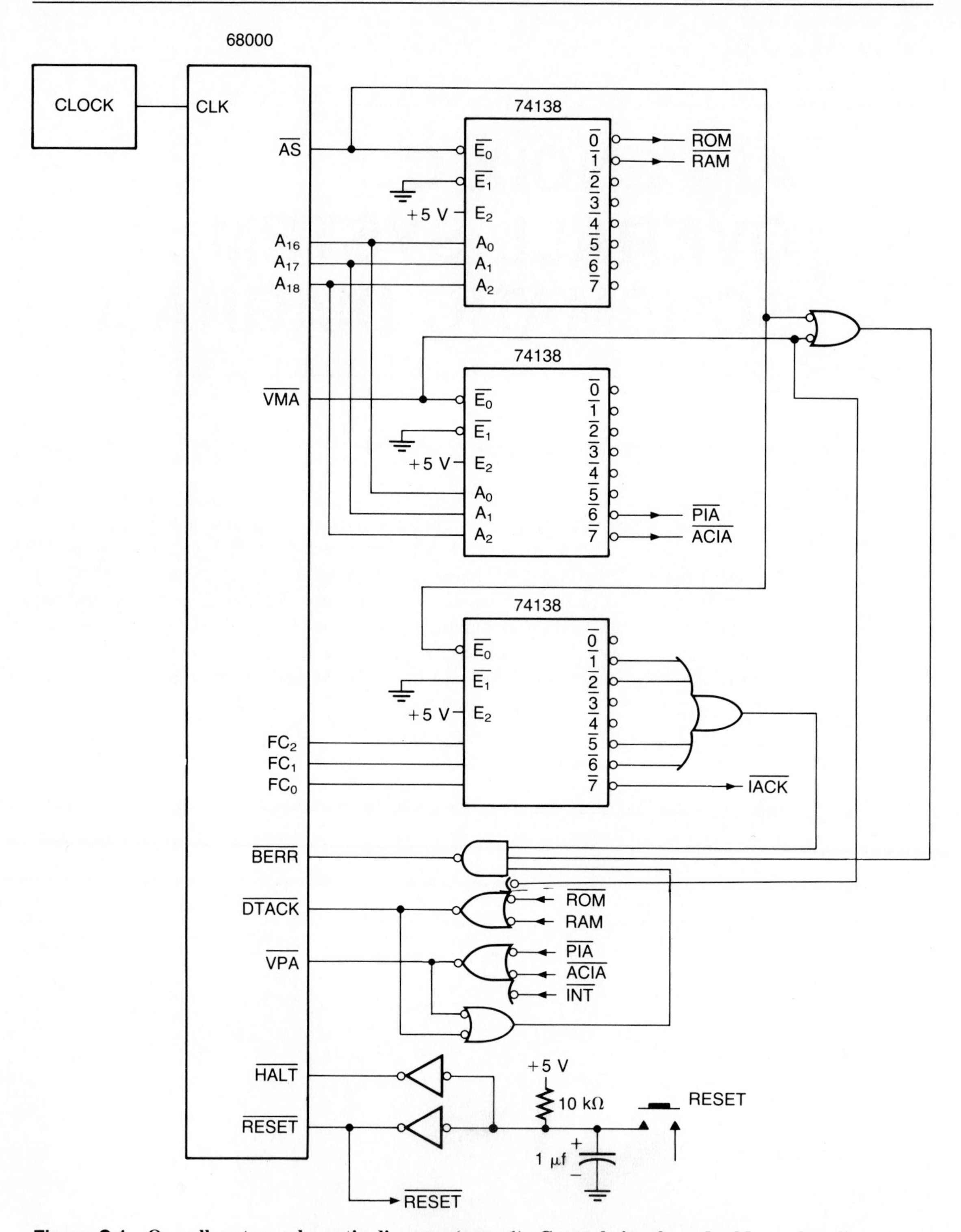

Figure C.1 Overall system schematic diagram (page 1). Control signals and address decoding.

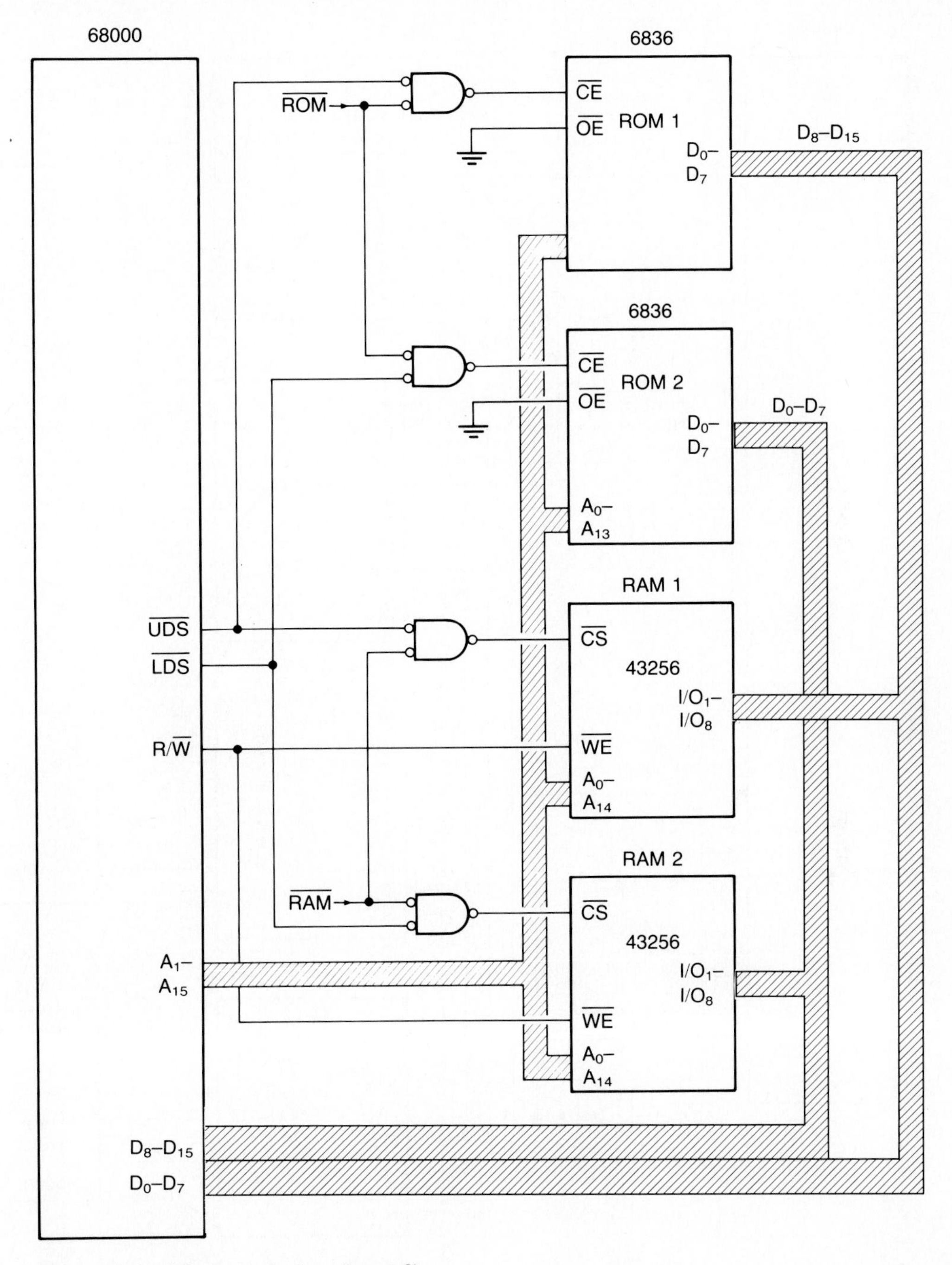

Figure C.1 Memory devices (page 2)

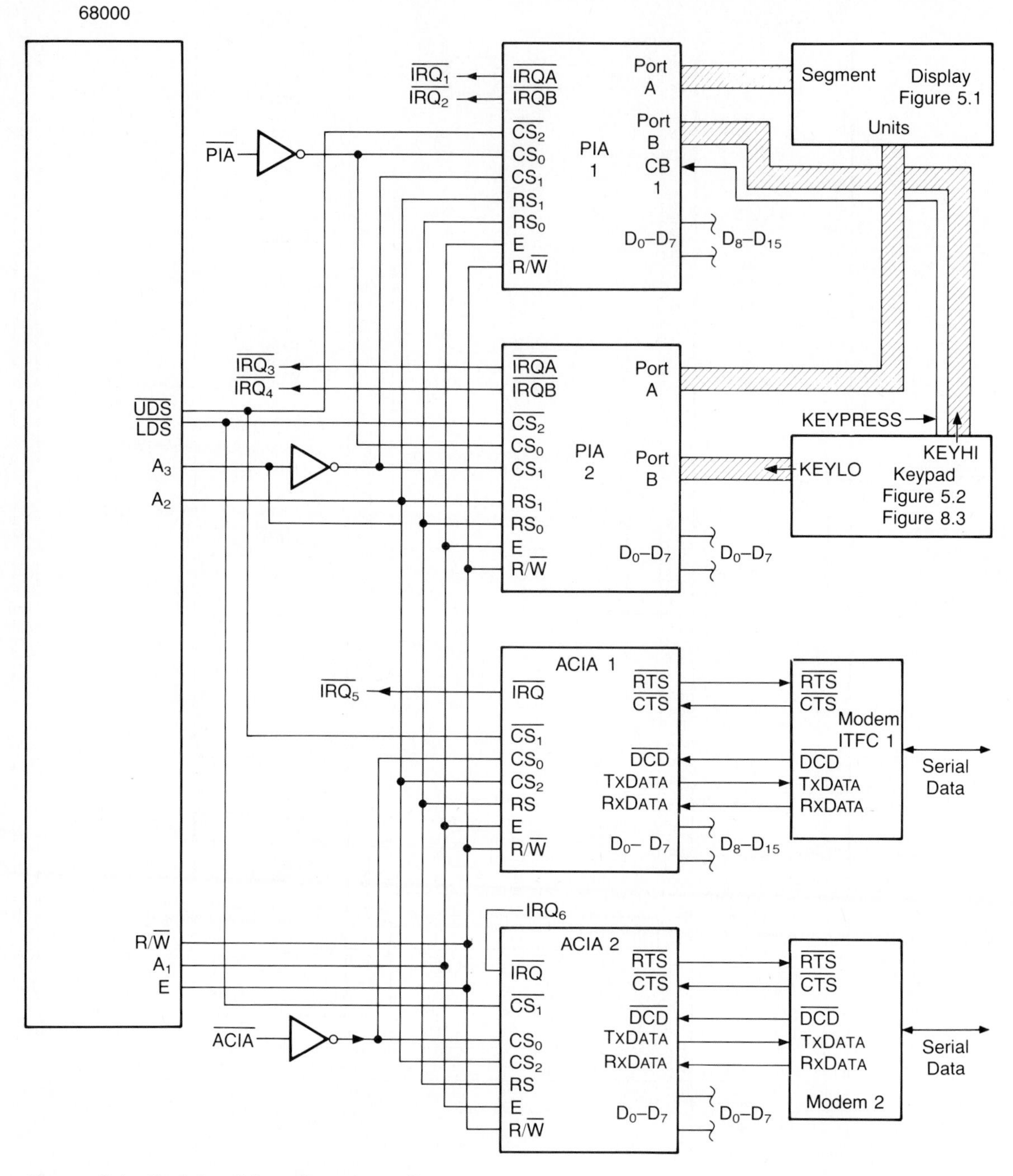

Figure C.1 **Peripheral interfaces (page 3)**

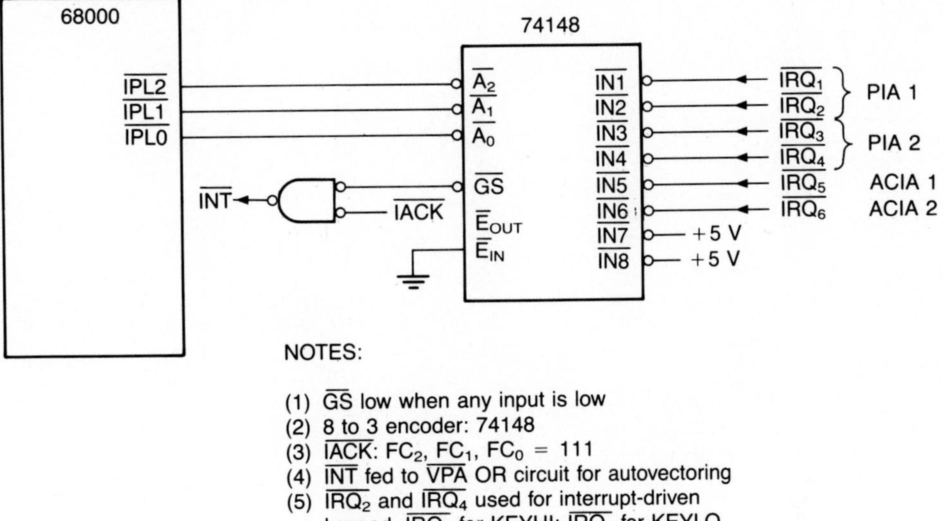

Figure C.1 **Interrupt priority circuit (page 4)**

APPENDIX D DEVICE DATA SHEETS

Device data sheets are reproduced courtesy of Signetics Corporation (pages 510–515), Intel Corporation (pages 516–526), NEC Electronics Inc. (pages 527–531), and Motorola, Inc. (pages 532–551).

1-OF-8 DECODER/DEMULTIPLEXER — 54/74 SERIES "138"

54S/74S138
54LS/74LS138

LOGIC SYMBOL

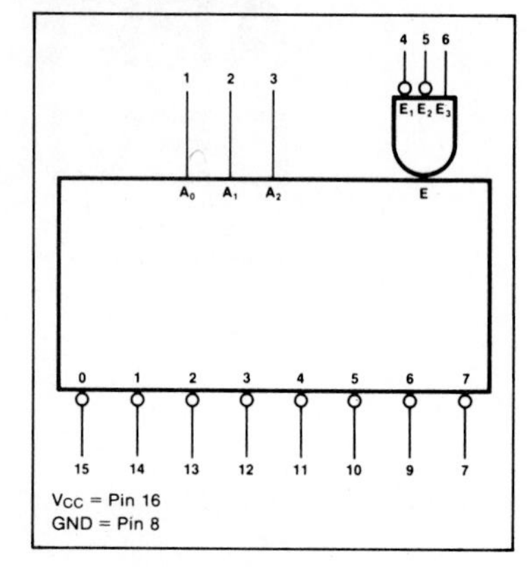

DESCRIPTION

The "138" is a HIGH speed 1-of-8 Decoder/Demultiplexer. The "138" is ideal for HIGH speed bipolar memory chip select address decoding. The multiple input enables allow parallel expansion to a 1-of-24 decoder using only three "138" devices; or to a 1-of-32 decoder using four "138" devices and one inverter.

FEATURES

- **Demultiplexing capability**
- **Multiple input enable for easy expansion**
- **Ideal for memory chip select decoding**
- **Direct replacement for Intel 3205**

PIN CONFIGURATION

Pin	Name	Pin	Name
1	A_0	16	V_{CC}
2	A_1	15	$\overline{0}$
3	A_2	14	$\overline{1}$
4	$\overline{E}_1$	13	$\overline{2}$
5	$\overline{E}_2$	12	$\overline{3}$
6	E_3	11	$\overline{4}$
7	$\overline{7}$	10	$\overline{5}$
8	GND	9	$\overline{6}$

ORDERING CODE (See Section 9 for further Package and Ordering Information)

PACKAGES	COMMERCIAL RANGES V_{CC} = 5V ± 5%; T_A = 0°C to 70°C	MILITARY RANGES V_{CC} = 5V ± 10%; T_A = −55°C to +125°C
Plastic DIP	N74S138N • N74LS138N	
Ceramic DIP	N74S138F • N74LS138F	S54S138F • S54LS138F
Flatpak		S54S138W • S54LS138W

INPUT AND OUTPUT LOADING AND FAN-OUT TABLE [a]

PINS	DESCRIPTION		54/74	54S/74S	54LS/74LS
A_0-A_2	Address inputs	I_{IH} (μA) I_{IL} (mA)		50 -2.0	20 -0.36
$\overline{E}_1$, $\overline{E}_2$	Enable (Active LOW) inputs	I_{IH} (μA) I_{IL} (mA)		50 -2.0	20 -0.36
E_3	Enable (Active HIGH) input	I_{IH} (μA) I_{IL} (mA)		50 -2.0	20 -0.36
$\overline{0}$-$\overline{7}$	Decoder ouputs	I_{OH} (μA) I_{OL} (mA)		-1000 20	-400 4/8[a]

NOTE

a. The slashed numbers indicate different parametric values for Military/Commercial temperature ranges respectively.

1-OF-8 DECODER/DEMULTIPLEXER — 54/74 SERIES "138"

FUNCTIONAL DESCRIPTION

The "138" decoder accepts three binary weighted inputs (A_0,A_1,A_2) and when enabled provides eight mutually exclusive active LOW outputs ($\overline{0}$-$\overline{7}$). The device features three Enable inputs: two active LOW ($\overline{E}_1$,$\overline{E}_2$) and one active HIGH (E_3). Every output will be HIGH unless $\overline{E}_1$ and $\overline{E}_2$ are LOW and E_3 is HIGH. This multiple Enable function allows easy parallel expansion of the device to a 1-of-32 (5 lines to 32 lines) decoder with just four "138's" and one inverter.

The device can be used as an eight output demultiplexer by using one of the active LOW Enable inputs as the data input and the remaining Enable inputs as strobes. Enable inputs not used must be permanently tied to their appropriate active HIGH or active LOW state.

TRUTH TABLE

INPUTS						OUTPUTS							
$\overline{E}_1$	$\overline{E}_2$	E_3	A_0	A_1	A_2	$\overline{0}$	$\overline{1}$	$\overline{2}$	$\overline{3}$	$\overline{4}$	$\overline{5}$	$\overline{6}$	$\overline{7}$
H	X	X	X	X	X	H	H	H	H	H	H	H	H
X	H	X	X	X	X	H	H	H	H	H	H	H	H
X	X	L	X	X	X	H	H	H	H	H	H	H	H
L	L	H	L	L	L	L	H	H	H	H	H	H	H
L	L	H	H	L	L	H	L	H	H	H	H	H	H
L	L	H	L	H	L	H	H	L	H	H	H	H	H
L	L	H	H	H	L	H	H	H	L	H	H	H	H
L	L	H	L	L	H	H	H	H	H	L	H	H	H
L	L	H	H	L	H	H	H	H	H	H	L	H	H
L	L	H	L	H	H	H	H	H	H	H	H	L	H
L	L	H	H	H	H	H	H	H	H	H	H	H	L

NOTES

H = HIGH voltage level
L = LOW voltage level
X = Don't care

LOGIC DIAGRAM

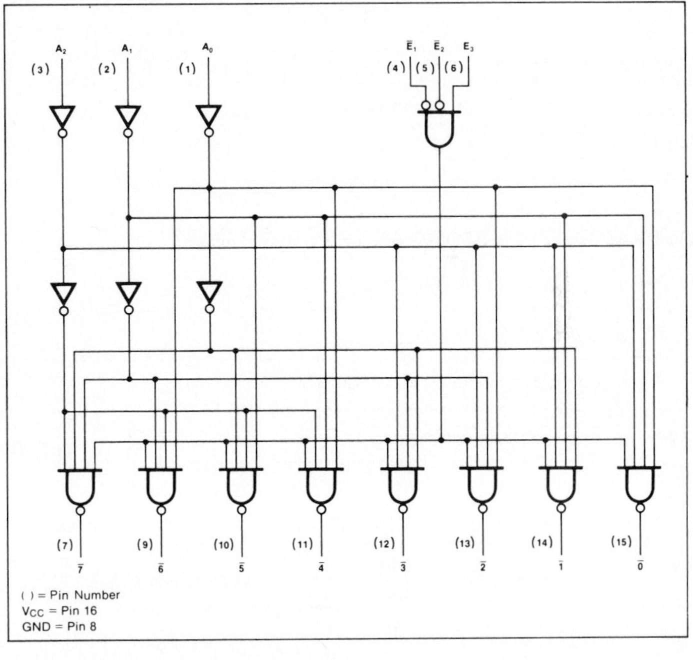

DC CHARACTERISTICS OVER OPERATING TEMPERATURE RANGE [b]

PARAMETER		TEST CONDITIONS	54/74		54S/74S		54LS/74LS		UNIT
			Min	Max	Min	Max	Min	Max	
I_{CC}	Supply current	V_{CC} = Max				74		10	mA

NOTE

b. For family dc characteristics, see inside front cover for 54/74 and 54H/74H, and see inside back cover for 54S/74S and 54LS/74LS specifications.

DUAL 1-of-4 DECODER/DEMULTIPLEXER — 54/74 SERIES "139"

54S/74S139
54LS/74LS139

DESCRIPTION
The "139" is a high speed Dual 1-of-4 Decoder/Demultiplexer. This device has two independent decoders, each accepting two inputs and providing four mutually exclusive active LOW outputs. Each decoder has an active LOW Enable input useable as a data input for a 1-of-4 demultiplexer. Each half of the "139" is useable as a function generator providing all four minterms of two variables.

FEATURES
- **Demultiplexing capability**
- **Two independent 1-of-4 decoders**
- **Multifunction capability**
- **Replaces 9321 and 93L21 for higher performance**

LOGIC SYMBOL

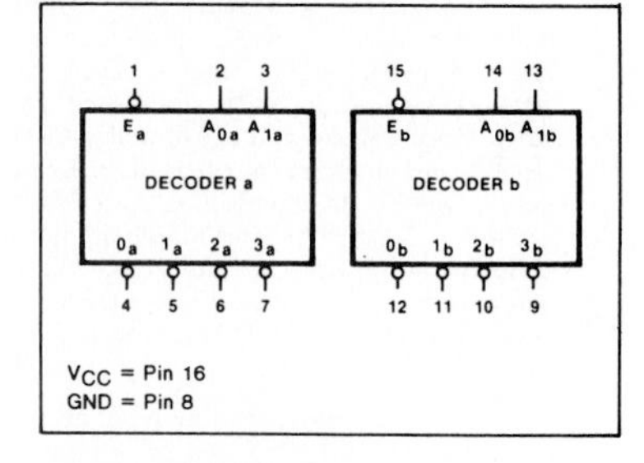

PIN CONFIGURATION

Pin	Name	Pin	Name
1	$\bar{E}_a$	16	V_{CC}
2	A_{0a}	15	$\bar{E}_b$
3	A_{1a}	14	A_{0b}
4	$\bar{0}_a$	13	A_{1b}
5	$\bar{1}_a$	12	$\bar{0}_b$
6	$\bar{2}_a$	11	$\bar{1}_b$
7	$\bar{3}_a$	10	$\bar{2}_b$
8	GND	9	$\bar{3}_b$

ORDERING CODE (See Section 9 for further Package and Ordering Information)

PACKAGES	COMMERCIAL RANGES $V_{CC}=5V\pm5\%$; $T_A=0°C$ to $+70°C$	MILITARY RANGES $V_{CC}=5V\pm10\%$; $T_A=-55°C$ to $+125°C$
Plastic DIP	N74S139N • N74LS139N	
Ceramic DIP	N74S139F • N74LS139F	S54S139F • S54LS139F
Flatpak		S54S139W • S54LS139W

INPUT AND OUTPUT LOADING AND FAN-OUT TABLE[a]

PINS	DESCRIPTION		54/74	54S/74S	54LS/74LS
A_0, A_1	Address inputs	I_{IH} (μA) I_{IL} (mA)		50 −2.0	20 −0.36
$\bar{E}$	Enable (Active LOW) inputs	I_{IH} (μA) I_{IL} (mA)		50 −2.0	20 −0.36
$\bar{0}$-$\bar{3}$	Decoder outputs	I_{OH} (μA) I_{OL} (mA)		−1000 20	−400 4/8(a)

DC CHARACTERISTICS OVER OPERATING TEMPERATURE RANGE[b]

PARAMETER		TEST CONDITIONS	54/74		54S/74S		54LS/74LS		UNIT
			Min	Max	Min	Max	Min	Max	
I_{CC}	Supply current	V_{CC} = Max, $V_{\bar{E}}$ = 0V				90		11	mA

NOTES
a. The slashed numbers indicate different parametric values for Military/Commercial temperature ranges respectively.
b. For family dc characteristics, see inside front cover for 54/74 and 54H/74H, and see inside back cover for 54S/74S and 54LS/74LS specifications.

8-INPUT PRIORITY ENCODER — 54/74 SERIES "148"

54/74148

DESCRIPTION

The "148" is an 8-Input Priority Encoder designed to accept eight parallel inputs and produce the binary weighted code of the highest order input. Cascading capability has been provided to allow expansion without the need for external circuitry.

FEATURES

- Code conversions
- Multi-channel D/A converter
- Decimal to BCD converter
- Cascading for priority encoding of "N" bits
- Input enable capability
- Priority encoding—automatic selection of highest priority input line
- Output Enable—active LOW when all inputs HIGH
- Group Signal output—active when any input is LOW

LOGIC SYMBOL

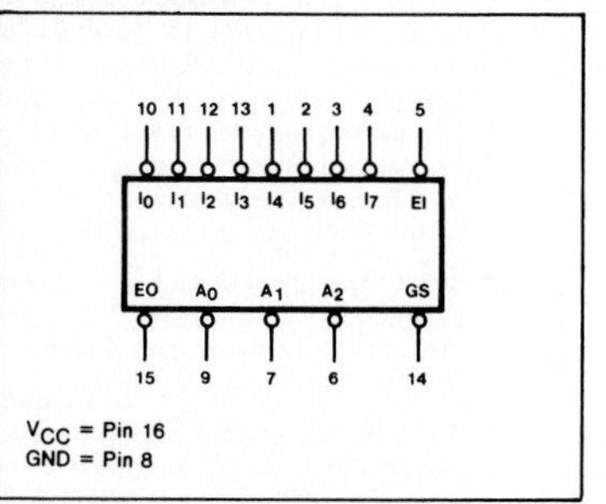

PIN CONFIGURATION

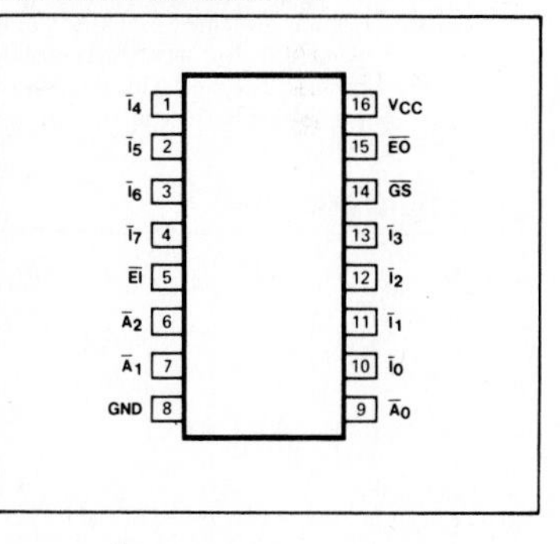

ORDERING CODE (See Section 9 for further Package and Ordering Information)

PACKAGES	COMMERCIAL RANGES $V_{CC}=5V\pm5\%$; $T_A=0°C$ to $+70°C$	MILITARY RANGES $V_{CC}=5V\pm10\%$; $T_A=-55°C$ to $+125°C$
Plastic DIP	N74148N	
Ceramic DIP	N74148F	S54148F
Flatpak		S54148W

INPUT AND OUTPUT LOADING AND FAN-OUT TABLE[a]

PINS	DESCRIPTION		54/74	54S/74S	54LS/74LS
$\overline{I_0}$	Priority (active LOW) input	I_{IH} (μA) I_{IL} (mA)	40 −1.6		
$\overline{I_1}$-$\overline{I_7}$	Priority (active LOW) inputs	I_{IH} (μA) I_{IL} (mA)	80 −3.2		
$\overline{EI}$	Enable (active LOW) input	I_{IH} (μA) I_{IL} (mA)	80 −3.2		
$\overline{A_0}$-$\overline{A_2}$	Address (active LOW) outputs	I_{OH} (μA) I_{OL} (mA)	−800 16		
$\overline{EO}$	Enable (active LOW) output	I_{OH} (μA) I_{OL} (mA)	−800 16		
$\overline{GS}$	Group Signal (active LOW) output	I_{OH} (μA) I_{OL} (mA)	−800 16		

NOTE

a. The slashed numbers indicate different parametric values for Military/Commercial temperature ranges respectively.

8-INPUT PRIORITY ENCODER — 54/74 SERIES "148"

FUNCTIONAL DESCRIPTION

The "148" 8-input priority encoder accepts data from eight active LOW inputs and provides a binary representation on the three active LOW outputs. A priority is assigned to each input so that when two or more inputs are simultaneously active, the input with the highest priority is represented on the output, with input line $\overline{I}_7$ having the highest priority.

A HIGH on the Input Enable ($\overline{EI}$) will force all outputs to the inactive (HIGH) state and allow new data to settle without producing erroneous information at the outputs.

A Group Signal output ($\overline{GS}$) and an Enable Output ($\overline{EO}$) are provided with the three data outputs. The $\overline{GS}$ is active level LOW when any input is LOW; this indicates when any input is active. The $\overline{EO}$ is active level LOW when all inputs are HIGH. Using the output enable along with the input enable allows priority coding of N input signals. Both $\overline{EO}$ and $\overline{GS}$ are active HIGH when the input enable is HIGH.

TRUTH TABLE

INPUTS									OUTPUTS				
$\overline{EI}$	$\overline{I}_0$	$\overline{I}_1$	$\overline{I}_2$	$\overline{I}_3$	$\overline{I}_4$	$\overline{I}_5$	$\overline{I}_6$	$\overline{I}_7$	$\overline{GS}$	$\overline{A}_0$	$\overline{A}_1$	$\overline{A}_2$	$\overline{EO}$
H	X	X	X	X	X	X	X	X	H	H	H	H	H
L	H	H	H	H	H	H	H	H	H	H	H	H	L
L	X	X	X	X	X	X	X	L	L	L	L	L	H
L	X	X	X	X	X	X	L	H	L	H	L	L	H
L	X	X	X	X	X	L	H	H	L	L	H	L	H
L	X	X	X	X	L	H	H	H	L	H	H	L	H
L	X	X	X	L	H	H	H	H	L	L	L	H	H
L	X	X	L	H	H	H	H	H	L	H	L	H	H
L	X	L	H	H	H	H	H	H	L	L	H	H	H
L	L	H	H	H	H	H	H	H	L	H	H	H	H

H = HIGH voltage level
L = LOW voltage level
X = Don't care

LOGIC DIAGRAM

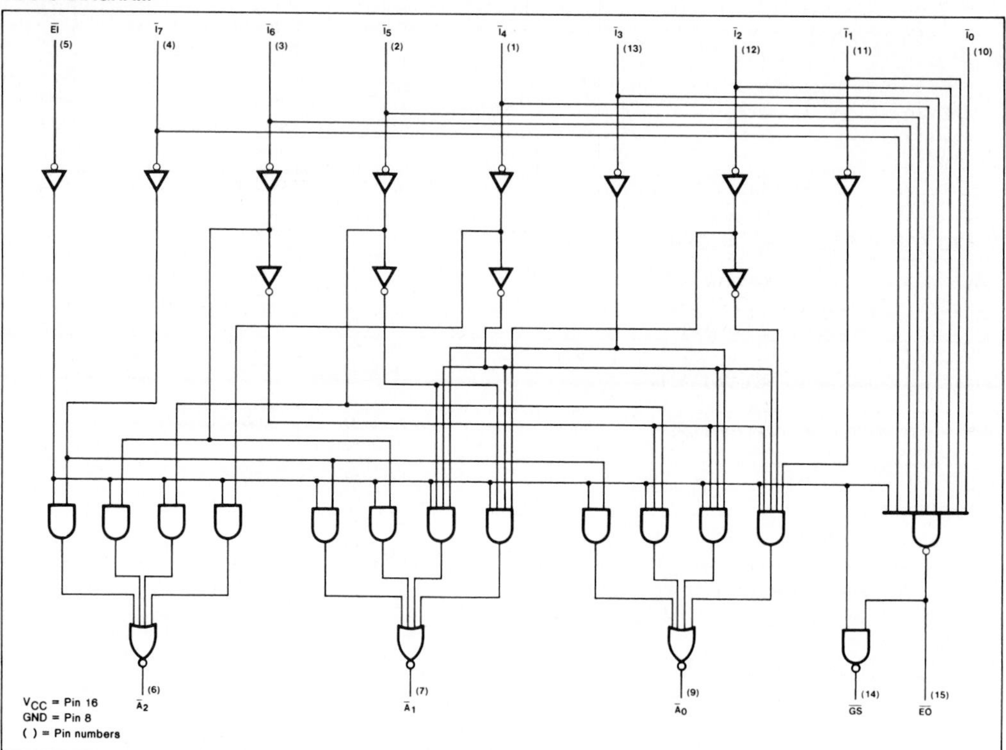

8-INPUT PRIORITY ENCODER — 54/74 SERIES "148"

DC CHARACTERISTICS OVER OPERATING TEMPERATURE RANGE[b]

PARAMETER		TEST CONDITIONS	54/74		54S/74S		54LS/74LS		UNIT
			Min	Max	Min	Max	Min	Max	
I_{OS}	Output short circuit current	V_{CC} = Max, V_{OUT} = 0V	−35	−85					mA
I_{CC}	Supply current	V_{CC} = Max, $V_7 = V_{EI}$ = 0V		60					mA
		V_{CC} = Max, inputs open		55					mA

AC CHARACTERISTICS: T_A=25°C (See Section 4 for Test Circuits and Conditions)

PARAMETER		TEST CONDITIONS	54/74		54S/74S		54LS/74LS		UNIT
			C_L = 15pF, R_L = 400Ω						
			Min	Max	Min	Max	Min	Max	
t_{PLH} t_{PHL}	Propagation delay $\overline{I}_n$ input to $\overline{A}_n$ outputs	Figure 1		19 19					ns ns
t_{PLH} t_{PHL}	Propagation delay $\overline{I}_n$ input to $\overline{EO}$ output	Figure 1		10 25					ns ns
t_{PLH} t_{PHL}	Propagation delay $\overline{I}_n$ input to $\overline{GS}$ output	Figure 2		30 25					ns ns
t_{PLH} t_{PHL}	Propagation delay $\overline{EI}$ input to $\overline{A}_n$ outputs	Figure 2		15 15					ns ns
t_{PLH} t_{PHL}	Propagation delay $\overline{EI}$ input to $\overline{EO}$ output	Figure 2		15 30					ns ns
t_{PLH} t_{PHL}	Propagation delay $\overline{EI}$ input to $\overline{GS}$ output	Figure 2		12 15					ns ns

NOTE

b. For family dc characteristics, see inside front cover for 54/74 and 54H/74H, and see inside back cover for 54S/74S and 54LS/74LS specifications

AC WAVEFORMS

WAVEFORM FOR INVERTING OUTPUTS

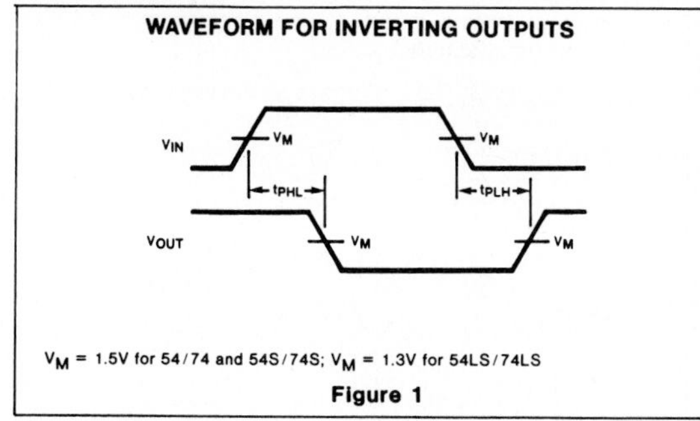

V_M = 1.5V for 54/74 and 54S/74S; V_M = 1.3V for 54LS/74LS

Figure 1

WAVEFORM FOR NON-INVERTING OUTPUTS

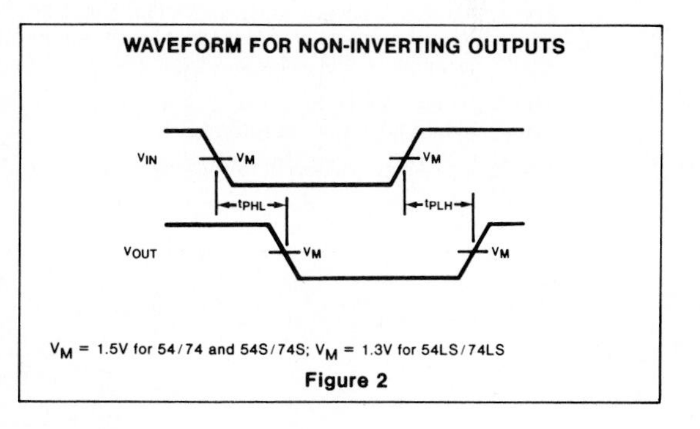

V_M = 1.5V for 54/74 and 54S/74S; V_M = 1.3V for 54LS/74LS

Figure 2

intel®

ADVANCE INFORMATION

27210
1M (64K x 16) WORD-WIDE EPROM

- **High-Performance HMOS* II-E**
  - **150 ns Access Time**
  - **Low 150 mA Active Power**
- **Complete Upgrade Capability**
  - **PGM "Don't Care" Status Allows Wiring in Higher Order Addresses**
- **Fast Programming**
  - **Quick-Pulse Programming™ Algorithm—8 Seconds Typical**
- **New Word-Wide Pinout**
  - **Clean, "Flow-Through" Architecture**
- **Standard EPROM Features**
  - **TTL Compatibility**
  - **Two Line Control**
  - **int$_e$ligent Identifier™ For Automated Programming**
- **40-Pin DIP and Compact 44-Lead PLCC(1) Packaging**
  (See Packaging Spec., Order #231369)

The Intel 27210 is a 5V only, 1,048,576-bit, Electrically Programmable Read Only Memory. It is organized as 64K-words of 16 bits each. It defines a new-clean memory architecture, oriented toward high-performance 16-bit and 32-bit CPUs, which simplifies circuit layout and offers a pin-compatible growth path to higher densities.

The 27210's unique circuit design provides for no-hardware-change upgrades to 4M-bits in the future. Since the $\overline{PGM}$ pin is a "don't care" state during read mode, direct connections to higher order addresses, A16 and A17, can be made without affecting the device's read operation. The 27210 will also be offered(1) in One-Time Programmable 40-pin plastic DIP and 44-lead PLCC—with the same 4M-bit upgrade path.

The 27210 provides the highest density and performance available to 16-bit and 32-bit microprocessors. Its by-16 organization makes it an ideal single-chip firmware solution in most microprocessor applications. The 27210's large capacity is sufficient for storage of operating system kernels in addition to standard bootstrap and diagnostic code. Direct execution of operating system software is made possible by the 27210's fast 150 ns access time, which yields no-WAIT-state operation in such high-performance CPUs as the 10 MHz 80286.

The 27210 is part of a three-product megabit EPROM family. Other family members are the 27010 and 27011. These two products have byte-wide organizations geared toward simple upgrades from lower densities. The 27010 is organized as 128K x 8 in a 32-pin DIP package which is pin-compatible with JEDEC-standard 28-pin 512K EPROMs. The 8 x 16K x 8 27011 utilizes page addressing, allowing "drop in" replacement of the 512 K-bit 27513 and continued no-hardware-change upgrades to 32 M-bits in the same JEDEC-compatible 28-pin site.

The 27210 shares several features with standard JEDEC EPROMs, including two-line output control for simplified interfacing and the int$_e$ligent Identifier™ feature for automated programming. It can also be programmed rapidly using Intel's Quick-Pulse Programming™ Algorithm, typically within 8 seconds.

The 27210 is manufactured using a scaled verison of Intel's advanced HMOS* II-E process which assures highest reliability and manufacturability.

*HMOS is a patented process of Intel Corporation.

NOTE:
1. Plastic dual-in-line package (P-DIP) and plastic leaded chip carrier (PLCC) will be available 2H '86.

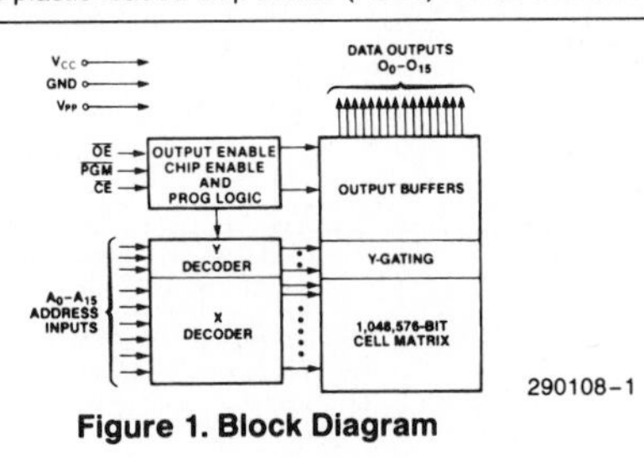

Figure 1. Block Diagram

Pin Names

A_0–A_{17}	ADDRESSES
$\overline{CE}$	CHIP ENABLE
$\overline{OE}$	OUTPUT ENABLE
O_0-O_{15}	OUTPUTS
$\overline{PGM}$	PROGRAM
N.C.	NO INTERNAL CONNECT
D.U.	DON'T USE

27210 P27210

4M	2M	27210	Pin	Pin	27210	2M	4M
V_{PP}	V_{PP}	VPP	1	40	VCC	V_{CC}	V_{CC}
$\overline{CE}$	$\overline{CE}$	$\overline{CE}$	2	39	$\overline{PGM}$	$\overline{PGM}$	A_{17}
O_{15}	O_{15}	O_{15}	3	38	NC	A_{16}	A_{16}
O_{14}	O_{14}	O_{14}	4	37	A_{15}	A_{15}	A_{15}
O_{13}	O_{13}	O_{13}	5	36	A_{14}	A_{14}	A_{14}
O_{12}	O_{12}	O_{12}	6	35	A_{13}	A_{13}	A_{13}
O_{11}	O_{11}	O_{11}	7	34	A_{12}	A_{12}	A_{12}
O_{10}	O_{10}	O_{10}	8	33	A_{11}	A_{11}	A11
O_9	O_9	O_9	9	32	A_{10}	A_{10}	A_{10}
O_8	O_8	O_8	10	31	A_9	A_9	A_9
GND	GND	GND	11	30	GND	GND	GND
O_7	O_7	O_7	12	29	A_8	A_8	A_8
O_6	O_6	O_6	13	28	A_7	A_7	A_7
O_5	O_5	O_5	14	27	A_6	A_6	A_6
O_4	O_4	O_4	15	26	A_5	A_5	A_5
O_3	O_3	O_3	16	25	A_4	A_4	A_4
O_2	O_2	O_2	17	24	A_3	A_3	A_3
O_1	O_1	O_1	18	23	A_2	A_2	A_2
O_0	O_0	O_0	19	22	A_1	A_1	A_1
$\overline{OE}$	$\overline{OE}$	$\overline{OE}$	20	21	A_0	A_0	A_0

290108–2

NOTE: Compatible Higher Density Word Wide EPROM Pin Configurations are Shown in the Blocks Adjacent to the 27210 Pins

Figure 2. Cerdip/Plastic(P) DIP Pin Configurations

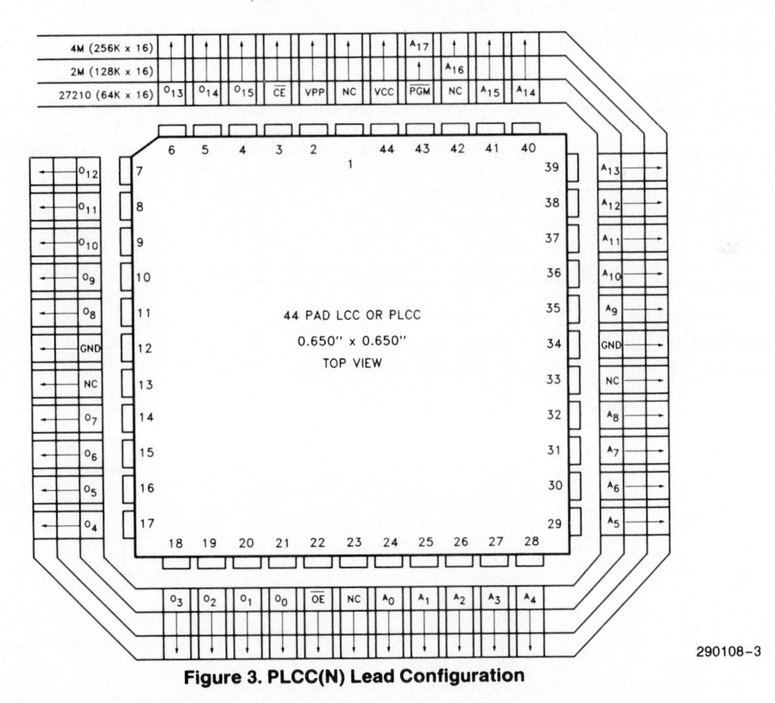

Figure 3. PLCC(N) Lead Configuration

NOTES:
1. P-DIP and PLCC will be available 2H '86.
2. Intel "Universal Site" compatible EPROM pin configurations are shown in the blocks adjacent to the 27210 pins.

intel 27210 ADVANCE INFORMATION

EXTENDED TEMPERATURE (EXPRESS) EPROMS

The Intel EXPRESS EPROM family is a series of electrically programmable read only memories which have received additional processing to enhance product characteristics. EXPRESS processing is available for several densities of EPROM, allowing the choice of appropriate memory size to match system applications. EXPRESS EPROM products are available with 168 ±8 hour, 125°C dynamic burn-in using Intel's standard bias configuration. This process exceeds or meets most industry specifications of burn-in. The standard EXPRESS EPROM operating temperature range is 0°C to 70°C. Extended operating temperature range (−40°C to +85°C) EXPRESS products are available. Like all Intel EPROMs, the EXPRESS EPROM family is inspected to 0.1% electrical AQL. This may allow the user to reduce or eliminate incoming inspection testing.

EXPRESS EPROM PRODUCT FAMILY

PRODUCT DEFINITIONS

Type	Operating Temperature	Burn-in 125°C (hr)
Q	0°C to +70°C	168 ±8
T	−40°C to +85°C	None
L	−40°C to +85°C	168 ±8

EXPRESS OPTIONS

27210 Versions

Packaging Options			
Speed Versions	Cerdip	PLCC*	Plastic*
-170/05	Q		
-200/05	T, L, Q		
-250/05	T, L, Q		
-170/10	T, L, Q		
-200/10	T, L, Q		
-250/10	T, L, Q		

*Available 2H '86

READ OPERATION

D.C. CHARACTERISTICS

Electrical Parameters of Express EPROM Products are identical to standard EPROM parameters except for:

Symbol	Parameter	T27210, L27210 Min	T27210, L27210 Max	Test Conditions
I_{SB}	V_{CC} Standby Current (mA)		50	$\overline{CE} = V_{IH}$, $\overline{OE} = V_{IL}$
I_{CC1}(1)	V_{CC} Active Current (mA)		170	$\overline{OE} = \overline{CE} = V_{IL}$
	V_{CC} Active Current at High Temperature (mA)		150	$\overline{OE} = \overline{CE} = V_{IL}$, $V_{PP} = V_{CC}$ $T_{Ambient} = 85°C$

NOTE:
1. The maximum current value is with Outputs O_0 to O_{15} unloaded.

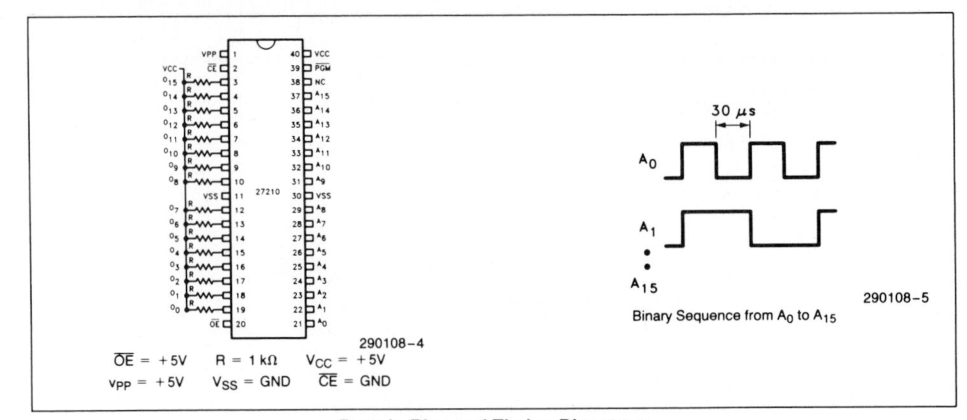

Burn-In Bias and Timing Diagrams

ABSOLUTE MAXIMUM RATINGS*

Operating Temperature During Read .... 0°C to +70°C

Temperature Under Bias .... −10°C to +80°C

Storage Temperature .... −65°C to +125°C

All Input or Output Voltages with Respect to Ground .... −0.6V to +6.25V

Voltage on A_9 with Respect to Ground .... −0.6V to +13.0V

V_{PP} Supply Voltage with Respect to Ground During Programming .... −0.6V to +14V

V_{CC} Supply Voltage with Respect to Ground .... −0.6V to +7.0V

**Notice: Stresses above those listed under "Absolute Maximum Ratings" may cause permanent damage to the device. This is a stress rating only and functional operation of the device at these or any other conditions above those indicated in the operational sections of this specification is not implied. Exposure to absolute maximum rating conditions for extended periods may affect device reliability.*

NOTICE: Specifications contained within the following tables are subject to change.

READ OPERATION

D.C. CHARACTERISTICS $0°C \leq T_A \leq +70°C$

Symbol	Parameter	Limits				Conditions
		Min	Typ(3)	Max	Units	
I_{LI}	Input Load Current			1	μA	$V_{IN} = 5.5V$
I_{LO}	Output Leakage Current			1	μA	$V_{OUT} = 5.5V$
I_{PP1}(2)	V_{PP} Load Current Read			1	μA	$V_{PP} = 5.5V$
I_{SB}	V_{CC} Current Standby			40	mA	$\overline{CE} = V_{IH}$
I_{CC1}(2)	V_{CC} Current Active			150	mA	$\overline{CE} = OE = V_{IL}$
V_{IL}	Input Low Voltage	−0.1		+0.8	V	
V_{IH}	Input High Voltage	2.0		$V_{CC}+1$	V	
V_{OL}	Output Low Voltage			0.45	V	$I_{OL} = 2.1$ mA
V_{OH}	Output High Voltage	2.4			V	$I_{OH} = -400$ μA

A.C. CHARACTERISTICS $0°C \leq T_A \leq +70°C$

Versions(5)	$V_{CC} \pm 5\%$	27210-150/05		27210-170/05 P27210-170/05 N27210-170/05		27210-200/05 P27210-200/05 N27210-200/05		27210-250/05 P27210-250/05 N27210-250/05		Unit
	$V_{CC} \pm 10\%$			27210-170/10		27210-200/10 P27210-200/10 N27210-200/10		27210-250/10 P27210-250/10 N27210-250/10		
Symbol	Characteristics	Min	Max	Min	Max	Min	Max	Min	Max	
t_{ACC}	Address to Output Delay		150		170		200		250	ns
t_{CE}	$\overline{CE}$ to Output Delay		150		170		200		250	ns
t_{OE}	$\overline{OE}$ to Output Delay		60		60		75		75	ns
t_{DF}(4)	$\overline{OE}$ High to Output Float	0	50	0	50	0	60	0	60	ns
t_{OH}	Output Hold from Addresses $\overline{CE}$ or $\overline{OE}$ Whichever Occurred First	0		0		0		0		ns

NOTES:

1. V_{CC} must be applied simultaneously or before V_{PP} and removed simultaneously or after V_{PP}.
2. The maximum current value is with Outputs O_0 to O_{15} unloaded.
3. Typical values are for $T_A = 25°C$ and nominal supply voltages.
4. This parameter is only sampled and is not 100% tested. Output Float is defined as the point where data is no longer driven—see timing diagram.
5. Packaging options: No prefix = Cerdip; Plastic DIP = P; PLCC = N. P-DIP and PLCC will be available in 2H '86.

CAPACITANCE(2) T_A = 25°C, f = 1MHz

Symbol	Parameter	Typ(1)	Max	Unit	Conditions
C_{IN}	Input Capacitance	4	6	pF	V_{IN} = 0V
C_{OUT}	Output Capacitance	8	12	pF	V_{OUT} = 0V
C_{VPP}	V_{PP} Input Capacitance		25	pF	V_{PP} = 0V

A.C. TESTING INPUT/OUTPUT WAVEFORM

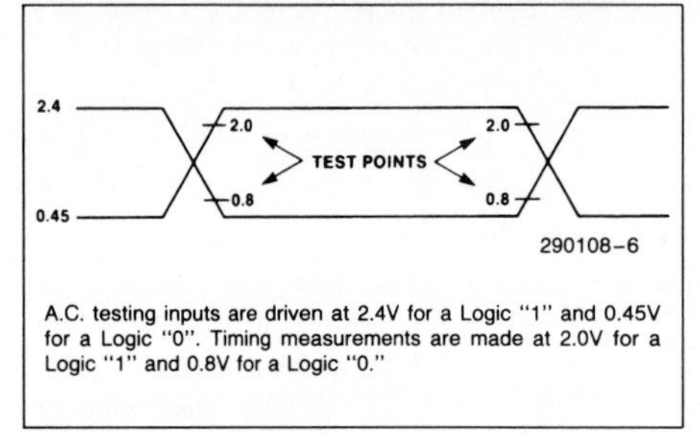

A.C. testing inputs are driven at 2.4V for a Logic "1" and 0.45V for a Logic "0". Timing measurements are made at 2.0V for a Logic "1" and 0.8V for a Logic "0."

A.C. TESTING LOAD CIRCUIT

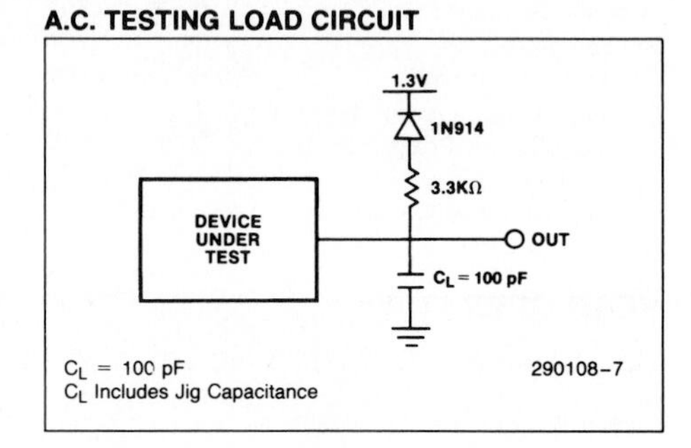

C_L = 100 pF
C_L Includes Jig Capacitance

A.C. WAVEFORMS

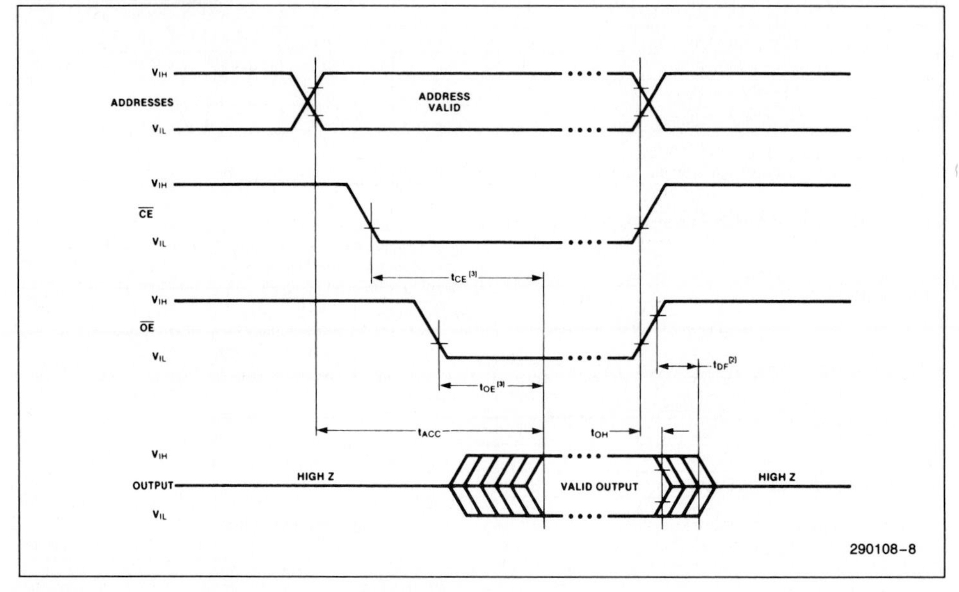

NOTES:
1. Typical values are for T_A = 25°C and nominal supply voltages.
2. This parameter is only sampled and is not 100% tested.
3. $\overline{OE}$ may be delayed up to t_{CE}—t_{OE} after the falling edge of $\overline{CE}$ without impact on t_{CE}.

27210 ADVANCE INFORMATION

DEVICE OPERATION

The modes of operation of the 27210 are listed in Table 1. A single 5V power supply is required in the read mode. All inputs are TTL levels except for V_{PP} and 12V on A_9 for int$_e$ligent Identifier.

Table 1. Modes Selection

Mode \ Pins		$\overline{CE}$	$\overline{OE}$	$\overline{PGM}$	A_9	A_0	V_{PP}	V_{CC}	Outputs
Read		V_{IL}	V_{IL}	X	X(1)	X	X	5.0V	D_{OUT}
Output Disable		V_{IL}	V_{IH}	X	X	X	X	5.0V	High Z
Standby		V_{IH}	X	X	X	X	X	5.0V	High Z
Programming		V_{IL}	V_{IH}	V_{IL}	X	X	(Note 4)	(Note 4)	D_{IN}
Program Verify		V_{IL}	V_{IL}	V_{IH}	X	X	(Note 4)	(Note 4)	D_{OUT}
Program Inhibit		V_{IH}	X	X	X	X	(Note 4)	(Note 4)	High Z
int$_e$ligent Identifier	Manufacturer(3)	V_{IL}	V_{IL}	X	V_H(2)	V_{IL}	V_{CC}	5.0V	0089 H
	Device(3)	V_{IL}	V_{IL}	X	V_H(2)	V_{IH}	V_{CC}	5.0V	00FFH

NOTES:
1. X can be V_{IL} or V_{IH}
2. $V_H = 12.0V \pm 0.5V$
3. A_1-A_8, $A_{10}-A_{15} = V_{IL}$
4. See Table 2 for V_{CC} and V_{PP} voltages.

Read Mode

The 27210 has two control functions, both of which must be logically active in order to obtain data at the outputs. Chip Enable ($\overline{CE}$) is the power control and should be used for device selection. Output Enable ($\overline{OE}$) is the output control and should be used to gate data from the output pins, independent of device selection. Assuming that addresses are stable, the address access time (t_{ACC}) is equal to the delay from $\overline{CE}$ to output (t_{CE}). Data is available at the outputs after a delay of t_{OE} from the falling edge of $\overline{OE}$, assuming that $\overline{CE}$ has been low and addresses have been stable for at least t_{ACC}-t_{OE}.

Standby Mode

EPROMs can be placed in standby mode which reduces the maximum current of the device by applying a TTL-high signal to the $\overline{CE}$ input. When in standby mode, the outputs are in a high impedance state, independent of the $\overline{OE}$ input.

Two Line Output Control

Because EPROMs are usually used in larger memory arrays, Intel has provided 2 control lines which accommodate this multiple memory connection. The two control lines allow for:

a) the lowest possible memory power dissipation, and

b) complete assurance that output bus contention will not occur

To use these two control lines most efficiently, $\overline{CE}$ should be decoded and used as the primary device selecting function, while $\overline{OE}$ should be made a common connection to all devices in the array and connected to the $\overline{READ}$ line from the system control bus. This assures that all deselected memory devices are in their low power standby mode and that the output pins are active only when data is desired from a particular memory device.

SYSTEM CONSIDERATIONS

The power switching characteristics of EPROMs require careful decoupling of the devices. The supply current, I_{CC}, has three segments that are of interest to the system designer—the standby current level, the active current level, and the transient current peaks that are produced by the falling and rising edges of Chip Enable. The magnitude of these transient current peaks is dependent on the output capacitive and inductive loading of the device. The associated transient voltage peaks can be suppressed by complying with Intel's Two-Line Control, and by properly selected decoupling capacitors. It is recommended that a 0.1 μF ceramic capacitor be used on every device between V_{CC} and GND. This should be a high frequency capacitor for low inherent inductance and should be placed as close to the device as possible. In addition, a 4.7 μF bulk electrolytic capacitor should be used between V_{CC} and GND for every eight devices. The bulk capacitor should be located near where the power supply is connected to the array. The purpose of the bulk capacitor is to overcome the voltage droop caused by the inductive effect of PC board-traces.

PROGRAMMING MODES

Caution: Exceeding 14V on V_{PP} will permanently damage the device.

Initially, and after each erasure, all bits of the EPROM are in the "1" state. Data is introduced by selectively programming "0s" into the desired bit locations. Although only "0s" will be programmed, both "1s" and "0s" can be present in the data word. The only way to change a "0" to a "1" is by ultraviolet light erasure (Cerdip EPROMs).

The device is in the programming mode when V_{PP} is raised to its programming voltage (See Table 2) and $\overline{CE}$ and $\overline{PGM}$ are both at TTL low. The data to be programmed is applied 16 bits in parallel to the data output pins. The levels required for the address and data inputs are TTL.

Program Inhibit

Programming of multiple EPROMS in parallel with different data is easily accomplished by using the Program Inhibit mode. A high-level $\overline{CE}$ or $\overline{PGM}$ input inhibits the other devices from being programmed.

Except for $\overline{CE}$, all like inputs (including $\overline{OE}$) of the parallel EPROMs may be common. A TTL low-level pulse applied to the $\overline{PGM}$ input with V_{PP} at its programming voltage and $\overline{CE}$ at TTL-Low will program the selected device.

Program Verify

A verify should be performed on the programmed bits to determine that they have been correctly programmed. The verify is performed with $\overline{OE}$ at V_{IL}, $\overline{CE}$ at V_{IL}, $\overline{PGM}$ at V_{IH} and V_{PP} and V_{CC} at their programming voltages.

int$_e$ligent Identifier™ Mode

The int$_e$ligent Identifier Mode allows the reading out of a binary code from an EPROM that will identify its manufacturer and type. This mode is intended for use by programming equipment for the purpose of automatically matching the device to be programmed with its corresponding programming algorithm. This mode is functional in the 25°C ±5°C ambient temperature range that is required when programming the device.

To activate this mode, the programming equipment must force 11.5V to 12.5V on address line A9 of the EPROM. Two identifier bytes may then be sequenced from the device outputs by toggling address line A0 from V_{IL} to V_{IH}. All other address lines must be held at V_{IL} during the int$_e$ligent Identifier Mode.

Byte 0 ($A0 = V_{IL}$) represents the manufacturer code and byte 1 ($A0 = V_{IH}$) the device identifier code. These two identifier bytes are given in Table 1.

INTEL EPROM PROGRAMMING SUPPORT TOOLS

Intel offers a full line of EPROM Programmers providing state-of-the-art programming for Intel programmable devices. The modular architecture of Intel's EPROM programmers allows you to add new support as it becomes available, with very low cost add-ons. For example, even the earliest users of the iUP-FAST 27/K module may take advantage of Intel's new Quick-Pulse Programming Algorithm, the fastest in the industry.

Intel EPROM programmers may be controlled from a host computer using Intel's PROM Programming software (iPPS). iPPS makes programming easy for a growing list of industry standard hosts, including the IBM PC, XT, AT and PCDOS compatibles, Intellec Development Systems. Intel's iPDS Personal Development System, and the Intel Network Development System (iNDS-II). Stand-alone operation is also available, including device previewing, editing, programming, and download of programming data from any source over an RS232C port.

For further details consult the EPROM Programming section of the Development Systems Handbook.

ERASURE CHARACTERISTICS (FOR CERDIP EPROMS)

The erasure characteristics are such that erasure begins to occur upon exposure to light with wavelengths shorter than approximately 4000 Angstroms (Å). It should be noted that sunlight and certain types of fluorescent lamps have wavelengths in the 3000-4000Å range. Data shows that constant exposure to room level fluorescent lighting could erase the EPROM in approximately 3 years, while it would take approximately 1 week to cause erasure when exposed to direct sunlight. If the device is to be exposed to these types of lighting conditions for extended periods of time, opaque labels should be placed over the window to prevent unintentional erasure.

The recommended erasure procedure is exposure to shortwave ultraviolet light which has a wavelength of 2537 Angstroms (Å). The integrated dose (i.e., UV intensity × exposure time) for erasure should be a minimum of 15 Wsec/cm^2. The erasure time with this dosage is approximately 15 to 20 minutes using an ultraviolet lamp with a 12000 μW/cm^2 power rating. The EPROM should be placed within 1 inch of the lamp tubes during erasure. The maximum integrated dose an EPROM can be exposed to without damage is 7258 Wsec/cm^2 (1 week @ 12000 μW/cm^2). Exposure of the device to high intensity UV light for longer periods may cause permanent damage.

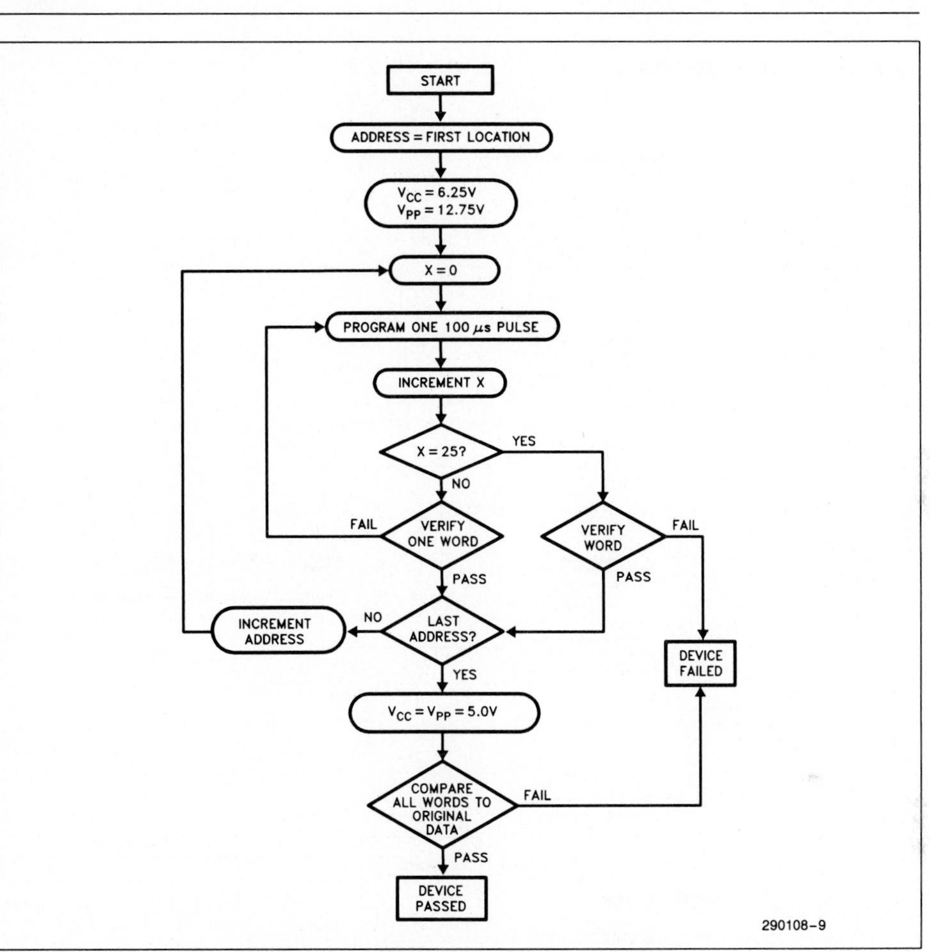

Figure 4. Quick-Pulse Programming™ Algorithm

Quick-Pulse Programming™ Algorithm

Intel's 27210 EPROMs can be programmed using the Quick-Pulse Programming Algorithm, developed by Intel to substantially reduce the throughput time in the production programming environment. This algorithm allows these devices to be programmed in under eight seconds, almost a hundred fold improvement over previous algorithms. Actual programming time is a function of the PROM programmer being used.

The Quick-Pulse Programming Algorithm uses initial pulses of 100 microseconds followed by a word verification to determine when the addressed word has been successfully programmed. Up to 25 100 μs pulses per word are provided before a failure is recognized. A flow chart of the Quick-Pulse Programming Algorithm is shown in Figure 4.

For the Quick-Pulse Programming Algorithm, the entire sequence of programming pulses and word verifications is performed at V_{CC} = 6.25V and V_{PP} at 12.75V. When programming of the EPROM has been completed, all data words should be compared to the original data with V_{CC} = V_{PP} = 5.0V.

In addition to the Quick-Pulse Programming Algorithm, 27210 EPROMs are also compatible with Intel's int$_{e}$ligent Programming Algorithm.

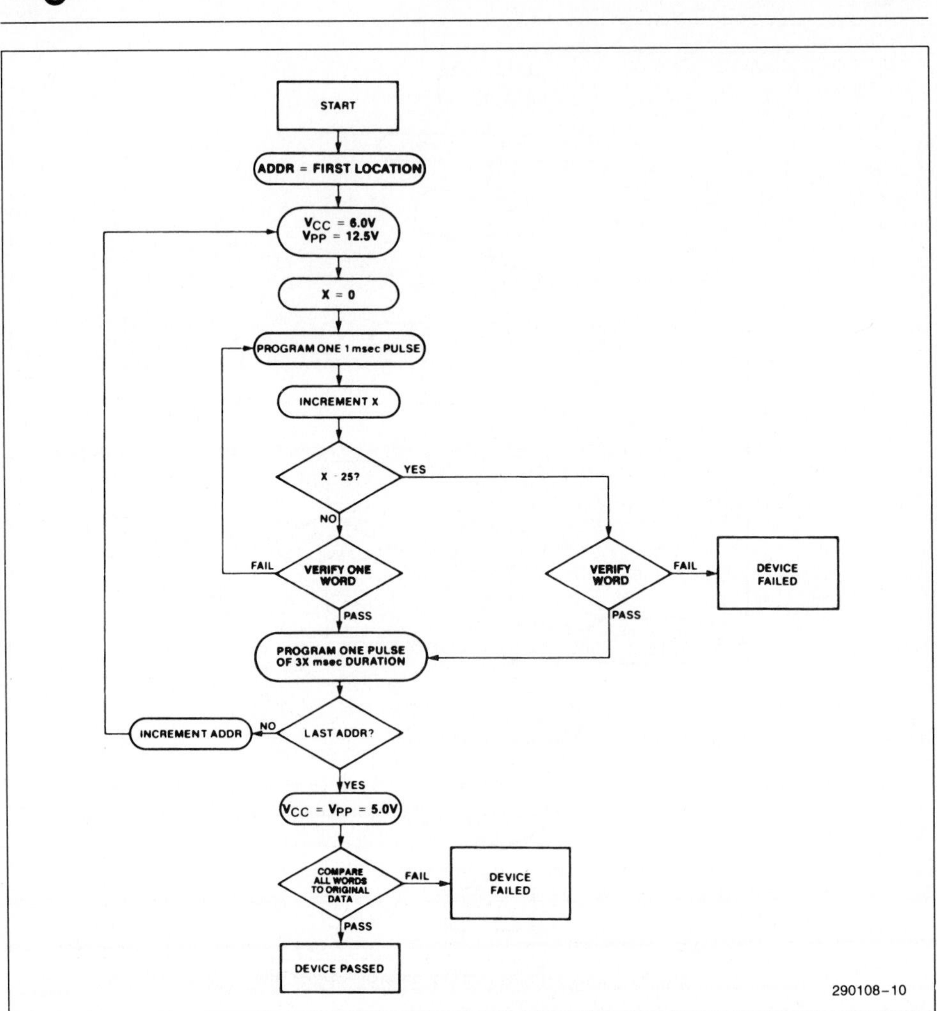

Figure 5. int$_e$ligent Programming™ Flowchart

int$_e$ligent Programming™ Algorithm

The int$_e$ligent Programming Algorithm has been a standard in the industry for the past few years. A flow-chart of the int$_e$ligent Programming Algorithm is shown in Figure 5.

The int$_e$ligent Programming Algorithm utilizes two different pulse types: initial and overprogram. The duration of the initial pulse(s) is one millisecond, which will then be followed by a larger overprogram pulse of length 3X msec. X is an iteration counter and is equal to the number of the initial one millisecond pulses applied to a particular location, before a correct verify occurs. Up to 25 one-millisecond pulses per word are provided for before the overprogram pulse is applied.

The entire sequence of program pulses and word verifications is performed at V_{CC} = 6.0V and V_{PP} = 12.5V. When the int$_e$ligent Programming cycle has been completed, all data words should be compared to the original data with V_{CC} = V_{PP} = 5.0V.

D.C. PROGRAMMING CHARACTERISTICS $T_A = 25°C \pm 5°C$

Table 2

Symbol	Parameter	Limits			Test Conditions (Note 1)
		Min	Max	Unit	
I_{LI}	Input Leakage Current (All Inputs)		10	μA	$V_{IN} = 6V$
V_{IL}	Input Low Level (All Inputs)	−0.1	0.8	V	
V_{IH}	Input High Level	2.0	$V_{CC}+1$	V	
V_{OL}	Output Low Voltage During Verify		0.45	V	$I_{OL} = 2.1$ mA
V_{OH}	Output High Voltage During Verify	2.4		V	$I_{OH} = -400$ μA
I_{CC2}(4)	V_{CC} Supply Current (Program & Verify)		160	mA	$\overline{CE} = \overline{PGM} = V_{IL}$
I_{PP2}	V_{PP} Supply Current (Program)		50	mA	$\overline{CE} = \overline{PGM} = V_{IL}$
V_{ID}	A_9 int$_e$ligent Identifier Voltage	11.5	12.5	V	$V_{CC} = 5V$
V_{PP}	int$_e$ligent Programming Algorithm	12.0	13.0	V	
	Quick-Pulse Programming Algorithm	12.5	13.0	V	
V_{CC}	int$_e$ligent Programming Algorithm	5.75	6.25	V	
	Quick-Pulse Programming Algorithm	6.0	6.5	V	

A.C. PROGRAMMING CHARACTERISTICS

$T_A = 25°C \pm 5°C$ (See Table 2 for V_{CC} and V_{PP} voltages.)

Symbol	Parameter	Limits				Conditions* (Note 1)
		Min	Typ	Max	Unit	
t_{AS}	Address Setup Time	2			μs	
t_{OES}	$\overline{OE}$ Setup Time	2			μs	
t_{DS}	Data Setup Time	2			μs	
t_{AH}	Address Hold Time	0			μs	
t_{DH}	Data Hold Time	2			μs	
t_{DFP}	$\overline{OE}$ High to Output Float Delay	0		130	ns	(Note 3)
t_{VPS}	V_{PP} Setup Time	2			μs	
t_{VCS}	V_{CC} Setup Time	2			μs	
t_{CES}	$\overline{CE}$ Setup Time	2			μs	
t_{PW}	$\overline{PGM}$ Initial Program Pulse Width	0.95	1.0	1.05	ms	int$_e$ligent Programming
		95	100	105	μs	Quick-Pulse Programming
t_{OPW}	$\overline{PGM}$ Overprogram Pulse Width	2.85		78.75	ms	(Note 2)
t_{OE}	Data Valid from $\overline{OE}$			150	ns	

*A.C. CONDITIONS OF TEST

Input Rise and Fall Times (10% to 90%) ...... 20 ns

Input Pulse Levels .... 0.45V to 2.4V

Input Timing Reference Level .... 0.8V and 2.0V

Output Timing Reference Level ...... 0.8V and 2.0V

NOTES:

1. V_{CC} must be applied simultaneously or before V_{PP} and removed simultaneously or after V_{PP}.
2. The length of the overprogram pulse (int$_e$ligent Programming Algorithm only) may vary from 2.85 msec to 78.75 msec as a function of the iteration counter value X.
3. This parameter is only sampled and is not 100% tested. Output Float is defined as the point where data is no longer driven—see timing diagram.
4. The maximum current value is with outputs O_0–O_{15} unloaded.

PROGRAMMING WAVEFORMS

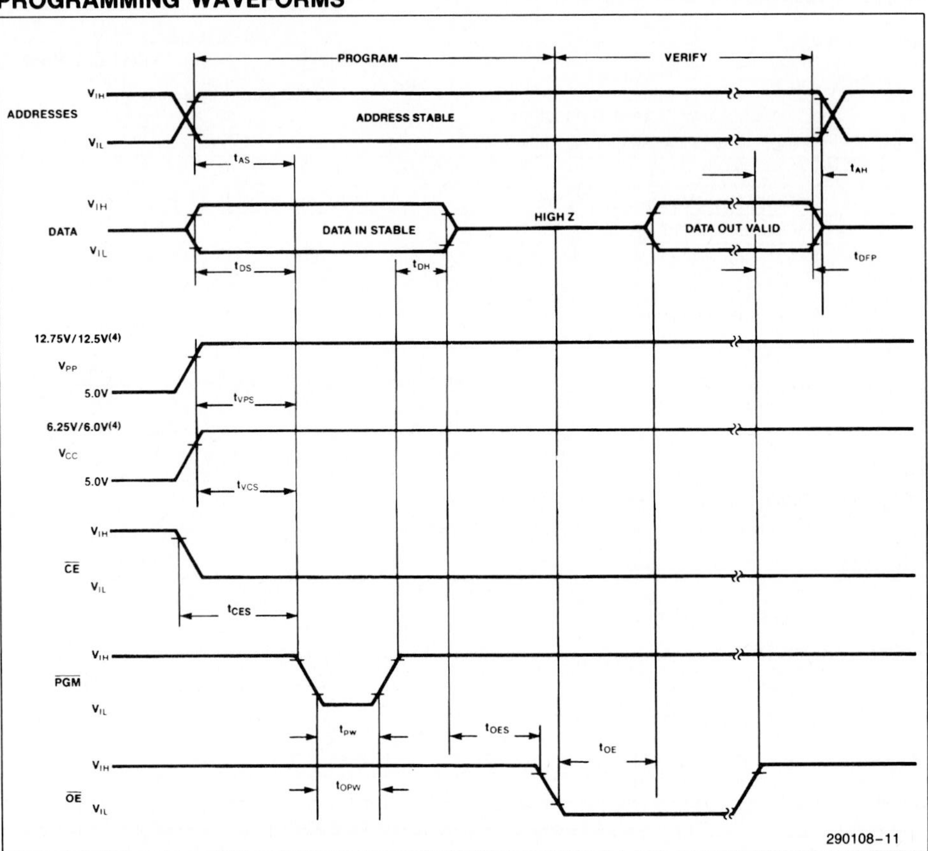

NOTES:

1. The Input Timing Reference Level is 0.8V for V_{IL} and 2V for a V_{IH}.
2. t_{OE} and t_{DFP} are characteristics of the device but must be accommodated by the programmer.
3. When programming the 27210, a 0.1 μF capacitor is required across V_{PP} and ground to suppress spurious voltage transients which can damage the device.
4. 12.75V V_{PP} & 6.25V V_{CC} for Quick-Pulse Programming Algorithm; 12.5V V_{PP} & 6.0V V_{CC} for int$_{e}$ligent Programming Algorithm.

NEC Electronics Inc.

μPD43256 32,768 x 8-BIT STATIC MIX-MOS RAM

Revision 1

Description

The μPD43256 is a high-speed, low-power, 32,768-word by 8-bit static MIX-MOS RAM fabricated with advanced silicon-gate MIX-MOS technology. The μPD43256 is a low standby power device using n-channel memory cells with polysilicon resistors. Furthermore, a novel circuitry technique makes the μPD43256 a high-speed and low operating power device which requires no clock or refreshing to operate.

Minimum standby power is drawn by this device when $\overline{CS}$ is at a high level, independently of the other inputs' levels.

Data retention is guaranteed at a power supply voltage as low as 2 V (μPD43256-10L/12L/15L).

The μPD43256C is packaged in a standard 28-pin plastic dual-in-line package.

The μPD43256G is packaged in a standard 28-pin plastic miniflat (SOP) package.

Features

- Single +5 V supply
- Fully static operation — no clock or refreshing required
- TTL-compatible — all inputs and outputs
- Common I/O using three-state output
- One Chip Select and one Output Enable input for easy application
- Data retention voltage
  - — μPD43256-10L/12L/15L: 2 V min
- Standard 28-pin plastic DIP and miniflat (SOP) packages

Performance Ranges

Device	Access Time	Cycle Time	Power Supply (Max) Active	Standby
μPD43256-10	100 ns	100 ns	70 mA	2 mA
μPD43256-12	120 ns	120 ns	70 mA	2 mA
μPD43256-15	150 ns	150 ns	70 mA	2 mA
μPD43256-10L	100 ns	100 ns	70 mA	100 μA
μPD43256-12L	120 ns	120 ns	70 mA	100 μA
μPD43256-15L	150 ns	150 ns	70 mA	100 μA

Capacitance

$T_A = 25°C$, f = 1 MHz

Parameter	Symbol	Limits Min	Typ	Max	Unit	Test Conditions
Input capacitance	C_{IN}			5	pF	$V_{IN} = 0 V$
Input/output capacitance	$C_{I/O}$			8	pF	$V_{I/O} = 0 V$

Pin Configuration

Pin	Name	Pin	Name
1	A_{14}	28	V_{CC}
2	A_{12}	27	$\overline{WE}$
3	A_7	26	A_{13}
4	A_6	25	A_8
5	A_5	24	A_9
6	A_4	23	A_{11}
7	A_3	22	$\overline{OE}$
8	A_2	21	A_{10}
9	A_1	20	$\overline{CS}$
10	A_0	19	I/O_8
11	I/O_1	18	I/O_7
12	I/O_2	17	I/O_6
13	I/O_3	16	I/O_5
14	GND	15	I/O_4

μPD43256

83-001401A

Pin Identification Table

No.	Symbol	Function
1-10, 21, 23-26	A_0-A_{14}	Address input
11-13, 15-19	I/O_1-I/O_8	Data input/output
14	GND	Ground
20	$\overline{CS}$	Chip select
22	$\overline{OE}$	Ouput enable
27	$\overline{WE}$	Write enable
28	V_{CC}	Power (+5 V)

Block Diagram

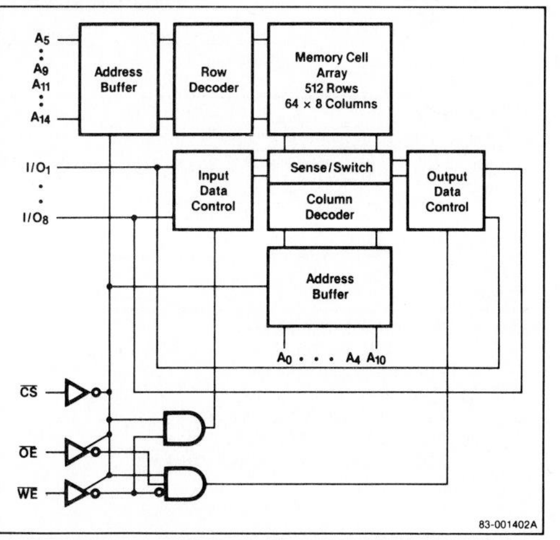

μPD43256 **NEC**

Absolute Maximum Ratings

Power supply voltage, V_{CC}	−0.5[1] to 7.0 V
Input voltage, V_{IN}	−0.5[1] to V_{CC} + 0.5 V
Output voltage, $V_{I/O}$	−0.5[1] to V_{CC} + 0.5 V
Operating temperature, T_{OPR}	0 to 70°C
Storage temperature, T_{STG}	−55 to 125°C
Power dissipation, P_D	1.0 W

Note: [1] −3.0 V min (pulse width 50 ns)

Comment: Exposing the device to stresses above those listed in Absolute Maximum Ratings could cause permanent damage. The device is not meant to be operated under conditions outside the limits described in the operational sections of this specification. Exposure to absolute maximum rating conditions for extended periods may affect device reliability.

Recommended DC Operating Conditions

T_A = 0 to 70°C

Parameter	Symbol	Limits Min	Typ	Max	Unit
Supply voltage	V_{CC}	4.5	5.0	5.5	V
Input low voltage	V_{IL}	−0.3[1]		0.8	V
Input high voltage	V_{IH}	2.2		V_{CC} + 0.5	V

Note: [1] −3.0 V min (pulse width 50 ns)

DC Characteristics

T_A = 0 to 70°C, V_{CC} = 5 V ± 10%

Parameter	Symbol	Limits Min	Typ	Max	Unit	Test Conditions
Input leakage current	I_{LI}			1	μA	V_{IN} = 0 to V_{CC}
I/O leakage current	I_{LO}			1	μA	$V_{I/O}$ = 0 to V_{CC} $\overline{CS} \geq V_{IH}$ or $\overline{OE} \geq V_{IH}$ or $\overline{WE} \leq V_{IL}$
Operating supply current	I_{CCA}		Note 1	70	mA	$\overline{CS} \leq V_{IL}$, Min Cycle $I_{I/O}$ = 0
Standby supply current	I_{SB}			Note 2	mA	$\overline{CS} \geq V_{IH}$
Standby supply current	I_{SB1}		Note 3	Note 3	mA	$\overline{CS} \geq V_{CC} - 0.2$ V
Output low voltage	V_{OL}			0.4	V	I_{OL} = 2.1 mA
Output high voltage	V_{OH}	2.4			V	I_{OH} = −1.0 mA

Notes: [1] μPD43256-10/10L: 35 mA typ
μPD43256-12/12L: 30 mA typ
μPD43256-15/15L: 25 mA typ
[2] μPD43256-10/12/15: 5 mA max
μPD43256-10L/12L/15L: 3 mA max
[3] μPD43256-10/12/15: 20 μA typ, 2 mA max
μPD43256-10L/12L/15L: 2 μA typ, 100 μA max

AC Characteristics

T_A = 0 to 70°C, V_{CC} = 5 V ± 10%

Parameter	Symbol	μPD43256-10/10L Min	Max	μPD43256-12/12L Min	Max	μPD43256-15/15L Min	Max	Unit
Read Cycle								
Read cycle time	t_{RC}	100		120		150		ns
Address access time	t_{AA}		100		120		150	ns
Chip select access time	t_{ACS}		100		120		150	ns
Output enable to output valid	t_{OE}		50		60		70	ns
Output hold from address change	t_{OH}	10		10		10		ns
Chip select to output in Lo-Z	t_{CLZ}	10		10		10		ns
Output enable to output in Lo-Z	t_{OLZ}	5		5		5		ns
Chip select to output in Hi-Z	t_{CHZ}		35		40		50	ns
Output enable to output in Hi-Z	t_{OHZ}		35		40		50	ns
Write Cycle								
Write cycle time	t_{WC}	100		120		150		ns
Chip select to end of write	t_{CW}	80		85		100		ns
Address valid to end of write	t_{AW}	80		85		100		ns
Address setup time	t_{AS}	0		0		0		ns
Write pulse width	t_{WP}	70		70		90		ns
Write recovery time	t_{WR}	5		5		5		ns
Data valid to end of write	t_{DW}	40		50		60		ns
Data hold time	t_{DH}	0		0		0		ns
Write enable to output in Hi-Z	t_{WHZ}		35		40		50	ns
Output active from end of write	t_{OW}	10		10		10		ns

AC Test Conditions

Input pulse levels	0.8 to 2.2 V
Input pulse rise and fall time	5 ns
Timing reference levels	1.5 V

Truth Table

$\overline{CS}$	$\overline{OE}$	$\overline{WE}$	MODE	I/O	I_{CC}
H	X	X	Not selected	Hi-Z	Standby
L	H	H	Not selected	Hi-Z	Active
L	L	H	Read	D_{OUT}	Active
L	X	L	Write	D_{IN}	Active

NEC **μPD43256**

AC Test Circuits

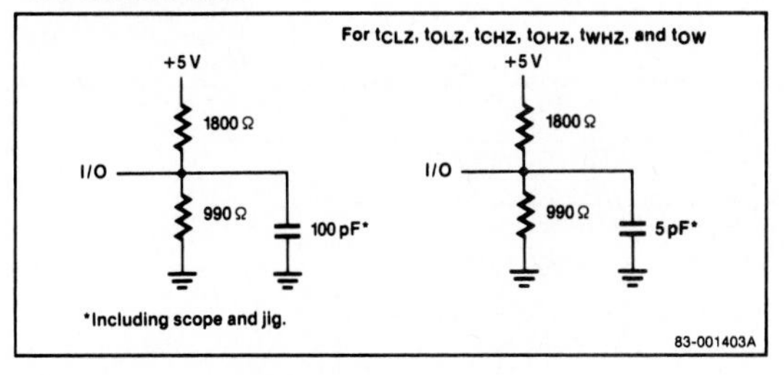

Timing Waveforms

Read Cycle No. 1 (Address Access) (Notes 1, 2)

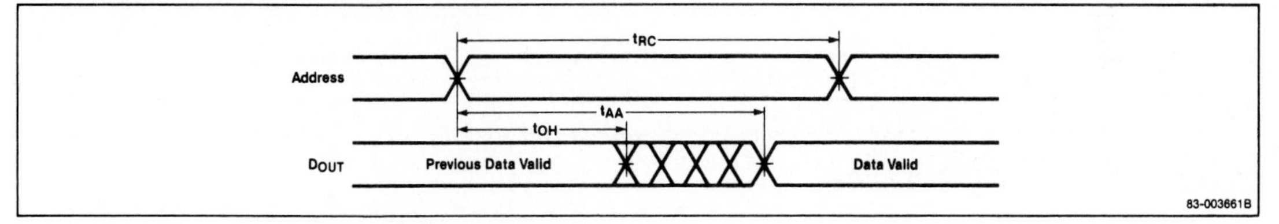

Read Cycle No. 2 (Chip Select Access) (Notes 1, 3)

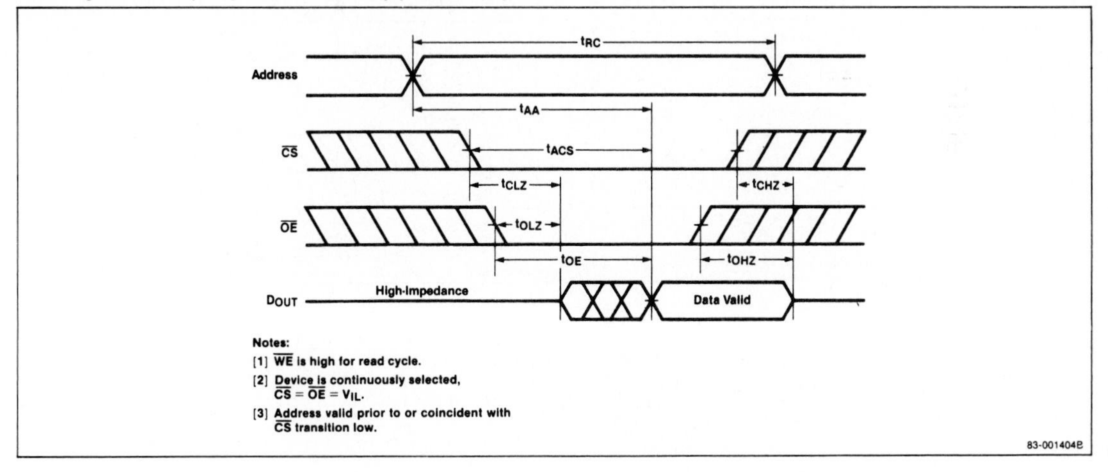

μPD43256 **NEC**

Timing Waveforms (Cont)

Write Cycle No. 1 ($\overline{WE}$ Controlled) (Notes 1, 2, 3)

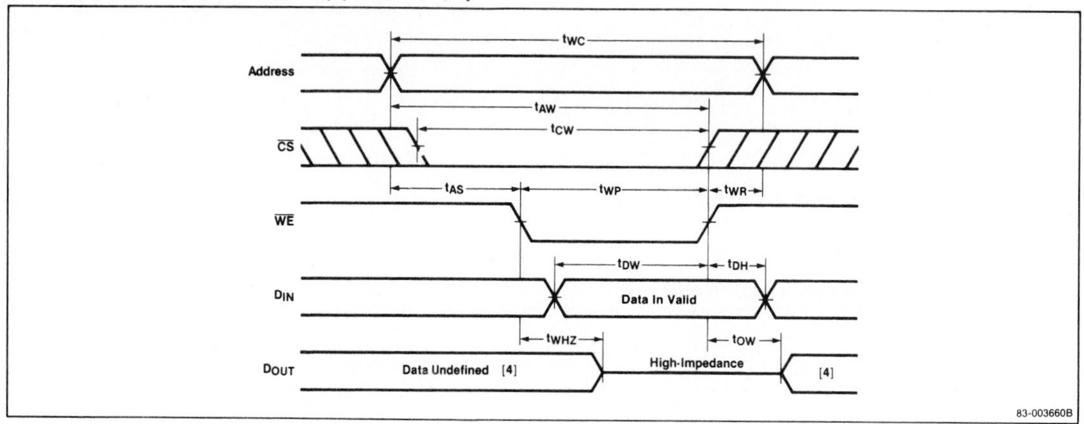

Write Cycle No. 2 ($\overline{CS}$ Controlled) (Notes 1, 2)

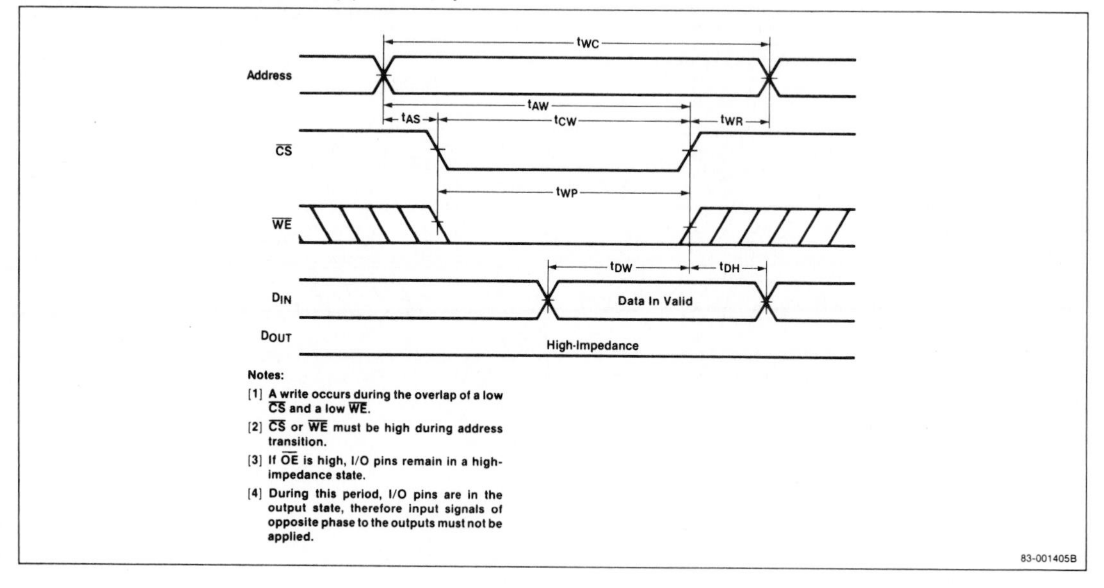

NEC **μPD43256**

Low V_{CC} Data Retention Characteristics

$T_A = 0$ to $70°C$ for μPD43256-10L/12L/15L

Parameter	Symbol	Limits Min	Limits Typ	Limits Max	Unit	Test Conditions
Data retention supply voltage	V_{CCDR}	2.0		5.5	V	$\overline{CS} \geq V_{CC} - 0.2\,V$
Data retention supply current	I_{CCDR}		1	50	μA	$V_{CC} = 3.0\,V$, $\overline{CS} \geq V_{CC} - 0.2\,V$
Chip deselection to data retention mode	t_{CDR}	0			ns	
Operation recovery time	t_R	t_{RC}			ns	

Data Retention Timing Chart

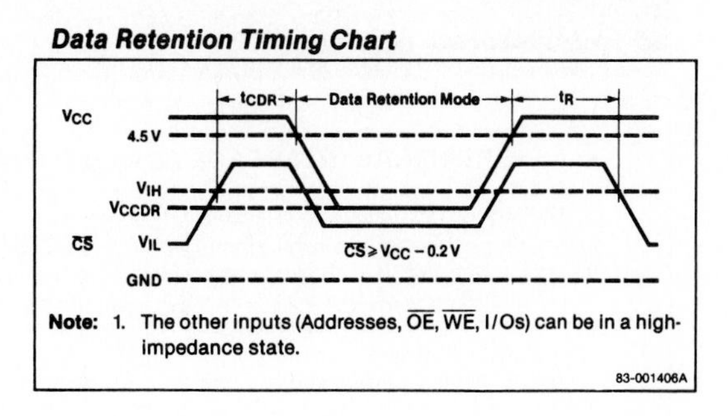

Note: 1. The other inputs (Addresses, $\overline{OE}$, $\overline{WE}$, I/Os) can be in a high-impedance state.

83-001406A

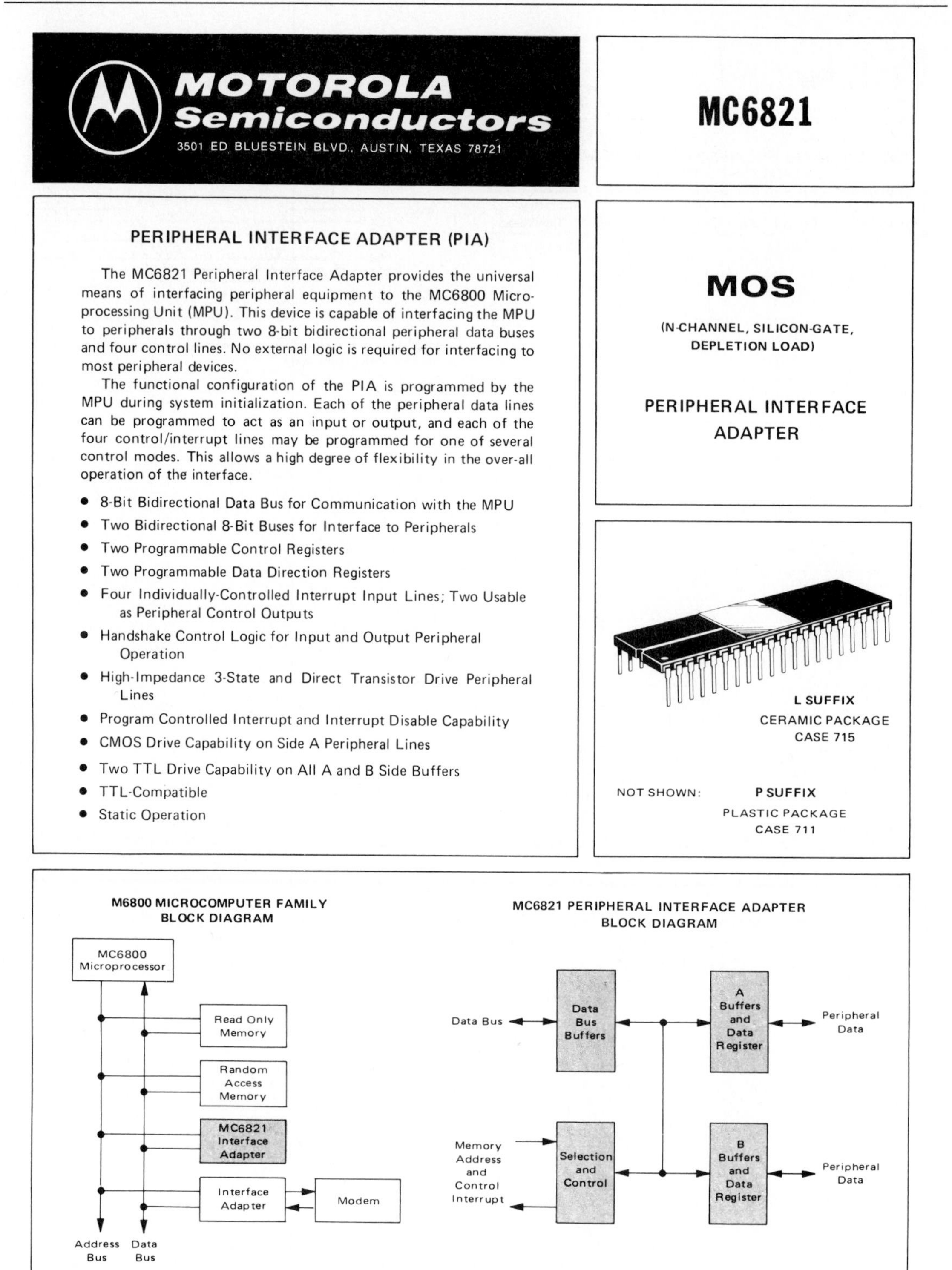

MC6821

MOS

(N-CHANNEL, SILICON-GATE, DEPLETION LOAD)

PERIPHERAL INTERFACE ADAPTER

PERIPHERAL INTERFACE ADAPTER (PIA)

The MC6821 Peripheral Interface Adapter provides the universal means of interfacing peripheral equipment to the MC6800 Microprocessing Unit (MPU). This device is capable of interfacing the MPU to peripherals through two 8-bit bidirectional peripheral data buses and four control lines. No external logic is required for interfacing to most peripheral devices.

The functional configuration of the PIA is programmed by the MPU during system initialization. Each of the peripheral data lines can be programmed to act as an input or output, and each of the four control/interrupt lines may be programmed for one of several control modes. This allows a high degree of flexibility in the over-all operation of the interface.

- 8-Bit Bidirectional Data Bus for Communication with the MPU
- Two Bidirectional 8-Bit Buses for Interface to Peripherals
- Two Programmable Control Registers
- Two Programmable Data Direction Registers
- Four Individually-Controlled Interrupt Input Lines; Two Usable as Peripheral Control Outputs
- Handshake Control Logic for Input and Output Peripheral Operation
- High-Impedance 3-State and Direct Transistor Drive Peripheral Lines
- Program Controlled Interrupt and Interrupt Disable Capability
- CMOS Drive Capability on Side A Peripheral Lines
- Two TTL Drive Capability on All A and B Side Buffers
- TTL-Compatible
- Static Operation

L SUFFIX
CERAMIC PACKAGE
CASE 715

NOT SHOWN: P SUFFIX
PLASTIC PACKAGE
CASE 711

M6800 MICROCOMPUTER FAMILY BLOCK DIAGRAM

MC6821 PERIPHERAL INTERFACE ADAPTER BLOCK DIAGRAM

ELECTRICAL CHARACTERISTICS (V_{CC} = 5.0 V ±5%, V_{SS} = 0, T_A = 0 to 70°C unless otherwise noted.)

Characteristic		Symbol	Min	Typ	Max	Unit
Input High Voltage		V_{IH}	V_{SS} + 2.0	–	V_{CC}	Vdc
Input Low Voltage		V_{IL}	V_{SS} – 0.3	–	V_{SS} + 0.8	Vdc
Input Leakage Current (V_{in} = 0 to 5.25 Vdc)	R/W, $\overline{Reset}$, RS0, RS1, CS0, CS1, $\overline{CS2}$, CA1, CB1, Enable	I_{in}	–	1.0	2.5	μAdc
Three-State (Off State) Input Current (V_{in} = 0.4 to 2.4 Vdc)	D0–D7, PB0–PB7, CB2	I_{TSI}	–	2.0	10	μAdc
Input High Current (V_{IH} = 2.4 Vdc)	PA0–PA7, CA2	I_{IH}	-200	-400	–	μAdc
Input Low Current (V_{IL} = 0.4 Vdc)	PA0–PA7, CA2	I_{IL}	–	-1.3	-2.4	mAdc
Output High Voltage		V_{OH}				Vdc
(I_{Load} = -205 μAdc)	D0–D7		V_{SS} + 2.4	–	–	
(I_{Load} = -200 μAdc)	Other Outputs		V_{SS} + 2.4	–	–	
Output Low Voltage		V_{OL}				Vdc
(I_{Load} = 1.6 mAdc)	D0–D7		–	–	V_{SS} + 0.4	
(I_{Load} = 3.2 mAdc)	Other Outputs		–	–	V_{SS} + 0.4	
Output Leakage Current (Off State) (V_{OH} = 2.4 Vdc)	$\overline{IRQA}$, $\overline{IRQB}$	I_{LOH}	–	1.0	10	μAdc
Power Dissipation		P_D	–	–	550	mW
Capacitance (V_{in} = 0, T_A = 25°C, f = 1.0 MHz)	D0–D7	C_{in}	–	–	12.5	pF
	PA0–PA7, PB0–PB7, CA2, CB2		–	–	10	
	Enable, R/W, $\overline{Reset}$, RS0, RS1, CS0, CS1, $\overline{CS2}$, CA1, CB1		–	–	7.5	
	$\overline{IRQA}$, $\overline{IRQB}$	C_{out}	–	–	5.0	pF
Peripheral Data Setup Time (Figure 1)		t_{PDSU}	200	–	–	ns
Peripheral Data Hold Time (Figure 1)		t_{PDH}	0	–	–	ns
Delay Time, Enable negative transition to CA2 negative transition (Figure 2, 3)		t_{CA2}	–	–	1.0	μs
Delay Time, Enable negative transition to CA2 positive transition (Figure 2)		t_{RS1}	–	–	1.0	μs
Rise and Fall Times for CA1 and CA2 input signals (Figure 3)		t_r, t_f	–	–	1.0	μs
Delay Time from CA1 active transition to CA2 positive transition (Figure 3)		t_{RS2}	–	–	2.0	μs
Delay Time, Enable negative transition to Peripheral Data Valid (Figures 4, 5)		t_{PDW}	–	–	1.0	μs
Delay Time, Enable negative transition to Peripheral CMOS Data Valid (V_{CC} – 30% V_{CC}, Figure 4; Figure 12 Load C)	PA0–PA7, CA2	t_{CMOS}	–	–	2.0	μs
Delay Time, Enable positive transition to CB2 negative transition (Figure 6, 7)		t_{CB2}	–	–	1.0	μs
Delay Time, Peripheral Data Valid to CB2 negative transition (Figure 5)		t_{DC}	20	–	–	ns
Delay Time, Enable positive transition to CB2 positive transition (Figure 6)		t_{RS1}	–	–	1.0	μs
Peripheral Control Output Pulse Width, CA2/CB2 (Figures 2, 6)		PW_{CT}	550	–	–	ns
Rise and Fall Time for CB1 and CB2 input signals (Figure 7)		t_r, t_f	–	–	1.0	μs
Delay Time, CB1 active transition to CB2 positive transition (Figure 7)		t_{RS2}	–	–	2.0	μs
Interrupt Release Time, $\overline{IRQA}$ and $\overline{IRQB}$ (Figure 9)		t_{IR}	–	–	1.6	μs
Interrupt Response Time (Figure 8)		t_{RS3}	–	–	1.0	μs
Interrupt Input Pulse Width (Figure 8)		PW_I	500	–	–	ns
Reset Low Time* (Figure 10)		t_{RL}	1.0	–	–	μs

*The Reset line must be high a minimum or 1.0 μs before addressing the PIA.

MAXIMUM RATINGS

Rating	Symbol	Value	Unit
Supply Voltage	V_{CC}	–0.3 to +7.0	Vdc
Input Voltage	V_{in}	–0.3 to +7.0	Vdc
Operating Temperature Range	T_A	0 to +70	°C
Storage Temperature Range	T_{stg}	–55 to +150	°C
Thermal Resistance	θ_{JA}	82.5	°C/W

This device contains circuitry to protect the inputs against damage due to high static voltages or electric fields; however, it is advised that normal precautions be taken to avoid application of any voltage higher than maximum rated voltages to this high impedance circuit.

BUS TIMING CHARACTERISTICS

READ (Figures 11 and 13)

Characteristic	Symbol	Min	Typ	Max	Unit
Enable Cycle Time	t_{cycE}	1.0	–	–	μs
Enable Pulse Width, High	PW_{EH}	0.45	–	–	μs
Enable Pulse Width, Low	PW_{EL}	0.43	–	–	μs
Setup Time, Address and R/W valid to Enable positive transition	t_{AS}	160	–	–	ns
Data Delay Time	t_{DDR}	–	–	320	ns
Data Hold Time	t_H	10	–	–	ns
Address Hold Time	t_{AH}	10	–	–	ns
Rise and Fall Time for Enable input	t_{Er}, t_{Ef}	–	–	25	ns

WRITE (Figures 12 and 13)

Characteristic	Symbol	Min	Typ	Max	Unit
Enable Cycle Time	t_{cycE}	1.0	–	–	μs
Enable Pulse Width, High	PW_{EH}	0.45	–	–	μs
Enable Pulse Width, Low	PW_{EL}	0.43	–	–	μs
Setup Time, Address and R/W valid to Enable positive transition	t_{AS}	160	–	–	ns
Data Setup Time	t_{DSW}	195	–	–	ns
Data Hold Time	t_H	10	–	–	ns
Address Hold Time	t_{AH}	10	–	–	ns
Rise and Fall Time for Enable input	t_{Er}, t_{Ef}	–	–	25	ns

FIGURE 1 – PERIPHERAL DATA SETUP AND HOLD TIMES
(Read Mode)

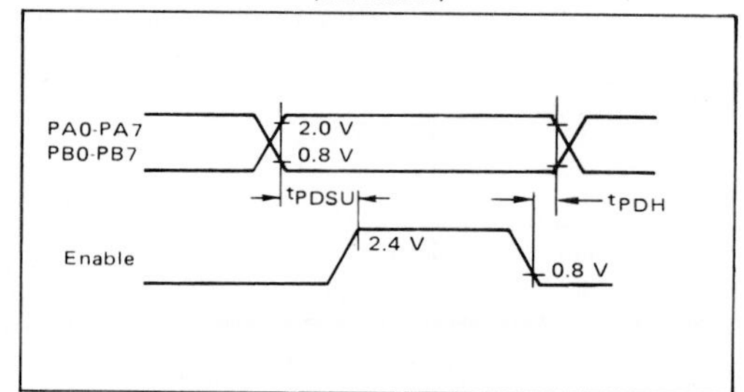

FIGURE 2 – CA2 DELAY TIME
(Read Mode; CRA-5 = CRA-3 = 1, CRA-4 = 0)

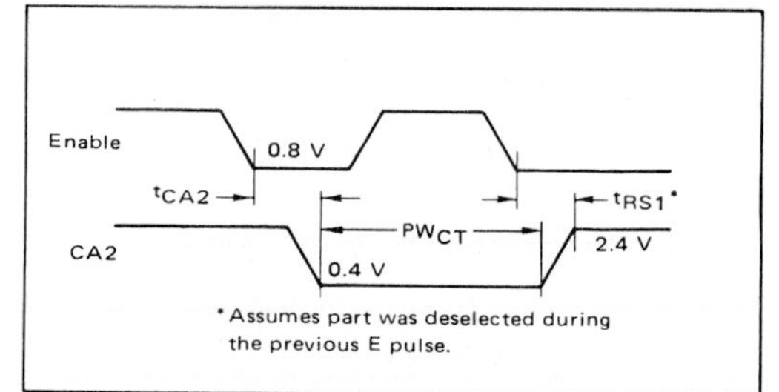

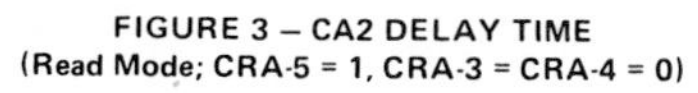

FIGURE 3 – CA2 DELAY TIME
(Read Mode; CRA-5 = 1, CRA-3 = CRA-4 = 0)

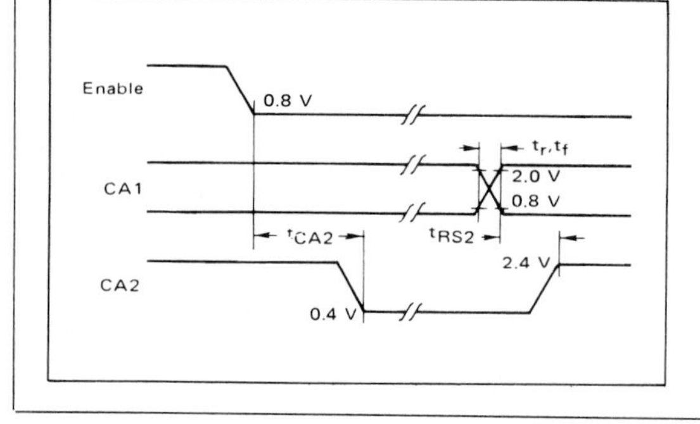

FIGURE 4 – PERIPHERAL CMOS DATA DELAY TIMES
(Write Mode; CRA-5 = CRA-3 = 1, CRA-4 = 0)

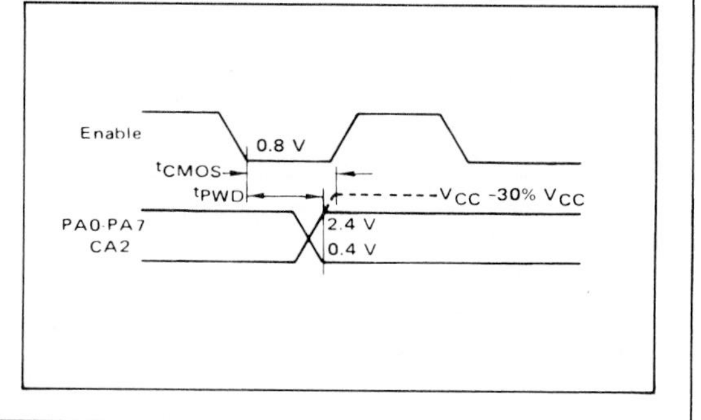

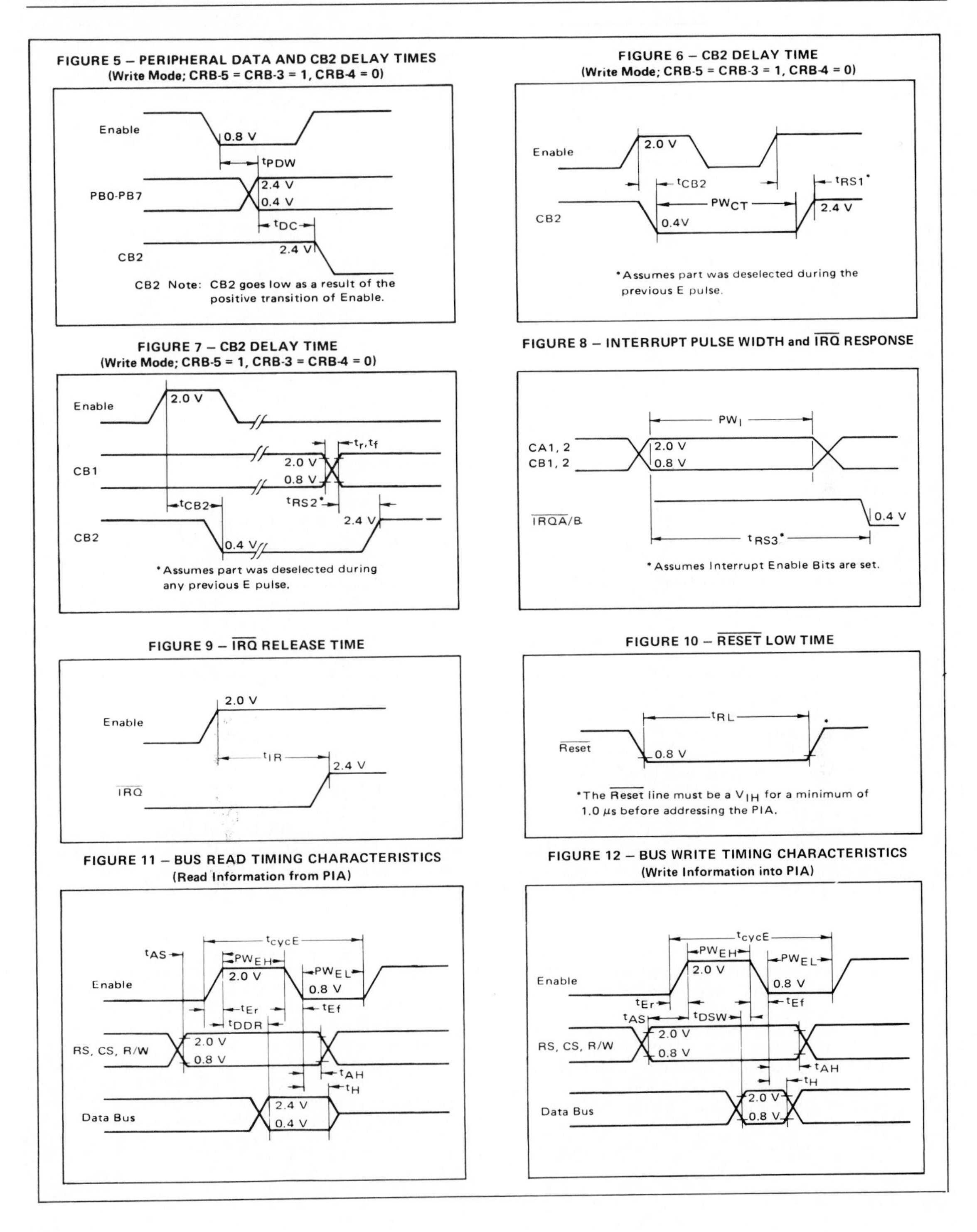
FIGURE 5 – PERIPHERAL DATA AND CB2 DELAY TIMES
(Write Mode; CRB-5 = CRB-3 = 1, CRB-4 = 0)
Enable
0.8 V
tPDW
PB0-PB7
2.4 V
0.4 V
tDC
CB2
2.4 V
CB2 Note: CB2 goes low as a result of the positive transition of Enable.
FIGURE 6 – CB2 DELAY TIME
(Write Mode; CRB-5 = CRB-3 = 1, CRB-4 = 0)
Enable
2.0 V
tCB2
tRS1*
PWCT
CB2
0.4V
2.4 V
*Assumes part was deselected during the previous E pulse.
FIGURE 7 – CB2 DELAY TIME
(Write Mode; CRB-5 = 1, CRB-3 = CRB-4 = 0)
Enable
2.0 V
tr, tf
CB1
2.0 V
0.8 V
tCB2
tRS2*
2.4 V
CB2
0.4 V
*Assumes part was deselected during any previous E pulse.
FIGURE 8 – INTERRUPT PULSE WIDTH and IRQ RESPONSE
PWI
CA1, 2
CB1, 2
2.0 V
0.8 V
IRQA/B
0.4 V
tRS3*
*Assumes Interrupt Enable Bits are set.
FIGURE 9 – IRQ RELEASE TIME
Enable
2.0 V
tIR
2.4 V
IRQ
FIGURE 10 – RESET LOW TIME
tRL
Reset
0.8 V
*The Reset line must be a VIH for a minimum of 1.0 µs before addressing the PIA.
FIGURE 11 – BUS READ TIMING CHARACTERISTICS
(Read Information from PIA)
tcycE
tAS
PWEH
PWEL
Enable
2.0 V
0.8 V
tEr
tEf
tDDR
RS, CS, R/W
2.0 V
0.8 V
tAH
tH
Data Bus
2.4 V
0.4 V
FIGURE 12 – BUS WRITE TIMING CHARACTERISTICS
(Write Information into PIA)
tcycE
PWEH
PWEL
Enable
2.0 V
0.8 V
tEr
tEf
tAS
tDSW
RS, CS, R/W
2.0 V
0.8 V
tAH
tH
Data Bus
2.0 V
0.8 V

FIGURE 13 – BUS TIMING TEST LOADS

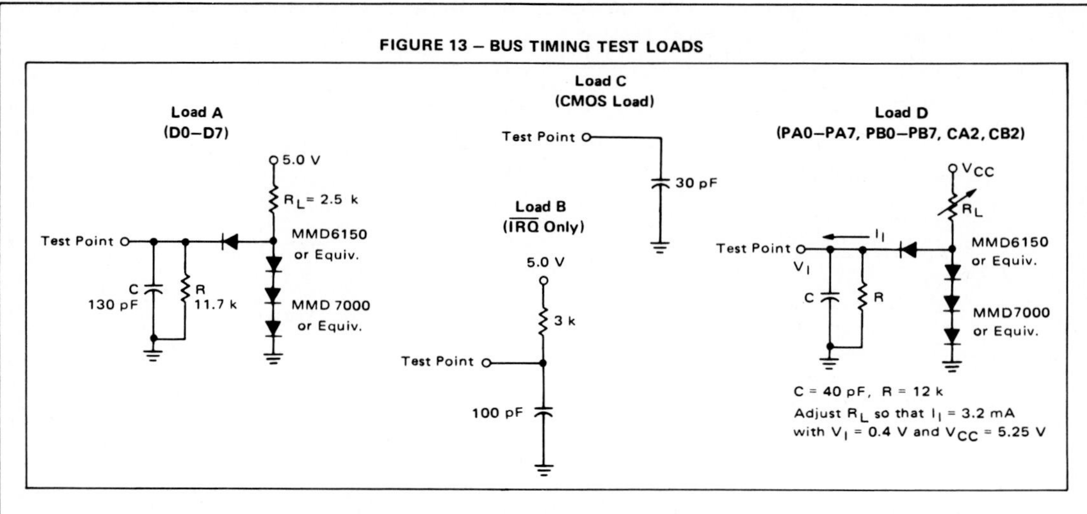

PIA INTERFACE SIGNALS FOR MPU

The PIA interfaces to the MC6800 MPU with an eight-bit bi-directional data bus, three chip select lines, two register select lines, two interrupt request lines, read/write line, enable line and reset line. These signals, in conjunction with the MC6800 VMA output, permit the MPU to have complete control over the PIA. VMA should be utilized in conjunction with an MPU address line into a chip select of the PIA.

PIA Bi-Directional Data (D0-D7) – The bi-directional data lines (D0-D7) allow the transfer of data between the MPU and the PIA. The data bus output drivers are three-state devices that remain in the high-impedance (off) state except when the MPU performs a PIA read operation. The Read/Write line is in the Read (high) state when the PIA is selected for a Read operation.

PIA Enable (E) – The enable pulse, E, is the only timing signal that is supplied to the PIA. Timing of all other signals is referenced to the leading and trailing edges of the E pulse. This signal will normally be a derivative of the MC6800 $\phi 2$ Clock.

PIA Read/Write (R/$\overline{W}$) – This signal is generated by the MPU to control the direction of data transfers on the Data Bus. A low state on the PIA Read/Write line enables the input buffers and data is transferred from the MPU to the PIA on the E signal if the device has been selected. A high on the Read/Write line sets up the PIA for a transfer of data to the bus. The PIA output buffers are enabled when the proper address and the enable pulse E are present.

$\overline{\text{Reset}}$ – The active low $\overline{\text{Reset}}$ line is used to reset all register bits in the PIA to a logical zero (low). This line can be used as a power-on reset and as a master reset during system operation.

PIA Chip Select (CS0, CS1 and $\overline{CS2}$) – These three input signals are used to select the PIA. CS0 and CS1 must be high and $\overline{CS2}$ must be low for selection of the device. Data transfers are then performed under the control of the Enable and Read/Write signals. The chip select lines must be stable for the duration of the E pulse. The device is deselected when any of the chip selects are in the inactive state.

PIA Register Select (RS0 and RS1) – The two register select lines are used to select the various registers inside the PIA. These two lines are used in conjunction with internal Control Registers to select a particular register that is to be written or read.

The register and chip select lines should be stable for the duration of the E pulse while in the read or write cycle.

Interrupt Request ($\overline{IRQA}$ and $\overline{IRQB}$) – The active low Interrupt Request lines ($\overline{IRQA}$ and $\overline{IRQB}$) act to interrupt the MPU either directly or through interrupt priority circuitry. These lines are "open drain" (no load device on the chip). This permits all interrupt request lines to be tied together in a wire-OR configuration.

Each Interrupt Request line has two internal interrupt flag bits that can cause the Interrupt Request line to go low. Each flag bit is associated with a particular peripheral interrupt line. Also four interrupt enable bits are provided in the PIA which may be used to inhibit a particular interrupt from a peripheral device.

Servicing an interrupt by the MPU may be accomplished by a software routine that, on a prioritized basis, sequentially reads and tests the two control registers in each PIA for interrupt flag bits that are set.

The interrupt flags are cleared (zeroed) as a result of an

EXPANDED BLOCK DIAGRAM

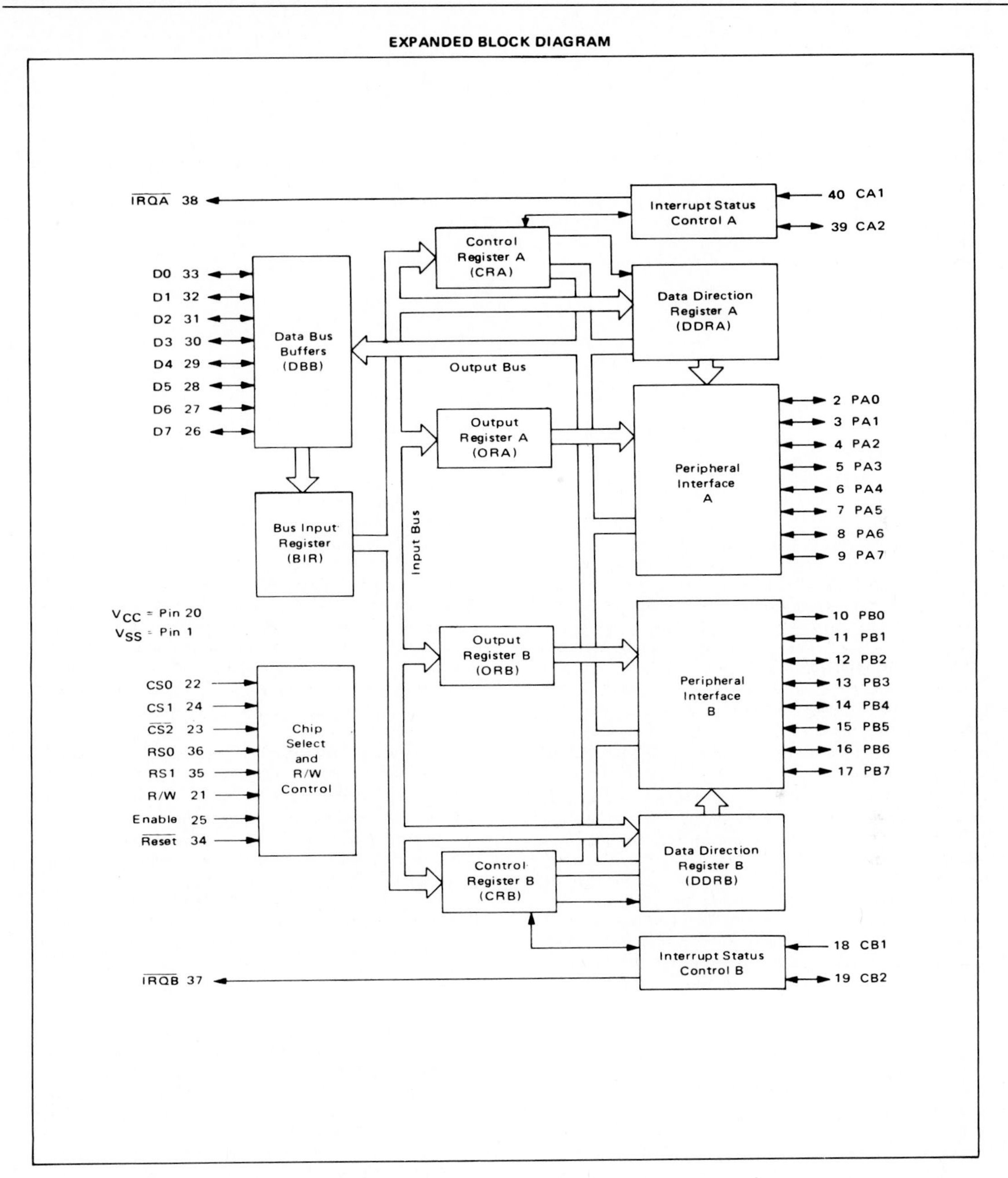

MPU Read Peripheral Data Operation of the corresponding data register. After being cleared, the interrupt flag bit cannot be enabled to be set until the PIA is deselected during an E pulse. The E pulse is used to condition the interrupt control lines (CA1, CA2, CB1, CB2). When these lines are used as interrupt inputs at least one E pulse must occur from the inactive edge to the active edge of the interrupt input signal to condition the edge sense network. If the interrupt flag has been enabled and the edge sense circuit has been properly conditioned, the interrupt flag will be set on the next active transition of the interrupt input pin.

PIA PERIPHERAL INTERFACE LINES

The PIA provides two 8-bit bi-directional data buses and four interrupt/control lines for interfacing to peripheral devices.

Section A Peripheral Data (PA0-PA7) – Each of the peripheral data lines can be programmed to act as an input or output. This is accomplished by setting a "1" in the corresponding Data Direction Register bit for those lines which are to be outputs. A "0" in a bit of the Data Direction Register causes the corresponding peripheral data line to act as an input. During an MPU Read Peripheral Data Operation, the data on peripheral lines programmed to act as inputs appears directly on the corresponding MPU Data Bus lines. In the input mode the internal pullup resistor on these lines represents a maximum of 1.5 standard TTL loads.

The data in Output Register A will appear on the data lines that are programmed to be outputs. A logical "1" written into the register will cause a "high" on the corresponding data line while a "0" results in a "low". Data in Output Register A may be read by an MPU "Read Peripheral Data A" operation when the corresponding lines are programmed as outputs. This data will be read properly if the voltage on the peripheral data lines is greater than 2.0 volts for a logic "1" output and less than 0.8 volt for a logic "0" output. Loading the output lines such that the voltage on these lines does not reach full voltage causes the data transferred into the MPU on a Read operation to differ from that contained in the respective bit of Output Register A.

Section B Peripheral Data (PB0-PB7) – The peripheral data lines in the B Section of the PIA can be programmed to act as either inputs or outputs in a similar manner to PA0-PA7. However, the output buffers driving these lines differ from those driving lines PA0-PA7. They have three-state capability, allowing them to enter a high impedance state when the peripheral data line is used as an input. In addition, data on the peripheral data lines PB0-PB7 will be read properly from those lines programmed as outputs even if the voltages are below 2.0 volts for a "high". As outputs, these lines are compatible with standard TTL and may also be used as a source of up to 1 milliampere at 1.5 volts to directly drive the base of a transistor switch.

Interrupt Input (CA1 and CB1) – Peripheral Input lines CA1 and CB1 are input only lines that set the interrupt flags of the control registers. The active transition for these signals is also programmed by the two control registers.

Peripheral Control (CA2) – The peripheral control line CA2 can be programmed to act as an interrupt input or as a peripheral control output. As an output, this line is compatible with standard TTL; as an input the internal pullup resistor on this line represents 1.5 standard TTL loads. The function of this signal line is programmed with Control Register A.

Peripheral Control (CB2) – Peripheral Control line CB2 may also be programmed to act as an interrupt input or peripheral control output. As an input, this line has high input impedance and is compatible with standard TTL. As an output it is compatible with standard TTL and may also be used as a source of up to 1 milliampere at 1.5 volts to directly drive the base of a transistor switch. This line is programmed by Control Register B.

INTERNAL CONTROLS

There are six locations within the PIA accessible to the MPU data bus: two Peripheral Registers, two Data Direction Registers, and two Control Registers. Selection of these locations is controlled by the RS0 and RS1 inputs together with bit 2 in the Control Register, as shown in Table 1.

TABLE 1 – INTERNAL ADDRESSING

RS1	RS0	Control Register Bit CRA-2	Control Register Bit CRB-2	Location Selected
0	0	1	X	Peripheral Register A
0	0	0	X	Data Direction Register A
0	1	X	X	Control Register A
1	0	X	1	Peripheral Register B
1	0	X	0	Data Direction Register B
1	1	X	X	Control Register B

X = Don't Care

INITIALIZATION

A low reset line has the effect of zeroing all PIA registers. This will set PA0-PA7, PB0-PB7, CA2 and CB2 as inputs, and all interrupts disabled. The PIA must be configured during the restart program which follows the reset.

Details of possible configurations of the Data Direction and Control Register are as follows.

DATA DIRECTION REGISTERS (DDRA and DDRB)

The two Data Direction Registers allow the MPU to control the direction of data through each corresponding peripheral data line. A Data Direction Register bit set at "0" configures the corresponding peripheral data line as an input; a "1" results in an output.

CONTROL REGISTERS (CRA and CRB)

The two Control Registers (CRA and CRB) allow the MPU to control the operation of the four peripheral control lines CA1, CA2, CB1 and CB2. In addition they allow the MPU to enable the interrupt lines and monitor the status of the interrupt flags. Bits 0 through 5 of the two registers may be written or read by the MPU when the proper chip select and register select signals are applied. Bits 6 and 7 of the two registers are read only and are modified by external interrupts occurring on control lines CA1, CA2, CB1 or CB2. The format of the control words is shown in Table 2.

TABLE 2 – CONTROL WORD FORMAT

	7	6	5	4	3	2	1	0
CRA	IRQA1	IRQA2	CA2 Control			DDRA Access	CA1 Control	
	7	**6**	**5**	**4**	**3**	**2**	**1**	**0**
CRB	IRQB1	IRQB2	CB2 Control			DDRB Access	CB1 Control	

Data Direction Access Control Bit (CRA-2 and CRB-2) – Bit 2 in each Control register (CRA and CRB) allows selection of either a Peripheral Interface Register or the Data Direction Register when the proper register select signals are applied to RS0 and RS1.

Interrupt Flags (CRA-6, CRA-7, CRB-6, and CRB-7) – The four interrupt flag bits are set by active transitions of signals on the four Interrupt and Peripheral Control lines when those lines are programmed to be inputs. These bits cannot be set directly from the MPU Data Bus and are reset indirectly by a Read Peripheral Data Operation on the appropriate section.

TABLE 3 – CONTROL OF INTERRUPT INPUTS CA1 AND CB1

CRA-1 (CRB-1)	CRA-0 (CRB-0)	Interrupt Input CA1 (CB1)	Interrupt Flag CRA-7 (CRB-7)	MPU Interrupt Request $\overline{IRQA}$ ($\overline{IRQB}$)
0	0	↓ Active	Set high on ↓ of CA1 (CB1)	Disabled — $\overline{IRQ}$ remains high
0	1	↓ Active	Set high on ↓ of CA1 (CB1)	Goes low when the interrupt flag bit CRA-7 (CRB-7) goes high
1	0	↑ Active	Set high on ↑ of CA1 (CB1)	Disabled — $\overline{IRQ}$ remains high
1	1	↑ Active	Set high on ↑ of CA1 (CB1)	Goes low when the interrupt flag bit CRA-7 (CRB-7) goes high

Notes:
1. ↑ indicates positive transition (low to high)
2. ↓ indicates negative transition (high to low)
3. The Interrupt flag bit CRA-7 is cleared by an MPU Read of the A Data Register, and CRB-7 is cleared by an MPU Read of the B Data Register.
4. If CRA-0 (CRB-0) is low when an interrupt occurs (Interrupt disabled) and is later brought high, $\overline{IRQA}$ ($\overline{IRQB}$) occurs after CRA-0 (CRB-0) is written to a "one".

Control of CA1 and CB1 Interrupt Input Lines (CRA-0, CRB-0, CRA-1, and CRB-1) – The two lowest order bits of the control registers are used to control the interrupt input lines CA1 and CB1. Bits CRA-0 and CRB-0 are used to enable the MPU interrupt signals $\overline{IRQA}$ and $\overline{IRQB}$, respectively. Bits CRA-1 and CRB-1 determine the active transition of the interrupt input signals CA1 and CB1 (Table 3).

TABLE 4 – CONTROL OF CA2 AND CB2 AS INTERRUPT INPUTS
CRA5 (CRB5) is low

CRA-5 (CRB-5)	CRA-4 (CRB-4)	CRA-3 (CRB-3)	Interrupt Input CA2 (CB2)	Interrupt Flag CRA-6 (CRB-6)	MPU Interrupt Request $\overline{IRQA}$ ($\overline{IRQB}$)
0	0	0	↓ Active	Set high on ↓ of CA2 (CB2)	Disabled — $\overline{IRQ}$ remains high
0	0	1	↓ Active	Set high on ↓ of CA2 (CB2)	Goes low when the interrupt flag bit CRA-6 (CRB-6) goes high
0	1	0	↑ Active	Set high on ↑ of CA2 (CB2)	Disabled — $\overline{IRQ}$ remains high
0	1	1	↑ Active	Set high on ↑ of CA2 (CB2)	Goes low when the interrupt flag bit CRA-6 (CRB-6) goes high

Notes:
1. ↑ indicates positive transition (low to high)
2. ↓ indicates negative transition (high to low)
3. The Interrupt flag bit CRA-6 is cleared by an MPU Read of the A Data Register and CRB-6 is cleared by an MPU Read of the B Data Register.
4. If CRA-3 (CRB-3) is low when an interrupt occurs (Interrupt disabled) and is later brought high, $\overline{IRQA}$ ($\overline{IRQB}$) occurs after CRA-3 (CRB-3) is written to a "one".

TABLE 5 – CONTROL OF CB2 AS AN OUTPUT
CRB-5 is high

CRB-5	CRB-4	CRB-3	CB2 Cleared	CB2 Set
1	0	0	Low on the positive transition of the first E pulse following an MPU Write "B" Data Register operation.	High when the interrupt flag bit CRB-7 is set by an active transition of the CB1 signal.
1	0	1	Low on the positive transition of the first E pulse after an MPU Write "B" Data Register operation.	High on the positive edge of the first "E" pulse following an "E" pulse which occurred while the part was deselected.
1	1	0	Low when CRB-3 goes low as a result of an MPU Write in Control Register "B".	Always low as long as CRB-3 is low. Will go high on an MPU Write in Control Register "B" that changes CRB-3 to "one".
1	1	1	Always high as long as CRB-3 is high. Will be cleared when an MPU Write Control Register "B" results in clearing CRB-3 to "zero".	High when CRB-3 goes high as a result of an MPU Write into Control Register "B".

Control of CA2 and CB2 Peripheral Control Lines (CRA-3, CRA-4, CRA-5, CRB-3, CRB-4, and CRB-5) — Bits 3, 4, and 5 of the two control registers are used to control the CA2 and CB2 Peripheral Control lines. These bits determine if the control lines will be an interrupt input or an output control signal. If bit CRA-5 (CRB-5) is low, CA2 (CB2) is an interrupt input line similar to CA1 (CB1) (Table 4). When CRA-5 (CRB-5) is high, CA2 (CB2) becomes an output signal that may be used to control peripheral data transfers. When in the output mode, CA2 and CB2 have slightly different characteristics (Tables 5 and 6).

TABLE 6 — CONTROL OF CA-2 AS AN OUTPUT
CRA-5 is high

CRA-5	CRA-4	CRA-3	CA2 Cleared	CA2 Set
1	0	0	Low on negative transition of E after an MPU Read "A" Data operation.	High when the interrupt flag bit CRA-7 is set by an active transition of the CA1 signal.
1	0	1	Low on negative transition of E after an MPU Read "A" Data operation.	High on the negative edge of the first "E" pulse which occurs during a deselect.
1	1	0	Low when CRA-3 goes low as a result of an MPU Write to Control Register "A".	Always low as long as CRA-3 is low. Will go high on an MPU Write to Control Register "A" that changes CRA-3 to "one".
1	1	1	Always high as long as CRA-3 is high. Will be cleared on an MPU Write to Control Register "A" that clears CRA-3 to a "zero".	High when CRA-3 goes high as a result of an MPU Write to Control Register "A".

PACKAGE DIMENSIONS

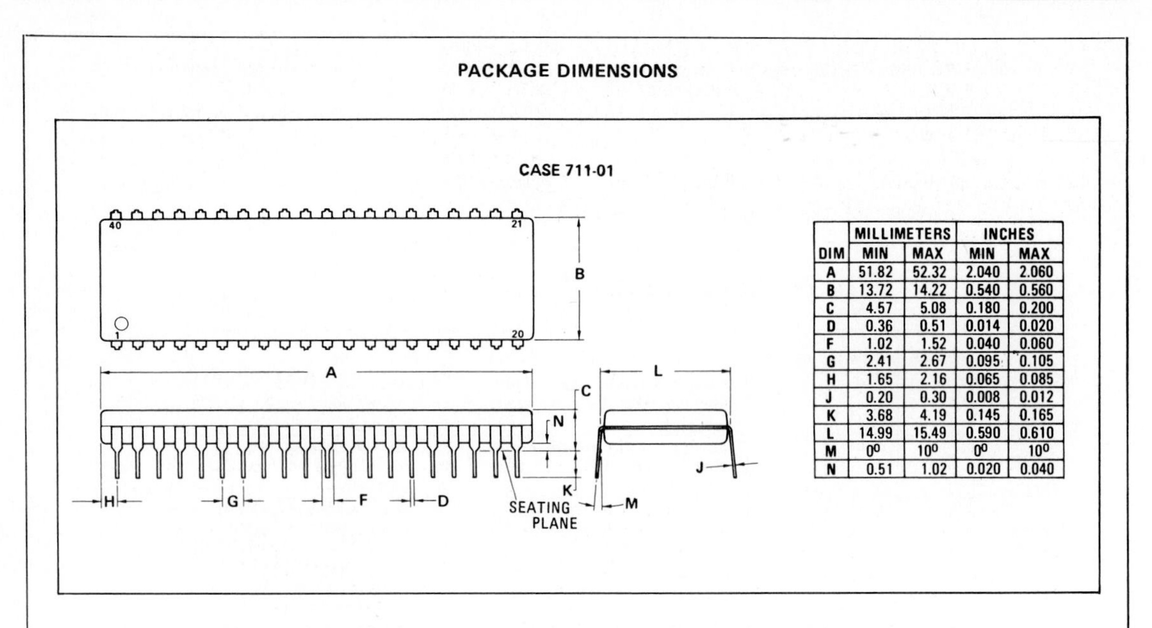

DIM	MILLIMETERS MIN	MILLIMETERS MAX	INCHES MIN	INCHES MAX
A	51.82	52.32	2.040	2.060
B	13.72	14.22	0.540	0.560
C	4.57	5.08	0.180	0.200
D	0.36	0.51	0.014	0.020
F	1.02	1.52	0.040	0.060
G	2.41	2.67	0.095	0.105
H	1.65	2.16	0.065	0.085
J	0.20	0.30	0.008	0.012
K	3.68	4.19	0.145	0.165
L	14.99	15.49	0.590	0.610
M	0°	10°	0°	10°
N	0.51	1.02	0.020	0.040

PIN ASSIGNMENT

Pin	Signal	Signal	Pin
1	V_{SS}	CA1	40
2	PA0	CA2	39
3	PA1	$\overline{IRQA}$	38
4	PA2	$\overline{IRQB}$	37
5	PA3	RS0	36
6	PA4	RS1	35
7	PA5	$\overline{Reset}$	34
8	PA6	D0	33
9	PA7	D1	32
10	PB0	D2	31
11	PB1	D3	30
12	PB2	D4	29
13	PB3	D5	28
14	PB4	D6	27
15	PB5	D7	26
16	PB6	E	25
17	PB7	CS1	24
18	CB1	$\overline{CS2}$	23
19	CB2	CS0	22
20	V_{CC}	R/W	21

PACKAGE DIMENSIONS

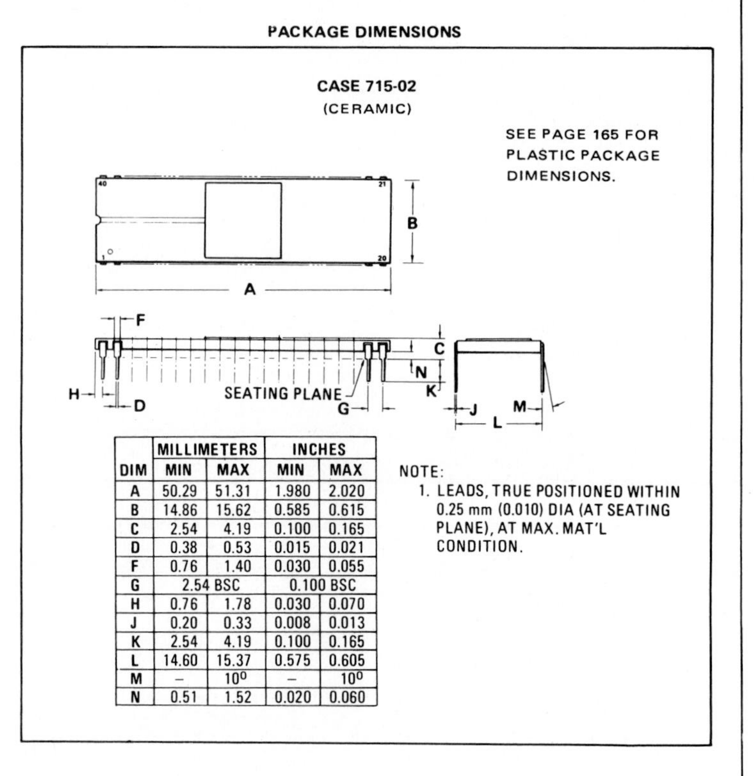

DIM	MILLIMETERS MIN	MILLIMETERS MAX	INCHES MIN	INCHES MAX
A	50.29	51.31	1.980	2.020
B	14.86	15.62	0.585	0.615
C	2.54	4.19	0.100	0.165
D	0.38	0.53	0.015	0.021
F	0.76	1.40	0.030	0.055
G	2.54 BSC		0.100 BSC	
H	0.76	1.78	0.030	0.070
J	0.20	0.33	0.008	0.013
K	2.54	4.19	0.100	0.165
L	14.60	15.37	0.575	0.605
M	–	10°	–	10°
N	0.51	1.52	0.020	0.060

NOTE:
1. LEADS, TRUE POSITIONED WITHIN 0.25 mm (0.010) DIA (AT SEATING PLANE), AT MAX. MAT'L CONDITION.

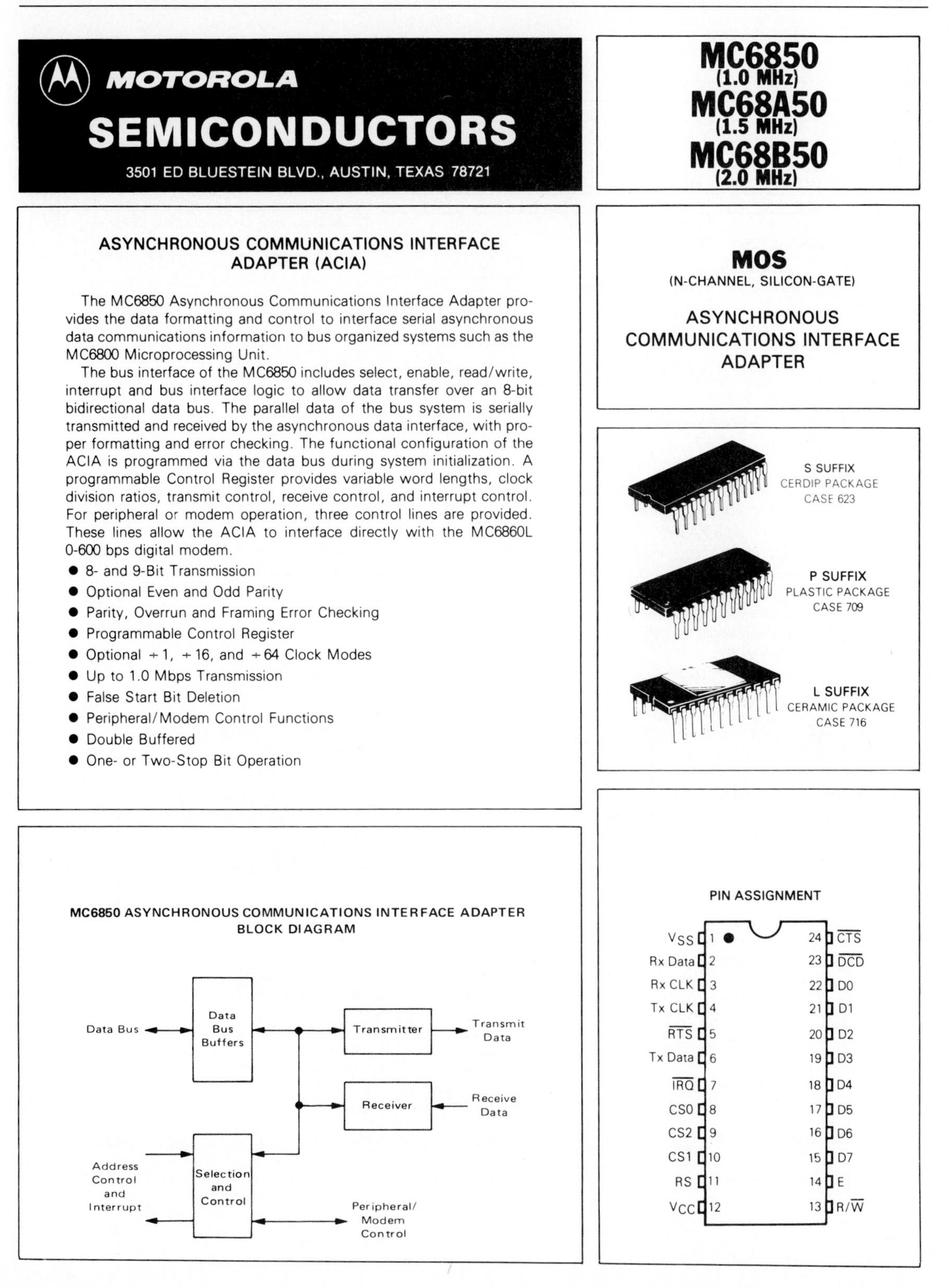

MOTOROLA

SEMICONDUCTORS

3501 ED BLUESTEIN BLVD., AUSTIN, TEXAS 78721

MC6850 (1.0 MHz)
MC68A50 (1.5 MHz)
MC68B50 (2.0 MHz)

ASYNCHRONOUS COMMUNICATIONS INTERFACE ADAPTER (ACIA)

The MC6850 Asynchronous Communications Interface Adapter provides the data formatting and control to interface serial asynchronous data communications information to bus organized systems such as the MC6800 Microprocessing Unit.

The bus interface of the MC6850 includes select, enable, read/write, interrupt and bus interface logic to allow data transfer over an 8-bit bidirectional data bus. The parallel data of the bus system is serially transmitted and received by the asynchronous data interface, with proper formatting and error checking. The functional configuration of the ACIA is programmed via the data bus during system initialization. A programmable Control Register provides variable word lengths, clock division ratios, transmit control, receive control, and interrupt control. For peripheral or modem operation, three control lines are provided. These lines allow the ACIA to interface directly with the MC6860L 0-600 bps digital modem.

- 8- and 9-Bit Transmission
- Optional Even and Odd Parity
- Parity, Overrun and Framing Error Checking
- Programmable Control Register
- Optional ÷1, ÷16, and ÷64 Clock Modes
- Up to 1.0 Mbps Transmission
- False Start Bit Deletion
- Peripheral/Modem Control Functions
- Double Buffered
- One- or Two-Stop Bit Operation

MOS

(N-CHANNEL, SILICON-GATE)

ASYNCHRONOUS COMMUNICATIONS INTERFACE ADAPTER

S SUFFIX
CERDIP PACKAGE
CASE 623

P SUFFIX
PLASTIC PACKAGE
CASE 709

L SUFFIX
CERAMIC PACKAGE
CASE 716

MC6850 ASYNCHRONOUS COMMUNICATIONS INTERFACE ADAPTER BLOCK DIAGRAM

PIN ASSIGNMENT

Pin	Name	Pin	Name
1	V_{SS}	24	$\overline{CTS}$
2	Rx Data	23	$\overline{DCD}$
3	Rx CLK	22	D0
4	Tx CLK	21	D1
5	$\overline{RTS}$	20	D2
6	Tx Data	19	D3
7	$\overline{IRQ}$	18	D4
8	CS0	17	D5
9	$\overline{CS2}$	16	D6
10	CS1	15	D7
11	RS	14	E
12	V_{CC}	13	R/$\overline{W}$

MAXIMUM RATINGS

Characteristics	Symbol	Value	Unit
Supply Voltage	V_{CC}	−0.3 to +7.0	V
Input Voltage	V_{in}	−0.3 to +7.0	V
Operating Temperature Range MC6850, MC68A50, MC68B50 MC6850C, MC68A50C, MC68B50C	T_A	T_L to T_H 0 to 70 −40 to +85	°C
Storage Temperature Range	T_{stg}	−55 to +150	°C

This device contains circuitry to protect the inputs against damage due to high static voltages or electric fields; however, it is advised that normal precautions be taken to avoid application of any voltage higher than maximum rated voltages to this high-impedance circuit. Reliability of operation is enhanced if unused inputs are tied to an appropriate logic voltage level (e.g., either V_{SS} or V_{CC}).

THERMAL CHARACTERISTICS

Characteristic	Symbol	Value	Unit
Thermal Resistance Plastic Ceramic Cerdip	θ_{JA}	 120 60 65	°C/W

POWER CONSIDERATIONS

The average chip-junction temperature, T_J, in °C can be obtained from:

$$T_J = T_A + (P_D \cdot \theta_{JA}) \qquad (1)$$

Where:

$T_A \equiv$ Ambient Temperature, °C

$\theta_{JA} \equiv$ Package Thermal Resistance, Junction-to-Ambient, °C/W

$P_D \equiv P_{INT} + P_{PORT}$

$P_{INT} \equiv I_{CC} \times V_{CC}$, Watts — Chip Internal Power

$P_{PORT} \equiv$ Port Power Dissipation, Watts — User Determined

For most applications $P_{PORT} \ll P_{INT}$ and can be neglected. P_{PORT} may become significant if the device is configured to drive Darlington bases or sink LED loads.

An approximate relationship between P_D and T_J (if P_{PORT} is neglected) is:

$$P_D = K \div (T_J + 273°C) \qquad (2)$$

Solving equations 1 and 2 for K gives:

$$K = P_D \cdot (T_A + 273°C) + \theta_{JA} \cdot P_D^2 \qquad (3)$$

Where K is a constant pertaining to the particular part. K can be determined from equation 3 by measuring P_D (at equilibrium) for a known T_A. Using this value of K the values of P_D and T_J can be obtained by solving equations (1) and (2) iteratively for any value of T_A.

DC ELECTRICAL CHARACTERISTICS (V_{CC} = 5.0 Vdc ±5%, V_{SS} = 0, T_A = T_L to T_H unless otherwise noted.)

Characteristic		Symbol	Min	Typ	Max	Unit
Input High Voltage		V_{IH}	V_{SS}+2.0	–	V_{CC}	V
Input Low Voltage		V_{IL}	V_{SS}−0.3	–	V_{SS}+0.8	V
Input Leakage Current (V_{in} = 0 to 5.25 V)	R/$\overline{W}$, CS0, CS1, $\overline{CS2}$, Enable RS, Rx D, Rx C, $\overline{CTS}$, $\overline{DCD}$	I_{in}	–	1.0	2.5	μA
Three-State (Off State) Input Current (V_{in} = 0.4 to 2.4 V)	D0-D7	I_{TSI}	–	2.0	10	μA
Output High Voltage (I_{Load} = −205 μA, Enable Pulse Width < 25 μs) (I_{Load} = −100 μA, Enable Pulse Width < 25 μs)	D0-D7 Tx Data, $\overline{RTS}$	V_{OH}	 V_{SS}+2.4 V_{SS}+2.4	 – –	 – –	V
Output Low Voltage (I_{Load} = 1.6 mA, Enable Pulse Width < 25 μs)		V_{OL}	–	–	V_{SS}+0.4	V
Output Leakage Current (Off State) (V_{OH} = 2.4 V)	$\overline{IRQ}$	I_{LOH}	–	1.0	10	μA
Internal Power Dissipation (Measured at T_A = T_L)		P_{INT}	–	300	525	mW
Internal Input Capacitance (V_{in} = 0, T_A = 25°C, f = 1.0 MHz)	 D0-D7 E, Tx CLK, Rx CLK, R/W, RS, Rx Data, CS0, CS1, $\overline{CS2}$, $\overline{CTS}$, $\overline{DCD}$	C_{in}	 – –	 10 7.0	 12.5 7.5	pF
Output Capacitance (V_{in} = 0, T_A = 25°C, f = 1.0 MHz)	RTS, Tx Data $\overline{IRQ}$	C_{out}	– –	– –	10 5.0	pF

SERIAL DATA TIMING CHARACTERISTICS

Characteristic		Symbol	MC6850 Min	MC6850 Max	MC68A50 Min	MC68A50 Max	MC68B50 Min	MC68B50 Max	Unit
Data Clock Pulse Width, Low (See Figure 1)	÷16, ÷64 Modes ÷1 Mode	PW_{CL}	600 900	– –	450 650	– –	280 500	– –	ns
Data Clock Pulse Width, High (See Figure 2)	÷16, ÷64 Modes ÷1 Mode	PW_{CH}	600 900	– –	450 650	– –	280 500	– –	ns
Data Clock Frequency	÷16, ÷64 Modes ÷1 Mode	f_C	– –	0.8 500	– –	1.0 750	– –	1.5 1000	MHz kHz
Data Clock-to-Data Delay for Transmitter (See Figure 3)		t_{TDD}	–	600	–	540	–	460	ns
Receive Data Setup Time (See Figure 4)	÷1 Mode	t_{RDS}	250	–	100	–	30	–	ns
Receive Data Hold Time (See Figure 5)	÷1 Mode	t_{RDH}	250	–	100	–	30	–	ns
Interrupt Request Release Time (See Figure 6)		t_{IR}	–	1.2	–	0.9	–	0.7	μs
Request-to-Send Delay Time (See Figure 6)		t_{RTS}	–	560	–	480	–	400	ns
Input Rise and Fall Times (or 10% of the pulse width if smaller)		t_r, t_f	–	1.0	–	0.5	–	0.25	μs

FIGURE 1 — CLOCK PULSE WIDTH, LOW-STATE

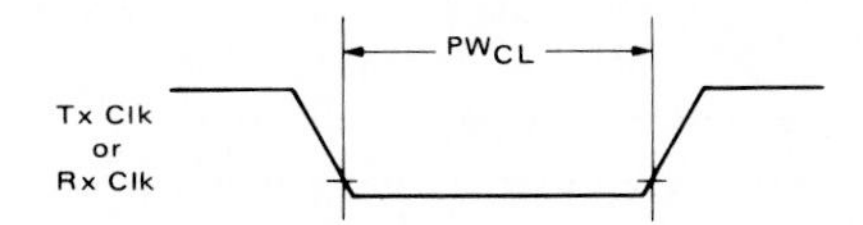

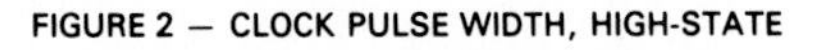

FIGURE 2 — CLOCK PULSE WIDTH, HIGH-STATE

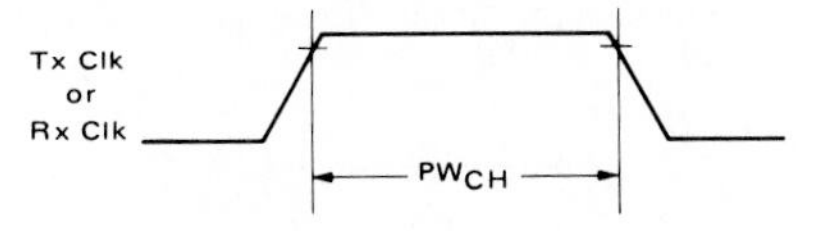

FIGURE 3 — TRANSMIT DATA OUTPUT DELAY

Tx Clk

t_{TDD}

Tx Data

FIGURE 4 — RECEIVE DATA SETUP TIME (÷1 Mode)

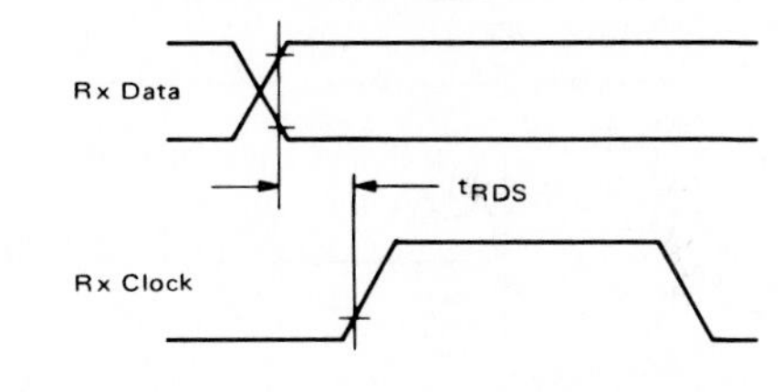

FIGURE 5 — RECEIVE DATA HOLD TIME (÷1 Mode)

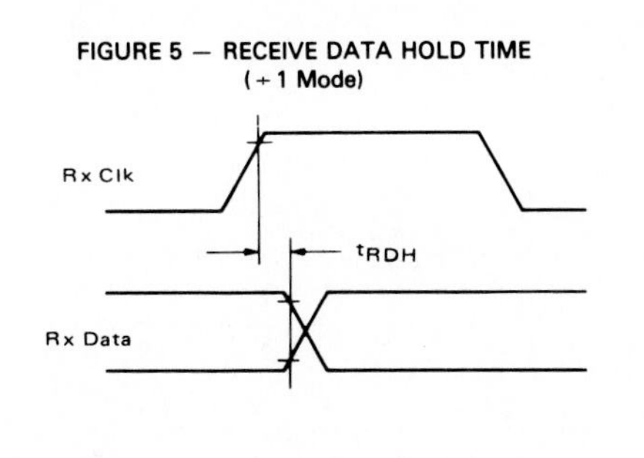

FIGURE 6 — REQUEST-TO-SEND DELAY AND INTERRUPT-REQUEST RELEASE TIMES

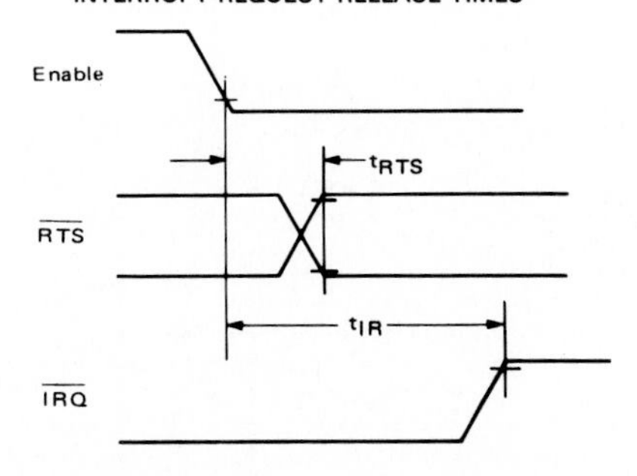

Note: Timing measurements are referenced to and from a low voltage of 0.8 volts and a high voltage of 2.0 volts, unless otherwise noted.

BUS TIMING CHARACTERISTICS (See Notes 1 and 2 and Figure 7)

Ident. Number	Characteristic	Symbol	MC6850		MC68A50		MC68B50		Unit
			Min	Max	Min	Max	Min	Max	
1	Cycle Time	t_{cyc}	1.0	10	0.67	10	0.5	10	μs
2	Pulse Width, E Low	PW_{EL}	430	9500	280	9500	210	9500	ns
3	Pulse Width, E High	PW_{EH}	450	9500	280	9500	220	9500	ns
4	Clock Rise and Fall Time	t_r, t_f	–	25	–	25	–	20	ns
9	Address Hold Time	t_{AH}	10	–	10	–	10	–	ns
13	Address Setup Time Before E	t_{AS}	80	–	60	–	40	–	ns
14	Chip Select Setup Time Before E	t_{CS}	80	–	60	–	40	–	ns
15	Chip Select Hold Time	t_{CH}	10	–	10	–	10	–	ns
18	Read Data Hold Time	t_{DHR}	20	50*	20	50*	20	50*	ns
21	Write Data Hold Time	t_{DHW}	10	–	10	–	10	–	ns
30	Output Data Delay Time	t_{DDR}	–	290	–	180	–	150	ns
31	Input Data Setup Time	t_{DSW}	165	–	80	–	60	–	ns

*The data bus output buffers are no longer sourcing or sinking current by t_{DHR}max (High Impedance).

FIGURE 7 — BUS TIMING CHARACTERISTICS

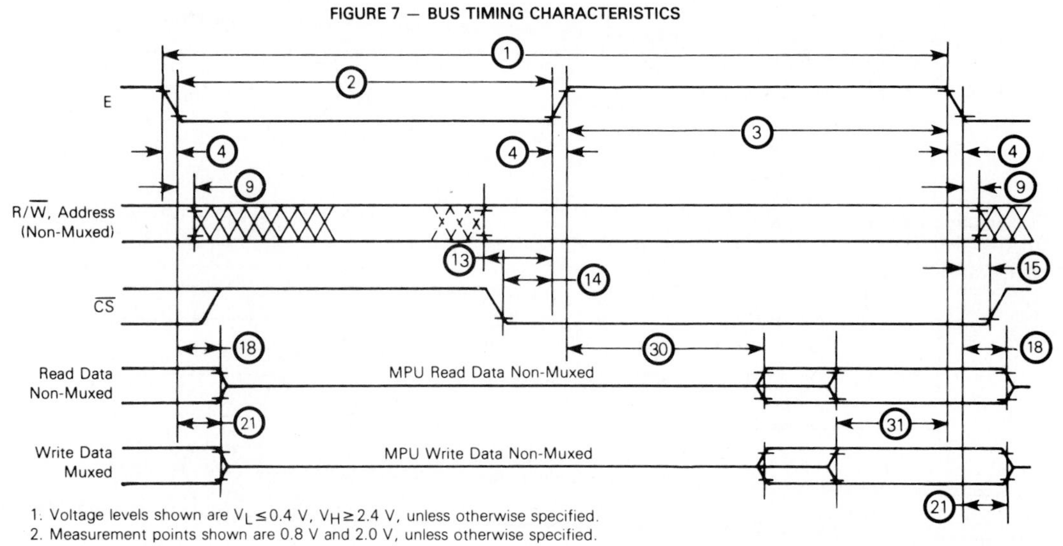

1. Voltage levels shown are $V_L \leq 0.4$ V, $V_H \geq 2.4$ V, unless otherwise specified.
2. Measurement points shown are 0.8 V and 2.0 V, unless otherwise specified.

FIGURE 8 — BUS TIMING TEST LOADS

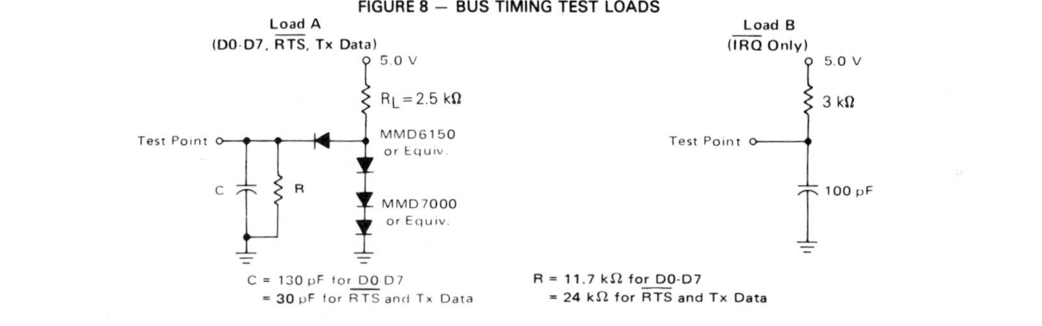

FIGURE 9 — EXPANDED BLOCK DIAGRAM

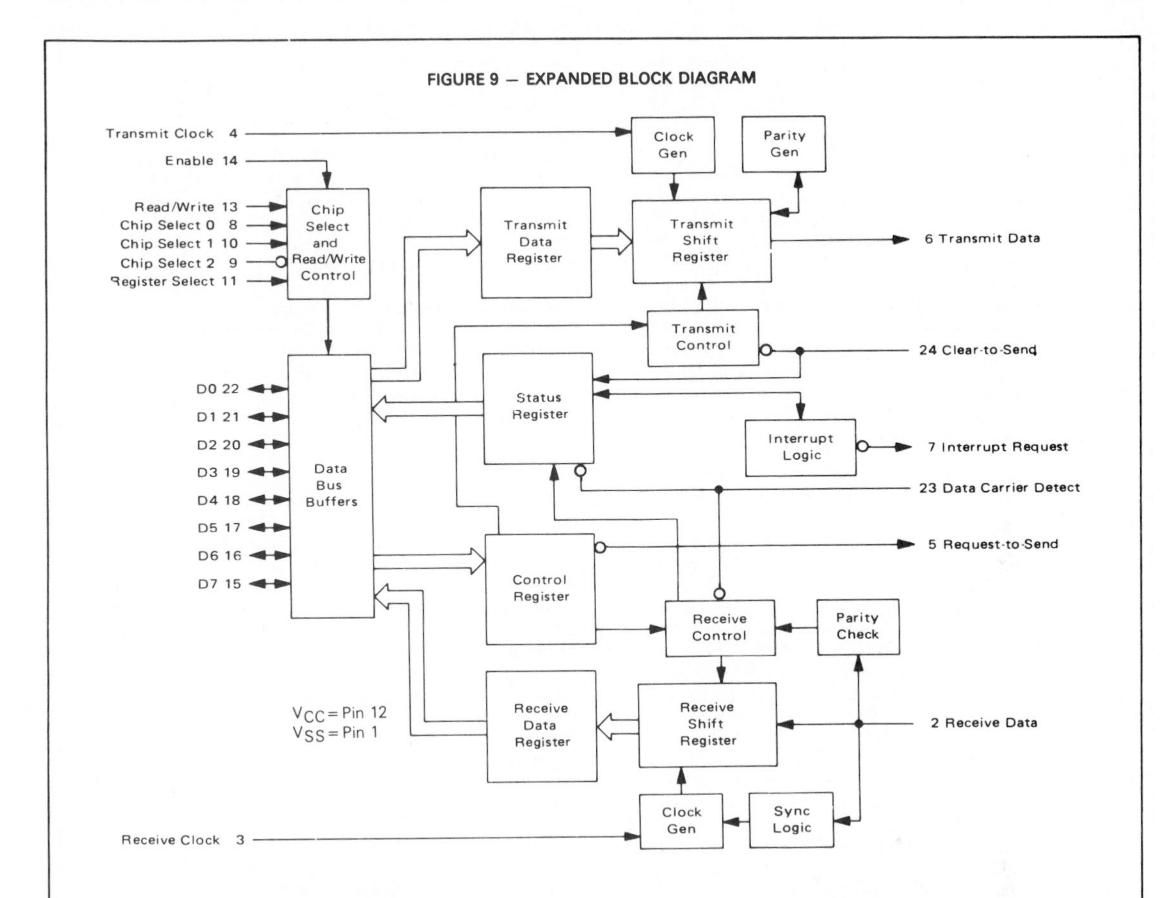

DEVICE OPERATION

At the bus interface, the ACIA appears as two addressable memory locations. Internally, there are four registers: two read-only and two write-only registers. The read-only registers are Status and Receive Data; the write-only registers are Control and Transmit Data. The serial interface consists of serial input and output lines with independent clocks, and three peripheral/modem control lines.

POWER ON/MASTER RESET

The master reset (CR0, CR1) should be set during system initialization to insure the reset condition and prepare for programming the ACIA functional configuration when the communications channel is required. During the first master reset, the $\overline{\text{IRQ}}$ and $\overline{\text{RTS}}$ outputs are held at level 1. On all other master resets, the $\overline{\text{RTS}}$ output can be programmed high or low with the $\overline{\text{IRQ}}$ output held high. Control bits CR5 and CR6 should also be programmed to define the state of $\overline{\text{RTS}}$ whenever master reset is utilized. The ACIA also contains internal power-on reset logic to detect the power line turn-on transition and hold the chip in a reset state to prevent erroneous output transitions prior to initialization. This circuitry depends on clean power turn-on transitions. The power-on reset is released by means of the bus-programmed master reset which must be applied prior to operating the ACIA. After master resetting the ACIA, the programmable Control Register can be set for a number of options such as variable clock divider ratios, variable word length, one or two stop bits, parity (even, odd, or none), etc.

TRANSMIT

A typical transmitting sequence consists of reading the ACIA Status Register either as a result of an interrupt or in the ACIA's turn in a polling sequence. A character may be written into the Transmit Data Register if the status read operation has indicated that the Transmit Data Register is empty. This character is transferred to a Shift Register where it is serialized and transmitted from the Transmit Data output preceded by a start bit and followed by one or two stop bits. Internal parity (odd or even) can be optionally added to the character and will occur between the last data bit and the first stop bit. After the first character is written in the Data Register, the Status Register can be read again to check for a Transmit Data Register Empty condition and current peripheral status. If the register is empty, another character can be loaded for transmission even though the first character is in the process of being transmitted (because of double buffering). The second character will be automatically transferred into the Shift Register when the first character transmission is completed. This sequence continues until all the characters have been transmitted.

RECEIVE

Data is received from a peripheral by means of the Receive Data input. A divide-by-one clock ratio is provided for an externally synchronized clock (to its data) while the divide-by-16 and 64 ratios are provided for internal synchronization. Bit synchronization in the divide-by-16 and 64 modes is initiated by the detection of 8 or 32 low samples on the receive line in the divide-by-16 and 64 modes respectively. False start bit deletion capability insures that a full half bit of a start bit has been received before the internal clock is synchronized to the bit time. As a character is being received, parity (odd or even) will be checked and the error indication will be available in the Status Register along with framing error, overrun error, and Receive Data Register full. In a typical receiving sequence, the Status Register is read to determine if a character has been received from a peripheral. If the Receiver Data Register is full, the character is placed on the 8-bit ACIA bus when a Read Data command is received from the MPU. When parity has been selected for a 7-bit word (7 bits plus parity), the receiver strips the parity bit (D7=0) so that data alone is transferred to the MPU. This feature reduces MPU programming. The Status Register can continue to be read to determine when another character is available in the Receive Data Register. The receiver is also double buffered so that a character can be read from the data register as another character is being received in the shift register. The above sequence continues until all characters have been received.

INPUT/OUTPUT FUNCTIONS

ACIA INTERFACE SIGNALS FOR MPU

The ACIA interfaces to the M6800 MPU with an 8-bit bidirectional data bus, three chip select lines, a register select line, an interrupt request line, read/write line, and enable line. These signals permit the MPU to have complete control over the ACIA.

ACIA Bidirectional Data (D0-D7) — The bidirectional data lines (D0-D7) allow for data transfer between the ACIA and the MPU. The data bus output drivers are three-state devices that remain in the high-impedance (off) state except when the MPU performs an ACIA read operation.

ACIA Enable (E) — The Enable signal, E, is a high-impedance TTL-compatible input that enables the bus input/output data buffers and clocks data to and from the ACIA. This signal will normally be a derivative of the MC6800 $\phi 2$ Clock or MC6809 E clock.

Read/Write (R/$\overline{W}$) — The Read/Write line is a high-impedance input that is TTL compatible and is used to control the direction of data flow through the ACIA's input/output data bus interface. When Read/Write is high (MPU Read cycle), ACIA output drivers are turned on and a selected register is read. When it is low, the ACIA output drivers are turned off and the MPU writes into a selected register. Therefore, the Read/Write signal is used to select read-only or write-only registers within the ACIA.

Chip Select (CS0, CS1, $\overline{CS2}$) — These three high-impedance TTL-compatible input lines are used to address the ACIA. The ACIA is selected when CS0 and CS1 are high and $\overline{CS2}$ is low. Transfers of data to and from the ACIA are then performed under the control of the Enable Signal, Read/Write, and Register Select.

Register Select (RS) — The Register Select line is a high-impedance input that is TTL compatible. A high level is used to select the Transmit/Receive Data Registers and a low level the Control/Status Registers. The Read/Write signal line is used in conjunction with Register Select to select the read-only or write-only register in each register pair.

Interrupt Request ($\overline{IRQ}$) — Interrupt Request is a TTL-compatible, open-drain (no internal pullup), active low output that is used to interrupt the MPU. The $\overline{IRQ}$ output remains low as long as the cause of the interrupt is present and the appropriate interrupt enable within the ACIA is set. The IRQ status bit, when high, indicates the $\overline{IRQ}$ output is in the active state.

Interrupts result from conditions in both the transmitter and receiver sections of the ACIA. The transmitter section causes an interrupt when the Transmitter Interrupt Enabled condition is selected (CR5•$\overline{CR6}$), and the Transmit Data Register Empty (TDRE) status bit is high. The TDRE status bit indicates the current status of the Transmitter Data Register except when inhibited by Clear-to-Send ($\overline{CTS}$) being high or the ACIA being maintained in the Reset condition. The interrupt is cleared by writing data into the Transmit Data Register. The interrupt is masked by disabling the Transmitter Interrupt via CR5 or CR6 or by the loss of $\overline{CTS}$ which inhibits the TDRE status bit. The Receiver section causes an interrupt when the Receiver Interrupt Enable is set and the Receive Data Register Full (RDRF) status bit is high, an Overrun has occurred, or Data Carrier Detect ($\overline{DCD}$) has gone high. An interrupt resulting from the RDRF status bit can be cleared by reading data or resetting the ACIA. Interrupts caused by Overrun or loss of $\overline{DCD}$ are cleared by reading the status register after the error condition has occurred and then reading the Receive Data Register or resetting the ACIA. The receiver interrupt is masked by resetting the Receiver Interrupt Enable.

CLOCK INPUTS

Separate high-impedance TTL-compatible inputs are provided for clocking of transmitted and received data. Clock frequencies of 1, 16, or 64 times the data rate may be selected.

Transmit Clock (Tx CLK) — The Transmit Clock input is used for the clocking of transmitted data. The transmitter initiates data on the negative transition of the clock.

Receive Clock (Rx CLK) — The Receive Clock input is used for synchronization of received data. (In the $\div 1$ mode, the clock and data must be synchronized externally.) The receiver samples the data on the positive transition of the clock.

SERIAL INPUT/OUTPUT LINES

Receive Data (Rx Data) — The Receive Data line is a high-impedance TTL-compatible input through which data is received in a serial format. Synchronization with a clock for detection of data is accomplished internally when clock rates of 16 or 64 times the bit rate are used.

Transmit Data (Tx Data) — The Transmit Data output line transfers serial data to a modem or other peripheral.

PERIPHERAL/MODEM CONTROL

The ACIA includes several functions that permit limited control of a peripheral or modem. The functions included are Clear-to-Send, Request-to-Send and Data Carrier Detect.

Clear-to-Send ($\overline{CTS}$) — This high-impedance TTL-compatible input provides automatic control of the transmitting end of a communications link via the modem Clear-to-Send active low output by inhibiting the Transmit Data Register Empty (TDRE) status bit.

Request-to-Send ($\overline{RTS}$) — The Request-to-Send output enables the MPU to control a peripheral or modem via the data bus. The RTS output corresponds to the state of the Control Register bits CR5 and CR6. When CR6 = 0 or both CR5 and CR6 = 1, the $\overline{RTS}$ output is low (the active state). This output can also be used for Data Terminal Ready (DTR).

Data Carrier Detect ($\overline{DCD}$) — This high-impedance TTL-compatible input provides automatic control, such as in the receiving end of a communications link by means of a modem Data Carrier Detect output. The $\overline{DCD}$ input inhibits and initializes the receiver section of the ACIA when high. A low-to-high transition of the Data Carrier Detect initiates an interrupt to the MPU to indicate the occurrence of a loss of carrier when the Receive Interrupt Enable bit is set. The Rx CLK must be running for proper $\overline{DCD}$ operation.

ACIA REGISTERS

The expanded block diagram for the ACIA indicates the internal registers on the chip that are used for the status, control, receiving, and transmitting of data. The content of each of the registers is summarized in Table 1.

TRANSMIT DATA REGISTER (TDR)

Data is written in the Transmit Data Register during the negative transition of the enable (E) when the ACIA has been addressed with RS high and R/$\overline{W}$ low. Writing data into the register causes the Transmit Data Register Empty bit in the Status Register to go low. Data can then be transmitted. If the transmitter is idling and no character is being transmitted, then the transfer will take place within 1-bit time of the trailing edge of the Write command. If a character is being transmitted, the new data character will commence as soon as the previous character is complete. The transfer of data causes the Transmit Data Register Empty (TDRE) bit to indicate empty.

RECEIVE DATA REGISTER (RDR)

Data is automatically transferred to the empty Receive Data Register (RDR) from the receiver deserializer (a shift register) upon receiving a complete character. This event causes the Receive Data Register Full bit (RDRF) in the status buffer to go high (full). Data may then be read through the bus by addressing the ACIA and selecting the Receive Data Register with RS and R/$\overline{W}$ high when the ACIA is enabled. The non-destructive read cycle causes the RDRF bit to be cleared to empty although the data is retained in the RDR. The status is maintained by RDRF as to whether or not the data is current. When the Receive Data Register is full, the automatic transfer of data from the Receiver Shift Register to the Data Register is inhibited and the RDR contents remain valid with its current status stored in the Status Register.

TABLE 1 — DEFINITION OF ACIA REGISTER CONTENTS

Data Bus Line Number	Buffer Address			
	RS • $\overline{R/W}$ Transmit Data Register	RS • R/$\overline{W}$ Receive Data Register	$\overline{RS}$ • $\overline{R/W}$ Control Register	$\overline{RS}$ • R/$\overline{W}$ Status Register
	(Write Only)	(Read Only)	(Write Only)	(Read Only)
0	Data Bit 0*	Data Bit 0	Counter Divide Select 1 (CR0)	Receive Data Register Full (RDRF)
1	Data Bit 1	Data Bit 1	Counter Divide Select 2 (CR1)	Transmit Data Register Empty (TDRE)
2	Data Bit 2	Data Bit 2	Word Select 1 (CR2)	Data Carrier Detect ($\overline{DCD}$)
3	Data Bit 3	Data Bit 3	Word Select 2 (CR3)	Clear-to-Send ($\overline{CTS}$)
4	Data Bit 4	Data Bit 4	Word Select 3 (CR4)	Framing Error (FE)
5	Data Bit 5	Data Bit 5	Transmit Control 1 (CR5)	Receiver Overrun (OVRN)
6	Data Bit 6	Data Bit 6	Transmit Control 2 (CR6)	Parity Error (PE)
7	Data Bit 7***	Data Bit 7**	Receive Interrupt Enable (CR7)	Interrupt Request ($\overline{IRQ}$)

* Leading bit = LSB = Bit 0

** Data bit will be zero in 7-bit plus parity modes.

*** Data bit is "don't care" in 7-bit plus parity modes.

CONTROL REGISTER

The ACIA Control Register consists of eight bits of write-only buffer that are selected when RS and R/$\overline{W}$ are low. This register controls the function of the receiver, transmitter, interrupt enables, and the Request-to-Send peripheral/modem control output.

Counter Divide Select Bits (CR0 and CR1) — The Counter Divide Select Bits (CR0 and CR1) determine the divide ratios utilized in both the transmitter and receiver sections of the ACIA. Additionally, these bits are used to provide a master reset for the ACIA which clears the Status Register (except for external conditions on $\overline{CTS}$ and $\overline{DCD}$) and initializes both the receiver and transmitter. Master reset does not affect other Control Register bits. Note that after power-on or a power fail/restart, these bits must be set high to reset the ACIA. After resetting, the clock divide ratio may be selected. These counter select bits provide for the following clock divide ratios:

CR1	CR0	Function
0	0	÷1
0	1	÷16
1	0	÷64
1	1	Master Reset

Word Select Bits (CR2, CR3, and CR4) — The Word Select bits are used to select word length, parity, and the number of stop bits. The encoding format is as follows:

CR4	CR3	CR2	Function
0	0	0	7 Bits + Even Parity + 2 Stop Bits
0	0	1	7 Bits + Odd Parity + 2 Stop Bits
0	1	0	7 Bits + Even Parity + 1 Stop Bit
0	1	1	7 Bits + Odd Parity + 1 Stop Bit
1	0	0	8 Bits + 2 Stop Bits
1	0	1	8 Bits + 1 Stop Bit
1	1	0	8 Bits + Even parity + 1 Stop Bit
1	1	1	8 Bits + Odd Parity + 1 Stop Bit

Word length, Parity Select, and Stop Bit changes are not buffered and therefore become effective immediately.

Transmitter Control Bits (CR5 and CR6) — Two Transmitter Control bits provide for the control of the interrupt from the Transmit Data Register Empty condition, the Request-to-Send ($\overline{RTS}$) output, and the transmission of a Break level (space). The following encoding format is used:

CR6	CR5	Function
0	0	$\overline{RTS}$ = low, Transmitting Interrupt Disabled.
0	1	$\overline{RTS}$ = low, Transmitting Interrupt Enabled.
1	0	$\overline{RTS}$ = high, Transmitting Interrupt Disabled.
1	1	$\overline{RTS}$ = low, Transmits a Break level on the Transmit Data Output. Transmitting Interrupt Disabled.

Receive Interrupt Enable Bit (CR7) — The following interrupts will be enabled by a high level in bit position 7 of the Control Register (CR7): Receive Data Register Full, Overrun, or a low-to-high transition on the Data Carrier Detect ($\overline{DCD}$) signal line.

STATUS REGISTER

Information on the status of the ACIA is available to the MPU by reading the ACIA Status Register. This read-only register is selected when RS is low and R/$\overline{W}$ is high. Information stored in this register indicates the status of the Transmit Data Register, the Receive Data Register and error logic, and the peripheral/modem status inputs of the ACIA.

Receive Data Register Full (RDRF), Bit 0 — Receive Data Register Full indicates that received data has been transferred to the Receive Data Register. RDRF is cleared after an MPU read of the Receive Data Register or by a master reset. The cleared or empty state indicates that the contents of the Receive Data Register are not current. Data Carrier Detect being high also causes RDRF to indicate empty.

Transmit Data Register Empty (TDRE), Bit 1 — The Transmit Data Register Empty bit being set high indicates that the Transmit Data Register contents have been transferred and that new data may be entered. The low state indicates that the register is full and that transmission of a new character has not begun since the last write data command.

Data Carrier Detect ($\overline{DCD}$), Bit 2 — The Data Carrier Detect bit will be high when the $\overline{DCD}$ input from a modem has gone high to indicate that a carrier is not present. This bit going high causes an Interrupt Request to be generated when the Receive Interrupt Enable is set. It remains high after the $\overline{DCD}$ input is returned low until cleared by first reading the Status Register and then the Data Register or until a master reset occurs. If the $\overline{DCD}$ input remains high after read status and read data or master reset has occurred, the interrupt is cleared, the $\overline{DCD}$ status bit remains high and will follow the $\overline{DCD}$ input.

Clear-to-Send ($\overline{CTS}$), Bit 3 — The Clear-to-Send bit indicates the state of the Clear-to-Send input from a modem. A low $\overline{CTS}$ indicates that there is a Clear-to-Send from the modem. In the high state, the Transmit Data Register Empty bit is inhibited and the Clear-to-Send status bit will be high. Master reset does not affect the Clear-to-Send status bit.

Framing Error (FE), Bit 4 — Framing error indicates that the received character is improperly framed by a start and a stop bit and is detected by the absence of the first stop bit. This error indicates a synchronization error, faulty transmission, or a break condition. The framing error flag is set or reset during the receive data transfer time. Therefore, this error indicator is present throughout the time that the associated character is available.

Receiver Overrun (OVRN), Bit 5 — Overrun is an error flag that indicates that one or more characters in the data stream were lost. That is, a character or a number of characters were received but not read from the Receive Data Register (RDR) prior to subsequent characters being received. The overrun condition begins at the midpoint of the last bit of the second character received in succession without a read of the RDR having occurred. The Overrun does not occur in the Status Register until the valid character prior to Overrun has been read. The RDRF bit remains set until the Overrun is reset. Character synchronization is maintained during the Overrun condition. The Overrun indication is reset after the reading of data from the Receive Data Register or by a Master Reset.

Parity Error (PE), Bit 6 – The parity error flag indicates that the number of highs (ones) in the character does not agree with the preselected odd or even parity. Odd parity is defined to be when the total number of ones is odd. The parity error indication will be present as long as the data character is in the RDR. If no parity is selected, then both the transmitter parity generator output and the receiver partiy check results are inhibited.

Interrupt Request ($\overline{IRQ}$), Bit 7 – The $\overline{IRQ}$ bit indicates the state of the $\overline{IRQ}$ output. Any interrupt condition with its applicable enable will be indicated in this status bit. Anytime the $\overline{IRQ}$ output is low the $\overline{IRQ}$ bit will be high to indicate the interrupt or service request status. $\overline{IRQ}$ is cleared by a read operation to the Receive Data Register or a write operation to the Transmit Data Register.

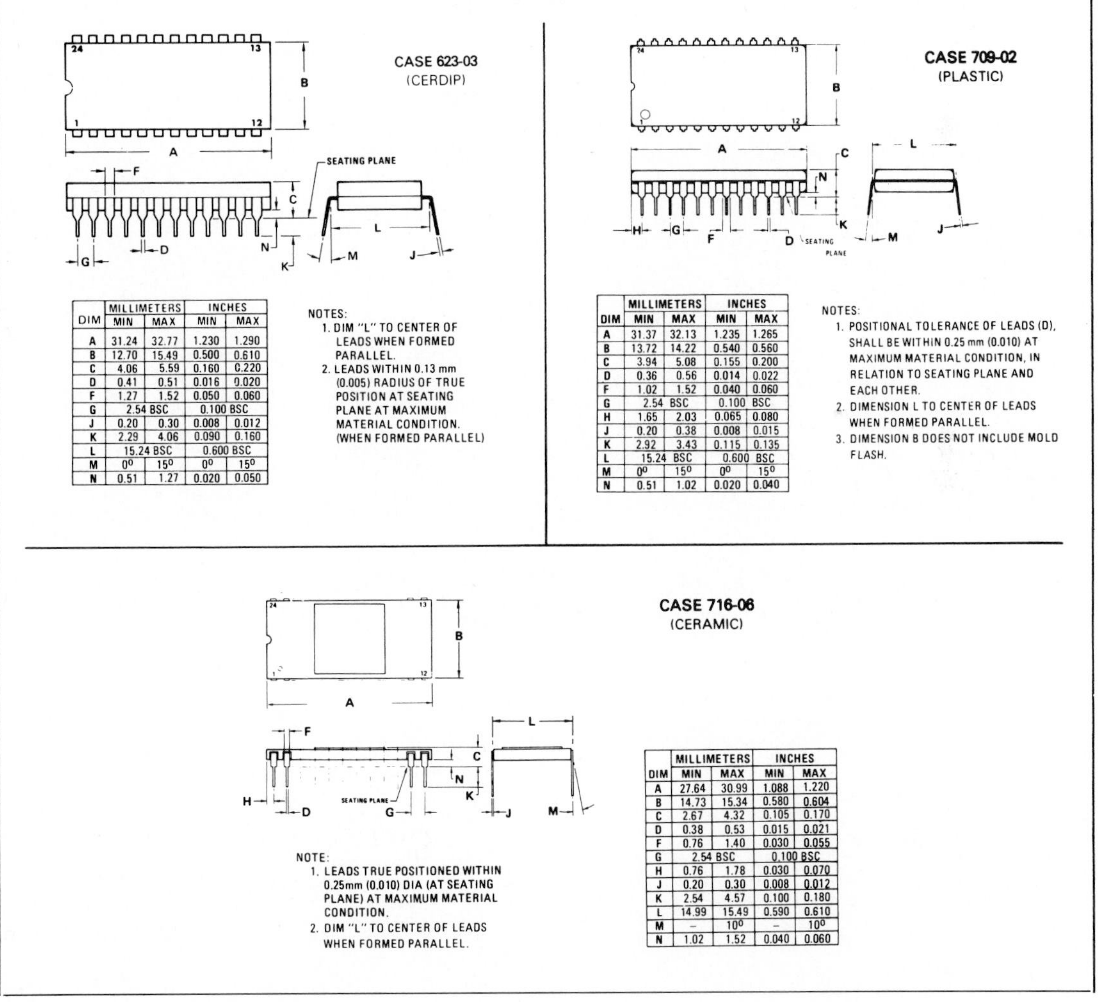

PACKAGE DIMENSIONS

CASE 623-03
(CERDIP)

DIM	MILLIMETERS MIN	MILLIMETERS MAX	INCHES MIN	INCHES MAX
A	31.24	32.77	1.230	1.290
B	12.70	15.49	0.500	0.610
C	4.06	5.59	0.160	0.220
D	0.41	0.51	0.016	0.020
F	1.27	1.52	0.050	0.060
G	2.54 BSC		0.100 BSC	
J	0.20	0.30	0.008	0.012
K	2.29	4.06	0.090	0.160
L	15.24 BSC		0.600 BSC	
M	0°	15°	0°	15°
N	0.51	1.27	0.020	0.050

NOTES:
1. DIM "L" TO CENTER OF LEADS WHEN FORMED PARALLEL.
2. LEADS WITHIN 0.13 mm (0.005) RADIUS OF TRUE POSITION AT SEATING PLANE AT MAXIMUM MATERIAL CONDITION. (WHEN FORMED PARALLEL)

CASE 709-02
(PLASTIC)

DIM	MILLIMETERS MIN	MILLIMETERS MAX	INCHES MIN	INCHES MAX
A	31.37	32.13	1.235	1.265
B	13.72	14.22	0.540	0.560
C	3.94	5.08	0.155	0.200
D	0.36	0.56	0.014	0.022
F	1.02	1.52	0.040	0.060
G	2.54 BSC		0.100 BSC	
H	1.65	2.03	0.065	0.080
J	0.20	0.38	0.008	0.015
K	2.92	3.43	0.115	0.135
L	15.24 BSC		0.600 BSC	
M	0°	15°	0°	15°
N	0.51	1.02	0.020	0.040

NOTES:
1. POSITIONAL TOLERANCE OF LEADS (D), SHALL BE WITHIN 0.25 mm (0.010) AT MAXIMUM MATERIAL CONDITION, IN RELATION TO SEATING PLANE AND EACH OTHER.
2. DIMENSION L TO CENTER OF LEADS WHEN FORMED PARALLEL.
3. DIMENSION B DOES NOT INCLUDE MOLD FLASH.

CASE 716-06
(CERAMIC)

NOTE:
1. LEADS TRUE POSITIONED WITHIN 0.25mm (0.010) DIA (AT SEATING PLANE) AT MAXIMUM MATERIAL CONDITION.
2. DIM "L" TO CENTER OF LEADS WHEN FORMED PARALLEL.

DIM	MILLIMETERS MIN	MILLIMETERS MAX	INCHES MIN	INCHES MAX
A	27.64	30.99	1.088	1.220
B	14.73	15.34	0.580	0.604
C	2.67	4.32	0.105	0.170
D	0.38	0.53	0.015	0.021
F	0.76	1.40	0.030	0.055
G	2.54 BSC		0.100 BSC	
H	0.76	1.78	0.030	0.070
J	0.20	0.30	0.008	0.012
K	2.54	4.57	0.100	0.180
L	14.99	15.49	0.590	0.610
M	–	10°	–	10°
N	1.02	1.52	0.040	0.060

ANSWERS TO ODD-NUMBERED PROBLEMS

Chapter 1

1.1 The PC is used to keep track of program instruction location in memory.

1.3 Instruction register

1.5 +32,768 and −32,738

1.7 (a) Set (b) Set (c) Reset (d) Reset

1.9 (a) $C0 (b) $C0 (c) $E1 (d) C2

1.11 Two

1.13 The CPU receives an active $\overline{\text{RESET}}$ and places its output lines into a high impedance state. The I bit in the status register is reset to 0. The address bus is cleared to $00 0000. After the RESET signal is removed, location $00 0000 is accessed and the data transferred to the PC.

1.15 $7612

1.17 Set the segment or unit code portion of $00 00A0 to $00.

1.19 Change BHI to BGT.

1.21 After determining the letter (A, B, C, D, or F) associated with the grade, take the original grade (unchanged by the grades program) and use DATAWORD from Problem 1.20 to display those values on units 3 and 4. The upper zeros of the grade number data in D0 will need to be blanked out. Next fetch the grade letter and convert it directly to its ASCII or 7-segment codes and use the subroutines developed to display the entire message.

1.23 Use SEARCH to input the data and the ABCD instruction to do the adding.

1.25 Insert a routine before the first SEARCH that loads the 7-segment codes for rEAdy. Modify SEARCH so that it calls the SHOW subroutine until a key is pressed (D0 =

$0000). Following MOVE to $00 8004 from D0 in the program for Problem 1.24, place the instructions that send the 7-segment codes for SEnd to locations $00 004–00 004A. Insert the JSR SHOW into the read loop, so SEnd displays until b15 goes low.

Chapter 2

2.1
```
$00 0120 12
$00 0121 34
$00 0122 AB
$00 0123 CD
```

2.3 16-bit data bus and 24-bit address bus

2.5 $\overline{UDS}$ and $\overline{LDS}$; FC_2, FC_1, FC_0 = 001; $\overline{AS}$, $\overline{DTACK}$, R/$\overline{W}$ = 0

2.7 (a) $\overline{AS}$ (b) $\overline{BR}$ (c) $\overline{BGACK}$ (d) $\overline{RESET}$ and $\overline{HALT}$ (e) $\overline{BERR}$
(f) $\overline{UDS}$ and $\overline{LDS}$ (g) $\overline{VMA}$ (h) $\overline{BG}$ (i) $\overline{DTACK}$ (j) $\overline{RESET}$

2.9 D_3 = $1234 3C2B

2.11 T = TRACE, a single-step execution of the user program.
S = supervisor mode, which indicates when an exception program is in process.

Chapter 3

3.1 (a) D0 = $1122 3344 (b) D3 = $5566 FACE (c) D1 = 9ABC DEDD
(d) A2 = $0000 2006 (e) D4 = $0012 0002 (f) A5 = $0000 2001
(g) A6 = $FFFF C345

3.3 D0 = $0000 0AFE D1 = $0000 4A63 D2 = $0000 0002
D3 = $FFFF FFFF D4 = $0000 2006 D5 = $0000 1010
A2 = $0000 200E

3.5
```
MOVE #52,D4
MOVEA.L #$2002,A2
MOVEA.L #$011006,A4
NEXT MOVE (A2)+,(A4)+        * Move a word of data
     DBF D4, NEXT            * Decrement counter
DONE BRA DONE
```

3.7 (a) D7 = $0000 CD10 (b) D6 = $AABB CCDD

3.9
```
        CLR.L DO            * Initialize holding register
        MOVE $2000,A0       * Initialize pointer
        MOVE $2002,D1       * Initialize counter
        SUBQ #$1,D1         * Adjust for DBF instruction
CHECK   CMP (A0)+,D0        * Check data sizes
        BLS KEEP            * D0 contents larger causes branch
        MOVE -2(A0),D0      * Replace D0 with larger value
KEEP    DBF D1, CHECK       * Loop counter
        MOVE D0, $2004      * Store results
DONE    BRA DONE
```

3.11 (a) D0.W + D3.W = CE00. The sum of these two positive numbers produces a negative result and sets the overflow flag.

(b) This division would produce a quotient larger than a word causing overflow (V) to be set.
(c) MULU causes the overflow flag always to be reset.
(d) The sum of these two negative numbers produces a positive result of \$6AD9 CC0F and sets the overflow flag.
(e) This division causes a quotient larger than 16 bits, setting the overflow flag.
(f) MULU causes the overflow flag always to be reset.

Chapter 4

4.1 Word (16 bits)

4.3 $\overline{\text{DTACK}}$

4.5 $\overline{\text{HALT}}$ is used as a wait signal. The length of the wait time is determined by the device holding the $\overline{\text{HALT}}$ asserted. Note: Wait states can also be inserted by withholding $\overline{\text{DTACK}}$.

4.7 D1 = \$3CB3 5A27. Overflow occurs.

4.9 There are several problems with this program:

1. The ABCD is an add with extend, but no provision is made to account for the initial condition of the X flag.

Solution: Add these instructions at the beginning of the program:

```
CLR D6
MOVE D6,CCR
```

2. There is a similar problem preceding the SBCD instruction.

Solution: Insert this instruction between ABCD and SBCD:

```
MOVE D6,CCR
```

3. The loop is never exited because the zero flag is not ever set by the SBCD instruction.

Solution: Add this instruction after the SBCD:

```
TST.B D0
```

The completed program is

```
     CLR D6
     MOVE D6,CCR
     MOVE #5,D0
     MOVE #1,D1
     MOVE #20,D2
MULT ABCD D2,D2
     MOVE D6,CCR
     SBCD D1,D0
     TST.B D0
     BNE.S MULT
DONE BRA.S DONE
```

Chapter 5

5.1
```
CLR ACONHI             * b2 = 0
MOVE #$FF,PORTAHI      * Makes ports outputs
MOVE #$1F1F,ACONHI     * Control register data
```

5.3 b_7 is set by receipt of a high to low transition on the CA1. It is reset by reading the data register of the appropriate side (MOVE PORTAHI,Dn).

5.5

From I/O	To I/O
CA1: Initiate input routine	CA2: 68000 read data; send more to port A
CB1: Initiate output routine	CB2: Data is in port B

5.7 b_0 = CA1 interrupt mask
b_1 = CA1 interrupt transition
b_2 = data/direction register access
b_6 = CA2 interrupt flag
b_7 = CA1 interrupt flag
The remaining flags are listed in order of analysis:
b_5 = selects CA2 as an input or output control lead
b_4 = selects between mode control and steady state for CA2 as an output lead or selects the interrupt transition for CA2 as an input lead.
b_3 = For CA2 as an input lead, b_3 acts as the interrupt mask. As an output selection, this bit's use depends on the condition of b_4. If b_4 selects the mode control, b_3 selects between pulse and handshake modes. If b_4 is set for steady state, then the steady state level of CA2 is selected by b_3.

5.9 Since there is no requirement for using the data ports, go directly to configuring the control register. A squarewave will be produced by controlling the state of CA2.

```
START MOVE.B #$34,ACONHI     * Selects steady state low on CA2
      BSR DELAY              * Controls how long CA2 is low
      MOVE.B#$3C,ACONHI      * Selects steady state high on CA2
      BSR DELAY              * Same time for CA2 to be high
      BRA START
```

Chapter 6

6.1 \$B6 = 1011 0110

6.3 \$61 = 0110 0001

6.5 111 1010 1100

6.7 73.17 kHz to 80.88 kHz

6.9
```
MOVE.B CONST2,D0        * Read status register
BTST.B #$04,D0          * Check DCD
BNE NOCARR
BTST.B #$20,D0          * Check overrun
BNE OVERRUN
BTST.B #$40,D0          * Check parity error
BNE PARITY
BTST.B #$10,D0          * Check framing error
BNE NOSTOP
MOVE.B SDATA2,SAVE      * Read data register
RTS
```

6.11 ACIA 1 data rate = 300 bps
ACIA 2 data rate = 75 bps

6.13

23	22	21	20	19	18	17	16	15	14	13	12	11	10	9	8	7	6	5	4	3	2	1	0
					1	1	1												R4	R3	R2	R1	L

R = Register Select

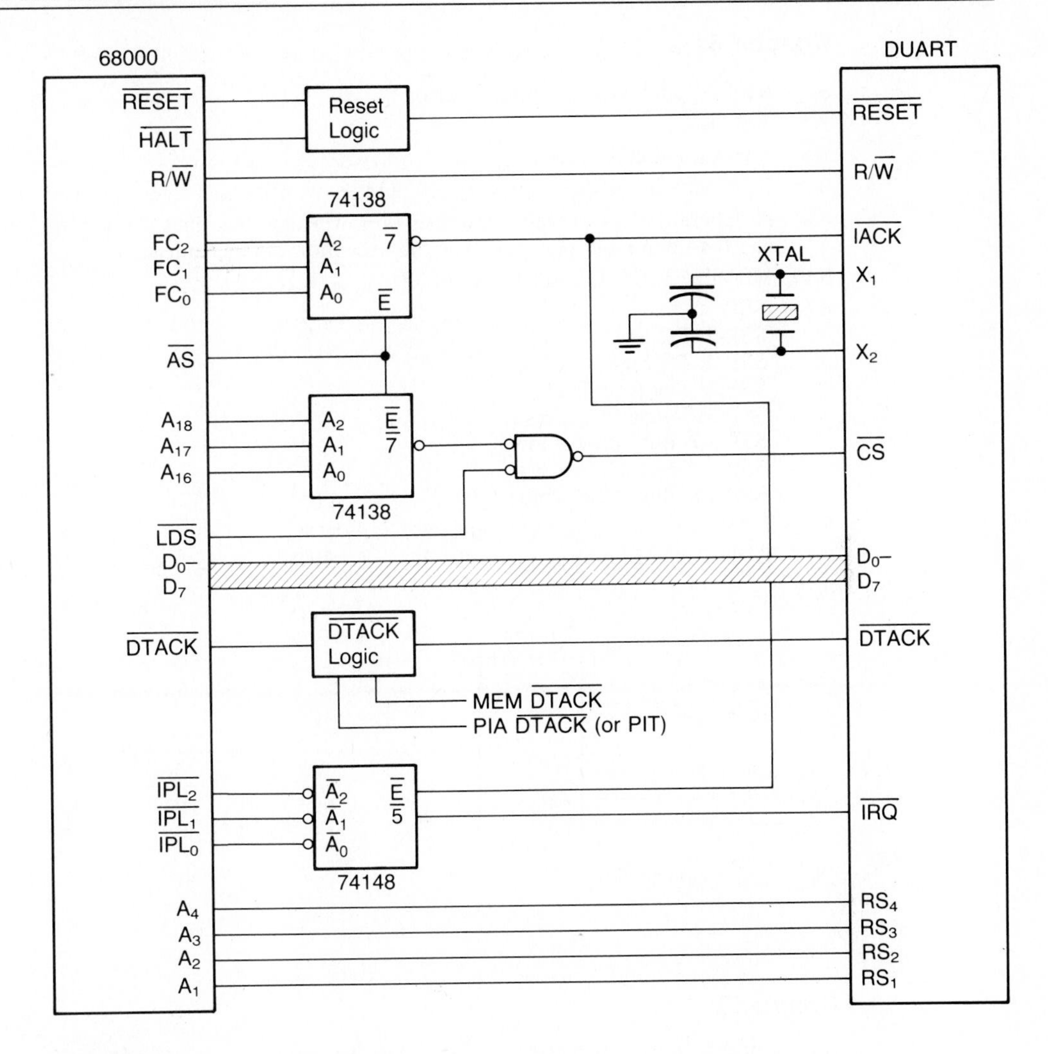

Chapter 7

7.1 .PC 1BB00 .US CC2000

7.3 MD 4000
4000 FE ED B0 0B 1C C4 DE ED 00 6B 22 BC 98 05 0E AC . . . . . . k". . . . . .

7.5 BT 10C000 10FFFE

7.7 COPY TESTDATA.LS,# E TESTDATA.LS

7.9 HOME DC.W $AB12,BB55,78DE,CC10

7.11 INCLUDE ADDBYTES

7.13 ASM BEGONE. SA,,,;R

7.15 (a) QUIT (b) CNTL A (c) F1 (d) CNTL S (e) OFF or BYE
(f) TM followed by a BREAK key (g) DF (h) DC %1001100110110101110

Chapter 8

8.1 RESET; address error and bus error

8.3 $00 0030

8.5 S is set and T is reset.

8.7 By comparing address lines A_1, A_2, and A_3 with its own interrupt level code (IPLs)

8.9 Peripheral places exception number on lower data bus lines. Peripheral returns a $\overline{\text{VPA}}$, and the microprocessor develops the exception number internally.

8.11 MOVEM.L D0–D2/A0/A3/A5,–(A7)

8.13 SSP
SSP-2: PC low
SSP-4: PC high
SSP-6: Status register
SSP-8: Instruction register
SSP-10: Bus address low
SSP-12: Bus address high
SSP-14: Bus cycle word

8.15 T flag is reset by the exception service process.

8.17 TRAP7 is exception number 39 (32 + 7). 39 × 4 = 156, which is $00 009C.

8.19

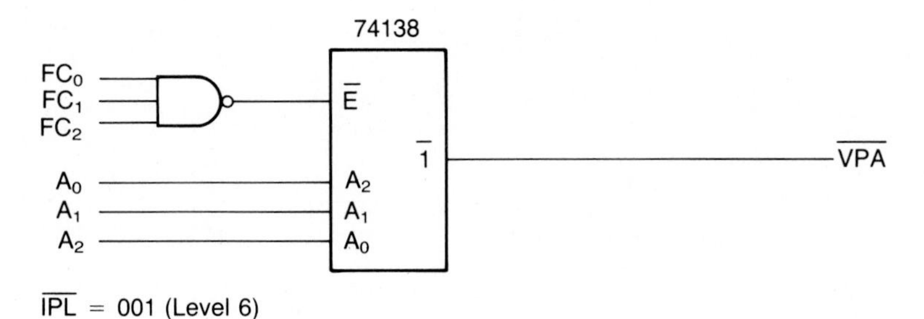

$\overline{\text{IPL}}$ = 001 (Level 6)

Chapter 9

9.1
```
MOVE.B #$01,ACONHI      * Enable CA1 interrupt
MOVE.B #$AD,CONST1      * Set up ACIA
```

Interrupt Program:

```
RING BSR RECV           * Read received data each RDRF = 1
CMPI #0,D0              * Break received?
BNE.S RING
MOVE COUNT,D1
XMT BSR SEND            * Send a character when TDRE = 1
DBF D1,XMT
RTE
```

9.3 (Guided by instructor/student preference)

9.5

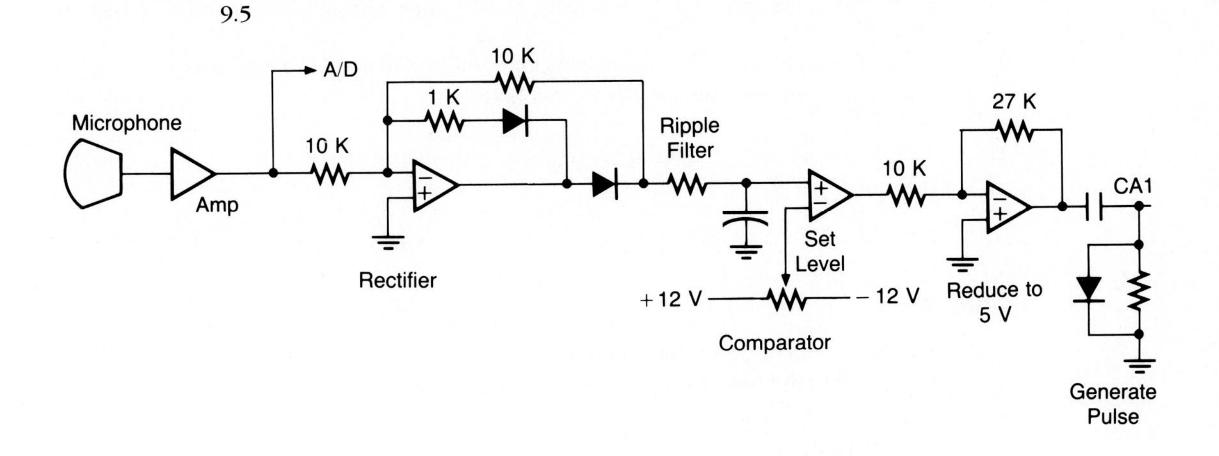

Chapter 10

10.1 Smaller address bus (20 vs. 23 bits)
Smaller data bus (8 vs. 16 bits)
One data strobe ($\overline{DS}$) instead of 2 ($\overline{UDS}$ and $\overline{LDS}$)
No $\overline{VMA}$
No $\overline{BGACK}$
IPL_0 and IPL_2 combined

10.3

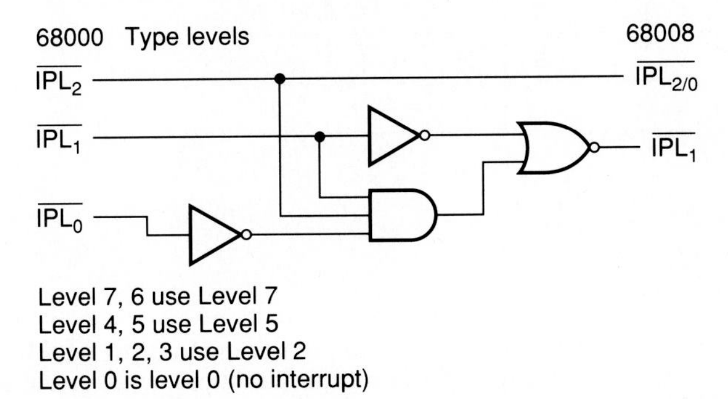

Level 7, 6 use Level 7
Level 4, 5 use Level 5
Level 1, 2, 3 use Level 2
Level 0 is level 0 (no interrupt)

10.5 Virtual memory provides a means of accessing memory locations beyond the physical memory of the system.

10.7 \$20 3CBA

10.9 CPU spaces are interrupt acknowledge (space code $F) and breakpoint (CPU space code $0).

10.11 The new exception is the format exception, and it has the lowest priority.

10.13 The breakpoint opwords are $4848 to $484F.

10.15 Bus error

10.17 Access is made where there is no physical or logical address.

10.19 $\overline{BR}$ and $\overline{BGACK}$

Chapter 11

11.1 The fully associative cache. The tags are checked simultaneously.

11.3 Byte, word, and long word. $\overline{DSACK0}$ and $\overline{DSACK1}$.

11.5 Byte 1; byte 2; byte 1; byte 2

11.7 $AA00 1DBE

11.9 D2 = $8702 3158

11.11 (a) D1 = $1234 5678 D2 = $ABCD EF89 D3 = $ABCD EF89
(b) D1 = $ABCD EF89 D2 = $ABCD EF89 D3 = $1234 5678

11.13

Address	b7	b6	b5	b4	b3	b2	b1	b0
0000 233E								
0000 233F								
0000 2340								X
0000 2341	X	X	X	X			0	0
0000 2342	0	0	0	0	0	0		
0000 2343								

Chapter 12

12.1 A coprocessor is an extension to a microprocessor and interacts with the main processor. A DMA device forces the main processor to go "off line" during a DMA operation. A coprocessor is dedicated to a specific set of tasks and does actual processing. A DMA provides for rapid transfer of data between memory and/or an I/O device, but doesn't actually perform any additional processing.

12.3 The BRA NEUTRAL is taken. The N flag in the main processor is cleared, indicating the initial sum is positive. As a result, the BMI will not occur. Similarly, the N flag in the coprocessor is set as a result of doubling the data ($5A2B × 2 = $B456, a negative number). As a result, the FBPL is not taken. This leaves the BRA at the end of the program segment to be executed.

12.5 It should reaccess the response register because b_{15} is set.

12.7 The number is negative (N flag set). There was a quotient value in excess of −$35 because the overflow exception flag is set and the quotient value is −$35. The resulting number is inexact. The accrued exception byte reflects both the overflow and the inexact exceptions.

12.9 SNAN, OPERR, UNFL and DZ

12.11 (a) 18.67 (b) 771.2 (c) 82820 (d) $3A.7F (e) $B.17A (f) 111.6

12.13
```
START   MOVEA.L #ANSWER,A0     * Initialize pointer
        MOVE.L D0,D1           * Copy data into D1
        SWAP D1                * Set D1 (magnitude)
        FCOS.D D0,FP0          * Find cosine of angle
        FMUL.D D1,FP0          * Find real value
        FMOVE.D FP0,(A0)+      * Save real
        FSIN.D D0,FP0          * Find sine of angle
        FMUL.D D1,FP0          * Find imaginary
        FMOVE.D FP0,(A0)+      * Save imaginary
DONE    BRA DONE               * End program
```

12.15 D3

12.17 Option 000: $0000 3010
Option 100: $0100 4600

12.19 $0001 0000

Chapter 13

13.1 None

13.3 Instruction cache—caches instructions during prefetch
Data cache—cache operand data
Address translation cache—contains translations from logical to physical memory addresses

13.5 HALT is an input pin only. Output halt functions now appear on STATUS.

13.7 Root pointer registers point to the top of the first table in a translation tree.

13.9 $6CC0 3468, $6CC0 346C, $6CC0 3460, $6C00 3440

13.11 Write-through means that both the cache and memory are updated during a memory write cycle. Used by the data cache.

13.13 $2000 0002

13.15 Search instruction cache—miss
Check TT0 and TT1—no transparent translation
Search ATC—miss
Begin translation table search—find translation address load ATC
Retry cycle—hit in ATC—access memory cache opword to instruction cache
Retry cycle—hit in instruction cache—move into pipeline

13.17 A long descriptor includes a limit value and a supervisor flag not included with the short descriptor. A table descriptor points to another table, a page descriptor points to the physical address sought.

13.19 $403F 8460

13.21 PMOVE.L #$80D2AA88,TC

13.23 Invalid descriptor (DT = 0); limit violation; BERR asserted; user write to supervisor space or write protected area

Chapter 14

14.1 FPO–FP7

14.3 Cache address register

14.5 LINE MOVE16

14.7 single: 8-bit exponent, 23-bit fraction, integer = 1
double: 11-bit exponent, 52-bit fraction, integer = 1
extended: 15-bit exponent, 64-bit fraction, explicit integer

14.9 −0.359375

14.11 (a) SIZx: number of bytes being transferred during the current cycle
(b) TT: transfer type defines external bus access type
(c) TM: transfer mode defines memory access areas
(d) $\overline{\text{LOCK}}$: indivisible read-write-modify cycle in process
(e) $\overline{\text{LOCKE}}$: last cycle of a read-modify-write cycle
(f) $\overline{\text{TS}}$: the start of an external transfer cycle
(g) $\overline{\text{TIP}}$: transfer cycle is in progress
(h) $\overline{\text{BR}}$: 040 is requesting use of the buses
(i) $\overline{\text{RSTO}}$: reset out—asserted by RESET instruction
(j) $\overline{\text{IPEND}}$: interrupt is accepted but not yet serviced

14.13 A dirty bit is a flag for each long word in a data cache line entry. It indicates that the data in the corresponding memory location is old and does not match the data in the cache.

14.15 Bus error
MMU translation error
virtual memory swap

14.17 There is no difference in size or content of these two stacks. The basic stack ($0) is built when the ISP is used and the throw away stack ($1) is made when the MSP is in use.

14.19 TT1, TT0 = 1, 1 TM = interrupt level
$A_0 - A_{31}$ = $FFFF FFFF

14.21 Attempting to fetch an opword or operand from an odd address; odd relative address value

14.23 A real number that has been rounded

14.25 BSUN: branch or set instruction testing a NAN
OPERR: operand error—no mathematical interpretation of an operand
OVFL: number too large for data type
UNFL: number too small for data type
INEX2: operation inexact result
INEX1: BCD input that is inexact

14.27 Condition code byte—flag indications of the results of floating point operation
Quotient—least seven bits and sign of quotient from a FDIV operation
Exception status—indicates when an FP exception is requested
Accrued exceptions—history of the exceptions during the current program tasks

BIBLIOGRAPHY

For additional reading and reference, consult the following texts and references.

Alisouskas, V. , and Tomasi, W. 1985. *Digital and Data Communications*. Englewood Cliffs, NJ: Prentice-Hall.

Bishop, R. 1979. *Basic Microprocessors and the 6800*. Hasbrouck Heights, NJ: Hayden Book Company.

Coffron, J. 1983. *Using and Troubleshooting the MC68000*. Englewood Cliffs, NJ: Prentice-Hall.

Dolhoff, T. 1979. *16-Bit Microprocessor Architecture*. Reston, VA: Reston Publishing Co.

Greenfield, J.D., and Wray, W.C. 1981. *Using Microprocessors and Microcomputers: The 68000 Family*. New York: John Wiley & Sons.

Hall, D.V. 1980. *Microprocessors and Digital Systems*. New York: McGraw-Hill.

Jaulent, P. 1985. *The 68000 Hardware and Software*. New York: Macmillan.

Kane, G.; Hawkins, D.; and Leventhal, L. 1981. *68000 Assembly Language Programming*. New York: McGraw-Hill.

Kane, G. 1981. *68000 Microprocessor Handbook*. New York: McGraw-Hill.

McNamara, J.E. 1977. *Technical Aspects of Data Communications*. Bedford, MA: Digital Corporation.

Osborne, A., and Kane, G. 1981. *Sixteen-Bit Microprocessor Handbook*. New York: McGraw-Hill.

Scanlon, L.J. 1981. *The 68000: Principles and Programming*. Indianapolis: Howard W. Sams.

Staugaard, A.C. 1981. *How to Program and Interface the 6800*. Indianapolis: Howard W. Sams.

Titus, J.; Titus, C.; Baldwin, J.; Hubin, W.; and Scanlon, L.J. 1982. *16-Bit Microprocessors*. Indianapolis: Howard W. Sams.

Young, P. 1985. *Electronic Communication Techniques*. Columbus, OH: Merrill.

Motorola Applications Notes:

AN 808: *Interfacing M6800 Peripheral Devices to the MC68000 Asynchronously* (A.J. Morales, 1981).

AN 810: *Dual Sixteen-Bit Ports for the MC68000 Using Two MC6821s* (J. McKenzie, 1981).

AN 817: *Asynchronous Communications for the MC68000 Using the MC6850* (C. Melea, 1981).

AN 818: *Synchronous I/O for the MC68000 Using the MC6852* (J. McKenzie, 1981).

AN 819: *Prioritized Individually Vectored Interrupts for Multiple Peripheral Systems with the MC68000* (R. Davis, 1981).

AN 854: *The MC68230 Parallel Interface/Timer Provides an Effective Printer Interface* (V.A. Scherer and W.G. Peterson, 1982).
AN 867: *A Higher Performance MC68000L12 System with No Wait States* (T. West, 1982).
AN 880: *An Evaluation Tool for the MC68451 MMU* (H. Scales, 1982).
AN 881: *Dual-Ported RAM for the MC68000 Microprocessor* (T. West, 1983).
AN 882: *The Bisync Protocol and a Bisync Data Link Between Two MC68000 MPU-Based Systems* (J. Vaglica, 1982).
AN 896A: *Serial I/O, Timer, and Interface Capabilities of the MC68901 Multifunction Peripheral* (G. Brown, 1984).
AN 897: *MC68008 Minimum Configuration System* (G. Brown and K. Harper, 1984).
AN 899: *A Terminal Interface, Printer Interface and Background Printing for an MC68000-Based System Using the MC68681 DUART* (K. Harper, 1984).
AR 208: *Design Philosophy Behind Motorola's MC68000* (T.W. Starnes, 1983).
AR 209: *Virtual Memory for the MC68010* (D. McGregor and D. Mothersole, 1983).
AR 211: *Built-In Tight-Loop Raises Microprocessor's Performance* (D. MacGregor and B. Moyer, 1983).
AR 214: *Microcoded Microprocessor Simplifies Virtual-Memory Management* (T.W. Starnes, 1983).
AR 217: *The Motorola MC68020* (D. MacGregor, D. Mothersole, and B. Moyer, 1984).
AR 219: *The MC68020 and System V/68* (B. Beims, 1984).
AR 220: *Multiprocessing Capabilities of the MC68020 32-Bit Microprocessor* (B. Beims, 1984).
AR 225: *Testing Approaches in the MC68020* (Kuban and Salick, 1985).
AR 227: *Product Development for the MC68020* (J. Reinhart, 1984).

Other Motorola Publications:
BR 176: *The MC68000 Family* (1984).
BR 265: *MC68881 HCMOS Floating Point Coprocessor* (1985).
DC 001: *Virtual Memory Using the MC68000 and the MC68451 MMU* (H. Scales, 1982).
EB 83A: *The Inter-Relationship Between Access Time and Clock Rate in an MC68000 System* (1982).
EB 97: *A Discussion of Interrupts for the MC68000* (B. Taylor, 1983).
MEX68KECB: *MC68000 Educational Computer Board* (1981).
MEX68KECB/D2: *MC68000 Educational Computer Board User's Manual, Second Edition* (1982).
M86KMASM/D9: *M68000 Family Resident Structured Assembler Reference Manual, Ninth Edition* (1985).
MCA1-1: *MC68000 16-Bit Microprocessor Course Notes* (1984).
MTT20-1: *MC68020 32-Bit Microprocessor Course Notes* (1984).
Motorola Systems News: MC68020—The 32-Bit Performance Standard (1984).

Magazine Articles

Johnson, T.L. "A Comparison of MC68000 Family Processors." *Byte* 11 (1986): 205–17.
Soloman, L. "Motorola's Muscular 68020." *Computers and Electronics* 122 (1984): 74–76.
Stockton, J., and Scherer, V. "Learn the Timing and Interfacing of the MC68000 Peripheral Circuits." *Electronic Design* 23 (1979): 58–64.
Zorpette, G. "The Beauty of 32 Bits." *IEEE Spectrum* 22 (1985): 65–71.

Data Manuals and Reference Books

M68000 16/32-Bit Microprocessor Programmer's Reference Manual, Fourth Edition. Motorola, Inc., 1984.

M68000 16-Bit Microprocessor Specification Book. Motorola, Inc., 1983.

MC68020 32-Bit Microprocessor User's Manual. Motorola, Inc., 1984.

MC68681 Dual Asynchronous Receiver/Transmitter (DUART). Motorola, Inc., 1985.

MC68881 Floating Point Coprocessor User's Manual. Motorola, Inc., 1985.

Memory Products. NEC Electronics, Inc., 1986.

Motorola 8-Bit Microprocessor and Peripheral Data Book. Motorola, Inc., 1983.

Motorola 16/32-Bit Microcomputer System Components. Motorola, Inc., 1984.

Signetics Logic—TTL Data Manual. Signetics Corp., 1978.

MC68030 Enhanced 32-Bit Microprocessor User's Manual, Second Edition. Prentice Hall. Copyright Motorola, Inc., 1989.

MC68040 32-Bit Microprocessors User's Manual. Motorola, Inc., 1989.

MC68000/20/30/40, MC68851/81/82 Programmer's Reference Manual. Motorola, Inc., 1989.

INDEX

R

S

T

U

V

W

African American Eras

Segregation to Civil Rights Times

African American Eras

Segregation to Civil Rights Times

Volume 2:
Communications and Media
Demographics
Education

U·X·L
A part of Gale, Cengage Learning

Detroit • New York • San Francisco • New Haven, Conn • Waterville, Maine • London

African American Eras:
Segregation to Civil Rights Times

Product Managers: Meggin Condino and Julia Furtaw

Project Editor: Rebecca Parks

Rights Acquisition and Management: Leitha Etheridge-Sims, Kelly Quin

Composition: Evi Abou-El-Seoud

Manufacturing: Rita Wimberley

Imaging: John Watkins

Product Design: Pamela Galbreath

Cover photographs reproduced by permission of Getty Images (photos of March on Washington and Ida B. Wells) and the Library of Congress (photo of Booker T. Washington).

Library of Congress Cataloging-in-Publication Data

African American eras. Segregation to civil rights times.
p. cm.
Includes bibliographical references and index.
ISBN 978-1-4144-3596-1 (set) -- ISBN 978-1-4144-3597-8 (v. 1) -- ISBN 978-1-4144-3598-5 (v. 2) -- ISBN 978-1-4144-3599-2 (v. 3) -- ISBN 978-1-4144-3600-5 (v. 4) 1. African Americans--History--1863-1877--Juvenile literature. 2. African Americans--History--1877-1964--Juvenile literature. 3. African Americans--Segregation--History--Juvenile literature. 4. Segregation--United States--History--Juvenile literature. 5. African Americans--Civil rights--History--19th century--Juvenile literature. 6. African Americans--Civil rights--History--19th century--Juvenile literature. 7. African Americans--Biography--Juvenile literature. 8. United States--Race relations--Juvenile literature.
I. Title: Segregation to civil rights times.
E185.6.A254 2010
973'.0496073--dc22

2010012405

Gale
27500 Drake
Farmington Hills, MI 48331-3535

ISBN-13: 978-1-4144-3596-1 (set) ISBN-10: 1-4144-3596-7 (set)
ISBN-13: 978-1-4144-3597-8 (Vol. 1) ISBN-10: 1-4144-3597-5 (Vol. 1)
ISBN-13: 978-1-4144-3598-5 (Vol. 2) ISBN-10: 1-4144-3598-3 (Vol. 2)
ISBN-13: 978-1-4144-3599-2 (Vol. 3) ISBN-10: 1-4144-3599-1 (Vol. 3)
ISBN-13: 978-1-4144-3600-5 (Vol. 4) ISBN-10: 1-4144-3600-9 (Vol. 4)

This title is also available as an e-book.
ISBN-13: 978-1-4144-3705-7 ISBN-10: 1-4144-3705-6
Contact your Gale, a part of Cengage Learning sales representative for ordering information.

Printed in Mexico
1 2 3 4 5 6 7 14 13 12 11 10

Table of Contents

chapter six *Education* 339

VOLUME 3

VOLUME 4

Reader's Guide

U•X•L *African American Eras: Segregation to Civil Rights Times* provides a broad overview of African American history and culture from the end of the Civil War in 1865 through the civil rights movement up to 1965. The four-volume set is broken into thirteen chapters. Each chapter covers a major subject area as it relates to the African American community. Readers have the opportunity to engage with history in multiple ways within the chapter, beginning with a chronology of major events related to that subject area and a chapter-specific overview of developments in African American history. They are next introduced to the men and women who shaped that history through biographies of prominent African Americans, as well as topical entries on major events related to the chapter's subject area. Primary sources provide a firsthand perspective of the people and events discussed in the chapter, and readers have the opportunity to engage with the content further in a research and activity ideas section.

The complete list of chapters is as follows:

- Activism and Reform
- The Arts
- Business and Industry
- Communications and Media
- Demographics
- Education

- Government and Politics
- Health and Medicine
- Law and Justice
- Military
- Popular Culture
- Religion
- Science and Technology

These chapters are then divided into seven sections:

Chronology: A timeline of significant events in the African American community within the scope of the chapter's subject matter.

Overview: A summary of major developments and trends in the African American community as they relate to the subject matter of the chapter.

Headline Makers: Biographies of key African Americans and their achievements within the scope of the chapter's subject matter.

Topics in the News: A series of topical essays describing significant events and developments important to the African American community within the scope of the chapter's subject matter.

Primary Sources: Historical documents that provide a firsthand perspective on African American history as it relates to the content of the chapter.

Research and Activity Ideas: Brief suggestions for activities and research opportunities that will further engage the reader with the subject matter.

For More Information: A section that lists books, periodicals, and Web sites directing the reader to further information about the events and people covered in the chapter.

OTHER FEATURES

The content of U•X•L *African American Eras: Segregation to Civil Rights Times* is illustrated with 240 black-and-white images that bring the events and people discussed to life. Sidebar boxes also expand on items of high interest to readers. Concluding each volume is a general bibliography of books and Web sites, and a thorough subject index that allows readers to easily locate the events, people, and places discussed throughout the set.

COMMENTS AND SUGGESTIONS

We welcome your comments on U•X•L *African American Eras: Segregation to Civil Rights Times* and suggestions for other history topics to consider. Please write: Editor, U•X•L *African American Eras: Segregation to Civil Rights Times*, 27500 Drake Rd., Farmington Hills, MI 48331-3535; call toll-free: 1-800-877-4253; or send e-mail via http://www.galegroup.com.

Chronology

1862 Congress passes the Homestead Act, which provides people with 160 acres of free land in the American West. The act, which applies to African Americans, sets the stage for the so-called "Colored Exodus" of the late 1870s and early 1880s.

1862 Freedmen's Hospital is founded in Washington, D.C., by the U.S. secretary of war. Its purpose is to meet the medical needs of African Americans, including newly freed slaves.

1865 **March 3** Congress passes the Freedmen's Bureau Act, creating the Bureau of Refugees, Freedmen, and Abandoned Lands for the purposes of helping former slaves obtain property, employment, and an education.

1865 **December 2** The Thirteenth Amendment to the United States Constitution, which formally outlaws slavery in the United States, is ratified.

1866 United States Congress commissions six all-black U.S. Army units. The units, which become known as the Buffalo Soldiers, serve the United States in the Indian wars and in the Spanish-American War, among others.

1868 Howard University College of Medicine is founded in Washington, D.C., as one of the few medical schools open to African Americans.

1868 **July 9** The Fourteenth Amendment is added to the Constitution. Its most important provisions declare that African Americans who

are born in the United States are citizens, that all persons are entitled to due process of law, and that no person shall be denied the equal protection of the laws.

1869 **May 1** Ebenezer Bassett becomes the first African American to serve the United States as a diplomat when President Ulysses S. Grant appoints him to be minister resident (the nineteenth-century equivalent of an ambassador) to Haiti.

1870 **February 3** The Fifteenth Amendment to the United States Constitution, which outlaws discrimination against voters based on race, is ratified.

1870 **February 23** Hiram Revels of Mississippi becomes the first African American to serve as a United States senator.

1875 **March 1** The Civil Rights Act of 1875, which forbids discrimination based on race for all public accommodations, is signed into law.

1877 **March** Under the Compromise of 1877, which enabled Republican candidate Rutherford B. Hayes to be elected president, the Republican Party agrees to withdraw the U.S. military from the South and end Reconstruction.

1883 **October 15** The Civil Rights Act of 1875 is declared unconstitutional by the Supreme Court, opening the way for Southern states to enact Jim Crow laws and institute policies of segregation.

1892 Journalist Ida B. Wells publishes *Southern Horrors: Lynch Law in All Its Phases,* the first comprehensive study of lynchings in the United States.

1895 **September 18** Booker T. Washington delivers a speech popularly known as the "Atlanta Compromise," in which he encourages African Americans to have patience and prove themselves worthy of equality to whites.

1896 **May 18** In the court case *Plessy v. Ferguson,* the United States Supreme Court rules that segregation is legal as long as blacks are provided "separate but equal" accommodations and facilities.

1899 **September 18** Scott Joplin publishes his first successful ragtime composition, "Maple Leaf Rag," which becomes the first instrumental sheet music to sell over one million copies.

1903 Author and activist W. E. B. Du Bois publishes *The Souls of Black Folk,* his landmark collection of essays about race relations in the United States.

1905 Madame C. J. Walker, a former employee of Annie Malone's Poro Systems, goes into business for herself, selling hair straighteners, creams, and other styling products designed specifically for African American women.

1906 **April** William Joseph Seymour begins the Azusa Street Revival, which is often credited as a key development in the growth of the Pentecostal faith. The Azusa Street Revival becomes the longest-running continuous revival in United States history.

1908 **December 26** Boxer Jack Johnson defeats Canadian heavyweight Tommy Burns in Australia to become the first African American world heavyweight champion.

1909 **February 12** Civil rights activists gather to form the organization that becomes known as the National Association for the Advancement of Colored People.

1910 Sickle-cell anemia is scientifically described for the first time. James Herrick, a Chicago physician, publishes a report describing the disease in Walter Clement Noel, a young black student from Grenada in the West Indies.

1919 **March** Oscar Micheaux's *The Homesteader,* the first feature-length film written and directed by an African American, premieres.

1920 Marcus Garvey, as part of his Back to Africa project, moves the headquarters of the Black Star shipping line to Liberia, a country in western Africa.

1925 Dancer and singer Josephine Baker arrives in Paris and quickly becomes the most popular American performer in Europe.

1927 **January 7** The first officially recorded game featuring the Harlem Globetrotters is played in Illinois.

1931 **November 7** The tragic death of Juliette Derricotte in a car accident in Dalton, Georgia, sparks a national outrage over segregated hospitals. Derricotte does not receive adequate medical care because the local hospital does not admit African Americans.

1936 **November 3** Seventy-six percent of African Americans who vote in the presidential election cast their vote for Franklin D. Roosevelt, a Democrat. The election marks the beginning of a major shift of African American voters away from the Republican Party.

1937 **April 25** The Brotherhood of Sleeping Car Porters becomes the first all-black union in American history to negotiate a labor

agreement with a major corporation when it enters into a collective bargaining agreement with the Pullman Company.

1940 **February 29** Actress Hattie McDaniel becomes the first African American to win an Academy Award for her portrayal of Mammy in the film *Gone with the Wind* (1939).

1941 **June 25** President Franklin D. Roosevelt signs Executive Order 8802, making it illegal for government agencies and private companies that do business with the government to refuse to hire African Americans.

1945 **November** John Harold Johnson launches the lifestyle magazine *Ebony,* which quickly becomes the most popular African American magazine in the nation.

1945 **November 1** Brooklyn Dodgers owner Branch Rickey signs Negro League baseball player Jackie Robinson to play in the major leagues, the first African American in modern major league baseball.

1952 After completing a jail sentence for burglary, Malcolm Little adopts the new name Malcolm X and becomes the leading spokesperson for the Nation of Islam.

1954 **May 17** The Supreme Court unanimously rules in *Brown v. Board of Education* that segregated schools are unconstitutional. The Court's ruling overturns its previous decision in *Plessy v. Ferguson* (1896) and marks the beginning of the end of legalized racial segregation.

1954 **October 30** The last racially segregated unit in the United States military is disbanded, completing the military's transition from completely segregated to completely integrated in just over five years.

1955 **August 28** Emmett Till, a fourteen-year-old African American boy from Chicago, Illinois, is taken from his uncle's house in Money, Mississippi, and murdered for allegedly whistling at a white woman.

1955 **December 1** Rosa Parks is arrested after refusing to give up her seat to a white passenger on a Montgomery, Alabama, city bus; the arrest leads to a year-long bus boycott and the eventual desegregation of Montgomery city buses.

1957 Martin Luther King Jr. and Ralph Abernathy co-found the Southern Christian Leadership Conference (SCLC), a group that teaches the use of nonviolent direct action to protest injustice and promote civil rights for African Americans.

1957 **August 13** A postal worker alerts the suburb of Levittown, Pennsylvania, that a black family has moved into the neighborhood. This incident sparks a wave of violence and terrorism against the family as white residents attempt to force them to leave the neighborhood.

1957 **September 24** President Dwight D. Eisenhower deploys a U.S. Army division to Little Rock, Arkansas, and federalizes the Arkansas National Guard in order to protect nine African American students and enforce integration at Little Rock Central High School. The students become known as the "Little Rock Nine."

1958 Dancer Alvin Ailey forms the Alvin Ailey American Dance Theater in New York City, one of the most influential modern dance companies in the country.

1959 **March 11** The play *A Raisin in the Sun* by Lorraine Hansberry premieres, becoming the first Broadway play written by an African American woman.

1960 **February 1** Four college students stage a sit-in at a Woolworth's lunch counter in Greensboro, North Carolina, launching a massive campaign that results in the desegregation of lunch counters throughout the city.

1960 **April 14** Berry Gordy Jr. founds the recording label Motown Records in Detroit, Michigan.

1962 **October 1** James Meredith is admitted to the University of Mississippi as its first African American student; the event leads to riots among white supremacists.

1963 **January 14** Newly elected Alabama governor George C. Wallace famously declares in his inauguration day address that he will support "segregation now, segregation tomorrow, segregation forever."

1963 **August 28** At least two hundred thousand demonstrators participate in the March on Washington for Jobs and Freedom; millions of viewers around the world are moved by the event and by Martin Luther King's "I Have a Dream" speech.

1964 **March** Malcolm X leaves the Nation of Islam and founds his own religious organization, the Organization of Afro-American Unity, built on a belief in world brotherhood.

1964 **May 25** The Supreme Court rules in *Griffin v. County School Board of Prince Edward County* that state governments in the South cannot close their public schools as a strategy for avoiding the

racial integration mandated by the Court's decision in *Brown v. Board of Education.*

1964 **July 2** The Civil Rights Act of 1964, which outlaws segregation based on race in virtually all instances, becomes law.

1964 **December 10** Martin Luther King Jr. is awarded the Nobel Peace Prize for his campaigns of nonviolent resistance.

1965 The federal programs Medicare and Medicaid are created, finally bringing an end to the long practice of segregated hospitals and medical discrimination.

1965 **February 21** Malcolm X, while attending a meeting in Harlem, is shot dead by three members of the Nation of Islam.

1965 **August 6** President Lyndon B. Johnson signs into law the Voting Rights Act of 1965, which prohibits all forms of racial discrimination in voting and the administration of elections. The Voting Rights Act and the Civil Rights Act of 1964 are widely regarded as the most important pieces of legislation enacted in the country in the twentieth century.

Era Overview

The American Civil War ended in 1865 with a victory for the North and freedom for African American slaves. The federal government quickly enacted several constitutional amendments aimed at establishing and protecting the rights of African Americans. The federal government was able to protect these rights during the period of time immediately following the Civil War known as Reconstruction. The presence of federal troops in the South during Reconstruction allowed African Americans to make impressive advances in a variety of fields. Numerous African Americans were elected to local and state offices as blacks exercised their right to vote. Schools were built throughout the South to help educate and train freed slaves for new careers in mainstream American society.

Federal troops remained in the South for more than a decade to enforce Reconstruction policies. Reconstruction ended in 1877 after the Northern Republicans agreed to withdraw federal troops from the South in exchange for Southern Democratic support of the Republican presidential candidate Rutherford B. Hayes (1822–93). The withdrawal of federal protection was devastating to the Southern black population. Many African Americans found themselves in circumstances as bad or worse than before the war. Southern blacks once again worked the plantation fields owned by whites—now as sharecroppers, trapped by debt just as they had once been bound by slavery.

To make matters worse, state governments throughout the South began to institute "black codes," which were laws aimed at restricting the rights of African Americans. These laws supported strict segregation throughout the

South. Separate public facilities—everything from schools to hospitals to water fountains—were created for blacks. These facilities were almost never equal in quality to those offered for white citizens. The constitutionality of this kind of segregation was tested in the United States Supreme Court case *Plessy v. Ferguson* in 1896. The Court ruled that "separate but equal" facilities for blacks did not violate the Fourteenth Amendment, which gave African Americans full citizenship. The "separate but equal" concept was used to justify segregation for many decades.

The South in particular saw increasing violence against blacks at the end of the nineteenth century, mainly in the form of lynching. Lynchings are executions held outside the bounds of the law, usually by large groups, and often without much proof that the victim had committed a crime. African American journalist and editor Ida B. Wells (1862–1931) exposed the racist motivations for lynchings in her pamphlets *Southern Horrors* (1892) and *A Red Record* (1895). She campaigned vigorously for stronger anti-lynching laws.

Even in the midst of segregation and lynchings, many African Americans in the South were furthering their educations thanks to schools such as the Tuskegee Institute in Alabama. Booker T. Washington (1856–1915), born into slavery, was the leader of the Tuskegee Institute and a persuasive voice in the struggle for African American acceptance by white society. Washington's philosophy was to provide African Americans with basic trade skills so they could prove their worth to whites as productive members of society. He felt that protests and demands for increased liberty would prove disastrous for blacks, and that the key component in defusing racial tension was time. Many African American intellectuals in the North, such as W. E. B. Du Bois (1868–1963), felt that blacks should not have to wait for whites to give them the rights they deserved. Du Bois was a driving force behind the creation of the National Association for the Advancement of Colored People (NAACP) in 1909. He utilized the organization's official publication, *The Crisis*, to rally African American readers in the struggle for civil rights.

In the first decades of the twentieth century, many African Americans began moving to growing cities in the North and Midwest in search of greater economic opportunities and freedoms. This became known as the Great Migration. By 1930, millions of African Americans had migrated to urban centers such as New York City, Detroit, and Chicago. One of the largest concentrations of African Americans in the North was found in Harlem, a neighborhood in New York City. The relative economic prosperity enjoyed by African Americans in the North, combined with the influx of cultural influences from the South, were key factors in the Harlem

Renaissance. The Harlem Renaissance was a flourishing of African American art and culture that began in the 1920s and continued into the 1930s.

Equally notable was the growing acceptance of African Americans by mainstream American society. While many white Americans reacted negatively to African American boxer Jack Johnson (1878–1946) when he won the heavyweight boxing title in 1908, Joe Louis (1914–81) was praised as an American hero when he secured the title in 1937. When Jackie Robinson (1919–72) debuted as the first African American baseball player in the major leagues in 1947, uncertainty among white fans quickly gave way to respect for his unquestionable talents.

African Americans made great strides in other fields as well. In 1939, Charles Drew (1904–50) invented a blood storage method that allowed for the creation of blood banks, thereby helping to save the lives of millions. Singer Marian Anderson (1897–1993) earned worldwide acclaim for her performances of both opera and traditional songs. And in 1940, Benjamin O. Davis Sr. (1877–1970) became the first African American to achieve the rank of general in the United States Army. In each case, however, these pioneers struggled against the widespread racism that still divided American culture. Drew resigned from the Red Cross in 1941 when the United States Army ordered that blood from black donors must be separated from blood donated by whites. In 1939, Anderson was barred from performing at Constitution Hall in Washington, D.C., because she was black. And while Davis served as a general in the U.S. Army during World War II (1939–45), the soldiers themselves were still divided on the battlefield according to their race.

One important factor in changing American perceptions about blacks was popular media. Even as the roles for African Americans in Hollywood films remained largely stereotypical, performers like Bill "Bojangles" Robinson (1878–1949) won over audiences with their talent and charm. Hattie McDaniel (1895–1952) became the first African American to win an Academy Award for her supporting performance as a servant in the 1939 film *Gone with the Wind.* Still, it would be fifteen years before another African American woman, Dorothy Dandridge (1922–65), would be nominated in the Best Actress category, and twenty-four years before Sidney Poitier (1927–) would become the first African American man to win an Oscar for Best Actor.

Even more significantly, the media played a key role in shaping American views on the growing struggle for civil rights for African Americans. The South was still as segregated in the 1950s as it had been in the 1800s, even as black Americans made important contributions to the worlds of art, science, and business. One of the most important challenges to segregation

occurred in 1954. That year the United States Supreme Court ruled in the case *Brown v. Board of Education of Topeka* that separate schools for blacks and whites are, by definition, unequal, and therefore against federal law. Soon after, segregation was challenged on other fronts as well; for example, Alabama resident Rosa Parks (1913–2005) earned fame when she refused to give up her seat to a white bus passenger in 1955. Her arrest sparked the Montgomery bus boycott, one of the first major protest campaigns of the modern civil rights era. In 1957, nine African American students in Little Rock, Arkansas, defied the state's governor and National Guard troops to attend a high school that had previously admitted only white students.

The leading voice of the civil rights struggle in the South was Martin Luther King Jr. (1929–68), a clergyman and advocate of nonviolence in the struggle for civil rights. King's influence resulted in peaceful demonstrations such as sit-ins and boycotts of businesses that supported segregation, and earned him the Nobel Peace Prize in 1964. However, the peaceful efforts of King and other activists were often met with violence by whites. Activists such as Medgar Evers (1925–63) and James Chaney (1943–64) were brutally murdered because of their efforts to help African Americans secure their constitutional right to vote. Across the South, African American churches were bombed, and peaceful demonstrators were beaten—often by police and at the direction of state and local government officials.

In 1963, more than two hundred thousand demonstrators participated in the March on Washington for Jobs and Freedom, aimed at securing basic civil rights for African Americans nationwide. Massive media coverage of the event helped sway millions of white Americans to support civil rights reform. The following year, the Civil Rights Act of 1964 was passed by the United States Congress. This legislation formally ended segregation and guaranteed civil rights for all Americans regardless of race. More than a century after Abraham Lincoln issued the Emancipation Proclamation to free African American slaves, the federal government finally followed through on guaranteeing their basic rights as Americans.

chapter four

Communications and Media

Chronology

1892 Journalist Ida B. Wells publishes *Southern Horrors: Lynch Law in All Its Phases,* the first comprehensive study of lynchings in the United States.

1905 May Robert Sengstacke Abbott publishes the first issue of the *Chicago Defender,* which grows into the largest and most influential African American newspaper in the United States.

1910 November *The Crisis,* the official magazine of the NAACP created and edited by W. E. B. Du Bois, publishes its first issue.

1917 The *Chicago Defender* launches a campaign encouraging African Americans in the South to move northward, helping to spur the Great Migration.

1919 March Oscar Micheaux's *The Homesteader,* the first feature-length film written and directed by an African American, premieres.

1927 December 4 Duke Ellington makes his debut as the leader of the orchestra at the Cotton Club, and his band's performances are broadcast live on the radio.

1928 March 19 Popular radio series *Amos 'N Andy,* featuring African American characters voiced by white creators Freeman Gosden and Charles Correll, premieres on Chicago radio station WMAQ.

1929 November 3 Jack L. Cooper, a journalist turned radio broadcaster, launches *The All-Negro Hour,* the first radio series created by and for African Americans, on WSBC in Chicago.

1931 Popular African American newspaper the *Pittsburgh Courier* launches a campaign to end the radio show *Amos 'N Andy* for its offensive depictions of African Americans; the newspaper ends its unsuccessful campaign after several months.

1931 Author George S. Schuyler publishes *Black No More,* widely regarded as the first successful satirical novel by an African American.

1935 February 22 The film *The Little Colonel,* featuring Shirley Temple performing a dance number with African American tap legend Bill "Bojangles" Robinson, premieres.

1939 April 9 Classically trained singer Marian Anderson performs a free, open-air Easter concert on the steps of the Lincoln Memorial for an audience of seventy-five thousand; millions of others across the country listen to the concert by radio broadcast.

1940 February 29 Actress Hattie McDaniel becomes the first African American to win an Academy Award

for her portrayal of Mammy in the film *Gone with the Wind* (1939).

1942 January 31 Singer and actress Lena Horne signs a seven-year contract with MGM Studios, becoming the first African American to be guaranteed "legitimate" roles that did not rely on black stereotypes.

1945 November John Harold Johnson launches the lifestyle magazine *Ebony,* which quickly becomes the most popular African American magazine in the nation.

1946 Broadcaster Jack L. Cooper debuts the first African American–produced radio talk show, *Listen Chicago.*

1947 November 24 Actress Hattie McDaniel becomes the first African American lead in a network radio series when she assumes the title role in *The Beulah Show.*

1948 Gordon Parks becomes the first African American staff photographer for *Life* magazine.

1948 Sports journalist Sam Lacy of the *Afro-American* becomes the first African American member of the Baseball Writers' Association of America.

1950 August 16 The Joseph L. Mankiewicz film *No Way Out* is released, marking the big-screen debuts of Sidney Poitier, Ossie Davis, and Ruby Dee.

1951 June 28 Popular radio series *Amos 'N Andy* premieres as a CBS television series with an all-black cast; despite strong ratings, the show ends after two seasons due to protests by the NAACP and others.

1951 November The weekly news magazine *Jet,* created by John Harold Johnson, debuts.

1954 October 28 The Otto Preminger film *Carmen Jones* is released, starring Dorothy Dandridge in the title role; for her performance, she becomes the first African American nominated for an Academy Award for Best Actress.

1956 February 6 The *Chicago Defender* becomes the first African American daily newspaper in the United States when it switches from a weekly to a daily format.

1956 November 5 Singer and musician Nat "King" Cole hosts the premiere episode of *The Nat King Cole Show,* a variety program featuring primarily African American guests and performers; the show—a ratings failure that could not land a corporate sponsor—lasts fourteen months.

1958 Louis E. Lomax becomes the first African American television

journalist, working for WNTA in New York.

1964 April 13 Actor Sidney Poitier becomes the first African American to win an Academy Award for Best Actor for his role in the film *Lilies of the Field* (1963).

1965 February 27 Actor and activist Ossie Davis delivers the eulogy at the funeral of Malcolm X.

1965 September 8 Actress Dorothy Dandridge is found dead from a drug overdose at the age of forty-two.

Overview

African Americans had wide-scale access to mass media for the first time in the years following the Civil War (1861–65). Through books, magazines, and newspapers, they were finally able to share ideas and information with others in an open forum. African Americans began to make their mark on the dominant forms of media in a relatively short period of time, sharing their perspectives and talents with a society from which they had been largely excluded for decades.

The earliest African American media successes were in journalism. Across the North, local newspapers appeared that focused on African American topics and news. Ida B. Wells (1862–1931) ran just such a newspaper in Memphis, Tennessee, until racial violence drove her to New York City. There she published the thoroughly researched *Southern Horrors: Lynch Law in All Its Phases* (1892), which was the first use of the media for an all-out attack against lynching. Wells worked for a time at the *New York Age,* a paper founded by editor and activist T. Thomas Fortune (1856–1928), before moving to Chicago.

Chicago and New York were home to the two most vibrant African American media cultures of the late nineteenth and early twentieth centuries. Conditions worsened throughout the South after the end of Reconstruction in 1877, when federal troops returned control of state governments to people who had largely opposed granting civil rights to African Americans. This prompted many African Americans to move northward to the large urban centers of New York and Chicago in search of prosperity. One man who saw opportunity in this was Robert Sengstacke Abbott (1870–1940), who created a newspaper aimed at African American readers in 1905. The newspaper—at first just a simple pamphlet assembled by Abbott in his landlord's apartment—was called the *Chicago Defender,* and it quickly grew to be the most successful African American newspaper in the country. In fact, Abbott discovered that the appeal of his paper was not limited to Chicago residents. The majority of his subscriptions were sold outside the city, and he had many readers in the South. Abbott used his influential paper to launch a campaign urging more African Americans throughout the South to move northward. In this way, the *Chicago Defender* played a key part in the Great Migration of African Americans during the 1920s and 1930s.

Another publication that exerted great influence among African American readers was *The Crisis.* It was launched in New York in 1910 as the official publication of the National Association for the Advancement of Colored People (NAACP), and was edited by the well-known civil rights

activist W. E. B. Du Bois (1868–1963). It offered news and opinions that readers would never see in the mainstream newspapers of the day, all in an effort to achieve its stated goal of revealing and defeating racial prejudice. In its early years, *The Crisis* proved more popular than the NAACP itself, and outsold many mainstream magazines.

Radio was another medium that saw great advances in both New York and Chicago. The first commercial radio programs began in 1920, but only a few featured African American entertainers. Duke Ellington (1899–1974) and his band were among the first black entertainers heard by radio listeners across the country, as their live performances from the Cotton Club in Harlem were aired on a chain of stations owned by the Columbia Broadcasting System (CBS). At the same time, a pair of white performers became famous for their portrayals of African American characters on a Chicago radio station. The show, called *Amos 'N Andy,* was a nightly series about two working-class African Americans newly arrived in Chicago from the South. Although the main characters were sympathetic, the performances by the creators—Freeman Gosden (1899–1982) and Charles Correll (1890–1972)—were highly exaggerated and played into white stereotypes about African Americans. Even as the show became highly popular across the country, some critics spoke out against it as demeaning and offensive.

Listeners looking for an authentic African American voice on the airwaves finally found one in 1929, when journalist turned broadcaster Jack L. Cooper launched *The All-Negro Hour* on a Chicago radio station. It was the first radio series to feature an African American host and behind-the-scenes crew. Cooper went on to create the first African American talk show, as well as the first public service radio show in the United States.

As radio was shaped by African American innovators, so too was the world of motion pictures—but at a much slower pace. Oscar Micheaux (1884–1951) was an early pioneer, becoming the first African American to write and direct a feature-length film in 1919 with *The Homesteader*. Despite Micheaux's relative success, Hollywood—where most movies were made—had a narrow view of the part played by African Americans in filmed entertainment. Roles were mostly limited to minor parts as servants. Hattie McDaniel (1895–1952) was an actress who made a living off such roles. Her talents earned her an Academy Award for Best Supporting Actress in the film *Gone with the Wind* (1939), making her the first African American to win an Academy Award. Still, it was not until Lena Horne (1917–2010) signed a film contract with MGM in 1942 that an African American woman would be guaranteed more prestigious roles. Horne

herself barely benefited from this promise; the studio did not yet know how to utilize an African American woman as a leading lady. But her victory helped actresses who followed in her footsteps like Dorothy Dandridge (1922–65), the first African American woman nominated for an Academy Award for Best Actress in 1955. Actors like Sidney Poitier (1927–) likewise expanded Hollywood's view of African American men, and in 1964 he became the first to win an Academy Award for Best Actor, for his role in the film *Lilies of the Field* (1963). However, it would be nearly forty years before another African American, male or female, would win an Academy Award for a leading role.

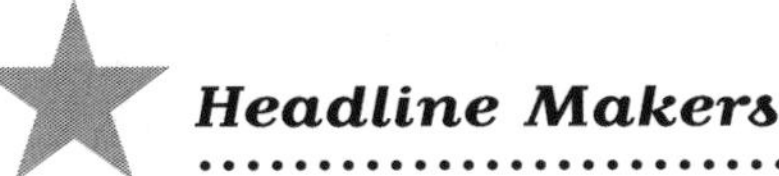

Headline Makers

DOROTHY DANDRIDGE (1922–1965)

Dorothy Dandridge was the first African American film actress to truly earn the title of "leading lady." She was the first African American to receive an Academy Award nomination for Best Actress, and was widely regarded as one of the most beautiful women in Hollywood. However, she encountered numerous racial barriers that prevented her from attaining the stardom that other actresses achieved. An accidental drug overdose brought her life to a tragic and premature end.

Dandridge was born in Cleveland, Ohio, to Cyril and Ruby Dandridge, the former a minister and the latter an aspiring stage performer. Her parents divorced just months after she was born. Dorothy had an older sister, Vivian, and Ruby encouraged the girls to perform together at churches and other events. The experience instilled in Dorothy a love for the stage at the tender age of three. The Dandridge daughters developed a reputation in African American Baptist church circles, and soon became known as the Wonder Children. They toured the South with their mother as their manager, performing mostly in churches. Ruby expanded the act with a piano player named Eloise Matthews, who also served as a chaperone and guardian for the girls.

Dorothy Dandridge.
AP Images

Ruby moved with the girls to Los Angeles, California, in 1930, after five years of touring. They continued to perform at local churches, but Ruby sought to expand their appeal. Noting the popularity of female singing trios like the Andrews Sisters, Ruby hired another girl named Etta Jones (1928–2001) to play the role of a third sister, and renamed the group the Dandridge Sisters. By 1935, the group was appearing in motion pictures. Dorothy's beauty and lighter complexion brought her the most interest from producers, and she appeared solo in several films. Dorothy's mother also found success on radio programs like *Amos 'N Andy*.

In 1940, the Dandridge Sisters recorded several songs for Columbia Records. Dorothy's

film career began to take off at the same time when she was cast in her first major role in the film *Four Shall Die*. She appeared in seven more films in 1941, including one with Harold Nicholas, whom she married in 1942. She gave birth to a daughter named Harolyn the following year. Harolyn was mentally disabled, and, unprepared to raise the child herself, Dandridge left her in the care of family friend Eloise Matthews. In addition to appearing in movies, Dandridge toured the country performing in nightclubs. She fought racism and segregation whenever she encountered it in her travels. For example, she refused to fulfill her engagement in one all-white, upscale hotel in Miami, Florida, until she was allowed to stay at the hotel.

Dandridge starred in the film *Bright Road* with Harry Belafonte (1927–) in 1953. The following year, she landed the title role in another film with Belafonte titled *Carmen Jones,* directed by Otto Preminger (1906–86). The film was an adaptation of the opera *Carmen* (1875). The sultry, seductive character of Carmen was a complete departure for Dandridge. Her success in the role earned her an Academy Award nomination for Best Actress, making her the first African American woman to receive such an honor. The film also led to an affair between the married Preminger and Dandridge, who had divorced her first husband in 1951.

Dandridge had made her mark as a leading lady and did not want to go back to accepting small roles portraying stereotypical black characters such as maids. Unfortunately, there were few leading roles for an African American woman in Hollywood. Dandridge did not star in another film until *Island in the Sun* in 1957. This was the first of several movies she starred in that dealt with interracial love, which was still largely forbidden throughout the South. The films brought controversy, but not much critical acclaim.

In 1959, Dandridge worked again with Preminger on the film adaptation of the musical *Porgy and Bess* (1959). Dandridge was cast as Bess in the film, and Sidney Poitier (1927–) played Porgy. Dandridge did not do her own singing for the film, which received mixed reviews and did poorly at the box office. Dandridge's performance fared much better as she was nominated for a Golden Globe Award for Best Actress.

Also in 1959, Dandridge married Jack Denison, a white nightclub owner. She attempted to resume her career as a stage entertainer, but failed to draw in audiences. Her husband's club began to fail, and Dandridge—who had been the highest-paid African American actress in the country—could no longer make ends meet. She divorced Denison in 1962 and filed for bankruptcy soon after. She lost her home and moved into a small apartment. Dandridge had to move her daughter Harolyn to a state institution since she was unable to continue paying Matthews to care for her.

Three difficult years later, Dandridge finally began to receive work again. She booked a nightclub engagement in New York, an American film project, and two more films to be shot in Mexico in 1965. These plans were cut short when she was found dead in her apartment on September 8, 1965, at the age of forty-two. Her death was determined to be caused by an accidental overdose of an antidepressant medication.

★ OSSIE DAVIS (1917–2005)

Ossie Davis was a pioneer in American entertainment, serving as one of the few African Americans in the 1950s and 1960s able to portray black characters that went beyond popular media stereotypes. He was not only an actor but also a successful playwright and film director—one of the first in Hollywood. In addition, Davis was a vocal activist for African American civil rights. Most of his accomplishments were carried out alongside actress and activist Ruby Dee (1924–), his wife of fifty-seven years.

Davis was born in Cogdell, Georgia, to Kince Charles Davis, a railway engineer, and Laura Cooper Davis. He was the first of five children. The family lived a modest life punctuated by threats from the local members of the Ku Klux Klan. The family moved to Waycross, Georgia, when Davis was still young, and there he remained through high school. After graduating, he received scholarships to two prestigious colleges, one of them being the Tuskegee Institute in Alabama. Davis could not cover the

Ossie Davis. *Herb Snitzer/ Michael Ochs Archives/ Getty Images*

remaining cost of tuition, however, and instead chose to work and save some money.

In 1935, Davis moved to Washington, D.C., where he could live with two aunts while he attended Howard University. His goal was to become a writer, but while he was at Howard, he became interested in a career as an actor. Davis left Howard after three years of college and moved to Harlem, where he joined an acting group called the Rose McClendon Players. In 1940, the American Negro Theater was formed in Harlem, and Davis became a member. The following year, he appeared in the play *Joy Exceeding Glory*.

His stage career was interrupted by World War II (1939–45). Davis was drafted into the U.S. Army, where he used his writing skills to create material for the U.S. Army newspaper and for motion picture shorts used to educate soldiers. After the war, Davis returned to Harlem and continued acting and writing plays. His first Broadway stage role was in a play titled *Jeb* (1946). It was during this production that he met Ruby Dee, whom he married two years later.

His first film role was a small part in *No Way Out* (1950), which was also the first film for fellow African American actor Sidney Poitier. Nearly ten years later, Davis would replace Poitier as the lead in the Broadway production of *A Raisin in the Sun* (1959). In the years between, Davis starred in several films and plays, including a television adaptation of the Eugene O'Neill play *The Emperor Jones* in 1955. He also continued writing, and achieved great success with his play *Purlie Victorious* (1961), starring both Davis and Dee. The play opened on Broadway to positive reviews, and spawned both a film adaptation in 1963 and a stage musical in 1970.

Davis was also a noted activist in the struggle for civil rights. He was a key part of the March on Washington for Jobs and Freedom in 1963 in Washington, D.C., the best-known and most effective civil rights demonstration of the era. Along with his wife, Davis served as master of ceremonies for the entertainment leading up to the march itself. Davis and Dee also grew close to Malcolm X (1925–65), an activist known for his aggressive verbal attacks against white America. Malcolm invited Davis and Dee to attend a meeting for his new organization on February 21, 1965, but they were unable to attend. At that meeting, Malcolm was shot dead by three militant Black Muslims. Less than a week later, Davis delivered the eulogy at Malcolm's funeral in Harlem.

Davis's career blossomed in the 1960s and 1970s. He went on to star in over forty films, and appeared on dozens of television shows. One show, *Ossie and Ruby!* (1980), featured Davis and his wife as the hosts.

In 1970, Davis became the second African American—after Gordon Parks (1912–2006)—to direct a major studio film with *Cotton Comes to Harlem.* Davis appeared in several films by director Spike Lee (1957–), including *Do the Right Thing* (1989) and *Jungle Fever* (1991). For Lee's biographical film about Malcolm X, Davis provided a new reading of the eulogy he gave at Malcolm's funeral. Many of Davis's awards were shared with his wife, including his induction into the National Association for the Advancement of Colored People (NAACP) Image Awards Hall of Fame in 1989, the Presidential Medal of the Arts in 1995, the Lifetime Achievement Award from the Screen Actors Guild in 2001, and the Grammy Award for Best Spoken Word Album in 2007. Davis died in 2005 while in Miami, working on a film.

★ RUBY DEE (1924–)

Ruby Dee was one of the most accomplished African American actresses of the twentieth century. She earned praise on the stage, in film, and on television over the course of a career that spanned several decades. She also became a successful author and activist, and is best known for her longtime marriage and professional partnership with actor Ossie Davis.

Ruby Dee. *The Library of Congress*

Dee was born Ruby Ann Wallace in Cleveland, Ohio, in 1924. She was the third child born to Edward Nathaniel Wallace and Gladys Hightower, who were both still teenagers at the time. Soon after, Gladys ran off with a preacher, leaving Edward alone with the children. He married another woman, Emma Amelia Bunson, a teacher who was thirteen years older and could not have children of her own. The family relocated to the African American neighborhood of Harlem in New York City, near where Emma attended school at Columbia. Ruby was raised by Emma, and did not know of her biological mother until she was eleven. The Wallace children were encouraged in their education as well as in their musical abilities. Ruby's academic skills enabled her to attend Hunter College High School, an institution for specially gifted students. It was there that she discovered a love of acting.

Dee attended Hunter College after graduating high school and joined the American Negro Theater (ANT). She appeared in several notable productions while still in college, including the Broadway musical *South Pacific* (1943) and the successful ANT production of *Anna Lucasta* (1944), in which she played the lead. She also met and married a singer named Frankie Dee in 1941. The marriage did not last, and in 1946 Ruby met her future husband, Ossie Davis, while performing with him in the Broadway production *Jeb*. The two married in 1948.

Dee's first big break in motion pictures came with *The Jackie Robinson Story* (1950), a biographical film about the baseball player that actually starred Robinson (1919–72) as himself and Dee as his wife. Dee and her real-life husband Davis often worked together on stage and in film. In 1959, Dee landed a role in Lorraine Hansberry's groundbreaking play *A Raisin in the Sun*. When the male lead, Sidney Poitier, left the production early in its run, Davis stepped in. In 1961, the two starred in *Purlie Victorious,* a play written by Davis that was later adapted into a film and a stage musical. Later films that starred both Dee and Davis include *Wattstax* (1973), *Do the Right Thing* (1989), and *Jungle Fever* (1991).

The couple also participated actively in the struggle for civil rights in the United States. They were an important part of the March on Washington for Jobs and Freedom in 1963, and worked closely with organizations like the Congress of Racial Equality (CORE) and the Southern Christian Leadership Conference (SCLC). They were close to both Martin Luther King Jr. (1929–68), and Malcolm X, two of the most influential civil rights leaders of the era. Dee was also a founding member of the Association of Artists for Freedom, a group that formed after the bombing of an African American church in Birmingham, Alabama, killed four young girls in 1963.

In the 1960s, Dee's television career blossomed, and she appeared in series such as *Play of the Week* (1960), *The Fugitive* (1963), and *Peyton Place*. She would continue to appear in dozens of series and films for television over the next several decades, and earned nine Emmy Award nominations—and one win—for her work. In 1980, she and Davis hosted their own series on public television titled *Ossie and Ruby!* It introduced viewers to various African American actors, musicians, and other entertainment professionals.

Like her husband, Dee was not limited to acting. She produced and co-wrote the film *Up Tight!* (1968), and also wrote several books. The first of these was a book of poetry for children titled *Child Glow and Other Poems* (1973). Other books include *My One Good Nerve* (1987) and *Two Ways to Count to Ten: A Liberian Folktale* (1988). With Davis, she wrote a

combined autobiography titled *With Ossie and Ruby: In This Life Together* (1998). She also wrote several plays, including *Zora Is My Name* (1983), about African American author Zora Neale Hurston (1891–1960).

Dee created a scholarship fund for African American actresses, and with Davis she founded the Institute of New Cinema Artists to train entertainment professionals. In 1988, Dee entered the Theater Hall of Fame. Over the course of her career, Dee won two Drama Desk awards for her stage work, two Image Awards from the National Association for the Advancement of Colored People (NAACP), and an Academy Award nomination for Best Supporting Actress for her role in *American Gangster* (2007). With Davis, she also shared numerous awards, including the Presidential Medal of the Arts, the Frederick Douglass Award from the National Urban League, and the Lifetime Achievement Award from the Screen Actors Guild.

★ T. THOMAS FORTUNE (1856–1928)

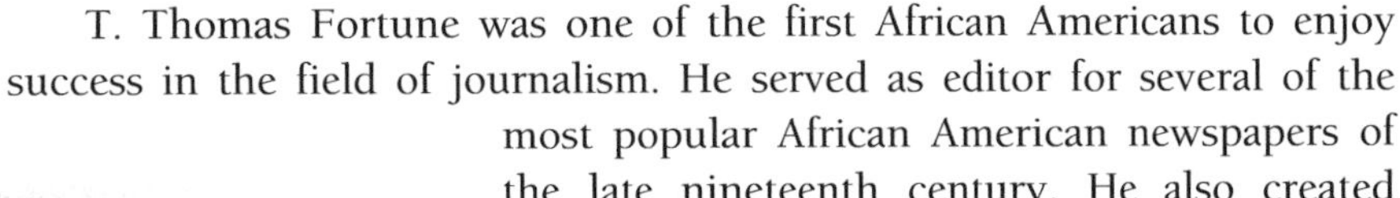

T. Thomas Fortune was one of the first African Americans to enjoy success in the field of journalism. He served as editor for several of the most popular African American newspapers of the late nineteenth century. He also created organizations intended to protect the rights of African Americans long before other activists such as W. E. B. Du Bois (1868–1963) formed similar groups. Despite his accomplishments, Fortune is not as well known today as other activists of the same period.

T. Thomas Fortune. *Fisk University Library. Reproduced by permission.*

An Early Love of Journalism

Fortune was born to slave parents Emanuel and Sarah Jane Fortune in Marianna, Florida. He was still a child when he obtained his freedom during the Civil War (1861–65). After the war, his father became a noted politician in the South. Fortune became interested in printing and the newspaper business at a young age while spending time at the office of the local paper, the *Marianna Courier*. The family moved to Jacksonville, Florida, at least partly in response to threats against Fortune's father by local white supremacists. There Fortune spent time at both the *Jacksonville Courier* and the *Union*.

He attended school for a time, but spent most of his young adult life working as a typesetter and as an employee of the post office. Fortune's lack of a formal education did not keep him from being admitted to Howard University in Washington, D.C., in 1875. From there, he got a job at an African American newspaper called the *People's Advocate.* At around this time he also married his girlfriend from back in Florida, Carrie Smiley. Returning to Florida, Fortune earned a living as a teacher, but was disappointed at the lack of progress made by African Americans in the South while he was in Washington. In fact, things seemed to be getting worse; with the end of Reconstruction in 1877, federal troops left the South and allowed the state governments to institute laws that greatly restricted the freedoms of African Americans.

Fortune moved to New York City in 1879, where he worked for the *Weekly Witness.* From there he was hired as an editor for the *New York Sun.* He went on to become editor and publisher of the *New York Globe* and then the *New York Age,* prominent African American newspapers. He also became an editor for the *New York Evening Sun,* making him one of the only African American editors at the time to work at a mainstream newspaper whose readership was mainly white.

Fights for Civil Rights

In 1884, Fortune published the book *Black and White: Land, Labor, and Politics in the South.* In it, he argued that the race issues of the South were largely issues of social class—of the working-class poor versus the wealthy landowners. However, Fortune also recognized that racist policies had been instituted across the South to keep African Americans from achieving equality. He used his influence as an author and editor to fight segregation and promote civil rights at every available opportunity.

In 1887, Fortune began to work on creating an organization for African Americans with the aim of securing equal rights and ending racial violence. The organization was called the National Afro-American League, and its first convention was held in Chicago in 1890. Fortune also used his power as editor and co-owner of the *New York Age* to publish anti-lynching articles by Ida B. Wells (1862–1931), an outspoken journalist whose own newspaper offices in Memphis were destroyed by white supremacists in 1892.

The National Afro-American League failed to bring about any significant reform in segregation laws. Its name was changed to the Afro-American Council in 1898, but the group did not gain enough support to survive. One of the most prominent members of the council was activist W. E. B. Du Bois, who later formed his own group called the Niagara Movement. The principles of the Niagara Movement were strikingly similar

to those of the earlier National Afro-American League. The work of the Niagara Movement would eventually lead to the creation of the National Association for the Advancement of Colored People (NAACP).

Fortune maintained strong relationships with both Du Bois and Booker T. Washington (1856–1915), one of the most influential African American leaders in the South. Although Fortune did not always agree with Washington's ideas, the two remained friendly and respectful of each other. They each wrote a chapter of *The Negro Problem* (1903), a book of essays about African Americans and their place in society. In 1907, Fortune—suffering from personal problems that may have been related to alcoholism—sold his share of the *New York Age* and gave up his position as editor.

Fortune wrote a few other works, including a book of poetry in 1905 and a book about African Americans in journalism in 1915. He also wrote articles for several newspapers, including the *New York Age,* in which Washington had bought a controlling interest. However, Fortune was no longer recognized as a leading voice of the African American struggle for equality. He became the editor of *Negro World,* a weekly newspaper created by the controversial black activist Marcus Garvey (1887–1940), in 1923. He remained editor of *Negro World* until his death in 1928, at the age of seventy-one.

★ SAM LACY (1903–2003)

Sam Lacy was one of the first African Americans to achieve success as a sportswriter. His writing often called attention to the unfair conditions and treatment experienced by African American athletes. Lacy was one of the earliest advocates of the integration of professional sports like baseball.

Lacy was born in Mystic, Connecticut, the son of an African American law researcher and his Native American wife. The family moved to Washington, D.C., when Lacy was young. Baseball was quickly earning its place as the national pastime, and Washington was home to one of the most popular teams, the Washington Senators. As a boy, Lacy spent a great deal of time at the local stadium watching the team practice. He often served as an assistant to the players, and later got a job selling refreshments in the stands during each game.

In high school, Lacy was a successful athlete in his own right. He played baseball, basketball, and football, and after graduating he played semiprofessional baseball. At the time, baseball was segregated. African

Sam Lacy. *AP Images*

American players could only play in the Negro leagues, which were not considered professional leagues. Although their skills were equal to those of white players, black players earned considerably less money and were often kept from playing in certain stadiums because of the color of their skin. Even black spectators had to watch games from a separate viewing area.

Lacy had greater hopes than just playing minor league baseball. He attended prestigious Howard University with the aim of becoming a sports coach. At the same time, he began working at a local African American newspaper, the *Washington Tribune*. Lacy realized that he could combine his athletic and writing talents by becoming a sportswriter.

As a journalist, Lacy often focused on how sports would benefit from integrating black and white players. He argued that not only were African American players being deprived of a chance to earn greater fame and higher salaries, but white spectators were also missing out on seeing some of the country's best players simply because they were black. Lacy covered the 1936 Olympic Games in Berlin, Germany, where African American athlete Jesse Owens (1913–80) earned four gold medals in track and field events—more than any other single competitor, regardless of ethnicity.

Lacy also actively fought for major league baseball to allow players from the Negro leagues to play alongside their white counterparts. Some major league owners and managers like Branch Rickey (1881–1965) and Bill Veeck (1914–86) favored integration, but the commissioner of the sport, Kenesaw Mountain Landis (1866–1944), ensured that baseball remained segregated. In 1940, Lacy moved to Chicago to work for the *Chicago Defender,* a popular African American newspaper. He had hoped that living in Chicago would earn him a chance to influence Landis, who also lived there. Despite promises that Lacy could address the commission regarding integration in 1943, the meeting took place without him.

Lacy moved to Baltimore and became a writer for the *Afro-American,* where he would remain for over fifty years. Landis died in 1944, and Happy Chandler (1898–1991)—a former governor and senator from Kentucky—took over as commissioner of baseball the following year. Chandler was open to the idea of integration, and soon Brooklyn Dodgers owner Branch Rickey asked Jackie Robinson to sign with his team as the first African American to play in the major leagues. Lacy was assigned to follow Robinson as he attended spring training and spent a short period of time in the minor leagues before making his major league debut. Lacy documented the segregation and racism that Robinson faced long before he ever appeared in a major league game—hardships that Lacy himself often faced, such as being excluded from hotels and denied access to stadiums. His stories may have been instrumental in strengthening public support for baseball's integration.

In 1948, Lacy became the first African American member of the Baseball Writers' Association of America (BWAA). However, he still suffered from discrimination and unequal treatment at stadiums across the country. In New Orleans, for example, he was once forbidden from entering the press box with the other sports journalists. He was told that he could report from atop the dugout, above where the players sat. Several white sportswriters, angered by the unfair policy, joined Lacy on the roof of the dugout.

Lacy covered many different sports, including the Olympic Games, but was best known for stories about baseball. As more African American players entered the major leagues, Lacy became their champion in the press, arguing for higher salaries and an end to segregated living arrangements for players. He continued writing for the *Afro-American,* despite lucrative offers from more famous publications. In 1998, after over sixty years as a journalist, Lacy was honored with the J. G. Taylor Spink Award from the BWAA, and was also inducted into the Baseball Hall of Fame. He died in 2003 at the age of ninety-nine.

★ HATTIE MCDANIEL (1895–1952)

Hattie McDaniel was one of the most famous African American actresses of the 1930s and 1940s. Even though her roles were generally limited to portraying housekeepers and servants, McDaniel earned praise for bringing depth and sympathy to parts that might otherwise be considered offensive or racist. In 1940, she became the first African American to win an Academy Award for her portrayal of the housekeeper Mammy in *Gone with the Wind* (1939). The role remains her best-known work.

A Child of the Stage

McDaniel was born in Wichita, Kansas, to Henry and Susan McDaniel. She was the youngest of thirteen children. Henry was a minister and born entertainer, and Susan was a singer. Together they raised their children in an environment filled with creativity and encouraged a love of the performing arts. The family moved to Denver, Colorado, when Hattie was just six. Although there were few African Americans in the area, Hattie and her siblings were accepted by the community much more readily than blacks in many other parts of the country.

Hattie McDaniel. *AP Images*

From an early age, Hattie showed a natural gift for singing, dancing, and acting. While Hattie was still in school, her father formed a troupe to perform minstrel shows. Minstrel shows originated as humorous portrayals of Southern black life, with the exaggerated African American characters played by white performers wearing black face makeup. By the early 1900s, however, many minstrel shows featured African American performers and offered more culturally authentic songs and dances. Henry McDaniel's minstrel troupe consisted of many of his children. He agreed to let Hattie drop out of school and join as a performer in 1910. McDaniel performed in her father's troupe for ten years, and also toured the region with other minstrel shows.

McDaniel had married a man named Howard Hickman in 1911, who died four years later from pneumonia. Her mother died in 1920, and her father passed away just two years later. During this period, McDaniel remarried;

her husband was a laborer named Nym Lankfard, and the couple had a stormy relationship until they were finally divorced in 1928.

McDaniel's career was more successful than her romantic life; she was becoming a notable singer and songwriter. She formed an act with musician George Morrison (1891–1974) in 1924 that toured the country. She also performed live on a Denver radio station in 1925, making her one of the first African American women to perform live on an American radio station. She sometimes had to work as a maid or laundry washer between entertainment jobs, but McDaniel's talents usually kept her employed. She decided to move to Hollywood to pursue a career in motion pictures in 1930.

Her brother Sam and her sister Etta had been living in Hollywood already, and both had enjoyed some success in radio and film. For the most part, the only roles open to African Americans at the time were as servants or as comical black stereotypes. McDaniel took these roles and made the most of them, eventually capturing the attention of famed director John Ford (1894–1973). He cast her in one of the main roles in his film *Judge Priest* (1934). McDaniel even had the opportunity to sing a duet with the film's star, famed humorist Will Rogers (1879–1935). She also improvised another song for the film.

Another important role came the following year in *The Little Colonel* (1935), a film featuring the child star Shirley Temple (1928–) and African American entertainer Bill "Bojangles" Robinson (1878–1949). Once again McDaniel was cast as a housekeeper, but her charisma and talent earned her positive reviews and several other film roles. She starred alongside famed African American actor and singer Paul Robeson (1898–1976) in the 1936 film adaptation of the musical *Show Boat*. She also starred in films with actors Clark Gable (1901–60), Katharine Hepburn (1907–2003), and Henry Fonda (1905–82). She earned a reputation for portraying humorously outspoken maids and servants, and was often credited with—or accused of—stealing attention away from the stars of the film.

McDaniel the Star

In 1939, McDaniel landed the role that defined her career. She was chosen to portray Mammy, Scarlett O'Hara's housekeeper in the film adaptation of *Gone with the Wind*. McDaniel and several other African Americans who starred in the film were unable to attend its premiere in Atlanta, Georgia, due to the harsh segregation laws still in place across the South. However, she received glowing reviews for her performance. She became the first African American nominated for an Academy Award as Best Supporting Actress in 1940. Not only was she nominated, but she

won. She received one of the most enthusiastic standing ovations ever seen in the history of the awards up to that time.

During the 1940s, McDaniel continued to star as servants and housekeepers opposite Hollywood superstars like Bette Davis (1908–89) and Humphrey Bogart (1899–1957). However, some civil rights activists began to protest the fact that Hollywood did not offer a wider range of roles for African Americans. Some even went so far as to criticize McDaniel and other actors for accepting such roles. McDaniel tried her best to inject sympathy and realism into her performances, but was not inclined to turn down work, regardless of the role. She reportedly said that she would much rather play the role of a maid for seven hundred dollars a week than be forced to work as a real one for seven dollars a week.

In 1945, McDaniel was part of an important lawsuit that helped influence segregation laws in the United States. McDaniel had purchased a home in an upscale neighborhood of Los Angeles. Several white residents sued to restrict the neighborhood from allowing black homeowners, and McDaniel organized a legal response to the suit. The judge presiding over the case sided with McDaniel and other black residents, refusing to allow racial restrictions in the neighborhood.

McDaniel married twice again during the 1940s, but neither marriage worked out. In addition to film roles, McDaniel landed the role of the title character of a radio series called *The Beulah Show* in 1947, in which she once again portrayed a housekeeper. However, McDaniel was now the star of the show, and became the first African American woman to play the lead in a major network radio series. McDaniel was diagnosed with breast cancer in 1952, and quickly became too ill to continue working. She died later that year at the age of fifty-seven.

★ GORDON PARKS (1912–2006)

Gordon Parks was a one-man revolution in the media arts. He was an accomplished photographer, author, poet, musician, filmmaker, and activist. He gained fame as the first African American staff photographer for *Life* magazine, and later earned recognition for his autobiographical novel *The Learning Tree* (1963). Parks was also the first African American to direct a major Hollywood motion picture.

An Unsettled Childhood and Early Career in Music

Parks was the youngest of fifteen children born to Jackson and Sarah Parks in Fort Scott, Kansas. According to Parks, he was stillborn, and the doctor who delivered him brought him back from death. His parents gave

Gordon Parks. *Alfred Eisenstaedt/Time & Life Pictures/Getty Images*

him the doctor's name—Gordon—as a token of gratitude. He was raised in a relatively poor farm family, but the house did have an upright piano that Gordon learned to play at a young age. The school Gordon attended was not segregated because the small town could not afford separate schools for white and black students. However, he grew up dealing with racism and prejudice on a regular basis.

Gordon's mother died when he was fifteen, and his father could not afford to look after him. He sent the boy to live with one of his older sisters in St. Paul, Minnesota. In Minnesota, he found people somewhat more tolerant of African Americans than he had in Kansas. The sister's husband made life difficult for Gordon, however, and eventually threw him out of the house. He lived on his own for a while, getting a job as a piano player and later as a busboy. His musical talent earned him a position as the only African American member of a traveling orchestra, and he began touring with the group.

The band's leader disappeared upon their arrival in New York City, leaving Parks stranded in an unfamiliar city. Parks signed up for the Civilian Conservation Corps, and while on leave he married his longtime girlfriend. After funding for the project stopped, he ended up looking for a job once again. He finally found steady employment working for a railway car company as a porter. During one trip, he picked up a magazine left by a passenger and saw a collection of photographs taken for the Farm Security Administration (FSA). The group was funded by the government to help poor farmers, and the photos were gritty, realistic portrayals of the farmers' difficult lives. Parks was inspired by the pictures, and soon after he bought a used camera at a pawn shop for $7.50. He began taking pictures wherever he went, and was encouraged when employees from Eastman Kodak assured him that his shots displayed skill and talent.

Talent for Photography Leads to Job with *Life*

Parks's pictures earned him a display in the window of Eastman Kodak, and he got an assignment taking fashion photos. This led him to Chicago, where he documented the city's people and places with an unblinking yet artistic eye. He earned a Julius Rosenwald Fellowship in 1942. He used the fellowship money to support his family while

continuing to perfect his photography skills. He traveled to Washington, D.C., where he worked with the FSA and then with the Office of War Information. One of Parks's most famous photographs, "American Gothic," (1942) was taken during the course of his work for the FSA. The image shows a black cleaning woman in front of an American flag with mop and broom in hand. The image is a parody (imitation for comedic or satiric effect) of artist Grant Wood's 1930 painting of the same name, which shows a white Iowa farmer and his wife.

Parks left the FSA in the mid-1940s and moved to Harlem in New York City. There he wrote two how-to books on photography, and worked as a freelance fashion photographer for *Vogue*. Parks's photoessay on a Harlem gang leader caught the attention of *Life* magazine, which offered him a staff position as a photographer and writer in 1948. He was first African American to hold such a position. For the next twenty years, Parks covered everything from sports and fashion to the civil rights movement. Parks completed one of his most noteworthy assignments in 1961. He traveled to Rio de Janeiro, Brazil, to record life in the slums of the glamorous city. Parks created a photoessay on the family of Flavio da Silva, a young boy suffering from disease and malnutrition and living in a one-room shack. *Life* readers were so moved by the piece, they donated enough money to pay for the boy's medical care and, eventually, buy his family a new home.

Makes History with *Shaft*

While working for *Life,* Parks also pursued other interests. He wrote a piano concerto in 1953, and he started writing poetry to accompany some of his photographs. In 1963, he published an autobiographical novel titled *The Learning Tree.* The book was a success, and Parks decided to create a film adaptation of it in 1969. He wrote the screenplay, served as producer, composed the score, and became the first African American to direct a major Hollywood motion picture. After that, he spent three years as editorial director for *Essence,* the first magazine aimed at young African American women.

Parks directed several more films, most notably *Shaft* (1971) and *Shaft's Big Score* (1972). Although he was sometimes criticized for offering an overly negative portrayal of African American communities in his photographs and films, he always favored positive role models. He was active in the civil rights movement, and remained close friends with Malcolm X until the activist was assassinated. Parks also later wrote the music and libretto (text) for a ballet honoring civil rights leader Martin Luther King Jr.

Parks returned to writing and photography in his later years. He wrote new volumes of his autobiography, and he wrote a biographical

novel about famed painter J. M. W. Turner (1775–1851) titled *The Sun Stalker* (2003). Parks passed away in 2006 at the age of ninety-three.

★ GEORGE SAMUEL SCHUYLER (1895–1977)

George Samuel Schuyler was a groundbreaking journalist and editorial writer who rose to fame in the 1920s and 1930s. His novel *Black No More* (1931) is generally considered to be the first satirical novel by an African American. A satirical novel is one in which the author uses humor to point out wrongs in society. Schuyler was known for criticizing African American intellectuals as often as he criticized whites. His staunch opposition to communism later in life alienated Schuyler from many African American readers, resulting in the loss of much of his popularity.

A Military Career of Highs and Lows

Schuyler was born in Providence, Rhode Island, the only child of George and Eliza Schuyler. His father was a chef who died while Schuyler was young. Schuyler was raised by his mother and stepfather in Syracuse, New York. His parents had worked as servants for wealthy families, and so they were considered among the highest class of African Americans in the city. This closeness to wealth led Eliza to raise Schuyler with the finest manners and education she could afford, with the hope that he would one day achieve great fame and wealth. Eliza also kept her son from interacting with poorer African Americans in the neighborhood, especially those migrating from the South. She felt that they were of a lower class and were generally uneducated, which would reflect poorly on all African Americans in the neighborhood.

George Samuel Schuyler. *© Hulton-Deutsch Collection/Corbis*

Schuyler was intelligent and did well in school, but he realized that his opportunities in Syracuse were severely limited because he was African American. Even with a high school diploma, he would likely find work only as a servant or as a railway porter like his stepfather. Schuyler chose to enlist in the U.S. Army at the age of seventeen instead of completing high school.

He served in the U.S. Army for three years, spending much of that time in Hawaii, which was not yet a state in the United States. The U.S. Army was strictly segregated, keeping African American soldiers in separate regiments. Even so, Schuyler was exposed to many different cultures and people in Hawaii. After spending several months working as a civilian there, he enlisted once again in the U.S. Army. Based on his performance and past record, he was chosen to become an officer. At around this time, he also began to publish humorous stories in military publications. Many of them focused on Southern characters similar to the soldiers with whom he worked.

Schuyler had attained the rank of first lieutenant by 1917. His military rank, however, did not shield him from racism. In 1918, while he was on leave from Fort Dix, a poor Philadelphia street worker refused to shine Schuyler's boots due to his race. Schuyler decided that he would rather leave the U.S. Army than defend people who would not recognize him as an equal. He hid out for several months, but eventually turned himself in and served nine months in a New York detention center for being absent without leave. After he had finished his sentence, a friendly officer who worked at the detention center helped Schuyler get a temporary civil service job.

Schuyler settled in New York City. He became involved with the socialist movement in 1921, and began writing articles for a socialist magazine called the *Messenger* in 1923. Socialists believed that the wealthiest members in a society had a duty to look after the needs of the least fortunate members of the society. They also fought for the rights of workers to earn a suitable wage. Schuyler saw socialism as a way for African Americans to achieve greater economic freedom and civil rights. He became known for a satirical column in the *Messenger,* which led to another weekly column for the *Pittsburgh Courier,* a popular African American newspaper.

A Sharp Turn Away from Socialism

After a few years with the *Messenger,* Schuyler became increasingly critical of those involved in the socialist movement, feeling that they were not truly interested in helping African Americans at all. He was equally critical of African American intellectuals who promoted the idea of the Harlem Renaissance, a flowering of African American art and culture during the 1920s. In 1926, Schuyler wrote an essay for the *Nation* called "The Negro-Art Hokum." In it, he argues that artistic works produced by African Americans should not be viewed as separate from broader American culture. Works created by Italian Americans or Irish Americans, for example, would not be considered as a separate kind of art. He thought this separation of black art from mainstream art would be harmful to

African Americans. The essay was criticized by many African Americans, most notably poet Langston Hughes (1902–67), who thought Schuyler was devaluing African American culture. While Schuyler was consistent in fighting and exposing white racism, his views on African American identity and culture often conflicted with those of leading African American public figures. Schuyler's brash, unapologetic essays found a welcoming home in *American Mercury,* a magazine founded and edited by H. L. Mencken (1880–1956), a man also famous for his brash, unapologetic writing. *American Mercury* published ten of Schuyler's articles between 1927 and 1933.

In 1928, Schuyler married a wealthy white woman from Texas named Josephine Cogdell. This led to the publication of his first book, *Racial Intermarriage in the United States* (1929). In it, he suggests that promoting interracial marriages would help bring about an end to racism by creating a nation of mixed-race children that benefited from the best qualities of each ethnic group. This was a shocking suggestion to many Americans, especially in the South, where interracial marriage was still against the law. Schuyler and his wife had a daughter named Philippa in 1931. She became a successful musician, but died in a helicopter crash in Vietnam in 1967.

Schuyler is best known for his 1931 satirical novel *Black No More,* in which a scientist invents a way for black people to change their skin color to become white. Schuyler offered humorous portrayals of contemporary leaders on both sides of the race issue, making fun of both white supremacists and African American civil rights activists at the same time. Also in the 1930s, Schuyler investigated the ruling forces in Liberia, a country on the western coast of Africa. The country was established in 1821 by the United States as a homeland for African American slaves who wished to return to Africa. However, Schuyler discovered that natives from the area had been driven out and even sold into slavery by Liberia's rulers. In addition to the articles he wrote on the subject for major mainstream newspapers, Schuyler published a novel inspired by his findings called *Slaves Today: A Story of Liberia* (1931). He also wrote dozens of short stories for the *Pittsburgh Courier,* using as many as seven different pen names.

At Odds with the Civil Rights Movement

Schuyler served as business manager for the National Association for the Advancement of Colored People (NAACP) from 1937 to 1944. Throughout the 1940s, Schuyler's work became increasingly conservative and focused on condemning socialism and communism. He continued to criticize civil rights leaders, believing that their liberal political philosophy veered too much toward communism. At the same time, Schuyler wrote

articles insisting that conditions for African Americans were improving rapidly even without the intervention of liberal activists. Schuyler's opinions were not shared by many African American readers, and his popularity suffered greatly. His obsession with attacking socialism and communism led him to harshly criticize Martin Luther King Jr. after King received the Nobel Peace Prize in 1964. The *Pittsburgh Courier* refused to print Schuyler's editorial, titled "King: No Help to Peace," and two years later, the paper ended Schuyler's column entirely.

Schuyler published his autobiography, *Black and Conservative,* in 1966. He continued writing columns for a smaller paper, the *Manchester Union Leader,* and for other conservative publications. At the time of his death in 1977 at the age of eighty-two, Schuyler and his earlier achievements as an African American author and journalist had been all but forgotten.

★ IDA B. WELLS (1862–1931)

Ida B. Wells was a journalist and activist best known for her crusade against lynching in the 1890s and early 1900s. Lynching is a form of mob violence in which a person accused of a crime, normally African American, is killed before being given a fair trial. Lynchings became common in the South after the Civil War as a way to intimidate the newly-free black population.

Ida B. Wells. *R. Gates/Hulton Archive/Getty Images*

A Strong Woman with a Strong Voice

Wells was born in Holly Springs, Mississippi, to slave parents. Her father, James Wells, was the son of a slave owner and a female slave, and received special treatment such as training in the field of carpentry. After the Civil War, James became a carpenter and leading black citizen of the town. Ida was given an opportunity to study at Shaw University, a school that had been newly created in Holly Springs for African American children. Ida's mother even attended school with her daughter in order to learn to read the Bible, since, as a slave, she had never received an education.

Both of Ida's parents fell victim to yellow fever and died when she was only sixteen. Her youngest brother, still an infant, also died from

the illness. As the oldest child, Wells took charge of caring for her five surviving siblings. She got a job as a teacher, and in 1881 moved to Memphis, Tennessee, where she continued her teaching career. She also began taking summer classes at Fisk University in Nashville, Tennessee.

In May 1884, Wells was traveling by train from Memphis when she was asked to move from the first-class section to the smoking car, despite the fact that she had a first-class ticket. The train conductor was trying to enforce a new rule of segregation, in which black passengers could be put in a separate car from whites as long as the quality of the other car was comparable. Wells refused to move, and the conductor forced her from her seat. When she returned to Memphis, she filed a lawsuit against the railroad for failing to provide her the first-class accommodations for which she had paid. She won her suit, but the decision was overturned by the state supreme court, which ruled that the train's smoking car qualified as a first-class substitute for black passengers.

Despite this setback, Wells continued teaching, and also began writing articles for local newspapers about the lawsuit and other local African American concerns. She became co-owner of one of these papers, the *Memphis Speech and Headlight,* later renamed the *Free Speech.* An article she wrote in 1891 criticized the Memphis school board for its failure to provide adequately for black students, and led to her dismissal as a teacher. With no other career options, she turned to the newspaper business full-time. Wells became quite popular among African American readers in Memphis, but her editorials brought her ridicule from some white residents.

Begins Her Crusade

In 1892, three African American men who ran a local grocery store in Memphis defended themselves from a group of white men who attacked them. In the chaos, three of the attackers—who turned out to be local police officers in disguise—were wounded by the store owners. The three store owners were arrested, but before they could be taken to trial, a lynch mob pulled them from the jail and killed them. Wells, who considered the store owners both colleagues and friends, wrote an angry and passionate editorial about the incident. She warned other African Americans in Memphis that if they had the means, they should leave, since it was clear that no one in authority would protect them from mob violence.

In the weeks after the incident, she continued to write angry editorials about the injustices of lynch mobs. Less than three months later, while Wells was away on a trip to Philadelphia, angry white residents in Memphis burned the *Free Speech*'s offices and printing equipment to the ground. Fearing that she would be killed if she returned to Memphis,

Wells decided to move to New York City. She secured a job as a reporter for the *New York Age,* a newspaper by and for African Americans. Outraged by the lynch mob violence that had occurred in her own Memphis community, Wells began to research similar crimes that were happening across the country, mostly in the South.

Her research resulted in a long article for the *New York Age,* which was later expanded and published as a pamphlet titled *Southern Horrors: Lynch Law in All Its Phases* (1892). In it, Wells describes her own experiences in Memphis, and also examines the common justifications for lynching that were given by whites. At the time, it was generally believed that a lynching only occurred as punishment against an African American man who had raped a white woman. However, Wells discovered that less than one-third of the lynchings in the South involved such a charge. Even when charges of rape were made, they were almost never proven. Wells noted that sometimes a white woman would engage in a consensual affair with a black man, but would lie and say that the man forced himself upon her to avoid disgrace when others found out about the affair. These assertions by Wells challenged the widely accepted image of the Southern white woman, and also challenged the idea that African Americans were largely unable to control their primitive urges.

The publication of *Southern Horrors* began Wells's crusade against lynching. She gave lectures throughout the northern United States as well as England on the horrors of lynching. Among the statistics she recited to her audiences was the fact that in 1894 more deaths by lynching occurred in the United States than deaths by court-ordered execution. Some audience members, especially in England, had a difficult time believing that Wells's horrific stories of mob murder could possibly be true, but she offered photographic evidence for any doubters. In 1895, she published *The Red Record: Tabulated Statistics and Alleged Causes of Lynching in the United States,* which documented lynchings that had occurred since the publication of *Southern Horrors.*

Also in 1895, Wells married lawyer Ferdinand Barnett and settled in Chicago, Illinois. Barnett owned the *Chicago Conservator,* an activist newspaper, and Wells became the editor. Married life—and the four children it brought—slowed her down a little, but Wells continued working to fight lynching and to secure better legal protections for African Americans. She also grew increasingly concerned about the state of African American women, who were perceived by the American public as immoral and uneducated. In 1896, she helped gather support for the founding of the National Association of Colored Women's Clubs (NACWC), an organization aimed at presenting a positive image of African American women and at

helping underprivileged black women. Wells also fought for the right of all women to vote, also known as women's suffrage.

In 1900, Wells wrote *Mob Rule in New Orleans,* the true story of a riot that broke out after an African American man shot and killed a police officer (in self-defense, according to Wells) and then escaped. Several thousand angry white residents marched through the streets, attacking any African Americans they came across. Over two dozen people were killed during the riots. Wells's account offers brutal descriptions of the various attacks on innocent black citizens, and finishes with a survey of lynch mob deaths by burning, which were on the increase across the South.

Wells continued her anti-lynching campaign for the rest of her life, and also continued to fight for women's suffrage. In 1913, she created the Alpha Suffrage Club, the first suffrage group for black women in Illinois. In 1919, she witnessed firsthand the Chicago race riot, in which nearly forty people—mostly African Americans—died. Wells died in 1931 at the age of sixty-eight. Her anti-lynching campaign played a key role in the steady decrease in the number of lynchings in the United States after 1892, the year she first wrote on the subject.

THE *CHICAGO DEFENDER* CHAMPIONS AFRICAN AMERICAN INTERESTS

African Americans relied primarily on local newspapers for information and entertainment in the late nineteenth and early twentieth centuries. Most mainstream papers, however, did not contain news and viewpoints of interest to African Americans. For this reason, enterprising black publishers began their own newspapers, containing both local news and items relevant to African Americans across the nation. One of the most successful of these was the *Chicago Defender*.

The *Chicago Defender* began in May 1905 as a four-page pamphlet created by Robert Sengstacke Abbott (1870–1940) in his landlord's kitchen. It started out with a circulation of just three hundred copies, and featured both reprinted material and news of local interest. Three other African American newspapers were already being printed in Chicago, so Abbott's business was a gamble. He overcame tough odds to achieve modest success by interacting with the people in his neighborhood, both to promote his paper and to discuss their own concerns, which helped him shape his news stories.

The paper was helped greatly by interest in the Brownsville Affair, a racial conflict that occurred in Texas in 1906. A regiment of African American soldiers was accused of killing a white bartender in the town of Brownsville, near their station at Fort Brown. Over 150 black soldiers were dishonorably discharged over the matter, even though evidence suggested that none of them could have committed the shooting. The *Defender* reported on the affair and the subsequent investigation by Congress, supporting the soldiers even as the government ruled against them. The paper's in-depth coverage of the incident and its aftermath brought it to national attention, and subscriptions increased dramatically.

A paperboy sells copies of the black newspaper *Chicago Defender* in 1942. *The Library of Congress*

Abbott added new sections to the paper, including sports columns, editorials, and sections on the arts and society. He also changed to a full-sized format in 1915. At the time, very few advertisers were willing to buy ad space in an African American newspaper. Abbott's profits

came almost exclusively from subscriptions. The majority of these subscribers were located outside of Chicago, and many lived in the South. Black railway workers helped in distributing the paper among Southern blacks. Beginning in 1917, Abbott launched a campaign to encourage African Americans in the South to move northward for better treatment and better economic opportunities. Abbott thought that if African Americans moved out of the South, their quality of life would improve and incidents of lynching would decrease.

By encouraging African Americans northward, Abbott and the *Defender* played a key role in the Great Migration, the movement of millions of African Americans from the South to urban centers in the North such as New York City and Chicago between 1910 and 1930. As the population of Chicago grew, so too did the circulation of the newspaper. By 1920, the *Defender* sold over two hundred thousand copies of each issue, making it the most successful African American newspaper of the era.

The paper was famous not only for supporting civil rights for African Americans, but also for supporting the rights of all workers. This made the paper especially relevant when the Great Depression, a time of extended economic troubles, began in 1929. About half of all able-bodied African American men were unable to find jobs during the worst times of the Depression, and circulation for the newspaper suffered. However, Abbott found ways to cut costs and continue in print. The *Defender* supported the causes of labor unions like the Brotherhood of Sleeping Car Porters, one of the earliest and most important African American labor unions in the country.

During World War II (1939–45), the newspaper's tradition of protest was tempered by the need for national unity. After Abbott passed away in 1940, his nephew John Sengstacke (1912–97) took over the paper. Sengstacke focused on stories that highlighted the achievements of African American soldiers during the war. However, the paper did not avoid discussing the fact that African Americans, who were being asked to risk their lives for the freedom of others in Europe and Asia, did not have access to their own full freedom back in the United States. At the time, African American soldiers were still kept separate from white soldiers. Sengstacke used the *Defender* to actively campaign for the integration of black and white soldiers in the armed forces. In 1948, President Harry Truman (1884–1972) signed an order for the integration of all branches of the United States military.

The *Defender* became the first African American daily newspaper in 1956. Continuing its tradition of fighting segregation and oppression, the *Defender* was one of the most important voices in the struggle for civil

rights during the 1950s and 1960s. It reported on the violent attacks suffered by civil rights workers across the South, as well as the advances being made through nonviolent demonstrations like sit-ins and boycotts. The newspaper celebrated its first century as a leading champion of African American interests in 2005.

W. E. B. DU BOIS FOUNDS *THE CRISIS* MAGAZINE

The National Association for the Advancement of Colored People (NAACP) was formed in 1909 as an organization focused on the fight for African American civil rights. Many other similar organizations had been created before it, yet none was able to survive and thrive. The NAACP would achieve great success where others had failed, and much of the credit for its success goes to *The Crisis,* the group's official publication, created by W. E. B. Du Bois.

Du Bois was one of the most influential African American intellectuals of the early 1900s. He had written *The Souls of Black Folk* (1903), an important collection of essays on race relations. He also cofounded the Niagara Movement, an early activist group whose principles formed the core of what would become the NAACP. In fact, the three people credited

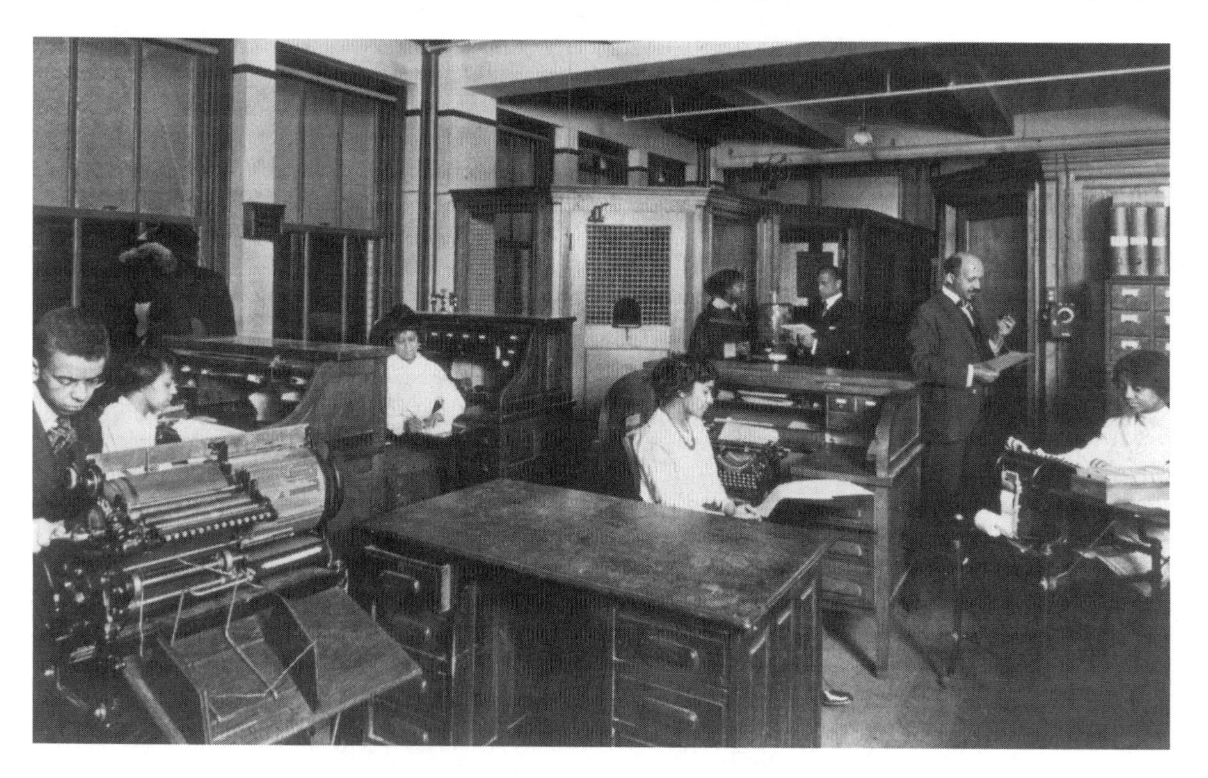

W. E. B. Du Bois (right) at the headquarters of *The Crisis* magazine in 1932. *Hulton Archive/Getty Images*

How The Crisis Got Its Name

Some sources have credited Du Bois himself with coming up with the name of the magazine. This theory is proven false, however, by the magazine's official history, which credits Mary White Ovington (1865–1951) and William English Walling (1877–1936), two of the organization's founders. While discussing poetry and the aims of the new magazine one day during the summer of 1910, Ovington mentioned that the poem "The Present Crisis" (1844) by James Russell Lowell (1819–91) held special meaning for her. The poem, written in the years before the Civil War, was originally intended to persuade readers of the need to abolish slavery. In it, Lowell argues that slavery is primitive and violates the spirit of equality upon which the United States is based. The poem ends by asserting that humankind should not "attempt the Future's portal with the Past's blood-rusted key." Upon hearing Ovington's remark, Walling observed that *The Crisis* would make an excellent title for the as-yet-unnamed magazine, and so it was decided.

with founding the NAACP had been among the Niagara Movement's only white members.

Du Bois was the only African American elected to serve as an executive officer when the NAACP was officially created. His title was director of publicity and research, and it was his job to oversee the creation of the group's official publication, *The Crisis*. The name reflected the urgent nature of the reforms that were needed to bring equality to African Americans.

Du Bois settled into his new office in New York City in the summer of 1910 and began work on the magazine. He was already an experienced editor, and had overseen the creation of two other periodicals, though both had ceased publication. Du Bois had learned valuable lessons from his experiences, and he had a clear vision for *The Crisis*. The other executive officers of the NAACP, confident of Du Bois's talents, were happy to let him do as he wished.

The magazine's most important function was to record news related to the struggle for African American civil rights. It also offered news about the struggles of other oppressed groups, including women (who did not yet have the right to vote in the United States). Every issue contained editorials by Du Bois and others reflecting contemporary thought on civil

rights and race relations. Du Bois was also intent on including the finest examples of art, poetry, and fiction created by African Americans. Following this basic structure, *The Crisis* began publication in November 1910. Du Bois stated in his first editorial for the magazine, "The object of this publication is to set forth those facts and arguments which show the danger of race prejudice, particularly as manifested today toward colored people."

The magazine began with a circulation of only one thousand copies, but grew steadily. By 1918, its circulation was over one hundred thousand copies per issue—greater than that of many prominent mainstream publications such as the *Nation* and *New Republic*. The magazine was more popular than the NAACP itself, and it served to draw in new members for the organization. It was without a doubt one of the most important publications in the fight for civil rights, and throughout the 1920s and 1930s was one of the premier African American arts publications. Its success proved that Du Bois's activist ideas—which had been at the heart of the failed Niagara Movement, as well as two unsuccessful magazines—had finally come of age.

OSCAR MICHEAUX FORMS FILM COMPANY

Oscar Micheaux (1884–1951) was a pioneer in a field not normally considered accessible to those without a great deal of money or education: filmmaking. The making of a film requires not only detailed technical knowledge but also expensive equipment and supplies. These barriers would have seemed insurmountable to almost anyone, regardless of race. Micheaux overcame all obstacles to become the first successful African American filmmaker.

Oscar Micheaux became the first African American filmmaker when he wrote, directed, and produced the 1919 movie *The Homesteader*. *AP Images*

Micheaux had grown up as part of a large working-class family in southern Illinois, far from the glamour of Hollywood. He grew up at a time when racial prejudice was perhaps at its worst in the United States. It was an era of strict segregation, and the rise of the Ku Klux Klan and mob violence against African Americans. Despite these obstacles, Micheaux firmly believed that African Americans could achieve the same levels of success as white people if they possessed the competence and initiative to make it happen.

Michaeux published his first novel, *The Conquest: The Story of a Negro Pioneer,* in 1913. The book was presented as fictional, but it contains many details of Micheaux's childhood and early adult life. The book describes his road to success. He grew up in a farm family, and moved to Chicago to work at various jobs before becoming a railway car attendant. This job allowed him to travel the country. It was during these travels that he learned of an opportunity that shaped his future. A portion of an American Indian reservation in South Dakota had been reclaimed by the federal government. Sections of it were being sold cheaply to selected applicants, called homesteaders, who agreed to maintain and work the land. Micheaux applied, and was not selected; however, some of the plots were made available again after the original owners failed to maintain them as the government required. Micheaux was able to purchase a large piece of land, and established his own large farm there.

For various reasons, his farming career was not successful. Micheaux recognized, however, that his story was unique, so he described his experience in *The Conquest.* He followed it with two more novels: *The Forged Note* (1915) and *The Homesteader* (1917). It was this third novel that attracted the attention of Noble Johnson (1881–1978), a famous African American actor, and his brother George. They had started their own film production company in 1915 in an effort to produce films for African American audiences, though the film directors were white.

At the time, films catered mostly to white audiences. The most successful film at the time was *The Birth of a Nation* (1915) by D. W. Griffith (1875–1948). The film offers an inarguably racist portrayal of events in the South during and after the Civil War, including the formation of the Ku Klux Klan. In it, a former slave is lynched by a mob after he makes romantic advances toward a white woman. The former slave was portrayed by a white actor in black makeup. Many African Americans, including the Johnsons, were outraged by the film and troubled by its huge success.

They believed that Micheaux's novel *The Homesteader* would make an excellent response to the negative portrayals of blacks in *The Birth of a Nation. The Homesteader* was about an African American farmer surviving the challenges of both nature and the mainly white society in which he lives. Micheaux was interested in directing the film adaptation himself, but since he had no experience, the Johnsons would not agree to a deal. Micheaux decided to form his own film production company, the Micheaux Book and Film Company, and direct the film himself.

Micheaux financed the film by selling shares of his company to farmers throughout the Sioux City, Iowa, area where he based his company. He also

shot the film there, since it required open expanses of farmland. Micheaux served many roles on the film, including producer, writer, and casting director. He was also the director of the movie, which made him the first African American film director. Micheaux released the film version of *The Homesteader* in 1919. He had hoped that he could also draw white audiences into the theater, but once the film was complete and screenings began, he realized that his viewers would be almost exclusively African American. Thus, the film and others that followed were classified as "race films"—not because of their content, but because of their audience. Micheaux also produced another film in 1919; *Within Our Gates* was a film about white Southerners taking unfair advantage of black laborers in the years immediately after the Civil War.

Micheaux continued to write, produce, and direct films for his independent production company for almost thirty years. In that time he created over forty films, only about a dozen of which are still known to exist. He achieved success where no other African American had, although he never worked with large budgets. His final film before his death in 1951, *The Betrayal* (1948), was the first film by an African American to premiere at a theater on Broadway. It was also the first film by an African American to receive a review in the *New York Times*.

DUKE ELLINGTON BROADCASTS ACROSS THE NATION

Radio was a brand-new medium in the 1920s. The first commercial radio stations had just begun operating across the country at the start of the decade. Station owners were still experimenting to discover what types of shows they could produce that appealed to listeners and were inexpensive. Live music became a popular format, and one of the first musicians to gain widespread popularity through the radio was jazz musician Duke Ellington (1899–1974).

Ellington and his band became the in-house orchestra for the Cotton Club, a popular nightclub for white customers in Harlem, in December of 1927. Ellington, still at the beginning of his career, was actually the fourth choice for the job. Other prominent African American bandleaders had already rejected offers to headline the club, with one important reason being that they knew their fellow black musicians would not be able to visit and listen to them play. At first, Ellington's style of music was not what club management expected from its house band. Rather than light, enjoyable versions of popular songs, Ellington and his band played aggressive and complex jazz that surprised and confused some listeners.

The unusual sound worked to the advantage of the club's management, since it appealed to white audiences' notion of the kind of wild, "jungle-like" music African Americans would create. The sound caught on, and a local radio station began to broadcast three live shows featuring the band each week: a late show on Monday nights and early evening shows on Wednesdays and Fridays. Radio listeners were soon visiting the club after hearing Ellington's band play on the radio, and business was booming. Ellington and his band had a special group of admirers who worked for the entertainment publication called *Variety*, and the band often received positive reviews there.

The radio broadcasts featuring Ellington were heard along the East Coast, but the club soon received an offer that promised even greater exposure. A broadcaster named William S. Paley approached the club's management and asked them to switch their radio show to his network of stations. Paley's plan was to expand his radio network into a nationwide system of stations that featured blocks of programming at the national level. This meant that Ellington and his orchestra would be heard all across the nation. The management of the club agreed, knowing the exposure would benefit both the club and the musicians. The network formed by Paley came to be known as the Columbia Broadcasting System, or CBS.

Ellington's broadcasts from the Cotton Club remained popular until 1931, when the bandleader and his group left the club and began touring to take advantage of their newfound nationwide fame. Cab Calloway (1907–94) became the orchestra leader in Ellington's absence, and enjoyed similar popularity when his band was featured on the Cotton Club's radio broadcasts.

RACISM OF *AMOS 'N ANDY* DRAWS CRITICISM

One of the most popular forms of entertainment in the United States during the late nineteenth and early twentieth centuries was the minstrel show. The minstrel show featured white stage performers who wore black face makeup and played African American characters. The portrayals were heavily based on stereotypes about African Americans; they were usually depicted as uneducated, lower class, and lazy, though also often good-natured and sympathetic. Even as minstrel shows declined in popularity from the 1920s onward, their legacy was clearly seen in *Amos 'N Andy*, one of the most popular radio shows of all time.

Amos 'N Andy was the creation of Freeman Gosden (1899–1982) and Charles Correll (1890–1972), two white performers who had made a name for themselves on a Chicago radio variety show called *Correll and Gosden*,

OPPOSITE PAGE

Duke Ellington sits at the piano as his band performs at the legendary Cotton Club in Harlem, New York, in the 1930s. *Frank Driggs Collection/ Hulton Archive/Getty Images*

the Life of the Party in 1925. Some of their most popular characters were African American in the minstrel tradition: sympathetic and humorous, but crudely stereotyped and voiced with a heavy Southern black dialect. The following year, the men created a serial comedy for radio station WGN. A serial comedy has the same characters in every episode, and each episode picks up the story where the last left off. The show was about two African American friends from Alabama who had recently moved to Chicago. It was called *Sam 'n' Henry*.

The show was a huge success, and Gosden and Correll left WGN with the hope that they could find a station willing to broadcast their program nationwide. WGN still owned the rights to *Sam 'n' Henry*, so the men created another show called *Amos 'N Andy*. *Amos 'N Andy* was remarkably similar to *Sam 'n' Henry*. In the show, Amos, voiced by Gosden, was a sensible, hardworking man who did his best to achieve success and at the same time keep Andy out of trouble. Andy, voiced by Correll, was always looking for a way to get something for nothing, and tended to avoid real work whenever possible. Between them, the men voiced dozens of other characters that appeared over the course of the show's run.

Amos 'N Andy debuted in 1928, appearing on seventy different radio stations before the end of its second year. Beginning in 1929, it was broadcast across the entire country on two different NBC radio networks. By the early 1930s half of all radio listeners in the United States tuned in to the show. Its popularity was so far-reaching that some restaurants would even halt service for the fifteen minutes the show aired each evening so that the waitstaff would not miss it.

Despite its popularity, the show did receive criticism from some African American activists. They complained that the show depicted African Americans as either lower-class dimwits or as devious conmen. A popular African American newspaper, the *Pittsburgh Courier*, launched a campaign to gather one million signatures of Americans protesting the show. The campaign faded away several months after it started, though they reportedly gathered two-thirds of their million-signature goal.

Gosden and Correll also drew controversy by starring in a film about the two characters called *Check and Double Check* (1930). The characters were African American so the white performers played the roles wearing black face makeup. The film was financially successful, but did not satisfy critics, civil rights activists, or even its creators.

Despite these early objections to the show and film, *Amos 'N Andy* remained popular throughout the 1940s. The show expanded its cast beyond Gosden and Correll, hiring African American actors to portray many of the characters that the two had previously voiced by themselves.

In the early 1950s, CBS produced a television adaptation of the show, even as the radio show continued to be a success. There was one main difference between the television show and the radio production: the roles of Amos and Andy were not played by Gosden and Correll, but by African American actors Alvin Childress (1907–86) and Spencer Williams (1893–1969). Both the network and the show's creators hoped to avoid the controversy caused by the earlier film.

Alvin Childress and Spencer Williams starred in the television version of *Amos 'N Andy* in the early 1950s. © *Bettmann/Corbis*

The casting switch did not keep controversy at bay. The general sentiment among African Americans had changed dramatically over the course of twenty years. The show's premiere in 1951 brought protests that far surpassed any earlier complaints. Few stopped to acknowledge the show's landmark status as the first television series to feature a cast made up entirely of African Americans. Instead, groups like the National Association for the Advancement of Colored People (NAACP) launched a campaign against the show and its portrayal of negative black stereotypes. The complaints were much the same as before, but at a time when African Americans were increasingly standing up against symbols of racism, the negative public opinion became impossible to ignore.

The television series version of *Amos 'N Andy* aired for two seasons, and ranked among the top twenty-five shows on television for both years. The overwhelming protest against the show forced its cancellation in 1953. Many local stations still aired the series in syndication for more than a decade, where it continued to draw audiences. The radio series finally ended in 1960, after more than thirty years of success.

JACK L. COOPER HOSTS RADIO PROGRAM FOR AFRICAN AMERICANS

The 1920s and 1930s were the age of radio, when families would gather together and listen to live music, dramas, and comedies that streamed magically through the air and into their homes. For all the entertainment and information the radio offered in the 1920s, it offered very few African American voices. Some African American musicians performed on a few stations, and many popular comedies featured African American characters. These characters, however, were offensive stereotypes by today's standards and usually voiced by white performers. But not until Jack L. Cooper took

to the airwaves in Chicago in 1929 did listeners hear a genuine African American voice hosting a weekly program created by and for African Americans.

Jack L. Cooper (1888–1970) was a man of varied talents. He was one of twelve children in his family, left to care for themselves and each other after their father died while Jack was still a baby. Early in his career he was a professional lightweight boxer and a semiprofessional baseball player. For a time, he was a vaudeville stage performer. He also worked as a journalist for several African American newspapers, including the *Freeman* in Indianapolis and the *Western World Reporter* in Memphis. He then moved to Chicago, where he worked as the drama editor for the *Chicago Defender,* one of the largest African American publications in the country.

It was in Chicago that Cooper made his biggest mark on mass media. He had already done some work in radio, performing on a comedy show in Washington, D.C., and working as an announcer on occasion. Like many African Americans, he was frustrated by the lack of African American voices and viewpoints on the various radio stations in town. In 1929, an African American community group approached the owner of radio station WSBC and suggested that the station offer a weekly program aimed at African Americans and featuring an African American announcer. The station already offered various programs targeted at other ethnic and cultural groups; the black population boom in Chicago made a black radio program a risk worth taking for station owner Joseph Silverstein.

On November 3, 1929, *The All-Negro Hour* premiered on WSBC with Cooper as the host. The cost of the airtime was paid by advertisers that Cooper himself had to find. The show featured a combination of live musical acts, comedy, dramatic performances, and news. It was the first radio show produced entirely by African American staff and technicians, just as Cooper was the first African American to host a weekly radio show on a major station. His on-air personality was exactly the opposite of what many listeners had come to expect from African Americans on the radio. Rather than speaking with an exaggerated dialect usually associated with poor Southern blacks, Cooper spoke clearly and articulately, displaying class and intelligence at least equal to that of white radio announcers.

Although the signal transmitted by the station was weak, the show became very popular among the African American residents within its broadcast range. What began as a once-weekly broadcast soon expanded to ten hours of programming each week. Cooper hosted *The All-Negro Hour* until it ended in 1935, and went on to host other programs for WSBC. Cooper sometimes announced play-by-play accounts of local sports events. He also hosted and created shows for other stations like WHFC.

Who Was the First Radio Disc Jockey?

Live music was the only kind of music heard on the nation's airwaves when Jack L. Cooper first began *The All-Negro Hour.* Sometimes, broadcasters transmitted shows live from clubs or music halls, as with Duke Ellington and his band's performances at the Cotton Club. In other cases, musicians had to bring their instruments into the radio's studio and perform live there. Vinyl records existed at the time, but no one thought to play recorded music on a commercial radio station until the early 1930s. The first person to do so would likely hold the title as the very first disc jockey, sparking a trend that quickly dominated radio. But who was it?

According to some sources, it was none other than Cooper. During one of his live broadcasts in either 1931 or 1932, the pianist who worked on the show walked out after a salary dispute. Cooper reportedly covered for the lack of live music by hooking up a portable record player, which would make him the first to broadcast existing musical recordings on the radio. Other sources credit a Los Angeles radio announcer, Al Jarvis (1909–70), with being the first disc jockey in 1932. Jarvis developed a program that relied solely on records called *Make Believe Ballroom.* The debate remains unsettled, though some are willing to concede that both men share the title, with both Cooper and Jarvis pioneering the technique in different regions of the country, each without knowledge of the other.

Recognizing the potential of the radio market, he became a producer for other radio shows. He bought the airtime and sold it to others for different types of shows. Many of these early programs centered on religion or local community affairs.

In 1938, Cooper created the first radio show that would qualify as a public service program. The show, called *Search for Missing Persons,* was dedicated to using the power of radio to track down friends and family members that loved ones had been unable to locate. At the time, the booming African American population in Chicago was a reflection of the Great Migration. The Great Migration refers to the spread of the African American population from the South to urban centers throughout the northern and midwestern United States from the 1910s to the 1930s. Millions of African Americans left the South due to poor economic opportunities and harsh treatment by local whites. This scattering of such

a large population led to many people losing touch with friends and relatives. Thanks to his sizable audience within the African American community, Cooper's show was an invaluable tool in helping tens of thousands of people reconnect over the course of twelve years.

Cooper created the first African American–produced talk show, *Listen Chicago,* in 1946. The show focused specifically on issues considered important to African American listeners, as Cooper interviewed intellectuals and other notable figures about topics that were scarcely, if ever, covered by other media outlets. His many successes opened doors for other African American radio personalities, and even led to an entire radio station in Memphis devoted to African American programming. The giant strides made by African American professionals in the radio industry throughout the 1940s and 1950s all sprang from Cooper's conviction that a large, untapped segment of America's citizens had been waiting for someone to respond to their needs.

"BOJANGLES" SHARES THE BIG SCREEN WITH SHIRLEY TEMPLE

African American performers enjoyed increasing success in mainstream entertainment during the 1930s. They appeared on Broadway, on the radio, and in films. Most of these productions, however, were intended specifically for African American audiences. Black performers in major Hollywood films almost always appeared in minor roles as servants or slaves. Seldom were they allowed to demonstrate the full range of their talents. Bill "Bojangles" Robinson (1878–1949) was the first African American performer to share a stage equally with a white star in a major film when he co-starred with Shirley Temple in the 1935 movie *The Little Colonel.*

Robinson was raised by his grandmother after his parents died when he was seven years old. He began performing at an early age, and spent many years tap dancing on the vaudeville circuit with a partner named George W. Cooper. At the time, African American entertainers were not allowed to perform solo for white audiences; they had to follow what was known as the "two-colored" rule, meaning two black performers on stage at all times. Robinson was the first African American performer to break that rule in 1916, and caused great controversy in the process. Still, his phenomenal dancing abilities drew large audiences. He was especially known for his "stair dance," in which he performed a routine on a staircase, changing his rhythm with every step he took. Though he was often named as one of the greatest dancers in the world and starred in numerous stage shows, Robinson was in his fifties by the time he appeared in his first film.

By contrast, Shirley Temple (1928–) had become a superstar almost overnight. She had starred in short films since the age of three, but four films released in 1934 revealed her budding talent to audiences across the nation: *Stand Up and Cheer!, Baby Take a Bow, Little Miss Marker,* and *Bright Eyes.* In *Bright Eyes,* she performed the song "On the Good Ship Lollipop," which became a huge hit and was recognized as Temple's signature song. The following year, at the age of six, she received a special Academy Award for her unique contributions to cinema.

Bill "Bojangles" Robinson dancing with child star Shirley Temple in the 1935 movie *The Little Colonel.* *AP Images*

In 1935, Temple and Robinson starred together in two of her most successful films, *The Little Colonel* and *The Littlest Rebel.* Robinson played a butler in the first film, and a slave in the second. Robinson's characters were stereotypes, but his interactions with Temple were filled with genuine affection and charisma. In *The Little Colonel,* Robinson teaches Temple how to do his famous stair dance. This is the first notable instance of an interracial duo performing a dance routine together in a major film. At a time when segregation was still more common than not, audiences were uniformly delighted by the routine.

Robinson starred alongside Temple in two more films, *Rebecca of Sunnybrook Farm* and *Just Around the Corner* (both 1938). Although both would go on to star in several more films, neither would achieve greater fame than they did performing with each other. In a later interview, after working with many of Hollywood's biggest talents, Temple still named Robinson her all-time favorite co-star.

LENA HORNE WINS MAJOR FILM STUDIO CONTRACT

African American film roles were limited in the 1930s. A few films were made with all-black casts, and these revolved around music and dance performances. In mainstream films, African American women were nearly always portrayed as domestics, such as cooks or maids, and were rarely significant to the film's story. The first notable success for an African American actress was in 1940, when Hattie McDaniel (1895–1952) received an Academy Award for her performance as Mammy in the film *Gone with the Wind* (1939). The second major success came two years later, when singer Lena Horne (1917–2010) signed a landmark contract with a major film studio.

Horne had been raised mainly by her grandmother in a middle-class neighborhood of Brooklyn in New York City. Her mother was a stage performer, and occasionally took Horne with her when traveling the country on show tours. Despite discouragement from other members of her family, Horne followed her mother's dream of a career on the stage. From the start, she proved more successful than her mother, landing a position on the chorus at the famed Cotton Club in Harlem, and performing on Broadway soon after.

Horne became a featured performer at the Cafe Society Downtown in New York City in 1941. It was here that she first began interacting with prominent civil rights activists. This began a lifetime commitment to the cause of civil rights, and played an important role in her Hollywood career. Horne moved to Hollywood to perform in a club on the Sunset Strip at the

beginning of 1942. At the same time, Walter White (1893–1955), the executive secretary for the National Association for the Advancement of Colored People (NAACP), approached Hollywood film studio executives in an effort to change how African Americans were portrayed in movies. The vast majority of roles for African Americans at the time were bit parts as servants or slaves, and the characters were usually written with stereotyped or exaggerated qualities that fit with white views of black people. The studios also dealt with African American actors unfairly in their business practices; white performers were often kept under contract with a studio and paid a steady salary, while African American actors simply waited for work to appear, often working odd jobs to make ends meet.

One studio executive in particular seemed open to White's ideas. This was Louis B. Mayer (1884–1957), who ran Metro-Goldwyn-Mayer (MGM) Studios. Mayer saw that with an increasing number of African Americans interested in going out to the movies, he could appeal to both black and white audiences with the right star. White had seen Horne perform at the Cafe Society Downtown, and thought that she offered the perfect combination of looks and talent for mass appeal. He recommended Horne to Mayer.

Horne met with executives at MGM, and sang a song for Mayer in his office. He was impressed, and expressed interest in hiring her for film work. That night, Horne reportedly called her father, who was in Pittsburgh, and asked him to come out to Hollywood and help her deal with the movie executives. She was just twenty-four, and did not want to be persuaded into doing something she might later regret. The next day, her father met with studio executives and told them that his daughter would not play a maid, a cook, a jungle native, or any of the other stereotypical roles usually available for African American women. They assured him that they had different plans in mind.

Horne felt convinced that the studio would not hire her. She was proved wrong within two weeks when she received word that the studio wanted to sign her to a seven-year contract to star in films. The terms of the contract specifically stated that she would be given "legitimate roles" rather than the types of roles usually offered to African Americans. Horne agreed, and became the only African American signed to a studio contract at the time.

Her first role for MGM was as a club singer in *Panama Hattie,* an adaptation of a Cole Porter musical. Horne appeared in over a dozen more films for MGM during her contract. Most of these were musicals in which Horne performed musical numbers not tied directly to the main story of the film. This was done so that the studio could remove her scenes before

The First African American Actress in Hollywood

More than twenty-five years before Horne came to Hollywood, another African American actress became the first to sign a film contract and to be featured in major motion pictures. Her stage name was Madame Sul-Te-Wan (1873–1959); her real name was Nellie Conley, and she had grown up surrounded by stage performers in Louisville, Kentucky. She became an actress and performed with touring companies across the East Coast. She moved to California with her husband and three children in 1915. Her husband abandoned the family there, and, desperate for work, she showed up on the set of film director D. W. Griffith's first blockbuster, *The Birth of a Nation.* She boldly approached Griffith and asked for a job, and he gave her one. She was hired to clean dressing rooms at the studio, but Griffith used her to play several roles in the film.

Sul-Te-Wan continued on the company payroll and appeared in Griffith's next film, *Intolerance* (1916). This was the start of a career that spanned more than forty years. Madame Sul-Te-Wan appeared in over forty films, including *Uncle Tom's Cabin* (1927), *King Kong* (1933), *Sullivan's Travels* (1941), and *Mighty Joe Young* (1949). In most of her appearances, she was uncredited, and the roles consisted mainly of servants, convicts, and slaves. Still, Sul-Te-Wan was an African American pioneer in the art of film, and after her death she finally received recognition for her landmark contributions.

showing the films in the South, where some white audiences were not yet ready to accept an African American singer appearing prominently in a mainstream film.

Horne appeared in only one leading role during her contract: in the film adaptation of *Cabin in the Sky* (1943), which featured an all-black cast and was not widely distributed in the South. According to Horne and others, the studio could not figure out how to utilize her in roles that expanded beyond the old, stereotypical Hollywood offerings. Although her light complexion and "Caucasian" features conformed to white standards of beauty at the time, she was clearly African American. The studio considered an African American leading lady a financial risk.

Some other African American performers resented the fact that Horne, new to Hollywood, received such a contract while they struggled for years

OPPOSITE PAGE
Lena Horne broke into movies in a small singing role in the 1942 movie *Panama Hattie.* *John Kobal Foundation/ Hulton Archive/Getty Images*

to move beyond the shallow, stereotypical roles they were offered. Later in life, Horne herself would view the MGM contract as an attempt by the NAACP to integrate the film industry. Although she never achieved leading-lady status in the film world, she paved the way for later performers like Dorothy Dandridge (1922–65) and Cicely Tyson (1933–). She also became the first African American woman to successfully break the stereotypes that had limited black contributions to American film.

MARIAN ANDERSON SINGS AT THE LINCOLN MEMORIAL

Marian Anderson (1897–1993), a gifted African American singer of classical training and worldwide fame, did not set out to create a defining moment in the struggle for African American civil rights. She simply wanted to perform for an audience in Washington, D.C. The discrimination she encountered while attempting to find a place to host her concert in 1939, however, resulted in a high-profile struggle against segregation. Anderson eventually performed her concert at the Lincoln Memorial on Easter Sunday in 1939 before a crowd of thousands. The concert was an early and notable victory in the civil rights movement.

Success in Europe, Segregation in the United States

Anderson rose from humble beginnings that were in stark contrast to the refined, upper-class environments usually associated with opera performers. She was the oldest of three sisters in a poor Philadelphia family. Her singing talent appeared at a young age, though she did not receive formal training until she was fifteen. Members of her church choir raised the money to pay for her lessons.

In 1925, the New York Philharmonic held a singing competition. Anderson claimed the top prize out of the three hundred singers who entered. She soon discovered that classical singing opportunities in the United States were limited for an African American woman. Opera houses like the Metropolitan in New York City refused to hire blacks as performers, regardless of their talents.

This discrimination led her to Europe, where she became an immediate success and remained for five years. She returned to the United States in 1935 and toured the country performing recitals, enjoying success and critical praise. In 1939, representatives from the prestigious African American school Howard University booked Anderson to perform in Washington, D.C., on April 9, Easter Sunday. They attempted to reserve Constitution Hall, a venue owned by a group known as the Daughters of the American Revolution (DAR). The group's members are descended directly from the original participants of the American Revolutionary War.

The hall was managed for the DAR by Fred Hand, who refused to accept the Anderson booking. He claimed that another concert was already booked but also mentioned that the hall did not permit African American performers.

University officials attempted to book the hall on a different date, but it soon became clear that Hand and the DAR would not allow Anderson to perform in the hall at all. There were no other large venues available in the city, so Anderson's concert was without a home. University officials were unwilling to accept racism as a reason for refusing to allow Anderson to perform in the hall. They mounted a publicity campaign to make sure people knew about the hall's "whites only" policy. One person who was surprised at the policy was First Lady Eleanor Roosevelt (1884–1962), who was herself a member of the DAR. The concert organizers then tried to secure the use of an auditorium at one of the city's largest high schools. The school in question was for white students only, and the request was denied.

Overwhelming Support Leads to a Landmark Event

Public support of Anderson and the concert promoters grew by the day. The National Association for the Advancement of Colored People (NAACP) was instrumental in organizing support for the singer. A group of protestors picketed the Board of Education to protest the superintendent's decision. Roosevelt, who had been privately outraged over the matter, finally spoke out: She announced that she was resigning as a member of the DAR. Roosevelt's resignation made the story front-page news. In fact, Anderson herself did not know about Roosevelt's resignation until she read about it in the newspaper. Still, the DAR refused to bow to public pressure.

Harold L. Ickes (1874–1952), the United States secretary of the interior, joined Roosevelt in criticizing the DAR's decision. Ickes's government position put him in charge of the National Park Service, which oversaw public lands that included national monuments. The Lincoln Memorial was one such monument, located very near Constitution Hall and featuring a large expanse that could easily accommodate a concert audience. Ickes, with Roosevelt's support, granted Anderson permission to stage her Easter Sunday concert on the steps of the Lincoln Memorial. The monument would serve as a symbol of the fight against racial intolerance that Anderson and virtually every other African American struggled with on a daily basis. The concert would be free of charge for anyone who wished to attend.

On Easter Sunday, an estimated seventy-five thousand people gathered in front of the steps of the Lincoln Memorial to watch Anderson perform. Over three hundred special invitees attended the event, including members of Congress, the presidential cabinet, and even Hollywood stars like

Marian Anderson singing at an Easter concert at the Lincoln Memorial on April 1, 1939, after being denied use of a hall by the Daughters of the American Revolution because of her race. *Thomas D. McAvoy/ Time & Life Pictures/ Getty Images*

Katharine Hepburn (1907–2003). Boy scouts, both white and African American, distributed programs among the audience members. Anderson appeared with only a pianist to accompany her, and opened the concert with her rendition of "America" (1832), popularly known as "My Country, 'Tis of Thee." In the second and third lines of the song, "Sweet land of liberty, / Of thee I sing," Anderson changed the words to "*To* thee *we* sing." The change suggests the frustration that she and other African Americans felt as citizens of a country that did not recognize them as equals.

Anderson also performed two classical pieces and several African American spirituals. When she concluded, the crowd erupted with enthusiastic applause. In addition to the audience gathered at the Lincoln Memorial, millions of listeners across the country were able to enjoy the concert through radio broadcasts. The event was a milestone in the struggle for civil rights. It exposed a large segment of the American population to Anderson's talents and suggested substantial support among white Americans to end segregation policies. The event also encouraged African Americans who had previously avoided taking an activist role; membership in the NAACP doubled in the year following the concert. Four years later, the DAR finally invited Anderson to perform in Constitution Hall.

EBONY AND *JET* BEGIN PUBLICATION

Many large cities offered newspapers aimed at African American readers during the 1930s. Some of the largest included the *Chicago Defender*, the *Pittsburgh Courier*, and the *Afro-American* in Baltimore. There were very few national magazines targeted toward African Americans, however. *The Crisis* was successful, but it was closely tied to the National Association for the Advancement of Colored People (NAACP) rather than being an independent publication. This glaring gap in the magazine business changed dramatically and permanently in 1942, thanks to John Harold Johnson (1918–2005).

Johnson was born in Arkansas City, Arkansas, a tiny, segregated town with few opportunities for African Americans. His father died in a sawmill accident when Johnson was just eight years old, and his mother later

remarried. Johnson was an excellent student, but the only school for African American children in Arkansas City ended at grade eight; with no other options available, Johnson repeated the eighth grade just to have an opportunity to continue learning. Johnson's mother knew that her son could not achieve his true potential in the segregated South. She saved up her money and moved the family north to Chicago, Illinois, in 1933.

Johnson attended DuSable High School, and proved to be an exceptional student. He became the president of his class, as well as the editor of the school newspaper. He finished school in just three years, and was awarded a tuition scholarship to the University of Chicago. Harry Pace (1884–1943), a successful African American businessman in Chicago, heard Johnson speak at an Urban League luncheon, and offered him a job at Supreme Life Insurance Company. Johnson eventually became Pace's assistant. One of Johnson's monthly duties as Pace's assistant was to search through magazines and newspapers and gather together the most important news stories relevant to African Americans. After collecting and organizing the articles, he presented them to Pace as a sort of news digest focused on African American issues. It occurred to Johnson that many other African Americans would be interested in a similar news digest, and the idea for his first magazine was born.

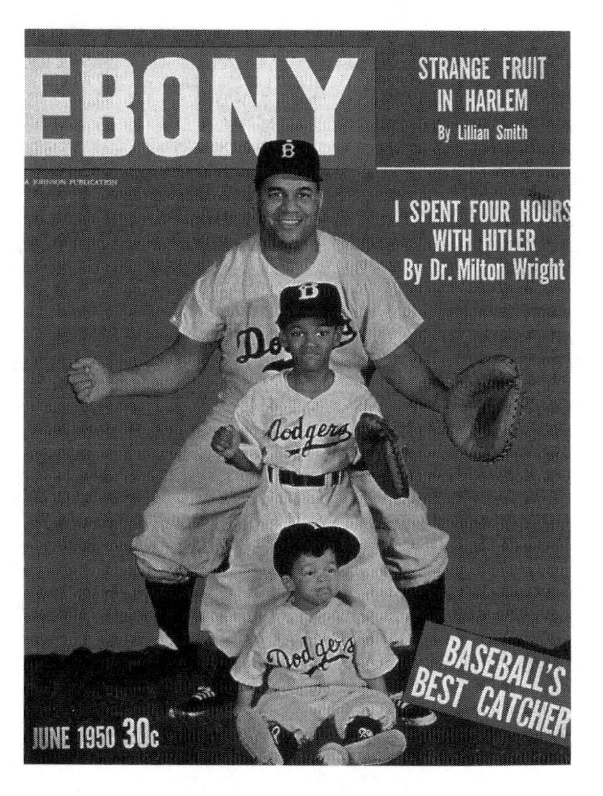

Cover of a 1950 *Ebony* magazine. *Ebony* was one of the first magazines to target African Americans. *Transcendental Graphics/ Getty Images*

Johnson had to use his mother's furniture as collateral for a loan to launch *Negro Digest*. This gave him enough money for the first issue, published in November 1942. The magazine was an immediate success, with monthly circulation reaching 50,000 in less than a year. This led him to consider ideas for additional magazines aimed at African American readers.

Noting the popularity of *Life* magazine, which featured photographs, interviews, and feature stories, Johnson decided to create his own version tailored to African Americans. The result was named *Ebony* after Johnson's wife pointed out that the ebony wood was renowned for its black color, its strength, and its beauty. The first issue of *Ebony* appeared on newsstands in November 1945, and quickly surpassed the success of *Negro Digest*. Johnson had difficulty convincing companies to purchase advertising space in the magazine at first. That situation changed quickly when executives realized the size

and potential buying power of the magazine's readership, with over four hundred thousand copies sold per issue by 1947.

The magazine's huge success prompted companies to tailor advertisements specifically for the magazine, featuring African American models in photo ads. The magazine's popularity also reopened American newsstands to African American publications, which had always been considered poor sellers and were previously ignored by many distributors. The success of *Ebony* helped African Americans to be viewed as mainstream consumers.

In 1951, Johnson phased out *Negro Digest* and launched another magazine, *Jet*, a digest-sized weekly news magazine. Although somewhat similar to *Negro Digest*, *Jet* did not rely on reprinted articles, and it also included more lifestyle content such as beauty and diet tips. Noting the shortage of African American–oriented magazines beyond his own, Johnson launched several more, including an entertainment news magazine and one aimed at children. By 1955, Johnson had created a magazine empire that sold over 2.5 million issues each month. As of 2009, *Ebony* and *Jet* remained the most popular African American magazines in the world, reaching a combined twenty million readers per issue.

THE NAT KING COLE SHOW DEBUTS ON TELEVISION

In the 1950s, African American entertainers were frequent guests on variety shows such as *The Jack Benny Program* and *The Ed Sullivan Show*. One of these frequent guests was popular African American singer Nat "King" Cole (1919–65). In 1956, he attempted to join the ranks of Benny and Sullivan with the creation of *The Nat King Cole Show*, a variety program that focused on performances by African American singers and musicians. It was the first time an African American hosted a major network variety show.

The talented singer and pianist was the perfect choice for this breakthrough for African Americans on television. He had achieved fame with his first hit song, "Straighten Up and Fly Right," in 1943. He performed jazz, but also had won over mainstream audiences with songs like "Mona Lisa" (1950) and "Unforgettable" (1951). He was one of the first African American musicians to achieve success in what was called popular or "pop" music. He appealed to both white and black audiences in an era when white audiences were leery of "race music," or music performed by African Americans.

Cole had already appeared on television many times as a performer, and his professionalism and charisma left little doubt about his ability to host a show. The National Broadcasting Company (NBC) gave Cole a

Singer Nat "King" Cole hosted his own variety show, *The Nat King Cole Show*, starting in 1956. *Michael Ochs Archives/Getty Images*

fifteen-minute slot on Monday nights. The first episode of *The Nat King Cole Show* aired on November 5, 1956. At the time, it was common for entertainment shows to only fill fifteen minutes of a network's prime-time schedule. On other nights of the week, performers like Dinah Shore (1916–94) and Jonathan Winters (1925–) occupied the same time slot as Cole.

The Nat King Cole Show featured regular performers like The Boataneers and the Randy Van Horne Singers, as well as special guests like Ella Fitzgerald (1917–96), Sammy Davis Jr. (1925–90), and Harry Belafonte (1927–). Cole also performed several songs on each show. Some episodes of the show were broadcast from different cities where Cole was performing, such as New York City, Miami, and Las Vegas. The show earned positive reviews in its early weeks. NBC moved it to a thirty-minute slot on the schedule eight months after it premiered.

The early success of the show did not attract corporate sponsors. The show only had one national sponsor, Carter Products, as well as regional sponsors in different areas of the country. Large corporations stayed away from *The Nat King Cole Show*. Corporate executives feared that whites might avoid their products if they advertised on a show hosted by an African American. NBC continued to air the program even without sponsorship, and Cole's guests agreed to appear for little or no money in order to save on production costs.

The show lasted for over a year, but it never achieved the ratings that Cole and the network wanted. Cole himself decided that it was time to end the show when programming executives told him that they planned to move the show to Saturday evenings, one of the least desirable time slots in the programming week. It would be more than a decade before an African American would host another network television show. Della Reese (1931–) tried to find mainstream success with the talk show *Della* in 1969, but the show did not find a wide audience. One year later, comedian Flip Wilson (1933–98) debuted *The Flip Wilson Show* (1970–74), which became a hit. The show won two Emmy Awards and was the country's second most popular television show in its first two years.

LOUIS E. LOMAX BECOMES THE FIRST AFRICAN AMERICAN NEWSCASTER

Television first achieved broad success in the United States in the years after World War II (1939–45). African American entertainers were an important part of television programming from the beginning. African American singers and actors appeared on variety shows and comedies as early as 1948, sometimes with great success and sometimes with great controversy. African Americans were not able to break into television in non-entertainment roles, however, until the late 1950s when Louis E. Lomax (1922–70) first appeared as a television journalist.

Lomax had already established himself as one of the most respected African American journalists of his time before making the transition to television. His original goal, however, was to become a teacher. He had risen from humble beginnings in Valdosta, Georgia, and attended college in Augusta, Georgia. From there he moved north and studied at both American University in Washington, D.C., and at Yale University, where he earned his doctorate. He spent some time teaching in the South, before deciding to pursue journalism.

Lomax worked as a reporter for the *Afro-American* in Baltimore and the *Chicago Defender,* two of the most popular African American papers in the

country. He reported on many civil rights issues in the South, including the 1955 murder of Emmett Till. Till was a fourteen-year-old African American boy who was beaten and murdered in Mississippi for allegedly flirting with a white woman. Lomax's stories about Till and other victims of racial violence called attention to the unacceptable conditions endured by African Americans throughout the South.

Lomax was hired as a television news reporter for WNTA in New York in 1958. He was the first black television news reporter. One of his fellow reporters was Mike Wallace (1918–), who later became famous as a correspondent on the CBS news show *60 Minutes.* One day, Lomax told Wallace about the Nation of Islam (NOI), its beliefs, and its prominent spokesmen, which included Malcolm X (1925–65) and Elijah Muhammad (1897–1975). Wallace was surprised to learn that the group, which called white people "devils" and advocated the complete separation of white and black society, was becoming popular across the country.

Working with Wallace, Lomax arranged for filmed interviews with key officials in the NOI, allowing them to explain their beliefs in their own words. These were the basis for a five-part documentary that aired in July 1959, hosted by Wallace, about the NOI. The documentary was titled *The Hate that Hate Produced,* and for millions of white viewers, was their first exposure to the teachings of the organization. Though Lomax had conducted the interviews, it was Wallace who framed the discussion with his own views about how the organization was potentially very dangerous. Lomax and Wallace knew that many would find the material shocking, and some speculated that exposing the group's extreme views would hurt its popularity. In fact, the documentary caused interest in the NOI to soar among African Americans, and the group grew substantially in the years following the documentary.

Lomax himself did not agree with the teachings of the NOI or its ideas about blacks forming a society separate from whites. Still, he and Malcolm X developed great respect for each other. They maintained a close relationship, and Lomax later wrote about the NOI and Black Muslim movement in his book *When the Word Is Given* (1963). He also wrote a book comparing the murders of Malcolm X and Martin Luther King Jr., titled *To Kill a Black Man* (1968).

Lomax began hosting a television show on KTTV in Los Angeles that ran for four years in 1964. He died in a tragic car accident in 1970, cutting his career short. Several other African Americans entered the field of television journalism after Lomax, including Mal Goode (1908–95), who became the first correspondent for a national news show in 1962. Bill Matney (1924–2001) became the second such correspondent in 1963. This

slow progress in the field of television journalism would lead, fifteen years later, to Max Robinson (1939–88) as the first African American to appear as a network news anchor.

SIDNEY POITIER WINS AN OSCAR FOR A LEAD ROLE

There were few opportunities for serious African American actors in the early days of Hollywood. Most film roles for black men were as butlers, convicts, or slaves. They were written to exaggerate negative stereotypes of African Americans. Acclaimed stage actor Paul Robeson (1898–1976) gave up on Hollywood in the early 1940s, frustrated by the stereotyped roles he was offered. It was not until the 1950s and 1960s that the domain of black actors in Hollywood films was redefined, largely due to the work of Sidney Poitier (1927–).

Poitier's parents lived in the Bahamas, and Poitier was raised there. He was born an American citizen because his mother gave birth to him while on one of the family's frequent trips to Miami, Florida. Poitier moved to the United States as a young man, and became a part of the American Negro Theater (ANT) in New York City. At a time when most roles for black performers involved singing and dancing, Poitier instead concentrated on dramatic acting. He landed a role in the Broadway revival of ANT's most successful production, *Anna Lucasta* (1944) in 1947. Three years later, he starred as the male lead in the film *No Way Out,* one of the first Hollywood films to feature a black actor in a leading dramatic role.

Poitier went on to star in several more leading roles. He starred opposite Tony Curtis (1925–) in the film *The Defiant Ones* in 1958. Poitier and Curtis, who played two fugitives shackled together as part of a chain gang, both received Academy Award nominations for Best Actor. This was the first time a black actor had been nominated for an Academy Award. (Performer James Baskett had received an honorary Academy Award in 1948 for his portrayal of Uncle Remus in the Disney animated film *Song of the South.*)

In 1963, Poitier starred in the film *Lilies of the Field* as an unemployed construction worker who ends up assisting a group of German nuns running a farm in Arizona. Unlike other films featuring a black man in a leading role, the film did not focus on racial conflict. Instead, the main theme was how people from radically different worlds can learn to respect and help each other. The film was a labor of love for director Ralph Nelson (1916–87), who was given such a small budget for the film that he could not pay his cast for rehearsal time. It was shot in less than two weeks outside of Tucson, Arizona, and premiered in November 1963.

Sidney Poitier with the Oscar he won for Best Actor in 1964. *AP Images*

The film received rave reviews, with Poitier's performance in particular earning praise. *Lilies of the Field* received five Academy Award nominations in 1964, including Poitier's second nomination as Best Actor. This time he won. Hattie McDaniel had been the first African American to win an Academy Award for her portrayal as Mammy in *Gone with the Wind,* but Poitier's win seemed to signal a change in how black performers were viewed in Hollywood. The roles available to Poitier were far from the traditional "black" roles like McDaniel's Mammy. His Academy Award win showed a growing acceptance of black performers as performers first and foremost, regardless of race.

Poitier continued as a leading man, starring in films like *Guess Who's Coming to Dinner* (1967) and *In the Heat of the Night* (1967). However, he was never again recognized with an Academy Award nomination for his film work. (He was given an honorary award for lifetime achievement in 2002.) Poitier himself noted that despite the strides he had made, it would mean nothing unless it opened the door for other black actors to achieve similar success. It would be six years before another African American actor earned a nomination for Best Actor, when James Earl Jones (1931–) was nominated for his role in *The Great White Hope* in 1970. It would be thirty-eight years until Denzel Washington (1954–) became the second black performer to win an Academy Award for Best Actor, earned in 2002 for his role in *Training Day*.

EDITORIAL FROM *THE CRISIS* (1918)

The NAACP publication *The Crisis* was one of the most important sources for African American perspectives on current events in the 1910s and 1920s. The editorial below, written by John Haynes Holmes, appeared in *The Crisis* in 1918. The United States was fighting in World War I (1914–18), and African Americans had contributed both their blood and their money to the war effort. Yet, they continued to experience racial discrimination in all areas of American society. The editorial expresses the frustration common among African Americans at the time, many of whom had fought in World War I to help defend freedoms that they themselves were not free to fully enjoy. Such an opinion from an African American would never have received publication in a mainstream newspaper or magazine at the time. Black newspapers and black publications like *The Crisis* remained the only print media forum for African American perspectives until mainstream newspapers began to cover the civil rights movement in the 1960s.

How precious the Negro is when society wants to use him!

How invaluable is his service when the cotton is waiting in the fields to be picked, rolled into bales, and transported to the world's markets! How indispensable is his loyalty, when the army is recruited for the great war to make "the world safe for democracy!" How welcome are his dollars, when a $5,000,000,000 Liberty Loan is floated by the government! Does anybody think of denying the black man the opportunity to do the work that nobody else will do? Has anybody urged that the black man be exempted from military service? Has any black man laid down his fifty dollars in a Liberty Loan booth, and been refused a bond? Just to suggest such possibilities is to reveal their **inherent** absurdity. In these and countless other directions, the Negro is usable, as a shovel is usable to dig a ditch or a truck to carry a burden; and society pays tribute to his worth.

But what happens when the Negro asks for **reciprocity** in this matter of service—seeks as a return for duties done the free exercise of privileges **conferred?**

What if he wants to use the public schools, the public libraries, the public parks, the theatres, hotels and railroads, public institutions and utilities generally, on an equal footing with other men? What if he buys real estate and builds a home in a neighborhood which will provide the best possible conditions for the rearing of his children? What if he enters not a Liberty Loan booth but a voting booth and seeks not to purchase a bond but deposit a ballot? This is different, is it not? The Negro is

Inherent
Contained within as an integral part

Reciprocity
A state of fair exchange between two parties

Conferred
Given as a gift

suddenly not so precious as before. On the contrary, he is worthless; or, worse than worthless, dangerous, and therefore not merely to be cast aside but **trodden** under foot.

Trodden
Stepped on or walked on

The situation is interesting, and, if the whites be wise, to be enjoyed while it lasts. For not always will the Negro, for all his patience and good cheer, be willing to recognize the **validity** of this onesided compact. Some day he will learn that rights are the complements of duties, and freedom the reward of service—at least in a democracy! And then will he insist upon knowing whether or not "a nation conceived in liberty and dedicated to the proposition that all men are created equal," has perished from the earth?

Validity
The quality of being fair, truthful, or legitimate

MOVIE REVIEW OF *THE LITTLE COLONEL* (1935)

African American actors struggled to get good roles in Hollywood from the beginning of the movie industry in the 1910s. They almost always had small parts playing slaves, servants, or criminals. African Americans began to see small improvements in the 1930s, beginning with the 1935 movie *The Little Colonel*. The movie starred white child star Shirley Temple and featured the aging African American dancer Bill "Bojangles" Robinson as the family butler. Though Robinson still played a servant, the role was far more substantial than most available to African American actors. The most famous scene of the movie shows Robinson demonstrating his famous "stair dance" to Temple, which was the first time an African American actor shared the screen equally with a white actor.

The significance of Robinson's role in the movie can be seen in the following review from *Variety*, a long-running publication containing entertainment industry news and features. The review highlights Robinson's contribution to the film even before mentioning the performances of veteran white actors such as Lionel Barrymore. His standout performance did not support greater variety in roles for African Americans, however. Four years after *The Little Colonel*, Hattie McDaniel would become the first black actress to win an Academy Award—but still while playing a servant to a white family.

• •

Cute tots have traditionally been tough subjects to fit with stories. **Little Colonel** is skillful **hokum** that will please in general, although the sophisticated minority may make a point of being superior to such sentimentality. Widely read book gives the film a head start, too.

Hokum
Mindless or unsophisticated entertainment

A southern colonel (Lionel Barrymore) is **embittered** when his daughter (Evelyn Venable) elopes with a northerner (John Lodge) and banishes them from the arc of his

Embittered
Containing feelings of bitterness

benevolence. Things go badly with the couple and the wife is forced to return with her small daughter (Shirley Temple) to occupy a cottage near her father's estate.

Benevolence
Generosity, goodness, or kindness

Of course, the child is the means of patching everything up, finale taking the form of a "pink party" given by the grandfather and photographed in Technicolor [by William V. Skall]. It's a gingerbread fade-out for a film loaded with sweetness and light.

Bill Robinson, vet colored hoofer from vaudeville, grabs standout attention here. Voice is excellent, he reads lines with the best of 'em, and his hoofing stair dance is ingeniously woven into the yarn. He plays the kindly and aging family butler. A strong point for the film is the youngster doing Robinson's stair dance with him.

Dressed in the bustled costumes of the 1880s, the **diminutive** miss is a fetching, beautiful and amiable infant. Her appeal is certain here and her acting range remains surprising.

Diminutive
Small in size or stature

Barrymore plays the colonel with scarcely one of his usual mannerisms, and with a zippy tempo, in contrast to the sidled-down technique he so often employs. Outside of the principals the other parts are incidental.

GORDON PARKS ON HIS YEARS AS A FEDERAL PHOTOGRAPHER (1964)

Gordon Parks is best known as the first African American staff photographer for *Life* magazine in 1948. Prior to this historic position, he worked for a federal government project to document the American experience in photos as part of the Historical Section of the Farm Security Administration (FSA). His photos highlighted the lives of the lower classes, with a focus on farmers. The job required him to travel across the country and photograph the living conditions and daily struggles of the poor.

Parks discussed his memories of working as part of the FSA with Richard Doud of the Smithsonian Institution in 1964. In the following excerpt, Parks talks about the significance of the work, as well as his own experiences, both good and bad, with those he encountered in his travels.

• •

Richard Doud: You may have come along too late in the game to really give me much idea on this but when this Historical Section was first established and the Information Division of what was then, I think, Rural Resettlement, it was primarily to keep a record of what the Resettlement Administration was doing and to take some pictures to show the public that, you know, it was helping the distressed areas and that sort of thing. But there was no real end goal as far as doing a survey of America or a picture of rural America or this sort of thing. What do you think were the main factors in this thing sort of ballooning the way it did and coming up with this tremendous file of photographs you'd never expect from a government agency? Why did this thing get so far out of government hands, so to speak, and become a really personal project?

Gordon Parks: I just really couldn't say because I wasn't there at that period. I can't be much help. But I can only surmise that the pictures themselves were so powerful and in touch with reality and to such an extent that they were their own power and they grew out of themselves. You know? Such pictures as they were making at that time anyone with any feeling about the sufferings of people were bound to look and notice and try in some way to absorb the message that they were preaching, and take it for them. I'd say that these things sort of grew out of themselves, their own strength. These men were all, I think, dedicated to enlightening the rest of the nation in a sense. Whether they knew it or not, it was that taking place. At least I ... the type of men that I've met there and have known since. They just had a marvelous chance to do it. I don't know at times it looked like an **indictment** of the government itself. But....

Indictment
An accusation of wrongdoing or a statement casting blame

Doud: Congress thought so, too, apparently.

Parks: But I think it served a great purpose. No country should be afraid to face its weaknesses.

Doud: I guess countries are like people in that respect. They shy away from their own shortcomings. Well what ... could you give me some idea of your more interesting assignments or the more memorable experiences in your time with the FSA or ... would that be hard to answer?

Parks: Well, I.... The whole thing was memorable to me. I ... it's very difficult to pick out any special projects or assignments that meant more than the others. I think the ... as I say before, of course, the great value that I received out of it was a great **humanitarian** feeling, brotherhood and so forth and so on. I do remember going once to Springfield, Massachusetts and going into a restaurant and that particular day there had been a lynching and I was, you know, just a little angry and I saw at the counter, sitting by the only stool left, a man who I knew would be a southerner who had the perforated shoes, the little straw hat and pinched lips and a mean look and I said, "Well, this is one day you're going to get it if you stop." I sat down by this guy and was ready to belt him if he said just one word and he turned around to me and said, "It's a nice day." And, "You're the nicest person I've seen in this place," he said. "Nobody talks to you in Springfield, you know." And so it was quite a shock to me and that is one favorable thing that caught me and then there were times when I was travelling in nights of sub-zero weather. I was up in New York state someplace and I went to a hotel, bitterly cold, twenty below zero or something like that. The guy refused me a room, just wouldn't talk with me and a young white boy came in with a **mackinaw** on and a guitar in his arm. He was standing behind me so the guy says, "Well, what do you want?" and he says, "Well, I'm after this man." And the hotel man said, "Well, we don't allow Negroes or Jews in this hotel, the management doesn't." And the other boy says, "Oh, if it's not good enough for him, it's not good enough for me." And he turned and walked back out in the cold. Well, he never stopped to apologize to me for this man, for his race or anything of this sort. It was the only hotel in town and I figured he

Humanitarian
Expressing kindness and sympathy toward other people

Mackinaw
A thick wool coat

must have gone somewhere but I thought the act was something I could never forget and I think I appreciated it even more because he didn't stop to talk. He showed me he believed in what he ... he didn't want to quarrel about it and off he went back into the cold and that was something that you don't forget, a thing like this. I often would like to have known who this guy was. A lot of people say that they thought maybe it was Pete Seeger, but I didn't know him but he was tall and he had a guitar under his arm. Anyhow, I just don't want to know who he was; to me he's better unknown.

Research and Activity Ideas

1. Many early films and radio shows reflect the times in which they were created. This means that early portrayals of African American characters might be considered offensive and racist by modern standards. Many works that have suffered from such controversy, such as the Disney film *Song of the South* (1946) and the film adaptation of *Porgy and Bess* (1959), are not available to modern audiences because they might be viewed as offensive. In your opinion, should older films and shows with potentially offensive portrayals be suppressed? Why or why not?

2. African American newspapers appeared in the late nineteenth century because mainstream papers did not adequately or fairly cover stories that interested African Americans. Using your library, the Internet, or other available resources, find an article from a major African American newspaper of the early twentieth century. The article should be about a significant news story, like Marian Anderson's 1939 concert at the Lincoln Memorial or the 1963 March on Washington. Then find an article covering the same event from a major mainstream newspaper, such as the *New York Times*. In a short report, compare and contrast the two news stories. Do they emphasize different facts, or does one leave out important details? Do the writers approach the subject with different attitudes?

3. Roles available for African American performers in the 1930s were limited to a handful of "types," such as the servant, the butler, or the slave. Compare the roles of African Americans in the 1930s to roles of other minorities in modern media. Are people of Middle Eastern or Asian descent limited to a handful of "types" as African Americans once were? Be sure to provide specific examples from modern media such as television shows and films.

4. Creations like Jack L. Cooper's radio show *The All-Negro Hour* and John Harold Johnson's magazine *Ebony* were designed to meet the needs of a specific audience. Working in a small group with other students, create an issue of a newspaper or a five-minute radio show that focuses on the news and views of a particular group in your area, such as fans of a local sports team or people who shop at a specific store. Present it for other students in your class.

5. Popular performers are often given the opportunity to influence others by speaking out on issues like racism. Ossie Davis and Ruby Dee, for example, were known for their vocal activism and support of the civil rights movement. Sometimes they can help bring about

positive change; in other instances, as with Paul Robeson and Lena Horne in the 1950s, they find themselves shut out by the media because of their opinions. Some argue that performers should not use their fame to influence audiences regarding issues that are outside their fields of expertise, like politics. Performers like Marian Anderson, for example, go out of their way to avoid controversy and focus purely on their work. In your opinion, should artists and performers use the media to speak out about issues important to them? Why or why not? Write a short essay supporting your view.

For More Information

BOOKS

Bogle, Donald. *Bright Boulevards, Bold Dreams: The Story of Black Hollywood.* New York: One World Books, 2006.

Bogle, Donald. *Toms, Coons, Mulattoes, Mammies, and Bucks: An Interpretive History of Blacks in American Films.* New York: Continuum International Publishing Group, 2001.

Dandridge, Dorothy, and Earl Conrad. *Everything and Nothing: The Dorothy Dandridge Tragedy.* New York: Perennial, 2000.

Davis, Ossie, and Ruby Dee. *With Ossie and Ruby: In This Life Together.* New York: Perennial, 2004.

Dingle, Derek T. *Black Enterprise Titans of the B.E. 100s: Black CEOs Who Redefined and Conquered American Business.* New York: John Wiley and Sons, 1999.

Fearn-Banks, Kathleen. *Historical Dictionary of African-American Television.* Lanham, MD: The Scarecrow Press, 2006.

Freedman, Russell. *The Voice that Challenged a Nation: Marian Anderson and the Struggle for Equal Rights.* New York: Clarion Books, 2004.

Gavin, James. *Stormy Weather: The Life of Lena Horne.* New York: Simon and Schuster, 2009.

Green, Adam. *Selling the Race: Culture, Community, and Black Chicago, 1940–1955.* Chicago: University of Chicago Press, 2007.

Green, J. Ronald. *With a Crooked Stick: The Films of Oscar Micheaux.* Bloomington: Indiana University Press, 2004.

Keiler, Allan. *Marian Anderson: A Singer's Journey.* New York: Scribner, 2000.

Lawrence, A. H. *Duke Ellington and His World.* New York: Routledge, 2001.

Mills, Earl. *Dorothy Dandridge: An Intimate Biography.* Los Angeles: Holloway House Publishing, 1999.

Parks, Gordon. *A Hungry Heart: A Memoir.* New York: Simon and Schuster, 2005.

Parks, Gordon. *Voices in the Mirror: An Autobiography.* New York: Harlem Moon, 2005.

Ruuth, Marianne. *Nat King Cole.* Los Angeles: Holloway House Publishing, 1992.

Vaillant, Derek. *Sounds of Reform: Progressivism and Music in Chicago, 1873–1935.* Chapel Hill: University of North Carolina Press, 2003.

Watts, Jill. *Hattie McDaniel: Black Ambition, White Hollywood.* New York: HarperCollins, 2005.

Williams, Oscar Renal. *George S. Schuyler: Portrait of a Black Conservative.* Knoxville: University of Tennessee Press, 2007.

PERIODICALS

Clayton, Edward T. "The Tragedy of Amos 'N' Andy." *Ebony,* vol. 16, no. 12 (October 1961): 66–73.

Robinson, Louie. "Dorothy Dandridge: Hollywood's Tragic Enigma." *Ebony,* vol. 21, no. 5 (March 1966): 71–82.

"Shirley Temple Recalls that Bias Experienced by 'Bojangles' Robinson Taught Her About Racism." *Jet,* vol. 95, no. 4 (December 21, 1998): 37.

WEB SITES

Schechter, Patricia A., Ph.D. "Biography of Ida B. Wells." Illinois Historical Digitization Projects, Northern Illinois University Libraries. http://dig.lib.niu.edu/gildedage/idabwells/biography.html (accessed on December 30, 2009).

Tennessee History Classroom. "Ida B. Wells." http://www.tennesseehistory.com/class/Ida.htm (accessed on December 4, 2009).

chapter five Demographics

Chronology

1847 **July 26** Joseph Jenkins Roberts declares Liberia an independent republic as opposed to a colony of the United States. Less than three months later he becomes Liberia's first president.

1862 Congress passes the Homestead Act, which provides people with 160 acres of free land in the American West. The act, which applies to African Americans, sets the stage for the so-called "Colored Exodus" of the late 1870s and early 1880s.

1877 The two-year migration of some fifty thousand African Americans from the American South to Kansas begins. These people, called "Exodusters," are in search of a new life, free from the oppression of the South after the end of Reconstruction.

1887 The town of Mound Bayou, Mississippi, is founded and becomes one of the nation's first all-black towns.

1891 Henry McNeal Turner, bishop of the African Methodist Episcopal Church, makes four visits to the west coast of Africa. Turner, a supporter of church mission work in Africa, goes on to support the so-called "Back to Africa" movement, encouraging African Americans to settle in Africa.

1910 Ninety percent of African Americans live in the South, 80 percent of which live in rural areas.

1914 Marcus Garvey founds the Universal Negro Improvement Association (UNIA), an organization that leads the struggle for independence for African descendents all over the world.

1915 Carter G. Woodson helps found the Association for the Study of Negro Life and History (ASNLH), which is dedicated to supporting the study of African American history, culture, and heritage.

1917 The number of African Americans moving out of the South as part of the Great Migration reaches a high point that continues over a three-year period. Historians estimate that somewhere between seven hundred thousand and one million African Americans move out of the South and into cities in the North, West, and Midwest during this time.

1918 Alain Leroy Locke becomes the first African American ever to receive a doctoral degree in philosophy from Harvard University in Cambridge, Massachusetts.

1920 The number of Caribbean-born persons living in the United States rises to almost 580,000 from 90,000 in 1880.

1920 Marcus Garvey, as part of his Back to Africa project, moves the headquarters of the Black Star shipping line to Liberia.

1922 Charles S. Johnson publishes *The Negro in Chicago: A Study in Race Relations and a Race Riot,* his landmark seven-hundred-page study that demonstrates that racial discrimination in housing and employment contributed to the Chicago race riots of 1919.

1922 Carter G. Woodson publishes *The Negro in Our History,* a landmark college textbook that examines patterns of migration, family composition, community structure, and other demographic characteristics of the African American community.

1925 Alain Leroy Locke publishes *The New Negro,* a collection of pictures, essays, stories, and poems by African American artists and authors that generates national interest in the ongoing Harlem Renaissance.

1927 June Angry white Southerners force Edward Franklin Frazier to leave Atlanta, Georgia, after he publishes an article arguing that racism is a form of paranoid, delusional mental illness.

1930 The United States Census shows that African Americans represent only 9.7 percent of the population. This is lowest percentage recorded by the national census, down from a high of 19.3 percent in 1790.

1939 Edward Franklin Frazier publishes *The Negro Family in the United States,* a landmark sociological study that exposes the influence of history, society, culture, and racism on the inner workings of African American families.

1940 The United States is home to more than seventy all-black towns with populations of ten thousand or more.

1940 The Second Great Migration begins. Over the course of a ten-year period more than 1.5 million African Americans leave the South for cities in the North, Midwest, and West.

1943 More than six thousand African Americans are arriving in Los Angeles every month as part of the Second Great Migration.

1948 The United States Supreme Court decides *Shelley v. Kraemer,* a landmark case in which the Court holds that racially restrictive housing agreements are not legally enforceable.

1950 United States Census statistics show Washington, D.C., has a population that is 35 percent African American. By the time the census is conducted again in 1960, the percentage has risen to 53 percent, making it the first major U.S. city with a majority black population.

1957 August 13 A postal worker alerts the suburb of Levittown, Pennsylvania, that a black family has moved into the neighborhood. This sparks a wave of violence and terrorism against the

family as white residents attempt to force them to leave the neighborhood.

1960 Forty percent of all African Americans live outside the South. Seventy-five percent live in cities.

1960 African Americans make up 28.9 percent of the population of Detroit, up from 1.4 percent in 1900.

1964 In California, Proposition 14 reverses fair housing rules prohibiting racial discrimination enacted by California state and local governments. The proposition is declared unconstitutional by the California Supreme Court in 1966.

Overview

The period from 1865 to 1965 saw a number of significant demographic shifts relating to the African American population in the United States. Many of these shifts involved large-scale migrations. One of the first migrations after the Civil War (1861–65) was known as the "Colored Exodus." It spanned the years from the late 1870s to the early 1880s. The peak year was 1879. The Colored Exodus was led by a former slave named Benjamin "Pap" Singleton (1809–92). More than fifty thousand African Americans moved from the South to Kansas. They were known as "Exodusters." Singleton's vision was that African Americans could create their own communities and economic opportunities, away from the racially oppressive culture of the South. The timing of the exodus was significant. It began the same year that the United States military withdrew from the South, marking the end of the Reconstruction era (1865–77). This meant that African Americans lost the physical protection of the military as well as many of the legal protections they had gained immediately following the Civil War.

A major demographic trend that began in the 1880s and continued into the early 1900s was the forming of more than one hundred all-black towns. These towns were formed in response to widespread racism in the South. They were founded throughout the South, but eventually the territory of Oklahoma became home to the most all-black towns. By 1940, there were more than seventy all-black towns with ten thousand residents or more.

Another demographic trend that lasted from the early 1800s into the 1900s was the so-called "Back to Africa" movement. The back to Africa idea was that free slaves should find a new home in Africa, the homeland of their ancestors. The most famous organization responsible for the back to Africa movement was the American Colonization Society (ACS). The ACS was an unlikely alliance of abolitionists, slave owners in the South who did not want free slaves living in the South, and African Americans who felt that they could not make a home in the United States. The ACS was successful in creating colonies in western Africa in the 1800s, including the settlement that would eventually become the independent nation of Liberia. Liberia's first president was a free black from the United States named Joseph Jenkins Roberts (1809–76).

The Back to Africa movement did not end in 1920. It was kept alive by a man named Marcus Garvey (1887–1940), who founded a group called the Universal Negro Improvement Association (UNIA). The purpose of the UNIA was to unite people of African descent all over the world. Garvey's

vision was that African peoples—whether in the United States, Africa, or the Caribbean—could work together to form a collective community and government. The organization strived to create economic opportunities for blacks that they could not find in white society. The group also championed emigration, or a permanent move by people out of a country, to Africa. While widespread emigration to Africa never occurred, the idea of blacks creating communities where they could govern themselves and create their own opportunities was influential and long-lasting.

A final demographic trend that spanned the nineteenth and twentieth centuries was a wave of immigration into the United States from the nations and islands of the Caribbean. This immigration was the result of several factors. The United States had increased its economic and military involvement in the Caribbean in the late 1800s. In addition, many Caribbean residents lived in poverty. Increased cultural interdependence and the lure of better economic prospects set the stage for immigration. From 1880 to 1920, the number of Caribbean-born persons living in the United States increased from 90,000 to 580,000. Many of these immigrants were of African descent. As a result, Caribbean immigration significantly reshaped U.S. demographics.

A major demographic shift known as the Great Migration took place in the early part of the twentieth century. The Great Migration was a large-scale movement of African Americans out of the South and into cities in the North, West, and Midwest. Nearly two million African Americans left the South between 1910 and 1930. Racial discrimination and limited economic opportunities created a desire to go elsewhere. The promise of new jobs in the country's new industrial centers lured African Americans to cities such as New York, Chicago, and Los Angeles. The Great Migration produced massive changes in American demographics. Most importantly, the number of African Americans living outside the South skyrocketed.

A similar demographic shift, prompted by the outbreak of World War II (1939–45), took place during the 1940s. It came to be known as the Second Great Migration. Once again, large numbers of African Americans left the South and moved to cities in other parts of the country. This time, they were drawn by the promise of industrial jobs created as part of the war effort. More than 1.5 million African Americans left the South in the 1940s alone. From 1940 to 1970, the total number of African American emigrants out of the South approached five million.

After the end of World War II, increasing numbers of Americans moved from city centers to suburban communities. The racial integration of these communities proved controversial. White and black families famously clashed in the suburban town of Levittown, Pennsylvania.

"Levittowns" (there was also a Levittown in New York) were suburban communities built by the company Levitt & Sons. Levitt & Sons was one of the first companies to perfect the quick, cheap, and efficient mass-manufacture of houses to meet the growing demand of soldiers returning from the war who sought housing for their growing families. In 1957, a crisis broke out in Levittown, Pennsylvania, when an African American family moved into the neighborhood. Unable to legally evict them from the neighborhood, angry white mobs turned to violence and acts of terrorism to try to force them to move. State authorities intervened to protect the family, but they decided to move away rather than face such harassment from their neighbors. Sadly, such events were common as the suburbs started to become integrated all over the country.

Headline Makers

EDWARD FRANKLIN FRAZIER (1894–1962)

Edward Franklin Frazier was a sociologist and professor who made important contributions to the study of African American culture and families. Frazier conducted pioneering studies that disproved stereotypes about African Americans that were common in the early twentieth century. He argued forcefully that the problems facing African American families were the result of racism and social, cultural, and economic forces. Frazier also published important books documenting the distinctive character and structure of African American families. He was a prolific writer, which means he wrote very frequently. Frazier was one of the leading African American academics of the first half of the twentieth century.

Edward Franklin Frazier was born on September 24, 1894, in Baltimore, Maryland, into difficult circumstances. His grandfather had been a slave, and his father never attended school. His father taught himself to read and write, and he encouraged Frazier to work hard in school. His father died when Frazier was eleven years old, but Frazier continued to heed his advice. He helped his mother support the family by working before school as a paper boy and after school at a grocery store.

Edward Franklin Frazier.
AP Images

Even so, he was such a good student that when he graduated from the Colored High School in Baltimore in 1912, he received a full scholarship to Howard University in nearby Washington, D.C.

Frazier was such a serious, talented student that his classmates at Howard nicknamed him "Plato," after the famous ancient Greek philosopher. Frazier worked hard on his studies and participated in extracurricular activities. He graduated *cum laude* (with honors) in 1916. He spent the next three years working as a teacher. In 1919, he resumed his education at Clark University in Worcester, Massachusetts. He received a master's degree in sociology in 1920. Sociology is the study of society, social institutions, and social relationships. Frazier viewed sociology as a way to help understand and solve the problem of racism in the United States.

A Talented Scholar Emerges

Frazier blossomed as a scholar and a writer over the next several years. He studied and taught in New York City and in Copenhagen, Denmark. In 1922, Frazier and his new wife, Marie Brown, moved to Atlanta, Georgia, so Frazier could accept a dual job as a professor of social work at Atlanta University and a professor of sociology at Morehouse College. He authored thirty-three academic articles during his time in Atlanta. The most famous of those articles was entitled "The Pathology of Race Prejudice," published in 1927. In it, Frazier argued that racism is a form of paranoid, delusional mental illness. The article so enraged white Southerners that Frazier and his wife were threatened with lynching (a lawless murder of an individual by a mob).

Frazier had begun work on a Ph.D. at the University of Chicago in 1923. A Ph.D. is the highest degree a student can earn at an American university. In June 1927, Frazier decided to leave Atlanta and move with his wife to Chicago so he could fully devote himself to his Ph.D. studies. The University of Chicago had an outstanding sociology department. Frazier immersed himself in his work and began to develop some of his central ideas. For example, he published an article in 1928 challenging common white stereotypes of African Americans. Frazier pointed to the significant differences between and among African Americans along lines such as geography, wealth, family, property, education, vocation, culture, and even language. How could a stereotype about all African Americans be true, Frazier asked, when African Americans were so different? This analysis of the demographic diversity of the African American community was a major scholarly insight.

Frazier also began to develop a sociological account of racism. In the early 1900s, there were some white professors and scholars who believed that science could prove that African Americans were inferior to white

Americans. As evidence, they often pointed to the fact that most African Americans lived in poverty. They also argued that African American families suffered from more problems than white families, which they said was evidence of natural inferiority. Frazier's academic work demonstrated that these so-called "scientific" studies were founded on racist thinking.

Produces Pioneering Work on Black Demographics

Frazier completed his Ph.D. at Chicago in 1931. His dissertation (a book-length work of original scholarship that is the final requirement for earning a Ph.D.) was titled *The Black Family in Chicago.* It was another attack on scientific justifications for racism. In it, Frazier analyzed the different ways that black families were organized and structured in different parts of the city of Chicago. Frazier built on this analysis in his famous 1939 book, *The Negro Family in the United States.* He argued that black family life had been influenced over time by a wide variety of sociological forces, including slavery, Reconstruction, urbanization, and a distinctively matriarchal (focused and centered on the mother) culture. Frazier's studies acknowledged that many black families faced serious problems, but he argued that these problems were the result of economic hardship and the socially damaging effects of racism, not some inborn inferiority. Frazier's study of the black family was a critical advance in academic demographics.

Frazier's final major contribution to the demographic analysis of African Americans in the United States was his analysis of the black middle class. He argued that the legacy of slavery and the continuing problems of racism and segregation had a distorting effect not only on African Americans' place within American culture, but also on the internal workings of African American culture itself. He identified a distinctive "black middle class" whose differences in behavior, values, and attitudes could largely be explained by the effects of racism.

Frazier enjoyed a long and successful career. He joined the faculty at Howard University in 1934 and remained there until his death in 1962. He authored eight books and more than one hundred academic articles. He dedicated his life and his work to the cause of debunking stereotypes about African Americans and advancing the cause of racial equality. His ideas remain influential to this day.

★ MARCUS GARVEY (1887–1940)

Marcus Garvey was a leader dedicated to the improvement of conditions for African descendents all over the world. He focused his efforts primarily in the Caribbean, the United States, and Africa. He

founded the Universal Negro Improvement Association (UNIA), an organization that would prove important for spreading Garvey's message of racial uplift and independence for African Americans. Garvey's teachings would be an inspiration to the black power movement in the United States and to the decolonization movement in Africa, both in the latter part of the twentieth century.

Marcus Garvey.
MPI/Getty Images

Early Life in Jamaica

Marcus Garvey was born in 1887 in St. Ann's Bay, Jamaica. His parents instilled in him a love of learning and reading. He attended elementary school and did well there. He became an apprentice to his godfather's printing business rather than attend secondary school. An apprentice is someone who works for someone with skill in a trade, typically working for free in exchange for learning a new skill. He learned the printing business, and moved to the much larger city of Kingston, Jamaica, at the age of sixteen. There, he was the youngest printer foreman in Jamaica. He also became active in politics and current issues.

Garvey left Kingston in 1910 and traveled the world. He moved first to Costa Rica, along with many other Caribbean natives seeking work in Latin America. He traveled all over Latin America for about four years. His travels, combined with his own experience growing up in the Caribbean, made him aware of the struggles and hardships suffered by people of African descent. (The Caribbean, especially Jamaica, was a major hub of the trans-Atlantic slave trade.) Garvey next moved to England. There, he studied law and worked for a journal. During this time, he more seriously began to ponder the situation of people of African descent throughout the world.

Garvey's commitment to helping people of African descent prompted him to return home to Jamaica in 1914. There, he founded the Universal Negro Improvement Association (UNIA) and African Communities League. The African Communities League was the governing body for the UNIA. The purpose of the organization was to provide opportunity for blacks all over the world. Garvey corresponded with famous African American educator Booker T. Washington (1856–1915), and he adopted Washington's

model for trade schools. He hoped that blacks would learn skills and trades, and provide humanitarian help and missionary assistance for "Mother Africa." Garvey did not find all of the support he needed for the UNIA in Jamaica, so he decided to go to the United States on the invitation of Booker T. Washington.

Brings UNIA to the United States

Garvey arrived in the United States in 1916 at the age of 28. He moved to New York City, where he found the spirit of the time was the right environment in which to grow the UNIA. It was in 1916 that the so-called "New Negro" movement began in the Harlem neighborhood of New York City. The New Negro movement was closely connected with the Harlem Renaissance, a period of great cultural and artistic achievement for African Americans. African Americans were becoming frustrated with the lack of equality and racial progress at home, despite the supposed commitment of the United States to the aims of equality and democracy that were fueling support for World War I (1914–18).

The UNIA grew in the years after Garvey came to the United States. Garvey was committed to a vision of African nationalism. This meant that he wanted blacks all over the world to have greater political self-determination and independence. Africa figured prominently in this equation since it was the origin of most blacks' ancestors. There was a growing feeling that political self-determination would not be possible in the United States. Garvey became interested in resettling thousands of blacks from the United States and the Caribbean to Africa.

Garvey soon turned his attentions to creating businesses and business opportunities for African Americans. The most famous of these was a shipping company called the Black Star Line. He hoped that the Black Star Line would allow blacks to find wealth in the international shipping trade. He sold shares of stock to raise the money to buy ships for the business. Garvey hoped the Black Star Line would further contribute to his resettlement dreams. He moved the headquarters for the Black Star Line to the African country of Liberia in 1920.

Troubles Later in Life and Garvey's Legacy

Garvey's Black Star Line was plagued by financial troubles. These were only magnified by the fact that the United States government suspected the Black Star Line of illegal activities. This was part of the reason that Garvey decided to move the Black Star Line to Liberia in 1920. Garvey returned to the United States in 1922, and was arrested and charged with mail fraud. Garvey decided to represent himself in his trial, a decision that proved to be unwise. He was sentenced to five years in prison. Garvey was

also caught up in a power struggle within the UNIA. There were fights over leadership and disputes between the American and African factions. This was caused in large part by a secret meeting between Garvey and the leader of the racist American organization the Ku Klux Klan, which supported the idea of relocating African Americans to Africa for reasons very different from Garvey's.

Garvey's prison sentence was commuted in 1927, and he was deported to Jamaica. He tried to keep his UNIA activities going, but the organization never regained the splendor it had known in its Harlem days. In 1928 he submitted a petition to the League of Nations (the forerunner of the United Nations) in Geneva, Switzerland, protesting the treatment of people of African descent throughout the world. In 1931, he founded the Edelweiss Amusement Company, which promoted Jamaican musicians and entertainers. Garvey moved to London in 1935, where he worked until his death, which followed two strokes, in 1940. He was buried in London. In 1964, his body was exhumed (dug up) and reburied in National Heroes Park in Kingston, Jamaica.

Garvey was in many ways controversial during his lifetime, but his legacy is a proud one. Some of his projects were failures, but he is credited with helping to promote the racial pride that would fuel the civil rights movement and black power movement in America. His teachings were also important in the decolonization movement in the later twentieth century, in which many African countries fought for and won their independence from colonial European powers.

James Theodore Holly. *Fisk University Library. Reproduced by permission.*

★ JAMES THEODORE HOLLY (1829–1912)

James Theodore Holly was the first African American bishop of the Episcopal Church. He was inspired by the plight of African Americans during the era of slavery, especially by the fact that even free blacks sometimes had to fear being captured and enslaved. This fear caused him to move to Canada, where he became a champion of emigration for African Americans. Emigration refers to a permanent move by people out of a country. He advocated African Americans emigrating from the United States to

Haiti, a country in the Caribbean. He organized a successful mission to Haiti, where he eventually became ordained as an Episcopal bishop.

James Theodore Holly was born on October 3, 1829, in Washington, D.C. He was the descendant of a freed slave. His great-great-grandfather, also named James Theodore Holly, was a Scotsman who lived in Maryland. He freed his slaves in 1772.

The Hollys' status as free blacks allowed them to enjoy a better standard of living than many blacks in the United States at the time, who were born into slavery. James Overton Holly, James Theodore's father, was a shoemaker and earned a living for the family by his trade. He was also intent on his children receiving an education. James Theodore Holly was educated in the private school of John H. Fleet, a black physician. His parents enrolled him in Fleet's school in 1837. He studied there until the family moved to Brooklyn, a neighborhood in New York City, in 1844.

The family opened its own shoe shop in Brooklyn in 1845. They also became very involved with abolitionism. Abolitionism, based on the word "abolish," meaning "put a stop to," refers to the movement to end slavery in the United States. Holly completed his training as a shoemaker in 1846. He also continued his formal education. The Holly family was Catholic, and James studied with Father Felix Varela of Transfiguration Church in New York City. He studied the classics and mathematics. He also gained writing experience. He published five articles on slavery in Frederick Douglass's publication, *North Star*.

The year 1850 was a turning point for James Theodore Holly. It was the year that the United States Congress passed the Fugitive Slave Act. The Fugitive Slave Act put all blacks in the United States, even free blacks, in danger. The law enforced a hefty fine on law enforcement officials who did not capture African Americans suspected of being fugitive, or runaway, slaves. Little was needed to prove that a person was a runaway slave. Once a person was captured and returned to a slave state, he did not have an opportunity to go to court or prove that he was free. This meant that a black person who had been free his whole life could be wrongly captured and forced into slavery.

The Fugitive Slave Act put free blacks like the Hollys at risk, so they decided to leave the United States. They moved to Windsor, Canada, in 1851. There, Holly became dedicated to helping fugitive slaves. He collaborated, or worked closely with, former slave Henry Bibb, who published a newspaper called *Voice of the Fugitive*. Holly put his formal education to work, helping to edit the newspaper. He also published an article in *Voice of the Fugitive,* urging African Americans to emigrate, or permanently move, to Canada. Henry Bibb held a conference in Toronto

in September 1851, which he called the North American Convention of Negroes. At that conference, he presented Holly's idea of forming a league of blacks in North America and the Caribbean to help blacks emigrate from the United States to Canada.

Holly quickly began working to make his dream of emigration a reality. In 1854, he attended the first Emigration Convention, which was held in the United States in Cleveland, Ohio. He and his family moved back to the United States permanently in 1856. They settled in New Haven, Connecticut, where Holly became the pastor of St. Luke's Episcopal Church. He also worked as a teacher. Preaching and teaching were both good forums for Holly to promote his vision of emigration. He made a series of lectures in favor of emigration, which were later published as "Vindication of the Capacity of the Negro Race for Self Governance and Civilized Progress."

Holly next began organizing and fundraising, trying to get together an actual trip for some African Americans to emigrate to Haiti. He exchanged letters with United States Congressman Francis P. Blair (1821–75) about his plans for emigration. He asked the congressman to help with funding for the trip. He also sought assistance from the Episcopal Church. He contacted the church's Board of Missions and asked for funding for his trip to Haiti.

Holly made his first trip to Haiti a reality in 1861. He gathered a group of 110 men, women, and children in New Haven for the first emigration to Haiti. Tragedy struck Holly in the first year. Many members of the group became ill because they did not have resistance to tropical diseases such as yellow fever and malaria. Fifty-two people from his group died in the first year. Sadly, this included his mother, his wife, and his two children.

The tragic first year of Holly's mission to Haiti did not stop him from promoting emigration. He kept fighting for funding to establish a permanent mission in Haiti. The Board of Missions of the Episcopal Church accepted sponsorship of his mission in 1865. Over the years, the mission grew. The Episcopal Church honored him in 1874 by making him missionary bishop of Haiti. He was the first African American bishop in the Episcopal Church. Some years later, he was recognized also as bishop of the Orthodox Apostolic Church of Haiti. He also managed to recover somewhat from the personal tragedies of his first year in Haiti. He remarried, to a woman named Sarah Henley. Sarah and James went on to have nine children. He became bishop of the Anglican Orthodox Episcopal Church of Haiti and, in 1897, took charge of the Episcopal Church in the Dominican Republic. He died in Haiti in 1911.

★ CHARLES S. JOHNSON (1893–1956)

Charles S. Johnson was a sociologist, educator, and magazine editor who dedicated his long, successful career to studying and identifying the distinctive characteristics, challenges, and history of the African American community in the United States. Johnson did groundbreaking research on the issue of race relations early in his career. He also edited a magazine called *Opportunity* that played an important role in the Harlem Renaissance, a flowering of African American arts that occurred in New York City in the 1920s. As a sociologist, Johnson wrote numerous books that provided key insights into the black community. His goal as a writer was to present social scientific insights in a way that would be accessible and useful to government officials as they framed public policy. He closed his career as the first African American president of Fisk University in Nashville, Tennessee.

A Sociology Career Begins in Chicago

Charles Spurgeon Johnson was born on July 24, 1893, in Bristol, Virginia, a small town in the southwest corner of the state. He was the oldest of five children born to Winifred Branch and the Reverend Charles H. Johnson. There was no school for African Americans in Bristol, so Johnson's father sent him more than three hundred miles away from home to study at the Wayland Baptist Academy, the high school affiliated with Virginia Union University, in Richmond, Virginia. Johnson also attended college at Virginia Union. He graduated from college with honors in 1916.

Charles S. Johnson. *Fisk University Library. Reproduced by permission.*

The poverty and discrimination faced by African Americans in Virginia spurred Johnson to study sociology. Sociology is the study of society, social institutions, and social relationships. Johnson believed sociology could provide research, concepts, and insights that would help analyze and solve the problem of racial discrimination. Therefore, he moved to Chicago in 1917 to pursue a degree in sociology at the University of Chicago. The University of Chicago had one of the best sociology departments in the country. Johnson earned a bachelor's degree from the University of Chicago in

1918. He then took a job with the Chicago Urban League, a civil rights organization dedicated to ending racial discrimination and securing economic opportunities for African Americans. Johnson served as the group's director of research.

One year later, a major race riot broke out in Chicago in which thirty-eight people were killed. The state of Illinois wanted to conduct a study to figure out what had caused the riot. Johnson's connections with the University of Chicago and the Chicago Urban League made him a good candidate to help write the study. The study was completed in 1922. It was titled *The Negro in Chicago: A Study in Race Relations and a Race Riot.* The study concluded that the riot was the result of deep-seated tension and unrest that were the products of racial discrimination in employment, housing, and other areas of public life. This was a controversial conclusion. Johnson bolstered (added strength to) his conclusion with more than seven hundred pages of research and analysis. The study was a landmark work of scholarship in the area of demographics.

Plays a Key Role in the Harlem Renaissance

After completing his Chicago study, Johnson took a job in New York. He became the Urban League's national director of research in 1922. One of his responsibilities was to edit the organization's magazine, the *Urban League Bulletin.* When Johnson took over, the *Bulletin* was a lightly regarded, little-read publication. Johnson had a more ambitious vision. He founded a new magazine called *Opportunity: A Journal of Negro Life* in 1923. In the first issue of *Opportunity,* Johnson explained that the magazine would be dedicated to the study of African American life and culture. He wanted to use the magazine as a tool for bringing sociological research and surveys to a wider audience. Under Johnson's leadership, the magazine took on the difficult issues confronting African Americans, especially discrimination in the areas of employment, health care, and housing.

Opportunity also took on a central role in the Harlem Renaissance. The Harlem Renaissance was an era of great creativity and productivity in African American culture, art, music, and literature that took place in the Harlem neighborhood of New York City. *Opportunity* magazine often published poems, essays, short stories, paintings, and drawings from young African American authors and artists as part of its mission to document African American culture. *Opportunity* thus helped build momentum and excitement within the community of black artists and intellectuals living in Harlem. The magazine also introduced their work to white patrons (people who provide financial support to artists) who helped finance the work of the Harlem Renaissance.

Fisk University Welcomes Its Future President

Johnson left the New York Urban League in 1928 to take a job at Fisk University, an historically black university in Nashville, Tennessee. Johnson became the chair (head professor) of the university's sociology department. He remained in that position until 1946. His leadership made the department a leading training ground for African American sociologists. Johnson also authored numerous books during his time at Fisk. His work made important contributions to the field of sociology by helping to disprove myths and stereotypes about African Americans. For example, his 1934 book *Shadow of the Plantation* disproved the belief held by many white Southerners that the plantation and sharecropping system was beneficial to African Americans.

All told, Johnson wrote, coauthored, or edited eleven books during his time at Fisk. The common theme uniting Johnson's works was the idea that the practice of racial segregation is not the natural order of things but a constructed social system. Racism was so entrenched (deeply rooted) and pervasive (commonplace and accepted) that many people at the time, especially whites, argued that fighting against it was futile. Johnson's academic research aimed to demonstrate that social patterns of race relations could be changed via social and political activism.

Johnson ended his storied (successful and acclaimed) career as the president of Fisk University. He was the first African American ever to hold that position. He was Fisk's president from 1946 to 1956. During that time, he scaled back his academic research and concentrated his efforts on raising money, enhancing Fisk's reputation, and strengthening the faculty and curriculum. Johnson died unexpectedly on October 27, 1956, at the age of sixty-three.

★ ALAIN LEROY LOCKE (1886–1954)

Alain Leroy Locke was one of the most famous and accomplished African American intellectuals of the early 1900s. He became famous for his promotion, interpretation, and criticism of various forms of African American art. Locke was one of the leading promoters of the Harlem Renaissance in the 1920s, which was a flowering of African American arts and culture in the Harlem neighborhood of New York City. Scholars consider his book, *The New Negro: An Interpretation* (1925), to be the definitive account of the Harlem Renaissance. Locke was also an accomplished scholar, author, and educator. He was the first African American ever to earn a doctoral degree in philosophy from Harvard University.

He developed a philosophy of cultural pluralism and also authored numerous works about art, music, and literature.

Education Opens Doors from a Young Age

Alain Leroy Locke was born on September 13, 1886, in Philadelphia, Pennsylvania. He was the only son born to Pliny and Mary Hawkins Locke. Both of his parents were very well educated. His mother was a career educator. His father earned a degree from the Colored Institute of Youth (now known as Cheyney University) in Philadelphia, the oldest historically black university in the United States. Alain received from them a love for, and belief in the value of, learning. He graduated second in his high school class in 1902, and he was first in his class at Philadelphia's two-year School of Pedagogy (the study of education). In 1904, he enrolled at Harvard University in Cambridge, Massachusetts, one of the most prestigious universities in the world.

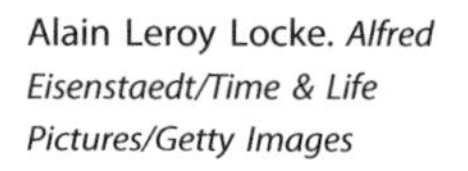

Alain Leroy Locke. *Alfred Eisenstaedt/Time & Life Pictures/Getty Images*

Locke was an outstanding student at Harvard. He was elected to Phi Beta Kappa, which is the best-known and most-respected honor society for college students. He graduated *magna cum laude* (with high honors) in 1907, with degrees in English and philosophy. Locke won the Bowdoin Prize, an award for an essay in English that is considered to be Harvard's highest academic award for undergraduates, and he became the first African American ever to be named a Rhodes scholar. Rhodes scholars receive two years of paid higher education in England. The Rhodes scholarship is very selective and a great honor. After Locke completed his Rhodes-sponsored studies in England, he traveled to France and Germany to study philosophy, art, and literature. He returned to the United States in 1912.

Locke was a professor of English, philosophy, and education at Howard University in Washington, D.C., from 1912 to 1916. Howard was the national center of African American educational and intellectual life in the early 1900s. While working at Howard, Locke became a believer in W. E. B. Du Bois's idea of the "Talented Tenth." The philosophy of the Talented Tenth holds that the most talented 10 percent of African Americans had a duty to use their talents and education to improve the fortunes of the black community. Locke left Howard to return to Harvard

to earn his Ph.D. in 1916. A Ph.D. is the highest degree a student can earn at an American university. In 1918, Locke became the first African American ever to earn a Ph.D. in philosophy from Harvard.

Launches His Career as a Critic and Thinker

Locke returned to Howard University after completing his Ph.D. He became the chair (head professor) of the philosophy department there. Locke's philosophical works centered on the idea of cultural pluralism. Many philosophers at that time believed that philosophy was an attempt to discover timeless principles that were universally true. Locke argued that all truths, values, and attitudes reflect the culture that produces them. He believed that concepts such as logic, facts, and even reality always reflect some part of the culture in which they are expressed. This philosophy came to be known as cultural pluralism or cultural relativism. Locke's key philosophical insight was into the importance of historical, social, and cultural context. Locke published numerous philosophical works on this topic throughout his career. He was a well-respected, influential philosophical thinker. He remained the chair of Howard's philosophy department until his retirement in 1953.

Locke also began to cultivate an interest in the arts during his early years as a Howard professor. He immersed himself in the literature, art, and theater of the African American community. Locke began to view the arts as a key medium through which he could put his ideas and views into practice. Consistent with his belief in the Talented Tenth, Locke viewed the arts as a way to advance the status and improve the self-image of African Americans. Consistent with his philosophy of cultural pluralism, Locke believed that African American art could reflect the beauty, insights, and values of African American culture in a way that would combat and undermine the myth of white superiority.

Locke's increasing interest in the arts coincided with the first Great Migration. The Great Migration lasted approximately from 1910 to 1930. About 1.3 million African Americans left the South and relocated to cities in the North, the West, and the Midwest in hopes of finding better work than was available in the South at the time. African Americans who previously had lived in the South took up residence in cities such as Washington, D.C., Philadelphia, Chicago, and especially the Harlem neighborhood of New York City. These former Southerners brought with them a rich variety of traditions in language, music, poetry, art, folklore, dance, and writing. They also tended to congregate in the same neighborhoods in their new homes in the North. In short, large numbers of African Americans from diverse backgrounds were living in close proximity and enjoying a feeling of optimism about their prospects. The

result was an exciting period of artistic creativity and vibrancy in the black community.

The New Negro Is Published

Nowhere was this trend more pronounced than in Harlem. In fact, African American culture, art, music, and literature flourished there to such a great degree during the 1920s that the period became known as the Harlem Renaissance. Locke was familiar with the Harlem Renaissance from his work as an art promoter in Washington, D.C. He began making regular trips to Harlem in the early 1920s. Simultaneously, Locke was trying to persuade Howard University to position itself as the national center of black theater and art. The university's all-white board of trustees refused, and Locke left Howard in 1924. He would return in 1928, but his time away from Howard allowed him to focus his energies on the ongoing Harlem Renaissance.

Locke moved to New York and became heavily involved in the arts. He served as a friend and advisor to young black artists, writers, musicians, and intellectuals. He asked for donations from white philanthropists. He also authored regular contributions to magazines and newspapers as an art critic. He was a key player in the Harlem Renaissance. In 1925, he decided to document what was happening in Harlem. He published a book entitled *The New Negro*. The book was a collection of poems, stories, essays, and pictures by African American authors and artists. Locke wrote a famous introduction to the book in which he said that the material it contained documented the arrival of a "New Negro" with a new psychology, enlivened spirit, and unlimited potential.

The New Negro did more than document the artistic, cultural, and intellectual excitement of Harlem in the 1920s. It also helped contribute to it. It emboldened African American thinkers and artists and inspired other African Americans to take up new callings. It inspired white patrons of the arts to increase their support for black artists. It also cemented Locke's place as the nation's leading authority on African American art and culture. It was the crowning achievement of his career.

Locke returned to his position at Howard University in 1928. He remained there for the next twenty-five years. He authored numerous books, articles, and essays on black art, music, and culture. He also continued to write about philosophy and cultural pluralism. Locke was the preeminent critic of and commentator on African American art in the first half of the twentieth century. In 1951, he received a grant to write a book about his life's work on African American culture. He planned to devote himself to this project full-time during retirement. Unfortunately, Locke's retirement did not even last one year. He died on June 9, 1954,

in New York City. He was sixty-seven years old. One of his colleagues ensured that his final project would be completed. *The Negro in American Culture* was published in 1956. It remains the definitive study of the contributions of African Americans to American art, culture, and society in the first half of the twentieth century.

★ JOSEPH JENKINS ROBERTS (1809–1876)

Joseph Jenkins Roberts was the first president of the nation of Liberia in West Africa. Liberia began as an American colony founded for the

Joseph Jenkins Roberts. *Apic/Hulton Archive/ Getty Images*

purpose of resettling freed slaves. Roberts was instrumental in Liberia's declaration of its independence. It was the first of many African nations that would go on to declare their independence from the colonial powers.

The Child of Free Blacks

Roberts was born in 1809 in Norfolk, Virginia. His parents were free. His mother was Amelia Roberts, who married James Roberts. James Roberts was a wealthy businessman who taught his stepsons, including Joseph Jenkins Roberts, his trade. He also made sure that they learned to read and write, even though African Americans did not have much access to formal education at the time. Joseph was what was then known as an "octoroon," meaning that he was only one-eighth black. During this time in American history, a person was considered to be black if he had even "one drop" of black blood.

Virginia in 1809 was not an easy place for free blacks to live. This is because it was governed by a strict set of so-called "black codes." Black codes were state and local laws that regulated the conduct of free blacks in the region. They were especially prominent in the American South, though there were black codes in the North, too. The purpose of black codes in the South was to suppress the rights of free blacks, to force them to move, and to prevent blacks from other places from moving to the state. They were prohibited from serving in certain professions where they might be able to help slaves revolt or try to win their independence. This was the major purpose of the black codes: In addition to harassing the free blacks, the laws were designed to protect the institution of slavery.

Move to Liberia

Roberts and his family got the opportunity to emigrate, or move from one country to another, to Liberia in 1829. They were enthusiastic at the chance to move to Africa. This was in part because of the oppressive conditions of living under the black codes in Virginia. Another factor in their decision to move was that they were devout Christians. They believed in missionary work and wanted an opportunity to convert the African people to Christianity. They set sail for Africa on the ship *Harriet* in 1829. It was an expedition organized by a group called the American Colonization Society (ACS).

The American Colonization Society had been founded in 1816 by Reverend Robert Finley (1772–1817). The purpose of the organization was to find a new home for freed slaves outside of the United States. Finley believed that the natural place for African American former slaves to make their home was in Africa. A diverse group of interests in the United States supported the ACS's movement to help freed slaves move to

Africa. Some were religious people who wanted freed slaves to assist in their mission work in Africa. Others thought it was a way for African Americans to get in touch with their roots. Some radical leaders, like Bishop Henry McNeal Turner (1833–1915), thought that free blacks would never be accepted by white society, and that they should make their own way in Africa. Other, more questionable interests also supported the ACS's mission. Some slave owners in the American South wanted freed slaves to leave the United States. They feared that the presence of free blacks would incite slaves to try to seek their own freedom.

Roberts and his family set sail for the American colony of Liberia. They settled in the city of Monrovia, named for President James Monroe (1758–1831). The family prospered in Liberia. They first established a successful shipping business. They exported products such as ivory from Africa to the United States. They also sold American products to the people in Liberia.

The family members thrived with the family business. One of Roberts's brothers, John Wright Roberts, became an ordained minister. He eventually became a bishop of the Liberia Methodist Church. Another of his brothers, Henry Roberts, returned to the United States to train as a physician. He returned to Liberia to be a doctor after graduating from medical school. Roberts himself became sheriff of the colony of Liberia in 1833. By 1841, after several years working in the government, Roberts became the first nonwhite governor of the colony of Liberia. Several years after that, Roberts asked the legislature to declare the colony's independence from the United States, which it did in 1847.

Serves as President of Liberia

On October 5, 1847, the new republic of Liberia elected Joseph Jenkins Roberts as its first president. Roberts devoted the first year of his presidency to foreign policy. He wanted the governments of the United States and Europe to recognize Liberia's status as an independent nation. Roberts also had to face the challenge of unifying a diverse population. Liberia had three hundred settlers from the United States, but there were also many native Africans who lived there as well. Some of the coastal tribes converted to Christianity and were receptive to Roberts's rule. Other tribes were not as receptive, and posed challenges to the new government.

Roberts won reelection to the presidency three times. He served a total of eight years. He finally lost election in 1855 to his vice president Stephen Benson, who had run against him. During the years following his presidency, Roberts served as a general in the Liberian Army and continued his foreign relations work as an ambassador to Great Britain and France. He also helped found Liberia College in Monrovia.

In 1872, Roberts returned to his office of president of Liberia. There was trouble brewing. The president at the time, Edward James Roye (1815–72), was planning to suspend the upcoming elections. Thus, the Republican Party of Liberia deposed him, or removed him from office. An election was held, and Roberts again was elected president. He served two more terms as president of Liberia. He died in February 1876, very shortly after the end of his presidency.

★ BENJAMIN "PAP" SINGLETON (1809–1892)

Benjamin "Pap" Singleton was known as the "Father of the Colored Exodus." The so-called exodus occurred in the late 1870s, after Reconstruction ended in the American South. The South was a hostile place for free blacks and former slaves. Singleton set out to help African Americans create new settlements and opportunities for themselves outside of the American South. He turned his focus to settlements in Kansas. During the period of 1877 to 1879, more than fifty thousand African Americans, inspired by Singleton's call, moved from the South to Kansas.

Benjamin "Pap" Singleton. *Kansas State Historical Society. Reproduced by permission.*

Benjamin Singleton was born on August 5, 1809. He was born a slave in Nashville, Tennessee. Because he was a slave, the exact details of his early life are not well known. It is known that he escaped from Nashville to New Orleans, Louisiana, because he did not want to be sold to another master. He was eventually captured, though, and returned to Nashville. He was sold to owners in Alabama and Mississippi. He escaped several more times, and several more times he was returned to slavery. He eventually escaped to freedom in the North. He lived for a time in Windsor, Ontario, which is in Canada. He also lived for a time in Detroit, Michigan. During his time in the North, Singleton helped other escaped slaves find a place to stay.

Singleton stayed in the North until the end of the Civil War in 1865. He then moved back to his hometown of Nashville. His experiences with slavery inspired him to help free blacks in the South, whose lives remained very hard after

the end of the war. There were few economic opportunities, and whites in the South, bitter over having lost the war, were often hostile to blacks. The federal government had to send troops to the South to keep the peace and to enforce Reconstruction laws. Singleton worked, as he had when he was a slave, as a cabinet maker. Now, he also made coffins. He was disturbed to find that many of the coffins he made were used for African Americans who had been killed by white vigilantes (people who take the law into their own hands). Singleton decided that African Americans needed a place where they could settle and earn a living away from the hostile white Southern culture.

Singleton quickly learned that Tennessee would not be the ideal place for his settlements. Land in Tennessee was expensive, and many white landowners were strongly opposed to his project. He found that white landowners, scared of the idea of blacks owning land and having independence, refused to sell him land at a fair price. Singleton eventually decided on Kansas as the place for his settlements.

Singleton found a partner to help him realize his dream of creating black settlements in the American West. His name was Columbus Jackson. Singleton and Jackson formed their first settlement in Cherokee, Kansas. The settlement was not successful. Undaunted, Singleton and Jackson formed a second settlement in Morris, Kansas. They spread the word about their new settlement to African Americans in Tennessee and other parts of the American South using posters.

Word traveled quickly. Many people were enthusiastic about the opportunity for a fresh start in the West. Between the years of 1877 and 1879, hundreds of African Americans moved to Singleton's new settlements. The timing of the exodus tied in directly with the end of Reconstruction in the South in 1877. The Radical Reconstruction era was a ten-year period between 1867 and 1877 when federal troops stationed in the South enforced the civil rights of African Americans. The withdrawal of troops led to widespread discrimination against African Americans designed to reduce them to as close a state to slavery as possible. Southern legislatures passed oppressive laws. Vigilante groups such as the Ku Klux Klan staged lynchings, terrorizing African Americans in the South. These events led to the migration of fifty thousand African Americans from the South to Kansas during 1879.

The blacks that followed Singleton's invitation to the West were known as "Exodusters." They were named for the Exodus in the Old Testament of the Bible. The Exodus, as recorded in the Bible, happened when the Israelites won their freedom from the Egyptians, and Moses led them out of Egypt. The Exodus was a powerful symbol of freedom from

bondage. For this reason, Singleton was known as the "Father of the Exodus" or the "Moses of the Colored Exodus."

Whites in the South were not entirely happy about the exodus. Many patrolled roads and rivers, trying to stop the migration. The United States Congress even investigated the exodus. Democrats and Republicans argued over it. Democrats accused Republicans of supporting the exodus for political gain. In 1880, Congress called Singleton to testify about the events of the exodus.

Singleton remained committed to the idea of African American migration and resettlement despite opposition, but his focus had changed by the latter part of the 1880s. He was no longer convinced that migration to Kansas was the answer to the problems facing African Americans. For some time, he supported migration of African Americans from the United States to the island of Cyprus in the Mediterranean Sea. He finally decided that Africa was the best place for the resettlement of African Americans. He formed the Trans-Atlantic Society in support of this goal in 1885. Singleton died in 1892, but his ideas lived on. Marcus Garvey (1887–1940) became the new champion of Singleton's dream of African settlement in the 1920s.

★ CARTER G. WOODSON (1875–1950)

Carter G. Woodson was a historian, author, and educator who made so many important contributions to the study of African American history that he earned the nickname "The Father of Black History." Woodson did most of his work for an organization he founded called the Association for the Study of Negro Life and History (ASNLH). Under Woodson's leadership, the ASNLH published an academic journal dedicated to African American history. Woodson also founded a publishing company that printed books on black history. He came up with the idea for Negro History Week, the forerunner to Black History Month. Woodson's work legitimized African American history as a serious topic for scholarship.

World Travels Spawn Interest in History

Carter Godwin Woodson was born on December 19, 1875, in rural Virginia. He was one of nine siblings. Woodson came from a humble background. His family was poor when he was growing up, so as a boy he worked on the family farm. He hired himself out to work on other farms as a teenager. When his family moved to West Virginia in the late 1880s, Woodson took work as a coal miner. Woodson's work to help support his family prevented him from attending school as a child. He was very

bright, however, and when he finally enrolled in high school at the age of twenty, he graduated in just two years.

Woodson had a thirst for knowledge and wanted to continue his education. He enrolled at Berea College in Berea, Kentucky, in 1897. Berea was a unique environment. It was the only racially integrated college that admitted both men and women anywhere in the South in the late 1800s. Woodson studied there for two years, and then he took a job as a teacher in West Virginia. He became principal of the high school from which he had graduated, but his hunger for learning eventually prompted him to return to Berea. He completed college in 1903, earning a bachelor's degree in literature.

Woodson had taken several courses at the University of Chicago during his last year as a student at Berea. While he was in Chicago, he became aware of a program that sent Americans overseas to teach in foreign schools. Woodson applied for the program, and later in 1903, he set sail for the Philippines. He worked as a teacher there for four years. He began his return home in 1907, but did not sail home directly. Instead, he traveled through India, Palestine, Greece, Italy, and France. Woodson spent a semester studying at a college in Paris, having become fluent in French during his time in the Philippines. Woodson would say later that his travels had deeply affected him, opening his eyes to the importance of history and culture.

Study of African American History Begins

Woodson earned a master's degree from the University of Chicago in 1908 upon his return to the United States. Interestingly, his master's thesis (the final paper a student has to write to earn a master's degree) was not about African American history. He wrote about European diplomacy in the 1700s. But Woodson's interest would turn to African American history soon enough.

Woodson next enrolled in a Ph.D. program at Harvard University. A Ph.D. is the highest degree a student can earn at an American university. Harvard is one of the most prestigious universities in the United States. In 1912, Woodson became just the second African American ever to earn a Ph.D. in history from Harvard (and he was the first person descended from slaves to do so). Woodson's dissertation (the final, book-length paper a student has to write to earn a Ph.D.) was about the events leading up to Virginia's secession from the Union before the Civil War.

Woodson took a job as a high school teacher in Washington, D.C., upon completing his Ph.D. He published his first book, *The Education of the Negro Prior to 1861*, in 1915. That same year, Woodson helped found the ASNLH. The ASNLH was a response to the reality that, in the early 1900s, most historians ignored African Americans in their studies of

OPPOSITE PAGE
Carter G. Woodson.
Hulton Archive/Getty Images

American history. They chose not to investigate topics of particular importance to African Americans, such as slavery and the Reconstruction era that followed the Civil War. They also assumed that African Americans had not played a major role in important events in history. Woodson wanted to use the ASNLH to call attention to the importance of studying African American history.

Becomes "The Father of Black History"

Woodson promoted the study of African American history in four major ways. First, the ASNLH began publishing an academic journal under his leadership entitled *Journal of Negro History* in January 1916. The journal was a tremendous innovation. It was the first periodical publication devoted exclusively to African American history. Second, Woodson founded a publishing company called Associated Publishers in 1921 that was dedicated to publishing books about African American history. The journal and the publishing company combined to provide a much-needed outlet for the distribution of the writings of African American historians. It also helped attract new scholars to the field.

Third, Woodson himself wrote a number of important books on African American history. Two of his most famous works are *A Century of Negro Migration* (1918) and *The Negro in Our History* (1922). Each work concentrated on demographic characteristics of and changes in the African American community, including patterns of migration, family composition, and community structure. *The Negro in Our History* was a college textbook. It was a groundbreaking work that was used in classrooms across the country for the next fifty years. Woodson wrote a total of nine books on African American history over the course of his career.

Fourth, Woodson came up with the idea for Negro History Week. The idea was a response to his belief that schoolchildren needed to learn at a young age about the contributions of African Americans to American history. During Negro History Week, schools at all levels would dedicate the second week of February to the study and appreciation of African American history. It was celebrated for the first time in February 1926. It was a tremendous success. In 1976, Negro History Week was renamed and expanded into Black History Month. It is held every year in February.

Woodson earned the nickname "The Father of Black History" for all his contributions to developing the field of African American history. It can truly be said that the field of African American history would not exist as it is today but for the contributions of Carter G. Woodson. Woodson died on April 3, 1950, at the age of seventy-four.

THE BACK TO AFRICA MOVEMENT GAINS MOMENTUM

The so-called "Back to Africa" movement lasted from about 1820 to about 1920. The movement was founded on the idea that African Americans and whites could not peaceably and freely coexist in the United States. Proponents, or supporters, of the Back to Africa movement favored the creation of settlements of African Americans in their ancestral homeland of Africa. Many different people supported the movement. Some were supporters of African American self-determination, who thought that there could be no opportunity for African Americans in the United States, even after the end of the Civil War. Others sympathized with Southern slave owners, who feared free blacks settling in the South and causing uprisings among the slaves. Still others were well-meaning abolitionists who could not imagine how blacks and whites could live together without conflict.

The idea of settlements in Africa of freed slaves from the United States was not new in 1820. Indeed, many prominent politicians and judges had argued that free blacks and whites could not coexist peacefully and equally in the United States. The first successful settlement of former slaves in Africa was established at Fourah Bay, Sierra Leone, in 1787. Sierra Leone was then a region in West Africa. The Fourah Bay settlement was an experiment of the British Crown. Several hundred freed slaves were sent to Fourah Bay to establish a colony there.

American Colonization Society

The American Colonization Society (ACS) was founded in 1816 in Washington, D.C. The group's full name was "The American Society for Colonizing the Free People of Color of the United States." The purpose of the ACS was to raise money and provide the means for free slaves to move to Africa to set up communities there. The ACS was inspired in part by the successful settlement at Fourah Bay in 1787. It was also inspired by a resettlement that took place in 1816, led by a man from New England named Paul Cuffe (1759–1817). Cuffe was a free man, a successful black shipper. He had learned about the settlement at Fourah Bay and successfully led nine families to settle in Sierra Leone.

The ACS drew a varied group of supporters, who formed some unlikely alliances. Some ACS supporters were Southerners who did not support the immediate abolition of slavery. These Southerners, many of them slave owners, did not want free blacks living in the South. They feared that the presence of free blacks would influence and inspire slaves to try to win their

William Lloyd Garrison Attacks the ACS

Despite the many supporters of the ACS, it was not without critics. Perhaps the most famous of these was the white abolitionist William Lloyd Garrison (1805–79). Garrison was a prominent abolitionist and journalist. He advocated for the immediate end of slavery. He also published an abolitionist newsletter called *The Liberator*. He called on other whites who opposed slavery to oppose it by nonviolent civil disobedience. He argued in favor of whites helping slaves escape from slavery. He also publicly burned a copy of the United States Constitution, which he called "a Covenant with Death, an Agreement with Hell" because of compromises over slavery that were written into the Constitution.

Garrison helped to found the New England Antislavery Society (NEAS) in 1832. Before the NEAS, the ACS was the most prominent abolitionist group in the United States. Garrison and the NEAS criticized the ACS as being aligned with the interests of slave owners. NEAS criticized ACS's slow approach to abolition and advocated instead for an immediate end to slavery. The ACS retaliated, or fought back, against the NEAS's criticism of it. The group accused the NEAS of being a low-class organization. This angered many Northern abolitionists, who withdrew their support from the NEAS.

own freedom. Other ACS supporters were Northerners, who may or may not have supported slavery. Many of these people did not think that free blacks and whites could happily and successfully coexist. Other Northerners who were abolitionists, or people who supported the end of slavery, favored the ACS's cause. They thought that a colonization movement might encourage Southerners to support the ending of slavery, if there was a place for the freed slaves to go. Finally, some African Americans were interested in colonization because they did not think they would ever find equality in the United States.

Early ACS Colonization Efforts in West Africa

The ACS launched its first actual attempt at colonization in 1820. It sent eighty-eight African Americans on a ship called the *Elizabeth,* bound for the western coast of Africa. The colonists landed at Sherbro Island, where they set up camp. The settlement at Sherbro Island failed, though.

About one-fourth of the colonists quickly died. Most died of diseases such as malaria and yellow fever. This was not uncommon in various colonization efforts throughout the nineteenth century. The African American colonists did not have resistance to these diseases.

The failure at Sherbro Island did not stop the ACS from trying again. It undertook a second attempt at colonization in 1821. This time, the location for the colony was to be Cape Mesurado. This location was selected by an American military lieutenant named Matthew Perry (1794–1858). The ACS had to negotiate with the chiefs of the tribes who lived near Cape Mesaduro. The chiefs eventually agreed to sell sixty square miles of land along the coast to the settlers. The settlers arrived in 1822, ready to populate the new settlement. They would face fierce resistance from the native populations, though. After some fighting with the natives, and with help from the U.S. and British militaries, the settlers were victorious. They called the new settlement Monrovia, after U.S. president James Monroe (1758–1831). The colony was called "Liberia," taken from the Latin word for "free."

The settlers at Liberia continued to struggle at first with the native people. Eventually, Liberia became a commonwealth. In 1847, Liberia declared its independence from the United States. Its first president was a man named Joseph Jenkins Roberts (1809–76), an African American who had migrated to Liberia with his family.

Church Efforts: Bishop Henry McNeal Turner

The ACS was the biggest supporter of African Americans settling in Africa in the 1800s. However, there were other people who sponsored resettlement efforts. One of these was Henry McNeal Turner (1834–1915), bishop of the African Methodist Episcopal Church. Turner was born before the end of slavery, but he was born free. He worked during his younger years as a traveling minister, and he witnessed firsthand the conditions of African Americans in the American South. What he saw convinced him that there was no place in the United States for African Americans to live free and equal to whites.

Turner was elected bishop of the African Methodist Episcopal Church in 1880. During his time as bishop, he made a major push for the church to increase missionary efforts in Africa. He made many trips to Africa himself. He took four trips to Africa in 1891 alone. He helped the church to establish schools there and to provide humanitarian support for native Africans. He also became a supporter of African American emigration, or a permanent move by people out of a country, to Africa. He lobbied the United States government to provide funding for emigration.

A Christian mission in Liberia, c. 1860. *Apic/Hulton Archive/Getty Images*

Garvey and the Black Star Line

After the turn of the twentieth century, another African American leader took up the cause of repatriation (returning to one's home country) to Africa. Marcus Garvey (1887–1940) founded the Universal Negro Improvement Association (UNIA) in 1914 with the aim of uniting and uplifting all Africans and people of African descent. To that end, Garvey and the UNIA founded the Black Star Line, a steamship company, in 1919. Garvey intended the Black Star Line to give blacks entry into international trade, and to provide blacks with comfortable and dignified sea transportation. The Black Star Line was part of Garvey's goal of helping people of African descent resettle in Africa.

The project was not a success. Garvey mishandled the company's finances and business dealings. He drastically overpaid for his ships and was cheated by engineers and representatives of the company. Garvey moved the headquarters of the Black Star Line to Liberia in 1920, in part because the U.S. federal government was investigating the company for illegal activities. Garvey was also interested in developing Liberia as an industrial base. Garvey's "Back to Africa" efforts were much criticized by African American leaders who wanted to push for integration rather than radical separation of whites and blacks. He outraged civil rights advocates by aligning himself with the white supremacist group the Ku Klux Klan. The Klan supported Garvey's plan to move African Americans back to Africa. Garvey was arrested for mail fraud in 1922. He was sent to prison in 1925. His sentence was commuted in 1927, and he was deported to Jamaica. While the idea of black separatism or nationalism would become popular again during the 1960s, organized "Back to Africa" movements in the United States ended in 1922 with the closing of the Black Star Line.

EXODUSTERS LEAVE THE SOUTH FOR KANSAS

In the late 1870s and early 1880s, a man named Benjamin "Pap" Singleton (1809–92) led more than fifty thousand African Americans from the American South to settle in Kansas. Singleton's idea was that African Americans needed a place to settle where they could create their own communities and economic opportunities. Singleton's settlements filled a need for African Americans living in the South after the Civil War. Freedom from slavery had not brought economic opportunity or freedom from persecution. Rather, many African Americans in the South after the Civil War worked as sharecroppers, barely less poor than they had been as slaves. Increasingly restrictive laws passed by Southern governments prevented blacks from rising above these conditions. All-white terrorist groups like the Ku Klux Klan also posed a constant threat to free blacks.

The migration westward led by Singleton came to be known as the "Colored Exodus." Singleton was known as the "Father of the Exodus" or the "Moses of the Colored Exodus." The people who moved to his settlements in Kansas were called "Exodusters." The book of Exodus in the Bible describes how the Israelites were held in captivity and enslaved by the Egyptians. The Israelites' leader, Moses, tried to convince the pharaoh (the monarch of Egypt) to free the Israelites. After God put a series of plagues on the Egyptians, the pharaoh let the Israelites go. Moses led them all out of Egypt, to the "promised land." The Exodus, then, was a powerful symbol of African Americans' experience with slavery in the United States. Various migration movements have been compared to the biblical Exodus.

Homestead Act Sets the Stage for the Exodus

Though the black exodus did not happen until the 1870s and 1880s, the stage for the migration was set more than a decade earlier. An event that made the great exodus possible was the passage of the Homestead Act by the United States Congress in 1862. The Homestead Act made it possible for people to become landowners for very little money in the American West. The law said that individuals who agreed to work a piece of land, to farm and improve it, could get 160 acres free in the American West. They would only have to pay a small filing fee, and the land belonged to them.

The purpose of the Homestead Act was to encourage people to migrate to the American West. As the country expanded westward during the 1800s, it gained millions of acres of new territory. The land was not cultivated, meaning that no one had farmed it. The idea of the Homestead Act was to give people free land so that they in turn would populate the Western states and start farming there.

The reason the Homestead Act was unique was that the only major qualification to become a homesteader under the act was that a person must not have taken up arms against the United States government. Anyone could qualify to take advantage of the Western land grants, even African Americans. This was very appealing to the African Americans who chose to move. African Americans could legally buy land, but white landowners in the South refused to sell land to blacks at fair prices. This made it very difficult for freed slaves to own land and prosper in the American South after the Civil War.

End of Reconstruction Prompts Blacks to Flee the South

Another major factor that led to the black exodus of the 1870s and 1880s was the official end of Reconstruction in the South. "Reconstruction" is the term used for the period directly following the Civil War in the South. It lasted from 1865 to 1877, and it was a period of rebuilding. It was also a period in which the federal government forced Southern states to make many legal and cultural changes. They had to abolish slavery and ratify the Fourteenth Amendment to the Constitution, which established that all people should enjoy equal protection of the laws. The year 1867 marked the passage of the first Reconstruction Act, of which there would be a total of three. The Reconstruction Acts forced the Southern states to give blacks the right to vote, and they placed the South under the control of the United States military. Military rule served to enforce the new rules the South would have to live by and provided some protection for freed African Americans.

The military withdrew from the South in 1877. This was the technical end of Reconstruction. At this time, conditions became much worse for African Americans in the South. Southern states began to pass so-called Jim Crow laws, which placed a series of restrictions on African Americans. Jim Crow laws required segregation in public places. They segregated schools, and African Americans often lacked access to education. The states often placed restrictions on voting that effectively deprived many former slaves of the right to vote. Many white landowners refused to sell land to African Americans. Many African Americans then had no choice but to work as sharecroppers. Sharecropping was a practice in which a white landowner would allow free blacks to farm part of his land. The landowner in turn would keep a substantial portion of the crops. In practice, this meant that many former slaves lived in crushing poverty. These conditions made for an unhappy existence for many African Americans in the South.

Singleton Leads Exodus from the South to Kansas

The black exodus was led and inspired by a man named Benjamin "Pap" Singleton. Singleton himself had been born a slave in Nashville, Tennessee. He escaped several times from slavery and was several times captured and returned. He eventually managed to escape across the border to Canada, where he helped other escaped slaves find a place to live. After the Civil War ended, Singleton returned to Tennessee. He was disheartened by what he found there. Life was not remarkably better than the life he had known as a slave. He began to imagine creating settlements for African Americans where they could make a life for themselves away from white society. He inquired about buying land in Tennessee, but found that either land was too expensive or white landowners refused to give him a fair price.

Singleton ruled out Tennessee as a good place to settle, so he turned his sights to Kansas. The Homestead Act of 1862 meant that there was a great deal of land available in Western territories for little or no money. Singleton founded his first settlement, along with his partner, Columbus Jackson, in Cherokee, Kansas. The settlement was not a success. They did not give up, though. They founded their

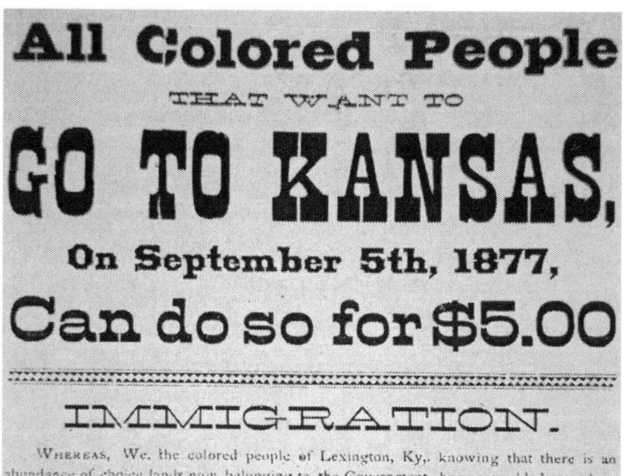

An 1877 poster urges African Americans to move to homesteads in Kansas. *Poster encouraging black exodus from the South to Kansas, 1877 (print), American School (19th century)/Private Collection/ Peter Newark American Pictures/The Bridgeman Art Library International*

second settlement in Morris, Kansas, which did succeed. Singleton advertised his settlements through posters and pamphlets that circulated rapidly throughout the South. Between 1877 and 1879, the settlements were a modest success. The largest and most famous settlement was Nicodemus, Kansas. Stories of these black communities circulated in African American communities throughout the South and inspired many others to try their luck in Kansas. Many heard the false rumor that the federal government had set aside Kansas for former slaves. In 1879 and 1880, about fifteen thousand African Americans migrated from the South to Kansas.

Several Exoduster towns grew and thrived during the 1880s, opening schools, churches, and even newspapers and baseball teams. But the towns lacked a connection to the expanding railroad lines. Without this key means of transportation, they were cut off from easy access to supplies and business opportunities that helped other pioneer towns succeed. Most Exoduster towns were abandoned by the turn of the twentieth century. Nicodemus declined steadily, dwindling from a peak population of 500 in the 1880s to just 16 in 1950.

AFRO-CARIBBEAN IMMIGRATION TO THE UNITED STATES INCREASES

A wave of immigration from the Caribbean region into the United States began in the late 1800s. This wave of immigration was caused by several factors. The United States had increased its military and economic presence in the region. At the same time, the Caribbean region's economy was suffering. The United States was a logical destination for people looking for economic opportunities. Many Caribbean immigrants had African ancestors, so this wave of immigration is often called Afro-Caribbean immigration. Hundreds of thousands of Afro-Caribbeans arrived in the United States in the late 1800s and early 1900s. The rate of immigration slowed after 1920, but increased again after World War II (1939–45).

Several Factors Trigger Immigration

Three primary factors came together to trigger a wave of immigration from the Caribbean in the late 1800s and early 1900s. First, the United States military occupied or intervened in the affairs of four different Caribbean islands in relatively quick succession. The American military occupied Cuba and Puerto Rico at the end of the Spanish-American War in 1898. The United States withdrew from Cuba in 1902, but it continued to exercise dominion (governmental control) over Puerto Rico into the twenty-first century. The United States intervened in Haiti in 1915 and in the Dominican Republic in 1916. It did not withdraw from the Dominican Republic and Haiti until 1924 and 1934, respectively.

Second, the United States became more economically involved in the Caribbean region in the late 1800s and early 1900s. The completion of the Panama Canal in 1914 created a shortcut route for shipping between the Pacific Ocean and the Atlantic Ocean, which sharply increased the volume of U.S. shipping traveling through ports in the Caribbean. At the same time, natural resources in the Caribbean were becoming increasingly important to the U.S. economy. Specifically, the sugar trade was booming. Sugar cane, the plant from which refined sugar is made, grew abundantly in the Caribbean. American companies began investing large amounts of money there. They purchased sugar plantations, built sugar mills, and shipped sugar north out of Caribbean ports.

Third, residents of the Caribbean faced significant economic hardship. Most of what are now the nations of the Caribbean were still European colonies in the late 1800s and early 1900s. That means a European nation had invaded and conquered the area and set up a government of its own choosing. Colonial governments were undemocratic, unfair, and exploitative to native peoples. Colonialism funneled wealth and resources away from colonies and into the colonizing nation. As a result, the vast majority of native Caribbean residents lived in extreme poverty. Few jobs were available. Those that were available required long hours of grueling work and paid low wages.

These three factors worked in combination to attract residents of the Caribbean to the United States. Culturally, the increased presence of Americans in the Caribbean made the United States seem less foreign and strange. It also became much easier to find transportation via ship to the United States. Economic interdependence prompted Americans to be more welcoming of Caribbean immigrants.

Afro-Caribbean Immigration Reshapes the Demographic Landscape

Immigration into the United States from the Caribbean had a distinctive demographic character. Hundreds of thousands of Africans were forcibly brought to the Caribbean as part of the trans-Atlantic slave trade from the 1600s to the middle of the 1800s. Many of these Africans continued on to the United States, but many others remained on Caribbean islands. Over the years, persons of African descent in the Caribbean came to make up a distinctive racial or ethnic group called Afro-Caribbeans. Centuries of slavery had left most Afro-Caribbeans living in poverty. As a result, many were eager to journey to the United States in search of better economic opportunities.

Afro-Caribbean immigration began in earnest in the late 1800s. According to the U.S. census, in 1880 there were only 90,000 Caribbean-born persons living in the United States. By 1920 that number had climbed to almost 580,000. Some Caribbean immigrants took work in agricultural

jobs in the South that had recently been abandoned by African Americans who had moved north as part of the first Great Migration. The majority continued on to cities in the North, especially New York City and Baltimore, Maryland.

Immigration from the Caribbean into the United States slowed down after 1920. Legislation passed in 1924 restricted the number of people who were allowed to immigrate into the United States from the Caribbean. The Great Depression began in the United States in 1929. This was a ten-year period of economic hardship that was widespread throughout the country. As a result, fewer jobs were available, and wages were not as good. Fewer Caribbean residents were willing to leave home with economic prospects in America looking dreary. The onset of World War II also slowed immigration from the Caribbean. The United States put severe restrictions on who was allowed to immigrate into the country during the war for security reasons. These restrictions were lifted once the war was over. The war also helped end the Great Depression.

Afro-Caribbean immigration into the United States therefore resumed at a very high rate following the conclusion of World War II. Some of these immigrants were motivated to leave by political unrest in their home countries. Others sought employment and better wages in the United States. These immigrants arrived in a variety of ways. Repressive (undemocratic and reliant on force) governments in Cuba and Haiti refused to let their citizens emigrate to the United States legally. Their citizens were forced to take dangerous, clandestine (secret and unauthorized) voyages on small boats. Afro-Caribbean immigrants from other nations arrived in more conventional ways via airplane or boat.

Immigration from the Caribbean has had a significant, lasting impact on the demographics of the United States. Afro-Caribbean immigrants have tended to settle in cities along the East Coast. Substantial immigrant populations and communities have developed, especially in Miami, Florida, and New York. These immigrant communities have transformed the political, business, culinary (food), and cultural makeup of these cities in particular and the nation as a whole. By 1970, there were more than 1.8 million Caribbean immigrants living in the United States. The wave of Afro-Caribbean immigration that began in the late 1800s continued into the twenty-first century.

AFRICAN AMERICANS FOUND ALL-BLACK TOWNS

The formation of all-black towns in the late 1800s and early 1900s was a noteworthy trend in American demographics. African Americans were driven to establish all-black towns by the terrible state of race relations in

the South during that time. Not only did African Americans have to live with racial segregation, but some towns in the South had gone so far as to become "whites-only." All-black towns were a logical response. The first all-black towns were established in the late 1880s. They were established all over the country, but Oklahoma soon became home to more all-black towns than any other state or territory. All-black towns remained in existence well into the mid-twentieth century.

Bad Conditions in the South Prompt a Search for an Alternative

Historians in the early twentieth century dubbed the period from 1880 to 1930 "The Nadir of the Negro," or simply the Nadir era. A nadir is the lowest possible point in an orbit or path. The Nadir era took its name from the fact that from 1880 to 1930, African Americans encountered the worst discrimination they had faced since the end of slavery. Ninety percent of African Americans lived in the South at the time, and racism in the South was everywhere. Jim Crow laws mandating racial segregation were passed throughout the South. Every aspect of Southern society was designed to reinforce the notion that African Americans were inferior to white Americans. The majority of African Americans worked as sharecroppers and lived in grinding poverty. They also lived in constant fear for their safety and their lives, as the practice of lynching African Americans (the unlawful killing of an individual by a mob) peaked during the Nadir era.

One feature of the Nadir era that was particularly noteworthy was the creation of so-called "sundown towns." A sundown town was a town that was deliberately organized in a way that gave it an all-white population. The name "sundown town" came from the fact that African Americans were only allowed to visit or pass through such towns during the day. Once sundown came, African Americans were not allowed within city limits. Sometimes this policy was enforced through restrictive housing covenants (laws that made it illegal for African Americans to buy or rent homes in a certain neighborhood). At other times, the policy was enforced through illegal intimidation and violence. Either way, the message was clear: African Americans were not welcome.

All-black towns were created as a response to the problems of the Nadir era in general and sundown towns in particular. The all-black town phenomenon should be analyzed against the backdrop of the Great Migration. The Great Migration was a large-scale movement of African Americans out of the rural South and into urban areas in the South and eventually the North, Midwest, and West from 1910 to 1930. African Americans living in the South faced constant discrimination. To many, the promise of freedom and equality held out at the end of the Civil War seemed like an impossibility. Many former slaves thus concluded that

Storefronts in the all-black town of Mound Bayou, Mississippi, in 1939. *© Corbis*

the only way they could escape from discrimination was to move away from communities of white Southerners.

All-Black Towns Offer an Alternative to Jim Crow

Prominent former slaves began organizing groups of African Americans and planning moves to new areas. More than one hundred all-black towns were established between the 1880s and 1920. The founding of the town of Mound Bayou, Mississippi, is a typical example. Mound Bayou was one of the very first all-black towns established anywhere in the United States. Mound Bayou was founded by Isaiah Montgomery (1847–1924), a former slave who had run a successful business for many years after the Civil War. He used some of the money he had made in business to purchase a large piece of land in a remote part of northwest Mississippi in the early 1880s. Montgomery and eleven other former slaves established Mound Bayou in the spring of 1887. As the need arose, they organized their own government, schools, churches, banks, and stores. The economy of the town was driven by farming, and the former slaves owned the land on which they farmed. There were no white towns nearby. Living in relative isolation enabled the residents of Mound Bayou to also live in relative autonomy.

All-black towns were founded throughout the South and Midwest, but Oklahoma eventually became the centerpiece of the all-black town

movement. This was largely because of the Land Run of 1889, in which the federal government allowed people to claim ownership of property in Oklahoma on a first-come, first-served basis. The free land attracted many African American settlers who could not afford to purchase land in their home states. Of the more than one hundred all-black towns founded nationwide, more than fifty of them were founded in Oklahoma. The most famous of these was Boley, Oklahoma. Boley became so well regarded that Booker T. Washington (1856–1915), the most prominent African American leader in the nation at the time, paid it a visit in 1904 and called it the most interesting and enterprising all-black town in the United States.

All-black towns were a significant feature of American demographics for many years. There were more than seventy all-black towns that had populations of ten thousand or more by 1940. The trend toward establishing such towns began to slow down, however, following World War II. The civil rights movement of the 1950s and 1960s created significant social momentum for racial integration. In addition, many all-black towns were economically dependent on small, family-owned farms. As agriculture shifted toward large-scale farms, many black families sold their farms and left their homes in all-black towns in favor of life in a city. Even so, there were still thirteen towns in the state of Oklahoma that were founded as all-black towns and still had all-black or nearly all-black populations as of 2004.

GARVEY FOUNDS THE UNIVERSAL NEGRO IMPROVEMENT ASSOCIATION

The Universal Negro Improvement Association (UNIA) was founded by Marcus Garvey in 1914. The organization aimed to unite people of African descent all over the world and help them find both political and economic independence. The concept was known as pan-Africanism. The organization and Garvey were controversial, but their influence was long-lasting. Garvey's pan-African ideas would go on to influence civil rights leaders for decades to come.

Marcus Garvey, the founder of the UNIA, was born in St. Anne's Bay, Jamaica, in 1887. At that time, Jamaica was a British colony. There were many people of African descent in Jamaica because Jamaica had been a major hub for the international slave trade. Garvey witnessed firsthand poverty and racism caused by colonial rule. In 1911, Garvey left Jamaica to travel. He eventually traveled to England, where he met many black intellectuals and activists. He learned about African history and culture and became committed to improving the position of people of African descent throughout the world.

A man stands outside a Universal Negro Improvement Association (UNIA) club in New York in 1943. *Gordon Parks/Hulton Archive/Getty Images*

Garvey returned to Jamaica in 1914 and founded the Universal Negro Improvement Association. He was inspired by the work of Booker T. Washington, an American who advocated for blacks to improve themselves through education. Washington argued that African Americans ought to improve their skills and work to better their own fortunes. He founded the Tuskegee Institute to carry out his mission. Garvey hoped to use the UNIA to establish a trade school in the model of Tuskegee. The lack of economic opportunities in Jamaica made it an unsuitable place to carry out his goals.

Garvey moved to the United States in 1916 and promptly established the American branch of the UNIA in 1917. He settled in New York City, but he traveled widely, speaking to large audiences of African Americans. He spoke on the topic of African Americans and African peoples all over the world seeking to better themselves. Garvey was a powerful speaker and charismatic person. He soon developed a large following among African Americans who were frustrated with a lack of economic opportunity and

with segregation in the era of Jim Crow laws. He started to publish *Negro World* magazine, which was a forum for his views.

The goal of the UNIA was "to unite all people of African ancestry of the world to one great body to establish a country and absolute government of their own." The motto of the UNIA was, "One God! One Aim! One Destiny!" Garvey aimed for African peoples all over the world to establish political and economic independence and self-determination. He hoped to start black-owned businesses to create opportunities for blacks where others did not exist. For instance, Garvey formed the Negro Factories Corporation to raise money for black-owned businesses like grocery stores, restaurants, laundries, and manufacturing plants.

Another of Garvey's business ventures was the Black Star Line. The Black Star Line was a steamship company with a number of purposes. One purpose was to allow blacks a way to enter into the lucrative international trade of goods. The Black Star Line would also provide a way for blacks to travel without the humiliating restrictions and segregation of other carriers during the Jim Crow era. The Black Star Line also fit in with Garvey's vision of emigration to Africa. Part of the so-called "Back to Africa" movement, Garvey imagined that blacks might successfully settle in their ancestral homelands.

Garvey's Back to Africa project with the Black Star Line was ultimately a failure. Garvey managed to purchase three steamships, but he paid too much for them, which was only the beginning of the Black Star Line's financial troubles. He also decided to move the Black Star Line's headquarters to Liberia, an African nation that had once been an American colony of free blacks. He moved the headquarters in 1920, in part because the U.S. federal government was investigating the Black Star Line for illegal activities. Garvey's arrest in 1922 for mail fraud only made the Black Star Line's financial troubles worse.

Despite the UNIA's message of racial uplift, the organization had tense relations with other black organizations at the time, such as the National Association for the Advancement of Colored People (NAACP). Garvey disliked these organizations because he did not support their goal of integration, or blacks and whites trying to live together. He did not believe that blacks would ever be fully accepted in the United States. Members of organizations like the NAACP did not like Garvey, and many considered him a fool. He also had controversial ideas about racial purity. He drew even more anger when it became public knowledge that he had held a secret meeting with a leader of the white supremacist group the Ku Klux Klan.

Controversy over Garvey's ideas became larger than life when he was put on trial for mail fraud. Facing a federal prison term, Garvey decided to

represent himself at his trial rather than use a lawyer. This proved to be unwise because he lost his case and served a sentence for fraud beginning in 1925. Garvey's decision to represent himself also made it seem to jurors and the public that his ideas were on trial. The focus was shifted away from whether he committed a crime to a sort of referendum, or popular vote, on Garvey's black nationalist ideas.

Garvey's jail sentence was commuted in 1927, and he was deported to Jamaica. Without Garvey's leadership, the UNIA in the United States lost its driving force. A new leader was elected, but many devoted Garvey supporters split off from the group to form "Garvey Clubs." Garvey himself operated a rival UNIA organization from abroad until his death in 1940. The original UNIA continued to operate out of New York City until 1941, when the headquarters moved to the country of Belize. Headquarters were moved back to New York in 1961, where the UNIA continued to operate into the twenty-first century.

GREAT MIGRATION TO NORTH OPENS NEW OPPORTUNITIES

The Great Migration was a large-scale movement of African Americans out of the South in the early twentieth century. The Great Migration lasted from roughly 1910 to 1930. During that time, nearly two million African Americans left the South and moved to cities in the North, Midwest, and West. They left because of the poor conditions and racial discrimination they encountered in the South. They were inspired to move by reports of available jobs and newfound economic opportunity in other parts of the country. The Great Migration produced a massive change in American demographics and had important consequences for American society and the economy.

Poor Conditions in the South Prompt an Exodus

African Americans were still overwhelmingly concentrated in the South as the nineteenth century came to a close. Fully 90 percent of all African Americans in the United States lived in the South as late as 1910. What is more, the vast majority of African Americans lived in rural areas. Only about 20 percent of African Americans living in the South lived in cities at that time. However, a combination of factors caused this distribution of the population to change.

First, African Americans faced a crippling lack of economic opportunity in the South. Most African Americans worked as share-croppers. Sharecropping is a system of farming in which a landowner rents land that he owns to a family and allows them to farm it. In exchange, he takes a percentage of the profits the family makes by selling their crops.

African American travellers from the South arrive in Chicago in 1918. During the Great Migration between 1910 and 1930, thousands of African Americans left their homes in the South for new opportunities in the North. *Chicago History Museum/ Hulton Archive/Getty Images*

White landowners in the South took huge percentages of sharecroppers' profits. The other jobs available to African Americans, such as domestic labor for women, also paid very little. As a result, most black families in the South were trapped in poverty.

Second, African Americans in the South faced racism in both legalized and illegal forms. Racism was legalized in the form of Jim Crow laws that mandated racial segregation. Racial segregation is the practice of forcing members of different races to live apart from one another. The purpose of segregation was to reinforce the idea that whites are superior to blacks and that blacks are second-class citizens. Illegal racism took the form of murders and violence. Lynching of African Americans had become commonplace in the South in the late 1800s and early 1900s. A lynching is an illegal killing (usually by hanging) of an individual by a mob. Thousands of African Americans were lynched in the South between 1880 and 1930.

Economic inequality and institutional racism pushed African Americans out of the South. Industrialization and the promise of new jobs drew them to the North, Midwest, and West. Industrialization occurs when a large number of businesses that previously operated on a small scale use new forms of energy and technology to produce and manufacture goods on a very large scale. Railroads, steelmakers, meatpackers, petroleum producers,

automobile manufacturers, and other industrialized businesses took root in cities across the North, West, and Midwest in the late 1800s. These new, large businesses needed workers. Many of them went so far as to send labor agents to the South to recruit African Americans to come work in their factories. Word of the employment opportunities available in the cities also traveled south through black churches and black newspapers.

Hundreds of Thousands Move North and West

The Great Migration was a two-stage process for many African Americans. First, families would leave sharecropping behind and make their way into cities in the South. Southern cities such as Memphis, Tennessee; New Orleans, Louisiana; and Birmingham, Alabama were hubs for railroads and steamboats. Migrating families' choice of their next destination was sometimes influenced by considerations of geography. For example, African Americans in rural Mississippi often took steamships north on the Mississippi River to St. Louis, Missouri. From there they might take a railroad to Chicago, Illinois, or Detroit, Michigan. Major railroad lines connected New Orleans to Los Angeles, California, and Atlanta, Georgia, to Philadelphia, Pennsylvania, and New York City.

The scope and scale of the Great Migration was truly monumental. Historians estimate that somewhere between 700,000 and one million African Americans migrated out of the South just during the three-year period spanning 1917 to 1920. African American migrants tended to congregate in certain cities, and the faces of those cities were transformed. Detroit's African American population increased from 6,000 to 120,000 during the Great Migration. In Cleveland, Ohio, the increase was from 8,500 to almost 72,000. Los Angeles saw its number of black residents spike from 7,600 to about 40,000. Big gains during the 1910–1930 era were realized in Chicago (44,000 to 234,000), New York City (100,000 to 328,000), and Philadelphia (84,000 to 221,000).

African Americans negotiated the logistics of the move out of the South in various creative ways. Sometimes entire families emigrated (left the region). Other times, only the husband in a family would go. Weekly black newspapers such as the *Chicago Defender* and the *Pittsburgh Courier* printed information about job availability in various cities. Porters, firemen, and other black railroad employees passed along helpful information to black passengers. African Americans who had completed the journey north or west would send letters to friends and family back in the South to tell them about neighborhoods, jobs, and housing. As a result, once a group of migrants from the same place in the South had settled in a new city in the North, other migrants from that region tended to be drawn to the same city.

The Great Migration Has Lasting Consequences

African Americans leaving the South often did not find ideal conditions in their new homes and workplaces. Cities in the North, Midwest, and West were growing so quickly that there was not enough housing. Living conditions in urban areas were often very bad. Many apartments lacked sanitation and clean water. There were more people who wanted apartments than there were available apartments, so landlords responded by charging high rents and requiring large groups of unrelated people to share one apartment. Racial discrimination was also a problem. Jim Crow laws and legalized segregation did not exist outside the South, but segregation was often just as much a way of life in the North. Homeowners and landlords simply refused to sell or rent to African Americans who tried to find a home outside of all-black neighborhoods.

Job opportunities also were not as plentiful and well-paying as many African Americans had hoped. Many employers practiced racial discrimination in hiring. As a consequence, the most desirable jobs in factories and on assembly lines were often unavailable to African Americans. Those employers who did hire black employees often required them to work in the dirtiest or most dangerous jobs. However, as white workers took factory and industrial jobs, they left behind jobs in stockyards, meatpacking facilities, mills, and labor yards. These jobs offered better conditions and higher wages than the jobs most African Americans had left behind in the South. In addition, the onset of World War I opened up many new industrial jobs to African Americans.

An African American female porter poses in uniform with her cleaning supplies in a New York City subway, c. 1917. Thousands of African Americans left the South for jobs in the North between 1910 and 1930. *Buyenlarge/Hulton Archive/Getty Images*

African American women in particular found the Great Migration to be a mixed blessing. In the South, the few African American women able to find paid work usually worked as domestic employees. Domestic employment takes place inside the house or home of a person who is not a member of the employee's family, such as employment as a nanny, butler, maid, or cook. Domestic employment is difficult and low-paying. Many black women left the South in hopes of finding work other than as domestic employees. They were largely disappointed. Many once again found themselves working as domestics. However, wages for domestic workers were substantially higher in the North than in the South—sometimes three times as high.

This increased earning power was a positive development for African American women.

Because of the Great Migration, the character of the rural and urban South changed significantly, and the character of many cities in the North, Midwest, and West was fundamentally and radically altered. By the end of the Great Migration, 30 percent of the nation's African American population lived outside the South. Indirectly, the Great Migration was a major factor in the civil rights movement. When 90 percent of all African Americans lived in the South, many whites in other parts of the country simply ignored the South's so-called "Negro Problem." But the Great Migration brought millions of white Americans into daily contact with African Americans. This increased the heterogeneity (the quality or state of having many different aspects or parts) of American cities and eventually helped give rise to broad-based support for civil rights and equality.

WORLD WAR II BRINGS SWIFT MIGRATION OUT OF THE SOUTH

A major demographic shift in the United States began in the 1940s. The beginning of World War II (1939–45) sparked a massive movement of African Americans out of the rural South that came to be known as the Second Great Migration. Some moved to cities in the South. Many others moved to cities in the North. They were drawn to urban areas by the promise of industrial jobs created by World War II. The war created a labor shortage as millions of industrial workers joined the military and went overseas. This labor shortage allowed African Americans to gain access to desirable jobs that had previously been unavailable to them because of racial discrimination. The movement of African Americans out of the South continued after the war ended. Between 1940 and 1970, approximately five million African Americans migrated out of the South.

The first Great Migration took place from about 1910 to 1930. During that time, almost two million African Americans left the South for points north and west. They were motivated to leave the South because of racial segregation and an overall lack of economic opportunity. They were drawn to cities in the North, Midwest, and West because of newly created industrial jobs. World War I (1914–18) accelerated this trend. The war increased demand for industrial production, and it also turned factory workers into soldiers, creating a need for new laborers.

Similar factors triggered the Second Great Migration. World War II began in Europe in the late 1930s. The United States did not join the war until 1941, but industrial production of military supplies for sale to American allies such as Great Britain and France began much earlier. At first, racial

A man protests racial discrimination in hiring during World War II. *John Vachon/Anthony Potter Collection/Getty Images*

discrimination prevented African Americans from getting the new jobs that were being created in wartime industries. That changed for two reasons.

First, black leaders took political action. A. Philip Randolph (1889–1979), a powerful union president, threatened to organize a march on Washington, D.C., in early 1941 to protest racial discrimination by

industrial companies. The threat of the march prompted President Franklin D. Roosevelt (1882–1945) to sign Executive Order 8802. An executive order is a document signed by the president that gives legally binding instructions on how to comply with the law. Executive Order 8802 made it illegal for industrial employers who were part of the war effort to refuse to hire African Americans.

Second, the entire American economy was transformed when the United States became an active participant in World War II. Hundreds of thousands of industrial workers were called overseas to serve as soldiers. This created a huge shortage of workers. Simultaneously, the war effort necessitated a vast increase in industrial production. The armed forces needed tanks, ships, planes, guns, bullets, and other supplies. The needs for supplies and labor were so acute that they largely drowned out old prejudices. Women made significant inroads into the industrial labor force for the first time. So, too, did African Americans. And just as industrialization had been geographically uneven two generations before, so too was the development of wartime industrial production facilities. The domestic war effort was concentrated in urban areas outside the South. By the early 1940s, the perception was that African Americans who were willing to leave the South were very likely to find industrial work elsewhere.

The decade of the 1940s saw the greatest migration ever of African Americans out of the South and into cities in the North, Midwest, and West. More than 1.5 million African Americans moved out of the South between 1940 and 1950. This was almost four times as many as had left between 1930 and 1940. California was a particularly attractive destination. About three hundred thousand African Americans moved to California between 1942 and 1945. The shipbuilding, aircraft assembly, munitions, and general manufacturing industries all had large factories in Los Angeles, San Diego, and the Bay Area (San Francisco and Oakland). At one point in 1943, more than six thousand African Americans were arriving in Los Angeles every month.

The movement of African Americans out of the South continued even after World War II concluded. Between 1950 and 1970, three million African Americans left the region. By the time the Second Great Migration concluded in 1970, the demographics of the United States in general, and its African American population in particular, had been fundamentally transformed. More than 80 percent of African Americans lived in cities in 1970. By contrast, 70 percent of the population as a whole lived in cities in 1970, and in 1910, nearly 80 percent of African Americans had lived in rural areas. In addition, only 53 percent of African Americans lived in the South in 1970, compared with 90 percent in 1910 and 70 percent in 1940.

INTEGRATION SPARKS VIOLENCE IN THE SUBURBS

In 1957, a violent incident in Levittown, Pennsylvania, symbolized a national struggle. Over the previous decade, increasing numbers of middle-class Americans had been moving from cities to suburbs. One of the first of these suburban communities was Levittown, Pennsylvania, built by Levitt & Sons builders. As in many of these communities, Levittown residents agreed that the neighborhood would be white-only. So when a black family moved into the neighborhood in 1957, violence erupted.

What is a Levittown?

The first Levittown was constructed on Long Island by the building company Levitt & Sons. After World War II, there was a need for more housing for soldiers returning from the war. Also, the mass production of automobiles made transportation easier. People no longer needed to live in cities near where they worked and shopped. William Levitt (1907–94) had a vision for a community outside of the city with its own roads, schools, and utilities. In 1947, he announced his plan to convert a former potato field into such a community. Levittown, New York, was a tremendous success. Levitt & Sons created a community with seventeen thousand homes over the course of three years. The neighborhood was popular with war veterans and their families.

The popularity of Levittown led to the spread of many similar communities. Some, like Levittown, Pennsylvania, were actually built by Levitt & Sons. These were called "Levittowns." There were also many similar suburban communities that were constructed during the postwar years. Levitt & Sons had, out of necessity, come up with a way to mass-produce houses cheaply to meet the growing demand of veterans returning from the war. These veterans had money to buy a house from the so-called G.I. Bill, the popular name for the Servicemen's Readjustment Act—a law that helped World War II veterans buy homes and get a college education when they returned to civilian life. This enabled millions of people to leave cities and become homeowners, something that had been out of reach in many parts of the country for all but fairly wealthy Americans. Levittown, Pennsylvania, was the second of the Levitt & Sons communities. Construction began soon after Levittown, New York, was completed.

"White Flight" and Restrictive Housing Covenants

The start of communities like Levittown coincided with a phenomenon called "white flight." White flight refers to the rapid movement of white Americans out of city centers and into suburban areas. This was in response to increasing numbers of African Americans moving to Northern cities in the early twentieth century. Many white Americans were uncomfortable with integrated living in city centers.

Supreme Court Rejects Restrictive Housing Covenants

The Supreme Court of the United States handed down a famous decision in 1948 that limited the reach of restrictive housing covenants. This was the same year that the Levittown community in Long Island, New York, was being established. The case was *Shelley v. Kraemer.*

The Shelleys were a black family who bought a house in St. Louis, Missouri. They were not aware when they bought the house that there was a restrictive covenant attached to the house. This covenant forbade the owner of the house to sell the property to African Americans. White families in the neighborhood filed suit in court to try to stop the Shelleys from moving into their house. The case went all the way to the United States Supreme Court.

The Supreme Court had to consider two questions in *Shelley v. Kraemer.* One was whether the Fourteenth Amendment to the United States Constitution prohibited restrictive housing covenants. The Fourteenth Amendment guarantees all people, regardless of their race, "equal protection of the law."

Levittowns had many strict community rules. Among these were rules about the racial makeup of the neighborhood. Levitt & Sons vowed that their communities were white-only, not for sale to African Americans. With the construction of the first Levittown in New York, the community had what was known as a restrictive housing covenant, which forbade sale or rent of Levittown homes to "any person other than members of the Caucasian race." These covenants were not legally enforceable by the time construction began in Levittown, Pennsylvania. But there were often still understandings or explicit agreements within neighborhoods that homes were not to be sold to nonwhite buyers.

Crisis in Levittown

In 1957, a black couple, Bill and Daisy Myers, bought a house in Levittown, Pennsylvania. That summer, they moved their family into their new home at 43 Deepgreen Lane. There were three small children in the family. Like many Levittowners, Bill and Daisy Myers had wanted to move to the suburbs so they would have more space for their growing family. They wanted a three-bedroom home.

The Supreme Court held that the Fourteenth Amendment did not stop private citizens from forming racially restrictive housing agreements. This is because the Fourteenth Amendment only regulates the conduct of states. The Court did hold, however, that these covenants could not be legally enforced. Courts were government entities and were part of the state. Thus, although the restrictive covenant in the Shelleys' neighborhood was legal, the neighbors could not obtain a court order preventing them from living there. Basically, the effect of *Shelley v. Kraemer* was to make these covenants unenforceable under the law.

The *Shelley v. Kraemer* decision would have important implications for the events that happened in Levittown, Pennsylvania. Even if Levittown was declared to be an all-white community, there was no legal way to stop black families from moving to the neighborhood. Sadly, this would not prevent racist residents of the neighborhood from resorting to violence to stop integration.

It was a mail carrier who alerted the neighborhood that a black family had moved into the neighborhood. On August 13, 1957, the postal worker ran down the streets shouting racial slurs, letting the neighbors know that an African American family was in their presence. Neighbors quickly gathered outside the Myers home. First, housewives congregated outside the home, some spitting at the family. By the end of the day, some three hundred people gathered outside the home. They threw rocks at the house and broke windows.

The violence would only continue and escalate. Crowds gathered every night. The Ku Klux Klan, an all-white terrorist organization, came to Levittown and set up a branch there. Crosses were burned in the front lawn of the Myers home (a common form of intimidation employed by the Ku Klux Klan). A cross was even burned on the lawn of some white neighbors who had defended the right of a black family to live in the neighborhood. The family one day discovered a bottle of gasoline with a cotton wick inserted outside the window of their infant daughter. The family had to flee their home three times in the first two weeks after moving in, for fear for their safety.

William and Daisy Myers in the kitchen of their home in Levittown, Pennsylvania, in 1957. The Myers endured violence and threats when they integrated the all-white community. *AP Images*

Local authorities were unwilling or unable to stop the violent attacks against Bill and Daisy Myers's family. Eventually, the state had to intervene. The governor sent in the state police to control the violence. Even still, the white neighbors had their way. After four years in Levittown, the Myers family moved to New York.

Primary Sources

MARCUS GARVEY EXPLAINS PAN-AFRICANISM (1921)

Marcus Garvey (1887–1940) was a prominent advocate of the philosophy of pan-Africanism in the early 1900s. Pan-Africanism is the idea that all people of African descent around the world, including African Americans, should unify into a global community. Garvey founded an organization to promote his pan-Africanist views called the Universal Negro Improvement Association. Below is the transcript of a radio address that Garvey gave in July 1921, titled "Explanation of the Objects of the Universal Negro Improvement Association." In the address, Garvey explains what pan-Africanism is and how his organization helps promote pan-Africanism.

Fellow citizens of Africa, I greet you in the name of the Universal Negro Improvement Association and African Communities League of the World. You may ask, what organization is that? It is for me to inform you that the Universal Negro Improvement Association is an organization that seeks to unite into one solid body the 400 million Negroes of the world; to link up the 50 million Negroes of the United States of America, with the 20 million Negroes of the West Indies, the 40 million Negroes of South and Central America with the 280 million Negroes of Africa, for the purpose of bettering our industrial, commercial, educational, social and political conditions.

As you are aware, the world in which we live today is divided into separate race groups and different nationalities. Each race and each nationality is **endeavoring** to work out its own destiny to the exclusion of other races and other nationalities. We hear the cry of England for the Englishman, of France for the Frenchman, of Germany for the Germans, of Ireland for the Irish, of Palestine for the Jews, of Japan for the Japanese, of China for the Chinese.

Endeavor
To strive to achieve or reach

We of the Universal Negro Improvement Association are raising the cry of Africa for the Africans, those at home and those abroad. There are 400 million Africans in the world who have Negro blood **coursing** through their veins. And we believe that the time has come to unite these 400 million people for the one common purpose of bettering their condition.

Course
To pass or travel quickly through

The great problem of the Negro for the last 500 years has been that of disunity. No one or no organization ever took the lead in uniting the Negro race, but within the last four years the Universal Negro Improvement Association has worked wonders in bringing together in one fold four million organized Negroes who are scattered in all parts of the world, being in the 48 states of the American union, all the West Indian Islands, and the countries of South and Central America and Africa. These 40 million

people are working to convert the rest of the 400 million scattered all over the world and it is for this purpose that we are asking you to join our ranks and to do the best you can to help us to bring about an **emancipated** race.

Emancipated
Freed from the restraint, control, or power of another

If anything praiseworthy is to be done, it must be done through unity. And it is for that reason that the Universal Negro Improvement Association calls upon every Negro in the United States to rally to its standard. We want to unite the Negro race in this country. We want every Negro to work for one common object, that of building a nation of his own on the great continent of Africa. That all Negroes all over the world are working for the establishment of a government in Africa means that it will be realized in another few years.

We want the moral and financial support of every Negro to make the dream a possibility. Already this organization has established itself in Liberia, West Africa, and has endeavored to do all that's possible to develop that Negro country to become a great industrial and commercial commonwealth.

Pioneers have been sent by this organization to Liberia and they are now laying the foundation upon which the 400 million Negroes of the world will build. If you believe that the Negro has a soul, if you believe that the Negro is a man, if you believe the Negro was **endowed** with the senses commonly given to other men by the Creator, then you must acknowledge that what other men have done, Negroes can do. We want to build up cities, nations, governments, industries of our own in Africa, so that we will be able to have the chance to rise from the lowest to the highest positions in the African commonwealth.

Endow
To provide or furnish with a gift or support

ALAIN LOCKE DESCRIBES THE HARLEM RENAISSANCE (1925)

Alain Locke (1886–1954) was a philosopher and writer who became famous for his work on the Harlem Renaissance. The Harlem Renaissance was a period of time during the 1920s when an exciting, vibrant community of African American businesspeople, intellectuals, artists, authors, and musicians came together in the Harlem neighborhood of New York City. Alain Locke was a central figure in the Harlem Renaissance. He published an important book in 1925 titled *The New Negro: An Interpretation.* The book was an anthology, which means it was a collection of shorter works by various authors. *The New Negro: An Interpretation* inspired a generation of African American authors and helped legitimize African American writing.

Alain Locke wrote an introductory foreword to *The New Negro: An Interpretation.* That foreword is reproduced below. In it, Locke comments on the ongoing demographic changes in the United States that are giving rise to a new genre of African American literature and art. Locke argues that African American art, writing, and music are the centerpiece of a new era in

African American culture. He also explains how this new literature, music, and art is both a product of, and a response to, ongoing demographic and social trends.

••••••••••••••••••••••••••

1. This volume aims to document the New Negro culturally and socially,—to register the transformations of the inner and outer life of the Negro in America that have so significantly taken place in the last few years. There is ample evidence of a New Negro in the latest phases of social change and progress, but still more in the internal world of the Negro mind and spirit. Here in the very heart of the folk-spirit are the essential forces, and folk interpretation is truly vital and representative only in terms of these. Of all the voluminous literature on the Negro, so much is mere external view and commentary that we may **warrantably** say that nine-tenths of it is about the Negro rather than of him, so that it is **the Negro problem** rather than the Negro that is known and **mooted** in the general mind. We turn therefore in the other direction to the elements of truest social **portraiture,** and discover in the artistic self-expression of the Negro today a new figure on the national canvas and a new force in the foreground of affairs. Whoever wishes to see the Negro in his essential traits, must seek the enlightenment of that self-portraiture which the present developments of Negro culture are offering. In these pages, without ignoring either the fact that there are important interactions between the national and the race life, or that the attitude of America toward the Negro is as important a factor as the attitude of the Negro toward America, we have nevertheless concentrated upon self-expression and the forces and motives of self-determination. So far as he is culturally articulate, we shall let the Negro speak for himself.

2. Yet the New Negro must be seen in the perspective of a New World, and especially of a New America. Europe **seething** in a dozen centers with emergent nationalities, Palestine full of a **renascent** Judaism—these are no more alive with the progressive forces of our era than the **quickened** centers of the lives of black folk. America seeking a new spiritual expansion and artistic maturity, trying to found an American literature, a national art, and a national music implies a Negro-American culture seeking the same satisfactions and objectives. Separate as it may be in color and substance, the culture of the Negro is of a pattern **integral** with the times and with its cultural setting. The achievements of the present generation have eventually made this apparent. Liberal minds today cannot be asked to peer with sympathetic curiosity into the darkened Ghetto of a segregated race life. That was yesterday. Nor must they expect to find a mind and soul bizarre and alien as the mind of a savage, or even as naïve and refreshing as the mind of the peasant or the child. That too was yesterday, and the day before. Now that there is cultural adolescence and then approach to maturity,—there has come a development that makes these phases of Negro life only an interesting and significant segment of the general American scene.

Warrantably
Correctly, rightly, or with good reason

The Negro Problem
A term used in the late 1800s and early 1900s by authors who were discussing race relations in the United States

Moot
To bring up for discussion or debate

Portraiture
Portrayal or the making of portraits

Seething
Suffering from violent internal excitement

Renascent
Reborn and rising again into being or vigor

Quicken
To enter a phase of active growth and development

Integral
Central or essential to

3. Until recently, except talent here and there, the main stream of this development has run in the special channels of "race literature" and "race journalism." Particularly as a literary movement, it has gradually gathered momentum in the effort and output of such progressive race periodicals as the *Crisis* under the editorship of Dr. Du Bois and more lately, through the quickening encouragement of Charles Johnson, in the brilliant pages of *Opportunity, a Journal of Negro Life*. But more and more the creative talents of the race have been taken up into the general journalistic, literary and artistic agencies, as the wide range of the acknowledgments of the material here collected will in itself be sufficient to demonstrate. Recently in a project of *The Survey Graphic,* whose Harlem Number of March, 1925, has been taken by kind permission as the **nucleus** of this book, the whole movement was presented as it is **epitomized** in the progressive Negro community of the American metropolis. Enlarging this stage we are now presenting the New Negro in a national and even international scope. Although there are few centers that can be pointed out approximating Harlem's significance, the full significance of that even is a racial awakening on a national and perhaps even a world scale.

Nucleus
The basic, core, essential part

Epitomize
To serve as the typical or ideal example

4. That is why our comparison is taken with those nascent movements of folk-expression and self-determination which are playing a creative part in the world to-day. The **galvanizing** shocks and reactions of the last few years are making by subtle processes of internal reorganization a race out of its own disunited and **apathetic** elements. A race experience penetrated in this way invariably flowers. As in India, in China, in Egypt, Ireland, Russia, Bohemia, Palestine and Mexico, we are witnessing the resurgence of a people: it has aptly been said,—"For all who read the signs aright, such a dramatic flowering of a new race-spirit is taking place close at home—among American Negroes."

Galvanize
To stimulate or excite into action

Apathetic
Lazy, unmotivated, or disinterested

5. Negro life is not only establishing new contacts and founding new centers, it is finding a new soul. There is a fresh spiritual and cultural focusing. We have, as the **heralding** sign, an unusual outburst of creative expression. There is a renewed race-spirit that consciously and proudly sets itself apart. Justifiably then, we speak of the offerings of this book embodying these ripening forces as culled from the first fruits of the Negro Renaissance.

Herald
To give notice or signal the approach of

Alain Locke
Washington, D.C.
November, 1925

Research and Activity Ideas

1. Nearly two million African Americans left the South and moved to cities in the North, Midwest, and West during the Great Migration. The Great Migration lasted from 1910 to 1930. Many of these African Americans left home in response to the promise of available industrial jobs in faraway cities. Some industrial companies actively recruited African Americans in the South to come work for them. Imagine that it is 1920 and you have been hired by an automobile manufacturer in Detroit, Michigan, to recruit African Americans in the South to come work for your company. You have decided to place an advertisement in a black weekly newspaper. Write a recruiting letter that you could use in the advertisement. What are some of the benefits of working for your company? What are some of the ways that life in Detroit might be better than life in the South?

2. The "Back to Africa" movement was a political movement that advocated for the creation of settlements of African Americans in their ancestral homeland of Africa. The idea behind the movement was that black and white Americans could not coexist peacefully in the same country. Some white supporters of the movement were sympathetic to African Americans and believed a return to Africa offered the best chance for a good life. Other white supporters of the movement were racists who believed the United States would be better off without any black residents. Write an essay in which you analyze the following question: How could people with such opposite views support the same position? Give at least two reasons why you think each side would have come to the conclusion it did.

3. African Americans began creating all-black towns in the late 1800s. This trend continued into the early 1900s. All-black towns were a refuge (a place of safety and escape) from the vicious racism African Americans then encountered on a daily basis. Imagine that you are an African American community leader (such as a minister or a teacher) in the South in the early 1900s. Write a speech in which you try to persuade a group of your fellow African American community members to leave their current homes and help you found an all-black town. Make at least three arguments about why their lives would be improved, be it for economic, social, or family reasons, by living in an all-black town.

4. Violence erupted in Levittown, Pennsylvania, in 1957, when an African American family moved into the previously all-white community. One of the reasons many white families had moved out of cities and into

Levittown was so they would only have white neighbors. When a black family moved into the community, white residents used intimidation, harassment, property damage, and violence to force the black family to leave. Have a group discussion with several of your classmates about the violence in Levittown. Why do you think the presence of one African American family in the neighborhood triggered such anger among the white residents? How much have racial attitudes and relations improved since 1957? What do you think the reaction would be today if a black family moved into a predominantly white neighborhood?

5. Carter Woodson (1875–1950) made a noteworthy contribution to the field of demographics by coming up with the idea for Negro History Week. Negro History Week was celebrated for the first time in February 1926 in public schools across the country with study and appreciation of African Americans' contributions to American history. In 1976 it was expanded into Black History Month, which is celebrated every February. Write an essay in which you address the following three questions: What are some reasons why it was important in 1926 to have a week specifically dedicated to the study of African Americans in history? Why do you think it was expanded into a full month fifty years later? What are some reasons why it is still important to celebrate Black History Month even today?

For More Information

BOOKS

Boehm, Lisa Krissoff. *Making a Way Out of No Way: African American Women and the Second Great Migration.* Oxford: University Press of Mississippi, 2009.

Garvey, Marcus, and Bob Blaisdell. *Selected Writings and Speeches of Marcus Garvey.* Long Island, NY: Dover Publications, 2005.

Gregory, James N. *The Southern Diaspora: How the Great Migrations of Black and White Southerners Transformed America.* Chapel Hill: University of North Carolina Press, 2005.

Kushner, David. *Levittown: Two Families, One Tycoon, and the Fight for Civil Rights in America's Legendary Suburb.* New York: Walker & Company, 2009.

Lemann, Nicholas. *The Promised Land: The Great Black Migration and How It Changed America.* New York: Vintage Books, 1992.

Painter, Nell Irvin. *Exodusters: Black Migration to Kansas After Reconstruction: The First Major Migration to the North of Ex-Slaves.* New York: W. W. Norton & Company, 1992.

Quarles, Benjamin. *The Negro in the Making of America*. New York: Macmillan Publishing Company, 1964.

Watkins-Owens, Irma. *Blood Relations: Caribbean Immigrants and the Harlem Community, 1900–1930 (Blacks in the Diaspora)*. Bloomington, IN: Indiana University Press, 1996.

WEB SITES

Lawrence, Jacob. *The Great Migration: A Story in Paintings*. http://www.columbia.edu/itc/history/odonnell/w1010/edit/migration/migration.html (accessed on December 28, 2009).

Locke, Alain. "The New Negro." Introduction to *The New Negro*. http://us.history.wisc.edu/hist102/pdocs/locke_new.pdf (accessed on December 28, 2009).

Universal Negro Improvement Association–African Communities League. *History*. http://www.unia-acl.org/ (accessed on December 28, 2009).

chapter six

Education

Chronology

1867 March 2 Congress passes a law creating Howard University, a college and medical school in Washington, D.C., for the education of African American students.

1881 April 11 Sophia B. Packard and Harriet E. Giles found the Atlanta Baptist Female Seminary, a school dedicated to providing educational opportunities for African American women, in Atlanta, Georgia.

1881 July 4 Booker T. Washington helps found the Tuskegee Normal and Industrial Institute, an all-black school that focuses on training teachers and teaching vocational skills, in Macon County, Alabama.

1884 The Atlanta Baptist Female Seminary is renamed the Spelman Seminary in honor of Laura Spelman Rockefeller, the wife of the philanthropist John D. Rockefeller. The school assumes its current name, Spelman College, in 1924.

1890 August 30 Congress passes the second Morrill Act, which requires states receiving federal funding for their land-grant colleges to allow African American students to attend those colleges.

1890 Kelly Miller begins a forty-four-year career at Howard University in Washington, D.C., when he is hired as a professor of mathematics.

1895 W. E. B. Du Bois becomes the first African American to earn a Ph.D. (the highest academic degree awarded by American universities) from Harvard University.

1895 September 11 Booker T. Washington delivers his famous "Atlanta Compromise" speech in Atlanta, Georgia, arguing that African Americans should pursue hard work and economic self-sufficiency rather than participation in the political process.

1903 September W. E. B. Du Bois publishes an article entitled "The Talented Tenth," in which he argues that the best way for the African American community to improve itself is for its smartest members to attain an education and use that education to lift up the fortunes of everyone around them.

1921 Benjamin Griffith Brawley publishes the famous scholarly book, *Social History of the American Negro*. The book remains in print for fifty years.

1926 Carter G. Woodson, a historian, chooses the week in February that falls between the birthdays of Frederick Douglass and Abraham Lincoln for the first-ever celebration of Negro History Week.

1926 June 20 Mordecai Wyatt Johnson becomes the first African American to

serve as the president of Howard University.

1928 Congress passes a law guaranteeing that Howard University will receive annual federal funding. The funding guarantee helps Howard grow in size from about one thousand seven hundred students in 1926 to more than six thousand students by 1960.

1930 May 10 Five historically African American fraternities and sororities form the National Pan-Hellenic Council on the campus of Howard University in Washington, D.C.

1937 November 7 Juliette Derricotte, a well-respected young educator, dies the day after she is involved in a car accident in Dalton, Georgia. Her death is partially the result of the local "whites-only" hospital's refusal to admit her for treatment.

1939 Psychologists Kenneth and Mamie Clark begin a famous series of experiments in which African American children use dolls and coloring books to express their ideas about race. These experiments demonstrate the psychological damage done by segregation. The United States Supreme Court refers to the Clarks' experiments in making its 1954 ruling in *Brown v. Board of Education.*

1940 July Benjamin Mays is named president of Morehouse College, an all-male, all-black college in Atlanta, Georgia. He serves as president of Morehouse for the next twenty-seven years.

1944 April 25 The United Negro College Fund is incorporated as a charitable organization dedicated to raising scholarship money to support African American students and historically black colleges and universities.

1946 February 28 Kenneth and Mamie Clark found the Northside Center for Child Development in the Harlem neighborhood of New York City to provide psychological services and guidance to emotionally disturbed children and their parents.

1950 June 5 The Supreme Court deals a major blow to the practice of racial segregation in the field of education with its decision in *Sweatt v. Painter.*

1954 May 17 The Supreme Court unanimously rules in *Brown v. Board of Education* that racial segregation in public schools is an unconstitutional violation of the Fourteenth Amendment.

1955 October 22 Willa Player becomes the first African American woman to head a four-year college in the United States when she is named president of Bennett College in Greensboro, North Carolina.

1957 Septima Clark founds the first citizenship school, designed to teach African American adults the reading skills they need to pass literacy tests and register as voters, on Johns Island, South Carolina.

1957 **September 24** President Dwight D. Eisenhower deploys a U.S. Army division to Little Rock, Arkansas, and federalizes the Arkansas National Guard in order to protect nine African American students integrating Little Rock Central High School. The students become known as the "Little Rock Nine."

1958 **February 11** Martin Luther King Jr. accepts Willa Player's invitation to speak on the campus of Bennett College after no other public or private organization in Greensboro, North Carolina, will allow him to use its facilities.

1960 **June 30** Mordecai Wyatt Johnson retires from the presidency of Howard University after thirty-four years of service, having overseen an increase in enrollment from two thousand to six thousand.

1962 **October 1** James Meredith, needing the protection of a military escort, becomes the first African American student to register and enroll in classes at the University of Mississippi.

1963 **June 11** Alabama governor George Wallace, fulfilling a campaign promise, literally stands in front of the door of the Forest Auditorium of the University of Alabama to block the entrance of two African Americans attempting to enroll. Federal marshals, the Alabama National Guard, and the state deputy attorney general confront him, and he stands aside.

1964 Almost two hundred citizenship schools, virtually all of which are run by teachers who trained under Septima Clark, are in operation throughout the South.

1965 **November 8** The Higher Education Act of 1965 designates 105 historically black colleges and universities (HBCUs) that are eligible to receive extra federal funding because of their historical and ongoing contributions to providing African Americans with educational opportunities.

Overview

African Americans encountered numerous challenges and also made significant progress in the field of education during the period between 1865 and 1965. African Americans had very little access to educational facilities and opportunities in the years immediately following the end of the Civil War (1861–65). But the number of colleges and universities that enrolled African American students gradually grew. Individual African Americans made important gains in the field of education and used their position as educators to contribute to society. The legal practice of racial segregation in education finally came to an end in the middle of the twentieth century, and African Americans fought for their right to integrated education.

Opening schools to serve the needs of the African American community was a top priority in the aftermath of the Civil War. The Freedmen's Bureau met that need at the elementary and secondary school level. Higher education posed more of a challenge. Progress in enabling capable African Americans to enroll in colleges and universities was slow. One of the most important advances in higher education for African Americans came in 1867, when Congress established Howard University in Washington, D.C. Howard was open to students of all races, but its students were almost all African American. Howard went on to become the center of African American intellectual life by the early 1900s.

Progress on the higher education front continued in the last two decades of the nineteenth century. The Atlanta Baptist Female Seminary was founded in Atlanta, Georgia, in 1881. The seminary enrolled an all-female student body. It thus filled an important gap in the higher education landscape for African Americans. It was eventually renamed Spelman College. Congress passed the second Morrill Act in 1890, which required states that operated agricultural and mechanical colleges to admit African American students to those colleges. The second Morrill Act greatly expanded the educational options available to African Americans, especially in the South.

Individual African Americans were also making their mark in the field of education. Anna Julia Cooper (1858–1964) published *A Voice from the South: By a Black Woman from the South* in 1892. Cooper's book was a passionate defense of the value of higher education for African Americans, especially African American women. W. E. B. Du Bois (1868–1963) made history in 1895 when he became the first African American to earn a Ph.D. (the highest academic degree awarded by American universities) from Harvard University. And Booker T. Washington (1856–1915) opened the

Tuskegee Normal and Industrial Institute in Alabama, where he put into practice his theory that African Americans should concentrate their energies on vocational, job-related training.

African American professors and scholars began to leave their mark on institutions of higher education as the nineteenth century gave way to the twentieth. Benjamin Griffith Brawley (1882–1939) became a prominent educator and author. He wrote or edited seventeen books over the course of his career, some of which were used even in the twenty-first century as textbooks in college courses. Kelly Miller (1863–1939) spent forty-four years as a professor at Howard University. Miller used his teaching and writing to argue against prejudiced, stereotypical beliefs about and views of African Americans. Juliette Aline Derricotte (1897–1931) was a promising young educator and academic in the early 1900s, whose career was tragically cut short when she died from injuries she sustained in an automobile accident. Her death—which was partially caused by the local "whites only" hospital's refusal to provide her medical care—caused outrage among African Americans nationwide.

African Americans continued to make important breakthroughs and advances in education in the early 1900s. Racial prejudice was rampant on college campuses at this time. This prejudice created a hostile environment for many African American students. African American students responded by starting fraternities and sororities (social and service organizations on college campuses). A group of five of these all-black fraternities and sororities came together in 1930 to form the National Pan-Hellenic Council (NPHC). The NPHC continued to be an important presence on college campuses into the twenty-first century.

African Americans fought back against the prejudice they faced in education through other avenues as well. Carter G. Woodson (1875–1950), a professional historian, came up with the idea for Negro History Week in 1926. Woodson wanted to push schools and colleges to recognize African Americans' positive contributions to American history. Negro History Week was an immediate success. It has been renamed and expanded into Black History Month. In the middle of the twentieth century, a group of African American university presidents joined together to raise money to support African American college students. The United Negro College Fund was founded in 1944, and it continues to raise millions of dollars each year to provide scholarships and financial assistance to historically black colleges and universities and African American students.

African Americans continued to blaze new trails in education as the twentieth century wore on, often rising to the highest levels of their profession. Benjamin Mays (1894–1984) served as the president of

Morehouse College (an all-black, all-men's college in Atlanta, Georgia), from 1940 to 1967. In 1955, Willa B. Player (1909–2003) became the first African American woman ever to head a four-year college when she became the president of Bennett College in Greensboro, North Carolina. Mordecai Johnson's election to the presidency of Howard University in Washington, D.C., in 1926 made him the first African American ever to hold the highest position at that esteemed university.

Perhaps the single most important moment in the history of African Americans' struggle for equal access to educational opportunities came in 1954. That year, the Supreme Court unanimously ruled in *Brown v. Board of Education* that racially segregated schools are unconstitutional. The Court relied heavily on research performed by two African American psychologists and researchers, Kenneth Clark (1914–2005) and Mamie Clark (1917–83), in its decision in *Brown*. The Court's decision touched off decades of struggle in the South. James Meredith (1933–) assumed a pivotal role in that struggle in 1962 when he became the first African American student ever to enroll at the University of Mississippi. And Septima Poinsette Clark (1898–1987) was a lifelong teacher who was so selfless in putting her skills as an educator to use in the service of racial integration and equal rights that she became known as the "queen mother" of the civil rights movement.

BENJAMIN GRIFFITH BRAWLEY (1882–1939)

Benjamin Griffith Brawley (1882–1939) was a prominent educator and author in the early twentieth century. Brawley used his scholarship and his teaching to explain and celebrate the African American experience and tradition. He built a reputation as an excellent teacher. Brawley was also a very productive author. He wrote or edited seventeen books over the course of his career. Some of the books he wrote continued to be used as textbooks in college courses in the twenty-first century. His best-known works were in the areas of literature, history, and African American studies.

Benjamin Griffith Brawley was born on April 22, 1882, in Columbia, South Carolina. His parents were Margaret Dickerson Brawley and Edward McKnight Brawley. His father was a Baptist preacher and a university professor. His father's professional duties required the family to move frequently during Brawley's youth. The young Brawley attended schools in various cities in the South, including Petersburg, Virginia, and Nashville, Tennessee. His mother made sure that the family's travels did not interfere with Brawley's education. She supplemented his schoolwork with education at home.

Benjamin Griffith Brawley. *Fisk University Library. Reproduced by permission.*

Brawley left his family when he was thirteen years old to continue his education in Atlanta, Georgia. He began attending the preparatory school at the Atlanta Baptist College. A preparatory school is a high school that is specifically designed to prepare its students to enroll at a college or university. Brawley was an excellent student. His best class was English. He was also involved in several extracurricular activities. He was captain of the football team and helped found a student-run literary journal. He enrolled in the Atlanta Baptist College after he graduated from high school. Atlanta Baptist College was renamed Morehouse College in 1913. It is an all-male school and one of the most famous historically black colleges in the United States. Brawley graduated from college in 1901, when he was just nineteen years old.

A Teaching Career Begins

Brawley found himself drawn to the profession of teaching. He spent his first year after graduating college as a teacher at a rural, one-room school in central Florida. He then moved back to Atlanta, where he had been hired to teach English and Latin at his alma mater (the college from which he graduated). Brawley spent the next six school years teaching in Atlanta. He published his first two books, both collections of poetry, during this time. He spent his summers in Chicago, Illinois, studying to earn a second bachelor's degree. He earned that degree in 1907 from the University of Chicago. Brawley spent the next year in Boston, Massachusetts, studying at Harvard University. He earned a master's degree from Harvard in 1908.

Brawley continued his career as a teacher at Atlanta Baptist College after graduating from Harvard. He spent two more years there before accepting a job offer from Howard University in Washington, D.C. Howard University was an all-black university that was the center of African American intellectual life in the first part of the twentieth century. Brawley was a professor of literature at Howard for two years. He met his wife, Hilda Damaris Prowd, while he was working at Howard. The two were married on July 20, 1912. Shortly after his marriage, Brawley accepted a position at Atlanta Baptist College as both a professor and the dean (the administrative head of the college).

Becomes a Scholar

Brawley spent the next eight years as a professor at Atlanta Baptist College. He continued to teach classes and also began writing scholarly works. He published five books between 1913 and 1919 on African American history, literature, and culture. Brawley briefly left academia (employment as a professor at a university) in the early 1920s. He traveled to Liberia, a small country in West Africa, in 1921. Liberia was a refuge (place of escape) for many African Americans who were former slaves. Brawley studied the schools, culture, and society in Liberia. He also was ordained as a Baptist minister in 1921. He spent a year serving as the pastor of a Baptist church in Massachusetts.

Brawley returned to working as a professor in 1922. He joined the faculty of Shaw University in Raleigh, North Carolina. His father was also a professor at Shaw. The previous year, Brawley had published the book that would become his most famous, *Social History of the American Negro.* Brawley stayed at Shaw for the next nine years. He published a literature textbook in 1925 designed for use in college courses. The textbook included the work of famous African American authors. Brawley also

Brawley's Writings

Benjamin Brawley was a prolific writer, which means he was an inventive and productive writer and wrote very frequently. Below is a list of all seventeen of the works that Brawley authored or edited over the course of his lifetime.

- *A Toast to Love and Death* (poems), 1902.
- *The Problem, and Other Poems* (poems), 1905.
- *A Short History of the American Negro,* 1913.
- *History of Morehouse College,* 1917.
- *The Seven Sleepers of Ephesus: A Lyrical Legend* (poems), 1917.
- *The Negro in Literature and Art in the United States,* 1918.
- *Women of Achievement,* 1919.
- *A Short History of the English Drama,* 1921.
- *A Social History of the American Negro,* 1921.
- *A New Survey of English Literature: A Textbook for Colleges,* 1925.
- *Doctor Dillard of the Jeanes Fund,* 1930.
- *History of the English Hymn,* 1932.
- *Early Negro American Writers* (edited volume), 1935.
- *Paul Laurence Dunbar, Poet of His People,* 1936.
- *Negro Builders and Heroes,* 1937.
- *The Negro Genius: A New Appraisal of the Achievement of the American Negro in Literature and the Fine Arts,* 1937.
- *The Best Stories of Paul Laurence Dunbar* (edited volume), 1938.

developed a reputation as a gifted, dedicated teacher during his time at Shaw.

Brawley finished his career at Howard University. He returned to Howard in 1931. He remained active as a scholar, writing or editing five more books in the 1930s. His scholarship centered on the African American experience, which he argued was an important but underappreciated part of American history and literature. His work catalogued and explained the accomplishments and contributions of African American artists and scholars. Brawley suffered a stroke in 1939. He died after a short illness on February 1, 1939, at the age of fifty-six.

★ KENNETH CLARK (1914–2005)

MAMIE PHIPPS CLARK (1917–1983)

Kenneth Clark and Mamie Phipps Clark were famous psychologists and educational researchers whose work helped advance the civil rights movement. The Clarks worked together as a husband-and-wife research team. Their most famous research was an experiment involving dolls that measured children's attitudes toward race. The results of this experiment helped convince the Supreme Court of the United States to rule in *Brown v. Board of Education* that segregated schools are unconstitutional. Both Kenneth and Mamie Clark conducted other valuable research over the course of their careers. They also co-founded a children's development center in the Harlem neighborhood of New York City.

Two Natural Born Academics Come Together

Kenneth Bancroft Clark was born on July 24, 1914, in the Panama Canal Zone. The Panama Canal Zone is a small strip of land surrounding the Panama Canal that was technically part of the United States from 1903

Kenneth Clark teamed up with his wife Mamie in a series of psychological experiments that proved that segregation in education had a negative impact on African American children. *William E. Sauro/New York Times Co./Getty Images*

to 1979. Clark, his sister, and his mother moved to New York when he was four years old. He attended school in Harlem during the Harlem Renaissance, a period during which African American cultural, intellectual, artistic, and political life flourished in New York. Clark had a very positive experience in school. He was an outstanding student and decided to attend Howard University in Washington, D.C., after graduation. It was at Howard that he would meet his future wife.

Mamie Clark was born Mamie Katherine Phipps on October 18, 1917, in Hot Springs, Arkansas. She was one of two children born to Harold and Katie Phipps. Clark graduated from high school in 1934. The country was in the grip of the Great Depression at that time, so her family did not have enough money to pay for her to go to college. Plus, there were very few options in the South for African Americans who wanted to attend college. However, Mamie Clark was able to win a scholarship to Howard University, the leading all-black university in the nation at the time.

Kenneth and Mamie Clark met in a class on abnormal psychology. Mamie was a student in the class, and Kenneth, who was a graduate student at the time, was a teaching assistant. They were drawn to each other immediately. Kenneth convinced Mamie to change her major from mathematics to psychology, and the two began working together. Kenneth earned his master's degree in 1936 and then enrolled in a doctoral program in psychology at Columbia University in New York. In 1940, he became the first African American to earn a Ph.D. from Columbia. Mamie followed a similar path. She earned a B.S. in psychology from Howard in 1938 and a master's degree in 1939. In 1940, she joined Kenneth at Columbia, and in 1944, she too earned a Ph.D. in psychology.

Research Helps End Segregation

The Clarks performed a series of famous experiments in the late 1930s and 1940s that later helped bring about the end of racial segregation in education. The experiments were designed to measure the effects of racism—specifically, racial segregation in education—on African American children. They published five articles between 1939 and 1950 documenting the results of their research.

The Clarks performed a series of experiments on two groups of African American children. One group attended segregated schools in Washington, D.C. The other group attended racially integrated schools in New York. In one experiment, the Clarks used dolls to measure children's attitudes about race. They gave the children a white doll and a black doll and asked them to pick one with which to play. They found that most African American children preferred to play with the white doll. When the Clarks asked the children to describe each of the dolls, they used positive words such as

"good" and "pretty" to describe the white doll and negative words such as "bad" and "ugly" to describe the black doll. The Clarks also asked the children to color in a picture of a child with the color of their own skin. The African American children consistently chose a color that was a lighter shade than their own skin.

The Clarks used the results of their experiments to argue that racial segregation had negative effects on African American children's self-esteem and self-image. The Clarks explained that the existence of racial segregation sent a powerful message to these children that being African American was negative and abnormal. Their conclusion was that African American children internalized (came to accept a belief as their own) the idea that African Americans were inferior to white Americans at a very young age. Their research also showed that this belief made it harder for African American children to do well in school.

The Clarks' research was groundbreaking and important. It was published in major academic journals and well received in the academic community. Perhaps more importantly, it later helped the National Association for the Advancement of Colored People (NAACP) convince the Supreme Court to declare that segregation in education was unconstitutional in the landmark case of *Brown v. Board of Education* (1954). The Supreme Court relied on the Clarks' research to support its conclusion that segregation did permanent damage to African American children. The Clarks thus contributed to one of the most important civil rights victories of the twentieth century.

Fighting for Integration

Although their work helped to prohibit legalized segregation in schools, the Clarks saw all around them the effects of segregation imposed by culture and society. The Clarks favored integration of schools, so that all students would experience more diversity of culture and viewpoints. In 1958, one of the most famous students in the country who attended an integrated school—Minnijean Brown (1941–), one of the first nine African American students to attend a previously all-white school in Little Rock, Arkansas—was expelled from school after an incident with a white student. The Clarks took her in and arranged for her to attend school near their own residence, in a suburb just outside New York City.

Frustrated by the continued segregation, Kenneth Clark started an organization called Harlem Youth Opportunities Unlimited (HARYOU) in 1962. The organization was wildly successful, but because of the intense political maneuvering and massive amounts of funding involved in the organization's educational projects, Clark found himself driven out of his own organization. He used the experience as inspiration to write *Dark*

Ghetto: Dilemmas of Social Power (1965), a book that examines the challenges faced by poor urban African Americans.

The Clarks Enjoy Long, Productive Careers

Both Kenneth and Mamie Clark remained active and successful in the field of psychology for many years after their research on segregation. The Clarks co-founded the Northside Center for Child Development in Harlem in 1946. The center provided psychological services for emotionally disturbed children and guidance to their parents. Mamie Clark served as the executive director of the Northside Center for thirty-four years, until her retirement in 1980. She and Kenneth co-authored several articles on personality development in children based on research they conducted at the center.

In 1966, Kenneth Clark was selected as a member of the New York State Board of Regents, an organization that effectively serves as the head of education in the state. Kenneth was the first African American to serve in such a position. He became the first African American to serve as the president of the American Psychological Association (APA) in 1970. Kenneth retired from his last job as a professor five years later. The Clarks founded a consulting firm in 1975 that helped large corporations recruit and retain minority employees. Mamie Clark died of cancer in 1983 when she was sixty-five years old. Kenneth Clark remained active in the consulting business well into the 1990s. He received an Outstanding Lifetime Achievement Award from the APA in 1994. Kenneth Clark died in 2005 at the age of ninety at his home in New York.

SEPTIMA POINSETTE CLARK (1898–1987)

Septima Poinsette Clark was a lifelong teacher who put her skills as an educator to use in the service of the civil rights movement. Clark organized a series of "citizenship schools" in the South in the 1950s and 1960s. These citizenship schools provided basic education and training to African American adults who had been unable to attend school. Graduates of the citizenship schools used their newly acquired reading skills to register as voters and participate in the civil rights movement. Clark's energy and leadership enabled her to play a role not just as an organizer but as an inspiration. She became known as the "queen mother" of the civil rights movement.

Training to Teach

Septima Poinsette Clark was born on May 3, 1898, in Charleston, South Carolina. She was the second of eight children born to Peter and

Victoria Poinsette. Clark's father had been born into slavery. He worked as a cook aboard ships that sailed between New York and Miami, Florida, when Clark was growing up. Her mother worked out of the family's home as a laundress (someone who earns money by washing other people's clothes). Victoria Poinsette had grown up on the island of Haiti. She had learned to read and write as a child and was proud of her education. Clark's father also firmly believed in the value of education.

Septima Clark, seated, with Bernice Robinson in 1986. Both women taught at the nation's first citizenship school on Johns Island. *© Karen Kasmauski/Corbis*

Clark attended private school for most of her education. She spent her middle school and high school years at the Avery Normal School. "Normal school" was the nineteenth-century term for a school that trained future teachers in the standards, or "norms," of education. Clark's teachers at Avery encouraged her to attend college after she finished high school. Her family, however, did not have enough money to pay for college.

Clark began working as a teacher in 1916, the same year she graduated from high school. Clark's hometown of Charleston had passed a law that made it illegal for African Americans to teach in public schools. As a result, Clark had to begin her teaching career elsewhere. She moved to Johns Island, South Carolina, a small town about eleven miles west of Charleston. She spent three years teaching at an all-black school in Johns Island. The school was small and lacked resources. Clark's single class included students who were in the fourth through the eighth grades. Clark focused on teaching her students how to read and write.

An Activist Is Born

Clark began to use her skills as a teacher for political purposes while she was in Johns Island. She joined the National Association for the Advancement of Colored People (NAACP) and began teaching adult reading classes in the evening. Clark returned to Charleston in 1919. She taught at the Avery Normal School for a year. She also helped the NAACP successfully lobby to repeal the law that prohibited African Americans from teaching in public schools. She taught at public schools in Charleston for the next several years. She also married a man named Nerie Clark in 1920. The couple had two children. One died as an infant. Clark's husband then died in 1924, and she was left to raise her son by herself.

Clark moved back to Johns Island in 1926 and spent three more years working as a teacher there. She then moved to Columbia, South Carolina, the state capital. She worked as an elementary school teacher in Columbia for seventeen years. She joined numerous civic and political organizations and stepped up her level of involvement in the NAACP. For example, she helped organize a lawsuit on behalf of African American teachers in the Columbia school district who were earning only half as much in salary as their white counterparts. Clark also used this time to further her own education. She used her summers to earn first a bachelor's degree and then a master's degree.

Clark's involvement with the NAACP eventually forced her to leave her home state. In 1954, the South Carolina legislature passed a law that made it illegal for anyone who worked in a public school to belong to the NAACP. Clark refused to drop out of the organization, so she was fired from her job. Clark relocated to Tennessee. She took a job in June 1956 as the director of education at the Highlander Folk School just outside of Chattanooga. Highlander was not a traditional school. It was a place where activists, community leaders, and political organizers would go to work together. Clark was invigorated by her work at Highlander. She began to take a more active role in the fledgling civil rights movement.

Citizenship Schools Help Register Voters

Clark decided to use her position at Highlander to help African Americans in her home state of South Carolina register to vote. She founded a citizenship school in Johns Island in 1957. South Carolina, as well as many other states in the South, had a law in place at the time that prohibited anyone who could not read from voting. The law was enforced by requiring anyone who wanted to register to vote to read a portion of the South Carolina state constitution and explain what it meant. Southern states had long used segregation to force African Americans to attend inferior schools and deny them access to a good education. As a result, many African Americans failed the literacy tests and could not vote. Many literate blacks also "failed" the literacy tests while illiterate whites were allowed to register and vote without problems. The tests were required, given, and scored selectively and unfairly with the purpose of keeping African Americans from voting.

Clark's citizenship school taught adults who wanted to register to vote how to read. Clark used innovative methods. Instead of using textbooks, the school started out using materials from daily life such as newspapers and mail-order catalogues. After students had learned the basics, they started reading the state constitution. The Johns Island citizenship school was an immediate success. Fourteen students had learned to read and

registered to vote within three months. Their success inspired Clark to create more citizenship schools throughout the South. She used her position at Highlander Folk School to recruit and train teachers. Within four years Clark had trained more than eighty teachers, and citizenship schools had opened in four states.

The success of the citizenship schools prompted a backlash from whites in the South. The state of Tennessee forced Highlander to close in 1961. Clark moved to Atlanta, Georgia, and went to work for the Southern Christian Leadership Conference (SCLC), the civil rights organization headed by Martin Luther King Jr. (1929–68). She continued to train teachers for citizenship schools in Atlanta at a school called the Dorchester Center. Dozens of future leaders of the civil rights movement learned under Clark at the Dorchester Center. Clark and the organizations she worked with eventually trained more than ten thousand citizenship-school teachers. Clark's central role in this process earned her the nickname of the "queen mother" of the civil rights movement.

Clark retired from the SCLC in the summer of 1970. However, she remained active in the civil rights movement for the entirety of her adult life. She was also active in the women's rights movement in the 1970s. In the late 1970s she got involved in South Carolina politics and won a seat on the Charleston School Board. She died on December 15, 1987, at the age of eighty-nine.

Juliette Aline Derricotte. *Fisk University Library. Reproduced by permission.*

★ JULIETTE ALINE DERRICOTTE (1897–1931)

Juliette Aline Derricotte was a promising young educator and academic in the early part of the twentieth century. She was the first woman ever to serve as a trustee of Talladega College. Talladega College is the oldest historically black college in the state of Alabama. A trustee is a person who helps manage a college's curriculum, finances, and administration. Derricotte was well on her way to a successful career in education when she died in a tragic car accident at the age of thirty-four. Derricotte's death created a national outrage in the African American community because she was refused admission to a "whites-only" hospital after the car accident.

A Love of Education Comes Early

Juliette Aline Derricotte was born on April 1, 1897, in Athens, Georgia. Her father, Isaac Derricotte, was a cobbler (a person who makes and repairs shoes and other leather goods). Her mother, Laura Hardwick Derricotte, was a seamstress (a person who sews clothing). Derricotte was the fifth of nine children. She grew up in the South during the era of racial segregation. Segregation is the practice of using the law to force members of different races to live apart from each other. Segregation functioned to deny African Americans many opportunities that were available to white Americans.

Segregation left a lasting influence on Juliette Derricotte. Derricotte was bright and inquisitive as a young girl. She was excited to go to school. She told her mother that she hoped to attend the Lucy Cobb Institute, an elite private school in Athens. The Lucy Cobb Institute did not admit African American students, however. Derricotte was discouraged when she learned that segregation would prevent her from attending the school of her choice. She responded to the disappointment by vowing to spend her life fighting against racial discrimination.

Derricotte did not let segregation prevent her from getting an education. She attended all-black public schools in Atlanta, Georgia. At first, it appeared that her family would not have enough money to pay for her to go to college after she finished high school. Eventually, they were able to send her to Talladega College in Talladega, Alabama. Talladega College is an historically black college. Derricotte was excited to be in college, studying alongside other African American students. She was disappointed, however, by the fact that almost all of the professors at Talladega were white.

Derricotte's time at Talladega strengthened her desire to dedicate her life to the service of others. Her friendliness, warmth, and intelligence made her a natural student leader. She used her position of leadership to speak out on behalf of causes in which she believed. Derricotte continued to be a student leader even after she graduated from college in 1918. She took a summer course with the Young Women's Christian Association's (YWCA) Training School in New York. She spent the next eleven years travelling to college campuses to work with student leaders on behalf of the YWCA.

A Life of Service Ends Too Soon

Derricotte also served on the general committee of the World Student Christian Federation (WSCF), an organization that brought together Christian student groups from around the world. Derricotte traveled to England in 1924 as an American delegate to the WSCF conference. She was one of only two African Americans selected to represent the United States. She traveled to India in 1928 for another WSCF event. Derricotte's

travels exposed her to students, cultures, and problems from around the world. She learned that discrimination was a global problem that took many forms. She decided to dedicate herself to studying what she called the "fundamental and eternal puzzles of economics, race, and religion."

Derricotte wanted to use what she learned to inspire future generations of student leaders. She earned a master's degree in religious education from Columbia University in New York in 1927. Her master's degree made it easier for her to seek employment at a university. Derricotte left her job at the YWCA in 1929 and took a job at Fisk University, an historically black university in Nashville, Tennessee. Derricotte worked as the dean of women. Her job was to oversee campus life and encourage students to become involved in activities. Derricotte was an immediate success. She was very popular with her students. During this time, Derricotte also became the first woman ever to serve as a trustee of Talladega College in Alabama.

Derricotte never got the chance to finish what she started at Fisk. On November 6, 1931, Derricotte and three Fisk students took a car trip from Nashville to Athens, Georgia. They were involved in a car accident in Dalton, Georgia, a small town about thirty miles southwest of Chattanooga, Tennessee. Derricotte and one of the students were seriously injured. Their injuries were severe enough that they should have been immediately hospitalized. However, the local hospital was segregated and refused to admit Derricotte and the student because they were African Americans. The student died that night, and Derricotte died the next day. She was just thirty-four years old.

African American leaders across the country were outraged by Derricotte's death. The National Association for the Advancement of Colored People (NAACP) alerted newspapers across the country of the circumstances of how she died. The NAACP hoped that it could increase support for racially integrated hospitals by publicizing the consequences of segregation. Derricotte's death became well known by the African American community. However, it did not result in hospital desegregation. Instead, honorary memorial services were held across the country. The African American community could only mourn the premature loss of a promising young leader.

★ MORDECAI WYATT JOHNSON (1890–1976)

Mordecai Wyatt Johnson was the first African American to serve as president of Howard University in Washington, D.C. Howard was the nation's leading center of African American intellectual life in the first half

Mordecai Wyatt Johnson.
Alfred Eisenstaedt/Time & Life Pictures/Getty Images

of the twentieth century. It was a major milestone for Johnson to attain the position of president. Howard University thrived under Johnson's leadership. Johnson was also a clergy member and an engaging, well-respected public speaker. He was a leading figure in the African American intellectual community in the early to mid-1900s.

Religion and Education Are Guiding Influences

Mordecai Wyatt Johnson was born on January 12, 1890, in Paris, Tennessee, a small town in the northwest corner of the state. His father, Reverend Wyatt Johnson, was a former slave who worked in a mill. His mother, Caroline Freeman Johnson, was a domestic employee, which

means she did paid work in another family's home. Johnson demonstrated a commitment to education from a young age. He finished middle school in Paris and then moved to Nashville, Tennessee, in 1903 to attend high school, but the school was destroyed in a fire in 1905. After briefly attending school in Memphis, Tennessee, Johnson moved to Atlanta, Georgia, to finish high school at the Preparatory Department of Atlanta Baptist College. A preparatory department is a high school that is affiliated with a college.

Johnson attended college at Atlanta Baptist College. Atlanta Baptist College, which was renamed Morehouse College in 1913, was an all-male, all-black college. Johnson was a standout student in college. He was a star athlete in three sports, a member of the debating team, and a singer in the glee club. A glee club is a musical group, similar to a choir, that specializes in singing short songs. Johnson also excelled academically. He was such a good student that the college offered him a position on its faculty upon his graduation in 1911. Johnson accepted a position as a professor of English and economics.

After a year of teaching, Johnson decided to pursue his own education as well as his interest in the ministry. He earned a second bachelor's degree from the University of Chicago in 1913. He then earned a bachelor of divinity degree from the Rochester Theological Seminary in Rochester, New York, in 1916. Johnson was deeply influenced by his time in Rochester. He accepted a position as the pastor of the First Baptist Church in Charleston, West Virginia, in 1917. He spent nine years as a pastor in Charleston. He organized the city's first branch of the National Association for the Advancement of Colored People (NAACP). During his time at First Baptist Church, Johnson took a leave of absence to earn a master's degree from the Harvard University Divinity School in 1921.

Becomes President of Howard University

In 1926, at the age of thirty-six, Johnson was selected to be the next president of Howard University in Washington, D.C. Johnson was the eleventh president in the all-black university's history, but he was the first African American to hold the position. At the time, all the other leading African American colleges and universities had white presidents. Johnson's appointment as president of Howard was viewed as a test of whether an African American was capable of leading an elite university. Johnson's selection turned out to be the beginning of an era at Howard. He would serve as Howard's president for the next thirty-four years, transforming the university along the way.

Howard was ill-suited for the prominent role it played in African American culture and society when Johnson became president. Howard

was one of the only universities in the country that would provide professional training to African Americans. More than 95 percent of the nation's African American lawyers attended Howard, as did nearly half of all African American doctors and dentists. Yet when Johnson took over, the university had inadequate facilities and materials and paid such low salaries that it had trouble retaining its best faculty.

Johnson made it his mission to improve Howard so that the quality of education it provided was on a par with the university's prominent position in the African American community. Johnson's first order of business was to secure guaranteed funding for the university. Previously, Howard University received funding from Congress on an erratic (irregular and unscheduled) basis. Congress frequently did not give the university enough money, and year-to-year variations in funding made efficient administration of the university impossible. Johnson lobbied Congress to pass a law guaranteeing annual funding for the university. Congress finally did so in 1928. It was a major victory for Johnson. All of Johnson's other goals for Howard required a regular, adequate supply of money. The NAACP awarded Johnson its Spingarn Medal "for the highest achievement of an American Negro" in 1929 in recognition of this landmark achievement.

Howard Thrives Under Johnson's Leadership

Johnson also oversaw a major transformation of the facilities at Howard University. During his time as president, twenty new buildings were added to the campus. These additions increased the estimated value of the campus from $3 million to $34 million. One of the centerpiece additions was a new, expanded library. Johnson also commissioned a new building for the medical school. As the size of the campus increased, so too did student enrollment. Howard went from having just under two thousand students when Johnson took over as president to just over six thousand by the time he retired.

Johnson also devoted his time and energy to improving the quality of the professional schools. When Johnson became president of Howard, the university's law school only offered night classes, and most of the classes were taught by lawyers from around Washington, D.C. Johnson transformed the law school into an elite center of scholarship and advocacy training. Most of the prominent leaders of the legal side of the civil rights movement, including future Supreme Court justice Thurgood Marshall (1908–1993), attended law school at Howard. The medical school enjoyed similar advances.

Johnson's time as president of Howard was not without controversy. Some faculty members felt Johnson was overly controlling and would not

listen to the ideas of others. Johnson also became the target of criticism from outside the university in the 1950s because he sometimes spoke favorably of the Communist government in the Soviet Union. These criticisms could not detract from the fact that Johnson's tenure as president of Howard was incredibly successful. He retired from Howard University in 1960 after thirty-four years as president. The university honored him in 1973 by naming a building on campus after him. Johnson died in 1976 at the age of eighty-six.

★ BENJAMIN MAYS (1894–1984)

Benjamin Mays served as the president of Morehouse College in Atlanta, Georgia, for more than twenty-five years. Morehouse is an all-male historically black college. Mays worked to make Morehouse one of the most highly regarded historically black colleges in the nation. Mays helped Morehouse expand its campus, improve its faculty, and attract better students. He also was an inspiration to thousands of students. The great civil rights leader Martin Luther King Jr. attended Morehouse College and described Mays as his "spiritual mentor."

Benjamin Mays.
AP Images

Benjamin Elijah Mays was born on August 1, 1894, in Greenwood County, South Carolina. He was the youngest of seven children born to Hezekiah and Louvenia Carter Mays. Both of Mays's parents were former slaves. Mays grew up in a time and place where African Americans faced constant, serious racial discrimination. His parents worked as renters, farmers who owned their own animals and tools but rented the land on which they farmed. They earned little money. The center of the family's life was their church, Mount Zion Baptist. Fellow congregation members at the church recognized that Mays was particularly intelligent. They encouraged him to pursue an education.

Southern states provided limited public schools for African Americans in the late 1800s and early 1900s. Mays thus attended various schools supported by the Baptist Church in South Carolina to further his education. His education was frequently interrupted by his duties on his family's farm. Even so, Mays was a

bright and dedicated student. He graduated as the valedictorian of the High School Department at South Carolina State College in 1916. Mays wanted to earn a college degree at a university that admitted both black and white students. There were no such universities in the South at the time. He thus enrolled in Bates College in Lewiston, Maine. He graduated from Bates with honors in 1920.

A Career in Education Begins

Mays spent one year in Chicago after graduating from Bates. He completed several semesters of graduate school at the University of Chicago. He accepted a position as a professor of mathematics at Morehouse College in 1921. He spent the next three years as a professor at Morehouse. Mays was ordained as a Baptist minister during his time in Atlanta. He returned to Chicago in 1924 to finish graduate school. He earned his master's degree in 1925. After spending a year as a professor of English at the State College of South Carolina, he moved to Tampa Bay, Florida, to work as the executive secretary for Florida's National Urban League. The National Urban League is a civil rights organization that, at the time, was largely devoted to fighting racial discrimination in employment.

Mays soon began focusing his professional energy on the academic study of religion. He became the national student secretary of the Young Men's Christian Association (YMCA) in 1928. He accepted an offer to direct a major national research study of African American churches in the United States in 1930. The results of the study were published in 1933 as *The Negro's Church*. Mays continued to write on the role of the church in African American society, and on the role of the African American church in American society, for many years. In 1934, Mays was named the dean of the School of Religion at Howard University in Washington, D.C.

Mays served as the dean of the School of Religion for six years. He built a sterling (highest quality) reputation with his work at Howard. Howard was already the nation's leading institution of African American education and scholarship. Mays improved the faculty and facilities of the seminary so much that it received a Class A rating from the American Association of Theological Schools. It became just the second all-black seminary to be rated Class A. Mays's work attracted the attention of another prominent all-black institution of higher education. Mays was named president of Morehouse College in 1940.

Morehouse Thrives Under Mays's Leadership

Mays was the president of Morehouse for twenty-seven years. Morehouse was already a highly regarded African American college when Mays became president. Mays worked to make Morehouse one of the best

colleges in the nation. He raised money to pay higher salaries to professors so that Morehouse could hire top-quality professors. He recruited the best students to enroll at the school. Mays also reached out to the best and brightest Morehouse graduates, encouraging them to return to Morehouse as professors after they earned advanced degrees.

Mays's approach paid dividends. Morehouse was granted full membership in the Southern Association of Colleges and Secondary Schools in 1957. This membership placed Morehouse, in Mays's words, on an "equal footing" with the best white colleges in the South. Mays achieved another major victory in 1967 when Morehouse was granted a chapter of Phi Beta Kappa. Phi Beta Kappa is the oldest and most famous academic honor society in the United States. It only has chapters at very good schools. Under Mays's leadership, Morehouse had become a truly elite university. In 1965, *The Alumnus,* a magazine for Morehouse graduates, described Mays as the most effective president in Morehouse history.

One of Mays's most treasured accomplishments was the relationship he built with Martin Luther King Jr. King attended Morehouse College from 1944 to 1948, graduating with a degree in sociology. Mays counseled and mentored King while King was a student at Morehouse and while King was attending seminary. After King's tragic death in 1968, Mays delivered the eulogy (a funeral speech honoring a person's life) at King's funeral.

Mays retired from Morehouse College in 1967. During his retirement, he wrote an autobiography entitled *Born to Rebel* and numerous other books and articles. He also remained active in education, serving as president of the Atlanta Board of Education from 1969 to 1981. He died on March 21, 1984, at the age of eighty-nine.

★ KELLY MILLER (1863–1939)

Kelly Miller was one of the leading intellectuals in the African American community in the late 1800s and early 1900s. Miller spent forty-four years as a professor at Howard University in Washington, D.C. He was a teacher, a scholar, an administrator, and a writer. He left his mark on several generations of students and professors at Howard. He also used his position at Howard to become a nationally syndicated columnist in the African American press. Miller dedicated his career to the issue of race. His scholarship and writing worked toward rooting out and eliminating racist prejudices and stereotypes against African Americans.

Kelly Miller was born on July 23, 1863, in Winnsboro, South Carolina. He was the sixth of ten children born to Kelly and Elizabeth

Roberts Miller. Young Miller attended a local school that was founded by the Freedmen's Bureau as part of Reconstruction. Reconstruction was the period following the end of the Civil War during which the federal government took an active role in promoting civil rights for African Americans living in the South. The Freedmen's Bureau was a government agency that established schools and provided education during Reconstruction. Miller was a very talented student, and his teachers took notice. His talents eventually earned him a scholarship to the Preparatory Department at Howard University in Washington, D.C. Miller graduated from high school and enrolled in college at Howard.

Becomes a Professor

Miller's intelligence and work ethic enabled him to land a job with the federal government while he was in college. Miller graduated from Howard in 1886, and he continued to work for the government after graduation. At the same time, he undertook a course of study in mathematics and science at the U.S. Naval Observatory. Miller was able to gain admission to graduate school at Johns Hopkins University in Baltimore, Maryland, after one year of studying at the Observatory. Johns Hopkins was one of the most selective universities in the nation at the time. Miller was the first African American to be admitted to the university. Miller studied at Johns Hopkins from 1887 to 1889. Financial difficulties prevented him from completing a graduate degree there.

Miller returned to Washington, D.C., in 1889 and taught high school for one year. Howard University hired him as a professor of mathematics in 1890. He taught mathematics at Howard for the next seventeen years. During this time Miller was also developing an interest in sociology. Sociology is the study of society, social institutions, and social relationships. It was a new academic discipline in the late 1800s. Miller believed sociology could provide valuable insights into the problem of racism in the United States. He worked hard to bring sociology into the curriculum at Howard. He taught sociology classes alongside his mathematics classes from 1895 until 1917. He taught sociology exclusively from 1917 until his retirement in 1934.

Miller's scholarship focused on race and African Americans' position in the United States. Many white people in the early 1900s believed that African Americans were naturally and genetically inferior to whites. Miller used his scholarship to argue that these beliefs were prejudices and stereotypes, not facts. Miller drew on the insights of sociology to contend that racial discrimination was a feature of American society that could be changed. Miller published numerous well-known pamphlets, articles, and books that disputed prevalent (common and widely held) stereotypes of African Americans.

An Intellectual Leaves His Mark

Miller also helped transform Howard University into a hub of African American intellectual life in the early 1900s. Miller served as the dean of the College of Arts and Sciences at Howard from 1907 until 1918. Miller focused on updating the college's curriculum. Previously, the curriculum had concentrated on providing a classical education in Roman and Greek language, literature, and culture. Under Miller's leadership, the college began offering more classes in mathematics, natural science, and sociology. This modernization of the curriculum helped Howard become the leading training ground for intellectuals and leaders of the African American community.

Miller used a syndicated newspaper column to expand his influence beyond Howard University. A syndicated newspaper column is a written work of opinion that appears in more than one newspaper. Miller wrote a weekly column that was syndicated in more than one hundred weekly African American newspapers. Historians estimate that his column was read by more than five hundred thousand people each week during the 1930s. Miller used his column to argue that African Americans should improve their economic fortunes through agriculture and improve themselves through education.

Miller retired from Howard University in 1934 after forty-four years as a professor. He began working on an autobiography, but he was not able to complete it. He suffered a heart attack just before Christmas in 1939. He died one week later on December 29, 1939, at the age of seventy-six.

★ WILLA B. PLAYER (1909–2003)

Willa B. Player dedicated her life to education. Player became the first African American woman ever to head a four-year college when she was named president of Bennett College in Greensboro, North Carolina, in 1955. She used her position as president of Bennett to support the civil rights movement. Player was a role model for, and an inspiration to, students at Bennett for more than thirty years. After her retirement from Bennett, she joined the federal Department of Education and oversaw a program that provided support to small, growing colleges. Player enjoyed a long, distinguished career as an educator.

Education Is an Early Passion

Willa Beatrice Player was born on August 9, 1909, in Jackson, Mississippi. She was the youngest of three children born to Clarence E. and Beatrice D. Player. The Player family moved to Akron, Ohio, in 1916.

The family was very involved in the Methodist Church. This involvement helped Player continue her education after she finished high school. Very few African American women had the opportunity to go to college at this time. However, Player was accepted to a Methodist college: Ohio Wesleyan University in Delaware, Ohio.

Player encountered some racial discrimination at Ohio Wesleyan. She and the two other African American students at the school were not allowed to live in the campus dormitories. But Player made sure to take advantage of the opportunity to get an education. She graduated from college in 1929 and immediately enrolled in a master's program at Oberlin College in Oberlin, Ohio. She earned her master's degree in 1930. She then accepted a job as a professor of Latin and French at Bennett College, a women's college in Greensboro, North Carolina.

Player was an effective and well-liked professor at Bennett. She used her time off to continue her own education. She received a Fulbright fellowship in 1935. A Fulbright fellowship is a very prestigious award that pays for its recipients to pursue academic studies overseas. Player used her Fulbright to study French at the University of Grenoble in France. When she returned from France, she became Bennett College's director of admissions. Her administrative responsibilities continued to grow after she earned a doctorate of education from Columbia University in New York in 1948. She became the college's coordinator of instruction in 1952, and in 1955 she became a vice president.

Willa Player made history on October 22, 1955, when she was named president of Bennett College. She became the first African American woman to serve as the president of a four-year college in the United States. Many women viewed Player's appointment as president of Bennett as a major symbolic victory for women in general and African American women in particular. Her attainment of the highest-ranking position at Bennett showed women everywhere that they could realistically aspire to the very highest levels of their chosen professions. Player downplayed the significance of her appointment and focused on making Bennett a better college.

Player focused on the importance of a liberal arts education during her time as president. She often explained that she believed that the primary purpose of a college education was to learn how to think. Instead of teaching students a particular set of facts, Player argued, a college education should teach students how to ask questions and how to think about problems in a critical, logical, and organized way. Player also defended the value of the single-gender education that students received at Bennett College. She contended that many female college students were

OPPOSITE PAGE
Willa Player
Courtesy of Bennett College

more confident and performed better when they did not feel that they were competing with men.

The Civil Rights Movement Finds an Ally

Player also used her position as president of Bennett College to lend support to the civil rights movement. In 1958, Martin Luther King Jr. wanted to come to Greensboro, North Carolina, to make a speech. The civil rights movement was still in its early stages, and there was a very real possibility of a violent response to King's appearance. The local government, which practiced racial segregation, would not allow King to use any public facilities. Local white-owned businesses did not want to be associated with an African American leader. Player defied local opinion and allowed King to make his speech on the Bennett College campus.

King's speech at Bennett College was hugely important. Player herself called it one of the most important moments in the college's history. The speech helped turn North Carolina into a hotbed of civil rights activity. The speech was particularly inspiring to Ezell Blair Jr., a student at nearby North Carolina Agricultural and Technical University. After hearing King's speech, Blair and three of his friends went into the Woolworth's in downtown Greensboro, sat down at the "whites only" lunch counter, and demanded service. This was the first in one of the most famous series of "sit-ins" in the civil rights movement.

Bennett students were soon involved in numerous sit-ins, and Player threw the full weight of Bennett College's support behind their efforts. Students who participated in the civil rights movement were frequently taken to jail. Bennett visited them in jail to ensure that they were not being mistreated. She had professors come to the jail to provide instruction and give homework to the students there. Player even brought her students their exams so that they could keep up with their classes. Civil rights leaders would later say that Player's support of her students was critical to the success of the Greensboro sit-ins.

Player retired from the presidency of Bennett College on February 28, 1966. Retirement from Bennett did not mean the end of her career in education. Player took a job with the U.S. Department of Education. She served as the director of the Division of College Support. In that role she helped the government provide grants and other kinds of financial support to small, growing colleges across the country. Player worked in this capacity for eleven years. During that time the program she oversaw more than tripled in size. She retired from the Department of Education in April 1977.

Player spent her retirement years working as an educational consultant. She also remained very active in the Methodist Church. Bennett College

honored her in 1988 by naming an endowed chair in the humanities after her and designating her as president emerita (an honorary president). Player died on August 27, 2003, at the age of ninety-four.

BOOKER T. WASHINGTON (1856–1915)

Booker T. Washington was the nation's most prominent African American leader in the late 1800s and early 1900s. Washington was a vocal advocate of providing educational opportunities to African Americans. He believed that African Americans could use vocational and industrial training as a vehicle to improve themselves and find a place in American society. Washington also helped found the Tuskegee Normal and Industrial Institute in Alabama. The Tuskegee Institute became an important center of African American education by providing training to numerous future teachers. Washington also gained recognition as an author, a public speaker, and a political leader.

A Young Man Is Drawn to Education

Booker Taliaferro Washington was born on April 5, 1856, on a farm in Virginia. He was the second of three children born to Jane Ferguson, who at the time was held as a slave. Washington, his mother, and his siblings lived in slavery until Washington was ten years old. Once the Civil War was over and slaves had attained freedom, Washington and his family moved to West Virginia so that his stepfather could find work. The family was very poor because they had not been allowed to own any property or earn income while they were slaves. As a result, Washington had to work in the salt and coal mines alongside his stepfather to help support his family.

Booker T. Washington. *The Library of Congress*

His family's difficult circumstances meant Washington often did not have much time to attend school. When he did attend, Washington was an eager and talented student. He soon learned of a school in Virginia that poor students could attend provided they paid their way by working. Washington left his home in West Virginia at the age of seventeen and enrolled at

the Hampton Institute in Hampton, Virginia, in October 1872. He had very little formal schooling or education, but he thrived at Hampton. He studied there for three years. The founder of the Hampton Institute, Samuel Armstrong (1839–93), believed that the best way for African Americans to improve themselves was to learn vocational (job-related) skills they could use to earn a living. Washington absorbed this philosophy from Armstrong and would devote his life to putting it into practice.

Washington graduated from Hampton in 1875. He returned home to West Virginia and spent two years as a schoolteacher. He then moved to Washington, D.C., to study at the Wayland Seminary. Washington was deeply influenced by his time in the nation's capital. Washington was very religious, and he believed that living in cities made people less hard-working and more sinful. Washington combined this belief with what he had learned at Hampton. He came to believe that the path to virtue and self-improvement lay in living in rural areas and working in agriculture. Washington returned to the Hampton Institute after his year in Washington, D.C. He worked as a teacher there from 1879 to 1881.

The Tuskegee Institute Is Founded

Washington began his rise to national prominence in 1881, when he helped found the Tuskegee Institute. That year, the Alabama legislature had set aside two thousand dollars to establish an all-black school in that state. Alabama politicians asked Samuel Armstrong to recommend someone to run the new school. Armstrong recommended Washington. Washington was only twenty-five years old, but he nonetheless became the school's first president. The school opened under the name Tuskegee Normal and Industrial Institute on July 4, 1881. The school soon became known simply as the Tuskegee Institute.

Washington used the Tuskegee Institute to implement the ideas about education and African American improvement that he had developed at the Hampton Institute. The industrial branch of the school was a training center in practical, hands-on job skills. Washington wanted the Tuskegee Institute to equip its students with the job skills they needed to perform gainful work. The curriculum focused on industrial and agricultural skills. "Normal school" was the nineteenth-century term for a school that trained future teachers in the standards, or "norms," of education. There was a shortage of African American teachers in the South, so training as a teacher brought promising career prospects.

The Tuskegee Institute thrived under Washington's leadership. Only thirty students attended the school when it opened in 1881. By 1906, it was home to 1,590 students and 156 faculty members. The institute became the nation's most important center of African American education

and scholarship in the late 1800s and early 1900s. Washington served as the institute's president from its opening in 1881 until his death in 1915. Throughout that time, Washington preached a philosophy of self-reliance. He encouraged his students to learn a trade and earn a good living. Washington believed that African Americans could only earn respect from—and eventually equality with—whites by proving that they could be economically self-sufficient.

Becomes a National Leader

Washington developed a political philosophy and strategy that was very similar to his philosophy of education. Washington did not believe African Americans should concentrate on exercising their right to vote or otherwise participating in the political process. Washington believed it was up to African Americans to disprove the negative stereotypes whites held about them. He contended the best way to do so was to concentrate on being economically productive and hard-working. This philosophy appealed to many white Americans who did not support full political equality for African Americans. Washington's role as president of a major African American educational institution thus positioned him to become a major national political leader as well.

Washington's most famous explanation of his political philosophy came in a speech in 1895. The speech—which Washington gave at a business convention in Atlanta, Georgia—came to be known as the "Atlanta Compromise." It was called a compromise because it advocated a middle position on the issue of African American civil rights, designed to appeal to both whites and blacks. In his speech, Washington argued that African Americans deserved equal opportunities. However, Washington did not argue that African Americans lacked those opportunities because of racism. Rather, Washington suggested African Americans needed to earn equality for themselves through hard work and "common labor." The position that Washington took in his speech had both supporters and critics. It was particularly popular among white Americans. Many African Americans supported Washington and his views because they believed Washington's position was realistic and actually could be accomplished.

By 1900, Washington had become the most famous African American leader in the United States. He wrote a series of books in the early 1900s that explained his own background and further developed his philosophy of industrial education and economic self-reliance. His autobiography, *Up From Slavery,* became a best-seller. Washington made national headlines in 1901 when he had dinner with President Theodore Roosevelt (1858–1919). He was the first African American to eat dinner at the White House as a social guest of the president.

Washington was not without his critics. Another well-known black leader named W. E. B. Du Bois (1868–1963) was developing a competing vision of African American self-improvement. Du Bois criticized Washington for stressing the importance of vocational training. Du Bois believed that an exclusive focus on agricultural, industrial, and manual labor would permanently force African Americans into the lower classes. Du Bois also believed Washington was wrong to ignore voting and participation in the political process as an avenue of African American self-improvement. Washington and Du Bois eventually became political rivals.

Washington's prominent leadership role put him in great demand as a public speaker. His role as the president of the Tuskegee Institute also required him to travel the country extensively raising money to support the school. His public commitments were extensive and exhausting. In the fall of 1915, Washington became seriously ill while he was in New York for a public appearance. He returned to the Tuskegee Institute in Alabama, where he died on November 14, 1915, at the age of fifty-nine. His death brought a premature end to the life of the first truly national African American leader to emerge after the end of slavery.

HOWARD UNIVERSITY IS FOUNDED

The founding of Howard University in Washington, D.C., was one of the most important events in the history of African American education. Howard University was established with the express purpose of providing higher education to African American students. At the time of its founding in the 1860s, very few colleges or universities would enroll African American students. Howard thus helped to fill a major void in the educational landscape. By the early 1900s, Howard University had become the national center of African American educational and intellectual life.

Howard University has its origins in the end of the Civil War (1861–65) and the emancipation of slaves throughout the United States. The end of the Civil War meant that more than four million African Americans who previously had been held as slaves were newly freed and in search of employment. Many of these slaves migrated toward large cities in the South, especially Washington, D.C. Tens of thousands of African Americans streamed into Washington, D.C., in 1865 and 1866. Many of them were unable to find work. They lived in poverty on the streets. Many white residents of Washington, D.C.—especially those who were members of the churches that had led the fight for the abolition of slavery—felt they had a duty to help newly freed African Americans find a way to support themselves.

The idea for Howard University was first proposed at a meeting of the Missionary Society of the First Congregational Church of Washington, D.C., on November 19, 1866. A missionary society is a group that seeks to promote the beliefs of its religion. The Missionary Society of the First Congregational Church believed that newly freed African Americans needed leaders within their race who could provide education and religious guidance. They concluded that the natural way to produce these leaders was to open a seminary, or a theological college, that catered to African American students.

A Seminary Gives Way to a Full-Fledged University

The Missionary Society's vision for this new school evolved over the next several months. Some suggested that, in addition to a seminary, the institution should include a normal school. "Normal school" was the nineteenth-century term for a school that trained future teachers in the standards, or "norms," of education. Eventually, it was decided that a full-scale university was needed. At first, the school was to be limited to African American students. The society soon concluded that the school should be

The Importance of Historically Black Colleges and Universities

An historically black college or university (HBCU) is a college or university that was founded before 1964 for the primary purpose of educating African Americans. The term "historically black college or university" first became significant when Congress passed the Higher Education Act of 1965. The act formally recognized 105 colleges and universities as HBCUs. These institutions had long provided educational opportunities to African Americans that were otherwise unavailable. The act recognized and rewarded this contribution by making HBCUs eligible for extra federal scholarships, grants, and other funding.

HBCUs were critically important institutions in African American education between 1865 and 1965. During that time, they were the only place where African Americans could receive a college education. HBCUs provided a unique haven (place of escape) from racial discrimination. They offered a hospitable environment that bred confidence and leadership. Almost all of the most prominent leaders of the civil rights movement attended HBCUs. Some of the nation's most famous HBCUs include Fisk University (Nashville, Tennessee), Howard University (Washington, D.C.), Morehouse College (Atlanta, Georgia), Spelman College (Atlanta), and Tuskegee University (Tuskegee, Alabama).

open to students of all races and backgrounds, including women. At the time, it was relatively uncommon for a university to admit both male and female students.

The Missionary Society drafted a charter for this new school that was presented to Congress on January 23, 1867. Normally, colleges and universities are run by state and local governments. However, since Washington, D.C., is not part of any state, all of its affairs and operations were handled by Congress. The school was to be located in the northwest portion of the District of Columbia. Congress approved the charter on March 2, 1867, and Howard University was born.

Howard University was named after General Oliver Otis Howard (1830–1909). General Howard was the commissioner of the Freedmen's Bureau. The Freedmen's Bureau was a federal government agency founded

in 1865 to provide assistance to former slaves throughout the South. The bureau provided food, helped African Americans find jobs, protected former slaves from violent attacks by white Southerners, and promoted education. General Howard was especially dedicated to providing African Americans in the South with educational opportunities. Under his leadership, the Freedmen's Bureau opened more than four thousand schools for African American students and provided education to almost two hundred fifty thousand people. Howard's dedication to the cause of equality inspired Congress and the Missionary Society of the First Congregational Church to name the new university in his honor.

Howard University Becomes a Center of African American Intellectual Life

When Howard University opened its doors in 1867, it offered courses in six departments: agriculture, collegiate, law, medical, normal and preparatory, and theological. Today, the "collegiate department" would be called a college of liberal arts. Even though the university had been founded with African American students in mind, its first four students were white females who were the daughters of two of the university's founders. The university was very small in its early days. The trustees (the people who help manage a college's curriculum, finances, and administration) immediately began acquiring additional land and building the necessary facilities and buildings. By 1870, the university had two dormitories, a medical building, a hospital, and two classroom buildings. Howard awarded its first degree in 1872.

Financial stability was a major concern at Howard during its early years. Congress had chartered the university, but it had not provided it with a regular source of funding. Howard received important financial support from the Freedmen's Bureau in its first few years of existence. The university also benefited from private contributions and donations. Its expansion was aided by Congress's decision in 1879 to provide a special appropriation (an allocation of money). Over the next several decades, Congress provided financial gifts at irregular intervals that totaled approximately $1.2 million. A major breakthrough that was critical to the university's long-term survival came in 1928 when Congress amended Howard's charter to provide for an annual federal appropriation to help pay for building construction, faculty recruitment and salaries, and improvement and maintenance.

African Americans were eager to enroll at Howard, and the university grew steadily over the next several decades. More than 1,300 students were enrolled at Howard by 1910; less than two decades later, enrollment had topped 1,700. Once regular annual funding was in place, the university's

The graduating class from Howard University's law school in 1900. *Buyenlarge/Hulton Archive/ Getty Images*

enrollment exploded. More than 6,000 students were enrolled at Howard in 1960. Howard became the most prestigious, well-respected center of African American intellectual life in the 1920s, 1930s, and 1940s. Many leading African American academics, as well as most of the leading figures in the civil rights movement, were educated at Howard. The school's law department, in particular, graduated several prominent lawyers who fought important civil rights cases, including future Supreme Court justice Thurgood Marshall. Congress's decision to charter Howard University in 1873 fundamentally reshaped the educational landscape for African Americans nationwide.

THE TUSKEGEE INSTITUTE PROVIDES TRAINING AND OPPORTUNITY

The founding of the Tuskegee Normal and Industrial Institute (commonly referred to as the Tuskegee Institute) in Macon County,

Alabama, in 1881, was a crucial event in African American education. In its early days, it concentrated on training teachers and on providing practical, hands-on training that would enable its students to find work after graduation. The Tuskegee Institute's focus on teaching occupational skills was controversial. However, the institute provided its students with knowledge and skills they could use to improve their own lives and economic fortunes, as well as those of countless other members of the African American community.

A New School Is Born

The founding of the Tuskegee Normal and Industrial Institute was largely the work of a man named Lewis Adams (1842–1905). Adams was a well-known African American leader in Macon County, Alabama, in the late 1800s. Two Alabama politicians approached Adams in 1880. They were hoping to convince African American residents in the town where Adams lived to vote for them in an upcoming election. They wanted Adams to campaign on their behalf. Adams said that he would campaign for them if they would promise to provide funding for an all-black school in Macon County.

The two politicians agreed to Adams's request, and they were true to their word. The Alabama state legislature allocated two thousand dollars a year for the school, beginning in 1881. Adams's next task was to find an educator to run the school. Adams decided to seek the recommendation of Samuel C. Armstrong (1839–93). Armstrong was then the president of the Hampton Institute, a well-respected normal school for African Americans in Virginia. One of the best students in Hampton's history was a man named Booker T. Washington (1856–1915). Washington was only twenty-five years old in 1881, but Armstrong thought very highly of Washington and recommended him to Adams. Adams hired Washington to become the first president of the new school.

Booker T. Washington would go on to become the most famous African American leader in the United States in the late 1800s. His leadership as president of the Tuskegee Institute played a major role in his rise to prominence. Adams had initially envisioned that the Tuskegee Institute would be a normal school. "Normal school" was the nineteenth-century term for a school that trained future teachers in the standards, or "norms," of education. Washington wanted to add a component that created a training center in the industrial arts, which are practical, hands-on job skills. The new school opened on July 4, 1881. It opened under the name Tuskegee Normal and Industrial Institute.

Washington Implements His Philosophy

Washington faced a challenge in getting the new school up and running. The Alabama legislature had appropriated (set aside money for a

particular purpose) two thousand dollars to pay teachers at the school. However, the school did not have a building, desks, books, or other facilities and materials. Washington conducted classes for the first year in a rundown building behind a local Methodist church. He hired teachers and recruited students. He put the students to work building a new building for the school. He also traveled around the country raising money for the school. Washington raised enough money that the school was able to buy one hundred acres of land to use as a campus in 1882.

The curriculum at the Tuskegee Institute reflected Washington's particular philosophy of African American education. Washington believed that providing moral training and developing good character was a central task for an African American school. He thus had students at the Tuskegee Institute attend daily church services. He also required his male students and female students to study separately. Washington believed that manual labor was an important device for building character and moral fitness. Therefore, students at Tuskegee helped build the school buildings and worked in the fields growing the food that they ate.

Washington firmly believed in the value of practical education. The goal of the Tuskegee Institute was to provide its students with the trade skills they needed to find work once they graduated. The practical education focused on industrial and agricultural skills. Tuskegee also trained African Americans to

Students in a workshop at the Tuskegee Institute in 1902. The Tuskegee Institute emphasized vocational skills in its curriculum. *The Library of Congress*

become teachers. There had long been a shortage of African American teachers in the South. Being trained as a teacher was thus a promising way to prepare for a career. Washington preached a philosophy of self-reliance. He believed that preparing students for jobs was critical because it would enable them to earn a good living and better themselves.

Tuskegee Leaves Its Mark

The Tuskegee Institute undertook an important mission of community education and outreach. Tuskegee students would travel through the rural areas surrounding the school to speak with African American sharecroppers (farmers who worked on land owned by someone else). Many of these sharecroppers could not afford to attend school full-time themselves. The Tuskegee students would teach the sharecroppers the latest agricultural science and techniques. Many Tuskegee graduates continued this outreach work by founding their own small schools in rural parts of the South. This outreach work was hugely important. It was many sharecroppers' only access to education and scientific knowledge.

The Tuskegee Institute thrived under Washington's leadership. The school had just thirty students when it opened in 1881. By 1906, Tuskegee had 1,590 students and 156 faculty members. The institute became the most important center of African American education and scholarship in the late 1800s and early 1900s. For instance, the famous agricultural scientist George Washington Carver (1864–1943) worked at the Tuskegee Institute for almost fifty years. By 1915, Tuskegee had a larger endowment (funds that have been donated to a university to support its educational mission) than any other all-black college in the country.

The Tuskegee Institute remained a focal point of African American education and scholarship throughout the twentieth century. Robert R. Moton (1867–1940), who succeeded Washington as president of Tuskegee, helped convince the federal government to locate a veterans' hospital at Tuskegee in 1923. The hospital helped Tuskegee become a major center for training African American medical professionals. The school's reputation led the U.S. Army to select Tuskegee to be the site of its first training program for African American pilots in 1941. The Tuskegee Normal and Industrial Institute shortened its name to the Tuskegee Institute in 1937. In 1985, it became Tuskegee University.

THE NATION'S FIRST COLLEGE FOR BLACK WOMEN IS FOUNDED

The founding of the Atlanta Baptist Female Seminary in Atlanta, Georgia, on April 11, 1881, was a landmark moment for African American women in the history of education. The seminary, which was eventually

renamed Spelman College, is the nation's oldest historically black college for women. The founding of Spelman College was a response to the lack of educational opportunities available to African American women in the South during the Reconstruction era (1867–77). Spelman itself enrolled countless African American women who otherwise would not have had access to higher education. It also inspired the founding of other African American women's colleges around the country.

A Women's College Is Needed

The Atlanta Baptist Female Seminary was the brainchild (original idea) of the Woman's American Baptist Home Mission Society. The Woman's American Baptist Home Mission Society was a religious organization in the Northern states of the United States dedicated to ministering (giving aid and service) to people in need. Members of the society became concerned about the welfare of African Americans in the South after Reconstruction ended in 1877. Reconstruction was the period immediately following the Civil War during which the North was actively involved in running the governments of Southern states and promoting civil rights for African Americans. In 1879, the society sent two women, Sophia B. Packard (1824–91) and Harriet E. Giles (d. 1909), to the South to study the living conditions among recently freed slaves.

Packard and Giles were longtime educators, Baptist missionaries, and women's rights activists from Boston, Massachusetts. Giles and Packard believed that the ongoing economic and social changes caused by the Industrial Revolution (a period during the late 1800s of rapid industrial, technological, and economic growth and innovation) were creating new challenges for women. Their experience as educators at the Oread Institute—a women's college in Worcester, Massachusetts—had convinced them that education for women was the key to creating equal rights and equal opportunity in an industrialized society. They traveled to the South in 1879 with these ideas firmly in mind.

Packard and Giles were shocked and disappointed by what they found in the post-Reconstruction South. African Americans were being denied basic opportunities, civil rights, and social equality. Packard and Giles observed that African American women in particular faced tremendous obstacles and hurdles. Their background as experienced educators led them to conclude that the most effective way to help improve conditions in the South would be to found a school for women. They returned to Boston and sought donations to open a school. The First Baptist Church of Medford, Massachusetts (a small town just outside Boston), donated one hundred dollars, and the women headed south to Atlanta, Georgia, to open a school.

A Great School Emerges from Humble Origins

The Atlanta Baptist Female Seminary had modest beginnings. Packard and Giles opened their school in the musty, cramped basement of the Friendship Baptist Church in Atlanta on April 11, 1881. They had only eleven students. All were women, and most were former slaves. The students primarily aspired to learn to read and write well enough to read the Bible and write letters to their families.

A hugely important event in the seminary's history took place the next year in 1882. That year, Packard and Giles travelled to a church conference in Cleveland, Ohio, to solicit money for their school. One of the attendees at the conference was John D. Rockefeller (1839–1937), the richest man in the world at the time and one of the most generous philanthropists in American history. A philanthropist is a person who regularly donates money to charities and nonprofit organizations. Rockefeller made a generous donation to the school. He also promised Packard and Giles that he would continue to support the school so long as they remained dedicated to it.

Rockefeller's support enabled the Atlanta Baptist Female Seminary to thrive and grow. The next year, the seminary used the money Rockefeller had donated to relocate from the basement of the church to a nine-acre site that was a former military barracks. The seminary was able to expand its enrollment because the new location was larger and had newer, more modern facilities. Rockefeller's support also inspired other philanthropists to contribute money. The African American community in Atlanta was particularly generous. In 1884, in appreciation of Rockefeller's support, the seminary changed its name to the Spelman Seminary in honor of Rockefeller's wife, Laura Spelman Rockefeller (1839–1915). The seminary awarded its first degrees in 1887 when it conferred high school diplomas on a graduating class of six female students.

Spelman Seminary continued to grow over the next several years. In 1888, Spelman acquired a charter from the state of Georgia, and Sophia Packard became Spelman's first president. When Sophia Packard died in 1891, Spelman had more than eight hundred students and thirty teachers. Harriet Giles succeeded Packard as Spelman's president and continued to expand the school. The seminary first began offering post-secondary, college-level courses in 1897, and it awarded its first college degrees in 1901. It took its current name of Spelman College on June 1, 1924.

THE SECOND MORRILL ACT ESTABLISHES BLACK LAND-GRANT COLLEGES

The Morrill Act of 1890, also known as the second Morrill Act, provided an important boost to African American education by requiring

states with land-grant colleges to admit African American students to those colleges. The land-grant colleges had been created in response to the first Morrill Act of 1862. The Morrill Act of 1862 provided federal financial support to states that opened colleges that provided training in agriculture and mechanics. These colleges focused on practical, job-related skills that helped their students earn a good living after graduation. By forcing states in the South to admit African Americans to these colleges, the Morrill Act of 1890 helped improve the economic opportunities available to countless African Americans.

The First Morrill Act Opens New Universities

The Morrill Act of 1862 was a revolutionary federal law. It takes its name from Vermont representative Justin Smith Morrill (1810–98), who introduced it into Congress. Morrill believed that agriculture and mechanical trades such as engineering were the backbone of the American economy. Yet in the mid-1800s, very few universities in the United States offered courses or training in these subjects. Morrill believed this lack of coursework in agriculture and mechanics would hurt the U.S. economy in the long term. Morrill came up with a plan for the federal government to give the states an encouragement to create colleges that specialized in these subjects. Morrill's plan became the Land Grant Act of 1862, more commonly known as the first Morrill Act.

Justin Smith Morrill was the U.S. representative who introduced the Morrill Land Grant College Act (1862), which provided land grants to those institutions that taught agriculture and mechanical arts. © *Corbis*

Morrill called his plan the Land Grant Act because it centered on land grants. Land grants are gifts of real estate. In the mid-1800s, the federal government owned millions and millions of acres of undeveloped land throughout the United States. The first Morrill Act had the federal government deed (make a gift of real estate) some of this land to the states. Each state got thirty thousand acres per senator and twenty thousand acres per representative. Therefore, the biggest, most populous states got the most land. In turn, the states were required to sell this land to private investors. The states were then to take the money they made from selling the land and use it to support colleges that offered courses in agriculture and mechanics. The states were required to establish such a college within five years.

The first Morrill Act passed Congress in 1862 and was signed into law by President

Abraham Lincoln (1809–65) on July 2, 1862. The act was a huge success. The federal government deeded more than seventeen million acres of land to the states. Some states used the funds to set up new colleges. These new colleges were known as A&M (agricultural and mechanical) colleges. Other states gave the money to existing colleges or universities and had them add A&M courses to their existing curriculum. Many scholars credit the first Morrill Act with the democratization of higher education. Democratization is the process of making something more democratic. The first Morrill Act helped democratize colleges and universities by causing them to offer courses in fields of study that were not associated with classical education and had previously been neglected.

The Second Morrill Act Extends Land-Grant Benefits to African Americans

The first Morrill Act had limited benefits for African Americans. It passed during the Civil War, so initially it did not apply to any of the states that formed the Confederacy. Once those states rejoined the Union, they also received land and money under the Morrill Act. Most Southern states did not extend the benefits of the Morrill Act to their African American residents, however. This was consistent with their policies of racial segregation. Only three Southern states directed a portion of their Morrill Act funds to African American colleges or universities. This discriminatory treatment angered many residents in the North.

The Morrill Act of 1890 was designed to bring an end to racial discrimination at A&M colleges in the South. The first Morrill Act had been a one-time grant of financial support (in the form of land) for A&M colleges. The second Morrill Act provided each state that maintained an A&M college with annual financial support, but the money came with strings attached. The second Morrill Act provided that no state would receive annual federal funding unless its A&M colleges admitted African American students. The second Morrill Act allowed the states to maintain racially segregated A&M colleges. But if a state did not have any A&M college that African American students could attend, that state did not receive federal financial support.

The second Morrill Act worked. The Southern states were facing economic difficulties and needed federal money to fund their colleges. Some states in the South opened new, separate A&M colleges that catered to African American students. Other Southern states began providing financial support to extant (already existing) all-black colleges. Technically, these colleges were not supported by the proceeds from sales of granted land. However, they have the same legal status as the colleges that were funded by the 1862 land grants. As a result, all of these schools are known

The Seventeen Black Land-Grant Colleges

The passage of the Morrill Act of 1890 prompted the Southern states to either establish A&M colleges that catered to African American students or provide financial support to existing all-black colleges. These schools are still in operation today. The list below identifies each of these schools by the name by which it is currently known, the city and state where it is located, and the year in which it was founded.

Lincoln University—Jefferson City, Missouri (1866)

Alcorn State University—Lorman, Mississippi (1871)

South Carolina State University—Orangeburg, South Carolina (1872)

University of Arkansas at Pine Bluff—Pine Bluff, Arkansas (1873)

Alabama A&M University—Huntsville, Alabama (1875)

Prairie View A&M University—Prairie View, Texas (1876)

Southern University and A&M College—Baton Rouge, Louisiana (1880)

Tuskegee University—Tuskegee, Alabama (1881)

Virginia State University—Petersburg, Virginia (1882)

Kentucky State University—Frankfort, Kentucky (1886)

University of Maryland Eastern Shore—Princess Ann, Maryland (1886)

Florida A&M University—Tallahassee, Florida (1887)

Delaware State College—Dover, Delaware (1891)

North Carolina A&T University—Greensboro, North Carolina (1891)

Fort Valley State College—Fort Valley, Georgia (1895)

Langston University—Langston, Oklahoma (1897)

Tennessee State University—Nashville, Tennessee (1909)

as land-grant colleges. Seventeen black land-grant colleges were established across the South between 1890 and 1915.

Black land-grant colleges became important centers of African American education in the South. Many of these schools specialized in agriculture, veterinary medicine, and engineering. Those fields, in turn, have provided economic opportunity to thousands of African Americans who graduated from black A&M colleges. As of 2010, there were seventeen black land-grant colleges in the sixteen Southern states (see sidebar).

A VOICE FROM THE SOUTH CHALLENGES TRADITIONAL VIEWS

The publication of Anna Julia Cooper's book titled *A Voice from the South: By a Black Woman from the South* in 1892 helped advance the cause of equal opportunity in education for African Americans. Cooper (1858–1964) was a lifelong educator and writer. She was a passionate defender of the importance of education to African Americans in general and African American women in particular. *A Voice from the South* was an important contribution to these causes. It was published at a time when there was growing sentiment that African Americans should confine themselves to vocational training. It also helped launch the idea of black feminism.

An Educator Defends Her Craft

Anna Julia Cooper was born into slavery on August 10, 1858, in Raleigh, North Carolina. After the Civil War ended, Cooper earned admission to a teacher training school in Raleigh. Cooper was an exceptionally talented student. As a child, she taught her mother how to read. Cooper completed high school in Raleigh and won admission to Oberlin College in Ohio. Cooper earned both a bachelor's and a master's degree from Oberlin. She completed her studies there in 1884.

Cooper spent three years after graduation working as a teacher in Raleigh. Then, in 1887, she was hired by Washington High School in Washington, D.C. Washington High, later known as M Street High School and eventually as Dunbar High School, was an all-black high school that was respected across the country for its excellent academics. It was widely considered the best all-black high school in the country. Cooper worked at this high school for the next forty years. Just five years after she started there, she published *A Voice from the South: By a Black Woman from the South.*

A U.S. postage stamp honors educator and scholar Anna Julia Cooper. *AP Images*

A Voice from the South is a collection of essays, speeches, and papers that address three related topics. First, Cooper argues that education is the key to African Americans' ability to lift themselves out of poverty and overcome the legacy of slavery. Second, she argues that different forms of discrimination, whether based on race, sex, or economic class, interact with and reinforce each other. Third, Cooper argues that the movement to uplift African Americans in

general must be supplemented by a movement specifically concerned with the uplift of African American women.

A Voice from the South Has a Lasting Impact

Each of the three main arguments Cooper advanced in *A Voice from the South* had a lasting impact on education and scholarship. Her first argument, about the value of formal education to African Americans, was especially important in the field of education. Cooper's position stood in contrast to the views of Booker T. Washington. Washington was the nation's most famous African American leader in the late 1800s. Washington argued that African Americans should forgo classical education and concentrate instead on vocational (job-related) training.

Cooper argued forcefully and persuasively that the most gifted African Americans should have the same access to higher education as white Americans so that they could become leaders, teachers, and activists in their communities. Cooper's *A Voice from the South* is thus similar to W. E. B. Du Bois's famous 1903 article, "The Talented Tenth." At a time when many African Americans, especially in the South, were on the verge of losing access to higher education entirely, Cooper's work provided a much-needed intellectual justification for equal opportunity in education.

Cooper's second main argument—that different forms of discrimination are interconnected—was revolutionary and ahead of its time. Cooper argued that racism and sexism are not independent of each other. For example, a black man might face racial discrimination, and a white woman might face gender discrimination. Cooper argued that not only do black women face both racial and gender discrimination, but they also face unique forms of discrimination specific to them as black women. In other words, racism and sexism are not wholly distinct and separate from each other; they are interactive and mutually reinforcing. This idea was virtually unheard of at the time Cooper published *A Voice from the South.* Today it is a widely accepted view. Cooper's book made an important contribution to this area of thought.

Cooper's final point was closely related to her second one. Cooper contended that African American women needed to band together politically so that their unique concerns would be addressed. At the time, what is now known as the feminist movement was in its fledgling (newborn, immature, and inexperienced) stages. Women still could not vote and were barred from holding many jobs. Even before the feminist movement had attained major victories, Cooper predicted that it would largely address itself toward the concerns and needs of white women. This insight has prompted many scholars to call *A Voice from the South* the wellspring of black feminism.

W. E. B. DU BOIS EARNS A PH.D. FROM HARVARD

W. E. B. Du Bois (1868–1963) made history in 1895 when he earned a Ph.D. from Harvard University in Cambridge, Massachusetts. A Ph.D. is a doctoral degree. It is the highest degree a student can earn at an American university. Harvard was regarded in the late 1890s as the best university in the United States, and possibly the entire world. Earning the Ph.D. was a remarkable personal accomplishment for Du Bois. On a larger scale, it was an important symbolic victory for an African American to earn an advanced degree from such a prestigious university just thirty years after the end of slavery.

A Rising Star Arrives at Harvard

W. E. B. Du Bois grew up in a small town in western Massachusetts. He said later in life that he encountered very little racial discrimination as a child. He was a talented student, and his high school teachers encouraged him to take college preparatory classes. He hoped to attend nearby Harvard University after graduating from high school, but his family could not afford to send him there. Instead, he attended Fisk University, an all-black university in Nashville, Tennessee. He excelled as a student and was the editor of the school newspaper. Du Bois also said his time in Nashville gave him firsthand exposure to the ugly realities of racism and segregation. Du Bois graduated from Fisk in 1888.

During the late 1800s, it was not uncommon for African American students to earn two college degrees, one from an all-black university and one from a mostly white university. Du Bois was able to earn a scholarship and enroll as a student at Harvard after his graduation from Fisk. Once again, he was a star student. When he graduated in 1890, he was selected to be one of only five graduating students to give a speech at the graduation ceremony. Du Bois then chose to continue his education by pursuing a master's degree at Harvard. He received a fellowship from Harvard to financially support his continued studies.

Du Bois immediately began developing the intellectual and research interests that would lead him toward a Ph.D. He worked on his master's degree from 1890 to 1891. A master's degree is a post-collegiate degree typically awarded after two years of additional study. To earn a master's degree a student must write a thesis, which is a long paper analyzing or researching a narrow topic. Du Bois wrote his master's thesis on the history of the enforcement of fugitive slave laws. His work gained him notice among the Harvard faculty. He was widely regarded as a gifted and dedicated student who showed tremendous promise as a scholar.

The master's degree gave Du Bois two degrees, but he was determined to complete his Ph.D. In 1890, he read an article in the *Boston Herald* about

a scholarship fund for African Americans that was being administered by former U.S. president Rutherford B. Hayes (1822–93). In the article, Hayes lamented the lack of worthwhile African American candidates for higher education. Du Bois took offense to the statement and sent Hayes a letter. Hayes eventually met with Du Bois personally and was so impressed with him that he awarded Du Bois enough money to support two years of study overseas. Du Bois studied at the University of Berlin in Germany from 1892 to 1894.

Du Bois Earns His Ph.D.

Du Bois returned to Harvard from Germany in 1894 with a well-developed intellectual outlook and philosophy. His firm belief was that social science—a combination of philosophy, history, and psychology that would later be known as the discipline of sociology—held the key to ending racial discrimination and advancing the fortunes of African Americans. Relying heavily on research he had performed in Germany, Du Bois completed his Ph.D. the next year.

To earn a Ph.D., a student is required to write a very long paper, almost as long as a book, called a dissertation. Du Bois's dissertation was titled *The Suppression of the African Slave-Trade to the United States of America, 1638–1870.* Du Bois argued in his dissertation that the slave trade had not ended because of any sense of moral obligation or outrage by white Americans. Rather, Du Bois argued, the slave trade simply died out once it was no longer economically profitable to import slaves from Africa. His argument was insightful, original, and provocative. In fact, Harvard University later published his dissertation in the prestigious journal *Harvard Historical Series.*

The Ph.D. Du Bois received in 1895 was the first doctorate degree Harvard had ever awarded to an African American. His accomplishment was an inspiration to African Americans across the country. Du Bois used the training he received at Harvard to launch a long, influential career as an academic and civil rights activist. He also opened the door for other African Americans to follow in his footsteps at Harvard and other elite universities.

WOODSON ESTABLISHES NEGRO HISTORY WEEK

The first celebration of Negro History Week in February 1926 marked a turning point in how students across the United States were educated about American history and African Americans' contributions to American history. Negro History Week was the brainchild (original idea) of Dr. Carter G. Woodson (1875–1950). Woodson was a professional historian. He

dedicated his career to the study of African American history. Negro History Week was Woodson's effort to introduce the public, especially schoolchildren, to the contributions of African Americans to American history. Negro History Week was successful in its ambitions. It was subsequently expanded into Black History Month, which since 1976 has been commemorated every February.

A Historian Fights for Recognition

Carter G. Woodson was born in Virginia to parents who were both former slaves. His parents worked as sharecroppers and coal miners, so Woodson did not begin high school until the age of twenty. He graduated in two years, enrolled in an all-black college in Kentucky, and became a teacher. Woodson had a deep-seated belief in the uplifting potential of education, so he enrolled at the University of Chicago to earn a second bachelor's degree and a master's degree. From there, he moved to Harvard University, where he earned a Ph.D. (the highest academic degree conferred by American universities) in history in 1912. He was just the second African American ever to earn a Ph.D. from Harvard.

In the early 1900s, professional, academic historians largely ignored African Americans in their study of American history. This ignorance showed itself in two ways. First, historians largely considered historical issues unique to African Americans, such as the slave trade or plantation culture, unworthy of serious study. Second, these historians also ignored or distorted the contributions of African Americans to American history overall. Most historians rarely discussed African Americans in their work. When they did, they portrayed African Americans solely as the victims of white racism and oppression. The dogma (widely accepted ideas and beliefs) of the day did not recognize that African Americans had been positive actors who had made important contributions to American history.

Carter Woodson dedicated his life and his career as a historian to studying African American history and to promoting the idea that such study was a worthy, serious academic pursuit. He organized the Association for the Study of Negro Life and History in 1915. The association began publishing an academic journal entitled *Journal of Negro History* in January 1916. Woodson founded a publishing company dedicated to printing books about African American history in 1921. Woodson himself wrote numerous scholarly works on African American history, including a textbook, *The Negro in Our History* (1922), that was used in college classrooms throughout the country. Woodson's contributions to the field earned him the nickname "The Father of Black History."

An African American classroom, c. 1940. Carter Woodson established Negro History Week in part to help African American students take pride in their heritage. *Lass/Hulton Archive/ Getty Images*

Negro History Week Reaches Out to Students Everywhere

These efforts were praiseworthy, but they were mostly addressed to other academics and professional historians. Woodson believed schoolchildren needed to learn at a young age about the contributions of African Americans to American history. The idea for Negro History Week was his response to this need. Woodson proposed that schools at all levels dedicate the second week of February to the study and appreciation of African American history. He chose the second week of February because it coincides with the birthdays of two figures of tremendous importance in African American history: Frederick Douglass (1818–95), the famous former slave, abolitionist, and orator; and President Abraham Lincoln (1809–65), who had issued the Emancipation Proclamation and successfully led the Union through the Civil War.

The idea for Negro History Week was a success. It was first celebrated in February 1926. Woodson and his colleagues at the Association for the Study of Negro Life and History assembled educational kits to be distributed for the event. The kits contained booklets, posters, photos, and other materials. They were distributed to schools, colleges, libraries, and museums. Woodson and the association strongly encouraged these institutions to make use of the kits and participate in Negro History Week. Thanks to their efforts, Negro History Week was widely observed throughout the United States. Woodson

continued to promote Negro History Week over the years. Eventually, it became a settled, yearly event in schools everywhere.

Negro History Week has proven to be a long-lasting, important contribution to American education. Negro History Week was renamed and expanded into Black History Month in 1976. It is held every year in February. School programs, exhibits, essay and poetry contests, and dramatic performances are held every February to highlight African Americans' history and achievements. Black History Month is also used as a vehicle for encouraging African American children to take pride in their history, culture, and heritage.

Thanks to Negro History Week, schools in the United States rapidly transitioned from systematically neglecting the historical contributions of African Americans to having a week-long event devoted to recognizing and celebrating those contributions. Negro History Week and its successor, Black History Month, have had an immeasurable impact on the attitudes of American students toward African American history.

ALL-BLACK FRATERNITIES AND SORORITIES BAND TOGETHER

The National Pan-Hellenic Council (NPHC) was formed on May 10, 1930, on the campus of Howard University in Washington, D.C. The NPHC was formed by five fraternities and sororities whose members were entirely or largely African Americans. The formation of the NPHC was important for several reasons. It helped black fraternities and sororities grow, recruit new members, and combat feelings of loneliness and isolation among African American college students. The NPHC also provided a way for African Americans to network (meet other people for business purposes) after they had graduated from college. The NPHC grew over the years and was made up of nine member organizations as of 2010.

Fraternities and sororities are social and service organizations for college students. Fraternities are made up of male members, while sororities' memberships are all-female. Fraternities and sororities are often referred to as "Greek lettered organizations" because their names are three Greek letters. Fraternities and sororities have long served two important functions for their members. First, they function as a social outlet. That means they help their members meet people and make friends while in college. Second, they function as a networking tool. That means that members of Greek lettered organizations often do business with other members of their organizations once they have graduated from college.

In the early 1900s, African American college students were largely prohibited from joining fraternities and sororities. Social custom and legal

segregation made it impossible for African American college students anywhere in the United States to join Greek lettered organizations. Racial prejudice was so widespread on college campuses—even in the North—that it frequently drove African American students to drop out of school entirely. For example, all six African American students who attended Cornell University in Ithaca, New York, during the 1904–05 school year dropped out and did not return the next year.

African American Students Form Their Own Organizations

African American students responded to this discrimination by forming their own Greek lettered organizations. Seven African American students enrolled at Cornell during the 1905–06 school year. The experience of their predecessors led them to form a fraternity so that they could support each other in a hostile racial environment. Thus, the Alpha Phi Alpha fraternity was formed in 1906. Additional all-black sororities and fraternities formed on various other college campuses over the next twenty-five years.

The all-black sorority Alpha Kappa Alpha meets at Howard University in 1946. The sorority was a charter member of the National Pan-Hellenic Council (NPHC) in 1930. *Alfred Eisenstaedt/Time & Life Pictures/Getty Images*

The Divine Nine

Nine fraternities and sororities made up the National Pan-Hellenic Council (NPHC) as of 2010. These nine organizations collectively have been nicknamed "The Divine Nine." Below is a list of the names of each of the nine organizations, the type of organization it is, the date when it was founded, and the university where it had its first chapter. The five charter members of the NPHC (members that originally formed the NPHC) are marked with an asterisk (*).

1. **Alpha Phi Alpha** (Fraternity): December 4, 1906; Cornell University—Ithaca, New York
2. **Alpha Kappa Alpha*** (Sorority): January 15, 1908; Howard University—Washington, D.C.
3. **Delta Sigma Theta*** (Sorority): January 13, 1913; Howard University—Washington, D.C.
4. **Kappa Alpha Psi*** (Fraternity): January 5, 1911; Indiana University—Bloomington, Indiana
5. **Omega Psi Phi*** (Fraternity): November 17, 1911; Howard University—Washington, D.C.
6. **Phi Beta Sigma** (Fraternity): January 9, 1914; Howard University—Washington, D.C.
7. **Zeta Phi Beta*** (Sorority): January 16, 1920; Howard University—Washington, D.C.
8. **Sigma Gamma Rho** (Sorority): November 12, 1922; Butler University—Indianapolis, Indiana
9. **Iota Phi Theta** (Fraternity): September 19, 1963; Morgan State University—Baltimore, Maryland

Five of these all-black Greek lettered organizations decided to band together in 1930. They met on the campus of Howard University in Washington, D.C., which was the nation's leading center of African American intellectual life at the time. The five organizations (two fraternities and three sororities) formed the National Pan-Hellenic Council on May 10, 1930. "Pan" is a Greek word meaning "all," and "Hellenic" is a synonym for "Greek." The stated purpose of the NPHC was to "consider problems of mutual interest to its member organizations." In other words, the NPHC was an effort by its member organizations to share information and

experiences on how best to navigate racism and prejudice on college campuses and beyond.

The NPHC was a success and grew quickly. Two more organizations joined in 1931, one joined in 1937, and a ninth member was added in 1997. The nine Greek lettered organizations that made up the NPHC were nicknamed "The Divine Nine." The strength and experience of the NPHC helped its member organizations expand and add chapters on new campuses. NPHC organizations especially thrived at historically black colleges and universities (HBCUs) in the South. An HBCU is a college or university that was founded before 1964 for the primary purpose of educating African Americans. NPHC fraternities and sororities cultivated unique, distinctive social and membership traditions such as lining (singing hymns to popular tunes in call-and-response fashion) and stepping (a form of percussive dance).

The nine organizations that make up the NPHC also perform valuable functions for their members after graduation. Many Greek lettered organizations limit their activities to the time when their members are in college. NPHC organizations have long functioned as more general social, civic, and service organizations in which their members remain active long after graduation. This has helped counteract the effects of racism, especially in the South, which prevented African Americans from joining civic and service organizations such as the Rotary Club.

The NPHC continues to thrive. As of 2010, there were more than four hundred undergraduate chapters—and an equal number of alumni (graduates of a particular college) chapters—of the nine fraternities and sororities that make up the NPHC. All told, there were more than 1.5 million members of the organizations that make up the Divine Nine.

THE UNITED NEGRO COLLEGE FUND IS FOUNDED

The founding of the United Negro College Fund in 1944 was a key moment in the history of African Americans' effort to obtain equal educational opportunities. The United Negro College Fund (UNCF) is a charitable organization that raises money to support African American students and historically black colleges and universities (HBCUs). An HBCU is a college or university that was founded before 1964 for the primary purpose of educating African Americans. The UNCF has done more than any other private organization in the country to advance the goal of providing African American students with access to higher education. The UNCF has distributed more than two billion dollars in financial assistance and helped more than three hundred fifty thousand students attend college.

Secretary of State John Foster Dulles (center) shakes hands with Frederick Patterson at a United Negro College fund event in 1955. Benjamin E. Mays, president of Morehouse College in Atlanta, looks on. *© Bettmann/Corbis*

Necessity Generates a New Idea for Fundraising

The United Negro College Fund was largely the brainchild (original idea) of Frederick Douglass Patterson (1901–88). Patterson was a veterinarian and professor who had become the president of the Tuskegee Normal and Industrial Institute in Alabama in 1935. Tuskegee is a famous HBCU founded by Booker T. Washington (1856–1915). Patterson was chosen to be just the third president in Tuskegee's fifty-four-year history. Patterson wanted to make significant changes to the curriculum and facilities at Tuskegee. He also wanted to hire new faculty members. However, he was hamstrung (made less effective) by a lack of adequate funding. Patterson was concerned that budgetary shortfalls were preventing Tuskegee from improving as an institution.

Patterson slowly became convinced that Tuskegee was not the only HBCU facing serious financial difficulties. He conducted an informal survey of the presidents of other historically black institutions in the early 1940s. They all reported that they were facing similar, or even more dire, financial constraints. At the same time, Patterson was aware that many of the most talented, promising African American high school students were not able to attend college because their families could not afford to pay their tuition. Tuskegee and other HBCUs were not able to offer these students scholarships or tuition waivers despite their financial difficulties.

These difficulties inspired Patterson to come up with the idea that HBCUs should pool their efforts and work together to raise money. Patterson first presented his idea for a collective fundraising organization in the early 1940s in a column in the *Pittsburgh Courier*, a weekly African American newspaper that had a nationwide circulation of more than three hundred thousand readers. The response to his idea was overwhelmingly positive. Patterson called a meeting of the presidents of more than a dozen of the nation's most prominent HBCUs in 1943. It was at this meeting that the idea for the United Negro College Fund began to take shape.

HBCUs Band Together for Financial Support

The creation of the UNCF was driven by a key insight: HBCUs needed to expand their fundraising base. HBCUs were not as old as other universities, so they did not have as many alumni who donated money to support them. They also tended to receive less financial support than other universities. As a result, most HBCUs relied heavily on large donations from very wealthy philanthropists. Philanthropists are people who regularly donate money to charities and nonprofit organizations. Patterson believed that there were many Americans who supported equal opportunity in education who did not know their support was needed. Patterson proposed that the HBCUs form an organization that would raise money from donors other than the very wealthiest benefactors (people who make gifts).

This plan became a reality on April 25, 1944, when Patterson and Mary McLeod Bethune (1875–1955)—president of Bethune-Cookman College in Daytona Beach, Florida—incorporated the United Negro College Fund in New York. The UNCF was originally made up of twenty-seven member institutions, HBCUs located throughout the South. The UNCF was an immediate success. Its goal was to solicit support from donors of all kinds, but many of its early supporters, such as Franklin D. Roosevelt (1882–1945) and John D. Rockefeller Jr. (1874–1960), made large individual donations. The UNCF raised seven hundred sixty thousand dollars in its first year of operation. By 1945, the number of member institutions had grown to thirty-two. The UNCF was raising more than one million dollars per year by 1947.

The United Negro College Fund continued to grow over the next several years. The Supreme Court's 1954 decision in *Brown v. Board of Education* caused public attention to become acutely focused on the question of how to provide equal opportunity in education. The UNCF relied on advertising campaigns that showed happy, hardworking African American college students to solicit funds. The UNCF received a major boost in 1963, when President John F. Kennedy (1917–63) donated the prize money he received after winning the Pulitzer Prize for his famous

book *Profiles in Courage*. In 1972, the UNCF launched what is still one of the most famous and effective advertising campaigns in history when it began soliciting funds using the slogan "A mind is a terrible thing to waste."

The UNCF supports equality in higher education in two ways. First, it provides money directly to its member institutions, which are HBCUs that educate a large proportion of African American college students. Second, individual students can apply to the UNCF directly for scholarships and other forms of financial support. Students of all races are eligible to apply for UNCF scholarships, although a substantial majority are African American. The UNCF continues to operate in the twenty-first century. As of 2009, it provided more than one hundred million dollars in financial support to African American students and to its thirty-nine member institutions.

BROWN DECISION ENDS SEGREGATION

The Supreme Court ruled in the 1954 case of *Brown v. Board of Education* that racially segregated public schools are unconstitutional. This decision was a hugely significant victory for African Americans in the area of education. Segregation is the practice of using the law to force members of different races to live apart from each other. Racial segregation in education was detrimental to (very bad for) African American students. African American students received lower quality education in segregated schools. Segregation was also used to limit the political and economic opportunities available to African Americans in other areas of life. Once segregation in education was declared unconstitutional, the foundation for the entire system of racial segregation in the South began to crumble.

The Supreme Court had ruled in the 1896 case of *Plessy v. Ferguson* that segregation was not unconstitutional so long as the facilities provided to members of each race were equal. The legal doctrine that justified segregation thus became known as "separate but equal." In practice, racially segregated schools were separate but unequal. School boards in the South refused to provide African American schools with adequate funding, facilities, materials, supplies, or teachers. It was virtually impossible for an African American student to get a top-notch education in a public school in the South.

Earlier Cases Lay the Groundwork

The National Association for the Advancement of Colored People (NAACP) began bringing lawsuits challenging segregation in the 1930s and 1940s. The NAACP was the nation's leading civil rights organization at the time. These lawsuits helped lay the foundation for the Supreme Court's decision in *Brown*. In the 1938 case of *Missouri ex rel. Gaines v. Canada,* the

The Little Rock Nine and the "Stand in the Schoolhouse Door"

The 1954 *Brown* ruling by the U.S. Supreme Court made segregation in schools illegal, but it did not end the practice overnight, or even in a decade. Many school districts across the South argued that obeying the law presented challenges that could not be overcome immediately. New school zones needed to be defined. Facilities needed to be expanded and updated. Thus, schools were granted extensions and exceptions in many cases.

At the same time, there was open protest, sometimes violent, against school integration by many white people in the South. In two well-publicized cases, Southern governors openly defied U.S. law by supporting school segregation. Governor Orval Faubus (1910–94) of Arkansas ordered the Arkansas National Guard to prevent nine African American students from attending Little Rock Central High School in the fall of 1957. President Dwight D. Eisenhower federalized (put under federal control) the Arkansas National Guard and sent in an army division in order to protect the students and enforce the law. Governor George Wallace (1919–98) of Alabama fulfilled a campaign promise to resist integration by standing in the doorway of the auditorium at the University of Alabama to block the entrance of two African Americans wishing to enroll in June 1963. He was confronted by U.S. marshals, and stood aside. These incidents made national news, but they were not isolated. Across the South, African American students faced intimidation and resistance as they tried to claim their legal right to equal education.

Supreme Court ruled that since the state of Missouri provided its white residents with a law school, it was required to also provide a law school to its African American residents. This case helped establish the idea that the notion of equality in education had to be taken seriously.

A major breakthrough came in the 1950 case of *Sweatt v. Painter*. In that case, an African American student argued that the state of Texas's "colored" law school was inferior to the state's whites-only law school. The Supreme Court agreed. The "colored" law school had a smaller faculty, a smaller library, fewer classes, and fewer extracurricular options. The

Plaintiffs in the famous *Brown v. Board of Education* Supreme Court case in 1953.
Carl Iwasaki/Time Life Pictures/Getty Images

Supreme Court ruled that these differences meant the law school was not "equal" to the whites-only law school.

The Court offered two reasons for its decision that eventually would lend crucial support for its decision in *Brown*. First, it recognized that there are important "intangible" elements of a high-quality education. It referred to them as "those qualities which are incapable of objective measurement but which make for greatness in a law school." Second, the Court finally acknowledged that the idea of separate but equal is a fiction when it said that it was "difficult to believe that one who had a free choice between these law schools would" not choose to attend the whites-only school. This was virtually an admission that segregated schools were always unequal. The time was ripe for a full-scale challenge to segregation in education.

Segregation Is Declared Unconstitutional

Brown v. Board of Education began in 1950 in Topeka, Kansas. That year, the local school board told Oliver Brown, an African American, that his

daughter would not be able to attend the elementary school that was a mere four blocks from their home because that school was only for white students. Brown's daughter would have to travel across town to attend the "colored" elementary school. Brown wanted to challenge this decision. The NAACP agreed to take Brown's case. By 1952, Brown's case, along with several other similar cases from around the country, had reached the Supreme Court.

The *Brown* case was a new kind of challenge to segregation in education. In previous cases such as *Sweatt v. Painter,* the NAACP had argued that segregated schools did not satisfy the equality component of "separate but equal." This had allowed the Supreme Court to rule in favor of the NAACP and African American students without confronting the constitutionality of segregation itself. The NAACP's lead attorney, Thurgood Marshall (1908–93), shifted course in *Brown* and challenged the very idea that segregated schools ever could be equal.

Marshall's strategy worked. The Supreme Court acknowledged that the schools under consideration in *Brown* "have been equalized, or are being equalized, with respect to buildings, curricula, qualifications and salaries of teachers, and other 'tangible' factors." The Court's use of the phrase "tangible factors" was a reference to its recognition in *Sweatt v. Painter* that true equality in education depends on intangible factors. For the first time ever, the Supreme Court was ready to acknowledge that segregation in education was a tool of racism and oppression.

The Supreme Court justified its decision with a three-step argument. First, the Court emphasized the importance of education, calling it "the very foundation of good citizenship." Second, the Court called attention to psychological studies and other evidence that showed that segregation in education "generates a feeling of inferiority" among African American children. Third, the Court reasoned that this sense of inferiority made it harder for African American children to learn and do well in school. As a result, the Court concluded that racial segregation in education was forbidden by the Fourteenth Amendment to the U.S. Constitution, which guarantees that all persons will receive "the equal protection of the laws."

Brown's historical significance extended beyond the area of education. It marked the beginning of the end of segregation in all areas of public life in the South. But putting the vision of desegregation outlined in *Brown* into practice proved to be a slow process. The decision triggered a wave of resistance among some white Southerners that the Supreme Court was largely powerless to counteract. It took more than twenty years for most schools in the South to desegregate. Nonetheless, *Brown* was the critical first step. It is widely regarded as one of the greatest decisions in the Supreme Court's history.

JAMES MEREDITH INTEGRATES "OLE MISS"

On October 1, 1962, James Meredith (1933–) became the first African American student ever to enroll at the University of Mississippi. The University of Mississippi, commonly referred to as "Ole Miss," was the flagship (best, largest, and most important) university of the state that was leading the fight against desegregation. Meredith was able to enroll there only after fighting to do so for more than a year and a half. Meredith's struggle to gain admission was marked by violence and required the intervention of the federal government. Meredith's enrollment at Ole Miss was an important victory for racial integration and civil rights.

The Supreme Court's landmark 1954 decision *Brown v. Board of Education* had declared that racial segregation in education was unconstitutional. That decision did not state how and when Southern states were required to end segregation, however. The next year, in the case known simply as *Brown II,* the Supreme Court ruled only that school desegregation in the South needed to take place "with all deliberate speed." Public schools and universities throughout the South relied on this vague language to avoid desegregating by claiming that they were making preparations to do so. As a result, most educational institutions did not desegregate unless an African American student brought a lawsuit requiring them to do so.

James Meredith was born and raised in the small town of Kosciusko, Mississippi. He excelled in school from a young age but had limited educational opportunities as an African American growing up in rural Mississippi in the 1930s and 1940s. He moved to Florida as a teenager in high school and was a standout student. Meredith joined the U.S. Air Force after graduating from high school. He served in the U.S. Air Force for nine years. Afterwards, he returned to his home state and enrolled at the all-black Jackson State College in Jackson, Mississippi.

Meredith Decides to Apply to the University of Mississippi

Meredith took enough classes to graduate from Jackson State. However, he wanted to apply to and graduate from the University of Mississippi. "Ole Miss" was the best university in Mississippi. Despite the Supreme Court's ruling in *Brown v. Board of Education,* Ole Miss was still an all-white school when Meredith applied for admission in January 1961. The university denied his admission when Meredith informed the university that he was an African American. Meredith decided to contact the National Association for the Advancement of Colored People (NAACP) for assistance. The NAACP was the nation's leading civil rights organization at the time and had extensive experience in bringing lawsuits over segregated education.

Meredith's NAACP lawyer, Constance Baker Motley (1921–2005), filed a lawsuit on his behalf in the summer of 1961. The lawsuit should have

been an easy victory in Meredith's favor. Ole Miss had clearly denied Meredith admission because he was African American, and this decision was clearly unconstitutional in light of *Brown v. Board of Education*. However, the first court to hear Meredith's lawsuit ruled in favor of the university. It took a full year for Meredith's case to be heard by a federal appeals court. On June 25, 1962, the Fifth Circuit Court of Appeals finally ruled that the university was required to admit Meredith.

The court's decision caused public outcry and anger in Mississippi. The governor of Mississippi, Ross R. Barnett (1898–1987), was a strong supporter of racial segregation. Barnett issued a proclamation on September 13, 1962, ordering the university to ignore the Fifth Circuit's decision and deny Meredith admission. One week later, a Mississippi court held a sham trial and convicted Meredith (who was not present for the trial) of the crime of false voter registration. The Mississippi legislature then passed a law prohibiting anyone who had been convicted of that crime from registering at Ole Miss.

Violence and Riots Precede Meredith's Enrollment

Word got out that Meredith nonetheless planned to register at Ole Miss. Students and local residents in Oxford, the city where Ole Miss is located, decided to take matters into their own hands. Thousands of people gathered outside the registrar's office at Ole Miss. They promised to use any means necessary, including violence, to prevent Meredith from enrolling at the university. Meredith was not deterred. He twice tried to register at Ole Miss in late September 1962. Both times, Governor Barnett came to Ole Miss and physically blocked Meredith from registering. Barnett continued to refuse to let Meredith register at Ole Miss even after the Fifth Circuit Court of Appeals specifically ordered him to stop doing so.

Earlier that year, Meredith had written a letter to recently elected President John F. Kennedy (1917–63), asking the federal government to support him in his effort to integrate Ole Miss. Kennedy was initially reluctant to get involved. The threat of riots, combined with Governor Barnett's open defiance of a federal court's decision, changed his mind. On September 30, 1962, Kennedy issued an executive order (a document signed by the president that gives legally binding instructions on how to comply with the law) authorizing the use of military force to require Ole Miss to enroll Meredith.

The Ole Miss campus was teetering on the edge of chaos. More than five hundred U.S. marshals and military members arrived on the Ole Miss campus on the evening of September 30 to escort Meredith as he registered for classes. A crowd of nearly one thousand people had gathered, and riots broke out that night. Two people were killed and hundreds were injured

James Meredith (center) attends class at the University of Mississippi accompanied by U.S. marshalls in 1962. *Buyenlarge/Hulton Archive/ Getty Images*

in the violence. Meredith, accompanied by his military escorts, finally registered as an Ole Miss student on October 1, 1962.

Meredith's integration of Ole Miss was an important victory for the civil rights movement. The government of the state of Mississippi had used all of the legal tools at its disposal to try to prevent the integration of Ole Miss. The people of Oxford had challenged the rule of law with violence and rioting. Meredith's courage and determination in the face of this response sent a message to whites throughout the South that racial integration could not be prevented or deterred by lawlessness and violence.

Primary Sources

W. E. B. DU BOIS RECOUNTS HIS TIME AT HARVARD (1960)

W. E. B. Du Bois (1868–1963) was a famous African American intellectual and political leader. Du Bois made history in 1895 when he became the first African American ever to earn a doctoral degree from Harvard University in Cambridge, Massachusetts. Harvard is one of the most prestigious universities in the world. Du Bois's receipt of a Ph.D. from Harvard was viewed as a breakthrough and symbolic moment within the African American community.

The following is an excerpt from an article written by Du Bois entitled "A Negro Student at Harvard at the End of the Nineteenth Century." The article was originally published in the Spring 1960 issue of *Massachusetts Review*. The excerpt describes Du Bois's time and experience at Harvard. Du Bois earned not only a Ph.D. but a master's degree and a bachelor's degree at Harvard. Du Bois describes what it was like to be one of only a very small number of African Americans attending Harvard in the late 1880s and early 1890s.

Harvard University in 1888 was a great institution of learning.... Seldom, if ever, has any American university had such a galaxy of great men and fine teachers as Harvard in the decade between 1885 and 1895.

To make my own attitude toward the Harvard of that day clear, it must be remembered that I went to Harvard as a Negro, not simply by birth, but recognizing myself as a member of a segregated **caste** whose situation I accepted but was determined to work from within that caste to find my way out....

Caste
A division or group of people created within a society

Venerable
Deserving of respect, worship, or veneration

Jamaican Maroons
Runaway slaves who revolted against their British owners in the 1700s

I hoped to pursue philosophy as my life career, with teaching or support. With this program I studied at Harvard from the Fall of 1888 to 1890, as undergraduate. I took a varied course in chemistry, geology, social science and philosophy. My salvation here was the type of teacher I met rather than the content of the courses....

When I arrived at Harvard, the question of board and lodging was of first importance. Naturally, I could not afford room in the college yard in the old and **venerable** buildings which housed most of the well-to-do students under the magnificent elms. Neither did I think of looking for lodgings among white families, where numbers of the ordinary students lived. I tried to find a colored home, and finally at 20 Flagg Street, I came upon the neat home of a colored woman from Nova Scotia, a descendant of those black **Jamaican Maroons** whom Britain deported after

solemnly promising them peace if they would surrender. For a very reasonable sum, I rented the second story front room and for four years this was my home. . . .

Following the attitudes which I had adopted in the South, I sought no friendships among my white fellow students, nor even acquaintanceships. Of course I wanted friends, but I could not seek them. My class was large, with some 300 students. I doubt if I knew a dozen of them. I did not seek them, and naturally they did not seek me. I made no attempt to contribute to the college periodicals, since the editors were not interested in my major interests. Only one organization did I try to enter, and I ought to have known better than to make this attempt. But I did have a good singing voice and loved music, so I entered the competition for the **Glee Club.** I ought to have known that Harvard could not afford to have a Negro on its Glee Club traveling about the country. Quite naturally I was rejected.

I was happy at Harvard, but for unusual reasons. One of these circumstances was my acceptance of racial segregation. Had I gone from **Great Barrington high school** directly to Harvard, I would have sought companionship with my white fellows and been disappointed and **embittered** by a discovery of social limitations to which I had not been used. But I came by way of Fisk and the South and there I had accepted color caste and embraced eagerly the companionship of those of my own color. This was, of course, no final solution. Eventually with them and in mass assault, led by culture, we Negroes were going to break down the boundaries of race; but at present we were banded together in a great crusade and happily so. Indeed, I suspect that the prospect of ultimate full human **intercourse** without reservations and annoying distinctions, made me all too willing to **consort** now with my own and to disdain and forget as far as was possible that outer, whiter world.

In general, I asked nothing of Harvard but the **tutelage** of teachers and the freedom of the laboratory and library. I was quite voluntarily and willingly outside its social life. . . . I found friends, and most interesting and inspiring friends, among the colored folk of Boston and surrounding places. Naturally, social intercourse with whites could not be entirely forgotten, so that now and then I joined its currents and rose or fell with them. I escorted colored girls to various gatherings, and as pretty ones as I could find to the **vesper exercises,** and later to the class day and commencement social functions. Naturally we attracted attention and the **Crimson** noted my girl friends; on the other part came sometimes the shadow of insult, as when at one reception a white woman seemed determined to mistake me for a waiter.

In general, I was encased in a completely colored world, self-sufficient and **provincial,** and ignoring just as far as possible the white world which conditioned it. This was self-protective coloration, with perhaps an inferiority complex, but with belief in the ability and future of black folk.

My friends and companions were taken mainly from the colored students of Harvard and neighboring institutions, and the colored folk of Boston and surrounding towns. With them I led a happy and inspiring life. There were among them many

Glee Club
A musical group, similar to a choir, that specializes in singing short songs

Great Barrington High School
The racially integrated high school Du Bois attended in Massachusetts

Embittered
Made to feel bitter

Intercourse
Interaction, exchange, and dialogue

Consort
To associate or keep company with

Tutelage
Teaching and mentoring

Vesper Exercises
Evening religious worship

The Harvard Crimson
Harvard's student-run newspaper

Provincial
Limited and narrow in scope or outlook

educated and well-to-do folk; many young people studying or planning to study; many charming young women. We met and ate, danced and argued and planned a new world.

Obsequious
Marked by extreme, exaggerated deference and respect

Toward whites I was not arrogant; I was simply not **obsequious,** and to a white Harvard student of my day, a Negro student who did not seek recognition was trying to be more than a Negro. The same Harvard man had much the same attitude toward Jews and Irishmen.

I was, however, exceptional among Negroes in my ideas on voluntary race segregation; they for the most part saw salvation only in integration at the earliest moment and on almost any terms in white culture; I was firm in my criticism of white folk and in my dream of a Negro self-sufficient culture even in America.

This cutting off of myself from my white fellows, or being cutoff, did not mean unhappiness or resentment. I was in my early manhood, unusually full of high spirits and humor. I thoroughly enjoyed life. I was conscious of understanding and power, and conceited enough still to imagine, as in high school, that they who did not know me were the losers, not I. On the other hand, I do not think that my white classmates found me personally objectionable. I was clean, not well-dressed but decently clothed. Manners I regarded as more or less **superfluous,** and deliberately cultivated a certain **brusquerie.** Personal **adornment** I regarded as pleasant but not important. I was in Harvard, but not of it, and realized all the irony of my singing "Fair Harvard." I sang it because I liked the music, and not from any pride in the Pilgrims. . . .

Superfluous
Unnecessary and extraneous

Brusquerie
A habit of being brusque, or short and abrupt of manner

Adornment
Manners and social graces

With my colored friends I carried on lively social intercourse, but necessarily one which involved little expenditure of money. I called at their homes and ate at their tables. We danced at private parties. Thus this group of professional men, students, white collar workers and upper servants, whose common bond was color of skin in themselves or in their fathers, together with a common history and current experience of discrimination, formed a unit which like many tens of thousands of like units across the nation had or were getting to have a common culture pattern which made them an interlocking mass; so that increasingly a colored person in Boston was more neighbor to a colored person in Chicago than to the white person across the street. . . .

In the general social intercourse on the campus I consciously missed nothing. Some white students made themselves known to me and a few, a very few, became life-long friends. Most of my classmates, I knew neither by sight nor name. . . . For the most part I do not doubt that I was voted a somewhat selfish and self-centered **"grind"** with a chip on my shoulder and a sharp tongue.

Grind
A slang term for a person who works all the time

Something of a certain inferiority complex was possibly a cause of this. I was desperately afraid of intruding where I was not wanted; appearing without invitation; of showing a desire for the company of those who had no desire for me. I should in fact have been pleased if most of my fellow students had wanted to associate with me; if I had been popular and envied. But the absence of this made me neither unhappy nor morose. I had my "island within" and it was a fair country. . . .

Harvard of this day was a great opportunity for a young man and a young American Negro and I realized it. I formed habits of work rather different from those of most of the other students. I burned no midnight oil. I did my studying in the daytime and had my day parceled out almost to the minute. I spent a great deal of time in the library and did my assignments with thoroughness and with **prevision** of the kind of work I wanted to do later. . . .

Prevision
Anticipation

In June 1890, I received my bachelor's degree from Harvard ***cum laude*** in philosophy. I was one of the five graduating students selected to speak at commencement. My subject was "Jefferson Davis." I chose it with deliberate intent of facing Harvard and the nation with a discussion of slavery as illustrated in the person of the president of the Confederate States of America. Naturally, my effort made a sensation. . . .

Cum Laude
Latin for "with honors."

A Harvard professor wrote to *Kate Field's Washington,* then a leading periodical: "Du Bois, the colored orator of the commencement stage, made a **ten-strike.** It is agreed upon by all the people I have seen that he was the star of the occasion. . . . One of the trustees of the University told me yesterday that the paper was considered masterly in every way. Du Bois . . . is an excellent scholar in every way, and altogether the best black man that has come to Cambridge."

Ten-strike
A casual term, taken from the best possible roll in the game of bowling, for a remarkably successful act

JAMES MEREDITH RECALLS INTEGRATING THE UNIVERSITY OF MISSISSIPPI (2002)

James Meredith (1933–) made national headlines in 1962 when he became the first African American ever to enroll at the University of Mississippi. At the time, Mississippi was a hotbed of resistance to racial integration. Meredith's decision to attend the University of Mississippi sparked riots in the state and required the intervention of U.S. marshals. Meredith dealt a devastating blow to racial segregation by integrating the flagship university of the state where segregation was most deeply entrenched.

Below is an excerpt from a public talk Meredith gave about his experiences at the University of Mississippi in 2002 to commemorate the fortieth anniversary of his enrollment there. Meredith discusses what motivated him to enroll at the university and a meaningful encounter he had with a black janitor after he registered.

• •

I'd already made the application 18 months before. And you got to remember, I went back to Mississippi in 1960 to fight a war; and to me, it was no different from World War One, World War Two, Korea, which all I'd had friends and relatives to die in, but that's what it was to me; it was a war. And I went back to Mississippi with plans devised to break the system of white supremacy. . . .

It was President Eisenhower who, at that time, was President, but more significantly, he had been the biggest general the United States had ever had, including Grant, and all of them. The biggest of all was Eisenhower. And Eisenhower had called the troops out in Arkansas to support the rights of citizenship for those Little Rock Nine. And, of course, as I said, I was a soldier. My mission was war, and I knew that I would have to have an armed force on my side bigger than Mississippi had on their side. The only force in the world bigger than Mississippi's armed force was the United States government. So once Eisenhower did that, I saw the opportunity of forcing the federal government to a position where they would have to support the citizenship rights of me and all of my kind. . . .

And I know you got a lot of goodhearted people in here, and they want to hear good-hearted, nice things. But I was at war, and what soldiers do, they kill enemies. And if there's anything that was really surprising to me about the whole Mississippi thing, was that I survived alive. I mean, if I had been me, I would have certainly killed me if I'd been on the other side. . . .

My mission was to force the federal government . . . into a position where they had to use organized violence to protect my rights of citizenship. And I had accomplished that days before. . . . I had already seen the soldiers assembling up in a militant Air Force base in Memphis, and so I knew what the deal was. I knew also that the Kennedy administration wanted very much to be reelected, and they didn't want to offend the South too badly, but they had already made the commitment. And I can't overstress the significance of using organized violence to protect me and my kind, and our rights of citizenship. . . .

[On the day that Meredith registered,] the marshals came in, and told me . . . that we could go register. So we went out to get in the car, and it was the same marshal's car we'd been riding in. All of the windows had been shot out, so we had to go back in the dorm and get some blankets to put over the seat, so the glass wouldn't cut us. And then we drove on over to the Lyceum Building, and, actually, at that time it was later that I learned that that's where all of the rioting had taken place, because there wasn't nobody there when I went over there but soldiers. So we went inside and registered. . . .

[After Meredith came out of the registrar's office, he saw a black man who was an employee of the University.] He was the only person in the building besides my group, and I thought it was odd, but as I walked past him, he had a broom under his arm, and he twisted his body and touched me with the broom. Well, he looked at me, and the message was very clear: "We are watching out for you. You don't have to worry about nothing." And you got to understand, at least 60 percent of the people working at the University of Mississippi were black. So that was a very, very important occasion in my life.

A lot of people don't realize, change is something that no one ought to experience. When I made application for Ole Miss, at least 95 percent of every black in

the state was against it, particularly if they were middle class or **high echelon** black. I mean, because they felt threatened. So, for the most part, I didn't have any way to know how the ordinary people like this janitor felt. But when he did that thing, that was probably the most important thing to me that had occurred to that point.

High Echelon
Upper class

Research and Activity Ideas

1. Numerous African American teachers, researchers, and activists made their mark in the field of education between 1865 and 1965. This chapter introduces ten notable educators who made tremendous contributions to and achieved notable firsts in their field. Write an essay in which you explain which of the African American educators you read about in this chapter you admire the most and why. Explain why you are especially impressed or inspired by this person's accomplishments. What kind of obstacles did this person have to overcome? How did his or her work benefit others?

2. In 1895, W. E. B. Du Bois became the first African American ever to receive a Ph.D. from Harvard University. Du Bois's accomplishment was viewed as an important milestone by the African American community. A Ph.D. is the highest academic degree awarded by American universities, and Harvard is widely regarded as one of the best universities in the United States. Work together with several of your classmates and have a group discussion about the reasons why other African Americans viewed Du Bois's receiving a Ph.D. from Harvard as an important event. What kind of symbolic value did Du Bois's achievement have? Why might that symbolic value have been particularly important to African Americans living in the 1890s?

3. W. E. B. Du Bois's 1903 essay "The Talented Tenth" was a direct response to Booker T. Washington's so-called "Atlanta Compromise" speech of 1895. Both men were addressing the pressing issues of how to "elevate" the African American population in the generations immediately following the Civil War while changing the negative attitudes most whites held about African Americans. Find Du Bois's essay and Washington's speech on the Internet. Then, use your library and the Internet to find out more about the social and economic challenges facing African Americans at the beginning of the twentieth century. Following the examples of Washington and Du Bois, write your own speech or essay addressing these challenges. Imagine you are writing in the first decade of the twentieth century. You can choose to support either Du Bois or Washington, or you can propose an entirely different solution. Make sure to support your argument with careful reasoning and facts.

4. James Meredith became the first African American ever to enroll as a student at the University of Mississippi (commonly known as "Ole Miss") on October 1, 1962. Many white residents of Mississippi

strongly believed that African Americans should not be able to attend Ole Miss. Imagine that you are addressing a crowd of white Mississippi residents in 1962. Prepare a speech in which you explain why it is important for African Americans such as James Meredith to be able to attend Ole Miss. Your speech should include at least one reason why African Americans will benefit, and at least one reason why whites will not be harmed.

5. The Supreme Court's famous 1954 decision *Brown v. Board of Education* brought an end to the practice of legal racial segregation in public schools. Specifically, the Supreme Court said in *Brown* that segregated schools deny African American children "the equal protection of the laws." The court relied heavily on research performed by Kenneth and Mamie Clark showing the negative psychological effects that segregated schools had on African American children. Write an essay that answers the following question: How does the Clarks' research support the Supreme Court's decision in *Brown*? How does their research strengthen the conclusion that laws requiring segregated schools are a form of unequal treatment?

For More Information

BOOKS

Brawley, Benjamin G. *History of Morehouse College.* New York: Cosimo Classics, 2009.

Doyle, William. *An American Insurrection: James Meredith and the Battle of Oxford, Mississippi, 1962.* New York: Anchor Books, 2003.

Gasman, Marybeth. *Envisioning Black Colleges: A History of the United Negro College Fund.* Baltimore: Johns Hopkins University Press, 2007.

Lanker, Brian. *I Dream a World: Portraits of Black Women Who Changed America.* New York: Stewart, Tabori, and Chang, 1999.

Lewis, David Levering. *W.E.B. Du Bois, 1868–1919: Biography of a Race.* New York: Henry Holt and Company, 1993.

Lewis, David Levering. *W.E.B. Du Bois, 1919–1963: The Fight for Equality and the American Century.* New York: Henry Holt and Company, 2000.

Logan, Rayford Whittingham. *Howard University: The First Hundred Years, 1867–1967.* New York: New York University Press, 1969.

Norrell, Robert J. *Up from History: The Life of Booker T. Washington.* Cambridge, MA: Belknap/Harvard University Press, 2009.

Patterson, James T. *Brown v. Board of Education: A Civil Rights Milestone and Its Troubled Legacy.* New York: Oxford University Press, 2001.

Samuels, Albert L. *Is Separate Unequal? Black Colleges and the Challenge to Desegregation.* Lawrence: University Press of Kansas, 2004.

Watson, Yolanda L., and Sheila T. Gregory. *Daring to Educate: The Legacy of the Early Spelman College Presidents.* Sterling, VA: Stylus Publishing, 2005.

WEB SITES

Brown Foundation for Educational Equity, Excellence and Research. *Brown v. Board of Education—Background Summary.* http://brownvboard.org/summary/index.php (accessed on December 8, 2009).

National Pan-Hellenic Council, Inc. *About Us.* http://www.nphchq.org/about.htm (accessed on December 8, 2009).

United Negro College Fund. *Historically Black Colleges and Universities (HBCUs)—An Historical Overview.* http://www.uncf.org/aboutus/hbcus.asp (accessed on December 8, 2009).

Where Do I Learn More?

BOOKS

Barbeau, Arthur E., Florette Henri, and Bernard C. Nalty. *The Unknown Soldiers: African American Troops in World War I.* Cambridge, MA: Da Capo Press, 1996.

Bartleman, Frank. *Azusa Street.* New Kensington, PA: Whitaker House, 1982.

Bogle, Donald. *Bright Boulevards, Bold Dreams: The Story of Black Hollywood.* New York: One World/Ballantine, 2005.

Brophy, Alfred, and Randall Kennedy. *Reconstructing the Dreamland: The Tulsa Race Riot of 1921, Race Reparations, and Reconciliation.* London: Oxford University Press, 2002.

Cooper, Anna Julia. *A Voice from the South.* Xenia, OH: Aldine Printing House, 1892.

Du Bois, W. E. B. *The Souls of Black Folk.* Ed. Henry Louis Gates Jr. New York: W. W. Norton, 1999.

Hill, Laban Carrick. *Harlem Stomp! A Cultural History of the Harlem Renaissance.* New York: Little Brown, 2003.

Katz, William. *The Black West: A Documentary and Pictorial History of the African American Role in the Westward Expansion of the United States.* New York: Harlem Moon, 1971.

Large, David Clay. *The Nazi Games: The Olympics of 1936.* New York: W. W. Norton, 2007.

Lemann, Nicholas. *The Promised Land: The Great Black Migration and How It Changed America.* New York: Vintage, 1992.

Love, Spencie. *One Blood: The Death and Resurrection of Charles R. Drew.* Chapel Hill, NC: University of North Carolina Press, 1996.

Malcolm X, with Alex Haley. *The Autobiography of Malcolm X.* New York: Grove Press, 1965.

Robeson, Paul. *Here I Stand.* Boston: Beacon Press, 1958.

Robinson, Jackie, and Alfred Duckett. *I Never Had it Made: An Autobiography of Jackie Robinson.* New York: Putnam, 1972.

Thomas, Vivien T. *Partners of the Heart: Vivien Thomas and His Work with Alfred Blalock: An Autobiography.* Philadelphia: University of Pennsylvania Press, 1998.

Till-Mobley, Mamie. *Death of Innocence: The Story of the Hate Crime that Changed America.* New York: One World/Ballantine, 2003.

Ward, Geoffrey C. *Unforgivable Blackness: The Rise and Fall of Jack Johnson.* New York: Vintage, 2006.

Wells, Ida B. *Southern Horrors and Other Writings: The Anti-Lynching Campaign of Ida B. Wells, 1892–1900.* Ed. Jacqueline Jones Royster. New York: Bedford/St. Martin's, 1996.

Wood, Sylvia. *Sylvia's Family Soul Food Cookbook: From Hemingway, South Carolina, To Harlem.* New York: William Morrow Cookbooks, 1999.

Woodward, C. Vann. *The Strange Career of Jim Crow.* New York: Oxford University Press, 1955.

WEB SITES

African American Odyssey: A Quest for Full Citizenship. Available at the Library of Congress Web site. http://memory.loc.gov/ammem/aaohtml/exhibit/aointro.html (accessed on March 1, 2010).

Chicago Defender. http://www.chicagodefender.com (accessed on March 1, 2010).

Documenting the American South. http://docsouth.unc.edu/index.html (accessed on March 1, 2010).

Duke Ellington: Celebrating 100 Years of the Man and His Music. http://www.dellington.org/ (accessed on March 1, 2010).

Jazz. Available on the PBS Web site at http://www.pbs.org/jazz/ (accessed on March 1, 2010).

Madame C. J. Walker Official Web site. http://www.madamcjwalker.com (accessed on March 1, 2010).

Mintz, S. "America's Reconstruction: People and Politics after the Civil War." *Digital History.* http://www.digitalhistory.uh.edu/reconstruction/index.html (accessed on March 1, 2010).

The Nation of Islam Official Web site. http://www.noi.org/ (accessed on March 1, 2010).

Rosa and Raymond Parks Institute for Self Development. http://www.rosaparks.org/about.html (accessed on March 1, 2010).

Tuskegee University. http://www.tuskegee.edu/ (accessed on March 1, 2010).

Index

Boldface type indicates entries; *Italic* type indicates volume; (ill.) indicates illustrations.

B

C

D

F

G

I

M

N

O

Q

R

T

U

V

X

Y

Z

BOCA RATON PUBLIC LIBRARY, FLORIDA
3 3656 0511575 1